Zoogeography:

THE GEOGRAPHICAL DISTRIBUTION OF ANIMALS

Frontispiece. The world. Double orthographic projection. Heavy broken line is course of Mid-Atlantic Ridge.

ROBERT E. KRIEGER PUBLISHING COMPANY
MALABAR, FLORIDA

Zoogeography:

THE GEOGRAPHICAL DISTRIBUTION OF ANIMALS

PHILIP J. DARLINGTON, JR.

Museum of Comparative Zoology

Harvard University

Original Edition 1957
Reprint Edition 1980, 1982

Printed and Published by
ROBERT E. KRIEGER PUBLISHING COMPANY, INC.
KRIEGER DRIVE
MALABAR, FLORIDA 32950

Printed in the United States of America

Library of Congress Cataloging in Publication Data

Darlington, Philip Jackson, 1904-
Zoogeography.

Reprint of the edition published by Wiley, New York.
Includes bibliographies.
1. Zoogeography.
[QL101.D3 1980] 591.9 79-26913
ISBN 0-89874-109-2

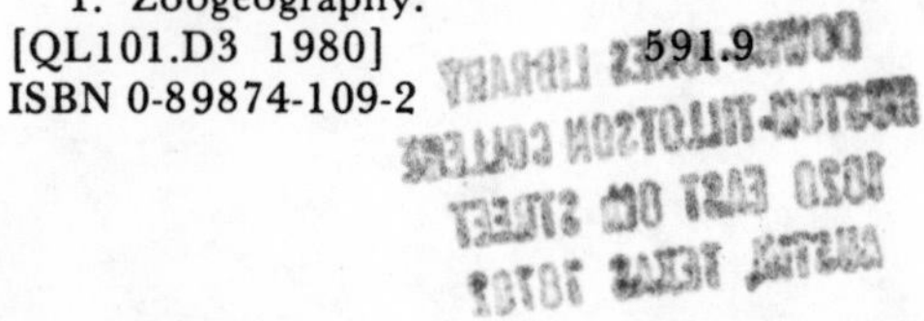

To the late Thomas Barbour,

a great naturalist and a great director

Preface

ABOUT THE BOOK. This book is a basic zoogeography. It is an attempt to amass facts, put the facts together, and discover or rediscover the principles of geographical distribution of animals over the world. It is, however, concerned only with the land and fresh water, not with the sea; and it is concerned principally with vertebrates, because they are the best-known, geographically most significant animals. Within these limits, the book will try to answer four questions:

What is the main pattern of animal distribution?
How has the pattern been formed?
Why has the pattern been formed?
What does animal distribution tell about ancient lands and climates?

This book is a geographical, not an ecological, zoogeography. This is not said apologetically. The ecological approach to animal and plant geography is fashionable now, but books written from that point of view, like Hesse, Allee, and Schmidt's excellent *Ecological Animal Geography,* are ecologies at heart and hardly geographies at all. I do not mean to make little of ecology but just to say that it is not zoogeography. Darwin said this long ago—that animal distributions cannot be accounted for simply in terms of climates and local physical conditions. For example (as someone has said), mammals browse on bushes in South Africa, South America, and Australia, but the mammals and the bushes are very different in each place, and ecology does not explain the differences. Ecology can contribute much to zoogeography; but zoogeography can contribute to ecology, too.

Geographical zoogeography has had a distinguished history. The distribution of plants and animals in South America was one of the things that first suggested evolution to Darwin, as he says in the first sentence of the *Origin*. Geographical zoogeography led Wallace, too, to Darwin's theory of evolution before Darwin published it, creating one of the most dramatic incidents in the history of science. Zoogeography will probably not again contribute to anything so mind-shaking as the idea of organic evolution, but it still has much to offer. Animal distribution can still tell biologists new things about evolution. It can still help geologists reconstruct the world of the past. And it is fascinating in its own right.

About methods and words

Zoogeography is concerned with a multitude of facts of its own, and it draws also on ecology, evolution, and geology. This vast field can be reduced to the limits of one book in two ways. Everything can be put at one superficial level, by means of selected examples, limited discussion, and arbitrary generalizations, but this is likely to result merely in uncritical rehashing of old ideas with not much chance of developing new ones. Or certain important things can be put in as much detail as space allows, and other things put very briefly. This is the method I have followed. I have described and discussed the distribution of land and fresh-water vertebrates as fully as I can, and I have cut many other subjects to bare references or just statements of my point of view—readers must know my point of view in, for example, geology and evolution in order to understand and criticize my zoogeography. Even so, I have had to keep my treatment of vertebrates within limits. I have continually had to restrain myself from research on details and be satisfied with good enough.

This has been true of my treatment of the literature, too. The literature that bears on zoogeography includes not only primarily zoogeographical items but also all that is published on the classification, habits, and history of all the animals concerned, and much that is published as general geography, climatology, ecology, and evolution. I have set no single standard and no deadline for search of all this literature, but have tried to give references sufficient to each case, sometimes just one recent reference to provide readers with an entry to other literature.

Nor have I been able to say much about the histories of zoogeographical ideas. I have concentrated on animal distribution, the thing itself, rather than on ideas about it, and have often just stated as facts things which now seem obvious but which have been worked out laboriously by generations of zoogeographers. If I have not always given credit where it ought to be given, I am very sorry. My excuse is necessity. If I had tried to trace histories of ideas and to give credit in detail, I would probably not have finished this book.

I write laboriously. I have to put what I want to say in whatever words come, and then criticize and revise it over and over again, trying always to say more exactly what I mean in fewer and better words. I have been helped and reassured in trying to write simply by a book called *Plain Words: Their ABC* (Sir Ernest Gowers, 1954, published by Alfred A. Knopf), which is designed for officials in England but which can help scientists avoid jargon, too.

When I was a young man, more given to argument than I am now, I was told that an author has the right to define words in his own way and that readers are to blame if they do not take the trouble to understand. But I do not want to be misunderstood, not even if it is the readers' fault. I have therefore tried to use ordinary words in ordinary ways, as well as to write simply and clearly. I have indeed defined some words, but so far as possible I have used the ordinary definitions given in my *Webster's Collegiate Dictionary.* For place names, I have followed *Webster's Geographical Dictionary* so far as possible.

Zoogeographical words are defined elsewhere (pp. 23–25), but a few general words and phrases are worth mention here. "As demonstrated almost conclusively," "according to the best scientific opinion," "in my carefully considered judgment," etc., are favorite phrases with scientific writers, but they usually mean no more than "probably," and that is the word I shall usually use here. I shall also say "I think" and "I guess" without circumlocution. The only other word to be mentioned now is "etc.," if it is a word. "Etc." has been called the track of a lazy man, but it is also the track of a man cramped for space. Many times in this book I have had to cut a sentence short without giving all the details, and I have used "etc." to show where details have been left out.

About the author

Who should write a zoogeography? No one, if personal familiarity with the whole subject is required. No one can know personally all animals and all parts of the world and all related subjects. But zoogeographies should be written, even if no one is fully qualified to write them. Perhaps a writer may partly make up for his deficiencies if he says plainly what his qualifications and limitations are.

I have been lucky. Part of the luck was to grow up and be educated near great museums; to be at home first in the old Boston Society of Natural History and later in the Museum of Comparative Zoology (the M.C.Z.) at Harvard; and to know such men as M. L. Fernald, W. M. Wheeler, and Thomas Barbour, all of them taxonomists or museum men as well as teachers.

At first I was (and still am) deeply interested in the distribution of life in North America, and I have done my share of collecting there. I first saw the tropics in 1926, as a graduate student twenty-two years old, when I spent three months at Harvard's Atkins research station at Soledad in central Cuba. That experience made such an impression on me that I have returned four times to the West Indies and have collected over all four of the main islands there. Perhaps no other islands except the Philippines show so well the geographic patterns of island life. In 1928–1929 I interrupted my graduate work to spend a year as resident entomologist for the Colombia Division of the United Fruit Company, inland from Santa Marta on the north coast of South America. Here I saw a variety of tropical American country—arid plains, the outskirts of the swamps and rain forest of the Magdalena delta, and the diversely forested slopes, *páramos,* and snow peaks of the Sierra Nevada de Santa Marta—at all seasons. In 1931–1932 I collected in Australia with a party from the Museum of Comparative Zoology and again saw a variety of country at all seasons. Finally, as an army entomologist during World War II, I spent two years moving from Australia to New Guinea, across the equator and across Wallace's Line to the Philippines, and to Japan.

As a boy I "watched" birds, and if I had had a license to collect them I might have become a professional ornithologist. I did become an entomologist, but (as I have said) this book is not about insects. I have systematically collected for the Museum not only insects and other invertebrates but, in the West Indies, also frogs and reptiles; in South America, birds and some other vertebrates; and in Australia, mammals, birds, reptiles, and amphibians almost more than insects at some seasons. Of all the main classes of vertebrates, the fishes are the only ones with which I cannot claim some useful field acquaintance. This is a serious gap, and I have sweated for it, trying to familiarize myself with fishes from the literature. Thanks to hard work on my part and especially to aid from George S. Myers and other ichthyologists, what I shall have to say about fresh-water fishes is, I suspect, finally no worse and perhaps better than any other part of this book.

I have the additional good fortune to be a taxonomist. Classification of animals is my profession. It is true that the animals I classify are beetles and not vertebrates, but my work has given me a first-hand knowledge of what classifications are, how they show relationships and distributions, and what their limitations are. This kind of knowledge is almost indispensable in zoogeography.

So I have been lucky in many ways, and one particular stroke of luck has tied all the rest together. In 1932 I was appointed to the staff of the Museum of Comparative Zoology and have worked there under the directorship first of Thomas Barbour and then of Alfred S. Romer. There has been no better place in the world to carry on zoogeographic studies than this Museum under these men. I planned this book at the Museum almost twenty years ago and have worked on it continually as other things permitted. I have had time for the work, facilities, help, and encouragement, and perhaps these are my greatest qualifications as a zoogeographer.

Acknowledgments

I am indebted more than I can acknowledge in detail to Thomas Barbour, Alfred S. Romer, and many past and present staff members of the Museum of Comparative Zoology. Some of them and some other specialists are named in the particular chapters with which they have helped me most. Of course these persons are not responsible for my errors, or for my conclusions. I am especially indebted to Ernest E. Williams for reading my final chapters and discussing them with me. He has an ability to be usefully critical without being hot about it. I am indebted to my wife for many kinds of help: typing when I needed it most, comments on my English, a bystander's comments on many parts of the book, and finally proofreading. Much of the material of the book was presented as "Biology 21: Zoogeography" in the summer term of Harvard in 1947. Some of the material on cold-blooded vertebrates appeared in condensed form in the *Quarterly Review of Biology* in 1948; I am indebted to the editors of that journal and to the Williams and Wilkins Company for permission to use parts of it again, but I have usually rephrased and extended it rather than repeated it. Finally I am indebted to Mrs. T. K. Searight for accurate typing of the manuscript, to Miss Patricia Washer for inking and lettering of illustrations, and to Dr. Karl P. Schmidt for careful reading of the manuscript.

PHILIP J. DARLINGTON, JR.

July 1957

Contents

chapter 1

Introduction

*B*rought together here under separate headings are some general facts about the world and the distribution of climate and vegetation on it, a partial outline of geological time, something about zoogeography in general and certain particular aspects of it, something about animal dispersal, a very short history of zoogeography, some definitions, and a list of working principles.

The world and maps

Zoogeography should begin with an understanding of certain simple facts about the world.

The world is a sphere. This fact is fundamental, but it is too often taken for granted and then forgotten. Zoogeographers should constantly remember that the world is round, and their maps should show it.

Maps can help or hinder understanding of the world. Wallace's maps were poor; he was a good zoogeographer in spite of them. He used the Mercator projection (Fig. 1), which gives an idea of the zonation of climate but which exaggerates areas toward the poles and gives no feeling of the position of the continents on the round earth. Other zoogeographers have used maps which make the world look like an orange peel, or a starfish, or the side of an onion, but it seems to me that all of these fancy projections do more harm than good, simply because they do not show the world as it is.

The best map of the world is a globe. A transparent globe would be best of all. The best flat map for use in zoogeography is the orthographic, which is simply a picture or diagram of the world as it would look from an infinite distance. The orthographic projection is rather scorned by cartographers, for neither distances nor areas can be measured on it and it can show only half the world in one continuous view,

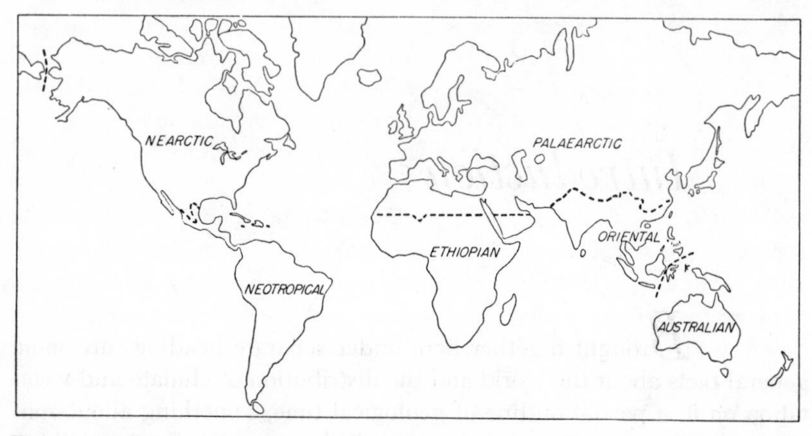

Fig. 1. Wallace's (1876) Mercator projection of the world. Simplified outline, with continental faunal regions as named and bounded by Wallace; *cf.* Frontispiece and Figure 46.

but no other flat map allows the eye so well to appreciate distances, areas, climatic zones, and broad geographic patterns in relation to the round earth. These things cannot be measured on an orthographic map but they can be seen and understood, and it is the understanding that matters. Construction of this projection in the equatorial plane is not difficult and is worth the time of any zoogeographer to learn. Instructions for it are given in textbooks of cartography. The objection that only half the world can be shown in one view can be overcome in some cases by superimposing two maps, of opposite sides of the world, to make what is virtually a picture of a transparent globe (see Frontispiece). The outline maps in this book are orthographic and have been constructed by myself. I have made them simple, almost diagrammatic, to show broad patterns rather than details, for the patterns are more important than details.

Land and oceans

Land covers only about 29 per cent of the earth's surface. It is not symmetrically arranged. There is more than twice as much land north of the equator as south of it. There is about twice as much in the eastern hemisphere (the Old World) as in the western one (the New World). And if the Pacific Ocean is taken as the center of a hemisphere and is compared with the opposite half of the world, the inequality is still greater: most of the land is in the opposite hemisphere, very little in the Pacific one. One possible explanation of this is that the moon came from the earth, from the Pacific side, and took most of the crust from that side with it (see p. 609).

In spite of the small amount of land and its uneven distribution, most of the land forms a single, more or less continuous system (see Frontispiece). All the main continents (except Antarctica) are connected or nearly so or are linked by archipelagos so that, as Wallace says (1876, Vol. 1, p. 37), it would probably be possible to travel over the whole system of continents without ever being out of sight of land. Moreover, very recently, perhaps only 10,000 years ago, the land was still more nearly continuous: Asia and North America were connected across the Bering Sea; Sumatra, Java, and Borneo were joined to Asia; Celebes was separated from Borneo by perhaps only 25 miles of water; and other water gaps between Asia and Australia were fewer and narrower than now. The main system of continents is arranged so that, north of the tropics, there are large areas of land which are nearly connected; within the tropics, large areas which are separated from each other; and south of the tropics, smaller areas which are very widely separated from each other.

Oceans are concerned in the distribution of land animals in several ways. They are barriers. They modify land climates, as described below. Ocean currents sometimes carry land animals to islands or perhaps even from one continent to another. And oceans and animals in the oceans tell things about the history of land. For example, the occurrence of closely related marine animals on opposite sides of Central America tells of a recent seaway interrupting the land; and fossil coastal-marine faunas tell of the distribution of climates and perhaps even of land itself in the past. Study of the ocean bottom tells or may tell things about the origin and history of the continents and about ancient land bridges. There are therefore many reasons why zoogeographers interested in land should know something about oceans too. Fortunately there are good, recent reference books on oceans, including Sverdrup, Johnson, and Fleming's (1942) *The Oceans,* Ekman's (1953—translated from the 1935 edition) *Zoogeography of the Sea,* and Kuenen's (1950) *Marine Geology.* I have found the last especially useful.

Climate

Climate is principally a matter of temperature and rainfall.

The temperature of the earth's surface is determined almost entirely by the amount of heat received from the sun, and the distribution of temperature over the earth is determined primarily by the earth's

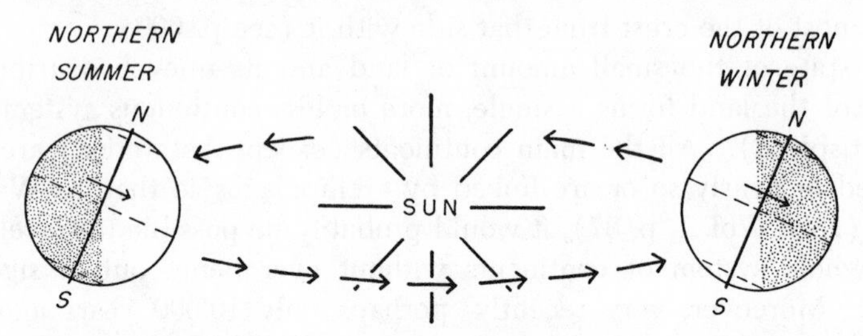

Fig. 2. The earth's motions in relation to the sun, to show how inclination of the axis of rotation (N–S) causes alternation of summer and winter in the higher latitudes. Diagrammatic: neither scale nor perspective is true.

shape and motions (Fig. 2). The greatest amount of heat falls on the tropics. The tropics are by definition the zone on which the sun's rays fall perpendicularly. The limits of the zone are therefore the northernmost and southernmost latitudes at which the sun is directly overhead, at the extremes of the earth's annual cycle. These latitudes are 23° 27′ north and south of the equator. In general, the tropics are hot throughout the year: maximum temperatures, at least in the wet tropics, are lower than those reached on hot summer days in many parts of the north-temperate zone, but the tropical heat is maintained within a few degrees at all seasons and by night and day. North and south of the tropics the heat received from the sun diminishes approximately with the curve of the earth, and becomes increasingly seasonal. In the temperate zones annual cycles of summer and winter appear, and the winters increase in severity with distance from the tropics; and still farther around the curve of the earth, across the Arctic and Antarctic Circles (which are about 23° 30′ from the poles), there are periods in summer when the sun never sets, and in winter when it never rises. The periods increase in length toward the poles, becoming six months of day and six of night at the poles themselves.

Alternation of seasons and severity of winters in the north may limit animal distributions more than mean temperature does.

On mountains, temperature falls with altitude, not because less heat is received from the sun (the sun's rays burn intensely at high altitudes) but because the air becomes thinner and therefore colder. The decrease of temperature with altitude can be compared with the decrease northward (Fig. 3). If conditions of snow and ice at 16,000 feet on mountains near the equator (*e.g.*, on the Carstensz Mountains of New Guinea—Umbgrove 1949, Pl. IX) are comparable to those

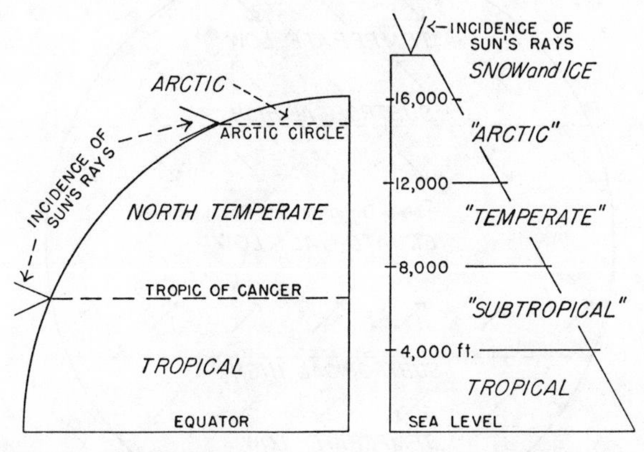

Fig. 3. Comparison of decrease of temperature northward on the earth and upward on a tropical mountain.

at sea level at 80° North, then it might be said that 200 feet of altitude has the same effect on climate as 1° of latiude northward. There is some truth in this comparison, but it is a partial truth and a misleading one. The effect of moving 1° north would be different at different places between the equator and the pole. The effect would be very small in the tropics and would gradually increase across the temperate zone, as the surface of the earth curved more and more away from the sun. But this is a minor criticism. The main criticism is that the decrease of temperature upward and the decrease northward are inherently different and have very different effects on life. Biogeographers know this but have not sufficiently emphasized it. They customarily call the successive altitudinal zones on tropical mountains "subtropical," "temperate," and "arctic." This invites misunderstanding. It is one of the misconceptions that, I think, might have been avoided if zoogeographers had used proper maps, which showed climate in relation to the round earth, instead of the queer distortions they do use. Altitudinal climates in the tropics are not the same as the northern climates for which they have been named. Climate in the north-temperate zone is cool and very seasonal; climate at high altitudes in the tropics is cool but nearly uniform throughout the year. In the north, a mean temperature of 50° F. may include extremes of from 100° to zero. High on a tropical mountain the mean may be the same, but the temperature may never fall to freezing.

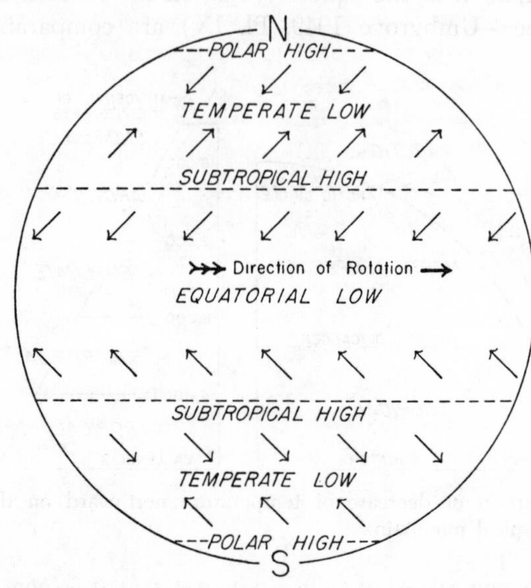

Fig. 4. Diagram of pressure and wind zones of one hemisphere. *Cf* Figure 8B.

That is one of the reasons why palms that thrive in the "temperate" zone on tropical mountains cannot exist in most parts of the north-temperate zone.

Rainfall, too, tends to occur in zones. Rain depends on wind, which picks up water vapor from oceans or other wet surfaces and carries it to where it condenses and falls. The basic zonation of rainfall is therefore determined by the earth's pattern of atmospheric pressure and wind (Fig. 4). There are zones of high pressure just outside the tropics in both the northern and southern hemispheres; winds blow away from the high-pressure zones, taking water with them; and that leaves deserts. Most of the great deserts of the world, in North Africa, southwestern Asia, and southwestern North America, and in Southwest Africa, Australia, and southwestern South America, lie in the high-pressure zones near or across the edges of the tropics. The deserts of east-central Asia (the Gobi, etc.) are exceptions; they owe their dryness to (among other things) distance from the sea. Winds (trade winds) tend to blow into the tropics from the subtropical high-pressure zones, bringing water with them; and that makes tropical rain and rain forest. And winds tend to blow north and south from the high-pressure zones into cool temperate regions, meeting other more variable winds and bringing variable rain.

This basic zonation of rainfall is modified, distorted, and interrupted in many ways. The size and form of land masses affect rainfall locally. On any given piece of land, rain is usually heaviest near the sea, on the windward side, and on mountains. Where the land is irregular and winds come mostly from one direction, rainfall may vary greatly in short distances. Inland from Santa Marta, on the north coast of South America, I could walk in one day from semi-desert, where acacias and giant cactus were the conspicuous plants, through intermediate scrub, to lowland rain forest, and then up the slope of the Sierra Nevada through deciduous forest (where rapid runoff robs the trees of part of the rain that falls) to permanently green cloud forest. Such abrupt differences of vegetation, reflecting local differences of rainfall, are common in the tropics. In many places in the tropics rainfall is seasonal, and the annual dry seasons may be so pronounced that forests lose their leaves, most insects disappear, and some birds migrate to feeding grounds hundreds of miles away.

The effect of oceans on land climates involves both temperature and rainfall. In general, oceans moderate extremes of temperature on nearby land. In special cases ocean currents carry warm (or cool) climates far beyond expected limits. For example the Gulf Stream carries warm water from the Gulf of Mexico diagonally across the Atlantic to the British Isles and Europe, warming the land and allowing given plants and animals to extend much farther north than they do in eastern North America. I have been north in Quebec as far as a road goes, beyond the northernmost farms, in thin spruce woods impoverished by cold, but the point I reached was barely level with the *southern* tip of England. The ocean and the Gulf Stream make the difference. For an opposite example, the Humboldt or Peruvian Current carries cold water from the south far up the west coast of South America, and has enabled a penguin to reach the Galapagos Islands on the equator. Oceans make rainfall, too. The water they give up by evaporation falls as rain wherever winds take it, especially where the winds first come against land and mountains.

All this is much too simple. Climate is a very complex thing. It is treated in more detail in such books as Kendrew's (1949) *Climatology* and Geiger's (1950) *The Climate Near the Ground,* which are themselves just short texts. Kendrew's (1937) *The Climates of the Continents* treats climate more geographically. Tannehill's (1952*a*) *Weather around the World* is a still shorter but very readable book.

The factors that make climate a partly zonal, partly local, very complex thing have existed ever since the surface of the earth has consisted of land and water. The earth has always been a revolving sphere with an equator which receives much heat from the sun and poles which receive little, so climates must always have been somewhat zonal, although the zonal climates may sometimes have been less differentiated than now. Moreover, the earth's axis has probably always been inclined so that, as the earth passes around the sun, northern and southern lands have always had alternately warm and cold seasons which increase the effect of zonation of climate on the distribution of life. The details of the relation of land to water and wind must have differed very much at different times in the past, as different pieces of land have changed their shapes and heights, but they must always have made a diversity of local climates. Every existing continent is partly wet and partly dry in a complicated pattern, and every continent has probably always been wet in some parts and dry in others. Geologists sometimes speak of uniform, world-wide climates, but they should not be taken literally. Neither temperature nor rainfall can really have been uniform over the world.

Vegetation

Vegetation, the plant cover of the world, is distributed primarily according to climate, especially temperature and rainfall, modified by the nature and history of land and by other things. The distribution of the most important vegetations of the world is shown diagrammatically in Figure 5. The distribution is zonal, but the vegetation zones are modified, distorted, and interrupted much as the temperature and rainfall zones are. In the tropics, the most obvious vegetation is rain forest, but it is not uniformly distributed. The largest areas of it are in Amazonian South America, the East Indies, and West Africa, but innumerable smaller strips and patches of it are very widely scattered in the tropics. Full-scale rain forest grades into other, poorer types of lowland forest, including tropical deciduous forest; and successive types of lower, wet, evergreen (but usually not coniferous) forest occur at increasing altitudes on wet tropical mountains, until (often at something like 12,000 feet altitude) the forest gives way to grass, and then (often at something like 16,000 feet) the grass gives way to rocks

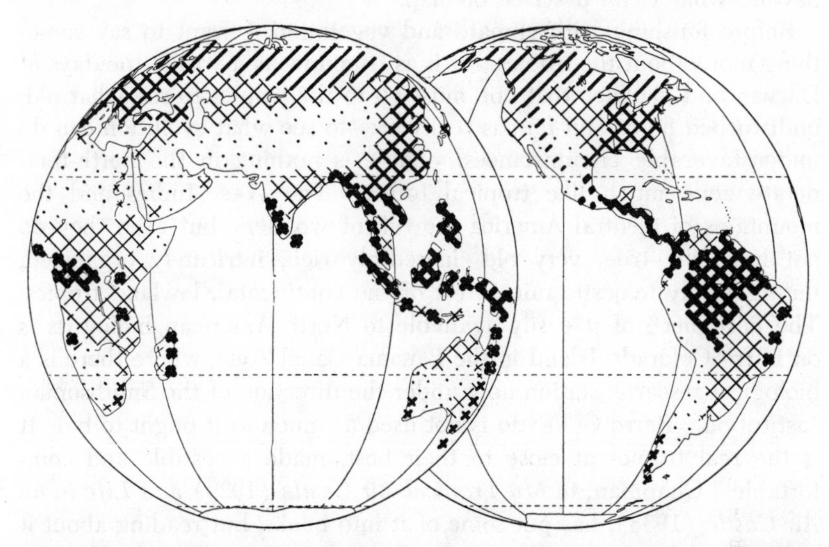

Fig. 5. Distribution of main vegetations of the world, extremely oversimplified; to emphasize features most important to animals. Density of vegetation is roughly proportional to density of patterns on the map. Heavy crosshatching indicates tropical rain forest; medium crosshatching, heavy temperate forest; diagonal hatching, northern coniferous forest; light crosshatching, light woodlands, grasslands; stippling, subdesert and desert. Sources consulted (not cited in other reference lists) include E. Milne-Redhead 1954, *Proc. Linnean Soc.* (*London*), **165,** p. 27, Figure 2 (vegetation map of Africa); anon. 1950, *The Australian Environment,* 2nd ed., Melbourne, Commonwealth Scientific and Industrial Research Organization, pp. 88–89, Figure 25 (vegetation map of Australia); anon. 1949, *Proc. National Inst. Sci. India,* **15,** p. 358 (map of climate of India in relation to rainfall distribution); Verdoorn *et al.* 1945, *Plants and Plant Science in Latin America,* Waltham, Mass., Chronica Botanica Co., endpaper (vegetation map of South and Central America).

and snow. There are also large areas of scrub and savanna in many places in the tropics, including great areas of thorn scrub and seasonally dry grassland in Africa. Sparse desert vegetation covers large areas near the edges of the tropics. Farther north, there is first a zone in which deciduous forest and grassland tend to cover alternate, irregular blocks around the mid-latitudes; then a more nearly uniform and continuous zone of northern coniferous forest; and finally a zone of arctic tundra. Southward, the south-temperate parts of Africa, Australia, and South America have complexly mixed vegetations. The complexity of the whole pattern over the world is, of course, far beyond what I can describe or map.

Before finishing with climate and vegetation, I want to say something more about the tropics. It is as true now as it was in the days of Darwin and Wallace that one must go to the tropics to see what old, undisturbed floras and faunas really are—to see what evolution can do under favorable circumstances. There is nothing in the north-temperate zone much like tropical forest. The West Indies and the mountains of Central America are full of wonders, but even they do not have the true, very old, incredibly rich, intricately organized, unforgettably majestic rain forest of the continental, lowland tropics. The best piece of it easily available to North American biologists is on Barro Colorado Island in the Panama Canal Zone, where there is a biological research station now under the direction of the Smithsonian Institution. Barro Colorado is not used as much as it ought to be. It is the real tropics at close to their best, made accessible and comfortable. Chapman, in *My Tropical Air Castle* (1929) and *Life in an Air Castle* (1938), has put some of it into books, but reading about it is not enough. It must be seen and felt. Any young naturalist who thinks he can understand the world and living things and evolution without experiencing the tropics is, I think, deceiving himself, to his own great injury.

Geological time

The earth has existed for two or three or four billion years or more. Simple plants (algae) probably appeared on it more than a billion years ago. Simple, boneless animals probably appeared considerably more than half a billion years ago. The oldest, jawless, fish-like vertebrates appeared perhaps 400 million years ago, but it was still many million years before the first amphibians appeared on land, and many million more before vertebrates were numerous enough on land and left enough known fossils in enough parts of the world to be geographically significant. I do not know how far back into the past zoogeographers may eventually hope to see the pattern and significance of animal distributions, but here I shall not try to see beyond about 200 million years, and it is hard enough to see anything clearly at that distance. This, therefore, is as far back as I shall go with the geological timetable: back through the Pleistocene, Tertiary, and

Mesozoic, to the Permian (Fig. 6). This timetable needs no detailed explanation here. Zeuner (1950) describes and explains it. The dating of the table is comparatively unimportant. It does not matter very much whether the Permian was really 200 million years ago or much more or much less. What does matter is the sequence. This sequence of eras, periods, and epochs of the past is the scale on which all the geological and biological events of the time concerned are placed in relation to each other.

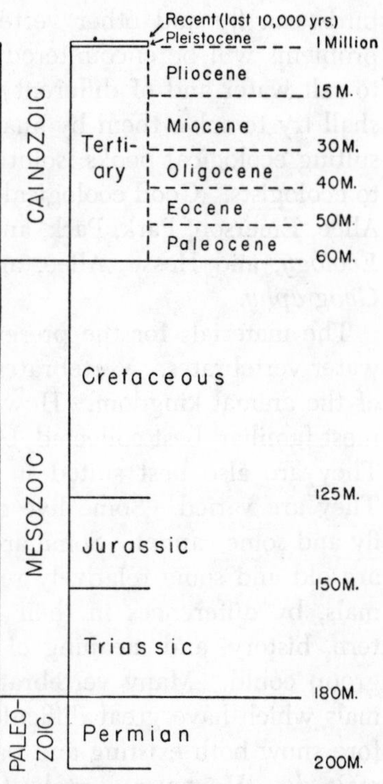

Fig. 6. Diagram of the last 200 million years of geological time.

Zoogeography: limits, levels, and materials

Zoogeography is concerned with the distribution of all animals over the whole world. Something like 1,000,000 different species of animals are known; for comparison, there are only about 200,000 words in the text (without the family lists, etc.) of this book. And the world is enormous and complex. Obviously a zoogeographer must set limits to his work. I have already said that this book is limited to the land and fresh water and to vertebrates. It is also limited to, or rather mainly concerned with, broad patterns rather than local ones.

A distinction between geographical distribution and local distribution was made by Wallace. Actually it is convenient to distinguish three levels: geographical distribution in the broadest sense, over the whole world; regional distribution, in selected segments of the world; and local distribution in Wallace's sense, which includes species geography and ecology. Of course I shall be concerned to some extent with all these levels, even with the lowest one (the presentation of detailed facts in Chapters 2 to 6), but the aim is to find and understand the broadest patterns. Species geography—the geographical distribution of species in relation to each other and to ecology and evolution—is not considered here. The evolutionary aspects of it are treated by Mayr (1942). Ecology is not formally considered here simply because it is not zoogeography, as I have said. The only ecology necessary to begin with here is a set of obvious generalities: that animals need space and food to live; that land animals have trouble crossing water, and water animals crossing land; that bats and most birds can fly and other vertebrates cannot; etc. Special ecological problems will be encountered (*e.g.*, the relation of fresh-water fishes to salt water and of different vertebrates to northern climates), but I shall try to solve them by analyzing distributions rather than by consulting ecological books; solutions obtained in this way may be useful to ecologists. Good ecological reference books do exist. They include Allee, Emerson, Park, Park, and Schmidt (1949), *Principles of Animal Ecology,* and Hesse, Allee, and Schmidt (1951), *Ecological Animal Geography.*

The materials for the present work are, of course, land and fresh-water vertebrates. Vertebrates comprise only 3 per cent or 4 per cent of the animal kingdom. However, they are the best-known animals: most familiar, best collected, best classified, with the best fossil record. They are also best suited in other ways for zoogeographical work. They are varied. Some live on land and some in water. Some can fly and some cannot. Some are cold- and some warm-blooded. Some are old and some relatively recent in their origins. Such varied animals, by differences in their distributions, tell more about the pattern, history, and meaning of animal distribution than a less varied group could. Many vertebrates are strictly land or fresh-water animals which have great difficulty crossing salt water and which therefore show both existing and past land connections as well as any animals do. However, vertebrates fail in some local situations. They apparently do not persist indefinitely in small, isolated areas and therefore do not show the histories of some old islands (*e.g.*, New Caledonia) and some other small, special areas (*e.g.*, the cold southern tip of South America) as well as some invertebrates do.

Everybody knows something about vertebrates. The best formal introduction to them for zoogeographical purposes is probably Romer's (1945) *Vertebrate Paleontology,* which describes vertebrates, classifies them, and traces their evolution and fossil record.

There are five main classes of vertebrates: fishes, amphibians, reptiles, birds, and mammals. (The fishes really form several classes, but most fresh-water fishes belong to one, Osteichthyes.) Each of the five main classes is treated in a separate chapter (Chapters 2 to 6). Each chapter begins with a short account of the classification, fossil record, and pertinent characteristics of the animals concerned and then describes and discusses their distribution under the following boldface headings:

Limits of distribution of . . .
Zonation of . . .
Radial distribution of . . .
Regional distribution of . . .
Transitions and barriers in distributon of . . .
Dominance and competition in relation to distribution of . . .
Summary: the pattern of distribution of . . .
History of dispersal of . . .

The general sequence of each chapter, as the headings suggest, is from description of existing distributions and of their relation to climate, barriers, etc.; to the effect of dominance and competition (and thus of evolution) on the distributions; to synthesis of main patterns; to the events and movements that have made the patterns. Special subjects are inserted under special headings, e.g., **What are fresh-water fishes?**, and **Geographical effects of bird migration.** Following the text of each of these five chapters is a formal classification and list of families, with each family's distribution—the families are the units of classification that best show how vertebrates are distributed over the world. These family lists contain the detailed facts on which the whole book rests. The lists include all existing families of land and fresh-water vertebrates. What fossils are included is stated at the beginning of each list.

For large and widely distributed families I have tried to give the following information:

Technical and English names of the family
Its region of occurrence in general terms
Its more precise limits, especially northward and southward, in transition areas, and on islands
Its main pattern of distribution within the limits
Its subfamilies (if any), numbered and discussed consecutively
Numbers of genera and species
Distributions of important or unusual genera and species (but unusual details are not emphasized)
Pertinent facts about the animals in question: their characteristics, habitats, habits, relationships, etc.
Pertinent fossil record

However, I have not always followed this outline exactly; and small families are treated much more briefly.

Distributions are very complex things, difficult to describe simply. Many details of the geographical occurrence even of vertebrates are not known, and known distributions are often confused by doubtful classifications or by doubts about whether certain animals in certain places are native or introduced by man. I have done my best to simplify complex distributions fairly and to be frank about the doubts. The difficulty of summarizing the distributions of even well-known small families and single species is suggested by some of the families near the beginning of the list of fresh-water fishes, *e.g.*, the Polyodontidae, Lepisosteidae, Amiidae, and Lepidosirenidae.

Dispersal

"Dispersal" means two things in zoogeography. The whole history of movement of a group of animals is the group's dispersal. In this sense, dispersal includes losses of ground as well as gains, for loss and gain are both involved in histories of movement. In a narrower sense, dispersal is movement of individual animals (eggs, young, or adults), especially their outward scattering into new places. There are two general cases of this to be considered: dispersal in continuous, favorable areas, and dispersal across barriers.

All vertebrates can disperse fast enough so that they would spread over the whole earth in a short time (most of them in a small fraction of a million years) if favorable habitats were continuous. Dispersal rates and distances are therefore, themselves, not very important in vertebrate distribution. Distributions are limited mainly by other things, especially by ecological and physical barriers.

The extent to which animals disperse across physical barriers has been bitterly argued about by zoogeographers. Often the method of argument has been to lay down a law to begin with, find a few cases that seem to support it, and then say that the law applies to all cases. This was done even by Wallace (1880, p. 235), who ruled that amphibians and terrestrial mammals never (or hardly ever?) reach oceanic islands across barriers of salt water, and that islands on which these animals occur must be continental. This was, I think, the worst mistake Wallace ever made, as a zoogeographer. But I do not want to start laying down the law myself. I want to find how animals really are distributed in relation to barriers, and then reach conclusions about them. Now, however, I want to consider briefly two agents of dispersal that may carry land animals across barriers: rafting and wind.

Rafting, zoogeographically, is the dispersal of land animals across water on floating objects. The objects may be more substantial than "raft" suggests. When a flooded river undercuts its banks, brings down whole hillsides, and carries tangled masses of trees out to sea, many land animals may be taken along. This happens most often in the wet tropics, where cloudbursts are common and where hillsides are very steep, not cut back by recent glaciation.

How more substantial rafts may be formed on fresh water is suggested incidentally by Brewster (1924, pp. 190–191) in describing a nesting site of American Bitterns on Lake Umbagog in Maine and New Hampshire. He says,

> . . we came upon a small floating island . . . circular in shape, about twenty yards across and densely covered with bushes . . . with here and there a muddy or grassy space. Near its center a few green and vigorous young larch trees rose to a height of fifteen or twenty feet. Interlacing roots of trees, shrubs, and grasses had combined with decaying vegetable matter, alluvial soil, and driftwood, to form a great, soggy raft, buoyant enough to rise and fall with the Lake, but so thin and flexible that big waves, rolling in under it, procured corresponding and very noticeable undulations of its surface. This was sufficiently elevated to appear comparatively dry until trodden on, when it gave beneath our steps, like that of a quaking bog, until the water came half-way to our knees. The island was then firmly anchored—probably by a few strong roots—to solid ground some five or six feet below, but we were told that not long before this it had come drifting across the flooded meadows from their western end. About ten years later it again broke loose from its moorings and journeyed still farther eastward, until stopped by a group of stubs.

A raft much like the one described by Brewster, 100 feet square with trees 30 feet high, evidently tied together by the roots of living plants, was seen in the Atlantic off the coast of North America in 1892, and is known to have drifted at least 1000 miles (Powers 1911). Rafts like this might be formed and floated into the sea anywhere in the world where shallow lakes or marshes drain into the sea without too rough a river passage. They do not require great rivers. And, because they are likely to be discharged during floods, they are likely to carry land animals trying to escape the floods. It seems to me that such rafts might carry almost any land animals almost any distances where ocean currents, winds, and climates are favorable.

Old arguments about rafting were mostly concerned with whether or not land animals could conceivably be rafted, but that was a rather futile question. The question should be, does the occurrence of ani-

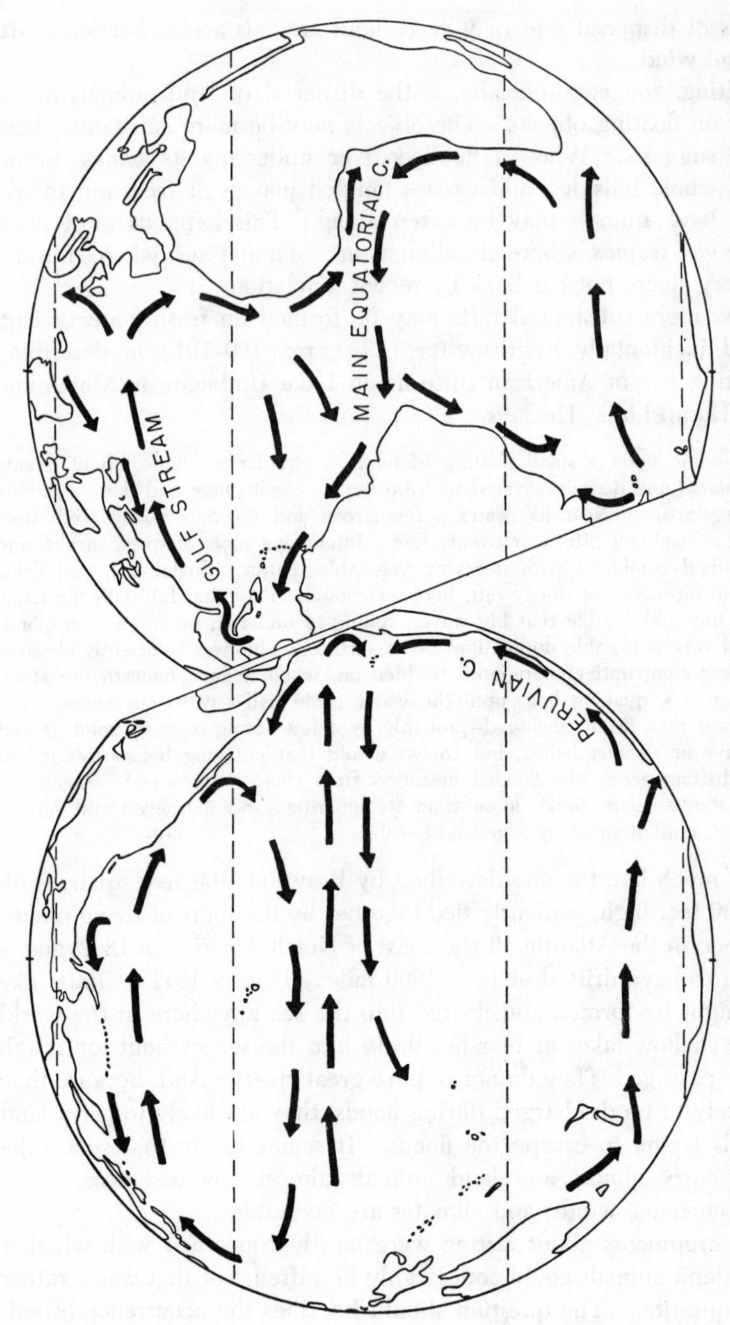

Fig. 7. Principal surface currents of the Pacific and Atlantic Oceans. Simplified diagram. *Cf.* Sverdrup *et al.* 1942, Map VII.

mals, especially on islands, indicate that rafting has occurred? If so, adequate rafts must have existed in the past, as they do now.

Direction of rafting presumably depends partly on ocean currents (Fig. 7). The Main Equatorial Current of the Atlantic is worth special consideration. It is an intercontinental current, which crosses the tropical Atlantic from West Africa to South America. Guppy (1917, p. 80), using data from the drift of marked bottles, estimated that only about twelve weeks would be needed for passage in this current from the Gulf of Guinea to Brazil. Much floating drift is carried by it, some coming from the Niger and Congo Rivers. Some of the drift is thrown up on the coast of Brazil and some (after a longer passage) on the West Indies. This current is important because, whether or not it has carried land animals, it may have carried the larvae of shallow-water marine invertebrates. Schuchert (1932) considered this possibility and decided against it, but his figures for distance and rate of drift seem to be wrong. He took the distance as 4000 miles (p. 903), but it may be much less than that; the shortest distance from Africa to South America is less than 2000 miles. If Guppy's estimate of time of crossing (twelve weeks) and Schuchert's estimate of time of larval stages (up to three months in some cases) are correct (I am not sure they are), the larvae of some marine invertebrates may be carried from Africa to South America by the Equatorial Current.

The other agent of dispersal to be considered now is wind.

Men instinctively underestimate the power of wind. Some zoologists today find it as hard to believe that winds and rising air currents can lift and carry organisms as people once found it to believe that heavier-than-air machines could fly. The difficulty is that a man's experience, upon which his instinctive judgment is based, is that of a large animal, exposed for the most part to moderate winds, blowing horizontally near sea level. The effect of stronger winds on smaller animals is far beyond anything a man can experience. The power of wind over small animals depends on the animals' ratio of body surface to weight. Surface varies as a square, weight as a cube. Decrease in size therefore decreases weight much more than surface. Very small animals have many times more surface for their weight than large ones do and are therefore much more at the mercy of wind. That is what a man cannot judge instinctively. Moreover, the force and carrying power of wind increase (roughly) with the square of the wind's velocity. It can be calculated that the effect of a wind of 100 miles an hour on an animal weighing 1 ounce (an adult house mouse weighs about 1 ounce) is something like 224 (14×16) times

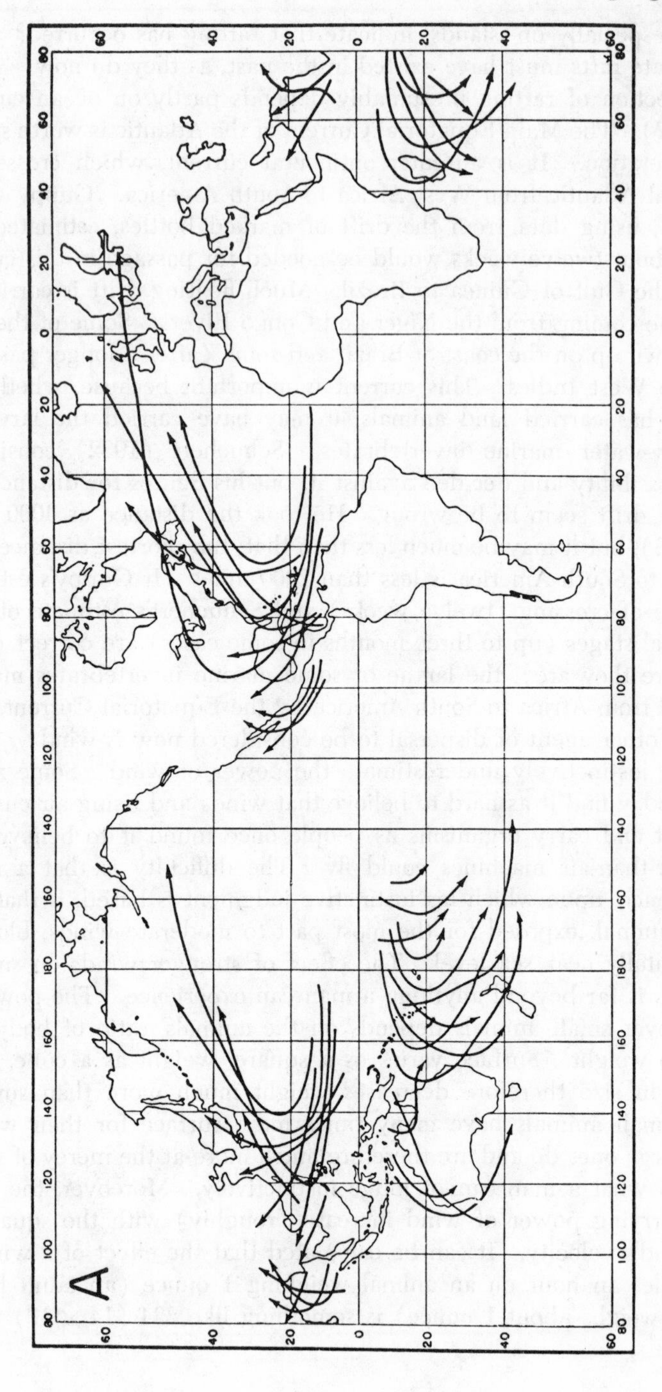
A

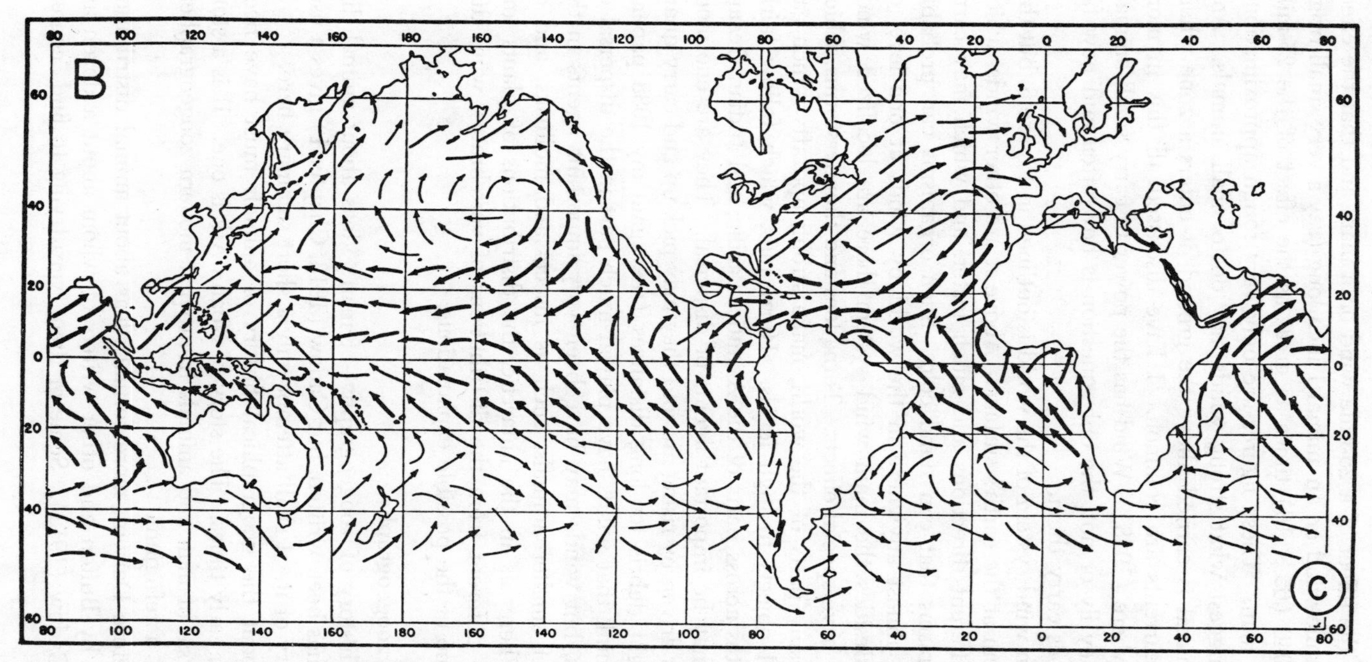

Fig. 8. (A) Typical tracks of tropical storms (hurricanes, etc.); (B) prevailing surface winds of the oceans in July. From Tannehill 1952*a*, *Weather Around the World*, Figures 7 and 24. Copyright Princeton University Press. Used by permission of Ivan Ray Tannehill and Princeton University Press.

as great as the effect of a 25-mile wind on a man; and that the effect of the 100-mile wind on an animal 1 inch long (say a very small frog) is about 1040 (65×16) times as great as the effect of the 25-mile wind on a man. These figures are only very rough approximations, but they suggest what strong winds may do to small animals. And once an animal is carried above the ground, it enters a zone where rising air currents are common. I have discussed all this in more detail elsewhere (1938). Wind has the power to carry small animals whether they fly or not; the only question is how often and how far it really does carry them.

The main wind system of the world is outlined in Figure 8, but this is the system at the earth's surface. Above it are other winds, which blow in different directions and much faster, and which may carry small organisms that can stand cold. Such organisms can probably be carried almost anywhere in the world by winds at one level or another. Besides the main wind system there are localized winds which may have special effects. Tornadoes and thunderstorms, which occur in many parts of the world, make strong updrafts which can carry small objects to high levels. Hurricanes, which follow fairly definite paths across parts of the Atlantic, Pacific, and Indian Oceans (Fig. 8), may be important agents of dispersal. They are enormous whirlwinds driven by rising air, and they may pick up and carry quantities of light debris for long distances (see again my 1938 paper). I do not think that winds have played much part in the dispersal of vertebrates, but winds may have been very important agents in the dispersal of smaller animals, such as insects and mollusks, and of plants. Where, as in the antarctic, the distributions of plants and insects are different from the distributions of vertebrates, wind dispersal is one of the possible explanations.

History of zoogeography

No full history of zoogeography—of man's ideas about animal distribution—has been written, but Gadow (1913, Chapter I) gives a useful summary of it. I shall attempt only a short summary here.

Ideas about the geographical history of animals must have been common in early times. The story of Noah's Ark is one. It is a story of redispersal of man and animals from what modern zoogeographers would call a refugium.

Even before Darwin, more formal ideas about animal distribution were held by Buffon and other writers of whom about a dozen are listed by Gadow (1913). Some of them were trying to find "centers of creation." This was Sclater's purpose when, in 1857 (published 1858), two years before publication of *The Origin of Species,* he divided the world into six regions according to the distribution of birds. This was the paper in which Sclater said that, if he could show the existence of separate centers of creation in different parts of the world, he could avoid "the awkward necessity of supposing the introduction of the red man into America by Bering's Straits, and of colonizing Polynesia by stray pairs of Malays floating over the water like cocoa-nuts, and all similar hypotheses." This gives a hint of the kind of controversy that was, I suppose, raging in those days. Nevertheless, Sclater analyzed the distribution of birds so well that his regions still stand with little change.

To Darwin and Wallace the pattern of distribution of animals in undisturbed parts of the world was one of the proofs of evolution; and the acceptance of the idea of evolution made modern zoogeography possible. Darwin kept to the evolutionary point of view. His two chapters on geographical distribution in *On the Origin of Species* are still one of the best discussions of the evolutionary principles of the subject. Wallace was more concerned with existing distributions. In 1876 he produced the first and only full-scale effort to show in detail how animals are distributed now. These two men, each great in his way, represent two points of view which persisted among later zoogeographers.

After Wallace, several generations of conventional zoogeographers worked to accumulate further details of the existing distribution of animals and to divide the world into faunal regions, almost to the exclusion of everything else. To some of them, "Wallace's Line," which is just one supposed boundary between two faunal regions, was the most important thing in zoogeography. I do not mean to belittle this work, which led to real advance toward understanding at that time. It led to a steady increase of factual knowledge, which was necessary before any sort of zoogeography could advance. Something more of the history of this work and of ideas about faunal regions is given in Chapter 7.

At the same time another group of zoogeographers may be said to have followed Darwin, but too often without his understanding. They were trying to work out the geographical histories of animals, but to do it they moved continents and built long land bridges with little regard for evidence or probability. I do not mean to belittle moving continents or making land bridges, if the evidence calls for it, but it ought to be done with reason. Darwin himself was annoyed by persons who, even in his time, made land bridges "as easily as a cook makes pancakes." Gadow, who was a good zoologist and whose little book, *The Wanderings of Animals* (1913), is excellent in some ways, made this error and moved continents beyond reason or necessity; and Scharff, Arldt, and others were worse. The last two authors did assemble books full of details, but the books were compiled uncritically and are full of errors and guesses passing as facts. These books are still sometimes cited as sources of "facts" by zoogeographers who do not know that the books were uncritical and unreliable to begin with. Matthew, in *Climate and Evolution* (1915, reprinted 1939), did much to counteract the more irresponsible historical zoogeographers. Regardless of the correctness of his conclusions, he set a good example in presenting evidence and trying to reconstruct the history of animal dispersal and the history of the continents by the evidence. A synthesis of this sort—putting Wallace and Darwin together, so to speak, and adding geology and other things in reasonable proportions—ought to be the purpose of modern zoogeographers. Of some, it is.

Recently there have been two not entirely justified tendencies among zoogeographers. One is to make zoogeography a subdivision of ecology. To bring ecology into zoogeography is a very good thing, but zoogeography is not ecology. The other tendency is to glorify historical zoogeography—the past evolutions and movements of animals—and almost to despise the study of present "static" distributions. This seems to me to be a great mistake. It is true that present distributions cannot be understood without reference to the past, but it is equally true that the past cannot be understood without reference to the present. It is present patterns which show how animal distributions really are related to climates and barriers, etc. Unless we know this, how can we understand what past distributions tell of past climates and barriers? It seems to me that, in zoogeography, the present and past are equally important and dependent on each other. I shall have something more to say about this in Chapter 7.

The two most recent books on zoogeography in English are *Plant and Animal Geography* (Newbigin 1936) and *Zoogeography of the Land and Inland Waters* (Beaufort 1951). They are both too short to cover the subject adequately and are mainly compilations of selected conventional ideas rather than reassessments of the subject.

This short history is more critical than I intended it to be. There have been many good zoogeographers, including some of those mentioned above (Beaufort and others). Most of them have been concerned with special groups of animals or special parts of the world or special situations, but their combined contributions cover the whole world and every important aspect of animal distribution. Some of them (but only a few of all zoogeographers) are mentioned elsewhere in this book.

Zoogeographical definitions

This is not a glossary of unfamiliar words. I have tried to avoid unfamiliar words, and those that I have had to use can be found in dictionaries or are defined in the text. This is (mostly) a list of familiar words that might be misunderstood. Readers might not know that these words need to be defined and might not look them up in a separate glossary. That is why I have put them here, where readers will, I hope, have to see them.

Autochthonous. Native in the sense of having originated (evolved) in the place in question.

Barrier. (1) A physical obstruction, such as a mountain range, which individual animals cannot or do not ordinarily cross. (2) A climatic or ecological factor, such as temperature or rainfall, which varies gradually from place to place, and which does not obstruct movement of individuals, but which does prevent successful establishment and spreading of species.

Competition. The struggle for places in the world; the struggle for existence. In this sense competition is any interaction among animals, no matter how complex or indirect it may be, that is or may be disadvantageous to any of them. Some zoologists use "competition" in a restricted way and they may object to my use of it, but I can find no other satisfactory word, and there is precedent (beginning with Darwin) for my use of it.

Complementary. Occurring in different areas which together make a whole area. The term is applied here to dominant, ecologically equivalent, presumably competing groups of animals that occupy complementary areas. This geographical relationship I have called *complementarity.* The latter word is not in dictionaries, but I can find no other word to express the meaning. It is, I think, the only word that I have invented here, except one or two proper names.

Derivative. Derived or descended from something different; comparatively late in evolutionary sequence. The opposite of this word is *primitive.*

Differentiation. (1) The process of becoming different. (2) The difference between different animals or different faunas caused by evolution and other processes. This is the ordinary English meaning of the word applied to zoogeography. The word has other, special, meanings in other branches of science.

Dispersal. (1) The total geographical movements of groups of animals. (2) The movements of individuals (eggs, young, or adults), especially their outward scattering (see p. 14).

Dominant. Conspicuously successful; successful in competition (see p. 552). This is the ordinary English meaning of the word. It has a special, different meaning in genetics.

Endemic. Confined to; occurring nowhere except in the place in question. This word has other meanings and should be used cautiously. I would not use it at all except that a very useful noun is formed from it, *endemism,* the existence of endemic forms. Synonyms: *exclusive; peculiar; precinctive.*

Feral. Secondarily wild, after domestication; implies not native.

Introduced. Brought by man, either intentionally or accidentally; not native.

Migration. (1) Non-recurrent directional movement (p. 549). (2) Recurrent seasonal movement, as of birds (p. 243).

Native. Occurring naturally; not introduced by man. This is the general term which includes the following special ones: *endemic, autochthonous, relict.* Synonym: *indigenous.*

Primitive. First; original; early in a given evolutionary sequence. The opposite of primitive is *derivative.*

Radiation. (1) Evolution of many different forms, with many different adaptations, from one ancestor. (2) Spreading in many directions from one center.

Range. The particular area occupied by a group of animals. Ranges often have to be shown on maps as if they had fixed, continuous boundaries, or as if they were solid blocks. Zoogeographers sometimes make the mistake of thinking of them that way, as if ranges were blocks which could be taken from their places and moved about and fitted together like building blocks. They are not. They are areas inhabited by living populations, and they are as complex, unstable, and dependent on their particular environments as living populations are.

Relict. A survivor; a plant or animal that (1) persists locally after extinction of it or its relatives elsewhere (a geographical relict) or (2) continues to exist after extinction of most other members of its group (an evolutionary or phylogenetic relict). The word is also used as an adjective.

Specialized. Adapted to a special, limited way of life. The opposite of this word is *unspecialized,* not *primitive. Unspecialized* and *specialized* concern animals' relation to the environment. *Primitive* and *derivative* concern animals' phylogenetic relation to each other. An unspecialized animal can be derivative, and a specialized one can be primitive. To use these words properly is not an affectation. It is necessary to clear thinking.

Unspecialized. See *specialized.*

Working principles

Animal and plant geographers from Wallace (1876) to Cain (1944) have tried to formulate the principles of their subject, but it seems to me that they have not been very successful, partly because they have usually mixed two kinds of principles. There obviously are two kinds. Rules of procedure and initial premises are working principles and should be understood before work is begun. General facts established by the work are derived principles and should not be finally formulated until the work is finished. For example, it is or should be a working principle of zoogeography that animals are constantly tending both to multiply and spread and to die and lose ground. Failure to take this as a working principle has invalidated the conclusions of some zoogeographers. On the other hand, the effect of barriers in limiting animal distributions is something to be found out by zoogeographical work and stated at its end as a derived principle. To assume at the beginning that such things as mountains and deserts are necessarily important barriers exaggerates their importance and prevents a real understanding of them.

At the risk of being pedantic I am going to suggest, number, and discuss what seem to me to be the important working principles of zoogeography.

1. *The first suggested working principle of zoogeography is to formulate working principles before work is begun.* Zoogeography is a complicated subject, set with pitfalls, and it should be approached with forethought. Moreover, a zoogeographer's readers ought to know what his working principles are in order to evaluate his conclusions.

2. *The second suggested working principle of zoogeography is to work with facts rather than opinions, so far as possible, and to understand the nature of the "facts" worked with.* Myers (1938, p. 340) remarks that "Generally speaking, zoogeographers, like Gaul, are divided into three parts—those who build bridges, those who do not,

and the proponents of continental drift." This is true, too true; and it might be added that each part defends its opinions stubbornly. There ought to be a fourth part, and I should like to belong to it, of zoogeographers who take facts and analyze them carefully and, instead of forming and defending fixed opinions, try to state alternative possibilities fairly.

Other peoples' opinions about doubtful matters are not a very good basis for zoogeography. When a zoogeographer falls back on the weight of opinion, he is likely to be short of facts. Also he may be wrong. For example, in 1935 Schuchert counted the opinions of fourteen qualified specialists and found the majority of them, thirteen to one, in favor of a land bridge origin of the plants and animals of the West Indies. But now most students of West Indian animals have turned against land bridges. Regardless of the truth in this matter, which is discussed in Chapter 8, a majority of zoogeographers either in 1935 or later has been very wrong. One must, of course, form opinions, but let them be based on facts and not on other persons' opinions, and let them not be defended too obstinately.

Notice that the second suggested working principle of zoogeography is to work with facts rather than opinions, *so far as possible.* The raw material of zoogeography is factual, a matter of what animals occur where. However, the "facts" may be mixed with errors. For example, in 1905 Ogilby described a snapping turtle from New Guinea, making a new genus for it, and for more than forty years its presence there was accepted as fact by zoogeographers. But no more specimens of it were found, and in 1947 Loveridge and Shreve showed that the turtle did not come from New Guinea and was not a new genus but was based on a mislabeled North American specimen. So there is no snapping turtle in New Guinea after all. It is easy to see how the mistake might have been made. Ogilby did receive other turtles from New Guinea. Perhaps a shipment came in one evening, and Ogilby unpacked the specimens and spread them out on a table unlabeled, for a preliminary gloat, moving that old stuffed American turtle onto the floor to get it out of the way. And perhaps the janitor found it that night, thought, "That careless curator again!" and put it back on the table with the New Guinea lot, with which it was labeled in the morning. These details are imaginary, but because of mistakes like this there are thousands of wrongly labeled specimens of different kinds of animals among the millions of correctly labeled ones in museums. Most of the errors do not cause so much trouble. They are too obvious or too unimportant. Most of the really important occurrences of animals in unexpected places have been confirmed by repeated collecting.

A much greater difficulty is that we do not yet know enough about relationships in some groups of animals to separate fact from opinion. The following fragments from the history of opinion about, for example, the New Zealand frog (*Leiopelma,* actually two or three closely related species all confined to New Zealand) should make any responsible zoogeographer think. That any frog should be on New Zealand is astounding. Before it was formally named and described, Darwin (1859; 1950 reprint, p. 333) heard of its presence there and doubted it; but the frog is on New Zealand. It was formally described by Fitzinger in 1861 as most closely related to a Chilean frog; but it is not really related. A more cautious taxonomist would have talked of similarity rather than relationship. Taxonomists (including myself) too often say "related" when they should say "similar to." In 1876 Wallace, probably following Steindachner, put the New Zealand frog in the Bomburatoridae, a family he supposed occurred otherwise only in Europe and temperate South America; but this family proved to be unnatural—the different frogs placed in it were really not related to each other. In 1911 Wallace, probably following an early arrangement of Boulenger, put the New Zealand frog in the Discoglossidae, a family then known otherwise only in Europe and China; but this was not a true relationship either. In 1910 Boulenger put the frog in the Leptodactylidae, as did Metcalf in 1923. This seemed to be a good arrangement geographically, for leptodactylids are abundant in Australia and South America; but the frog is not a leptodactylid. Dunn, in 1925, corrected Metcalf, saying that the New Zealand frog was "absolutely known" to be a discoglossid; but, as I have said, it is not one. Dunn was right in so placing it in the imperfect classification of that time, but his language was a little stronger than the final facts justified. Dunn should not be especially blamed for this. Zoogeographers (including myself) often make overemphatic statements about things that are not quite settled. Finally Noble, in 1931, after a painstaking study of frog anatomy, decided that *Leiopelma* did not belong in any of the families in which it had been placed but that it was almost uniquely primitive, one of the two most primitive existing frogs. The other, *Ascaphus,* not discovered until 1889, occurs on almost the other side of the world, in northwestern North America. This distribution is perhaps more remarkable than any of the supposed ones in which the New Zealand frog had been concerned; but it seems to be true.

One may ask how, after such a history, present opinion about the New Zealand frog can be taken as fact. The answer is that, up to Noble's time, increasing knowledge of frog anatomy and improvements in the general system of classification of frogs continually forced changes in the position of *Leiopelma*. But since Noble's work, new facts discovered by study of *Leiopelma, Ascaphus,* and frogs in general have only strengthened his conclusions. Perhaps there is no such thing as final fact in a case like this, but the many details now known make it exceedingly probable that *Leiopelma* and *Ascaphus,* although not directly related, are the only survivors of an ancient group of frogs, and they are usually put together in a family of their own.

Unfortunately the relationships of many other animals, including many other frogs, many snakes, some birds and mammals (especially some songbirds and rodents), and very many invertebrates, are still as unsettled as those of *Leiopelma* were in Wallace's time. There is not much that zoogeographers can do about this until taxonomists improve the poor classifications. In the meantime zoogeographers should work as much as possible with the best-known groups of animals, remember that classifications are imperfect, and proceed cautiously. Taxonomic experience is useful, almost essential, to a zoogeographer. An experienced taxonomist knows what classifications are and what other taxonomists mean when they talk about relationships. Also (but this is less important) a taxonomist can sometimes suspect unsound classification even of animals with which he is not familiar, much as an engineer might suspect an unsound bridge even if he has not built it himself.

But all this is the dark side of the matter. The bright side is that the facts now known of the geographical occurrence and classification at least of vertebrates are good enough. Even eighty years ago, when known facts were fewer and classifications were poorer, Wallace was able to reach sound general conclusions about the geographical distribution of animals. Today we at least have the factual materials to do much better than he.

3. *The third suggested working principle of zoogeography is to define and limit both the work to be done and the factual material to be worked with.* The necessity of setting limits to single projects within the vast field of zoogeography has already been discussed. This book, for example, is limited in its purpose to the consideration of broad patterns of distribution, and in its material to land and freshwater vertebrates, especially the families of them.

4. *The fourth suggested working principle of zoogeography is to present the selected material fully and fairly.* This is likely to require thought and experiment as well as good intentions. Tables, diagrams, and text have to be adapted to different cases. Outline maps can and should be standardized for clarity and ease of comparison. Everything should be designed to present the facts as fully, clearly, and fairly as possible.

5. *The fifth suggested working principle of zoogeography is to remember that animals are living things, which are constantly evolving and multiplying in some places, spreading into other places, and dying out in others, and thus forming new geographical patterns.* All zoogeographers know this, but not all think about it.

All animal distributions are products of complex equilibriums. At the bottom is the equilibrium between rate of reproduction and rate of death of individuals. In stable populations, reproduction and death are balanced, opposite, equal forces. No one would say that, in this balance, reproduction alone is important and death is something which can be ignored. Change in either rate of reproduction or rate of death can change the balance and move it equally well in either direction, toward increase or decrease in size of populations. Increase and decrease in size of populations are connected with gain and loss of ground by species. And gain and loss of ground by many species become spreading and receding of genera, families, orders, and whole faunas. Even at the highest level an equilibrium exists which, although it is very complex, depends ultimately on the balance between reproduction and death in populations, and the whole complex system can be moved equally well in either direction, toward either the spread or the recession of even major groups and faunas (see Chapter 9).

In stating "principles," zoogeographers almost always stress the power of animals to multiply and spread, but often minimize and sometimes forget their liability to die and lose ground. This is a serious mistake. Darwin (1859; 1950 reprint, pp. 95–97) is one of the very few zoogeographers who have thought this matter out thoroughly—I mention him not for the weight of his authority but because what he says is so obviously true. He thought that extinction must be a very common thing. He thought that evolution of new forms and extinction of old ones must tend to balance each other, that there is a limit to the number of species of animals that can exist, and that as new species evolve and multiply old ones *must* become rare and disappear. His discussion is more detailed than this and is well worth reading.

I have said something (1948, pp. 1–2) about this balance at the level of families and faunas rather than of species. During the evolution of animals upon the earth, many faunas have succeeded each other. The succession of faunas has involved on the one hand evolution, radiation, and spreading of new dominant groups of animals, and on the other recession and extinction of most species of previously dominant groups. These two processes are related, as Darwin said. There is a limit to the number of animals the earth can support. The total number probably increases gradually as animals become able to enter really new environments or to live in new ways, but the total increase is probably a small fraction of the sum of replacements that occur. When an old fauna disappears and is replaced by a new one, there is probably usually little change in the total of existing life. It is therefore likely that appearance of new species is almost balanced by extinction of old ones, and that for almost every dominant group of animals that spreads an old one disappears or is reduced to a few relicts.

This proposition cannot be proved but it can be illustrated. According to Simpson (1940, p. 158), just before North and South America were united toward the end of the Pliocene they had respectively about 27 and 29 families of land mammals. With one or two exceptions, they did not then have any families in common. After their union, the two continents exchanged mammals very extensively. In the Pleistocene they had 22 families in common, and further movements and many extinctions have occurred. Now, after all the shuffling, North America has only 23 families of land mammals, South America again 29. Each continent has gained some families but lost others; the total number of families on each is about what it was before.

It seems to me no more possible to have new groups of animals constantly spreading over the world without a nearly equal amount of recession of old groups than to have new generations of animals constantly being born without death of old generations. I think therefore that spreading and recession are about equally common and equally important. It might be argued that evolution and spreading are gradual processes, which continue for long periods, and that recession and extinction are rapid and soon completed, so that at any given time most groups of animals are spreading and few receding, but this may not be true. If the recession of some animals is due to the spread of others, the two processes may be equally slow.

I have gone into this matter at length because it is a vital one. Overstress on spreading and understress on recession of animals have led to the making of many unnecessary land bridges, simply because the bridgemakers could not believe that certain animals might have disappeared from wide areas of existing land. I am going to assume here that recession is as common as spreading, and that only careful analysis of situations and clues can tell (if anything can tell) whether particular groups of animals have been gaining or losing ground.

6. *The sixth suggested working principle of zoogeography is to understand and use fairly the clues to geographical histories of ani-*

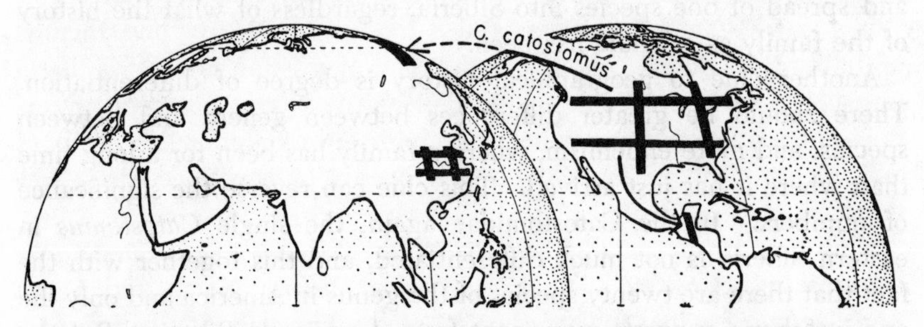

Fig. 9. Distribution of fresh-water fishes of the family Catostomidae.

mals. The clues are best explained by illustration. I shall illustrate them by means of a family of fresh-water fishes, the Catostomidae, which includes the suckers and buffalo fishes (Fig. 9).

The family Catostomidae is now chiefly North American: about 18 genera, 84 species occur in America south to Guatemala. Only 2 species of the family are known elsewhere. One belongs to the genus *Catostomus.* There are about 20 species of this genus in America. A single one of them, which ranges across northern North America from Maine to arctic Alaska, occurs also, as a slightly differentiated subspecies, in eastern Siberia. The other non-American species, with three subspecies, is in China. It is much differentiated, forming a very distinct genus of its own, *Myxocyprinus.*

Zoogeographers commonly use what I call "numbers clues." They find the place where the largest numbers of genera or species of a given group of animals now occur and take that as the place of origin of the group. In doing this they tacitly assume that genera and species usually increase in numbers at about the same rate everywhere. This can be true only if most animals are increasing and spreading most of the time. If recession and extinction are about as common as increase and spreading, as I think they are, numbers clues, used uncritically, will lead to wrong conclusions almost as often as not. Numbers clues used uncritically would put the place of origin of the family Catostomidae in North America, and that would probably be wrong. But in appropriate cases numbers are better clues. If a group of animals is dominant wherever it occurs, if it has left numerous fossils within but not outside the present range of the group, or if there is other evidence that the group is dominant, increasing, and spreading, then the numbers of genera and species in different places are more significant. In the dominant genus *Catostomus,* for example, numbers indicate an American origin of the genus and spread of one species into Siberia, regardless of what the history of the family as a whole has been.

Another clue to geographical history is degree of differentiation. There should be greater differences between genera and between species, and more endemism, where a family has been for a long time than where it has just arrived. This clue can reverse the significance of numbers. In the Catostomidae again, the single *Catostomus* in eastern Siberia is not much differentiated, and this together with the fact that there are twenty species of the genus in America and only the one in Siberia suggests movement from America to Siberia. But the catostomid in China is much differentiated, and this hints that, in spite of numbers of genera and species, it may be a relict from a time when catostomids were more numerous in eastern Asia.

Another clue is extent of area. A few zoogeographers suppose that the area occupied by a group of organisms increases directly with the group's age (p. 548). If this were so, the place of origin of a family would be the center of its range. If recession is as common as spreading, this clue is worthless by itself, but it has some value in connection with other clues. The fact that the single *Catostomus* in Siberia occupies a comparatively small area there increases the likelihood that it has recently come from America.

A fourth, more important clue is continuity of area. The Siberian *Catostomus* occupies an area as nearly as possible continuous with the area occupied in America, and this is consistent with its having reached Siberia recently. The Chinese *Myxocyprinus* also occupies a rather small area, but one widely separated from other existing catostomids, and this is consistent with its being a relict of an earlier Chinese catostomid fauna.

A fifth, very useful clue to family histories is the distribution of related, competing, and associated families. The closest relatives of the catostomids are the cyprinids, the carps and their allies. The two families are related and distributed in such a way as to suggest that catostomids were once numerous in eastern Asia, and that cyprinids have been derived from them and have replaced most of them there.

All the clues together suggest that catostomids originated in eastern Asia, moved primarily from Asia to North America, and (except for *Myxocyprinus*) have been replaced in Asia by cyprinids; that catostomids have radiated secondarily in North America; and that one *Catostomus* has returned to the near corner of Asia.

This apparent history of the catostomids illustrates a minor principle of animal distribution: that much differentiated forms (like *Myxocyprinus*) isolated from the main range of their groups, especially in or across areas inhabited by competing groups (in this case the Cyprinidae), are often not immigrants but relicts persisting where the first groups were once numerous but have been replaced. This suggests a further principle: that if groups of animals arise in certain places because of favorable conditions there, they are likely to begin recession in the same places because of the rise of later families responding to the same conditions. In other words, centers of evolution and dispersal are also likely to be centers of extinction. This is a corollary of the proposition that spreading and receding are related, nearly equal processes.

At this point I want to pause a moment. Taken together, the clues discussed thus far are significant, and I think they have probably told the truth about the catostomids. But clues like these are not infallible. I want to emphasize this by another example. *Esox* is a genus of fresh-water fishes, which includes the true pikes, pickerels, and muskallunge. The genus contains six species: four are confined to eastern North America, one is confined to northeastern Asia (the Amur River, etc.), and one (*E. lucius,* the Pike) occurs entirely around the northern part of the world—across North America from Labrador and New York to Alaska, and across northern Asia and Europe to the British Isles (Fig. 10). On the basis of these facts alone, almost any zoogeographer would say that *Esox* originated in eastern North America (five species, four endemic), then reached eastern Asia (two species, one endemic), and finally and recently reached Europe (one species, not endemic). The distribution of the Pike is almost continuous, as if it had spread recently. *Esox* is the only genus of its family, but the present distribution of related families, especially of the Umbridae, would fit well with a North American origin of the whole order, Haplomi, to which *Esox* belongs. Nevertheless, there are several well-preserved fossil species of *Esox* in the Oligocene and Miocene of Europe; and *Palaeoesox,* the earliest known member of the Haplomi, is in the Eocene of Europe, while neither *Esox* nor any related fish is known fossil in America. The geographical history of *Esox* has at least been more complex than the distribution of living species shows. In spite of all contrary clues, the Pike in Europe is perhaps more likely a survivor of a receding group than an immigrant from America. The reason for its survival may be

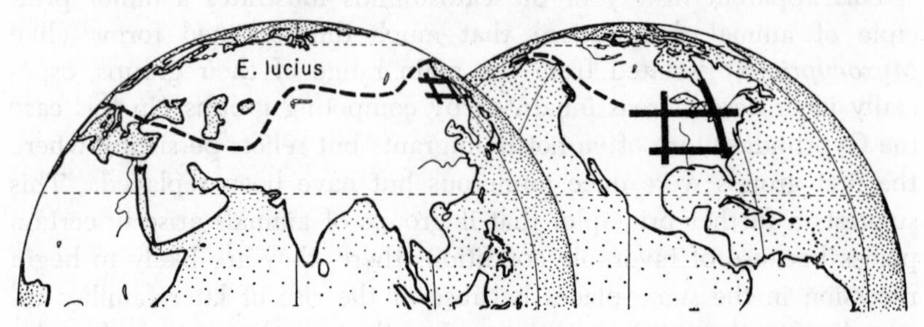

Fig. 10. Distribution of fresh-water fishes of the genus *Esox*. Broken line is approximate southern limit of *E. lucius* in Eurasia and western North America.

that it has unusual tolerance for cold; it is one of the few true fresh-water fishes that are widely distributed in the arctic. This probably enabled it to cross a late Siberia-Alaskan land bridge, but it perhaps crossed from Asia to North America rather than the reverse, and all American *Esox* apparently came from Eurasian stock in the first place.

The history of the Catostomidae has probably included at least two crossings of Bering land bridges, an early one from Asia to America, and a late one back to the near part of Siberia. The history of the Esocidae also has probably included two crossings, one early and one late, but both perhaps from Eurasia to America. The distributions of other animals that disperse more rapidly than fresh-water fishes may be products of still more complex movements. The possible complexity of geographical movements is in itself a reason for distrusting simple clues in working out the histories of families.

There are, of course, other, special, clues that help to reveal histories in special cases. For example, when a family's range includes a recently glaciated area, a movement into that area is obvious. One clue I do *not* trust is the distribution of primitive forms, which are supposed by some persons to mark the place of origin of a family, and by others, the periphery of its range. Probably they may mark either, or both (p. 554). If the existing representatives of a group of animals can be arranged in several true phylogenetic lines, and if the lines all lead back from different directions to one place, that is likely to be the place of origin of the group, but this method of tracing geographical histories is very difficult and often misused.

What I have been trying to say in the last few paragraphs is that the clues are only clues, not proofs, and that they should be used judiciously.

The best clues, of course, are fossils—the right fossils in the right places—but even fossils must be interpreted judiciously. The fossil record of many animals, even of many vertebrates, is very incomplete or confused by error. Fossils may show the evolution of a group of animals very well but still not show the geographical history of the group; one fossil can show a given stage in evolution, but many are required to show distribution at a given time. Poor fossil records can be very misleading. For example, parrots are known fossil in North America and Europe in the mid-Tertiary but nowhere else until the Pleistocene. Obviously this does not mean that parrots originated in the north and did not enter the tropics until recently; but equally unjustified conclusions have been drawn in other similar cases. In many cases, as in that of the parrots, the distribution of existing forms is a better guide to the past than a poor fossil record is.

7. *The seventh and last suggested working principle of zoogeography is to try working hypotheses when facts fail, remembering always that the hypotheses are not facts.* This procedure is useful, for example, in attempting to trace the geographical histories of animals with poor fossil records. The procedure is to work out probabilities independently in each of a series of cases, then put the cases together and look for a common pattern. If one is found, it will strengthen the probabilities in the separate cases. This is what I shall do for the different classes of land and fresh-water vertebrates: analyze their distributions separately, then look for common patterns.

The idea is sometimes encountered that the histories of certain animals, for example fresh-water fishes, and what they tell about land bridges and drifting continents can only be guessed about, not proved, and that because they cannot be proved they should not be written about at all. This idea seems to me to show a lack of understanding of scientific method. Science proceeds by defining problems and trying to solve them. There is no way of telling beforehand whether a particular problem at a particular time is soluble or not. We may suspect that it is not, but we are bound to try. To decide beforehand that it is insoluble makes it so. No one should say that because he has failed no one else should try, for no one can investigate a complex problem so thoroughly that nothing is left to be said. Not only the problem itself but its relation to other problems must be discussed and rediscussed; and even if all the attempts fail, they may show what work must be done to succeed. I shall in due course in this book set up a system of hypotheses about the history of animal dispersal and the world of the past. If I set it up properly, it will be useful at least for a while. I think that, in fact, the hypothetical history will have much truth in it, although it will be neither wholly true nor the whole truth. To be honest, I should add that guesses about the past have a fascination for me and for other zoogeographers. To leave the guesses out of zoogeography would be to leave half the fun out of it. But the guesses should be careful ones. They should be careful statements of probabilities and possibilities, not just flights of fancy, and not fixed ideas to be defended unreasonably.

So much for suggested working principles. Except for numbers 5 and 6, they are merely working principles of science applied to zoogeography. All together they come to this: decide what you want to do; select and understand the factual material to do it with; plan how to put the facts together to get the truth out of them; go carefully; and keep going.

REFERENCES

The lists of references at the ends of this and succeeding chapters serve two purposes: to give the principal sources of material used and to give readers an entry to further literature. The lists are as short as these purposes allow. Many of the works listed have bibliographies which lead as far back into the past or as deeply into complex subjects as readers may want to go. For example, the following list does not pretend to cover the history of zoogeography, but Gadow (1913) gives additional references which lead into the history of the subject. For another example, G. G. Simpson has made many important contributions to zoogeography, but only two are listed here. One is listed because it is referred to in the text. The other is not referred to in the text but is listed because it is recent and opens the way to Simpson's earlier zoogeographical work. The lists are arranged alphabetically, by authors, and chronologically under each author. Articles in journals are indicated by name of journal, volume number (in boldface), and page numbers. Citations in the text refer to these lists by authors' names, dates, and sometimes page numbers.

Allee, W. C., A. E. Emerson, O. Park, T. Park, and K. P. Schmidt. 1949. *Principles of animal ecology.* Philadelphia and London, W. B. Saunders.

Beaufort, L. F. de. 1951. *Zoogeography of the land and inland waters.* London, Sidgwick and Jackson; New York, Macmillan.

Brewster, W. 1924. The birds of the Lake Umbagog region of Maine. Part 1. *Bull. Mus. Comparative Zool.*, **66**, 1–209.

Cain, S. A. 1944. *Foundations of plant geography.* New York and London, Harper and Bros.

Chapman, F. M. 1929. *My tropical air castle.* New York and London, Appleton.

———. 1938. *Life in an air castle.* New York and London, Appleton-Century.

Darlington, P. J., Jr. 1938. The origin of the fauna of the Greater Antilles, with discussion of dispersal of animals over water and through the air. *Quarterly Review Biol.*, **13**, 274–300.

———. 1948. The geographical distribution of cold-blooded vertebrates. *Quarterly Review Biol.*, **23**, 1–26, 105–123.

Darwin, C. 1859, 1950. *On the origin of species . . . .* London, Murray; reprinted 1950, London, Watts and Co.

Ekman, S. 1953. *Zoogeography of the sea.* London, Sidgwick and Jackson. This is a translation of the 1935 German edition; about half the original maps and figures are omitted.

Gadow, H. 1913. *The wanderings of animals.* Cambridge, England, Cambridge U. Press; New York, G. P. Putnam's Sons.

Geiger, R. 1950. *The climate near the ground.* Cambridge, Mass., Harvard U. Press.

Guppy, H. B. 1917. *Plants, seeds, and currents in the West Indies and Azores.* London, Williams and Norgate.

Hesse, R., W. C. Allee, and K. P. Schmidt. 1951. *Ecological animal geography,* 2nd ed. New York, John Wiley; London, Chapman and Hall.

Kendrew, W. G. 1937. *The climates of the continents.* London, Clarendon Press.

———. 1949. *Climatology.* London, Oxford U. Press.

Kuenen, P. H. 1950. *Marine geology.* New York, John Wiley; London, Chapman and Hall.

Matthew, W. D. 1915, 1939. Climate and evolution. *Ann. New York Acad. Sci.*, **24,** 171–318; reprinted (1939) as *Special Pub. New York Acad. Sci.*, **1.**

Mayr, E. 1942. *Systematics and the origin of species.* New York, Columbia U. Press.

Myers, G. S. 1938. Fresh-water fishes and West Indian zoogeography. *Smithsonian Rep.*, **1937,** 339–364.

Newbigin, M. I. 1936. *Plant and animal geography.* London, Methuen.

Powers, S. 1911. Floating islands. *Popular Science Monthly,* **79,** 303–307.

Romer, A. S. 1945. *Vertebrate paleontology,* 2nd ed. Chicago, U. of Chicago Press.

Schuchert, C. 1932. Gondwana land bridges. *Bull. Geol. Soc. America,* **43,** 875–915.

———. 1935. *Historical geology of the Antillean-Caribbean region.* New York, John Wiley.

Sclater, P. L. 1858. On the general geographical distribution of the members of the Class Aves. *J. Proc. Linnean Soc.* (London), *Zool.*, 2, 130–145.

Simpson, G. G. 1940. Mammals and land bridges. *J. Washington Acad. Sci.*, 30, 137–163.

———. 1953. *Evolution and geography.* Eugene, Oregon, Oregon State System Higher Education.

Sverdrup, H. U., M. W. Johnson, and R. H. Fleming. 1942. *The oceans.* New York, Prentice-Hall.

Tannehill, I. R. 1952. *Hurricanes.* Princeton, New Jersey, Princeton U. Press.

———. 1952*a*. *Weather around the world,* 2nd ed. Princeton, New Jersey, Princeton U. Press.

Umbgrove, J. H. F. 1949. *Structural history of the East Indies.* London and New York, Cambridge U. Press.

Wallace, A. R. 1876. *The geographical distribution of animals.* London, Macmillan, 2 vols.

———. 1880. *Island life.* London, Macmillan.

———. 1911. *Island life,* 3rd ed. London, Macmillan.

Zeuner, F. E. 1950. *Dating the past,* 2nd ed. London, Methuen.

chapter 2

Fresh-water fishes

No part of this book has given me more trouble in the beginning or more satisfaction in the end than that on fishes. I should not have been able to complete it at all without the patient assistance of Dr. George S. Myers, whose own (1938) discussion of fresh-water fishes and of their distribution should be read by every zoogeographer. I have had useful information also from Dr. Carl L. Hubbs, especially in a long letter following the publication of my (1948) paper on the geographical distribution of cold-blooded vertebrates.

Fresh-water fishes are uniquely significant in zoogeography. Although they live in water, those that are confined to fresh water are as closely bound to land masses as are any animals. On land they are almost inescapably confined to their own drainage systems and can pass from one isolated stream basin to the next only by slow change of the land itself or by rare accident, so that they disperse over the world slowly and are likely to preserve old patterns of distribution. Compared to most other cold-blooded animals (reptiles, amphibians, and invertebrates), fishes are fairly well known in classification, evolution, present distribution, and powers of dispersal. And compared to the still better-known warm-blooded vertebrates (mammals and birds), the fishes presumably have more generalized distribution patterns, showing more clearly the effects of climate. If there is such a thing as a general pattern of animal distribution, the fresh-water fishes ought to show it perhaps better than any other animals, and they do in fact show broad patterns which are well defined and very informative. There is not space in this book to discuss all the classes of vertebrates as thoroughly as I should like, and I have therefore chosen to treat the fishes in somewhat more detail than any others, because they illustrate general patterns and principles so well.

The main system of classification of fishes used here, down to and including the orders, is Romer's (1945). Berg's (1947) classification has been referred to also in some cases. The families were originally taken from Regan (1929) because of the importance of his geographical work on fishes, but a good many changes and additions have been made. These classifications differ in details, but the differences are not very important zoogeographically. Fresh-water fishes are much less numerous than marine ones. Many strictly marine groups are not even mentioned here, and many fossil groups of unknown habits or unknown geographical significance are omitted, so that the classification of fishes at the end of this chapter is much abridged in places. The reader who is not familiar with fishes should consult Romer's text (1945) to learn the main features of fish evolution and classification.

For zoogeographical purposes fresh-water fishes may be grouped in several different ways. Cyclostomes and shark-like fishes are relatively unimportant in fresh water. The important fresh-water groups are all bony fishes, and of these it is useful sometimes to contrast the non-teleosts (archaic bony fishes) with the teleosts (higher bony fishes), and at other times to contrast the single dominant fresh-water order Ostariophysi with all other orders, lumped as non-ostariophysans. These groupings are designed to match primitive against derivative forms and non-dominant against dominant ones and to reveal their different patterns of distribution. The groups are compared in Table

TABLE 1. MAIN TAXONOMIC GROUPINGS OF FRESH-WATER FISHES FOR ZOOGEOGRAPHIC PURPOSES

Grouping	
Cyclostomes and shark-like fishes: not important	
Bony fishes	
Non-teleosts	Non-ostariophysans
Teleosts	
Other than Ostariophysi	Non-ostariophysans
Ostariophysi	Ostariophysi

1, and their content is given in more detail in the classification and list of families at the end of this chapter.

What are fresh-water fishes?

Fishes occur in almost all fresh waters, even in the most adverse climates and on the most remote islands. All the fishes that occur in fresh water are fresh-water fishes in a sense, but their zoogeographic significance varies with their salt tolerance and behavior. This is obvious and has long been understood in a general way, and various attempts have been made to classify fresh-water fishes according to their relation to the sea (*e.g.*, Boulenger 1905, pp. 414–415) and to account for some details of their distribution by passage of salt-tolerant forms through the sea (*e.g.*, Regan 1912, in his discussion of fishes and Wallace's Line), but Myers (1938) made possible a real advance in understanding of fish geography by proposing an ecological classification of fresh-water fishes based simply on their ability or inability to tolerate salt water.

Fishes that are strictly confined to fresh water Myers put in a *primary division.* Those that live chiefly in fresh water but have a little (not too much) salt tolerance he put in a *secondary division.* And fishes that occur in fresh water but have much salt tolerance may be put in a *peripheral division;* this term is not from Myers but from Nichols (1928, pp. 6–7). This brief preliminary statement of Myers' arrangement does not do it justice. The three divisions are very complex groups, which cut across taxonomic lines and which must be discussed before they can be understood and defined more accurately.

The primary division, of strictly fresh-water fishes, contains only bony fishes. The primitive, non-teleost groups that belong here are the African polypterids, the Chinese and North American paddlefishes, the North American bowfin, and the Australian and African-South American lungfishes. Of teleosts, there are many, of diverse origin, some or the other of them occurring on every habitable continent except perhaps Australia, where there is only one (doubtfully) primary-division teleost, an osteoglossid. In North America the primary-division teleosts are the mooneyes, many Ostariophysi (see below), the pikes and mud minnows and their allies, the trout perches and pirate perches, the centrarchid basses and sunfishes, and the true perches including the darters. A complete list of primary-division families is included in the taxonomic list of fish families at the end of this chapter.

About seven-eighths of all the primary-division fresh-water fishes of the world belong to one order, Ostariophysi. This order is distinguished by a chain of "Weberian ossicles" connecting the air bladder with the inner ear. The chain is thought to carry vibrations to the inner ear and to increase the sensitivity of hearing, and it is thought that ability to hear well may be especially useful under fresh-water conditions (Regan 1922, p. 204), and may account at least in part for the unique success of the Ostariophysi in fresh water. It is a curious and important fact that, although ostariophysans dominate most fresh waters, there are very few of them in the sea. Out of nearly 40 families and 5000 species of them, only 2 families, the ariid and plotosid catfishes, with not over 150 species, are marine, and they are probably derived from fresh-water ancestors. To the Ostariophysi belong the hordes of characins, gymnotid eels, carps, minnows, loaches, suckers, and catfishes, some or the other of which swarm in the fresh waters of every continent except Australia (paraphrased from Myers 1938, p. 345).

The secondary division, fishes usually confined to fresh water though a few enter the sea, contains only bony fishes, and not many even of them: of primitive non-teleosts, only the North and Central American gar pikes; and of teleosts, only the cyprinodonts and cichlids. These groups all live chiefly in fresh water. However, gar pikes enter the sea along the Gulf of Mexico coast. Many cyprinodonts do not seem to be much inconvenienced by salt water. One of them, a *Mollienisia*, enters the sea freely and multiplies in brine pools about Manila Bay, where it was accidentally introduced; a *Rivulus* has been reported from tide pools on Curaçao; and some *Fundulus* and *Cyprinodon* live permanently in sea water along coasts beyond the reach of estuaries. The *Challenger* expedition even caught a *Fundulus* in a mid-Atlantic pelagic haul. The cyprinodont genus *Aplocheilus* is presumed to have unusual salt tolerance because of its occurrence on islands, in the Orient, beyond the reach of most fresh-water fishes. Of cichlids, most can survive several hours or days in the sea, and one species (a *Tilapia*), collected in brackish water in Mozambique, survived for months in sea water at the New York aquarium. It is evident that many species of the secondary division might survive short sea journeys, and their distributions show that some of them have done so. (Much of the preceding part of this paragraph is from Myers 1938, p. 345.) Besides these, parts of some other families now placed in the primary division may have to be transferred to the secondary division when their salt tolerances are better known. This may be true even of a few Ostariophysi, especially of certain cyprinids and catfishes which will be mentioned again.

The peripheral division, of very salt-tolerant fresh-water fishes, and of fishes that are apparently derived from very salt-tolerant groups, contains a great number of forms of which only a few can be mentioned here. They include one family of cyclostomes and a few shark-like fishes, one family (the north-temperate sturgeons) of non-teleost bony fishes, and an almost endless diversity of teleosts. Even a few Ostariophysi belong here, the ariid and plotosid catfishes, which are chiefly marine but some of which enter or have become resident in fresh water, especially in isolated places.

The habits of peripheral fishes are diverse. Many are migratory. Of these, some (lampreys, sturgeons, salmon, and trout, etc.) breed in fresh water but spend much of their lives in the sea, except that some species and genera of each of the groups named have become permanently resident in fresh water. Others (anguillid eels, galaxiids, some gobies, etc.) breed in the sea but spend much of their lives in fresh water, and among these, too, certain forms (of galaxiids and gobies but not of anguillids) have become resident in fresh water. Some other migratory fishes move regularly between fresh and salt water for purposes other than breeding; and still others, for example some sticklebacks, appear to live indifferently in both. Migratory fishes can be classified in complex ways according to their behavior (Myers 1949), and the details are important in studies of fish biology, but they are much less important in zoogeography. One of the advantages of Myers' arrangement of fresh-water fishes in a primary and a secondary division is that the zoogeographically less important complications of their life histories are set aside and emphasis is placed simply on the extent to which the fishes can tolerate salt water and may have dispersed through it.

Besides the migratory and indifferent forms there are many fishes living in fresh water which have obviously been derived locally from migratory or fully marine ancestors. Fishes of this sort occur in fresh water in all parts of the world but are most conspicuous in remote places such as Australia, where they make up the main part of the fresh-water fish fauna, and on remote islands. Examples in North America are those lampreys, sturgeons, salmon, and trout (and the related whitefishes and graylings) that now live entirely in fresh water; the "White Bass" and "Yellow Bass" (sea bass family), Fresh-water Sheepshead (marine drum family), and Burbot (codfish family);

and certain cottids and atherinids. Further details will be found in the formal list of families of fresh-water fishes.

Finally, a very large number of marine fishes of many different families run up rivers or enter fresh water occasionally or locally. This happens in every part of the world, and examples will be found scattered through almost any good faunal work on fresh-water fishes. A good series of examples is shown by the family Atherinidae (see list of fish families). All fishes known to occur in both fresh and salt water in North and Middle America are listed by Gunter (1942), and his list gives an idea of the diversity of behavior of these fishes. Other papers bearing on the subject are by Breder (1934) and Myers (1949 and 1949*a*), and the last author gives additional references.

The whole subject of the relation of peripheral fishes to fresh and salt waters is much too complex to discuss further here. The important thing is that all the diverse fishes of the peripheral division either can pass through the sea or are probably derived rather recently from fishes that could do so, so that their present distributions are likely to be partly the result of dispersal through the sea.

This arrangement of fresh-water fishes in three divisions according to salt tolerance is simple in theory but very complicated in fact and open to endless argument about details. The salt tolerances of many fishes are unknown, and it is and probably always will be impossible to draw sharp lines between the divisions. For the present most fishes are grouped by families, on the assumption that all the members of one family probably have about the same salt tolerance. This may often be true, but it is not always so.

Many species even of primary-division families can tolerate brackish water, and a few can enter the sea. In the Ostariophysi, for example, besides the two families of catfishes that are actually marine, various characins, carps, suckers, and fresh-water catfishes enter brackish estuaries (Myers 1938, p. 344, footnote 4; and other sources); a Japanese carp (cyprinid) which is widely distributed in rivers is frequently taken also in the sea (Jordan 1905, Vol. 1, p. 256; Okada 1955, p. 93); a small British Columbian cyprinid sometimes occurs in the sea and has reached Vancouver and Nelson Islands, which the strictly fresh-water cyprinids of the same region have not done (Carl and Clemens 1948, p. 77); and some species of the Oriental genus *Mystus* (or *Macrones*) of the fresh-water catfish family Bagridae live in the lower parts of tidal rivers and enter or live partly in salt water (Weber and Beaufort 1911–1951, Vol. 2, p. 345; H. M. Smith 1945). Still other species of primary-division families are distributed in such a way as to suggest that they may have crossed narrow ocean barriers to the Philippines or Australia. These cases will be considered in due course.

As to the salt tolerances of ancient fishes, paleontologists and physiologists agree that the common ancestor of all fishes probably lived in fresh waters: the earliest fossil fish-like vertebrates are in fresh-water deposits (Romer 1945, pp. 44–45), and all fishes possess kidneys of a sort that probably originated in fresh water (H. W. Smith 1932).

Sharks and rays and their allies of the class Chondrichthyes probably originated in the sea, from an ancestor which had left the fresh-water environment. They have been ocean dwellers since the beginning of their known history (Romer 1945, p. 60), and their flesh is impregnated with urea, apparently as an adaptation to life in salt water (H. W. Smith 1932). Several persons have transferred marine sharks and rays from sea water to dilute salt water or fresh water with the "inevitable" result that the animals absorbed water, swelled up, and died, apparently because they could not make physiological adjustments rapidly enough (H. W. Smith 1932, p. 15). Nevertheless some sharks do enter fresh water in nature, and two species of *Carcharias* (a primarily marine genus of sharks) are believed to live only in fresh water, one in Lake Nicaragua, the other in the Zambezi (Regan 1929); sawfishes habitually ascend tropical rivers from the sea; some stingrays enter fresh water and others have become resident there (Garman 1913, pp. 415–427); and there are some other records of selachians occurring or living in fresh water. These facts show that a very long history of existence in one kind of water does not necessarily destroy the tolerance of fishes to other waters. The facts suggest, too, that the results of experiments made with a few species of fishes may be misleading. Laboratory experiments apparently do not always show the real tolerances in nature of even the species experimented with. The results of such experiments should not be applied too sweepingly to whole orders or families or to their ancient ancestors.

The earliest bony fishes probably inhabited fresh water (Romer 1945, p. 75). Polypterids and lungfishes have probably maintained continuous lines in fresh water since their origin, although at least a side line of lungfishes made a short-lived invasion of the sea in the Devonian (Schaeffer 1952, p. 101). However, the ancestors of most existing fresh-water fishes have probably at some time passed through the sea. Teleosts may have originated in the sea (Romer 1945, p. 104). If so, some of them returned to fresh water, and one, perhaps derived from a marine herring of the order Isospondyli, developed an improved hearing apparatus and evolved and diversified in the fresh waters of the great continents to form the order Ostariophysi. Certain ostariophysan catfishes then re-entered the sea. And finally some of the marine catfishes went back into fresh water in such remote places as Australia and Madagascar. If this history is correct, the endemic catfishes in Australian rivers are at the end of a line of teleost evolution which began in fresh water, went into the sea, back to fresh water, into the sea, and back to fresh water again.

All this should make it clear that Myers' primary division of fresh-water fishes is a heterogeneous assemblage of unrelated orders and families. A few species of even the greatest fresh-water groups can enter the sea, and the tolerances and habits of many species, especially in the tropics, are unknown. The ancestry of most primary-division fishes is not such as to suggest that they have always lived in fresh water. On the contrary, many of them, including the Ostariophysi, were probably derived from marine fishes long ago. That the fishes of the primary division are and long have been confined to fresh water is proved less by their nature and evolutionary history than by their distribution, which will be described. This should be remembered. Certain striking features of their distribution help to prove that these fishes are confined to fresh water, and when an occasional species is found which breaks out of the usual distribution pattern, as for example the Australian osteoglossid does, an immediate possibility is that it is not really a strictly fresh-water fish.

Myers (1949*a*, p. 317) mentions that some ichthyologists doubt the reality of his divisions. They are indeed somewhat artificial, there are probably no sharp lines between them, and there is plenty of doubt about details. But the divisions are based on a real and very important fact, that fresh-water fishes vary in their ability to pass through salt water and that their distributions show it. It seems to me that these divisions, or others like them, are absolutely necessary to a real understanding of fresh-water fish geography.

For practical purposes, and in the light of the preceding discussion, Myers' divisions may be slightly redefined as follows. The *primary division* consists of families or other groups of fresh-water fishes that ordinarily cannot or do not enter the sea (although exceptional species of some of the groups do so) and that have probably been confined to fresh water so long that their present distributions are the result of dispersal through fresh water, even though their remote ancestors may have lived in the sea. The *secondary division* consists of families or other groups of fishes that occur chiefly in fresh water but that can enter the sea and survive there for a limited time, or are recently descended from forms that could do so, so that their present distributions may be and apparently often are partly the result of dispersal along coast lines or across narrow ocean gaps. The *peripheral division* consists of families or other groups that, although found in fresh water, are somehow closely connected with the sea or have been so recently derived from it that their present distributions may be and often are largely the result of dispersal through the sea.

These divisions are designed for the study of distribution of fresh-water fishes over the whole world. Other classifications may be more useful for local studies in some cases, for some groups of fishes that are certainly peripheral so far as the whole world is concerned may be confined to fresh water in certain places. This is true of some northern salmonids which, though derived from migratory forms, are now confined to local fresh waters and show the histories of drainage systems. Another example may be found in the Atherinidae. This family as a whole is certainly a peripheral one. There is no reasonable doubt that it reached Australia through the sea. A number of Australian atherinid genera, however, forming the subfamily Melanotaeniinae, may now be strictly confined to fresh water and may be limited in the direction of the Solomons and Moluccas by old salt-water barriers, as if the Melonotaeniinae were primary-division fishes. This case is given as a hypothetical one, for there is some doubt about the fact of it (see Atherinidae in list of fish families).

Limits of distribution of fresh-water fishes

The limits of distribution of fresh-water fishes are outlined in Figure 11.

Northward, a few primary-division fishes extend more or less widely above the Arctic Circle. Of non-ostariophysans, the Pike (*Esox lucius*) reaches all the main arctic river systems of Eurasia (except western and northern Scandinavia), and parts of arctic Alaska, the Mackenzie delta, and the Anderson River, which is entirely above the Arctic Circle in western Canada. The Blackfish (*Dallia*) occurs in the central and western part of the "head" of Alaska north at least to the Seward Peninsula, in arctic and Bering drainages of the eastern tip of Asia (the Chukotski region), and on St. Lawrence Island in the northern Bering Sea. It is the only primary-division fish specially adapted and limited to an arctic habitat. It lives in shallow water on the tundra, and is a small (up to 8 inches), very fat fish, exceptionally

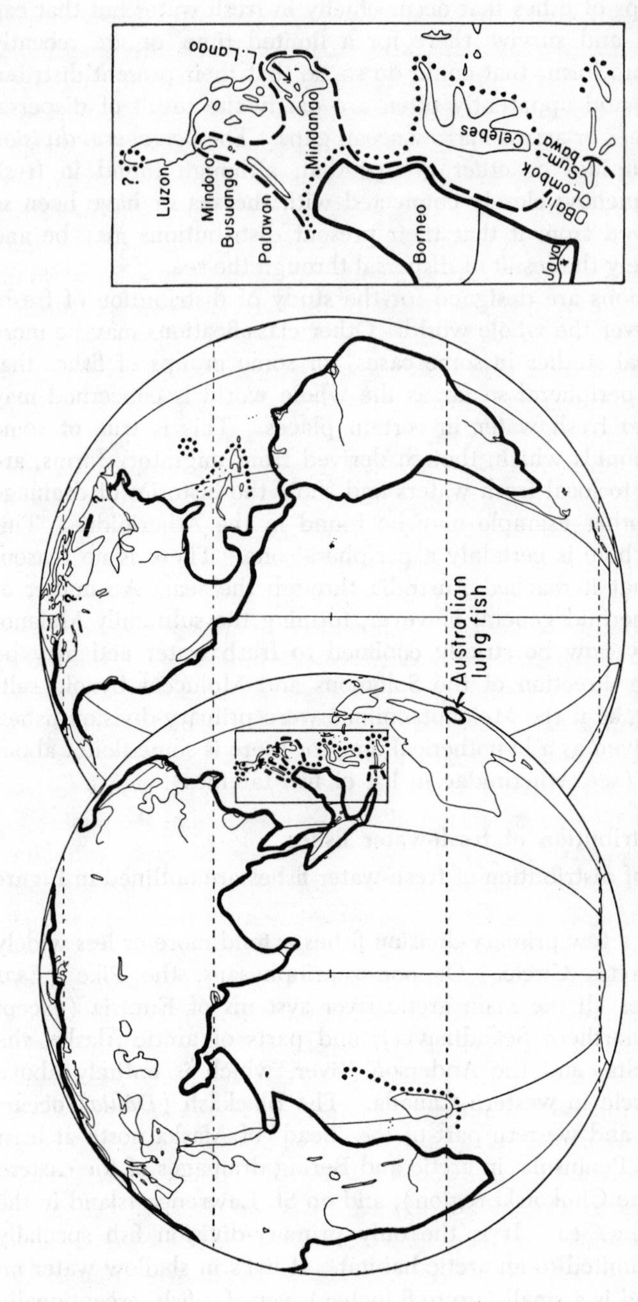

Fig. 11. Limits of distribution of primary- and secondary-division fresh-water fishes. Redrawn from *Quarterly Review of Biology*, 23, 1948, p. 7, Figure 1. Solid lines show limits of primary-division fishes, or, in East Indies, limit of main primary-division fish fauna; broken lines, limits of primary-division stragglers (mostly small cyprinds) in East Indies; and dotted lines, limits of secondary-division fishes on islands.

able to withstand cold. Individuals kept apparently solidly frozen for weeks, then eaten by dogs and so thawed out, are sometimes vomited up alive! That this fish, uniquely adapted to the tundra, is confined to such a small area, mostly south of the Arctic Circle, while the less specially adapted Pike and some other strictly fresh-water fishes are much more widely distributed in the arctic and go farther north, is a noteworthy fact. Two percids extend into the arctic in Eurasia, and the American Pike Perch reaches the Mackenzie delta. The Trout Perch (Percopsidae) crosses the Arctic Circle in western Canada. Of Ostariophysi, the Northern Sucker (*Catostomus catostomus*) extends above the Arctic Circle in western North America and eastern Siberia; several cyprinids extend into the arctic in Eurasia; and at least one American cyprinid (*Couesius*) reaches the Mackenzie delta.

Some of the fishes listed above reach true arctic environments, but others probably just follow large, north-flowing rivers, such as the Mackenzie in Canada and the Yenisei and Lena in Siberia, into the arctic but do not occur in the tundra country behind the main river banks. There are some arctic fresh-water habitats, notably stagnant ponds on the tundra, which are not inhabited by any primary-division fishes except (within the limits of its small range) by *Dallia*, and there are areas of even the continental arctic where there are no primary-division fishes. There may be such areas in parts of northern Quebec and Labrador. The Pike, the Northern Sucker, and one cyprinid (*Couesius*) extend north in the Labrador Peninsula to 56° or 57° N. in the George River, but in some other parts of the peninsula no primary-division fishes are known. However, they are probably not barred by existing climate. More likely they have been unable to get back to remote northeastern drainage systems from which they were eliminated by late Pleistocene ice. Peripheral fishes, especially salmonids and their allies, swarm in many arctic fresh waters both within and north of the limits of primary-division fishes, but even the peripheral forms decrease in variety in the farthest north. Berg (1950, pp. 18–19) says that, in Eurasia, a charr or trout (*Salvelinus*, a genus of peripheral Salmonidae) is the northernmost fresh-water fish, reaching Novaya Zemlya and the New Siberian Islands above Siberia. In America, the same genus extends still farther north, having been found in fresh water at least to 82° 34′ N. on northern Ellesmere Island, near the final northern limit of land. These details of distribution of fresh-water fishes in the arctic are mostly from Berg (various works) and Wynne-Edwards (1952).

Southward, some groups of African primary-division fishes reach the tip of South Africa, but many others do not (for details see p. 61). Those that fail to reach South Africa are probably stopped more by habitat limitations than by coolness. Primary-division fishes are absent in most of Australia but not because of cold climate; peripheral forms occur south through Tasmania and New Zealand. In South America, a catfish of the family Pygidiidae reaches 47° 30′ S., some 500 miles short of Cape Horn (Eigenmann 1918, p. 269). It may be the most southern primary-division fish in the world, but peripheral fishes occur in fresh water probably as far south as there is any on South America and Tierra del Fuego.

Within their northern and southern limits, faunas of true fresh-water fishes, of the primary division, are confined to the continents, *excepting Australia,* and to certain continental islands. No fact in zoogeography is more striking than this, and it is this fact which proves that many groups of fishes have long been confined to fresh water. The geographical limits of continental fish faunas are shown on Figure 11. These limits are passed by very few fishes indeed, except by such as are known to tolerate salt water or to be carried by man.

Both North and South America are rich in fresh-water fishes, and a number of the South American groups reach the continental island of Trinidad, but not one fish of the primary division is native to the West Indies proper, although some salt-tolerant secondary-division and peripheral ones are (Myers 1938, pp. 356–357). The northern island of Newfoundland, too, although a continental island lying close to the mainland, apparently lacks all primary-division fishes, although there are peripheral ones there (Frost 1940). True fresh-water fishes have probably been unable to reach this island since the Pleistocene.

Africa has a great fauna of both ancient and modern families of fresh-water fishes, but not one fish of the primary division is native on Madagascar, although a few secondary-division forms and, of course, peripheral ones have reached the island.

Europe and Asia have given one part or another of their primary-division fish faunas to the continental British Isles (Berg 1932, and many other authors), Japan (Mori 1936, pp. 10–14; Okada 1955), Formosa and Hainan (Mori 1936, pp. 47–51), and Ceylon (Day 1889, Munro 1955, and references in Hora's papers), and most of the primary-division fish families of southeastern Asia extend to Sumatra, Java, and Borneo. Here, however, true fresh-water fishes stop abruptly (Fig. 11), as has long been known (Regan 1912). Some of them may stop short of the outer ends of Java (Dammerman 1929, pp.

15–16) and North Borneo; not much is known about this. The rich Oriental fish fauna is well represented on Java as a whole, but little of it reaches even the first of the Lesser Sunda Islands. Various ichthyologists from Regan (1912) to Beaufort (1951, p. 180) have implied that a significant part of the Javan fish fauna reaches Bali, or at least that Bali has significantly more fresh-water fishes than Lombok. This is a mistake. The only primary-division fishes known and apparently native on Bali are two or three small cyprinids of the genera *Rasbora* and *Puntius,* and both these genera occur also on Lombok, and *Rasbora* is on Sumbawa also. *Fresh-water fishes do not mark Wallace's Line at its southern end.*

Farther north, many Oriental primary-division fishes reach Borneo, but not one reaches Celebes, except as (probably) carried by man. A very few Bornean fishes reach the southern Philippines (Herre 1928). They are mostly small cyprinids, which occur through Balabac, Palawan, and the Calamian Islands to Mindoro, and through the Sulu Archipelago to and through the whole of Mindanao (Fig. 11). Several genera are involved, but especially *Rasbora* and *Puntius* (*Barbodes*). There are endemic species localized along both the routes described, and in Lake Lanao on Mindanao there is a cyprinid fauna of at least four endemic genera and seventeen species, all apparently derived from one ancestor (Herre 1933). Besides the cyprinids, there is an endemic genus and species of silurid catfish on Palawan and the Calamian Islands and an endemic species of clariid catfish on Mindanao. This completes the list of known, probably native, primary-division fishes of the Philippines.

The only Oriental fishes (excepting a genus of cichlids in Ceylon and peninsular India) now placed in the secondary division are cyprinodonts. Of these, the genus *Aplocheilus* (including *Oryzias*) is notable for the islands it has reached. It is present on the Riu Kiu and Andaman Islands, where primary-division fishes are absent or very few; it occurs on Lombok and Timor; it has reached Celebes, where it has evolved endemic species (Aurich 1935) and probably also the endemic family Adrianichthyidae (see list of fish families); and a species has been described from one locality on northern Luzon (Herre and Ablan 1934) but, since cyprinodonts are unknown elsewhere in the Philippines, it may not be native there. In view of this distribution and of the known fact that many cyprinodonts are salt tolerant, there can be little doubt that *Aplocheilus* has dispersed partly through the sea.

To return for a moment to the small cyprinids, two genera of them, *Rasbora* and *Puntius* (*Barbodes*), repeatedly occur on islands beyond the limits of other primary-division fishes. Both genera cross Wallace's Line on the Lesser Sundas, and both go most of the way up both routes of immigration of cyprinids into the Philippines. In fact one species, *Puntius binotatus,* which occurs in fresh water from Singapore through Sumatra, Java, and Borneo, has gone beyond the limits of most fresh-water fishes in all three directions, to Bali and Lombok, to Palawan, and to Mindanao, where it is the apparent ancestor of the endemic cyprinids of Lake Lanao. This pattern, like that of *Aplocheilus* but on a smaller scale, suggests that these cyprinids have more than usual salt tolerance, though not so much as *Aplocheilus,* and that they have been able to cross narrow ocean barriers.

In the preceding discussion I have ignored certain primary-division fishes, particularly the labyrinth fishes *Anabas testudineus* and *Ophicephalus striatus* and the catfish *Clarias batrachus,* which are common food fishes, which live in a variety of fresh and brackish waters, which are very tenacious of life and can live out of water for considerable periods if kept damp, and which are often carried about alive by man. These fishes occur beyond the limits of most fresh-water fishes on the Lesser Sunda Islands, the Philippines, and Celebes, etc. They have probably been carried to these places by man.

In Australia, far outside the limits of the great continental faunas, are two fishes of supposedly primary-division fresh-water families. The osteoglossid *Scleropages leichhardti,* a large fish 2 or 3 feet long, inhabits rivers of parts of tropical Queensland and southern New Guinea. Another species often placed in the same genus occurs in the eastern Orient. This genus spans the gap between Asia and Australia in a way that no other living primary-division fishes do. The osteoglossids have been thought of as ancient since Regan's time (1911) but, known first in the early Tertiary, they are no older as fossils than the Ostariophysi or some other existing fresh-water fishes, and their relationships do not suggest great antiquity. We do not know to what extent they can tolerate salt water. The African osteoglossid, *Clupisudis,* enters brackish water, and a supposed osteoglossid is fossil in marine strata in the English Eocene. Whether *Scleropages* reached Australia through river systems over an ancient land connection or partly through the sea I shall not try to say. It is surely a case which should be interpreted with caution.

The Australian lungfish, *Epiceratodus forsteri,* is known to live only in the middle parts of two small river systems, the Mary and the

Burnett, in subtropical southeastern Queensland. It is not directly related to the living lungfishes of Africa and South America but to fossil forms which appeared in the Triassic and occurred over most of the world in the Mesozoic. It is a relict of a family much more ancient than the osteoglossids. It is the only undoubted relict of an ancient fresh-water family outside the limits of the great continental fish faunas. But we know so little of the tolerances of fishes, especially of ancient ones, that we cannot say for sure that even this one must have reached Australia entirely through fresh water.

Within the general limits of distribution of primary-division fishes, each separate group of them is still more limited. No family is cosmopolitan or even nearly so. The Osteoglossidae and Nandidae might be called tropicopolitan (except that nandids do not reach the Australian Region), but their representatives are somewhat localized within the tropical regions; and it is not quite certain that they have dispersed entirely through fresh water. The family Cyprinidae, which extends over almost the whole of tropical and temperate Eurasia, Africa, and North America, but which does not reach the Australian Region or South America, has the widest essentially continuous distribution of any primary-division family. A few primary-division genera, such as *Barbus* (Cyprinidae), *Clarius* (catfishes), *Anabas* and *Ophicephalus* (labyrinth fishes), and *Mastacembelus* (spiny eels), extend over most of the length of Africa plus most of the width of southern Asia. *Esox* is almost continuously distributed across northern Eurasia and northern and eastern North America. A few other primary-division genera are common to Eurasia and North America but are more discontinuous in distribution. The most widely distributed species of strictly fresh-water fish is probably the Pike (*Esox lucius*), which occurs across the whole width of northern Eurasia and North America.

Zonation of fresh-water fishes

Within the limits described above, many fresh-water fishes range along climatic zones more than along continuity of land. The north-temperate zone and the tropical zone of the world have great, distinct assemblages of primary- and secondary-division families. The arctic and south-temperate zones lack well-defined faunas of strictly fresh-water families but have many peripheral fishes. These fishes, too, are zoned with climate. Peripheral salmonids and their allies dominate arctic fresh waters; they occur southward through most of the north-temperate zone, but nowhere enter truly tropical waters. In the tropics, gobies are the commonest peripheral fishes, although they are not exclusively tropical; and many other tropical marine families occur in fresh water to some extent. In the south-temperate zone, galaxiids and their relatives are the most conspicuously zoned peripheral fishes. The genus *Galaxias,* with forty or more species, all living in or at least entering fresh water, occurs around the world in the cool southern hemisphere, on southern Australia and Tasmania, Lord Howe Island, New Zealand, southern South America, and the southern tip of Africa (see list of fish families for further details). Lampreys, too, are zoned, in the southern hemisphere as well as in the north; in their southern zone, they enter fresh water in southern Australia, Tasmania, New Zealand, and south-temperate South America, but not in South Africa.

Many marine fishes, too, are zoned with climate: the great shore-fish faunas of the tropics around the world are remarkably similar in many respects, and are relatively distinct from the temperate faunas immediately north and south of them (Myers 1940). Some marine families are characteristic of both north- and south-temperate zones but are absent or poorly represented in the tropics (see Myxinidae, Gadidae, and Cottidae in the list of fish families—and see also Hubbs 1952), and others are characteristic of either northern or far southern seas. That so many peripheral and marine fishes, which can disperse by sea, tend to be confined to single climatic zones shows that climate has a profound influence on their distribution. Zonal climate apparently has a profound influence on the distribution of true fresh-water fishes too, although its importance is not always recognized.

To analyze the zonation of true fresh-water fishes, it is best to divide them into two groups suggested earlier in this chapter. In one group go all except the order Ostariophysi, in the other the Ostariophysi alone.

It is mainly the non-ostariophysans that are zoned with climate (Fig. 12). Sixteen families of them are tropical: thirteen occur in Africa; nine, in the Orient (six shared with Africa); and three, in South America (all shared with Africa and two also with the Orient). The seven peculiar African families all belong to ancient or generalized orders and include some very archaic forms. The three peculiar Oriental families are higher, spiny-rayed fishes presumably more recent in origin. As compared with Africa, the Orient is notable for its lack of archaic fresh-water fishes. The South American families are notable for their small number, for the fact that all of them occur also in the Old World tropics, and for their phylogenetic diversity: they include one group of archaic non-teleosts, one of Isospondyli, and one of spiny-rayed fishes. Although these sixteen families form a zonal fauna,

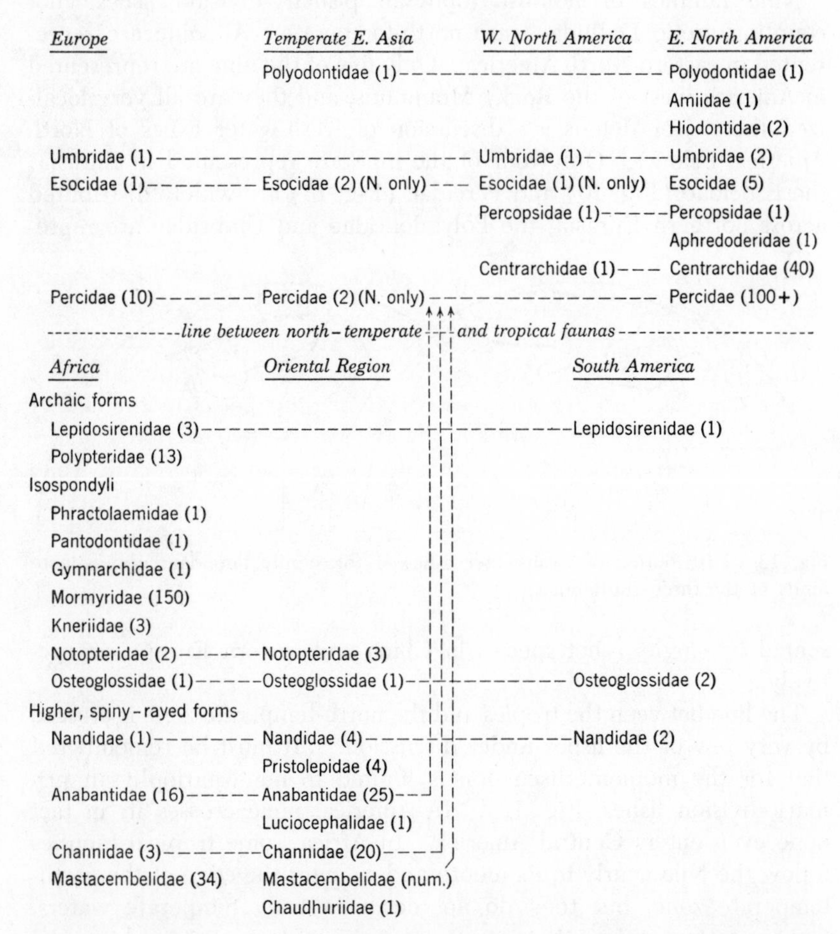

Fig. 12. Zonation of non-ostariophysan primary-division fresh-water fishes. Slightly modified from *Quarterly Review of Biology*, **23**, 1948, p. 9, Table 2. Numbers in parentheses are numbers of species but are only approximations in the larger families. Arrows indicate tropical Oriental families that extend into temperate eastern Asia. (A small percid that reaches western Mexico is not listed; it has nothing to do with the more northern percids that are discontinuously distributed.)

which is almost confined to the tropics and some parts of which go around the world, it is plain that the fauna differs profoundly on the different continents, and the differences presumably reflect differences in the histories of the continents as well as in the evolution and dispersal of different groups of fishes.

Nine families of non-ostariophysan primary-division fishes (not counting arctic Dalliidae) are north-temperate. All nine are represented in eastern North America. Only five of the nine are represented in America west of the Rocky Mountains, and they are all very localized there (for details see discussion of fresh-water fishes of North America, below). Only four of the nine are represented in Eurasia: the Esocidae (Fig. 10) and Percidae (Fig. 13) are widely distributed across northern Eurasia; the Polyodontidae and Umbridae are repre-

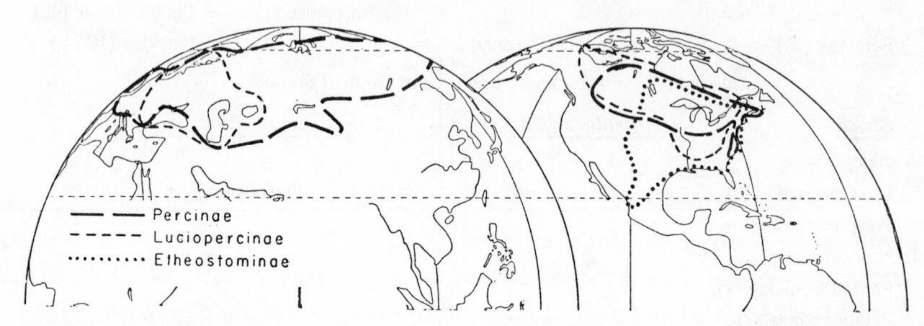

Fig. 13. Distribution of fresh-water fishes of the family Percidae: approximate limits of the three subfamilies.

sented by single, relict species in China and eastern Europe, respectively.

The line between the tropics and the north-temperate zone is crossed by very few of the fishes under discussion. (It must be remembered that for the moment discussion is limited to non-ostariophysan primary-division fishes, Fig. 12). In America, none crosses it; in fact none even enters Central America. In Africa, some tropical families follow the Nile nearly to its mouth and so enter the edge of the north-temperate zone, but they do not occur in truly temperate waters. Only in eastern Asia is there any overlapping of temperate and tropical groups. Of three temperate families in eastern Asia, none reaches the tropics and all are in fact either relict (Polyodontidae in China) or confined to the far north (Esocidae in the Amur and northward, Percidae only north of the Amur). But of the nine tropical Oriental families, three enter the temperate zone: the Anabantidae and Mastacembelidae reach North China, and the Channidae (Ophicephalidae) reach the Amur. These families lack endemic temperate genera; there is little endemism even of species near their northern limits;

and their temperate ranges are much smaller than and continuous with their great ranges in the Old World tropics. All this indicates that, in eastern Asia, temperate families have withdrawn and tropical families have pushed northward into the temperate zone.

The non-ostariophysan primary-division fishes have had diverse origins. The lungfishes and polypterids may have evolved continuously in fresh water, but all the other main groups have probably come independently from the sea, although when and where they did so is unknown. Some of them have apparently had complicated dispersals in fresh water, but others may simply have persisted at and near their places of origin. Fossils show two significant things. First, they show that some families have existed in the north-temperate zone since the Eocene and during the Tertiary have tended to withdraw from Eurasia and western North America and to survive in eastern North America; the distribution of some of the survivors (Fig. 12) confirms the withdrawals. The other fact shown by fossils is that two families have withdrawn from still greater areas. The Osteoglossidae, whatever their origin, have withdrawn from the north-temperate zone and survived in the tropics during the Tertiary. And ceratodontid lungfishes, which occurred in fresh water over most of the world in the Mesozoic, have been reduced to one species in two small Australian rivers.

The fresh-water fishes of the secondary division, all of which are non-ostariophysan bony fishes, have zonal distributions too but overlap climatic boundaries to some extent. Gar pikes now occur in temperate eastern North America and part of Central America, and on Cuba. They are fossil in temperate Asia and Europe as well as in North America in the late Cretaceous and early Tertiary, and in tropical India probably in the early Tertiary. They are therefore an old, mainly north-temperate group, but they once occurred in the tropics in Asia, and they have pushed farther into the American tropics than any other northern fresh-water fishes have done. However, they have not reached South America and do not meet the main South American fish fauna. Cyprinodonts are characteristic of the tropics of the world excepting the Australian Region, but they occur also north and south of the tropics well into the temperate zones, especially in America. Cichlids are chiefly in the African and American tropics, although a few occur in neighboring warm-temperate areas, and a few are on southern India and certain islands (see list of families for details).

Radial distribution of fresh-water fishes

The distribution of the order Ostariophysi is less zonal than that of the non-ostariophysans. Characins and most catfishes are mainly tropical but are not fully zoned. Cyprinids and allied fishes have a pattern of distribution which is not primarily zonal at all, although there is a tendency toward zonation of certain genera or groups of genera of less than family rank. The main pattern of distribution of the cypriniform families is radial rather than zonal; they lie along continuities of land outward from one main distribution center. The center is tropical Asia—the Oriental Region. Homalopteridae are confined to that region. Cobitidae are most numerous and diverse there, but a few of them extend over almost the whole of Eurasia and into northern Africa. Catostomidae (suckers), which now occur mostly in North America, do not themselves have a radial pattern of distribution, but they fit into the radial pattern of the cypriniforms as a whole. Cyprinidae are very diverse in tropical Asia; a good many extend over the whole of temperate Eurasia; comparatively few stocks reach Africa, but some of them extend over the whole of that continent; and a few stocks reach North America and extend over most of that continent too. This is the existing pattern of distribution of cypriniforms. Their fossil record (summarized by Romer 1945, p. 583) is too scanty to prove much, but is consistent with the fishes having formed the existing pattern by evolution in and movements outward from the Orient, chiefly during the Tertiary (Fig. 14).

Stated briefly, the difference between the distribution patterns of the non-ostariophysan primary-division fishes and of the Ostariophysi is this. The non-ostariophysans form a chiefly zonal pattern (Fig. 12): all of the families are confined, or nearly so, to either the tropics or the north-temperate zone; two of the tropical families occur around the world, in all the principal tropical regions of fresh-water fishes; and several north-temperate families occur entirely around the northern hemisphere or did so in the past. The Ostariophysi are much less zonal in distribution: they are much more numerous in species and somewhat more numerous in families than the non-ostariophysans, but no family follows a single climatic zone around the world. Some of the groups are partly zoned, but the pattern of distribution of the most dominant group, the cypriniforms, is primarily radial, and its radiation appears to have been from a center in tropical southeastern Asia. Over a longer period of time, the whole of the Ostariophysi may have radiated from the Old World tropics (Fig. 14). This follows if the South American forms are derived from the Old World, and if, within the Old World, movement has been from the tropics into temperate areas, as it seems to have been.

The zonal and radial patterns are not completely distinct. Although the non-ostariophysans are chiefly zoned, a few of them, espe-

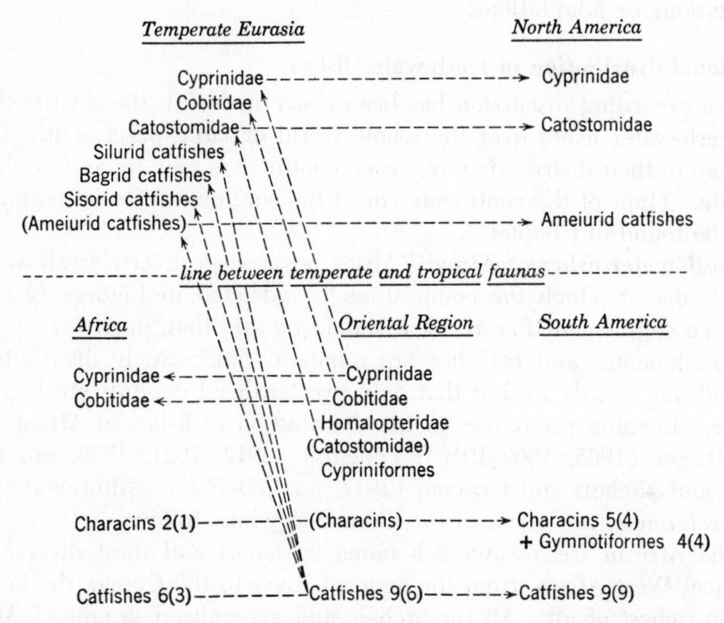

Fig. 14. Radiation of fresh-water fishes of the order Ostariophysi: present distribution and apparent chief lines of dispersal of the families. From *Quarterly Review of Biology,* 23, 1948, p. 11, Table 3. Names in parentheses indicate assumed former occurrence of groups in regions where they no longer exist. Numbers are numbers of families of characins and tropical catfishes in given regions; numbers in parentheses are numbers of endemic families.

cially the labyrinth fishes (Anabantidae and Channidae), have incipiently radial distributions. And although the Ostariophysi have a primarily radial distribution, some of them are partly zoned or localized in single zones. Moreover, the zonal pattern of the non-ostariophysans includes many localizations, discontinuities, and other indications of retreat (fossils). And the radial pattern of the Ostariophysi includes more continuity, and other indications of spreading. All this suggests that the radial pattern may mark an early stage in the dispersal of dominant groups of fresh-water fishes, and zonation a later stage in their history. Zonation is a kind of localization, and it may be a stage through which formerly dominant but retreating groups pass on their way to further localization and final extinction. This suggests a correlation between dominance and pattern of distribution which will be considered in more detail later.

Of course, not all zoned groups need have the same history. Some (perhaps salmonids?) may always be zoned because of their special limitations or adaptations.

Regional distribution of fresh-water fishes

The preceding discussion has been concerned with the distribution of fresh-water fishes over the whole world or large parts of it. The pattern of their distribution over each continent is now to be described briefly. Maps of the continents (or of the continental faunal regions) will be found in Chapter 7.

Fresh-water fishes of Africa. Africa possesses a diverse fresh-water fish fauna of which the composition is indicated in Figures 12 and 14. To supplement Figure 14, it should be said that, in Africa, cyprinids, characins, and catfishes are numerous and widely distributed, though not evenly so, but that only two localized cobitids are known there. Leading references to the distribution of fishes in Africa are Boulenger (1905, 1909–1916); Pellegrin (1912; 1921; 1933, pp. 10–19); and Nichols and Griscom (1917, pp. 739–752). Additional special references are given in the discussion which follows.

The African fresh-water fish fauna is richest and most diverse in tropical West Africa, from the Senegal River to the Congo, the latter being richest of all. All the archaic and generalized groups of African fresh-water fishes (see Fig. 12) are best represented in the west-tropical area, and some are confined to it. Much of this fauna extends through the Chad Basin to the upper Nile, and part of it follows the Nile down to its lower reaches or even to its mouth (Boulenger 1907). Many fishes of the Niger, Lake Chad, and the Nile are identical or closely related, suggesting recent connections between these water systems. The fishes of the Congo, though of the same general groups, are more distinct and have probably been isolated for a longer time. The headwaters of the Nile and affluents of the Congo are actually connected at flood season now, but the connection is probably a recent one and comparatively few fishes seem to have dispersed through it. The Congo and the Zambezi seem to have exchanged some fishes, but not their whole faunas; the Zambezi has a moderately rich fish fauna but lacks polypterids and some other tropical West African groups.

The east-flowing rivers of tropical Africa, other than the Zambezi, are relatively small and have a still more limited fish fauna, again lacking polypterids and other characteristic West African groups. This lack is partly made up by the presence of many species (but few genera) of Cyprinidae, which are comparatively few in West Africa, and which are thought to have reached eastern Africa from the north (from Asia), as a cobitid near Lake Tana has evidently done.

The great lakes of east-central Africa have highly endemic fish faunas dominated by cichlids, although a few other fishes are present (Brooks 1950). Lake Nyasa alone is credited with something like 178 species of cichlids, of which 174 are endemic, in 20 endemic genera.

Southward, south of the Zambezi, which itself has a somewhat depauperate fauna (see above), fresh-water fishes become progressively fewer (details from Barnard 1943, 1948). One genus of characins, but apparently only one, reaches the Limpopo and slightly beyond. Secondary-division cyprinodonts reach south of the Zambezi but fall far short of the tip of the continent. Secondary-division cichlids reach the vicinity of East London but not the actual Cape region. Two primary-division families of catfishes, Clariidae and Bagridae, reach the Orange River or even slightly beyond, but not the final southernmost drainages. The latter, the small rivers that drain the truncated tip of Africa, are reached by only two truly fresh-water fish families: Anabantidae and Cyprinidae. The Anabantidae are discontinuously distributed. The tropical forms do not go south of the Zambezi, but two endemic species (possibly forming a separate genus, *Sandelia*) are isolated on the southern tip of the continent. The Cyprinidae are more or less continuously distributed but have evolved an endemic group of species in the extreme south. That the cyprinids, which are comparatively recent in Africa, are the only strictly fresh-water fishes that range continuously to the southern tip of the continent is a fact to be noted. There are also several peripheral fishes in South African fresh waters, including a galaxiid. The latter, which is placed in an endemic subgenus, is common along most of the southern truncation of the continent and north to the Olifants River on the west. It is a variable species with two intergrading forms, a mountain and a lake (Vlei) form. It apparently does not enter the sea, as some of the Australian and South American members of the family do.

Northward in Africa, the great area of the Sahara Desert is inhabited by a few fresh-water fishes (Pellegrin 1921, 1937). Of primary-division groups there are a few cyprinids and two species of catfishes of the genus *Clarias,* and of secondary-division ones there are a few cyprinodonts and three genera of cichlids. All these families are widely scattered in the Sahara in oases, and some species, the same species that occur above ground, have been found in underground waters too (see references given under Cichlidae in list of fish families). These fishes are probably relicts from a time not long ago when the Sahara was wetter than now. Their geographical relationships are mixed. At least some of the cyprinids and cyprinodonts seem to be northern (Mediterranean) forms. The clariid catfishes and the cichlids are primarily tropical.

Finally, the northwestern corner of Africa (Pellegrin 1921), including the Atlas Mountains and the coastal area north of them, has a fresh-water fish fauna very different from that of the main part of Africa. The species are few, presumably because fresh water is limited and in part seasonal. This region has, of primary-division fishes, only Cyprinidae, some with European and some with "eastern" relationships, and one European cobitid. It has also a few secondary-division cyprinodonts, a trout in the mountain streams and a stickleback near the coast, both, of course, representing northern peripheral families, and an eel and other peripheral fishes. Catfishes, cichlids, and all other typical tropical African fresh-water fishes are absent.

The pattern of distribution of African fresh-water fishes is, then, this: great richness, together with a diversity of ancient stocks equaled nowhere else in the world, from the west-tropical area to the Nile, and richest of all in the Congo; moderate poverty in the east, partly compensated for by cyprinids of northern (Asiatic) origin; local radiation especially of cichlids in the great lakes; progressive poverty southward to the Cape; still greater poverty in the Sahara, but a few fishes still present in isolated waters; and, with tenuous transition, isolation in the northwestern corner of the continent of a very different, very limited fresh-water fish fauna closely related to that of Europe. This pattern is obviously determined partly by local climates, apparently partly by size of drainage systems (the larger the rivers, the more, and the more diverse, the fishes), and partly by the histories of different drainage systems.

Of the origin and dispersal of the older groups of African fresh-water fishes, little is known. The more recent cyprinids and cobitids have apparently come from Asia. Some of the spiny-rayed groups are apparently also of Asiatic origin (see Fig. 12).

Fresh-water fishes of Eurasia. The composition of the fresh-water fish fauna of Asia and Europe is indicated in Figures 12 and 14, but the figures do not fully show how unevenly the fishes are distributed. There is no single, good study of the distribution of fishes over this whole area, but parts of it are covered by Pellegrin 1923 (Syria); Kosswig 1952 (Turkey), Hora 1934 (Afghanistan); Day 1889 and Hora in various papers (India, etc.); H. M. Smith 1945 (Thailand); Weber and Beaufort 1911–1951 (Sunda Islands, etc.); Nichols 1943 (China); Mori 1936 (eastern Asia and Japan); Berg 1936 (Eurasia except the tropics), 1932 (Europe), and other papers culminating in a final edition of his *Fresh-water fishes of Union of Soviet Socialist Republics* (1948–1949, 3 volumes, in Russian, but with a table in Volume 3, pp. 1196–1233, giving distributions of every species by river basins, with the names of both fishes and rivers in Roman type); and many other authors, of whom a few are cited in the following discussion.

The main part of tropical Asia with the main islands of the Sunda shelf (Sumatra, Java, Borneo) has a large fresh-water fish fauna, which is extremely rich in cyprinids and their allies and moderately rich in spiny-rayed fishes but which lacks all the archaic and almost all the isospondyl families of Africa. The fauna as a whole is relatively uniformly distributed through the *eastern* part of the Oriental Region (Indochina, etc.) to Java and Borneo. Its distribution evidently dates from a recent land connection between the islands and the mainland. In fact the distribution of fresh-water fishes on the islands follows in some detail the old river systems of which the lower courses are now drowned by the sea (Fig. 59, p. 489). This main Oriental fauna includes a great variety of lowland fishes and also some families and parts of families specialized for life in mountain torrents. Many, but not all, of the lowland groups extend westward more or less continuously into peninsular India (Hora 1944). The torrent fishes cannot do this now for lack of a continuous series of mountain streams, but nevertheless some of them are isolated in the hills of southern India. Hora (1949) lists fifteen such species in half a dozen families in peninsular India, widely separated from their relatives or from other populations of the same species farther east. He suggests that these fishes invaded India not much before the Pliocene (too late for fishes of this sort to reach Ceylon), and that they crossed India via the Satpuras and other hills when the latter were more continuous and wetter than now. Many torrent fishes occur in northern (not peninsular) India, in streams on the south face of the Himalayas, but their origin (from the east) is easier to understand.

On the north face of the Himalayas and elsewhere in high central Asia the fish fauna is much more limited and consists primarily of a peculiar group of cyprinids (Schizothoracinae), one genus of cobitids (*Nemacheilus*), and one genus of sisorid catfishes (*Glyptosternum*), all able to live in mountain torrents and all apparently derived from the east. That is, they have apparently dispersed from east to west behind the Himalayas, not from south to north across them (Hora 1937, and other authors). Some other fishes reach the central highlands of Asia from one direction or another but are not so widely distributed there.

In eastern Asia, from the tropics northward through China, there is a steady change of fishes. Some of the tropical families do not reach the temperate zone at all. Homalopterids and sisorid catfishes reach the Yangtze Kiang, in which occur also the temperate-relict paddlefish and Chinese sucker and, in the upper reaches, salmonids. These fishes do not all occur together, but they are all in the same river system. Other tropical families extend for varying distances farther north, and silurid and bagrid catfishes and the labyrinth family Channidae reach the Amur, where they occur with many temperate fishes including a pike and peripheral lampreys, sturgeons, salmonids, cottids, sticklebacks, and *Lota.* This brief statement is, of course, a very great simplification of a complex situation. The south-north transition of main groups of fresh-water fishes is complex, and there is also an east-west transition or perhaps interdigitation (Nichols 1943, p. 5) of the Chinese fishes with those of the interior Asiatic highlands. The pattern is still further complicated by a great diversity of Cyprinidae, there being within this family a complex transition from southern to northern genera. Under these circumstances it is not surprising that ichthyologists argue about where to draw faunal lines. Berg (1936) considered that most of China is dominated by tropical fishes and should be included in the tropical Oriental Region. Mori (1936) considered that only southernmost China should be included in the Oriental Region, the main part of it being Palearctic (Old World north-temperate). He put the dividing line at the southern Nan Shan (Nan Ling) Mountains not far north of the Tropic of Cancer, and Nichols (1943, p. 5), too, finds a not very well-marked faunal line for fishes near the line between the tropical and north-temperate zones. Just how the line is drawn is unimportant compared to the main fact, that the Chinese fresh-water fish fauna is a great, complex, transitional one, the transition covering an enormous area from within the tropics to the Amur River, which is at the latitude of British Columbia.

Northern Asia, north of the Amur and north of the central highlands, has a fish fauna consisting of some cyprinids, a few cobitids, the Pike, a couple of percids, and, especially in the far north, many salmonids and other northern peripheral fishes. One of the latter, a cottid, has radiated in Lake Baikal to form two endemic families (Comephoridae and Cottocomephoridae) with nine genera and sixteen or eighteen species (see list of fish families). Northern Europe shares the greater part of the widely distributed, far-northern Asiatic fish fauna. Much of it reaches England and some, but less, Ireland.

Farther south in Europe are a few, but only a few, other fishes. A big catfish, *Silurus glanis,* which occurs east to the Aral region, ranges west to the Rhine. In southeastern Europe, especially in rivers entering the Black and Caspian Seas, are some endemic or relict percids (several genera and about ten species) and an umbrid minnow, the latter confined to parts of the Danube and Dniester Rivers. The European fresh-water fish fauna is dominated by cyprinids, of which some are wide-ranging but others endemic or relict, especially in southeastern Europe, in the Balkans. Mediterranean Europe has a few secondary-division cyprinodonts. A catfish of the Asiatic genus *Parasilurus* is isolated in Greece. But southern Europe apparently has no primary-division fishes derived from the south, from Africa. This poor European fauna becomes poorer still in Spain and southern Italy, where the only primary-division fishes are a number of cyprinids and two cobitids (which are cypriniforms)—no catfishes, no Pike, no percids. This lack is compensated for by the presence of various peripheral fishes of Mediterranean (warm-temperate) groups.

The southwestern corner of Asia, nearest Africa, from Asia Minor to Arabia and Iran, is not very hospitable to fishes and has only a limited variety of them. Most are cyprinids, of which a few brook-living species of the widely distributed (African-Asian) genera *Barbus* and *Garra* occur even in isolated waters in southern Arabia (Trewavas 1941). There are in southwestern Asia also a few cobitids and cyprinodonts and a very few other primary- and secondary-division fishes of diverse relationships: African cichlids reach Syria and Palestine and have evolved an endemic genus and several species there; a very few catfishes of widely distributed tropical groups are in Syria and the Euphrates-Tigris system, and a (tropical) mastacembelid is in the Euphrates-Tigris. The big European catfish and the Pike reach northern Asia Minor. A little farther east, in Afghanistan and probably also in southern Iran, various Indian fishes appear; and in northern Afghanistan are a (northern) trout and peculiar groups of cyprinids and catfishes from central Asia. It will be seen that dry southwestern Asia is a region of double transition of fish faunas, the transition being from African to Oriental forms in one direction, and from tropical to northern forms in another. The details of the double transition are not yet very well known.

The main pattern of distribution of fresh-water fishes in Eurasia is, then, this: richness, in spite of absence of archaic groups, in the main tropical part of Asia and on the recent continental islands of the Sunda shelf; poverty and specialization in the central highlands; transition involving many groups over a great area through eastern Asia from the tropics north to the Amur; wide east-west distribution of special northern groups, with evolution of endemic families (of peripheral origin) in Lake Baikal; poverty in Europe, increasing westward and in Spain and southern Italy, but with a few additional endemic and/or relict groups in southeastern Europe; and poverty and two-way transition in southwestern Asia.

As to movements of Eurasia fishes, cyprinids and their allies and perhaps certain other fishes seem to have evolved in and spread north and west from the Asiatic tropics, and other fishes, particularly the older north-temperate forms, seem to have withdrawn before them. This is suggested by the relationships and distribution of the cypriniforms themselves and by their obvious success in a great variety of climates and habitats; by the pattern of distribution of the whole fish fauna in eastern Asia, where north-temperate fishes seem to have withdrawn and tropical groups including cyprinids seem to have spread northward into the temperate zone; and by the small number of other fishes in temperate Eurasia, and by the relict distributions of some of them. The movements of fishes in temperate Asia and Europe during and after Pleistocene glaciation is a special subject not treated in detail here; references to it are given in Chapter 10.

Fresh-water fishes of North America. The composition of the fresh-water fish fauna of North America, too, is indicated in Figures 12 and 14. This fauna consists of a larger variety of non-ostariophysans than occur in temperate Eurasia but of a smaller basic variety of Ostariophysi, although the three ostariophysan families that do occur in North America are numerous in species there. (Two additional South American families of Ostariophysi just reach the southern edge of North America.) Known North American fresh-water fishes were all included in the check list of Jordan *et al.* (1930). There is no single good summary of the distribution of fishes over North America (one is much needed), but there are a number of good local faunal works, of which I have found Hubbs and Lagler (1947) especially useful.

The main North American fresh-water fish fauna is concentrated in the east-central part of the continent (see Fig. 12), in the Mississippi River and its tributaries. The much smaller, east-flowing rivers of the Atlantic coast are all poorer in fishes than the Mississippi, and their poverty becomes acute in the northeast. In New Hampshire (Carpenter and Siegler 1947), for example, the native primary-division fishes are only two *Esox,* three small centrarchids, three percids, and, of Ostariophysi, fourteen small cyprinids, three catostomids, and two ameiurid catfishes; the Pike, two black basses, three other centrarchids, the Pike Perch, etc., are introduced. In Nova Scotia (Livingstone 1953), native, strictly fresh-water fishes are still fewer: only one percid, eight cyprinids, one catostomid, one catfish, and one (secondary-division) cyprinodont; all twelve species occur on the isthmus which connects Nova Scotia to the mainland, but fewer than half of them reach the southern tip of Nova Scotia or northern Cape Breton Island, and some isolated Nova Scotian streams have no strictly fresh-water fishes. The sparsity of fishes northeastward is probably partly a result of the failure of some of them, especially the larger ones, to return to the more isolated coastal rivers since the retreat of Pleistocene ice.

To the southwest and west the great fish fauna of the Mississippi diminishes rapidly. Of the non-ostariophysan primary-division families, only the Centrarchidae and Percidae reach Mexico, finding their southern limits about on the line of the tropics there. All three North American families of Ostariophysi go south slightly or considerably beyond this, however, and so do secondary-division groups. Their limits are given in more detail in discussion of the fresh-water fishes of Central America (pp. 72ff.).

In the western United States, west of the continental divide, which follows the Rocky Mountains, there are only three known native species of primary-division non-ostariophysans, each representing a different eastern American family, each forming an endemic genus, and each confined to a different drainage system. They are *Novumbra hubbsi* (Umbridae) in the Chehalis River of Washington; *Columbia transmontana* (Percopsidae) in the Columbia River of Oregon and Washington; and *Archoplites interruptus* (Centrarchidae) in the Sacramento and San Joaquin river system in California. A fourth species, *Rafinesquiellus pottsii,* representing still another family (Percidae), reaches west-flowing rivers in Mexico. The Pike (Esocidae) reaches western drainage only in Alaska. The Trout Perch (Percopsidae) reaches at least the upper part of the west-flowing Yukon. And the Blackfish, *Dallia,* is, of course, confined in America to west-draining areas in Alaska. Of Ostariophysi west of the divide, there are catostomids and a number of small cyprinids but no native catfishes except two or three in western Mexico. Southward in the west there are also some secondary-division cyprinodonts. Some of the cyprinids and cyprinodonts occur in relict waters in even the most arid parts of the southwest.

Northward, too, the rich primary-division fauna of east-central North America diminishes progressively. A rather small part of it enters the arctic along the Mackenzie; a still smaller part occurs in truly arctic habitats on the tundra; and primary-division fishes apparently entirely fail to reach some remote tundra regions, including, perhaps, parts of the Labrador peninsula. Only three primary-division species are widely distributed in Alaska, although at least two more may enter Alaska along the Yukon. The three are the Pike, the Northern Sucker, and *Dallia.* All three, but no more-southern primary-division American species, occur also in Asia across Bering Strait, and this suggests a recent, cold land bridge there. Most of the arctic fresh-water fishes of North America are, of course, peripheral ones much like those of Asia and Europe, and they show transition or relationships in both directions. For example, several species of salmon are common to northwestern North America and northeastern Asia, and the Atlantic Salmon (*Salmo salar*) is common to northeastern North America and Europe. Of primary-division fishes, none (except the circumpolar Pike) is common to North America and Europe, but there are relationships between some eastern North American and European forms, including *Umbra,* some percids, and some cyprinids. Berg (1932, p. 178) has discussed this in more detail.

The main pattern of distribution of fresh-water fishes in North America is, then, this: richness and diversity in the east, especially in the largest river system, the Mississippi; some reduction of richness along the Atlantic coast, and increasing poverty northeastward; progressive poverty, at least of primary-division groups, southwestward and through Mexico; limitation west of the continental divide to cyprinids, catostomids, cyprinodonts, and a very few, scattered representatives of other eastern families (with some peripheral fishes chiefly near the coast); and transition northward to an arctic fauna of a few primary-division and many peripheral fishes closely related toward northern Asia and (in the case of some northeastern peripheral forms) northern Europe.

As to origins and movements, much of the existing North American fresh-water fish fauna is apparently part of a former, Tertiary fauna which was common to Europe, northern Asia, and North America (Berg 1932, p. 178). Most of the non-ostariophysan primary-division families of this fauna (Fig. 12) have died out in western North America (they have declined or died out in Europe and especially in Asia, too) and survived or even radiated, as the percids have done, in eastern North America. There are also some important non-ostariophysans (*e.g.*, perhaps the centrarchids) in eastern North America which may have originated there. Finally the three families of Ostariophysi in North America seem to be immigrants from eastern Asia. Just when they arrived is uncertain: ameiurid catfishes and catostomids may have come in the early Tertiary, cyprinids in the Miocene. The effect of the Pleistocene on North American, as on Eurasian, fishes is visible in several ways noted in Chapter 10.

Fresh-water fishes of South America. The South American fresh-water fish fauna is remarkable on the one hand for its small number of ancestral stocks and on the other for richness and endemism in certain groups of Ostariophysi. Of non-ostariophysan primary-division fishes (Fig. 12), only three families are known in South America, and they total only five known species there. Of Ostariophysi (Fig. 14), South America has only characins, gymnotid eels (which are derived from characins), and catfishes (no cypriniforms), but they total nearly 2000 known species in South America, with many endemic genera and about 17 endemic families, plus one family (Characidae) which is shared with Africa. South America has also a number of secondary-division cyprinodonts and cichlids and some peripheral fishes. This fauna has long attracted the attention of ichthyologists, and the main features of its distribution have been worked out especially by Eigenmann (1909, 1910, 1927, 1942 [with Allen], etc.).

The main fresh-water fish fauna of South America centers in the Amazon and its tributaries. This is by far the richest river system in the world in number of fishes. Known species of Amazonian fishes of all kinds now number well over 1000 and may approach 2000 when the river is fully investigated (Myers 1947–1949, Part 2, [p. 13]). Of catfishes alone the Amazon system is known to have more than 450 species, compared with less than 20 species from the whole of the much better known Mississippi system (Myers)! Characins almost equal the catfishes in variety and far surpass them in numbers of individuals, and the Amazon possesses also its share of gymnotid eels, all three families of non-ostariophysan primary-division fishes that occur in South America, various secondary-division cyprinodonts and cichlids, and some peripheral fishes. Many groups of Amazonian fishes but not all, for example not the lungfish, reach the Guianas and the Orinoco; and many but again not all, for example not the osteoglossids or nandids, reach the Paraguay-Paraná system. The headwaters of these rivers come close to the Amazon system and may even be connected with it. The somewhat smaller and more isolated Magdalena near the northwestern corner of South America, though still a fine river, has a more limited fish fauna (Miles 1947), which lacks, for example, all three of the non-ostariophysan primary-division families. The known fishes of the Magdalena number 149, as follows: 58 characins, 6 gymnotid eels, 67 catfishes, 5 cyprinodonts in 2 families, 4 cichlids, and 9 peripheral fishes in 7 families. The eastern coastal rivers of Brazil, south of the Amazon and north of the Paraná, are for the most part rather small. Their fish faunas are limited but are related to that of the Amazon.

The narrow western watershed of tropical South America, west of the Andes, has a very uneven fish fauna. There are fair numbers of true fresh-water fishes on the Pacific watershed of Colombia and Ecuador; the fishes on the two sides of this part of the Andes are in fact more alike than those on the two sides of the Rocky Mountains. Some of the western Colombian forms have apparently come around the northern end of the mountains from the east. Those of the western slope of Ecuador have apparently been derived from the Amazon either across the Andes or before the Andes were formed (Eigenmann 1927, pp. 5–6; see also Eigenmann and Allen 1942). Below Ecuador, from Peru southward, the narrow, dry Pacific watershed west of the Andes has few fresh-water fishes.

The high Andes themselves have a limited, specialized fauna of torrent fishes consisting chiefly, at highest altitudes (sometimes up to 14,000 feet) in the equatorial zone above the Amazon, of one genus of characins (*Lebiasina*) and two of catfishes (*Astroblepus* and *Pygidium*) (Myers 1947–1949, Part 4, p. 12). This fauna follows the Andes for nearly their whole length, from their northern end south into Chile and Patagonia, with some local changes in composition: *e.g.*, the genus *Hatcheria*, related to *Pygidium*, is endemic to parts of Chile and northern Patagonia. At somewhat lower altitudes in the equatorial Andes other fishes appear, living in moderately swift water, and below 2000 or 3000 feet the number and variety increases rapidly and there is complex transition to the lowland fauna. The latter is extremely complex ecologically as well as very rich (Myers 1947–1949, Part 4). One portion of the high Andes, the Altiplano of southern Peru, including Lake Titicaca (over 12,000 feet), possesses an endemic subfamily of cyprinodonts, the Orestiinae, of four genera and about twenty species, most in Lake Titicaca itself (Brooks 1950, p. 162). There is only one other fish, a catfish, in the Titicaca basin.

Much of the tropical lowland fish fauna goes south to the mouth of the Paraguay-Paraná system, following these rivers into the south-temperate zone (Eigenmann 1909, p. 230) as some African fishes do the Nile into the north-temperate. Except for this, there is a steady reduction of the fresh-water fish fauna southward, beginning even in southeastern Brazil, and below Buenos Aires the poverty of fresh-water fishes becomes acute. Of primary-division fishes, only a few characins and catfishes reach Chile and Patagonia. Some of them are closely related to and obviously derived from widely distributed tropical South American forms, but one of the genera of catfishes is supposedly the most primitive living American catfish and is placed in a family of its own, Diplomystidae, confined to parts of central Chile and central Argentina, and another catfish, *Nematogenys*, is supposedly the most primitive living pygidiid and is confined to part of Chile (Eigenmann 1927, p. 13). The geographical history of these genera is unknown, but it is a reasonable guess that they are relicts of comparatively primitive groups which have been replaced in the tropical part of South America. The farthest south reached by any characin may be 41° 18′, by *Cheirodon*, which lives in quiet lowland waters (Eigenmann 1927, p. 14). Hardy, mountain-and-stream-living catfishes of the family Pygidiidae go much farther south than this, the genus *Hatcheria* reaching at least 47° 30′ S. (Eigenmann 1918, pp. 269, 283). Beyond them are only peripheral fishes, which occur to the southern tip of the continent: lampreys, galaxiids, and *Aplochiton*. These are all "antarctic" peripheral groups which occur also in New Zealand and southern Australia, and which in South America become progressively fewer northward as the tropical fresh-water forms do southward. According to Eigenmann (1927, p. 15) the antarctic fish fauna as a whole extends north only to Valdivia, one *Galaxias* and *Aplochiton* continue north to Concepcion, and only lampreys reach Valparaiso. The antarctic fishes go less far north on the Atlantic side of South America. In the warmer parts of south-temperate South America are a few peripheral fishes of other groups: a mugilid, atherinids, and serranids.

The main pattern of distribution of fresh-water fishes in South America is, then, this: central richness (in spite of poverty of ancestral stocks) greatest in the Amazon, somewhat less in adjacent river systems north and south, and still less in the more isolated Magdalena; moderate poverty in western drainages of Colombia and Ecuador; progressive poverty southward beginning in southeastern Brazil but interrupted by the south-flowing Paraguay-Paraná system; poverty and specialization of torrent fishes in the Andes, and radiation of certain cyprinodonts mostly in Lake Titicaca; and transition in the far south to a completely different fauna, of antarctic peripheral fishes.

The origin of the South American fishes will be considered later. The present richness of the fauna is obviously due almost entirely to evolution on and movements within the continent. There has been much exchange of fishes among the large rivers, but the Amazon, the largest of all, still has much the largest and most diverse fish fauna; there is here the same general correlation between size of rivers and both number and diversity of fishes which has been noted in Africa and North America. In general, fishes seem to have spread from the Amazon in all possible directions (Eigenmann 1909, p. 371). The northern Andes seem not to have been much of a barrier; a number of fishes have got around or across them. Southward from the Amazon the fish fauna changes mostly by subtraction (this phrase is from Eigenmann and Allen 1942, p. 62), and the subtraction continues until, south of the La Plata region, only a few characins and catfishes are left, and then only catfishes, and finally no strictly fresh-water fishes at all. Eigenmann tried to see under this pattern evidence of a former division of South America into separate parts, a main Archamazonia and a southern Archiplata (1909, p. 370), or an Archibrazil and an Archiguyana (Eigenmann and Allen 1942, p. 35), each with its independent fish fauna, but I cannot see it. There seems to be only one main South American fauna of true fresh-water fishes, which radiates from the Amazon, and parts of which are extended or subtracted or localized or specialized in different places about as would be expected in the diverse habitats of a continent like South America.

Fresh-water fishes of Central America. Although Central America is neither a continent nor a principal faunal region, it is a transition area of such importance that it deserves separate treatment. Some earlier discussions of the distribution of fishes in this area are by Regan (1906–1908, pp. XII–XXXII), Eigenmann (1909, pp. 309–310), and Myers (1938, pp. 350–351). All Central American fishes known at that time are included in the list of Jordan *et al.* (1930). The fresh-water fishes of Panama, which is a critical area, are treated by Hildebrand (1938). And papers by Buen (1946) and Miles (1947) on the fishes of Mexico and of the Rio Magdalena, though not directly concerned with the Central American fauna, are important in connection with its nature and origin. A map of Central America will be found in Chapter 7, p. 457.

The fresh-water fish fauna of Central America is very depauperate. It completely lacks all the non-ostariophysan primary-division families (Fig. 12) of both North and South America. The disappearance of these fishes actually begins on the continents far short of Central America itself. Of the nine families of them in temperate North America, only two (Centrarchidae and Percidae) reach Mexico. Each of these families is represented by several species in the Rio Grande system, but only one species of each goes farther south, about to the edge of the tropics, the centrarchid east and the percid west of the continental divide. Of the three families (only five species) of non-ostariophysan primary-division fishes of South America, none reaches even the Magdalena or the smaller rivers of northwestern South America.

The Ostariophysi do better, but even they are few in Central America, and they too are reduced in numbers even on the adjacent edges of the continents. Of North American Ostariophysi, the cyprinids go south, in decreasing numbers, to the Rio Balsas in southern Mexico, within the tropics but short of the Isthmus of Tehuantepec. Catostomids go south to Guatemala, also in decreasing numbers: two North American genera stop in the northern edge of Mexico; a third (*Moxostoma*) stops in south-central Mexico; and a fourth, *Carpiodes,* alone reaches Guatemala. Ameiurid catfishes disappear progressively southward in a similar manner, only *Ictalurus* reaching Guatemala. Of South American Ostariophysi, there is some loss of main groups even in the Magdalena. A good many do reach Panama, but very few go farther. In fact, few of those that enter eastern Panama from South America reach even western Panama. Hildebrand (1938, pp. 221, 247) reports 19 genera, 32 species of characins in eastern and central Panama, but only 5 genera, 9 species of them in western Panama; and against 7 species of loricariid catfishes, some of them common, in central Panama, only one is known in the western part of the country. Still fewer South American ostariophysans range farther into Central America, and they reach different limits. Of characins, *Bramocharax* appears to be endemic in Lake Nicaragua; and of South American genera, *Roeboides* reaches Guatemala; *Brycon,* Guatemala and Yucatan; and *Astyanax,* tributaries of the Rio Grande in Texas. The single gymnotid eel that is known to range beyond central Panama reaches Guatemala. Of South American fresh-water catfishes, apparently only pimelodids chiefly of the genus *Rhamdia* go far into Central America, reaching in fact to the Rio Panuco on the east coast of Mexico. I give these details with some trepidation, for they are probably incomplete and perhaps in some cases wrong, but it is only by giving them that the progressive poverty of Central American primary-division fishes can be made clear. These fishes are not only few in number but very poor in endemic genera in Central America.

The poverty in strictly fresh-water groups is partly made up by disproportionate abundance in salt-tolerant, secondary-division ones. Cichlids are much more numerous in Central than in South America, and the area of their abundance is almost exactly complementary to that of the primary-division fishes: the cichlids become numerous in western Panama (Hildebrand 1938, p. 221), where the poverty of primary-division fishes first becomes acute, and, although several genera of them reach Mexico, they taper off there northward, only a single one reaching the Rio Grande and none coming into contact with the full North American fish fauna. The cichlids in Central America show considerable endemism of genera as well as species, although most or all of them seem to be derived from the South American genus *Cichlasoma* (Regan, Myers). Cyprinodonts too are disproportionately numerous, with a high degree of endemism, in Central America (Myers 1938, pp. 350–351). Gar pikes, which are, of course, now primarily North American, go south through eastern Mexico and Central America to Lake Nicaragua. There are also a fair number of peripheral fishes in Central American fresh waters. The composition of the fresh-water fish fauna of the more isolated parts of Central America is exemplified by the fishes of the Yucatan Peninsula (Hubbs 1936), which include only one species (with two subspecies) of characin and one species (with several localized subspecies) of pimelodid catfish, and no other primary-division forms, but, in the secondary division, thirteen species of cyprinodonts and six of cichlids, and several peripheral fishes.

The pattern of distribution of fresh-water fishes in Central America is, then, this: great poverty in strictly fresh-water groups (no primary-division non-ostariophysans, few Ostariophysi), the poverty being progressive, North American groups becoming fewer through Mexico and reaching only to Guatemala, and South American groups becoming fewer with distance from South America, although a very few of them extend along the whole length of Central America to Mexico, and one of them reaches the Rio Grande; and relative abundance in certain salt-tolerant groups, of which the area of abundance is complementary to that of the strictly fresh-water ones, and among which there is more endemism than among the strictly fresh-water forms.

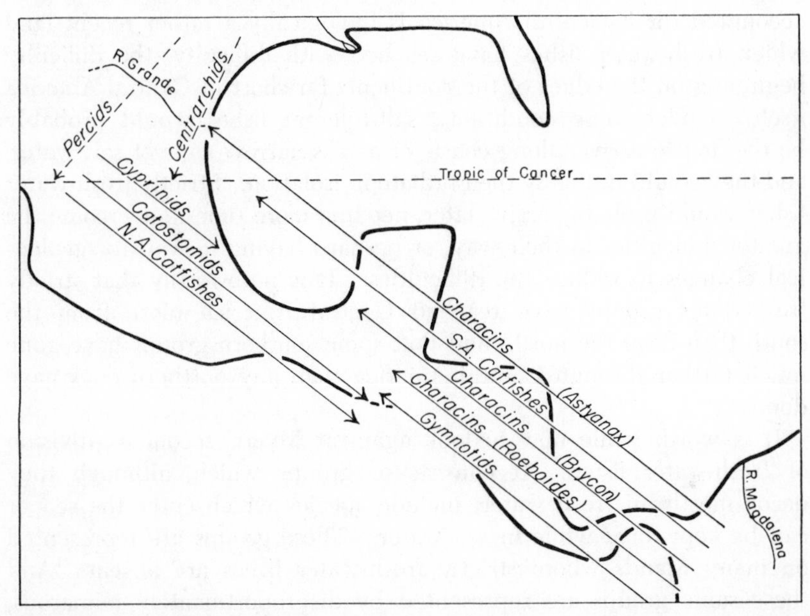

Fig. 15. Transition of strictly fresh-water fish faunas in Mexico and Central America. Broken arrows show extensions of two North American non-ostariophysan primary-division families into Mexico. Solid arrows show extensions of Ostariophysi from North and South America. The arrows show approximate distances the fishes have moved into the transition area, but the arrow heads do *not* mark exact localities reached.

Considered more broadly, Central America is the meeting place of the fringes of two great fish faunas, one north-temperate and the other tropical (Fig. 15) but the transition of faunas involves a much wider area than Central America alone. The North American fauna is best developed in the Mississippi, and subtractions from it occur progressively westward and southward into and through Mexico to the final limit (of strictly fresh-water forms) in Guatemala; and the South American fauna is best developed in the Amazon, and subtractions from it begin even in the other main rivers of South America and become obvious in the Magdalena and acute in Panama, although a small fraction of the fauna goes through Central America and reaches the edge of North America. And in the absence of all except a very few strictly fresh-water fishes, there has evolved in Central America and part of Mexico a characteristic, endemic fauna derived from salt-tolerant groups.

The nature of the Central American fresh-water fish fauna is well accounted for if Central America is biologically a rather recent land which fresh-water fishes have reached with difficulty, the difficulty beginning on the edges of the continents far short of Central America itself. Under these conditions, salt-tolerant fishes would probably be the first to arrive, along coasts or across narrow gaps of salt water, and they would probably then radiate in isolation. Strictly fresh-water fishes would probably arrive later, needing more time to overcome the greater difficulties in their way, or perhaps having to wait for geological changes to reduce the difficulties. It is noteworthy that strictly fresh-water groups have reached Central America more from the south than from the north, and that some southern groups have gone much farther through Central America than any northern ones have done.

It is worth while now to look again at Myers' secondary-division of fresh-water fishes. It consists of groups which, although they occur mostly in fresh water, include species which enter the sea or can be kept for a while in sea water. These groups are represented on many islands where strictly fresh-water fishes are absent. And these same groups are represented by disproportionately numerous, highly differentiated forms in Central America. These facts go together. They show important things about the dispersal of the fishes and about the history of some of the lands where the fishes occur. We may, if we wish, doubt the "reality" of Myers' divisions, but how can we doubt their usefulness in zoogeography?

It is worth while to look again also at the contrasting distributions of the non-ostariophysan primary-division fishes and the Ostariophysi. Of the non-ostariophysans, there are a number of families in North America, and some are old and some are abundant there. Of the Ostariophysi, there are only three families in the main part of North America, and at least some of them are probably rather recent there. But all of the ostariophysan families go farther south in North America than any non-ostariophysan primary-division fishes do. (These three families of Ostariophysi seem to have come from Asia and passed right through an older, non-ostariophysan fish fauna in temperate

North America.) This pattern is mirrored in the south: of South American primary-division fishes, only Ostariophysi reach Central America. This is one part of a general difference already noted: that over the world as a whole the non-osteriophysans are primarily zoned, apparently usually not spreading much and often retreating, while the ostariophysans are radiating and spreading. This is a very significant difference. To reveal it may justify even such a cumbrous phrase as "non-ostariophysan primary-division fishes."

Fresh-water fishes of Australia. Besides the lungfish and osteoglossid already discussed, Australia has many peripheral fresh-water fishes. They are included in McCulloch's (1929–1930) check list of Australian fishes, and their nature and distribution are briefly described by the same author (1927).

The fishes best represented in Australian fresh waters are lampreys, galaxiids etc., catfishes (of marine families), perch-like fishes (of marine families), gobies, and atherinids. Some other marine groups are represented by only one, or very few, fresh-water Australian species. There are also several anguillid eels; and one widely distributed synbranchid and one amphipnoid reach Australia. Finally there are several introduced salmonids and cyprinids, some of the latter now being pests.

Some of the native fresh-water (but peripheral) fishes are very widely distributed over Australia. Some others, notably the lampreys and the galaxiids etc., are zoned, confined to the cool southern part of the continent. There is also some localization of species and genera, but the details are not important here.

The fresh-water fishes of New Guinea, except perhaps the osteoglossid, are all peripheral, and the New Guinean fish fauna as a whole is closely related to that of Australia. In fact, the southern rivers of New Guinea have fishes (and other aquatic animals) of the same species that occur in the rivers of part of the north coast of Australia (Whitley 1943), suggesting (as on the Sunda shelf of Asia) the existence in the Pleistocene of a great river system of which the lower part is now under the sea (see Fig. 59, p. 489). The rivers of northern New Guinea have a related but somewhat different fish fauna, including endemic genera (Whitley 1943).

There is no mystery about the origin of most Australian fresh-water fishes. Except (perhaps) for the lungfish and osteoglossid, they have come directly from or through the sea.

Transitions and barriers in distribution of fresh-water fishes

What happens where different, adjacent fresh-water fish faunas meet tells much about the factors that control fish distribution. Usually there is no sharp line of separation but an area of transition or overlapping with progressive subtractions from both faunas, in opposite directions. The extent of the transition varies with the factors concerned.

Zonal climatic differences help to separate major fish faunas in several cases, but in only one case is zonal climate the only obvious main factor involved. This is in eastern Asia, where tropical Oriental and temperate Eurasian faunas meet in a great transition involving many fishes in an area nearly 2000 miles wide, from the tropics to the Amur River. Farther west in Asia, the zonal climatic difference is reinforced by mountain and desert barriers, and there is only a little mixing of tropical and temperate fish faunas; and between tropical Africa and temperate Europe the zonal climatic difference is reinforced by a desert and a salt sea, and there is no meeting at all of strictly fresh-water fishes, unless perhaps within the family Cyprinidae. In America, tropical and north-temperate fish faunas are separated not only by zonal climatic differences but by a narrow land bridge which was interrupted by salt water until a very few million years ago, and by a partial barrier of dry country in southwestern North America. Nevertheless a few strictly fresh-water fishes of South and North America meet and overlap in Guatemala and Mexico.

In two other cases zonal climatic barriers help to separate fish faunas which themselves are climatically alike. The tropical faunas of Africa and Asia are separated by southwestern Asia, which is warm-temperate rather than tropical, and also relatively dry. This compound barrier has been very effective, for the tropical African and Oriental fish faunas are very different as wholes. Nevertheless a few tropical fishes occur in the barrier region now and a few have evidently trickled through it in the past. The temperate faunas of Asia and North America are separated by a barrier of arctic climate. They are now separated also by salt water, but that is relatively recent; climate has apparently been the main barrier. When Asia and North America were most recently connected across Bering Strait, the bridge was apparently so cold that only arctic fishes crossed it. Longer ago, however, the bridge was evidently warmer and many fishes crossed it.

Deserts are certainly barriers to fishes, but their effect is probably less than is usually supposed. Nowhere are two different, major fish faunas separated by deserts alone, without help of other factors. Some fishes occur in isolated waters in deserts, including the Sahara, showing that the deserts were once wetter than now and less of a barrier to fishes; and ways are often open to fishes around or across deserts, for example the Nile across the deserts of northern Africa. However, although deserts alone may not be very important barriers to fishes during long periods of time, they are important in combination with other factors.

Mountains too are barriers to fishes over short periods of time but are less important over long periods. There are always ways around or across them. Many fishes have got around the mountains of central Asia on the east; that the lowland fish faunas of southern and northern Asia are so different is an effect of climate more than of the mountain barrier. Fishes have got both around and across the northern Andes. The Rocky Mountains of North America are probably the most effective existing mountain barrier to dispersal of fresh-water fishes, but cyprinids and catostomids as well as cyprinodonts are widely distributed on both sides of them; at least three other fishes have crossed them in the United States; and the continental divide has been crossed by at least two more fresh-water fishes in the far north, and two more in northern Mexico. That the fish faunas east and west of the Rockies are so different is obviously owing only partly to the mountain barrier, and partly to habitat limitations to the west.

The only barriers that are fully effective against fresh-water fishes seem to be those of salt water. Moderate gaps of salt water, such as now separate the West Indies from North and South America, Madagascar from Africa, and Celebes from Borneo, may completely stop all strictly fresh-water fishes for a long time. In the very few cases in which supposedly strictly fresh-water species have got across narrow salt-water barriers, there may be doubt about the fishes' salt tolerances.

In general it seems that, excepting salt water, no barriers completely stop the dispersal of fresh-water fishes. What such barriers as zonal climate, deserts, and mountains do is to slow up rates of dispersal and limit or select the numbers and kinds of fishes that disperse. The effect is greatest when two sorts of barriers are combined, as when zonal climatic differences are reinforced by barriers of arid country, but even such combinations do not usually stop dispersal of fresh-water fishes completely. A few fishes get across even the compound barriers in the course of time, and the ones that do so are likely to be either dominant fishes (*e.g.*, Cyprinidae) or fishes with special adaptations and perhaps some salt tolerance (*e.g.*, Cyprinodontidae).

Dominance and competition in relation to distribution of fresh-water fishes

I want now to return to the concept of dominance and its relation to patterns of distribution of fresh-water fishes.

Dominant animals are conspicuously successful ones. It will be remembered that Ostariophysi are the dominant fresh-water fishes, far outnumbering all other primary-division fishes together. Cyprinids and their allies are the predominant Ostariophysi.

The family Cyprinidae is by far the largest family of fresh-water fishes in numbers of genera and species, and individuals are often very numerous too. Within their area of greatest abundance, cyprinids dominate fresh-water fish faunas to an extent that no other fresh-water fishes do anywhere. Cypriniforms, chiefly the Cyprinidae themselves, rule in the whole of tropical and temperate Eurasia. In China, for example, which includes an area from within the edge of the tropics through much of the width of the north-temperate zone, Nichols (1943) finds 268 species of Cyprinidae, 83 of other cypriniforms, 65 of fresh-water catfishes, and only 15 of non-ostariophysan primary-division fishes (plus 2 secondary-division cyprinodonts and 73 peripheral fishes). In Europe, Berg (1932, p. 174) finds 94 species of Cyprinidae, 10 of other cypriniforms, only 2 of fresh-water catfishes, and 13 of non-ostariophysan primary-division fishes (plus 4 secondary-division cyprinodonts and 121 peripheral fishes, many of which occur in fresh water only occasionally).

The family Cyprinidae is fairly diverse in structure, and the cypriniforms as a whole are still more diverse. Structural diversity is not necessarily a criterion of dominance, but it must increase as and if a dominant group radiates.

Cyprinids have also very wide tolerances or great powers of adaptation to different fresh-water environments. Cyprinidae are dominant in both the tropics and the north-temperate zone, and in this they are unique. No other family of fresh-water fishes is strikingly successful in both zones. Moreover, cyprinids have entered and crossed inhospitable places that have been barriers to almost all other fresh-water fishes and have reached many relatively isolated places within the general limits of distribution of the family. In some places Cyprinidae and a few other cypriniforms are the only primary-division fishes. This is the case in Spain and southern Italy and in the temperate northwestern corner of Africa, in southern Arabia, in parts of western North America (the catostomids are cypriniforms), and perhaps in parts of South Africa. In some of these places cyprinids are accompanied by cyprinodonts, but the latter are secondary-division fishes and as a group are much less numerous and less successful than cyprinids. Cyprinodonts seem to owe their presence in isolated and inhospitable places to special characteristics including perhaps small size and ability to withstand not only saline but possibly alkaline water, while the presence of cyprinids in these places is only one detail of a general dominance.

Finally, fossils and other evidence (see again Figs. 12 and 14) suggest that, in temperate Eurasia, spread of cyprinids has been correlated with retreat and disappearance of some other fishes, and it may be guessed that direct competition has occurred and that the cyprinids have had the best of it.

Cyprinidae, then, are very numerous in species and individuals, fairly diverse in structure, and have wide tolerances (or are very adaptable) and are uniquely successful in both tropical and temperate climates and in a great variety of habitats, including relatively unfavorable and isolated ones, so that their distribution is nearly continuous over a great area, in a great range of climate, and across difficult places that have been barriers to most other fresh-water fishes. And cyprinids have apparently competed with and forced withdrawal of some other fishes in some places. All these characteristics go together to make a picture of a very dominant group, with what I have called a radial pattern of distribution. It is among such dominant animals that clues to places of origin and directions of dispersal are most trustworthy, and the clues show that cyprinids and cypriniforms have radiated from a main center in tropical Asia.

The whole order Ostariophysi, to which the cypriniforms belong, is dominant, comprising about seven-eighths of all primary-division fresh-water fishes. Ostariophysi are overwhelmingly dominant in tropical and temperate Eurasia; less so in Africa and North America; but more so in South America, where there are perhaps 1800 or 2000 known fresh-water Ostariophysi to only five species of non-ostariophysan primary-division fishes (plus considerable numbers of secondary-division and some peripheral ones). The distribution of the whole dominant order Ostariophysi falls into a pattern of apparent radiation from the Old World tropics: the Ostariophysi of the north-temperate zone seem to have come from the Old World tropics, and those of South America are part of an originally very limited and probably derived fauna, so that the tropics of the Old World are left as the apparent center of radiation of the order (Fig. 14).

Excepting the cypriniforms and the family Cyprinidae, the separate groups of Ostariophysi do not have obviously radial distributions. The characins and their allies, though numerous in the regions where they occur and very diverse in South America, are confined to Africa

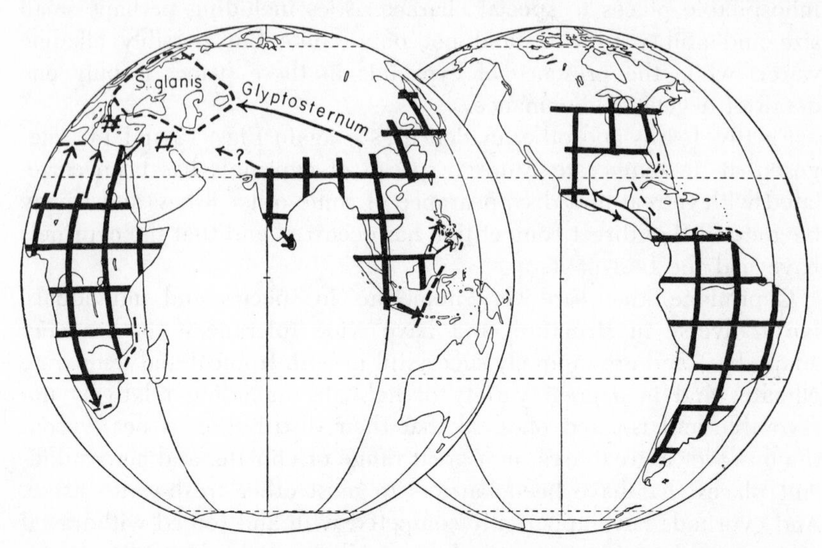

Fig. 16. Distribution of fresh-water (primary-division) catfishes. See text and list of families. Heavy crosshatching indicates principal catfish faunas; arrows, extensions of single genera or small parts of faunas, including extension of *Glyptosternum* across central Asia; light crosshatching, isolated areas with one or more catfishes; broken lines, some limits reached by fresh-water catfishes including the limits of distribution of *Silurus glanis* in Europe etc.

and America and are almost confined to the tropics there. The catfishes are worth more detailed consideration.

Catfishes are numerous, diverse, and widely distributed (Fig. 16). However, they nowhere dominate fish faunas to the extent that cyprinids sometimes do. Catfishes are very numerous only in the tropics. (Even the marine families are chiefly tropical.) In the north-temperate zone, there are fair numbers of them (but nothing like a tropical diversity) only in China and eastern North America. There are few of them in southwestern Asia, only two localized species in Europe, none across northern Asia above the latitude of the Amur River and none anywhere in or very near the arctic, none native in most of western North America, and none in some smaller areas where cyprinids occur (*e.g.*, the northwestern corner of Africa and part of Arabia). Catfishes are therefore not only fewer but also less evenly and less continuously distributed than cyprinids in the north-temperate zone. Besides not going so far north as cyprinids, catfishes do not go quite so far south in Africa, but they go farther south than any other true fresh-water fishes in South America, where catfishes attain their greatest dominance, and where they are almost unbelievably varied in structure and adaptation.

The interrelationships of different catfish families are not well understood, so their evolution and dispersal can only be guessed at. If the Ostariophysi originated in the Old World tropics, catfishes probably began there, but they must have reached South America long ago. If the ancestors of the South American catfishes came through the north, they have disappeared there, which is reasonable enough. Present distributions suggest that several later groups of catfishes have extended from tropical Asia into the north but have not done well there. The ancestors of the Ameiuridae may have moved from tropical into temperate Asia and then to North America. If so, they later died out in Asia and western North America. The Ameiuridae are supposed to be related to Old World forms, and the family is doubtfully fossil in eastern Asia; there is still more doubtful evidence (a painting) of a supposed existing species in China. The Siluridae have perhaps extended more recently from the tropical Orient into northern Asia and Europe. They may now be retreating in Europe, for the two existing European species are isolated and probably relict (see list of fish families). Finally, the Bagridae, which reach the Amur, and the Sisoridae, which reach central Asia, are presumably still more recent invaders, whose fate is not yet indicated.

The catfishes as a whole might be said to have a pattern of distribution that is partly zonal and partly radial. They are primarily tropical, but several groups of them have made minor radiations into the north-temperate zone, from tropical Asia.

Of other fresh-water fishes, the labyrinth fishes come nearest a pattern of radial distribution. They are most numerous and diverse in tropical Asia. Two of the Asiatic genera, *Anabas* and *Ophicephalus*, extend to Africa, and the same two genera extend north in eastern Asia, reaching southern Manchuria and the Amur River respectively. However, labyrinth fishes seem to be absent in western Asia and northern Africa, and this discontinuity suggests that the main (moderate) radiation of these fishes is over and that they are beginning to retreat.

Dominance and the radial patterns of distribution which are in some cases connected with it have been discussed in detail because of their importance. It is very important in zoogeography that some groups of animals are more successful than others, at least at certain times and in certain parts of the world, and that the more successful

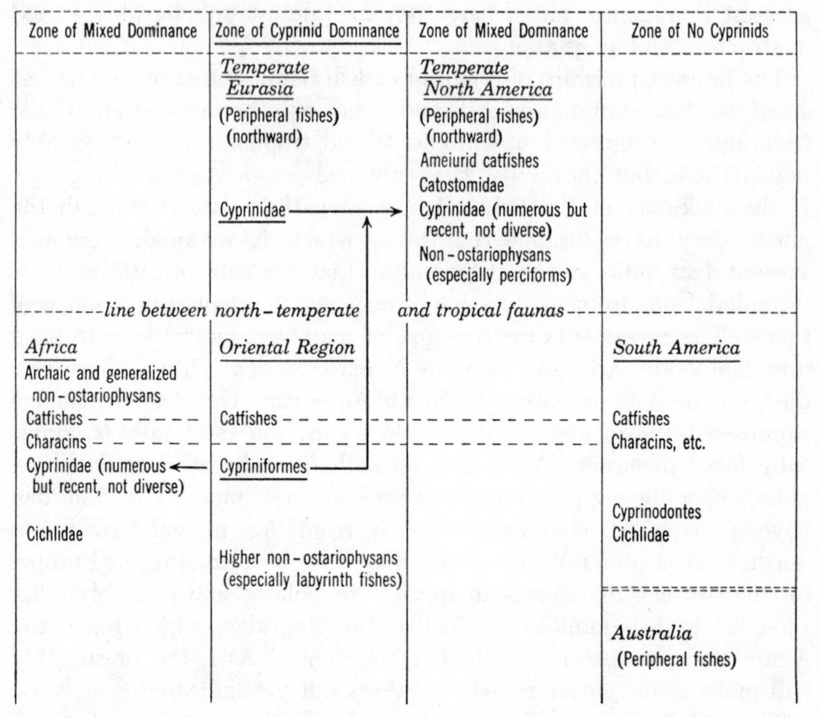

Zone of Mixed Dominance
Zone of Cyprinid Dominance
Zone of Mixed Dominance
Zone of No Cyprinids
Temperate Eurasia
(Peripheral fishes) (northward)
Cyprinidae
Temperate North America
(Peripheral fishes) (northward)
Ameiurid catfishes
Catostomidae
Cyprinidae (numerous but recent, not diverse)
Non-ostariophysans (especially perciforms)
line between north-temperate and tropical faunas
Africa
Archaic and generalized non-ostariophysans
Catfishes
Characins
Cyprinidae (numerous but recent, not diverse)
Cichlidae
Oriental Region
Catfishes
Cypriniformes
Higher non-ostariophysans (especially labyrinth fishes)
South America
Catfishes
Characins, etc.
Cyprinodontes
Cichlidae
Australia
(Peripheral fishes)

Fig. 17. Dominant groups of fresh-water fishes of main continental areas. From *Quarterly Review of Biology,* 23, 1948, p. 12, Table 4.

groups sometimes have special patterns of distribution which override the climatic zones and most kinds of barriers. This fact is as well exemplified by fresh-water fishes, especially by the Cyprinidae, as by any animals.

Dominance is relative, and it implies competition. There are various evidences that fishes do compete with each other. Introduction by man of certain fishes into certain waters sometimes causes reduction or elimination of other fishes. A similar elimination of some fishes by others evidently occurs naturally. The presence of the Pike reduces the numbers of some other fishes in some waters, and the presence of suckers reduces the numbers of trout. The full extent of this sort of thing in nature can only be deduced indirectly. The apparent history of fresh-water fishes in temperate Eurasia, where cyprinids have spread and some other fishes apparently retreated, suggests competition of whole groups. And the complementary distributions of dominant fishes over the world (Fig. 17), especially the general complementarity of cypriniforms and other fishes, suggest competition on the broadest lines. It looks as if the main pattern of distribution of fresh-water fishes is partly a result of competition among great groups of fishes in great areas during long periods of time. The competition has probably been very complex in detail and has probably involved long-term adaptability as well as short-term adaptations.

The qualities and adaptations that make some fresh-water fishes dominant—more successful than others over great areas and in diverse habitats—are not easy to ascertain. All Ostariophysi possess the Weberian apparatus (p. 42) which may sharpen hearing and may account for the order being dominant in fresh water but not in the sea. At least this is a plausible explanation, whether or not it is correct. I know no such simple explanation of the dominance of cyprinids among Ostariophysi. The moderate dominance of the labyrinth fishes may or may not be due partly to their possession of a supplementary air-breathing organ.

Why southeastern Asia should have been the center of radiation of cyprinids, some catfishes, and labyrinth fishes is a separate question. I shall try to answer it later.

Summary: the pattern of distribution of fresh-water fishes

Starting with the fact that fishes occur in almost all fresh waters over the world, the broadest pattern of their distribution is one of limitation and partial complementarity. Strictly fresh-water, primary-division fishes are almost limited to the continents and recent continental islands, excepting Australia. Somewhat salt-tolerant, secondary-division fishes tend to complement the strictly fresh-water ones on islands which are not too remote and in other places, notably Central America, which have been hard for strictly fresh-water fishes to reach. And very salt-tolerant, peripheral fishes, although they occur almost everywhere, are relatively abundant in the far north and far south and on Australia and remote islands where strictly fresh-water fishes and even secondary-division ones are few or absent.

Within their limits, strictly fresh-water fishes form several other broad patterns. The non-ostariophysans tend to be zoned: most families of them follow one or the other of the two main climatic zones around the world or are localized in one zone, and many secondary-division and peripheral fishes and some marine ones are zoned too. The Ostariophysi tend to be radial rather than zonal in distribution: the cypriniforms follow continuities of land from a main center in southeastern Asia, and the whole order Ostariophysi fits into an apparent pattern of radiation from the Old World tropics. However, the zonal and radial patterns are connected by intermediate, apparently transitional cases.

Another broad pattern of fresh-water fish distribution is differentiation of faunas on different continents. The fauna of Africa is a complex mixture of archaic fishes and higher ones, but with relatively few cypriniforms. That of tropical Asia lacks all archaic fishes but is very rich in cypriniforms and fairly rich in higher, spiny-rayed groups, especially labyrinth fishes. That of temperate Eurasia is dominated by cypriniforms. That of North America is fairly diverse; it includes two families of cypriniforms, but they do not dominate the fauna as they do in temperate Eurasia. That of South America is basically limited and entirely lacks cypriniform but is richest of all in endemic groups of other Ostariophysi. And that of Australia almost lacks strictly fresh-water fishes. These differences fall into one main pattern: there is a central area of cyprinid dominance in tropical and temperate Eurasia; next, mixed faunas, including some cypriniforms, on the adjacent continents of Africa and North America; and finally, highly endemic faunas, without cypriniforms, on the more isolated continents of South America and Australia. Figure 17 shows this pattern graphically.

In spite of the great differences in their composition, the fresh-water fish faunas of the different continents have a common internal pattern of distribution. The main part of each fauna is concentrated in the largest rivers and largest favorable areas, and progressive subtractions from it occur in smaller rivers and smaller and less favorable, outlying areas. This pattern is strongly marked in Africa, in Eurasia as a whole, in North America, and in South America. The main fish faunas of adjacent, connected continents are usually separated from each other by unfavorable areas in which comparatively few fishes occur. However, some members of each fauna usually extend into the unfavorable areas and overlap each other, forming transitions in which there is progressive subtraction from each fauna, in opposite directions.

To put all this still more briefly, the main pattern of distribution of fresh-water fishes consists of at least five subpatterns. They are

(1) geographical limitation of strictly fresh-water fishes, and partial complementarity with salt-tolerant ones; (2) zonation; (3) radiation; (4) differentiation of continental faunas within a world-wide pattern in which complementarity of dominant groups is conspicuous; and (5) concentration of each main fauna in a large and favorable area, with subtraction marginally, but with the fringes of adjacent faunas overlapping across most kinds of barriers. These subpatterns are related among themselves. The broadest limitation and complementarity contain or involve the other patterns. Complementarity is partly zonal and is involved in differentiation of continental faunas. Zonation and radiation are connected, and both contribute to the differentiation of continental faunas. And the concentration of main faunas in large and favorable areas, with subtraction marginally, strengthens other patterns, even though it blurs their edges. All this is complex, but that is unavoidable. Animal distribution is complex.

Within this main pattern are an infinite number of local patterns made by local climates, local barriers, local changes in drainages, and local evolutions and extinctions of different fishes. A few of these local patterns have been noted, but most are of necessity beyond the scope of this book.

The place of relicts in the pattern of distribution of fresh-water fishes must be stated for later reference. There are no ancient relicts of cyclostomes or shark-like fishes in fresh water. Existing fresh-water members of these classes appear to have come rather recently from the sea. Of bony fishes, all the non-teleosts are phylogenetic relicts. They are few. In the primary division of fresh-water fishes are only the bichirs, with two genera, thirteen species, in Africa; paddlefishes, with one genus and species in China and another genus, one or two species in east-central North America; the Bowfin (*Amia calva*) in eastern North America; lepidosirenid lungfishes, with one genus, three species in Africa and another genus, one species in South America; and ceratodontid lungfishes, with one species localized in Australia. In the secondary division is only the family Lepisosteidae, with one genus, four or five species in eastern North America, Central America, and Cuba. Of these fishes, only the Australian lungfish is geographically isolated, and it is the only one that is strikingly restricted in present distribution. All the others live with great faunas of true fresh-water fishes, some in the full tropics, some in the north-temperate zone, and all are more or less widely distributed within the regions in which they occur (see list of fish families for details). Of all the main regions in which faunas of true fresh-water fishes occur, only tropical Asia lacks relict non-teleosts. Some of the relicts, notably the lungfishes, may be protected from competition with most other fishes by special adaptations and modes of life, but others are not. *Amia calva*, for example, although the only relict of an order which dominated Jurassic seas and was once widely distributed in fresh water in the north-temperate zone, exists in open and successful competition with other large, predaceous fishes in the present, rich fresh-water fish fauna of eastern North America. These facts do not agree well with the idea, current among biologists, that phylogenetic relicts usually occur in isolated or protected places.

How has the pattern of distribution of fresh-water fishes been formed? The pattern of distribution of fresh-water fishes seems to have been formed mainly by movement of the fishes themselves, by a complex process of evolution, dispersal, limitation, retreat, and extinction over the available parts of a complex world. There is plain evidence, in patterns of distribution and occurrence of fossils, that some groups of fishes have spread very extensively in geologically recent times and that others have retreated extensively. Another indication of motion is that adjacent fish faunas usually are not sharply separated but overlap at the edges in a way suggesting movement in opposite directions of some members of each fauna, so that the pattern of fish distribution is blurred like that of a moving kaleidoscope rather than sharp like that of a motionless jigsaw puzzle.

The most important recent event in the dispersal of fresh-water fishes seems to have been the rise of cypriniforms in southeastern Asia and their radiation from there, accompanied by withdrawal of various other fishes from large parts of Eurasia. An even more important, earlier event was the rise of the Ostariophysi as a whole in fresh water and their radiation probably from the Old World tropics. These two events, imposed on a pattern otherwise primarily zonal, seem to have formed the most significant features of the present pattern.

I have just answered tentatively, for fresh-water fishes only, two of the four questions asked at the beginning of the preface: what is the main pattern of distribution?; how has the pattern been formed? Now, for later reference and to introduce some important principles, I shall give tentative answers, still for fishes only, to the other questions: why has the pattern been formed?; what does it tell about ancient lands and climates?

Why has the pattern been formed? It has been formed because of the nature of fishes and the nature of the world, by complex interaction of the two. This is self-evident, but the details are not all easy to understand. What is there about fishes and about the world that accounts for the five observed subpatterns of fresh-water fish distribution?

The virtual limitation of primary-division fishes to continental areas is due to the lack of salt tolerance of these fishes and to the position of old barriers of salt water. That some fresh-water fishes are limited by salt water is not surprising. It is much more surprising that so many other fishes can tolerate presence or absence of salt. That some groups of fishes can do this, and have thus been able to spread beyond the limits of the strictly fresh-water groups and to reach and evolve in fresh water on isolated lands, has produced the broad pattern of partial complementarity of strictly fresh-water and salt-tolerant forms.

The zonal subpattern of distribution of fresh-water fishes has been formed because whole groups of the fishes have become adapted to single climates, tropical, temperate, or arctic, which form zones around the world. The importance of climate itself is shown by the many peripheral and marine fishes that are zoned too. The zonation of strictly fresh-water fishes is accentuated by barriers, for the tropical and north-temperate zones of the land happen to be separated in most places by areas of arid country and in some places also by mountains or other obstacles to dispersal of fishes. Such barriers are absent only in eastern Asia. Judging by what happens there, it may be guessed that, in the absence of barriers, many fresh-water fishes would still be zoned but that there would not be sharp lines between the zonal faunas.

The radial subpattern of fresh-water fish distribution has been formed because certain dominant groups of fishes have evolved in certain places, especially in tropical Asia, and have spread from there in whatever directions continuity of land has allowed, with relatively little regard for climatic zones or for most kinds of barriers, except salt water.

Differentiation of continental fish faunas has been due to the very complex factors which have caused spreading or receding, survival, evolution, or extinction of different fishes on different continents. The main pattern of differentiation—of cyprinid dominance in Eurasia, mixed dominance on adjacent continents, and absence of cyprinids on more remote continents—is due to whatever causes led to the rise of cyprinids in southeastern Asia.

Finally, concentration of main fish faunas in large, favorable areas suggests a connection between extent of favorable area and size and nature of faunas. This is the first hint of what will prove to be a very important principle of animal distribution.

In summary, it might be said that the broadest pattern of distribution of fresh-water fishes has been determined by three things: old barriers of salt water, zonation of climate, and the nature and evolution of the fishes themselves, with area playing an as yet undetermined role.

What does the pattern tell about ancient lands and climates? North of the tropics, fossils and the general distribution of living fishes show that there has been a distinct fresh-water fish fauna around the world in the north-temperate zone since early in the Tertiary. This suggests that climatic zones, oriented as now, have existed at least since the early Tertiary. And it shows that Eurasia and North America were connected in the Tertiary and that the connection was warm enough to allow exchange of many temperate-zone fishes. Finer details of the distribution of fishes tell many things about changes of climate and of drainage systems in the Pleistocene. At this time Asia and North America were still or again connected, across Bering Strait, but the connection was cold. It was crossed by arctic fishes but not by more southern ones.

South of the tropics, in the south-temperate zone, the distribution of fresh-water fishes tells of no great changes in the world. The primary-division fishes of South Africa and southern South America have evidently come southward from the tropics independently on the two continents. There are no primary-division fishes in southern Australia. So, no primary-division fishes show relationships between the southern extremities of the continents. The galaxiids and other peripheral fishes that do occur around the southern hemisphere have, I think, dispersed by sea, although some persons still doubt it (*e.g.*, Stokell 1945).

Within the tropics, fresh-water fishes have a pattern of distribution so clear and significant that the fishes may well tell more about the histories of the continents than do any other living animals.

Fresh-water fishes suggest no great changes in the relation of the Old World continents to each other.

The fish faunas of Africa and tropical Asia are very different as wholes, but share about ten primary-division families and even a few genera. These two faunas are fairly effectively separated now and they seem to have been so for a long time, except for occasional exchange of a few fishes across southwestern Asia. A few fishes, especially some cyprinids, have been exchanged by this route recently. That earlier exchanges, too, were across southwestern Asia and not across a more direct land connection is indicated by the complete absence of strictly fresh-water fishes on Madagascar. The differences between the fish faunas of Africa and Asia suggest a profound difference in the histories of the two continents. The fauna of Africa is a complex mixture of archaic and more recent fishes, the latter apparently mostly received from Asia. That of tropical Asia consists mostly of more recent fishes of groups which seem to have originated and radiated in tropical Asia. The history of Africa seems to have favored survival of old fishes; of tropical Asia, evolution of new ones. The finer details of distribution of fresh-water fishes in these parts of the world tell many things about the histories of rivers and lakes, but this cannot be gone into at length here. Hora (several papers) has published interesting material on the histories of the fresh-water fishes and rivers of India; and Beaufort (1951, pp. 76–79, 80ff., maps on pp. 86, 104–109) has recently discussed from this point of view the histories of some of the lakes and rivers of India, the Greater Sunda Islands, and Africa.

Fresh-water fishes show no great change in the relation of Australia to other continents. The Australian osteoglossid probably came from Asia, and the Australian lungfish may have come from that direction as well as any other. These fishes are not enough to prove a former land connection, although the lungfish suggests one.

Before concluding this discussion of what fresh-water fishes tell of ancient lands and climates, I want to consider the origin of the South American fish fauna.

Origin of South American fresh-water fishes (a digression). South America has a fish fauna which raises greater difficulties. If we knew how fresh-water fishes reached South America, we would know much about the histories of all the continents since the Cretaceous. We do not know, but we can make some careful guesses, and they are well worth making. It is best to begin by getting firmly in mind the position of South America on the globe and its relation to other continents (see Frontispiece).

South America has the largest of all fresh-water fish faunas, but it is derived from few ancestors. South American primary- and secondary-division fishes belong to only seven basic groups (Table 2). All seven basic groups are represented also in Africa, but only part of them in Asia. Asia does have an osteoglossid, two genera of nandids, and catfishes, but no lepidosirenids, no characins, only one genus of

TABLE 2. THE BASIC GROUPS OF SOUTH AMERICAN FRESH-WATER FISHES

Primary division, non-ostariophysans

1. Osteoglossidae: one family, also in Africa, Orient, Australia (perhaps salt-tolerant)
2. Nandidae: one family, also in Africa, Orient
3. Lepidosirenidae: one family, also in Africa

Primary division, Ostariophysi

4. Catfishes: nine families; other families in Africa, Orient, etc.
5. Characiformes, with derived Gymnotiformes: nine families; characins also in Africa

Secondary division

6. Cichlidae: one family, also in Africa to southern India
7. Cyprinodontes: two principal families; Cyprinodontidae widely distributed, but dominant South American and African genera related; Poeciliidae shared with North America

cichlids, which apparently reached southern India rather recently from Africa, and only a few cyprinodonts, which are apparently not directly related to South American ones. Moreover, there seem to be no direct relationships between South American fresh-water fishes and existing, truly North American groups except in the salt-tolerant Cyprinodontes. The whole of the South American fresh-water fish fauna is primarily African in its relationships. The relationships are old. Only four primary- and two secondary-division families are now the same in Africa and South America, and no genus is the same. A further, important fact is that, although all the main groups of true fresh-water fishes in South America are represented in Africa, there are additional groups in Africa which are not represented in South America. This statement of the case is too simple, and some details of it may be questioned, but it is essentially true. The case has been stated somewhat differently and in more detail by Myers (1938, pp. 348–349), and the general relationship of South American to African fishes was well known to earlier authors including Eigenmann (1909, especially pp. 363ff.).

The unbalance of the South American fresh-water fish fauna goes beyond the fewness of its ancestors. The fauna is overwhelmingly dominated by Ostariophysi, which number something like 1800 or more known species in South America, a little less than half being characins and gymnotid eels, and more than half catfishes (but the characins are said to surpass the catfishes in number of individuals). The three other groups of primary-division fishes in South America total only five known species there. This, of course, does not include the cyprinodonts and cichlids, which are present in fair numbers but which, because of the known salt tolerance of some of them, are less important zoogeographically. A noteworthy detail is that the three families of non-ostariophysan primary-division fishes in South America differ extraordinarily among themselves. The lungfish is a primitive non-teleost, and is a large fish adapted to seasonally fluctuating water. The two osteoglossids are soft-rayed teleosts of the order Isospondyli, and are very large fishes which must live in large and permanent bodies of water. And the two nandids are higher, spiny-rayed teleosts, and are small and can live in small bodies of water. These differences in evolutionary level and in ecological requirements must be taken into account in considering how the fishes reached South America.

The limitation and unbalance of the South American fresh-water fish fauna suggest that it is a derived one, descended from a few immigrants which somehow reached South America from Africa, or from a fauna like that which as a whole is now confined to Africa. In South America, the Ostariophysi, both catfishes and characins, have radiated in isolation, like the cichlids in African lakes, or the cichlids and cyprinodonts in Central America, or some other fresh-water fishes elsewhere, but on a much larger scale. The fewness of ancestors and extent of radiation suggest that the fishes reached South America with difficulty. How they did so is the question.

The Ostariophysi are, I think, the key to the question of the origin of the South American fresh-water fishes, because of their numbers in South America, because (excepting two families of catfishes) they are as strictly confined to fresh water as any fishes, because they are dominant fishes over most of the world, and because for this reason clues to their history are likely to be relatively clear and dependable. However, it should be remembered that even Ostariophysi vary in dominance, and that some of them have retreated from some places.

Fresh-water fishes—Ostariophysi—might conceivably have reached South America in several different ways. They might have come through the sea, if they had enough salt tolerance in ancient times, or across a direct land connection from Africa which no longer exists, or by way of existing northern continents where the fishes no longer exist. The choice among these possibilities, if one can be made, must be made chiefly by clues from existing distributions, for the fishes concerned have not left a good fossil record.

Characins and fresh-water catfishes might have reached South America through the sea if they had a marine origin, or if some of the early fresh-water forms were salt tolerant. An actual origin in the sea is unlikely. Ostariophysi seem primarily adapted to fresh water and probably originated in it. Dispersal by sea of salt-tolerant members of early fresh-water groups is possible but not probable. If these fishes could cross even moderate widths of sea, they should be in Madagascar and perhaps Australia and other isolated places, but they are not, except ariid and plotosid catfishes, which are secondarily able to live in the sea. The complete absence of other Ostariophysi in remote places is significant, for these fishes are dominant and adaptable and should survive in fresh water on any reasonably large, stable, isolated land that they have ever been able to reach. A further argument against marine dispersal of these fishes is the presence in South America of other fresh-water animals, including certain mollusks and frogs, which show the same African relationships as the fishes but can hardly have dispersed by sea. That salt-tolerant cichlids and cyprinodonts may have reached South America from Africa through the sea is more probable. These fishes have reached Madagascar, though not Australia.

As to a direct connection, the possibility that Africa and South America were formerly in contact and have drifted apart has long intrigued students of fish distribution. At first sight the facts seem consistent with this possibility. All the true fresh-water fishes of South America seem to have old relationships toward Africa, and fresh-water fishes are just the animals which ought to preserve old geographical relationships. But a closer look changes the picture. South American fresh-water fishes do not represent the whole African fish fauna or any simple part of it. Particularly they do not represent the old part for, although a lungfish is present in South America, the ancient polypterids are not, and most of the presumably old African groups of Isospondyli are absent in South America too (Fig. 12). South American fresh-water fishes represent a fraction of the African fish fauna which is extraordinarily mixed in its evolutionary levels and ecological requirements, as has already been pointed out. This is not what would be expected from continental drift but rather what might result if a few fishes reached South America with difficulty, perhaps in different ways or at different times. Moreover, theories of continental drift usually involve all the continents, not just South America and Africa, and nowhere else in the world does the distribution of strictly fresh-water fishes suggest drift. The absence of primary-division fishes on Madagascar is a notable fact inconsistent with the usual drift theory. It seems to me (but some zoogeographers disagree) that the distribution of existing groups of fresh-water fishes over the world shows no evidence of drift.

There is a moral here. The relationships of South American fresh-water fishes are almost wholly African. Superficially this fact suggests continental drift. But better understanding of the situation changes its significance. The moral is obvious.

The possibility that fresh-water fishes reached South America across a long land bridge from Africa might be dismissed with the comment that fresh-water fishes are about the last sort of animals to get across a land bridge. They are. But they do cross land bridges, and fortunately a bridge now exists which shows the effect of bridge crossing on a fish fauna. The bridge connects South and North America. It is, from the beginning of the Isthmus of Panama to the Isthmus of Tehuantepec, about 1300 miles long, or about two-thirds as long as a hypothetical bridge from Africa to South America. Moreover, the existing American bridge is rather narrow, mountainous, and unstable and was broken during most of the Tertiary. The present complete connection was established late in the Pliocene, only a few million years ago.

The present distribution of fresh-water fishes on this bridge has been described under "Fresh-water fishes of Central America" (pp. 72ff.). The whole length of the bridge has been crossed by one genus of characins and one of fresh-water catfishes. A few other Ostariophysi have got part of the way across it. But no strictly fresh-water non-ostariophysans have even begun to cross it. Of salt-tolerant fishes, however, several cyprinodonts and one genus of cichlids have crossed the whole bridge, and gar pikes have got about halfway across.

The fishes that have crossed or are crossing this bridge, then, are only a few dominant Ostariophysi, extremely few considering the numbers of them in North and South America, and a few salt-tolerant fishes. The latter differ very much among themselves, except for their salt tolerance. The gar pikes are primitive non-teleosts and are very large fishes. The cyprinodonts are teleosts, not Isospondyli but still below the evolutionary level of the spiny-rayed families, and are small fishes which can live in small and sometimes seasonal waters. And the cichlids are spiny-rayed, and of moderate size and more or less ordinary requirements. These fishes have filtered across a land bridge, and they make up a fauna which as a whole (although some of the groups are different) is very much like the one from which the South American fauna seems to have been derived: a few dominant Ostariophysi plus a few other fishes extraordinarily mixed in their evolutionary levels and ecological requirements.

Derivation from Africa across a direct land bridge narrow and difficult enough to have had a strong selective or filtering effect, permitting passage of only a few dominant plus a few slightly salt-tolerant groups of fresh-water fishes, followed by radiation of the dominant Ostariophysi, accounts very well for the South American fresh-water fish fauna—if the lungfish, osteoglossids, and nandids as well as cyprinodonts and cichlids were slightly salt tolerant. But there is still another way in which fresh-water fishes may have reached South America, and the existence of a reasonable alternative makes it necessary to withhold judgment on the direct bridge. Perhaps the evidence of other animals will decide which explanation is correct. Strictly fresh-water fishes are about the last sort of animals to get across a bridge. If they get across at all, other animals should get across faster and in greater number and should show beyond question that the bridge existed. This is obvious, and it is what did happen in Central America.

The remaining possibility is that fresh-water fishes—Ostariophysi—reached South America through the north, over existing continents where the fishes no longer exist. This might have happened either if the fishes originated in the north and invaded the tropics of the Old and New Worlds, or if they originated in the Old World tropics and moved through the north to South America. These two possibilities are alike in requiring little change of land, but the evolutionary implications and dispersal patterns are different, and so is the expected fossil record. If the fishes rose to dominance in the north, they should have left an extensive fossil record there, but if they merely filtered through the edges of northern lands, the record might be scanty.

Matthew, in *Climate and Evolution* (1915, pp. 297–299; 1939, pp. 127–130), tried to apply his thesis of northern origins to fresh-water fishes, but the fishes do not fit the thesis. Matthew's statement of the supposed evidence of fossil fishes is almost wholly a list of errors, as Myers (1938, pp. 351–352, with footnote 18) has shown. There is hardly a fish fossil that really supports Matthew's ideas. Among existing fresh-water fishes, northern groups have not pushed into the tropics; on the contrary, tropical groups seem to have pushed northward. Moreover, tropical fresh-water fishes do not have the jumbled pattern of relationships to be expected of relicts, but a definite pattern of evolution and dispersal within the tropics. The pattern, of course, is one of survival of old fishes in Africa, evolution and radiation of new dominant groups in tropical Asia, and original limitation followed by independent radiation in South America.

The facts just given as arguments against a northern origin of tropical fresh-water fishes are much more consistent with a filtering of them through the north. I have discussed this elsewhere (1948, pp. 13–15), but the subject requires rediscussion along somewhat different lines.

If tropical fresh-water fishes filtered through the north under anything like existing conditions, they presumably went by way of eastern Asia, a Bering bridge, North America, and a Central American bridge to South America. It is therefore important to find whether this route has been used by fishes recently. Parts of it have. Some tropical fishes have started along the route. Labyrinth fishes, for example, have radiated (moderately) in tropical Asia and have spread over almost the whole of the accessible Old World tropics, and two widely distributed tropical genera have pushed north in eastern Asia to southern Manchuria and the Amur region, although labyrinth fishes are otherwise absent in the north-temperate zone. These fishes have not reached North America, but they might do so if the climate were a little milder and if land were continuous across Bering Strait. Cyprinids have followed the route much farther, probably beginning in tropical Asia, spreading through temperate Eurasia and North America, and re-entering the edge of the tropics in southern Mexico; but they can hardly be said to have filtered through the north, for they are dominant there as well as in the Old World tropics. This, however, is a special case; cyprinids are the *only* fresh-water fishes conspicuously successful both in tropical and in northern climates.

The catfishes and characins that reached South America long ago may have begun like the labyrinth fishes and then passed through North America as the cyprinids have done, but without becoming numerous in the north. This is consistent with what is known of the fishes.

Catfishes are primarily tropical. Some existing groups of them extend into the north-temperate zone, but they apparently do not usually do well there (see p. 82) and may not survive there long. Earlier catfishes may have extended into the north-temperate zone long enough to pass through it to South America without leaving much record of their passage.

Characins are now confined to Africa and America and are nearly confined to the tropics there. They do not occur in tropical Asia and are not known fossil there. However, cypriniforms are supposed to be derived from characins, and the origin and eventual dominance of cypriniforms in tropical Asia implies the former presence of characins there and accounts for their disappearance. Characins were apparently in tropical Asia once, and they may have extended northward in eastern Asia and through North America to South America, as catfishes may have done. These fishes may have passed right through an old north-temperate fish fauna, just as the three existing groups of North American Ostariophysi have done more recently (p. 76).

The other, non-ostariophysan fresh-water fishes of South America may have come through the north, if catfishes and characins did so, or they may have come in some other way, perhaps at least partly through the sea, either directly from Africa or along indirect coast lines. They are comparatively unimportant and are not worth further discussion now.

If Old World tropical fishes reached South America through eastern Asia and North America, they came not from a primarily African fauna but from a fauna which was once common to Africa and tropical Asia, which has survived more or less entire in Africa, but which has been largely replaced in Asia by evolution of later dominant fishes. This pattern—of survival in Africa and evolution of new, dominant groups in Asia—*is* the pattern of distribution of fresh-water fishes in the Old World tropics.

All this seems to me to make the derivation of South American fresh-water fishes through the north reasonably likely. The route followed need not have been the exact one described, but any route consistent with the state of Eurasia and North America when the fishes dispersed. They probably did so in the Cretaceous, for South America became isolated from all other continents about the beginning of the Tertiary. The record of the Ostariophysi and of most other fresh-water fishes in the Cretaceous is virtually blank over the whole world. The absence of Cretaceous fossil characins etc. in the north is, therefore, only one detail of a general absence of any record of these fishes at that time. The absence of a pertinent fossil record allows and even forces us to try to reconstruct the history of the fishes by other means, as I have done.

Of the various ways in which fresh-water fishes might have reached South America—from or through the sea, or by direct contact of Africa and South America or a direct land bridge, or from or through the north—only two seem likely for the strictly fresh-water forms. One is by direct but narrow bridge (not broad contact) from Africa; the other, through (not from) the north-temperate zone. Both routes might be thought of as filter bridges, the filtering being done in the first case by the difficulty of crossing a narrow land bridge, and in the second by cool or seasonal climate high above the tropics (see Frontispiece). In either case, it must be supposed that certain things have happened since the fishes reached South America. In the first case a long bridge of land must have disappeared into the Atlantic. In the second, certain groups of fishes must have disappeared from at least parts of the north-temperate zone; and some change of land must be supposed in this case too, at least the existence and later interruption of a probably late Cretaceous land connection between North and South America. So far as the fishes alone are concerned, I see no way to decide between these two possibilities now.

What does the pattern tell about ancient lands and climates (conclusion)? The distribution of fresh-water fishes seems to tell of a world not very different in the past from now, with no great changes in the arrangement of climatic zones or of continents, but with the north-temperate zone warmer in the Tertiary and colder in the Pleistocene than now, with land connections in the north, at least across Bering Strait, in the Tertiary and Pleistocene, and with a land bridge connecting South America with either Africa or North America in the Cretaceous. It tells of no close union of Africa and South America, nor does it tell of continental drift anywhere within the time of the fishes concerned, but their time may have been only since the late Cretaceous. The fishes also tell something of the history of different continents. They suggest that Africa has been relatively stable, that southern Asia has been unstable or has somehow favored the evolution of new dominant groups of fresh-water fishes, that South America was both well watered and isolated for a long time after existing fresh-water fishes reached it, and that Australia has been more isolated for a longer time.

After the preceding pages had been written and rewritten several times, it struck me that the apparent history of the Ostariophysi fitted with the history of the Tethys Sea. From the Mesozoic into the Tertiary the Tethys Sea, the great ancestral Mediterranean, of which the present Mediterranean is a fragment, stretched across the whole of southern Europe and Asia to the East Indies. It was not a stable sea. It lay in a geosyncline in which spectacular mountain building was going on, mountain building which eventually, in the Tertiary, produced the Alps and Himalayas. The region was evidently constantly changing, with changing shore lines, changing barriers and bridges, changing archipelagos, and presumably an immense, changing frontage between salt- and fresh-water habitats. This may well have been the place where, in the Cretaceous, the ancestor of the Ostariophysi entered fresh water, and where characins and early catfishes rose to dominance. And the eastern (southern Asiatic) part of the Tethys region, where favorable conditions may have been more widespread or may have lasted longer, was apparently the place where, perhaps about the beginning of the Tertiary, characins were converted into cypriniforms and the place from which the latter radiated.

History of dispersal of fresh-water fishes

The following is a brief, partly hypothetical history of dispersal of existing fresh-water fishes (Fig. 18).

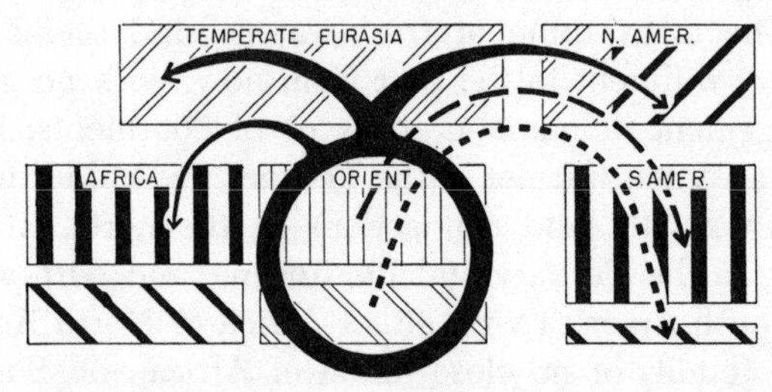

Fig. 18. Apparent history of dispersal of fresh-water fishes. Simplified diagram to show how existing distribution may have been produced by the following situation and events: zonal distribution of old non-ostariophysan fish faunas (diagonal bars) in the main Old World tropics and the north-temperate zone; rise of primitive Ostariophysi (characins and catfishes—vertical bars) in the main Old World tropics; movement of fractions of the old tropical fauna to South America, perhaps via the edge of the north-temperate zone (broken arrows), followed by multiplication of primitive Ostariophysi in South America; and radiation from the tropical Orient of dominant Cypriniformes (solid circle and arrows), causing extinction (white bars) of many older fishes in the Orient, temperate Eurasia, and western North America, and survival (solid bars) of the old tropical fish fauna in Africa and (modified) in South America and of the old north-temperate fauna in eastern North America.

I assume the existence, in the Cretaceous, of a variety of non-ostariophysan fresh-water families, independently derived from the sea, and mostly zoned with climate, some in the Old World tropics and some in the north-temperate zone. (Some of the non-ostariophysans may have entered fresh water later, but they cannot be distinguished here.)

During the Cretaceous, Ostariophysi originated in fresh water, probably in the Old World tropics and very likely in the Tethys region, and split into characins and catfishes. And part, but only part, of the resultant Old World tropical fauna of non-ostariophysan fishes plus primitive Ostariophysi (characins and catfishes) reached South America, either from Africa along a direct land bridge or from tropical Asia by way of North America.

Later, about the beginning of the Tertiary, catostomids evolved, supposedly from characins in southeastern Asia (the eastern Tethys region), and they and ancestral ameiurid catfishes spread through eastern Asia to North America. Then cyprinids evolved in turn, apparently from catostomids, radiated explosively in southeastern Asia, flooded the whole of Eurasia, and reached Africa and North America. At the same time some higher, spiny-rayed groups, especially labyrinth fishes, radiated from southeastern Asia to a lesser degree, reaching Africa and temperate eastern Asia but not beyond. The newly dominant fishes, especially the cyprinids, replaced many older fishes. In tropical Asia they replaced characins, cichlids, and most old freshwater families; and in temperate Eurasia they replaced most catostomids and several non-ostariophysan families, notably (in eastern Asia) percids. The destruction of older fishes in both tropical and temperate Eurasia, and also in western North America and Central America, was furthered by various geological and climatic changes, so that the old tropical fresh-water fish fauna of the Cretaceous survived as a whole only in Africa, and in a partial and derivative form in South America. Of course this is just the simplest outline of the main events of a history which must have been (if it occurred) very complex.

This discussion of fresh-water fishes has been long and complicated, and I want to end it by emphasizing certain important things. It is important that many fishes have zonal distributions, limited by climate. It is also important that some dominant fishes are not limited by climate but have radial distributions which override the climatic zones and most other barriers. It looks as if dominant fishes may first radiate with little regard for climate and become zoned later. The most dominant existing fresh-water fishes seem to have radiated from the Old World tropics; the north-temperate zone may have acted as a filter bridge for the passage of some of them to South America. A relation between size of rivers and number and diversity of fishes has been noticed. That there are the most fishes where there is the most space for them may seem too obvious to emphasize, but the effects of space (area) will turn out to be neither simple nor obvious but very important.

LIST OF FAMILIES OF FRESH-WATER FISHES AND PRIMITIVE FISH-LIKE VERTEBRATES

The general nature of this and the other lists of families, at the ends of succeeding chapters, is described in Chapter 1 (p. 13); and the classification and nature of fresh-water fishes are discussed at the beginning of the present chapter. The following list includes all existing families of fresh-water fishes of the primary (1st) and secondary (2nd) divisions but only the more important families of peripheral (per) fishes. If all marine groups of which one or more members live in or occasionally enter fresh water were included, the list would be increased by something like 100 families! I have, however, tried to make the list full enough to suggest the diversity of relationships of fresh-water fishes to marine ones. Fossils are mentioned only when they tell something about the history of existing fresh-water families. A dagger (†) before a name indicates that the group is extinct.

CLASS AGNATHA, primitive, jawless, superficially eel-like vertebrates: an ancient class of which only the *order Cyclostomata* (cyclostomes) survives. There are 2 existing groups or suborders of them, of which one, containing the Myxinidae or hagfishes, is entirely marine, in seas chiefly of the cool parts of the northern and southern hemispheres, rarely in the tropics (*e.g., Myxine circifrons* Garman, in the Gulf of Panama, in deep water). The other suborder contains the following single family.

(per) Petromyzonidae, lampreys: cool zones of both northern and southern hemispheres, but not the tropics: north coast of Africa, most of temperate and arctic Eurasia, Japan, western North America from the arctic to central Mexico (to Rio Lerma etc. just within the edge of the tropics), eastern North America from Labrador to Florida including many inland waters (Great Lakes, much of the Mississippi drainage, etc.), western Greenland, and Iceland; and southern Australia, Tasmania, New Zealand, and much of south-temperate South America (but not South Africa); about 8 genera (northern and southern genera are all different), 24 species; none wholly marine, many living in the sea but entering fresh water to breed, and some resident in fresh water in the northern hemisphere but apparently not in the southern one. The class Agnatha includes the earliest fossil vertebrates, but the cyclostomes themselves are unknown as fossils.

CLASS CHONDRICHTHYES, shark-like fishes, including rays: numerous since the Upper Devonian; now and probably always chiefly marine, but a few sharks enter rivers; 2 species of *Carcharias* (a primarily marine genus of sharks) are believed to live only in fresh water, one in Lake Nicaragua, the other in the Zambezi (Regan 1929); sawfishes

(Pristidae) habitually ascend tropical rivers in many parts of the world; and some marine Trygonidae (stingrays) enter fresh water, and the following related family is almost confined to fresh water.

(per) Potamotrygonidae, river rays: confined to South America, chiefly in the following river systems: Magdalena, Orinoco, Amazon, La Plata; some 3 genera, 14 species (Garman 1913).

CLASS OSTEICHTHYES, bony fishes: originated in the Devonian, probably in fresh water. Different groups since then have had diverse histories. A few existing non-teleosts may have evolved and dispersed entirely in fresh water, but the teleosts probably originated in the sea, in the Jurassic (Romer 1945, p. 104), and various groups have returned to fresh water, some of them so long ago that they are now as closely confined to fresh water as any fishes. For zoogeographic purposes I have grouped all the primitive, relict non-teleosts against the derivative, dominant teleosts. In the text, the teleosts are divided in turn, the dominant order Ostariophysi being contrasted with all the others (Table 1, p. 40).

(*Non-teleostei*), archaic bony fishes: several unrelated groups which are relict in both an evolutionary and a geographical sense.

(*Superorder Chondrostei*), ganoids in part

Order Polypterini

(1st) Polypteridae, bichirs: west and central tropical Africa to Lake Tanganyika and the whole Nile (not Lakes Victoria or Nyasa, not eastern drainages, not South Africa); 2 genera: *Polypterus,* with about 12 species, in the entire range of the family; and eel-like *Calamichthys,* one species, confined to West Africa, in swampy estuaries from the delta of the Niger to the mouth of the Congo (Pellegrin 1923*a,* p. 39). Polypterids are rather large, predaceous fishes. They can breathe air to some extent, and this may account for their survival in regions of seasonal drought (Romer). Their fossil scales have been found in Egypt, in the Eocene, but the family is unknown either Recent or fossil outside of Africa. It is apparently derived from palaeoniscoids, which were once numerous and widely distributed over the world and which had a long history (from the Devonian) in fresh and salt water.

Order Acipenseroidei

(per) Acipenseridae, sturgeons: cool parts of the northern hemisphere: most of temperate and arctic Eurasia (not North Africa), Iceland, Japan, west coast of North America from Alaska to California (and perhaps part of Mexico), Hudson Bay and Great Lakes and Mississippi drainages, east coast of North America from the St. Lawrence River (and probably Labrador) to Florida; about 5 genera, more than 20 species; very large fishes, none fully marine, some entering fresh water only to breed, some resident in fresh water. The existing genus *Acipenser* is fossil back at least to the Eocene in Europe and North America.

(1st) Polyodontidae, paddlefishes: now localized in China and east-central North America; 2 living genera and species: *Psephurus gladius,* often said to be confined to the Yangtze River of central China but listed also from the Hwang Ho of northern China by Mori (1936), whether correctly or not I do not know; and *Polyodon spathula* in the "Mississippi River system from the Missouri River of South Dakota to Pennsylvania and New York (Lake Chautauqua); southward to western North Carolina, Mississippi, Louisiana and Texas. Recorded a very few times from the Great Lakes basin, and thought by some to have reached these waters *via* canals, but more likely the species was encountered on the way to natural extirpation in the north . . . Most often in the open waters of large, silty rivers . . ." (Hubbs and Lagler 1947, pp. 28–29). (There may be a second, undescribed species of paddlefish in the Mississippi, Myers 1949*a*, p. 319, footnote 3.) They are very large fishes. Fossil genera in Europe in the Upper Cretaceous and in western North America in the Upper Cretaceous of Montana (MacAlpin 1947) and Eocene of Wyoming suggest that the family formerly occurred around the whole north-temperate zone.

(*Superorder Holostei*), ganoids in part

Order Semionotoidea

(2nd) Lepisosteidae, gar pikes: eastern North America in "Mississippi River affluents from Montana through the Great Lakes basins (apparently excepting that of Lake Superior) to the St. Lawrence-Champlain watershed of Quebec and Vermont, descending the St. Lawrence River nearly to salt water" (Hubbs and Lagler 1947, p. 31), south to Florida and the east coast of Mexico; Central America south to Lake Nicaragua (records for Panama are probably wrong, according to Hubbs, in letter); and Cuba; (and a living species of the genus has been reported from China, but probably in error); one genus, 4 or 5 species; large, heavily armored, predaceous fishes, which live in fresh water but are known in some cases to enter the sea. The Alligator Gar (*Lepisosteus tristoechus*) enters salt water commonly in the "passes" of the Mississippi delta, and a specimen of this species was taken in the open sea ¾ mile off the Louisiana coast by Gunter (1942, p. 317). This may be the species that is on Cuba, although its identity is not quite settled. The existing genus is fossil in Europe (Upper Cretaceous to Miocene), temperate Asia (Eocene), and North America (Upper Cretaceous and later), and also in several localities in tropical India (probably about the beginning of the Tertiary) (Hora and Menon 1952–1953).

Order Amioidea

(1st) Amiidae, bowfins: now confined to east-central North America "From the Mississippi River system [but not Lake of the Woods] in Minnesota to the St. Lawrence-Champlain basin in Quebec and Vermont; southward, west of the Appalachians, to Florida and Texas; northward on the Atlantic slope to the Carolinas and to the Susquehanna River (recorded from Connecticut, probably the result of introduction). Throughout the Great Lakes region, though not in the drainage basin of Lake Superior (except its outlet, St. Marys River)" (Hubbs and Lagler 1947, p. 32). One existing species, *Amia calva,* a rather large, common, voracious inhabitant of lakes, rivers, and swamps, which competes openly and successfully with other predaceous fishes of similar size. According to Romer (1945, pp. 99–100), the order

Amioidea includes some four families of fishes which dominated Jurassic seas. The family Amiidae includes Jurassic and Cretaceous genera, at least some of which were apparently marine. And *Amia* is fossil in fresh-water deposits in Europe (Lower Eocene to Lower Miocene) and North America (Middle Eocene and later). So the group apparently once dominated the seas, then became confined to fresh water in the northern hemisphere, and then was reduced to the single surviving species in eastern North America.

(*Subclass Amphibioidea*) (Choanichthyes): the following 2 orders are now usually placed at the end of the fishes, as ancestral to the amphibians, but for zoogeographic purposes they are best included with the other primitive, relict non-teleosts. Amphibioidea appeared in the Devonian and were common in fresh water in the late Paleozoic (Romer 1945, p. 114), and some also entered the sea, but there now remain of them only the few listed below.

Order Crossopterygii, ganoids in part: now existing, so far as is known, only in the sea off the southeast coast of Africa. Crossopterygians are fossil in other parts of the world from the Devonian to the Cretaceous, but no later fossils are known, and living forms were unknown until recently; the order was supposed to have become extinct at the end of the Mesozoic. However, a living specimen, a coelacanth, named *Latimeria chalumnae* J. L. B. Smith (1939), was caught on December 22, 1938, in the sea some miles off East London, South Africa, by trawl net at a depth of about 40 fathoms. Another specimen was taken in the Mozambique Channel near Madagascar about Christmas time, 1952, and reported in the public press (*Time,* January 12, 1953, p. 60), and others have been obtained since then. It is not yet clear whether or not they are all one species. They are large fishes, running over 100 pounds. They evidently live at moderate depths where there are many other fishes; they do not owe their survival to retreat to very deep water. Early coelacanths were, in general, fresh-water types, but in the Triassic, when coelacanths were common and varied, they appeared in the sea, where the last of them survive (Romer 1945, p. 121).

Order Dipnoi, lungfishes: ". . . first seen in the Middle Devonian . . . . In the late Paleozoic and Triassic, lungfish were moderately abundant and varied in the fresh-water environment in which they arose and to which they were adapted" (Romer 1945, p. 122). In their time, they were widely distributed as well as moderately numerous, and a few even occurred in the sea (in the Devonian), but they are now reduced and localized as described below. They are, of course, large fishes which can breathe air at certain times of need.

(1st) Ceratodontidae: now localized in Australia: one living species (*Epiceratodus forsteri*), confined to the middle reaches of 2 small river systems, the Mary and the Burnett, in subtropical southeastern Queensland (Spencer 1926). The same genus is fossil elsewhere in eastern Australia back to the Upper Cretaceous. The closely related, but older, extinct genus †*Ceratodus* dates from the Triassic and was more or less cosmopolitan in the Mesozoic, presumably entirely in fresh water.

(1st) Lepidosirenidae: much of tropical Africa and South America; 2 genera. *Protopterus,* with 3 species, is in Africa, from the Senegal to the upper Nile, south to the Congo and the Zambezi, and around Lakes Tanganyika and Nyasa but apparently not Victoria. *Lepidosiren,* with one species, is in South America, widely distributed in the Amazon and Paraná systems. It is now known to be common, but secretive, even along the lower Amazon (Myers 1947–1949), and there are records for the lower Paraná at Resistencia, Argentina (Barrio 1943), and from below San Pedro in the Paraná delta, at about 34° S. (MacDonagh 1945). Fossil teeth of *Protopterus* have been found in Egypt (Oligocene) and East Africa (Lower Miocene), and a fossil (late Miocene) lepidosirenid is listed from the upper Magdalena Valley, where the family does not now occur, in northern South America (Savage 1951), but no fossils surely assignable to the family are known outside of Africa and South America.

(*Superorder Teleostei*), higher bony fishes. Teleosts apparently originated in the sea in the Jurassic (Romer 1945, pp. 102–104) and are now the dominant fishes in both the sea and fresh water. Different fresh-water groups have had diverse histories.

Order Isospondyli: the central, most primitive group of living teleosts; probably of marine origin and still very numerous in the sea. Some of the marine families enter fresh water or have fresh-water representatives. For example the Clupeidae (herrings, shads, etc.) are mostly marine, but some enter fresh water or run up rivers to spawn and a few are resident in fresh water; the related Elopidae (tarpons) and Chanidae are small marine families which habitually enter fresh water; and the Osmeridae (smelts) are chiefly marine, but a few spawn in rivers and a few are landlocked in lakes. Some other isospondyl families, listed below, are chiefly or strictly fresh-water groups. The suborders of this order are not agreed on by all ichthyologists, and the families placed together in a suborder may not be actually related to each other.

(*Suborder Clupeoidea*): various marine and peripheral groups (herrings, etc.) and 2 small African fresh-water families.

(1st) Kneriidae (including *Cromeria*): tropical Africa; at least 3 genera, 6 species.

(1st) Phractolaemidae: tropical West Africa; one genus and species.

(*Suborder Salmonoidea*)

(per) Salmonidae, salmon, trout, etc.: the northern hemisphere, from the highest arctic (p. 49) south to the cooler waters of the northwestern corner of Africa, the Caucasus, northern Afghanistan (Hora 1934) (but not India), western and northern China (but not the main part of China proper), Japan, and the mountains of Formosa; and the northwestern corner of Mexico (Buen 1946, p. 87); and introduced into cool waters in many other parts of the world; various genera, numerous species, none wholly marine, many entering fresh water to spawn, some confined to fresh water.

(per) Coregonidae, whitefishes: northern Eurasia and North America, chiefly the glaciated areas. All species spawn in fresh water and all but a few are confined to it. Coregonids do sometimes enter salt water, however (Gunter 1942, p. 313). Berg (1948–1949) includes this small family in the Salmonidae.

(per) Thymallidae, graylings: northern parts of Eurasia and North America; few species, all in fresh water; related to salmonids.

(per) Galaxiidae: southern hemisphere, in cool fresh waters, with some species entering the sea; rather small, in some cases trout-like fishes of several genera, the principal one being *Galaxias,* which includes numerous species (about 40 to 70, depending on how finely they are "split") on southern Australia, Tasmania, and adjacent islands; Lord Howe Island; New Zealand and adjacent and subantarctic islands; Tierra del Fuego, the southern tip of South America north to about latitude 38° S. in Chile, adjacent islands, and Falkland Islands; and (an endemic subgenus with 2 intergrading forms) the southern tip of Africa. Other, small genera (some of which are doubtfully distinct) are localized within the range of *Galaxias,* on Tasmania, New Zealand, and Chile; and an endemic genus on New Caledonia, just within the edge of the tropics, probably sets the northern limit of the family. (A supposed galaxiid from India was long ago shown not to be one.) The family was once supposed to be confined to fresh water, and all the species probably occur there, but it has long been known that some of them enter or even breed in the sea. One species, *Galaxias attenuatus,* which breeds in the sea, occurs in fresh water with only slight differentiation of races in southern Australia, New Zealand, and southern South America. Leading references on this zoogeographically interesting family are Boulenger 1902; Regan 1905; Eigenmann 1927, pp. 48ff.; Scott 1936–1941; Barnard 1943 (the African forms); Stokell 1945; and other references given by the latter. A fossil apparently referable to *Galaxias* itself has been found in an apparently fresh-water deposit of Pliocene age in New Zealand (Stokell 1945).

(per) Aplochitonidae: southeastern Australia, Tasmania, New Zealand, and southern South America south of Concepción; in fresh water but with some species entering or breeding in the sea; about 3 genera, few species; probably related to the Galaxiidae (Chapman 1944).

(per) Retropinnatidae: southern and eastern Australia, Tasmania, and New Zealand; one genus, apparently several closely related species; some living in fresh and others in salt water, but all apparently breeding in fresh water, thus reversing the cycle of the galaxiids (Stokell 1941 and 1949). The relationships of this family are doubtful.

(*Suborder Osteoglossoidea*)

(?1st) Osteoglossidae: endemic genera scattered in all the principal tropical regions, as follows. *Clupisudis* (*Heterotis*): tropical West Africa to the upper Nile (not below Aswan); one species, usually in fresh water but Boulenger (1907, p. 85) quotes Steindachner that it is not rare in brackish water at the mouth of the Senegal. *Scleropages:* parts of the Oriental and Australian Regions; 2 species, one in southern Siam, Indochina, the Malay Peninsula, Sumatra, and Borneo, and the other in southern New Guinea and eastern tropical Australia, in rivers of the Gulf of Carpentaria and eastern Queensland south to the Dawson River; not known to enter salt water (Longman, in letter). *Osteoglossum* and *Arapaima:* monotypic genera of tropical South America, widely distributed in the Amazon-Guianas-Orinoco area (exact limits undetermined) but not reaching the Magdalena or the Paraná system. As to whether or not this family really belongs in the primary division of fresh-water fishes, and how "ancient" it is, see text (p. 52). The living forms are large or very large fishes which require large streams or other permanent bodies of water for their existence. Fossil osteoglossids of extinct genera occur in fresh-water deposits in North America (Eocene of Wyoming), India (Eocene–Hora and Menon 1952–1953), and Sumatra (probably Eocene), and in Australia (in Queensland, probably Oligocene, supposedly the same genus that was in North America in the Eocene); and a supposed osteoglossid is in apparently marine Eocene clays in England. No fossil osteoglossid has yet been found in South America.

(1st) Pantodontidae, African fresh-water flying fishes: tropical West Africa; one genus and species.

(1st) Notopteridae: main regions of the Old World tropics; 2 genera: *Notopterus,* with one species in tropical West Africa, from the Gambia to the Congo, and 3 in the Orient, from India to Java and Borneo; and *Xenomystus,* with one species, in tropical Africa including the upper Nile; in fresh and sometimes brackish water.

(1st) Mormyridae: tropical Africa and the whole Nile; about 10 genera, 150 species.

(1st) Gymnarchidae: tropical Africa including the upper Nile; one genus and species.

(1st) Hiodontidae, mooneyes: east-central North America, from probably the upper Mackenzie system, western and southern tributaries of Hudson Bay, and the St. Lawrence-Champlain drainage, south through most of the Mississippi system; 2 genera, 2 or 3 species.

Order Ostariophysi: this, the dominant order of fresh-water fishes, is represented in all the principal regions of the world, but the different main groups of the order are very differently distributed, and the only ones that reach Madagascar and the Australian Region are secondarily marine catfishes (see below). The order contains about seven-eighths of existing primary-division fishes, probably not far from 5000 known species in nearly 40 families. It includes also 2 families of catfishes which are chiefly marine, though probably derived from fresh-water forms; some members of these families have re-entered fresh water, especially in remote places. Regan divided this order into 2 suborders, the more or less carp-like forms and the catfishes, and the former he divided again into 3 groups here called superfamilies. I have followed Regan's arrangement rather than Nichols' (1938) because the latter gives less distributional information. In Regan's time (1922) no fresh-water family of Ostariophysi was known fossil outside of its existing range. A few extra-limital fossils have been found since then, but in most cases their relationships or significance are still doubtful. Known fossils indicate that some specialized ostariophysan families existed early in the Tertiary, which suggests that the order originated and first dispersed during the Cretaceous (Regan 1922 and some later authors), although Cretaceous ostariophysan fossils are few and somewhat doubtful.

(*Suborder Cyprinoidea*), carp-like fishes

(*Superfamily Characiformes*), characins: tropical and subtropical parts of Africa and America; about 6 families. In Africa there are 2 families with perhaps 20 genera and 130 species; they inhabit tropical West Africa, the whole Nile, East Africa, and (a few) warm-temperate South Africa, but they do not reach the extreme south. In America there are 5 families (one shared with Africa) with nearly 100 genera and probably more than 800 known species. They are most abundant in the tropics of South America itself; some 19 genera reach Panama (Hildebrand 1938), but only a few range farther even into tropical Central America, and only one reaches the Rio Grande in southern Texas and New Mexico; only 3 occur south of Buenos Aires, and none is known within 700 or 800 miles of Cape Horn (the farthest south may be 41° 18′ S. reached by *Cheirodon*—Eigenmann 1927, p. 14). A few fossil characins are known within the present range of the group, and also recorded are a very few from North America in the Upper Cretaceous and early Tertiary (Hay 1929, pp. 718–719) and from France in the early Tertiary (Piton 1940, pp. 269–271), but (if they are characins) their significance is doubtful. The complete absence of characins both living and fossil in the tropics of Asia is noteworthy; but if cypriniforms evolved from characins in tropical Asia, the characins must have been there long ago. Characins are small to medium-sized fishes of diverse habits and include the vicious piranhas and many aquarium fishes. A list of families of characins follows.

(1st) Characidae: Africa and America (within the limits given above); about 5 genera, 70 species in the former, and about 75 genera, 600 species in the latter, one reaching the Rio Grande along the southern edge of the United States (American members of the family revised by Eigenmann and Myers 1917–1929).

(1st) Citharinidae: Africa; about 15 genera, 60 species.

(1st) Hemiodontidae: America; about 5 genera, 60 species.

(1st) Anostomidae: America; about 9 genera, 150 species.

(1st) Xiphostomidae: America; 3 genera, about 8 species.

(1st) Gasteropelecidae, flying characins: America, north to Panama and south to Santa Fé, northern Argentina; 3 genera, 8 species (Fraser-Brunner 1950).

(*Superfamily Gymnotiformes*), gymnotid eels (including knife fishes and the Electric Eel): South and Central America; 4 families with together about 17 genera and 35 species, most in the tropics of the South American continent, south to the La Plata east of the Andes, and to the west coast of Ecuador and Colombia; several species reaching Panama, but only one ranging farther into Central America, to Guatemala; eel-like fishes related to and probably derived from characins. The group has been revised by Ellis (1913). The families are:

(1st) Rhamphichthyidae, Sternarchidae, Gymnotidae, and Electrophoridae.

(*Superfamily Cypriniformes*)

(1st) Catostomidae, suckers, buffalo fishes: North America and eastern Asia. There are about 18 genera, 84 species in North America (Jordan *et al.* 1930), south to Guatemala, including some in the Pacific drainage of North America south to California. One American species, *Catostomus catostomus,* extends into eastern Siberia as a slightly differentiated subspecies; and an endemic genus, *Myxocyprinus,* with one rather large, apparently relict but specialized species, with 3 subspecies, is isolated in east-central China (Nichols 1943, pp. 58–61; Nelson 1948). The distribution of this family is discussed in the text (pp. 31–33, Fig. 9). Fossils suggest the presence of the family in northern China in the Eocene, although this is not certain (Nelson 1949), and have been supposed to show its presence in North America too in the Eocene (Boulenger 1905, p. 415; Regan 1922, p. 206; Gosline 1944, p. 216; Darlington 1948, p. 2), but the earliest sure records for North America may be Miocene (Hay 1929–1930, Romer 1945).

(1st) Cyprinidae, carps, minnows, etc.: all continents *except* South America and Australia; most numerous and diverse in Eurasia, especially in southeastern Asia and China; less numerous at least in proportion to the whole fish fauna and much less diverse (fewer basic stocks) in Africa and North America, but widely distributed on both, occurring through the whole of Africa to its southern tip, and through almost the whole of North America both east and west of the Rockies, from parts of the arctic (north to the delta of the Mackenzie) to Mexico (where cyprinids occur in decreasing numbers south to the Panuco and Balsas drainages—Buen 1946). The family is not yet recorded from Alaska but probably occurs there, for one species, *Couesius plumbeus,* reaches the upper Yukon (Wynne-Edwards 1952, p. 18). The salt tolerance of certain cyprinids and their occurrence in the East Indies with reference to Wallace's Line (many reach it, very few cross it) are discussed in the text (pp. 51 and 52). The family Cyprinidae is probably the largest of all fish families, with hundreds of genera and something like 2000 known species. Regan divided the family into the following 3 subfamilies, which have suggestive distributions: Cyprininae, most diverse in southeastern Asia but numerous also in temperate Eurasia and in Africa; Bariliinae, most diverse in southeastern Asia but a few also in temperate Eurasia and in Africa; and Leuciscinae, abundant in both temperate Eurasia and North America. The last subfamily is the only one that occurs in America according to Regan, and Nichols (1938) agrees, except that he puts one American genus (*Notemigonus,* of eastern and central North America) in another Old World subfamily, Abramadinae, not recognized by Regan. Nichols (1943, p. 10) later published a table of adaptations of principal groups of carp-like fishes, with distributions indicated. Fossils show the existence of cyprinids in Europe since the Eocene, but in North America only since the Miocene.

(1st) Cobitidae, loaches: Eurasia etc. and northern Africa; numerous in southeastern Asia, to Java and Borneo; fewer through temperate Eurasia to Spain, Portugal, and the British Isles; and one European species in Morocco in Palearctic North Africa, and at least one endemic species in Abyssinia. The family occurs mostly in rapid streams.

(1st) Homalopteridae, loaches: southeastern Asia: India and southern China to Java and Borneo; 17 genera, about 50 species (Hora 1932); in torrential mountain streams.

(*Suborder Siluroidea*), catfishes: in fresh water in most parts of the world (Fig. 16), but most numerous in the tropics, fewer and more localized in temperate areas, and absent in very cold places; and the catfishes of Madagascar, Australia, etc. are peripheral fresh-water forms derived from marine families. Of true fresh-water catfishes, Africa has 6 families and some 250 or more species, most in the tropical part of the continent. Some follow the Nile nearly to its mouth (Boulenger 1907), and *Clarias* occurs in relict waters in the Sahara (Braestrup 1947), but otherwise temperate North Africa lacks catfishes. A few fresh-water catfishes reach warm-temperate South Africa, though not quite its southern tip (Barnard 1943). At least 2 African species reach Syria, and there are at least 2 fresh-water catfishes in the Euphrates-Tigris. Tropical Asia has 9 families (3 shared with Africa) of fresh-water catfishes. Species are numerous from India, the Himalayas, and China to Java and Borneo. All Oriental families except monotypic Cranoglanididae reach these islands, but none is native on Celebes, although 2 stocks do reach the southern Philippines. Several of the tropical Asiatic families extend into parts of temperate Eurasia, as described in the text (p. 83) and under the separate families below. North America has an endemic family, Ameiuridae, which extends south only to Guatemala. South America has 12 fresh-water families of catfishes, which make up about half (something like 1000 species) of South American fresh-water fishes. Several of the South American families reach Panama or Panama and Costa Rica, but only one (Pimelodidae) extends through Central America to Mexico, and there is little or no generic endemism in Central America. Only 4 genera occur as far south as central Chile and central Argentina, but one of them is *Diplomystes*, supposedly the most primitive living catfish, forming a family of its own localized in part of south-temperate South America, and another is *Nematogenys*, supposedly the most primitive living genus of the Pygidiidae, localized in a small area in Chile. Two other families of catfishes, Ariidae and Plotosidae, although probably derived from fresh-water ancestors, are marine, except that some of them have returned to fresh water, especially in isolated places. These 2 families are apparently not closely related to each other; they probably represent two separate invasions of the sea. Authorities do not yet agree on the classification or relationships of catfishes, so I have discussed catfish distribution in general terms (above) and treated the separate families rather briefly (below). It is to be understood that families listed from "Africa" and "South America" occur only within the general limits given above of catfish distribution on these continents. Fossil catfishes are known back to the Eocene in both fresh- and salt-water deposits, but their relationships are disputed and they are of little use in working out the origins and dispersals of the families, some of which probably date from the Cretaceous (*cf.* Gosline 1944, p. 216).

(*Old World fresh-water catfishes*): mostly after Regan

(1st) Bagridae: numerous in Africa, at least one genus (*Macrones*) represented in Syria (Pellegrin 1923) and the Euphrates-Tigris, and numerous in southern and eastern Asia, southeast to Java and Borneo, and north in decreasing numbers to the Amur River; an immense family, which may be related to the North American ameiurids (Regan) or South American pimelodids (Myers); at least one widely distributed Oriental species, *Macrones gulio,* is partly marine (Weber and Beaufort, Vol. 2, p. 345), and other species of the same genus are more or less so (H. M. Smith 1945).

(1st) Clariidae: Africa, Syria (one African species), and the Oriental Region to Java and Borneo, with one endemic species on Mindanao in the southern Philippines; another large family. Some species, such as *Clarias batrachus,* are hardy food fishes which can live out of water for some time and are carried about by man, even across Wallace's Line. The genus *Heterobranchus,* with several species in Africa and one isolated on Borneo and Banka Island in the Orient, is fossil in India (Siwalik, Lower Pliocene) (Hora 1937*a*) (but see Gosline 1944, p. 219, footnote 4).

(1st) Schilbeidae: Africa and the Oriental Region.

(1st) Mochochidae and Amphiliidae: Africa.

(1st) Malopteruridae, electric catfishes: tropical Africa and the whole Nile; one species.

(1st) Siluridae: Eurasia; numerous in tropical Asia; north in decreasing numbers into temperate eastern Asia to the Amur River; absent north of the Amur and absent in much of central Asia, but one very large species (*Silurus glanis*) from the Aral region west to the Rhine, and an endemic species of the Asiatic genus *Parasilurus* isolated in the Achelous River in Greece. The family is otherwise absent in southern Europe and absent in western Europe west of the Rhine. In the Orient, silurids occur southeast to Java and Borneo, with one endemic genus and species (not 2 as formerly supposed—Miss Haig in MS., according to Myers in letter) on Palawan and the Calamian Islands in the southern Philippines.

(1st) Sisoridae: Oriental Region, chiefly in mountain streams; and one genus (*Glyptosternum,* with 2 species—Hora 1937) north of the Himalayas in central Asia west to the Aral region (Berg 1949, pp. 1220–1221).

(1st) Amblycepidae, Pangasiidae (one genus), Chacidae (one species), and Cranoglanididae (one species): families inhabiting all or part of the Oriental Region.

(*South and Central American fresh-water catfishes*): mostly after Gosline (1945). Note that the North American family Ameiuridae (below) also enters Central America.

(1st) Diplomystidae: central Chile (between Santiago and Valdivia, common—Eigenmann 1927, p. 13), and the Rio Negro of central Argentina; one supposedly primitive, relict genus, with 2 species.

(1st) Doradidae: South America and Panama; 45 genera, 131 species, of which only one is in Panama.

(1st) Ageneiosidae: South America and Panama; 2 genera, 27 species, but only one species in Panama.

(1st) Pimelodidae: South and Central America north to central Mexico (Rio Panuco—Buen 1946); 56 genera, 281 species. Several genera reach Panama, but *Rhamdia*, with numerous species, is the principal or only genus that goes farther north.

(1st) Helogeneidae: part of northern South America; one species.

(1st) Hypophthalmidae: part of South America; one species.

(1st) Cetopsidae: South America; 4 genera, 12 species.

(1st) Pygidiidae: South America and Panama; 27 genera, 135 species, but only one genus (*Pygidium*) reaches Panama. This family inhabits mountain streams as well as lowlands, and it occurs farther south than any other primary-division fresh-water fishes, reaching about 47° 30′ S. latitude (Eigenmann 1918, p. 269). Myers (1944) discusses the genera of this family.

(1st) Bunocephalidae: South America; 10 genera, 31 species.

(1st) Callichthyidae: South America and Panama; 8 genera, 50 species, of which only one reaches Panama.

(1st) Loricariidae, armored catfishes: South America and lower Central America; 49 genera, 415 species; a number of the genera reach Panama, and one reaches Costa Rica.

(1st) Astroblepidae: South America, in the northern Andes; one genus, 37 species, one reaching Panama.

(*North-temperate fresh-water catfishes*): see Bagridae, Siluridae, and Sisoridae, above; and the following:

(1st) Ameiuridae: North America east of the Rocky Mountains; one or more eastern species are introduced in the West, and a few occur naturally in western drainages in Mexico; south in decreasing numbers to Guatemala; (a supposed existing Chinese ameiurid is probably mythical); 10 genera, about 36 species (Jordan *et al.* 1930). *Rhineastes,* fossil in North America in the Eocene and doubtfully in eastern Asia in the Pliocene, may belong in this family (Myers 1938, p. 347, including footnote 8). *Ameiurus* is fossil in North America back to the Oligocene.

(*Salt-water catfishes and their derivatives*)

(per) Ariidae: warm seas; in coastal seas and estuaries throughout the tropics and (in decreasing numbers) into the temperate zones (*e.g.*, 2 species reach Cape Cod, Massachusetts). Some enter rivers in many parts of the world and a few are probably resident in fresh water; Gosline (1945) lists those that occur in fresh water in South and Central America. Fossil ariids have been reported back to the Eocene, but they are probably best considered doubtful.

(per) Doiichthyidae: southern New Guinea; one species, more or less related to and presumably derived from Ariidae, in fresh and brackish water.

(per) Plotosidae: Indo-Pacific seas; some enter rivers and some are resident in fresh water, in Madagascar, the Australian Region, etc. The genera

Copidoglanis and *Cnidoglanis* are apparently confined to rivers of Australia and New Guinea.

Order Apodes: an order of about 20 marine families, and

(per) Anguillidae, fresh-water eels: widely but discontinuously distributed; in fresh water in Europe, around the Mediterranean and Black Seas, east coast of Africa (but not west coast), Madagascar, Mascarenes, and Seychelles, much of southern and eastern Asia, Japan, Indo-Australian Archipelago, much of northern and eastern Australia, New Zealand, and many remote Pacific islands (but not Hawaiian or Galapagos Islands), eastern North America (but not western North America) and rarely northern South America, from southern Greenland and Labrador to Brazil and the West Indies (but not most of South America); one genus, about 16 species; all breed in the sea but all (at least the females) spend much of their adult lives in fresh water. The European and North American species breed in overlapping areas in the western Atlantic just north of the West Indies. The African species are thought to spawn in the Madagascan Deep (Barnard 1943). The breeding grounds of most of the other species are still unknown. The last revision of this family is by Ege (1939), who gives references to the pioneer work of Schmidt.

Order Mesichthys

(*Suborder Haplomi*) (Esocoidei of Berg 1936*a*)

(1st) †Palaeoesocidae: Middle Eocene of Germany; one well-preserved fossil genus (Berg 1936*a*).

(1st) Umbridae, mud minnows: localized in different areas in the north-temperate zone; 2 genera: *Umbra,* with 3 species, one in the middle and lower Danube and lower Dniester Rivers in eastern Europe (Berg 1932, p. 157); one from Manitoba through most of the Great Lakes region to Quebec and the Lake Champlain basin (one record from the Erie Canal in the Mohawk River watershed), south in the central basin of the continent to the upper Ohio system, Reelfoot Lake in northwestern Tennessee, northeastern Arkansas, Kansas, and possibly eastern Nebraska (Hubbs and Lagler 1947, p. 74); and one along the Atlantic coast in lowland streams and swamps from Long Island to North Carolina; and *Novumbra* (sometimes placed in a separate family), with one species in western United States, in the Chehalis River at Satsop, Washington State. They are small fishes, very tolerant of adverse conditions including oxygen deficiency and cold.

(1st) Dalliidae, blackfish: northern Alaska and extreme eastern Siberia (Chukot Peninsula–Berg 1949); one genus and species, in shallow water on the tundra.

(1st) Esocidae, pikes, pickerels, etc.: cool parts of the northern hemisphere; one genus, about 6 species, distributed as described in Chapter 1 (pp. 33–34, Fig. 10). They are large, predaceous fishes.

(*Suborder Cyprinodontes*) (Microcyprini): see Myers (1931) for arrangement of the principal families. Besides the families listed below, several additional smaller families, Jenynsiidae, Anablepidae, and Tomeuridae (Myers 1947–1949, Part 3, p. 9) in America, and Horaichthyidae (Hubbs 1941) in India, are often recognized in this suborder. They are all localized within the ranges of the larger families listed below. Cyprinodontes are mostly small fishes, some of which are salt tolerant and some of which live in very small bodies of water, including oases or relict waters in deserts. Some of them are aquarium fishes or are used for mosquito control and are often introduced outside their natural ranges.

(2nd) Amblyopsidae, North American cave fishes: southeastern and central United States; 3 blind genera in caves, and a fourth, eyed genus in both swamps and caves. Myers puts this family in the primary division, but I see no useful reason to separate it from the other, secondary-division families of the suborder.

(2nd) Cyprinodontidae, top minnows, killifishes: tropical and warm-temperate regions of the world, except the Australian Region; Mediterranean Europe, most of Africa but not quite to the southern tip (Barnard 1943), Madagascar, the Seychelles, southern Asia north to Japan and northern China and southeast to Timor, Celebes, Formosa etc., and perhaps on Luzon in the Philippines; and North, Central, and South America, from the St. Lawrence River and the south-central Californian coast south to the La Plata River, and on some West Indies and Bermuda (distribution mainly from Myers 1931, p. 248); many genera and species especially in America, but few in the Orient; most in fresh or brackish water but some in the sea. As a rule the marine forms stay near the shore, but the *Challenger* caught a *Fundulus* in the mid-Atlantic, and a species of this genus has reached Bermuda, where it occurs in more or less saline water, sometimes among mangroves. The dominant South American genera of this family are said to be related to African, not North American, forms (Myers 1938, p. 348). One group of 4 genera and about 20 known species, comprising the subfamily Orestiinae, is confined to high altitudes in a small part of western South America (the high "Altiplano" including Lake Titicaca) where it has undergone a notable radiation (Tchernavin 1944, Brooks 1950).

(2nd) Poeciliidae, viviparous minnows: America, "from Delaware, Illinois, and Arizona to western Ecuador and northern Argentina" (Myers 1938, p. 348), and on some of the West Indies; numerous genera and species, most in fresh water, but some have considerable salt tolerance.

(2nd) Goodeidae: localized in central Mexico, mostly in the Rio Lerma system; numerous genera and nearly 30 species (Hubbs and Turner 1939, Turner 1946); another notable example of radiation in a limited area.

(2nd) Adrianichthyidae: Celebes, in Lakes Poso and Lindu; 2 genera, 3 species; presumably derived from a salt-tolerant cyprinodont. A possibly ancestral fossil cyprinodont of doubtful age has been described from Celebes by Beaufort (1934).

(*Suborder Synentognathi*): a group of about 4 primarily marine families, including the following:

(per) Hemirhamphidae, halfbeaks: all tropical and some temperate seas, with many species entering or living in fresh water; various genera, many species. Most of the fresh-water forms go in a viviparous subfamily (or family), Dermogenyinae, which occurs from southeastern Asia to Celebes and the Philippines, and which is nearly confined to fresh water (Hubbs, in letter). One of the fresh-water genera, *Nomorhamphus,* with 2 species, is confined to Celebes.

(*Suborder Thoracostei*): a primarily marine group of about 8 or more families, including the following:

(per) Gasterosteidae, sticklebacks: cool northern hemisphere; a small family of about a dozen species of small fishes, some of which occur in both salt and fresh water (for some details see Myers 1930, pp. 99–100; Wynne-Edwards 1952, pp. 22–23).

(*Suborder Salmopercae*): North America; 2 apparently relict, phylogenetically isolated families. They are rather small, more or less perch-like fishes.

(1st) Percopsidae, trout perches: 2 genera and species: *Percopsis omiscomaycus* in northern and eastern North America from "the Yukon and Alberta to Hudson Bay and Quebec; throughout the Great Lakes–St. Lawrence system to Lake Champlain; south on the Atlantic slope (generally rare) from the Hudson River system to the Potomac River, and in the Mississippi Valley to West Virginia, Kentucky, Missouri, Kansas and South Dakota (rare and local southward)" (Hubbs and Lagler 1947, p. 79); and *Columbia transmontana* localized in western North America, in the Columbia River in Oregon and Washington State.

(1st) Aphredoderidae, pirate perches: central and eastern North America: one species (at least 2 subspecies) from "southeastern Minnesota, southern Wisconsin, southern Michigan and the southern tributaries of Lake Ontario southward, west of the Appalachian Mountains and foothills, to the Gulf Coast, as far as Texas" (Hubbs and Lagler 1947, p. 80); and on the Atlantic slope from New York to Florida. There are fossil genera in western United States (Eocene and Oligocene).

Order Acanthopterygii, spiny-rayed teleosts: this is an order of more than 20 suborders (Romer) and a great many families. More than half the suborders and a great many of the families are strictly marine or occur in fresh water to only an insignificant extent. About 20 other families are primarily marine but commonly enter or are represented in fresh water; and a few families are confined to fresh water. The following list includes chiefly the latter, with a few of the marine families that have noteworthy fresh-water representatives.

(*Suborder Percoidea*), perch-like fishes: this suborder contains about 50 marine families (some of which have fresh-water representatives) and also several important primary- and secondary-division fresh-water families.

(per) Serranidae, sea basses: a large marine family, including the groupers, jewfishes, etc. Besides the scores of strictly marine species there are a few that enter fresh water to breed and a few, including the North American "White Bass" (*Lepibema chrysops*) and "Yellow Bass" (*Morone interrupta*), that are resident in fresh water. The "Nile Perch" (*Lates niloticus*) and its allies, very large fishes widely distributed in fresh water in Africa, southern Asia, and Australia, probably belong in this family. See Hubbs and Lagler (1947, pp. 80–81) for further details of the North American members of the family.

(per) Sciaenidae, drums: another large marine family of which a few enter fresh water and a very few are resident in it. The latter include the Fresh-water Sheepshead (*Aplodinotus grunniens*) of North America, which occurs from Hudson Bay drainage to Guatemala (Hubbs and Lagler 1947, p. 95).

(1st) Centrarchidae, fresh-water basses, sunfishes: North America; several genera, about 24 species east of the Rockies, from southern Canada to Mexico, one species reaching the Rio Soto la Marina, almost to the Tropic of Cancer in eastern Mexico (Buen 1946); one endemic genus and species, *Archoplites interruptus,* isolated in California in the Sacramento and San Joaquin river systems; and some eastern species introduced west of the Rockies.

(1st) Percidae, true perches: Europe, northern Asia, and eastern North America, but absent in extreme eastern Asia and western drainages of North America, except that one species (*Rafinesquiellus pottsii*) reaches west-flowing streams in Mexico. Limits of distribution of family shown in Figure 13. Three subfamilies. (1) Percinae: Eurasia and North America. *Perca fluviatilis* occurs across Eurasia from the British Isles to the Kolyma River but not to Pacific drainages of extreme eastern Asia, and *Perca flavescens* (the Yellow Perch) occurs in eastern North America northwest to Lake Athabasca but not in western North America, except as introduced there. A third species of *Perca* is endemic in the Lake Balkash-Tarim region of central Asia; it is the only endemic percid in Asia. Three more genera, 7 species, occur in southern Europe, especially in drainages of the Black and Caspian Seas, and one of the species (*Acerina cernua*) extends west to England and east to the Kolyma, reaching the same eastern limit as *Perca fluviatilis.* (2) Luciopercinae: Eurasia and North America. *Lucioperca* (3 species) occurs in Europe from the Aral Region to the Elbe but not in northern drainages and not in Asia. The closely related *Stizostedion* (2 species, the Sauger and Pike Perch) occurs across North America, within the limits of percids, and sets the northern limits of the family there. (3) Etheostominae: eastern North America from southern Canada to Mexico. *R. pottsii* (above) reaches the Rio Mezquital, which crosses the Tropic of Cancer in western Mexico (Buen 1946). There are about 30 genera, more than 100 species (Jordan *et al.* 1930). They are for the most part very small fishes living in rapid water; I suppose they fill the same niches in eastern North America that torrent cypriniforms do in Eurasia. Fossil percids occur in Europe and North America in the Eocene and later.

(1st) Nandidae: disconnected tropical areas: one or 2 endemic genera in each of the 3 principal tropical regions (not the Australian Region), each genus with only one or 2 species: in (West) Africa, *Polycentropsis;* in the Orient, *Nandus* (India to Borneo but not Java, in fresh and brackish water) and *Badis* (India to Burma and perhaps beyond); and in South America, *Polycentrus* (Trinidad and Guiana) and *Monocirrhus* (Guiana and the Amazon lowlands). They are rather small fishes.

G. S. Myers, who has seen all of the genera both alive and dead, writes (in letter) that all except *Badis* certainly can go in one family, and that *Badis* is very close and can just as well go in it, too.

(1st) Pristolepidae: the Oriental Region, discontinuously distributed: Indochinese Subregion to Java and Borneo, with one of the eastern species occurring also in an isolated area in peninsular India and Ceylon (Hora 1944); one genus, about 4 species; much like nandids.

(2nd) Cichlidae: concentrated in the tropics of Africa and America, with a few in adjacent places; many genera, 600 or more species. In Africa, most are in the western and central tropical regions and especially in Lakes Nyasa, Tanganyika, etc., where great "species flocks" have evolved (Brooks 1950, Greenwood 1951), but a few occur to the lowest Nile and even (with a few other fishes) in isolated or underground waters in the northern Sahara (Boulenger 1907, pp. 465, 502, and 522; Braestrup 1947) though none reaches the extreme northwestern corner of Africa north of the Atlas Mountains, and a few go far south, though not quite to the Cape of Good Hope (Barnard 1943); a few are in Syria and Palestine, including an endemic genus in the Jordan Valley (Trewavas 1942); there are 3 endemic genera on Madagascar; and one endemic genus (related to a Madagascan one) with 3 species occurs in estuaries and coastal rivers of Ceylon and peninsular India. [This genus is said by Beaufort (1951, p. 65) to occur in Siam too, but this is probably an error, for no cichlid is included in H. M. Smith's (1945) work on the fresh-water fishes of Siam.] In America, cichlids are moderately numerous through much of South America (but none reaches the colder parts of Chile and Argentina) and relatively very numerous through Central America to southern Mexico; they occur in decreasing numbers north through Mexico (Buen 1946), with one species reaching the southern edge of Texas (the Central American and Mexican forms are supposedly derived from South American ones). The widely distributed South and Central American genus *Cichlasoma* occurs also on Cuba and Hispaniola and has been recorded, but probably in error, from Jamaica. The species of this family normally occur in fresh or brackish water, but a few occur also in strongly saline inland waters, and some can survive for considerable periods in full-strength sea water (Myers 1938). Cichlids are typically rather small, perch-like fishes. Pellegrin monographed the family in 1903, and his work, though outdated in many details, is still useful. Some fossil cichlids are known, but they are not of much use in working out the geographical history of the family.

(*Suborder Blennioidea*): a marine group of about 20 families, including:

(per) Brotulidae: a marine family, some members of which occur in deep water; the only known fresh-water representatives are blind forms found in caves in Cuba (2 endemic genera) and Yucatan (one endemic genus) (Hubbs 1938).

(*Suborder Anacanthini*): another marine group of several families, including the following:

(per) Gadidae, codfishes, etc.: a primarily marine family of several score marine species, ranging from arctic to antarctic seas, but with relatively few, mostly deep-water forms within the tropics (Hubbs and Lagler 1947, p. 79), and with one genus, *Lota* (the burbots or eelpouts), with one species but several subspecies, in fresh water around the colder parts of the northern hemisphere (see Wynne-Edwards for further details). *Lota* is resident in fresh water in much of its range, but in the far north it apparently occurs also in the sea and is said to be migratory. At least one other gadid sometimes enters rivers in northeastern North America (Gunter 1942, p. 314), and several genera occur in fresh water in northern Asia (Berg 1949, pp. 1222–1223).

(*Suborder Chaetodontoidea*): still another marine group of several families, including:

(per) Toxotidae, archer fishes: the Indo-Australian area from India to the Philippines, south to northern Australia, and east to the Solomons and New Hebrides; one genus, several species, "in littoral water of low salinity, in brackish water of estuaries and in rivers" (Weber and Beaufort).

(*Suborder Scorpaenoidea*): a group of 20 or more families, of which most are strictly marine, but:

(per) Cottidae, sculpins etc.: numerous genera and something like 300 species in northern seas (and a few in antarctic seas) and some resident in colder rivers and lakes in the northern hemisphere. See Hubbs and Lagler (p. 96) and Wynne-Edwards (p. 21) for further details.

(per) Comephoridae and Cottocomephoridae: confined almost entirely to Lake Baikal in Siberia (2 species reach the Yenisei and one is endemic in the Lena) but evidently derived by radiation of a marine cottid ancestor; 9 genera, 16 or 18 species (Berg 1949, pp. 1230–1233; Brooks 1950).

(*Suborder Gobioidea*), gobies, etc.

(per) Gobiidae (with Eleotridae, etc.): a very large family or group of families of small fishes which live especially along the coasts of all tropical and many temperate seas, often on coral reefs or in tide pools or along sandy or muddy shores—a few even spend much of their time out of water chasing insects over wet sand or mud. Many others live in fresh water, sometimes descending to the sea to spawn, sometimes not. Some of the fresh-water forms occur, often in swift streams, in parts of the world including Madagascar, Australia, the West Indies, and many other tropical islands like Luzon and the Hawaiian Islands where strictly fresh-water fishes are absent. For example, the genus *Rhyacichthys*, sometimes placed in a family by itself, is known from Java, Celebes, the Philippines, and the Solomons (all apparently one species). It inhabits swift streams, clinging to the rocks (as other stream-living gobies often do) and slipping around and under them when disturbed, but its distribution and the known habits of other gobies suggest that it can pass through the sea. (Much of this information about gobies is from Herre 1927). A number of endemic gobies inhabit the Black Sea and especially the Caspian.

(*Suborder Anabantoidea*), labyrinth fishes: a fresh-water suborder of about 3 families. They are mostly small or medium-sized fishes. They have supplementary air-breathing organs.

(1st) Anabantidae (including *Osphronemus*), climbing or walking perches, fighting bettas, etc.: Africa and the Oriental Region, north into eastern Asia. In Africa there is probably only *Anabas* (which occurs also in the Orient—but the African and Oriental forms are sometimes put in separate genera) with nearly 20 species from parts of the upper Nile system to Cape Town (but absent across northern Africa; and the tropical forms go south only to the Zambesi, the 2 Cape species forming a geographically isolated endemic group—Barnard 1943). The supposed West African *Micracanthus* is probably a mislabeled Oriental *Betta,* according to Myers (in letter). In the Oriental Region there are some 11 genera (including *Anabas*) and numerous species, some reaching Java and Borneo, and a few extending north to central China or (one species) to southern Manchuria and Korea. One of the Oriental *Anabas, testudineus,* is a common, hardy food fish which lives in all sorts of fresh and brackish water and which is regularly carried about alive by man. It extends beyond the limits given above, to the Philippines and to Celebes and Halmahera, but it is probably introduced in these places. A supposed endemic anabantid on Cagayan Sulu, a small island northeast of Borneo, is now thought to be the same as a common Bornean species (Myers), and may have been introduced.

(1st) Channidae (Ophicephalidae), snakeheads: Africa and the Oriental Region, north into eastern Asia; 2 genera, something like 20 species. The principal genus, *Ophicephalus,* includes about 3 species in Africa, from West Africa to the upper Nile system, and south to Delagoa Bay (J. L. B. Smith 1950), and a number in the Orient, from India etc. to Java and Borneo, north in decreasing numbers through China, one species reaching the Amur. The other genus, *Channa,* with probably only one species, is confined to southeastern Asia. *Ophicephalus striatus* occurs beyond the limits given above, on the Lesser Sunda Islands, Celebes and Halmahera, and the Philippines, but it has probably been introduced there. It is another common food fish which can live for a while out of water and is often carried by man.

(1st) Luciocephalidae: Orient: from the Malay Peninsula to Borneo; one species.

(*Suborder Mugiloidea*), mullets etc.: primarily a marine group, but even some Mugilidae habitually enter or live permanently in fresh water (Myers 1938; Schultz 1946, especially key pp. 379–381 and last text paragraph p. 382), and the following family is still more notable for the number of peripheral fresh-water forms it has produced.

(per) Atherinidae, silversides: coasts of all tropical and temperate seas, and many isolated fresh waters. Some genera are strictly marine. Others are chiefly marine but enter fresh water. For example, *Atherina* is a widely distributed marine genus of many species, but *A. caspia* of the Mediterranean Sea enters fresh water along the coast and has formed local fresh-water races there, and *A. tamarensis* is found both along coasts and in streams in Tasmania and South Australia. Other genera which occur in both salt and fresh water are found along the eastern coast of the United States and on parts of both eastern and western coasts of tropical America, and *Austromenidia* inhabits coastal seas, streams, and lakes of south-temperate South America south to the Straits of Magellan. But of 59 genera of the family recognized by Jordan and Hubbs (1919), at least half are restricted to fresh or rarely brackish water. For example, in the Old World, there are 2 endemic genera in Madagascar, one (now 2—Aurich 1935) in Celebes, and about 10 in Australia, New Guinea, and neighboring islands. In the New World, one fresh-water genus is confined to eastern coastal regions of the United States; 3 are localized in Mexico (one has recently been described from Cuba); 2 are in southeastern Brazil; and one is in Andean streams on the Pacific slope of Peru and Chile, to an altitude of 8000 feet or more. Note that in most cases, but not all, the fresh-water genera are in places where primary-division fishes are absent or relatively few. Jordan and Hubbs's classification has been modified by recent work (*e.g.*, by Schultz 1948), but the main pattern of distribution of the family is not changed. The subfamily Melanotaeniinae, with about 10 genera, has sometimes been said to be confined to the warmer part of Australia, New Guinea, and adjacent islands on the same continental shelf (Beaufort 1926, map on p. 103; 1951, pp. 154–155), but Jordan and Hubbs (p. 21) include *Telmatherina* of Celebes in the same group. The atherinids are rather small, smelt-like fishes.

(per) Phallostethidae: Siam and the Malay Peninsula to Borneo and Luzon; some 10 genera, 18 species (Herre 1942); some in salt or brackish and others in fresh water; "tiny and almost invisible [transparent] fishes" (Herre).

(*Suborder Opisthomi*): a fresh-water suborder of 2 families.

(1st) Mastacembelidae, spiny eels: Africa to the Orient; 2 genera: *Mastacembelus,* with numerous species, of which about 30 are in tropical Africa (but apparently not in North Africa, not in most of the Nile), one in the Euphrates-Tigris system of southwestern Asia, 12 in the Oriental Region southeast to Java and Borneo, and one in China north to the Pei-ho (Mori 1936); and *Rhynchobdella,* one species, in the Orient. (Most of these details are from Boulenger 1912; a doubtful record from the Oxus River has apparently not been confirmed.)

(1st) Chaudhuriidae: Orient; one species, known only from Inlé Lake, Burma (Regan 1919).

(*Suborder Synbranchii*), mud eels: a group of a few small families which, although they live chiefly in fresh water, are not to be considered true fresh-water fishes.

(per) Synbranchidae: the tropics; several genera in salt, brackish, and fresh water. *Synbranchus,* in fresh and brackish water, rarely in the sea, has one species in tropical West Africa, one in the East from India to New Guinea and Australia (McCulloch), and one in tropical America from Mexico and Cuba to Brazil. Related blind genera have been found in fresh water, but not far from the sea, in West Africa and Yucatan.

(per) Amphipnoidae: India and Burma, and also in Queensland, Australia (McCulloch 1929); one species, in fresh and brackish water.

(per) Indostomidae: Orient; known only from a lake in Burma; one species. The systematic position of this family is doubtful (Bolin 1936).

REFERENCES

For explanation of this and other chapter reference lists see end of Chapter 1.

Aurich, H. 1935. Mitteilungen der Wallacea-Expedition Woltereck, XIII and XIV, Fische I and II. *Zool. Anzeiger,* **112,** 97–107, 161–177.

Barnard, K. H. 1943. Revision of the indigenous freshwater fishes of the S. W. Cape region. *Ann. South African Mus.,* **36,** 101–262.

———. 1948. Report on a collection of fishes from the Okovango River, with notes on Zambesi fishes. *Ann. South African Mus.,* **36,** 407–458.

Barrio, A. 1943. . . . *Lepidosiren paradoxa* . . . [in Argentina]. *Revista Argentina de Zoogeografia,* **3,** 9–20.

Beaufort, L. F. de. 1926. *Zoogeographie van den Indischen Archipel.* Haarlem, Netherlands, de Erven F. Bohn.

———. 1934. On a fossil fish from Gimpoe (central-Celebes). *Verh. Geol.-Mijnb. Genoot. Ned. Kolonien, Geol. Ser.,* **10,** 180–181. This is cited from the *Zoological Record;* I have only a photostat.

———. 1951. *Zoogeography of the land and inland waters.* London, Sidgwick and Jackson; New York, Macmillan.

Berg, L. S. 1932. Übersicht der Verbreitung der Süsswasserfische Europas. *Zoogeographica* (Jena), **1,** 107–208.

———. 1936. [Division of the Palearctic Region according to fishes—in Russian with French explanation of map.] *Trud. Ist. Vsyesozan Geogr. Syezda,* **3,** 1–8, map. This is cited from the *Zoological Record;* I have only a photostat of this paper, received from Prof. Myers.

———. 1936*a*. [The suborder Esocoidei—in Russian with English summary.] *Bull. Inst. Recherches Biol. Perm,* **10,** 385–391.

———. (1940) 1947. *Classification of fishes both Recent and fossil* [in Russian and English]. Ann Arbor, Michigan, J. W. Edwards, reprinted 1947 by offset from 1940 Russian edition.

———. 1948–1949. [*Fresh-water fishes of Union of Soviet Socialist Republics.*] Leningrad; 3 vols. This is in Russian, with distribution table (Vol. 3, 1196–1233) giving names of fishes and rivers in Roman type.

———. 1950. *Natural regions of the U.S.S.R.* New York, Macmillan.

Bolin, R. L. 1936. The systematic position of *Indostomus paradoxus* . . . . *J. Washington Acad. Sci.,* **26,** 420–423.

Boulenger, G. A. 1902. The explanation of a remarkable case of geographical distribution among fishes. *Nature,* **67,** 84.

———. 1905. The distribution of African fresh-water fishes. *Nature,* **72,** 413–421.

———. 1907. *The fishes of the Nile.* (In) Anderson's Zoology of Egypt, London, Hugh Rees, Ltd., 2 vols., text and pls.

———. 1909–1916. *Catalogue of the fresh-water fishes of Africa in the British Museum.* London, British Mus., 4 vols.

Boulenger, G. A. 1912. A synopsis of the fishes of the genus *Mastacembelus. J. Acad. Nat. Sci. Philadelphia,* Ser. 2, **15,** 197–203.

Braestrup, F. W. 1947. Remarks on faunal exchange through the Sahara. *Videnskabelige Meddelelser Dansk naturhistorisk Forening,* **110,** 1–15.

Breder, C. M., Jr. 1934. Ecology of an oceanic fresh-water lake, Andros Island, Bahamas, with special reference to its fishes. *Zoologica* (Sci. Contrib. New York Zool. Soc.), **18,** 57–88.

Brooks, J. L. 1950. Speciation in ancient lakes. *Quarterly Review Biol.,* **25,** 131–176.

Buen, F. de. 1946. Ictiogeographía Continental Mexicana. *Revista Soc. Mexicana Hist. Nat ,* **7,** 87–138.

Carl, G. C., and W. A. Clemens. 1948. *The fresh-water fishes of British Columbia.* British Columbia Provincial Mus. Handbook No. 5.

Carpenter, R. G., 2d, and H. R. Siegler. 1947. *A sportsman's guide to the fresh-water fishes of New Hampshire.* Concord, New Hampshire, State Fish and Game Commission.

Chapman, W. M. 1944. . . . *Aplochiton zebra* Jenyns. *J. Morphology,* **75,** 149–165.

Dammerman, K. W. 1929. On the zoogeography of Java. *Treubia,* **11,** 1–88.

Darlington, P. J., Jr. 1948. The geographical distribution of cold-blooded vertebrates. *Quarterly Review Biol.,* **23,** 1–26, 105–123.

Day, F. 1889. Fishes. (In) *The fauna of British India, including Ceylon and Burma.* London, 2 vols.

Dymond, J. R., and V. D. Vladykov. 1934. The distribution and relationship of the salmonoid fishes of North America and North Asia. *Proc. Fifth Pacific Sci. Congress,* **5,** 3741–3750.

Ege, V. 1939. *A revision of the genus* Anguilla *Shaw* . . . . Dana Rep. No. 16.

Eigenmann, C. H. 1909. The fresh-water fishes of Patagonia and an examination of the Archiplata-Archhelenis theory. *Rep. Princeton U. Exp. Patagonia, 1896–1899,* **3,** Part 3, 225–374.

———. 1910. Catalogue of the fresh-water fishes of tropical and south temperate America. *Rep. Princeton U. Exp. Patagonia, 1896-1899,* **3,** Part 4, 375–511.

———. 1918. The Pygidiidae, a family of South American catfishes. *Mem. Carnegie Mus.,* **7,** 259–398.

———. 1927. The fresh-water fishes of Chile. *Mem. National Acad. Sci.* (Washington), **22,** No. 2.

Eigenmann, C. H., and W. R. Allen. 1942. *Fishes of western South America.* Lexington, Kentucky, U. of Kentucky.

Eigenmann, C. H., and G. S. Myers. 1917–1929. The American Characidae. *Mem. Mus. Comparative Zool.,* **43.**

Ellis, M. M. 1913. The gymnotid eels of tropical America. *Mem. Carnegie Mus.,* **6,** 109–204.

Fraser-Brunner, A. 1950. . . . Gasteropelecidae. *Ann. and Mag. Natural History,* Ser. 12, **3,** 959–970.

Frost, N. 1940. *A preliminary study of Newfoundland trout.* Newfoundland Department Nat. Resources, Research Bull. 9 (Fisheries).

Garman, S. 1913. The Plagiostomia (sharks, skates, and rays). *Mem. Mus. Comparative Zool.,* **36.**

Gosline, W. A. 1944. The problem of the derivation of the South American and African fresh-water fish faunas. *Anais Acad. Brasileira de Ciencias,* **16,** 211–223.

Gosline, W. A. 1945. Catálogo dos nematognatos de água-doce da América do Sul e Central. *Boletim Mus. Nacional* [Brazil], New Ser., No. 33.

Greenwood, P. H. 1951. Evolution of the African cichlid fishes: the *Haplochromis* species-flock in Lake Victoria. *Nature,* **167,** 19–20.

Gunter, G. 1942. A list of the fishes of . . . North and Middle America recorded from both freshwater and sea water. *American Midland Naturalist,* **28,** 305–326.

Hay, O. P. 1929–1930. *Second bibliography and catalogue of the fossil Vertebrata of North America.* Carnegie Inst. Washington Pub. 390.

Herre, A. W. C. T. 1927. *Gobies of the Philippines and the China Sea.* Philippine Bureau Sci., Monograph 23.

———. 1928. True fresh-water fishes of the Philippines. (In) Dickerson *et al.*, *Distribution of life in the Philippines,* 242–247. Bureau of Sci., Manila, Monograph 21.

———. 1933. The fishes of Lake Lanao: a problem in evolution. *American Naturalist,* **67,** 154–162.

———. 1942. New and little known phallostethids, with keys to the genera and Philippine species. *Stanford Ichthyological Bull.,* **2,** 137–156.

Herre, A. W. [C. T.], and G. L. Ablan. 1934. *Aplocheilus luzonensis,* a new Philippine cyprinodont. *Philippine J. Sci.,* **54,** 275–277.

Hildebrand, S. F. 1938. A new catalogue of the fresh-water fishes of Panama. *Pub. Field Mus. Nat. Hist., Zool. Ser.,* **22,** No. 4, 215–359.

Hora, S. L. 1932. Classification, bionomics and evolution of homalopterid fishes. *Mem. Indian Mus.,* **12,** 263–330.

———. 1934. Fish of Afghanistan. *J. Bombay Nat. Hist. Soc.,* **36,** 688–706.

———. 1937. Comparison of the fish-faunas of the northern and the southern faces of the Great Himalayan Range. *Rec. Indian Mus.,* **39,** 241–250.

———. 1937*a*. Geographical distribution of Indian fresh-water fishes and its bearing on the probable land connections between India and the adjacent countries. *Current Science* (Bangalore), **5,** 351–356.

———. 1938. On the age of the Deccan trap as evidenced by fossil fish remains. *Current Science* (Bangalore), **6,** 370–372.

———. 1938*a*. Changes in the drainage of India, as evidenced by the distribution of freshwater fishes. *Proc. National Inst. Sci. India,* **4,** 395–409.

———. 1944. On the Malayan affinities of the freshwater fish fauna of peninsular India . . . . *Proc. National Inst. Sci. India,* **10,** 423–439.

———. 1949. Dating the period of migration of the so-called Malayan element in the fauna of peninsular India. *Proc. National Inst. Sci. India,* **15,** 1–7.

Hora, S. L., and A. G. K. Menon. 1952–1953. Distribution of Indian fishes of the past and their bearing on the geography of India. *Everyday Science,* **1,** 26–37; **2,** 105–113.

Hubbs, C. L. 1936. Fishes of the Yucatan peninsula. (In) The Cenotes of Yucatan. *Carnegia Inst. Washington Pub.* 457, 157–282.

———. 1938. Fishes from the caves of Yucatan. *Carnegie Inst. Washington Pub.* 491, 261–295.

———. 1941. A new family of fishes. *J. Bombay Nat. Hist. Soc.,* **42,** 446–447.

———. 1952. Antitropical distribution of fishes and other organisms. *Proc. Seventh Pacific Sci. Congress,* **3,** 324–329.

Hubbs, C. L., and K. F. Lagler. 1947. *Fishes of the Great Lakes region.* Cranbrook Inst. Sci., Bull. 26.

Hubbs, C. L. and R. R. Miller. 1948. Correlation between fish distribution and hydrographic history in the desert basins of western United States. (In) The Great Basin, with emphasis on glacial and postglacial times. *Bull. U. Utah, Biol. Series,* **10**, No. 7, 18–166.

Hubbs, C. L., and C. L. Turner. 1939. Studies of the fishes of the order Cyprinodontes. 16. *A revision of the Goodeidae.* Misc. Pub. Mus. Zool., U. Michigan, 42.

Jordan, D. S. 1905. *A guide to the study of fishes.* New York, Henry Holt, 2 vols.

Jordan, D. S., B. W. Evermann, and H. W. Clark. 1930. *Check list of the fishes . . . of North and Middle America . . . .* Rep. United States Commissioner of Fisheries for 1928, Part 2.

Jordan, D. S., and C. L. Hubbs. 1919. *A monographic review of the family of Atherinidae or silversides.* Leland Stanford Junior U. Pub., U. Ser. (no No.).

Kosswig, C. 1952. Die zoogeographie der türkischen Süsswasserfische. *Pub. Hydrobiological Research Inst., Faculty Sci., U. Istanbul,* Ser. B, **1**, 85–101.

Livingstone, D. A. 1953. The fresh-water fishes of Nova Scotia. *Proc. Nova Scotian Inst. Sci.,* **23**, 1–90.

MacAlpin, A. 1947. . . . polyodontid fish from Upper Cretaceous of Montana . . . . *Contrib. Mus. Paleontology, U. Michigan,* **6**, 167–234.

McC(ulloch), A. R. 1927. Fishes of Australia. (In) *Australian Encyclopaedia,* 3rd ed., Vol. 1, 465–467. Sydney, Angus and Robertson.

McCulloch, A. R. 1929–1930. A check-list of the fishes recorded from Australia. *Australian Mus.* (Sydney) *Mem.,* **5.**

MacDonagh, E. J. 1945. Hallazgo de una *Lepidosiren paradoxa* en el detta del Paraná. *Notas del Mus. de La Plata,* **10**, 11–16.

Matthew, W. D. 1915, 1939. Climate and evolution. *Ann. New York Acad. Sci.,* **24**, 171–318; reprinted (1939) as *Special Pub. New York Acad. Sci.,* **1.**

Miles, C. 1947. *Los Peces del Rio Magdalena.* Ministerio de la Economía Nacional, Sección de Piscicultura etc., Bogotá, Colombia.

Mori, T. 1936. *Studies on the geographical distribution of freshwater fishes in eastern Asia.* Apparently privately published.

Munro, I. S. R. 1955. *The marine and fresh-water fishes of Ceylon.* Canberra (Australia), Dept. of External Affairs.

Myers, G. S. 1930. The killifish of San Ignacio and the stickleback of San Ramon, Lower California. *Proc. California Acad. Sci.,* Ser. 4, **19**, 95–104.

———. 1931. The primary groups of oviparous cyprinodont fishes. *Stanford U. Pub., U. Ser., Biol. Sci.,* **6**, 241–254.

———. 1938. Fresh-water fishes and West Indian Zoogeography. *Smithsonian Rep.,* **1937**, pp. 339–364.

———. 1940. The fish fauna of the Pacific Ocean . . . . *Proc. Sixth Pacific Sci. Congress,* **3**, 201–210.

———. 1944. [Rearrangement of Pygidiidae.] *Proc. California Acad. Sci.,* Ser. 4, **23**, 591–602.

———. 1947–1949. The Amazon and its fishes [in five parts]. *Aquarium J.* 1947, March pp. 4–9 (Part 1), April pp. 13–20 (Part 2), May pp. 6–13 and 32 (Part 3), July–Aug. pp. 8–19 and 34 (Part 4); 1949, Feb. pp. 52–61, March pp. 76–85 (Part 5).

Myers, G. S. 1949. Usage of anadromous, catadromous and allied terms for migratory fishes. *Copeia,* **1949,** 89–97.

———. 1949*a*. Salt-tolerance of fresh-water fish groups in relation to zoogeographical problems. *Bijdragen Tot de Dierkunde,* **28,** 315–322.

———. 1951. Fresh-water fishes and East Indian zoogeography. *Stanford Ichthyological Bull.,* 4, 11–21.

———. 1953. Paleogeographic significance of fresh-water fish distribution in the Pacific. *Proc. Seventh Pacific Sci. Congress,* **4,** 38–48.

Nelson, E. M. 1948. The . . . Weberian apparatus of the Catostomidae . . . . *J. Morphology,* **83,** 225–251.

———. 1949. The opercular series of the Catostomidae. *J. Morphology,* **85,** 559–568.

Nichols, J. T. 1928. Fishes from the White Nile . . . [with table of "world's fresh-water fish faunae"]. *American Mus. Novitates,* No. 319.

———. 1938. Classification of carp-like fishes. *Zoologica* (New York), **23,** 191–193.

———. 1943. *The fresh-water fishes of China.* Natural History of Central Asia (American Mus. Nat. Hist., New York), Vol. 9.

Nichols, J. T., and L. Griscom. 1917. Fresh-water fishes of the Congo Basin . . . [including a discussion of African fresh-water fish faunas, pp. 739–752]. *Bull. American Mus. Nat. Hist.,* **37,** 653–756.

Okada, Y. 1955. *Fishes of Japan.* Tokyo, Maruzen Co.

Pellegrin, J. 1903. . . . la famille des cichlidés. *Mém. Soc. Zool. France,* **16,** 41–399.

———. 1912. Les Poissons d'eau douce d'Afrique et leur distribution géographique. *Mém. Soc. Zool. France,* **25,** 63–83.

———. 1921. Les Poissons des eaux douces de l'Afrique du Nord Française: Maroc, Algérie, Tunisie, Sahara. *Mém. Soc. Sci. Nat. Maroc,* **1,** No. 2.

———. 1923. Étude sur les poissons . . . . *Voyage Zool. d'Henri Gadeau de Kerville en Syrie,* 4, 5–37.

———. 1923*a*. *Les Poissons des eaux douces de l'Afrique Occidentale (du Sénégal au Niger).* Paris, Émile Larose.

———. 1933. Les Poissons des eaux douces de Madagascar . . . . *Mém. Acad. Malgache Tananarive,* **14.**

———. 1937. Les Poissons du Sahara occidental. *Compte Rendu Association Française Avancement Sci.,* 60th Session, 337–338.

Piton, L. 1940. *Paléontologie du Gisement Éocène de Menat.* Clermont-Ferrand, Imprimeries Paul Vallier.

Regan, C. T. 1905. A revision of the fishes of the family Galaxiidae. *Proc. Zool. Soc. London,* **1905,** Part 2, 363–384.

———. 1906–1908. Pisces. *Biologia Centrali-Americana.*

———. 1911. The classification of the teleostean fishes of the order Ostariophysi. *Ann. and Mag. Nat. Hist.,* Ser. 8, 8, 13–32, 553–577.

———. 1912. Discussion on Wallace's Line. *Rep. . . . British Association Advancement Sci.,* **1911,** 433–435.

———. 1919. Note on Chaudhuria . . . . *Ann. and Mag. Nat. Hist.,* Ser. 9, **3,** 198–199.

———. 1922. The distribution of the fishes of the order Ostariophysi. *Bijdragen Tot de Dierkunde,* **22,** 203–208.

Regan, C. T. 1929. Cyclostomata. Selachians. Fishes. (In) *Encyclopaedia Britannica*, 14th ed.

Romer, A. S. 1945. *Vertebrate Paleontology*, 2nd. ed. Chicago, U. of Chicago Press.

Savage, D. E. 1951. Report on fossil vertebrates from the upper Magdalena Valley, Colombia. *Science,* **114,** 186–187.

Schaeffer, B. 1952. Rates of evolution in the coelacanth and dipnoan fishes. *Evolution,* **6,** 101–111.

———. 1952*a*. The evidence of the freshwater fishes [in regard to Mesozoic land connections across the South Atlantic]. *Bull. American Mus. Nat. Hist.,* **99,** 227–234.

Schultz, L. P. 1946. A revision . . . Mugilidae. *Proc. United States National Mus.,* **96,** 377–395.

———. 1948. A revision of six subfamilies of atherine fishes . . . . *Proc. United States National Mus.,* **98,** 1–48.

Scott, E. O. G. 1936–1941. Observations on fishes of the family Galaxiidae. *Papers and Proc. R. Soc. Tasmania,* **1935,** 85–112; **1937,** 111–143; **1940,** 55–69.

Smith, H. M. 1945. The fresh-water fishes of Siam, or Thailand. *Bull. United States National Mus.,* **188.**

Smith, H. W. 1932. Water regulation and its evolution in the fishes. *Quarterly Review Biol.,* **7,** 1–26.

Smith, J. L. B. 1939. A living fish of Mesozoic type. *Nature,* **143,** 455–456.

———. 1950. Two noteworthy non-marine fishes from South Africa. *Ann. and Mag. Nat. Hist.,* Ser. 12, **3,** 705–710.

S(pencer), W. B. 1926. Ceratodus. (In) *Australian Encyclopaedia,* 3rd ed., Vol. 1, 248–250. Sydney, Angus and Robertson.

Stokell, G. 1941. A revision of the genus *Retropinna*. *Rec. Canterbury Mus.,* **4,** 361–372.

———. 1945. The systematic arrangement of the New Zealand Galaxiidae, part 1. *Trans. and Proc. R. Soc. New Zealand,* **75,** 124–137.

———. 1949. A fresh-water smelt from the Chatham Islands. *Rec. Canterbury Mus.,* **5,** 205–207.

Tchernavin, V. V. 1944. A revision of the subfamily Orestiinae. *Proc. Zool. Soc. London,* **114,** 140–233.

Trewavas, E. 1941. Fresh-water fishes. (In) *British Mus. (Nat. Hist.) Exp. South-west Arabia 1937–1938,* **1,** No. 3, pp. 7–15.

———. 1942. The cichlid fishes of Syria and Palestine. *Ann. and Mag. Nat. Hist.,* Ser. 11, **9,** 526–536.

Turner, C. L. 1946. *A contribution to the taxonomy and zoogeography of the goodeid fishes.* Occasional Papers Mus. Zool., U. Michigan, 495.

Weber, M., and L. F. de Beaufort. 1911–1951. (Vol. 8 by Beaufort alone, Vol. 9 by Beaufort and W. M. Chapman.) *The fishes of the Indo-Australian Archipelago.* Leiden, E. J. Brill, 9 vols.

Whitley, G. P. 1943. The fishes of New Guinea. *Australian Mus. Mag.,* **8,** 141–144.

Wynne-Edwards, V. C. 1952. Freshwater vertebrates of the arctic and subarctic. *Bull. Fisheries Research Board Canada,* No. 94.

chapter 3

Amphibians

The late Dr. Thomas Barbour, Mr. Arthur Loveridge, and Mr. Benjamin Shreve have answered many questions of mine about existing amphibians. Professor Alfred S. Romer has helped me with ancient ones. Dr. Karl P. Schmidt and Professor E. R. Dunn have read an early stage of the manuscript of this chapter and made useful comments on it. And Dr. G. S. Myers in correspondence and Dr. Paulo Vanzolini in conversation have contributed many additional facts and ideas. I am deeply indebted to all of these persons.

Romer (1945, modified 1947) recognizes ten orders of amphibians. Seven of them, with about forty families, contain only long-extinct forms. Some of the latter were important in their time. They were the first land vertebrates and were dominant before the Age of Reptiles. The three existing orders of amphibians are the Anura (frogs and toads), Urodela (salamanders), and Apoda (caecilians). These orders are very distinct from each other and all are probably old. Probable ancestors of the Anura occur before the Mesozoic, in the Pennsylvanian (Romer 1947, p. 259), and Anura themselves are known in the Jurassic. Of salamanders, the earliest known fossil is in the Lower Cretaceous; there is no record of them back through the Jurassic and Triassic; but they are apparently derived from still earlier, pre-Mesozoic lepospondyls (Romer 1945, p. 161). No fossils surely referable to the Apoda are known, but this order too is probably derived, independently, from pre-Mesozoic lepospondyls (Romer).

The classification of families and genera of existing amphibians is not very satisfactory. The salamanders are in best order. The caecilians are simplified, legless, burrowing forms known to few people and poorly understood taxonomically. Frogs and toads are better known, but they have no external characters of much phylogenetic importance, and there has been parallelism in the evolution and adaptations of different groups, even in the evolution of the skeleton (Parker 1932). Because of this, the real relationships of some frogs and toads are still doubtful. I am not competent to decide the doubts, so I have used the last detailed classification of living amphibians (Noble 1931) with only a few changes, chiefly Parker's (1934, 1940). The more important doubtful points are mentioned in the proper places below.

I have been familiar with the terms Anura, Urodela, Apoda, and also Salientia (a term meaning approximately the same as Anura) for many years, but I still have to pause to identify these words when I meet them. They are confusing words and I intend to avoid them. But that raises a minor difficulty: there is no inclusive, common English word for Anura, for frogs and toads together. I shall get around this difficulty by using "frogs" in a broad sense, to include toads too. This is a natural extension of meaning, for toads are not a taxonomic unit but merely frogs with thickened and warty skin.

Amphibians are, of course, cold-blooded. Typical frogs and salamanders and also some caecilians are amphibious, with aquatic larvae (tadpoles) and terrestrial adults. In each order, however, certain forms have become entirely aquatic through their whole lives, and others entirely terrestrial. In the latter the larval stage is omitted, or passed on land in the egg capsule, or carried by the adults, so that the animals spend their whole lives out of water. Frogs particularly have diverse adaptations for this (Barbour 1926), especially on the East and West Indies and some other islands, including the Seychelles and New Zealand.

There is, in the living orders of amphibians, a correlation between numbers, taxonomic and ecological diversity, and extent and diversity of area occupied. Caecilians form one family of some 17 genera and 73 known species. They are all burrowing or aquatic as adults. And they are confined almost entirely to the wet tropics of Africa, the Oriental Region, and America. Living salamanders form 8 families, some 43 or more genera, and about 200 full species. Most of them are terrestrial or aquatic (in surface water) as adults, but a few occur in subterranean water and a few are arboreal. And they are mostly confined to the wet parts of the north-temperate zone; a few have entered the tropics; but no salamander occurs in the south-temperate zone. Frogs, including toads, form about 37 families and subfamilies, about 200 or more genera, and perhaps 2000 or more known species. They include a variety not only of amphibious, aquatic, and terrestrial forms, the latter occurring in many different land habitats, but also many arboreal ones independently derived from several different families. And they are very widely distributed in both main climatic zones, occur in arid and other unfavorable places as well as favorable ones, and range farther north, farther south, and much farther onto islands than caecilians and salamanders do.

Amphibians have scant tolerance for salt water. The frog *Rana cancrivora* and related species in the Orient live near the sea coast and by brackish estuaries; the adults enter sea water for short times apparently with impunity; and the tadpoles have been found living in salt water in crab holes even below high-tide line (Boulenger 1903, Pearse 1911, M. Smith 1927). There are some other records of salamanders and frogs being found in salt water, probably by accident but alive, and living more or less normally in brackish water (Schmidt 1951). Many persons including Darwin (1859; 1950 reprint, p. 333) have thought that contact with salt water must be immediately fatal to all amphibians, but this is not quite strictly true (see again Schmidt 1951). Some amphibians, at least some frogs, can stand an occasional dip in the sea or contact with salt spray, and this must improve their chances of being rafted across narrow ocean straits. The actual tolerances for salt water of different amphibians under different conditions are almost entirely unknown and are well worth investigation. But simple experiments to this end will not be final. Whether or not amphibians have crossed salt water is more likely to be shown by their distributions than by anything that can be learned in the laboratory, and their distributions, compared with those of fresh-water fishes, show beyond reasonable doubt that frogs have crossed moderate gaps of salt water many times.

Amphibians have scant tolerance for desiccation. They vary somewhat in this. Thorson and Svihla (1943) found that *Scaphiopus* (spadefoot toads) could stand to lose more water than *Bufo* (ordinary toads), which could lose more than *Hyla* (tree frogs), which could lose more than *Rana* (ordinary frogs), the latter being least able to stand drying. These experiments were made with only a few species, and the authors wisely do not try to extend the results. For here again there is a great difference between the adaptations and tolerances of the animals experimented with and the adaptability and occurrence of the groups in nature. Not only *Scaphiopus* and *Bufo* but also *Rana* and its derivatives sometimes occur in deserts (see text, p. 147), and even *Hyla* does so sometimes (see under Hylidae in list of families); in fact different frogs find a variety of ways to avoid desiccation (Buxton 1923, p. 96). The extent to which dry country is a barrier to amphibians is therefore another thing to be found out from their distributions, not decided in advance or by experiment.

Some amphibians have been carried about by man for food, or as curiosities or pets, or to control insects, or by accident. Pope and Boring (1940) mention commercial shipments of the Giant Salamander from Japan to China and say that there has been difficulty in determining the distribution of this species in China not only because it is carried for food but also because it is kept as a curiosity by royalty and other rich persons, and that this probably accounts for certain unexpected records, including one from the sea at Amoy. Boulenger (1936, p. 307) says of the Edible Frog, *Rana esculenta,* that it is in England confined to the fens of Norfolk and Cambridgeshire, where it was probably introduced in the Middle Ages by Italian monks, since some of the British specimens agree with an Italian variety. Whether or not this frog really was brought to England so early, it has certainly been introduced there a number of times more recently, its history in England being one of repeated introduction and, after a varying number of years, disappearance of the introduced colonies (M. Smith 1951, pp. 141–143). The Giant Toad, *Bufo marinus,* native to the mainland of tropical America, has been introduced in attempts to control insects on many of the West Indies, Bermuda, the Hawaiian Islands, and parts of tropical Australia, New Guinea, the Solomons, and other Pacific islands. Many more instances like these could be given. Perhaps even more important is the fact that small frogs are carried around by children of every race, without record. And perhaps most important of all is the fact that some frogs, including tree frogs which can hide in thatch on boats or in bunches of bananas, are often transported accidentally, especially in the tropics. In some cases the hand of man is easily detected, but in others it is not. Human agency should always be suspected when an amphibian turns up in an unexpected place. But all this concerns details and has little to do with the main pattern of amphibian distribution.

Limits of distribution of amphibians

The northern and southern limits of distribution of existing amphibians (Fig. 19) are all set by frogs. Typical frogs of the genus *Rana* go farthest north. *Rana temporaria* in Europe reaches the northern limit of land in Scandinavia at about 71° N., well above the Arctic

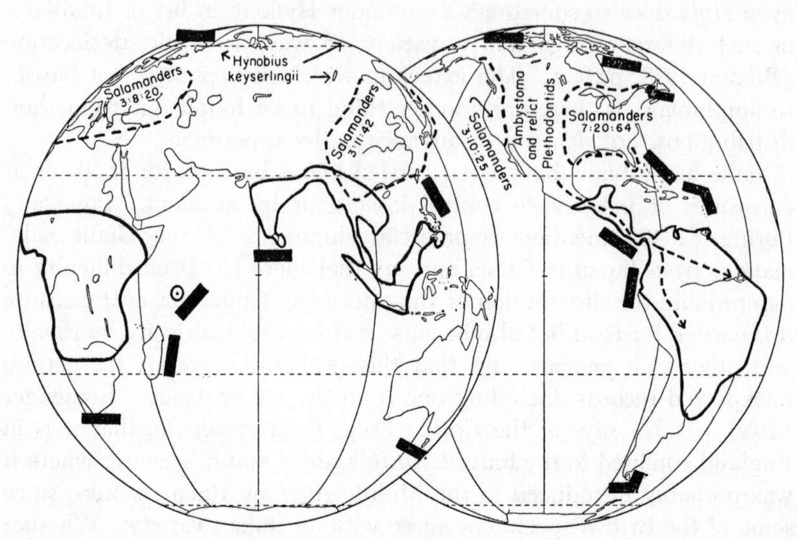

Fig. 19. Limits of distribution of amphibians. Heavy bars show approximate limits of frogs; solid lines, approximate limits of caecilians; broken lines, four principal areas of occurrence of salamanders in the north-temperate zone and extension of *Oedipus* to Central and South America. Numbers are numbers of families, genera, and species of salamanders in the areas indicated.

Circle, and this or related species cross the Circle at several points in Asia too. A subspecies of *Rana sylvatica* reaches or crosses the Circle in the west of North America and occurs nearly to the northern tip of the subarctic Labrador Peninsula in the east. Typical toads of the genus *Bufo* come close to the arctic in Europe, to about 65° N., but apparently reach only about 60° N. in western North America and a little above 55° in Labrador. Tree frogs of the genus *Hyla* reach only about 57° N. in Europe, perhaps not much above 52° in eastern Siberia, and to about 62° in western North America and 55° in the east, but another hylid genus, *Pseudacris,* which lives more in swamps and on the ground than in trees, approaches the Arctic Circle in western North America. Of salamanders, *Hynobius keyserlingii* of Siberia crosses the Arctic Circle to Verkhoyansk (Dunn 1923), which is one of the coldest places in the world in winter, with a January mean of −59° F. and a minimum of −90°, although the July mean is about 60° (*Webster's Geographical Dictionary*). This is an example of the fact noted by Wynne-Edwards (1952, p. 24), that summer rather than winter temperatures determine the limits of northern amphibians. Other salamanders are not known even to get near the Arctic Circle. They reach only about 63° N. in Europe and perhaps only about 59° and 53° in western and eastern North America respectively. Salamanders thus fall short of the northern limits of *Rana* and *Bufo* in Europe and of *Rana* and hylids in North America. These details are from Ekman (1922), Mertens and Müller (1940), Mills (1948), Schmidt (1953), Wynne-Edwards (1952), Logier and Toner (1955), Harper (1956), and other sources. The northernmost frogs (*Rana*) extend into regions where the subsoil is permanently frozen (Harper 1956, p. 99). How they survive the arctic winter is, I think, still unknown. Perhaps they hibernate under water deep enough to protect them from extreme cold.

Southward, frogs reach the tip of South Africa, Tasmania south of Australia, New Zealand, and the tip of South America and Tierra del Fuego. (Bridges 1948, p. 447, mentions small, unidentified frogs on Tierra del Fuego. I am indebted to Dr. C. T. Parsons for sending me this reference, after I had despaired of finding one to frogs so far south.)

The limits of distribution of amphibians on islands too (Fig. 19) are set by frogs. Salamanders are confined to mainland areas and continental islands such as the British Isles, Japan, the Riu Kius, and Formosa; a supposed record from Haiti in the West Indies is now discredited. Caecilians inhabit the continental tropics of Africa, Asia, and America, and occur on such continental islands as Ceylon, Sumatra, Java, and Borneo, and in America on Trinidad, but they do not occur on Madagascar or the West Indies proper and do not cross Wallace's Line. In these directions they are as sharply limited as strictly fresh-water fishes. A caecilian does reach the southern Philippine islands of Palawan and Mindanao, but so do a few cyprinid fishes. In only one place do caecilians notably surpass the limits of distribution of strictly fresh-water fishes, and that is on the Seychelles Islands in the Indian Ocean. This is the more remarkable because of the absence of caecilians on Madagascar. Frogs are present on all habitable continents and most continental islands, including all those mentioned above, and they are native also on some doubtful or biologically oceanic islands including Madagascar, the Seychelles, the whole Indo-

Australian Archipelago including the Philippines and Solomons and perhaps the Palau and Fiji Islands, and on New Guinea, Australia, Tasmania, New Zealand, and the West Indies. Additional details and references are given in Chapter 8. An interesting detail is that Newfoundland, although it is a large continental island near the mainland, seems to lack native amphibians as well as fresh-water fishes and reptiles. A negative fact like this is hard to prove, but at least I have been unable to find any records of amphibians from Newfoundland except of one frog (*Rana clamitans*) which has apparently been introduced in loads of hay from Nova Scotia (Mills 1948, p. 7; H. J. Squires in letter of October 24, 1951; C. H. Lindroth in letter).

Within the general limits of amphibians, a few genera especially of frogs are very widely distributed. *Rana,* in a broad sense (Fig. 21), inhabits all Africa, Eurasia, and North America and reaches the northern edge of Australia and the northern half of South America. *Bufo* (Fig. 22) is on all the continents except Australia. *Hyla* (Fig. 25) is world-wide except for a great gap in Africa and the Oriental tropics. But all other genera of frogs are much more limited in distribution. Of salamanders, *Triturus* is discontinuously Holarctic, but all other genera are much more localized. Of caecilians, no genus is known to extend beyond the limits of one region. No species of amphibians extends much beyond the limits of one region.

Zonation of amphibians

Within the limits described, two of the three existing orders of amphibians are strongly zoned (Fig. 19).

Caecilians are almost entirely tropical. They enter the edges of the temperate zones, to a very limited extent, only in eastern Asia and southern South America.

Salamanders are mostly north-temperate. The occurrence of one in tropical West Africa is very doubtful. Several do occur in the edge of the tropics in eastern Asia (see under Salamandridae in list of families). Ambystomid salamanders enter the edge of the tropics in Mexico, but only at rather high altitudes. And the plethodontid genus *Oedipus* or its components occur in the American tropics south to Bolivia; some species of this genus are on mountains, but others are in the full lowland tropics; one lives on the equator at the mouth of the Amazon (see under Plethodontidae in list of families). Although most salamanders are adapted to live only within a narrow range of cool temperature, the exceptions suggest that the animals are not inherently cold-adapted but may have as wide potential adaptability as frogs. This is a good example of the difference between adaptation and adaptability.

Frogs are very widely distributed, but the greatest numbers of them are in the tropics. Of 37 families and subfamilies of them here recognized, 27 are confined to the tropics or nearly so and 4 or 5 others are more tropical than temperate in distribution (Fig. 20). Of perhaps 2000 known species, about 80 per cent are tropical; a little more than 10 per cent, north-temperate; and probably less than 10 per cent, south-temperate.

No group of frogs is both confined to and distributed over the whole of a temperate zone. Pelobatinae are almost entirely confined to the north-temperate zone, entering the tropics only on the Mexican Plateau, but they are few in number and are absent in much of Asia, although fossil in Mongolia in the Miocene; and other subfamilies of Pelobatidae are isolated in the tropics in southeastern Asia, and on the Seychelles Islands. Discoglossidae occur chiefly in the north-temperate zone, but in the Old World only, and they are discontinuously distributed even there, occurring in Europe etc. and *eastern* Asia, and one genus of the family is isolated on one small island in the tropical Philippines. Finally, *Ascaphus* (one species), which is sometimes placed in a family by itself, sometimes brigaded with *Leiopelma* of New Zealand, is confined to cold brooks in northwestern North America. These three groups of frogs are the only ones of family or subfamily rank that come near to being peculiar to the north-temperate zone; all are primitive according to Noble (1931); and they all have relict distributions, either discontinuous or localized. A few other frogs that are primarily tropical enter the north-temperate zone in eastern Asia or eastern North America in such a way as to suggest local movements northward (see Fig. 20): at least five tropical Oriental genera besides *Rana* and *Bufo,* belonging to four families, reach central China, and two of them go still farther north, *Rhacophorus* to northern Honshu in Japan, and the brevicipitid *Kaloula* to Manchuria; and in America three tropical genera of leptodactylids and one of brevicipitids reach Texas and in one case southern Arizona too, and a second brevicipitid genus, *Microhyla,* extends north to Iowa, Indiana, etc. Representatives of the dominant, widely distributed genera *Rana, Bufo,* and *Hyla* complete the list of frogs of the north-temperate zone except for a few that just enter its southern edge.

In the south-temperate zone there is even less of a zonal frog fauna. The families Leptodactylidae and Hylidae are characteristic of the

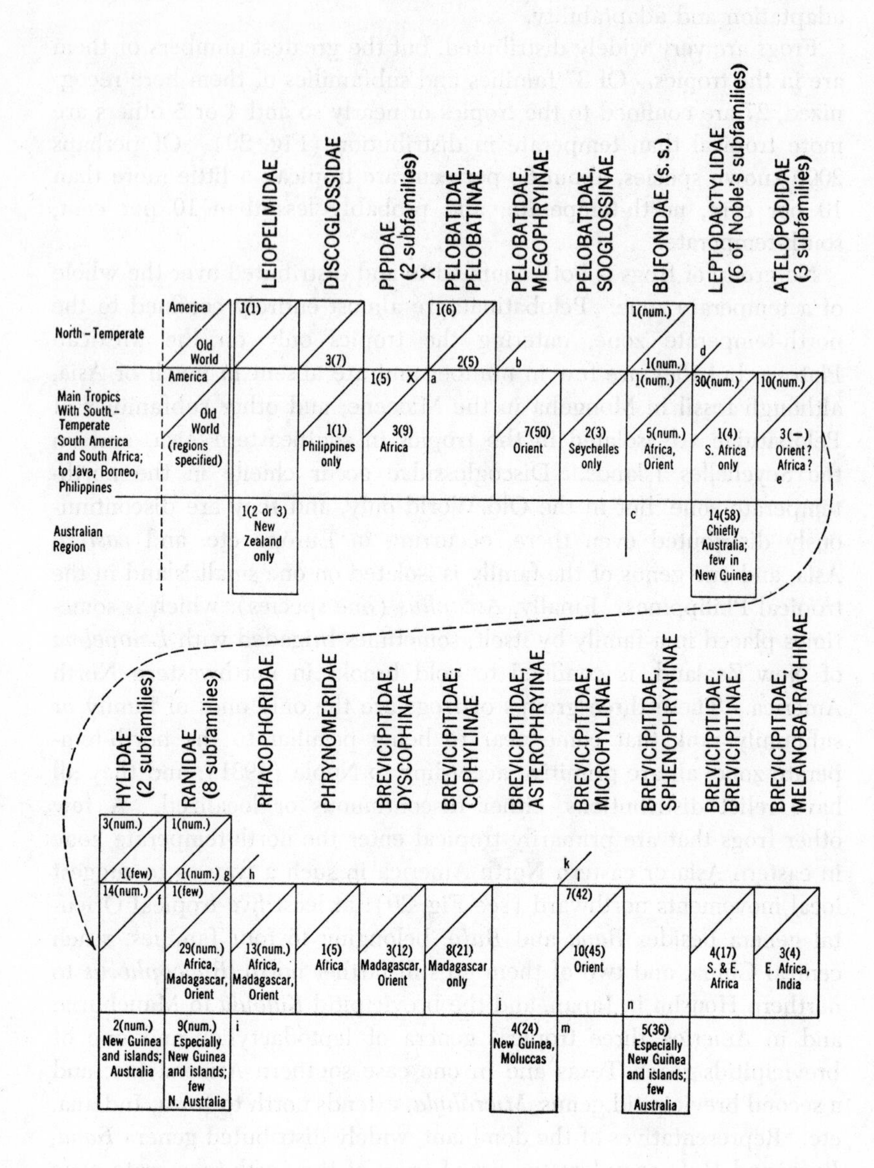
LEIOPELMIDAE
DISCOGLOSSIDAE
PIPIDAE (2 subfamilies) X
PELOBATIDAE, PELOBATINAE
PELOBATIDAE, MEGOPHRYINAE
PELOBATIDAE, SOOGLOSSINAE
BUFONIDAE (s. s.)
LEPTODACTYLIDAE (6 of Noble's subfamilies)
ATELOPODIDAE (3 subfamilies)
North – Temperate
America
Old World
Main Tropics With South-Temperate
South America and South Africa; to Java, Borneo, Philippines
Old World (regions specified)
Australian Region
1(1)
3(7)
1(6)
2(5)
b
1(num.)
d
1(num.)
1(5)
X
a
1(num.)
30(num.)
10(num.)
1(1) Philippines only
3(9) Africa
7(50) Orient
2(3) Seychelles only
5(num.) Africa, Orient
1(4) S. Africa only
3(–) Orient? Africa?
e
1(2 or 3) New Zealand only
c
14(58) Chiefly Australia; few in New Guinea
HYLIDAE (2 subfamilies)
RANIDAE (8 subfamilies)
RHACOPHORIDAE
PHRYNOMERIDAE
BREVICIPITIDAE, DYSCOPHINAE
BREVICIPITIDAE, COPHYLINAE
BREVICIPITIDAE, ASTEROPHRYINAE
BREVICIPITIDAE, MICROHYLINAE
BREVICIPITIDAE, SPHENOPHRYNINAE
BREVICIPITIDAE, BREVICIPITINAE
BREVICIPITIDAE, MELANOBATRACHINAE
3(num.)
1(num.)
1(few)
1(num.)
g
h
k
14(num.)
f
1(few)
7(42)
29(num.) Africa, Madagascar, Orient
13(num.) Africa, Madagascar, Orient
1(5) Africa
3(12) Madagascar, Orient
8(21) Madagascar only
10(45) Orient
4(17) S. & E. Africa
3(4) E. Africa, India
j
n
2(num.) New Guinea and islands; Australia
9(num.) Especially New Guinea and islands; few N. Australia
i
4(24) New Guinea, Moluccas
m
5(36) Especially New Guinea and islands; few Australia

southern barrier-limited continents of South America and Australia, but they are not restricted to south-temperate climates; in fact both families are primarily tropical in America. These facts suggest that frogs are primarily tropical animals, which frequently extend into the temperate zones locally or extensively and sometimes persist there for considerable times, but which do not form separate, dominant, primarily temperate faunas.

Radial distribution of amphibians

Two genera of frogs have strongly radial distributions. *Rana* (Fig. 21), which includes most of our ordinary, familiar north-temperate frogs, is very well represented in the Old World tropics. It occurs through all Africa and all the habitable parts of Asia; it has reached Madagascar; it goes down through the Indo-Australian Archipelago to New Guinea and (if *Hylarana* is included) to the northern corner of Australia; and it occurs through North America and (in decreasing numbers) through Central America to the northern half of South America (see list of families for further details). It belongs to the family Ranidae, which, excepting *Rana,* is confined to the warm parts of the Old World. (How many North American naturalists know that our common frogs are members of a great genus which has come from

---

Fig. 20. Distribution of families and subfamilies of frogs. Revised from *Quarterly Review of Biology,* **23**, 1948, p. 18, Table 7. Numbers such as 3 (7) are numbers of genera and species.

(X, x) An additional family, Rhinophrynidae, belongs here, with one species, localized in the northern edge of the American tropics.

(a) *Scaphiopus* enters edge of tropics in central Mexico.

(b) *Megophrys* enters edge of north temperate in China, and other genera occur in southwestern China.

(c) *Bufo* ranges to Celebes and Lombok.

(d) Three genera enter edge of north temperate, to Texas, New Mexico.

(e) See family list.

(f) *Hyla* enters north edge of tropics in southeastern Asia.

(g) A second genus (*Staurois*) enters edge of north temperate in China.

(h) *Rhacophorus* ranges north to central China and Japan.

(i) *Rhacophorus* ranges to Celebes, Timor.

(j) *Metopostira* does *not* reach the Sulu Islands; see family list.

(k) *Microhyla* (*Gastrophryne*) enters north temperate in eastern United States; a second genus (*Hypopachus*) reaches southern Texas.

(l) Two genera enter north temperate in eastern Asia: *Microhyla* reaches central China, Japan; *Kaloula,* Manchuria.

(m) *Kaloula* reaches Celebes and Lesser Sunda Islands; *Microhyla,* Bali.

(n) *Oreophryne* ranges west and north to Bali, Celebes, and Mindanao.

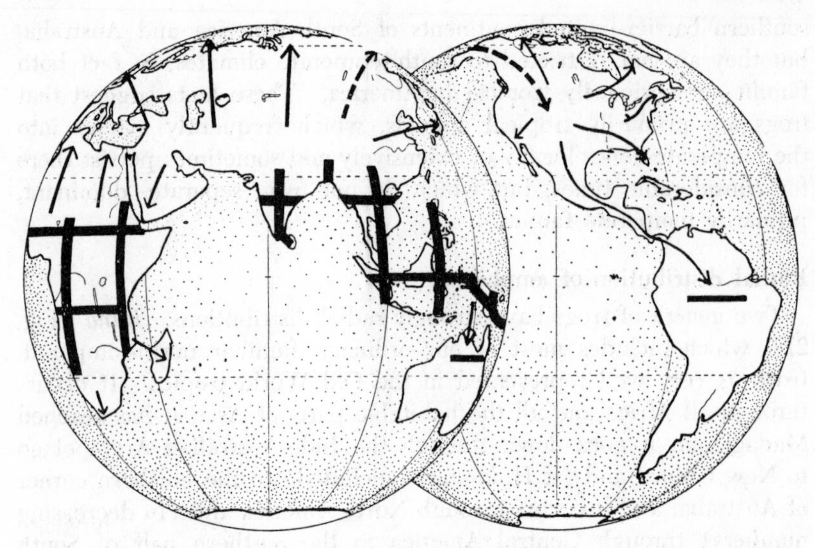

Fig. 21. Distribution of frogs of the family Ranidae. Heavy crossed bars show area of occurrence of other ranids with *Rana;* arrows, radiation of *Rana* (diagrammatic); bars, approximate limits reached by *Rana* in Australia and South America.

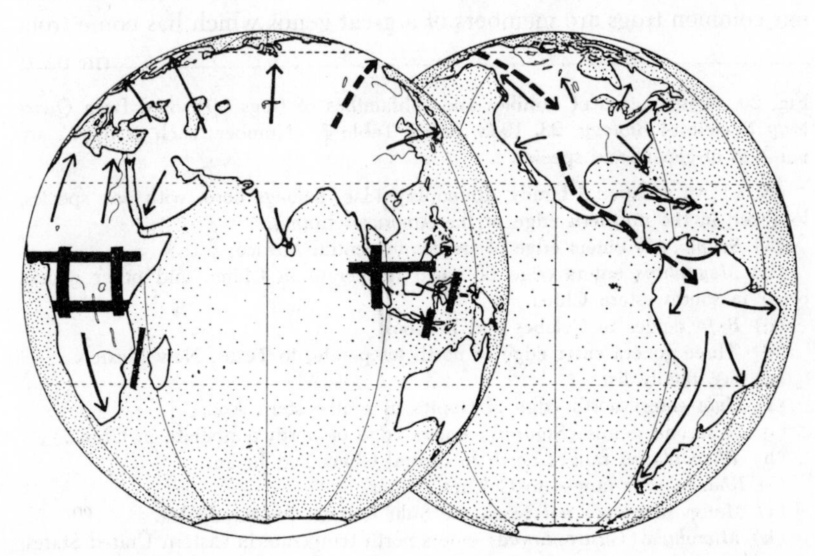

Fig. 22. Distribution of frogs (toads) of the family Bufonidae. Heavy crossed bars show areas of occurrence of other bufonids with *Bufo;* arrows, radiation of *Bufo* (diagrammatic); bars, limits reached by *Bufo* west of Africa and in the East Indies.

the Old World tropics and is just now sending spearheads into Australia and South America?) *Bufo* (Fig. 22), which includes the typical toads, occurs through all Africa, temperate and tropical Eurasia and islands to Celebes, etc. (but not the Australian Region), and North, Central, and all or almost all South America, and on some West Indies. It belongs to the family Bufonidae, which, too, in a strict sense and excepting *Bufo*, is confined to the Old World tropics.

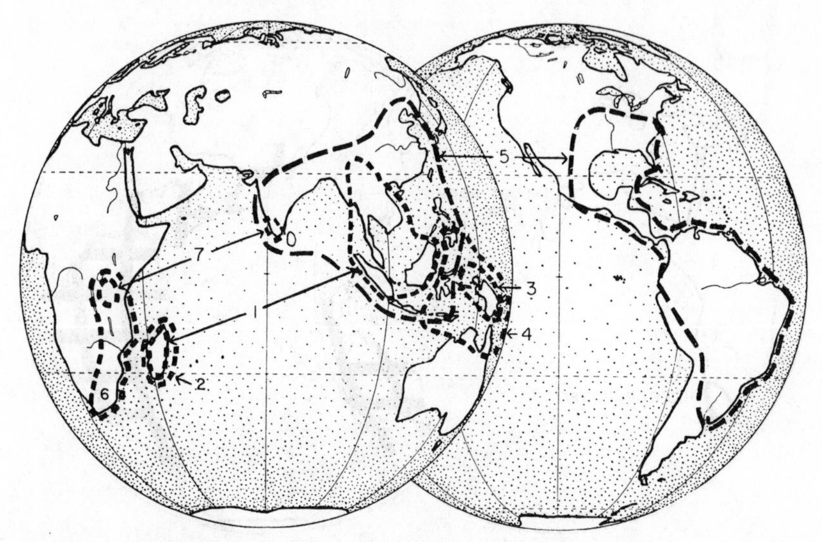

Fig. 23. Distribution of frogs (toads) of the family Brevicipitidae: approximate limits of distribution of the seven subfamilies, numbered as in the list of families.

(How many North American naturalists know that toads not very different from ours have hopped their way around the world and south to the farthest parts of Africa and South America?) The radial distributions of both these genera, and of their families, center on the main part of the Old World tropics.

Two other families of frogs that do not have obviously radial distributions are worth consideration now. One is the family Brevicipitidae (Microhylidae). This family (Fig. 23) is divided by Parker (1934) into seven subfamilies. All of them occur in the Old World tropics and most of them are localized or have discontinuous ranges there, the largest concentrations of them being now in Madagascar, tropical Asia, and the tropical part of the Australian Region. One of the subfamilies, Microhylinae, which occurs mainly in tropical Asia, extends northward into temperate eastern Asia, reaching Manchuria.

This is the only subfamily that occurs in America, where again it is mainly tropical. (See list of families for further details.)

The other family to be considered now is the Leptodactylidae (Fig. 24). Leptodactylids are numerous in Australia and Tasmania, and a few reach New Guinea; they are numerous also in South and Central America, and a few reach north to southern Texas and Arizona. An isolated genus, *Heleophryne,* in South Africa may belong to this

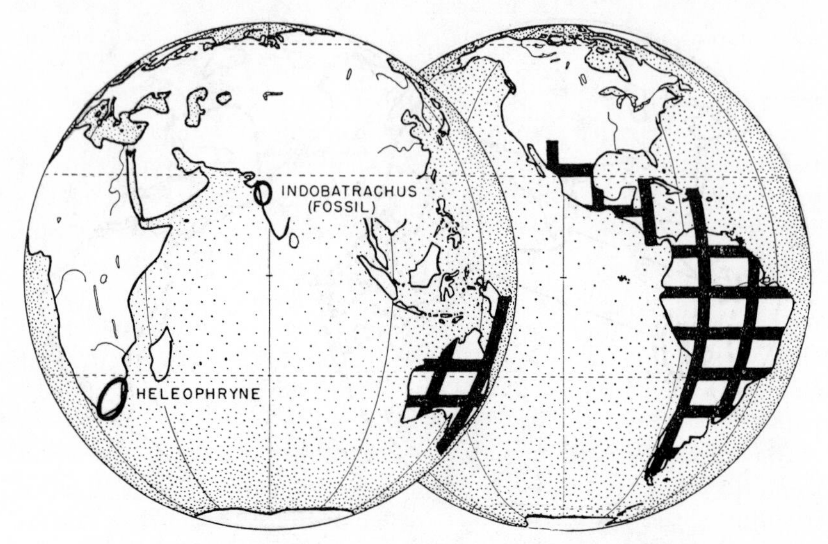

Fig. 24. Distribution of frogs of the family Leptodactylidae.

family, but it is doubtful. Except for *Heleophryne,* living leptodactylids are unknown outside the Australian Region and the warm parts of America. However, a fossil (Eocene) frog from Bombay in tropical India is put in this family; and the fact that bufonids, which are thought to be derived from leptodactylids, apparently originated in the main Old World tropics confirms the former presence of leptodactylids there.

These four families, Ranidae, Bufonidae, Brevicipitidae, and Leptodactylidae, form a significant sequence (Figs. 21 to 24). Something can be said of the history of each family considered alone. Ranidae have evidently radiated in the main Old World tropics, and *Rana* has spread from there over most of the rest of the world. Bufonidae and *Bufo* have probably had a generally similar history, of radiation in the main Old World tropics and spread of *Bufo* over much of the world. Brevicipitids have evidently radiated primarily in the Old

World tropics but seem to be retreating there now, and one subfamily has got through the north to the American tropics and reradiated there. And leptodactylids apparently occurred in the main Old World tropics once but have withdrawn from there and now survive in Australia, the warmer part of America, and perhaps South Africa. These histories are based on facts or obvious clues. However, something less obvious but more important than the separate histories emerges when the four families are considered together. They form an apparent sequence which suggests a common pattern of evolution and dispersal. It looks as if successive families have evolved in the main area of the Old World tropics, radiated from there, and then in the case of the older families retreated or become extinct there with the rise of later families, the older ones surviving chiefly in peripheral or barrier-limited areas. *Rana* and *Bufo* are near the height of this apparent cycle, brevicipitids farther along, and leptodactylids still farther along.

The four families just discussed are all primarily amphibious or terrestrial; none of the four is primarily arboreal, although they include aboreal derivatives. The two principal arboreal families of frogs, Hylidae and Rhacophoridae, form an independent geographical pattern.

Hyla (Fig. 25) is world-wide except for a gap in most of Africa and tropical Asia. Species of *Hyla* increase in numbers with distance from this gap. There are few across temperate Eurasia, more in temperate North America, and still more in Central and South America, especially in the tropics there; and there are many also, isolated, in New Guinea etc. and Australia. (How many North American naturalists know that our tree frogs and northern spring peepers are outliers of a genus which swarms in the rain forests of tropical America and, with marsupials, in New Guinea and Australia?) Most other hylids are in South and Central America, where *Hyla* is reinforced by many other genera (see list of families for further details). This distribution might be the result either of a primary radiation of hylids in tropical America and spread of *Hyla* from there (if a way was open to Australia) or of a primary radiation from some other center and secondary radiations in America and the Australian Region. If the latter, the history of hylids may have followed the pattern suggested by the ground-living families. Hylids may have radiated first in the main Old World tropics and later disappeared there; *Hyla* may have spread from there to the Australian Region and America and reradiated in these places; and additional hylids may have reached

America, perhaps before *Hyla* did so. *Hyla* would not need continuous forest for its dispersal. Most species of the genus are tree frogs, but some are rock dwelling or even terrestrial, and a few, though very little modified in structure, live in and can cross deserts (see again Hylidae in list of families).

There is a good possible reason for the disappearance of hylids in the Old World tropics: the rise of tree frogs of the family Rhaco-

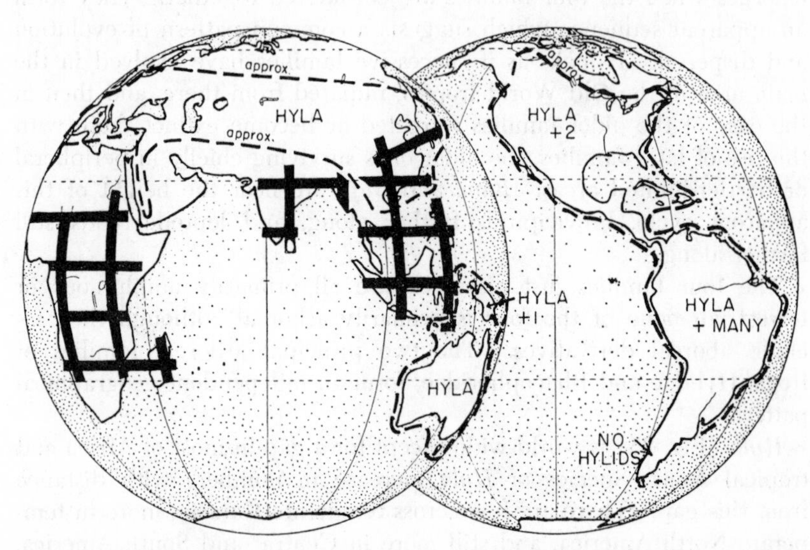

Fig. 25. Distribution of principal families of tree frogs: Rhacophoridae (heavy crossed bars) and Hylidae (broken lines). Numbers are numbers of genera of hylids in addition to *Hyla* in different areas.

phoridae (Fig. 25). Rhacophorids are numerous in the main part of Africa, Madagascar, and tropical Asia, etc. They reach southern China and Japan but are otherwise absent in the north-temperate zone, and they do not reach the Australian Region or America. They are dominant in an area almost exactly complementary to that now occupied by *Hyla.* Rhacophorids overlap the range of *Hyla* probably only in part of eastern Asia and on Timor, etc. (see list of families). Rhacophorids may be near the beginning of a cycle of radiation from the main Old World tropics, and they may be overrunning and replacing *Hyla* as they radiate.

The hypothesis of spread of successive groups of frogs from the Old World tropics has been set forth in different words in my *Quarterly Review* article (1948, p. 21). It is based on comparison of the distributions of families without regard to their phylogenetic position. It is important to find whether the distributions are correlated with phylogeny.

Frogs are derived from an amphibian which had ribs and a tail. Of existing frogs, (1) Leiopelmidae still have free ribs and are the only existing frogs with tail-wagging muscles—that is, as someone has said, they still have muscles to wag the tail they no longer have. (2) Discoglossidae have free ribs even as adults; Pipidae, only as larvae; and the family Rhinophrynidae is tentatively associated with these families because it has similar vertebrae and because its supposed larva is pipid-like. (3) Pelobatidae lack free ossified ribs but have other somewhat primitive characters. The remaining, "higher" frogs form two main series, differing in technical characters of the skeleton and musculature: (4) Leptodactylidae, Atelopodidae, Bufonidae, and Hylidae; and (5) Ranidae, Rhacophoridae, Phrynomeridae, and Brevicipitidae. The five groups numbered above are usually treated as suborders (Noble 1931; Romer 1945, p. 591; etc.) and are of course defined by various characters in addition to those mentioned. The suborders are named in the list of families. These five groups, although not necessarily related in any direct or simple way, apparently represent stages in frog evolution. Their distributions are correlated with their places on the evolutionary scale. The first, most primitive group survives only locally, in northwestern North America and in New Zealand. The families of the second group, though not so strikingly relict, are discontinuously distributed or localized and are nowhere dominant. The third group too is discontinuously distributed and is confined to certain parts of the world, but is subdominant in part of the Orient. The fourth and fifth groups both are nearly cosmopolitan, and both include radiating families and families which are dominant in various places; but the fourth group tends to predominate in peripheral or barrier-limited continents, and the fifth group, which Noble considers the highest of all, now predominates in the main area of the Old World tropics. In short, the most primitive groups of frogs are the most localized; successively higher groups are increasingly dominant; the next to the highest groups are dominant peripherally; and the highest of all are dominant primarily in the Old World tropics. This general correlation fits well enough with the idea of radiation of successive groups from the Old World tropics. The details that do not fit suggest complexities but do not (I think) spoil the main correlation.

As a general rule, I do not like to stress details in zoogeography, but one detail of frog distribution does seem especially significant in this connection. The presence of a very primitive frog on New Zealand and the complete absence of related frogs in Australia and South America suggest succession and replacement of frog faunas on the continents.

Fossils tell little about the geographical history of frogs. The leptodactylid *Indobatrachus,* in India in the Eocene, may be the only frog fossil of much geographical importance—if even it is beyond doubt. Some other fossils show that certain frogs were at certain places at certain times, but the details shown are mostly minor ones, and there is too much doubt about some of them. For example, a *Hyla* was described from the Lower Miocene of Europe by Noble (1928), who admitted the imperfection of the specimen but said, "I do not hesitate to refer the species to *Hyla.*" Parker (1929 and 1930) re-examined the actual specimen and decided that it was not a hylid. Yet the fossil is still commonly cited in the record of *Hyla* (Romer 1945, Darlington 1948, Schaeffer 1949). I do not know the rights of this matter; the point is that it is doubtful. Another fossil frog from the Oligocene-Miocene of Europe has been assigned to the Hylidae and to the existing genus *Amphignathodon,* a highly specialized genus now confined to a small area in South America, but Hoffstetter (1945) has shown that the European fossil is not *Amphignathodon* but the skull of a frog incorrectly associated with the jaw of a lizard! Schaeffer (1949) has done a useful thing in bringing together recent records of fossil frogs, but I doubt if the fossils mean as much as he implies they do. The fact is, I think, that frogs are among the animals of which the fossil record is still inadequate, and of which the distribution of existing forms is still the best guide to their geographical history.[1]

Regional distribution of amphibians.

Space cannot be taken here to treat the amphibian faunas of different parts of the world in much detail, but their general composition and relationships will be summarized.

The amphibians of the north-temperate zone have already been

[1] Metcalf (1923, 1940) has tried another way of tracing the evolution and dispersal of frogs, by means of their opalinid parasites. This is "host-parasite zoogeography." It appeals to zoologists and it may be useful in some cases, but this was not a successful example of it. Metcalf's methods and conclusions have been criticized in special papers by Noble (1925) and Dunn (1925).

mentioned (pp. 134–135): salamanders; a few primitive groups of frogs, all more or less localized; a few frogs of tropical groups which enter temperate eastern Asia and/or temperate North America; and (dominating the fauna) representatives of the three great frog genera *Rana, Bufo,* and *Hyla* (plus two other hylid genera in North America). The amphibian faunas of the two halves of the north-temperate zone, Eurasia and North America, are much alike as wholes but different in details, and amphibians are very unevenly distributed on each continent. Salamanders illustrate this.

Within the north-temperate zone, salamanders occur as follows (Table 3): in Europe and adjacent parts of southwestern Asia and

TABLE 3. SALAMANDERS OF FOUR PRINCIPAL AREAS IN THE NORTH TEMPERATE ZONE

All families are listed, and all genera that occur in more than one of the four areas. Numbers in parentheses after families are numbers of genera and species; after genera, numbers of species. The numbers in some cases are only approximations.

Europe, etc.	Eastern Asia	W. North America	E. North America
	Hynobiidae (5, 26+)		
	Cryptobranchidae (1, 1)	....	Cryptobranchidae (1, 2)
		Ambystomidae (3, 6)	Ambystomidae (1, 9)
		Ambystoma (4)	Ambystoma (9)
Salamandridae (6, 18)	Salamandridae (5, 15+)	Salamandridae (1, 4)	Salamandridae (1, 3)
Triturus (10)	Triturus (3)	Triturus (4)	Triturus (3)
	(Notophthalmus)	(Notophthalmus)	
			Amphiumidae (1, 2)
Plethodontidae (1, 1)	....	Plethodontidae (6, 15)	Plethodontidae (13, 38)
		Plethodon (5)	Plethodon (11)
		Aneides (3)	Aneides (1)
Hydromantes (1)	....	Hydromantes (2)	
Proteidae (1, 1)	....	....	Proteidae (1, 7)
			Sirenidae (2, 3)
3 Families (8, 20)	3 Families (11, 42)	3 Families (10, 25)	7 Families (20, 64)

North Africa, 3 families, 8 genera, about 20 full species; in a great area of western and northern Asia, only *Hynobius keyserlingii;* in eastern Asia including the eastern Himalayas etc., China, and Japan, 3 families, 11 genera, at least 42 species; in western North America, from the lower corner of Alaska through California, 3 families, 10 genera, 25 species; in west-central North America including the Great Basin, Rocky Mountains, and Great Plains, only one wide-ranging species of *Ambystoma* and a few localized, geographically isolated ambystomids and plethodontids; and in eastern North America, from the Ozarks, eastern Texas, etc., to the Atlantic, and from southern

Canada to Florida, 7 families, 20 genera, and about 64 full species. It will be seen that salamanders are concentrated in four widely separated areas in the north-temperate zone: Europe etc., eastern Asia, western North America, and eastern North America; and only the latter has a nearly complete representation of existing families. Many families and genera of salamanders have discontinuous ranges (Table 3), and fossils indicate that some of them once occurred far outside their present ranges: *e.g.*, the cryptobranchid *Megalobatrachus* (giant salamanders) and the salamandrid *Tylototriton*, both now confined to eastern Asia, are fossil in Europe in the Miocene. The geographical relationships of different groups run in every possible direction among the four main areas of distribution. Salamandrids link Europe and eastern Asia; *Hydromantes*, Europe and western North America; proteids, Europe and eastern North America; certain *Triturus*, eastern Asia and western North America; living cryptobanchids, eastern Asia and eastern North America; and certain plethodontids and *Ambystoma*, western and eastern North America.

The geographical pattern formed when an old, diverse, widespread fauna is depleted and withdraws into restricted areas is called a relict pattern. It has two main characteristics. The first is discontinuity: related forms tend to occur in widely separated areas, and fossils may occur far outside the ranges of the living forms. The second is diversity of geographical relationships: the geographical ties of surviving forms tend to be diverse and confused, reflecting chances of extinction and survival rather than directions of dispersal. The salamanders in the north-temperate zone illustrate this kind of pattern perhaps better than any other animals.

The tropical and south-temperate parts of the world are best considered together, so far as amphibians are concerned. Of caecilians it need be said here only that they are confined to separate areas, almost entirely within the tropics, in West Africa, East Africa, the Seychelles Islands, the Oriental Region, and America, and that the interrelationships of the forms in the different areas are uncertain. Salamanders extend only a little way into the northern edge of the tropics in the Old World, and the one group of them that is widely distributed in tropical America has evidently been derived rather recently from the north. Frogs (including toads) are abundant in the tropics and occur in a number of more or less separate areas of distribution.

The frogs of Africa south of the Sahara (Noble 1922, pp. 64–67; 1924) belong to seven full families and about forty or more genera—probably many more by modern standards. They are most numerous and diverse in the forests of tropical West Africa, but many of the genera are widely distributed over the African tropics. Only *Rana* and *Bufo* go far into the Sahara (Angel and Lhote 1938, table opposite p. 378); *Pyxicephalus*, recorded from the south-central Sahara by Guibé (1950) along with several species of *Bufo*, is simply a *Rana* with shortened legs and improved digging apparatus (Loveridge, in conversation). The temperate northwestern corner of Africa is inhabited by European amphibians, including at least three genera of salamanders, and frogs of the genera *Discoglossus*, *Hyla*, and *Pelobates* as well as *Rana* and *Bufo*. Temperate South Africa has various *Rana* and *Bufo*, a few frogs of other characteristic African groups, and two additional noteworthy genera. *Heleophryne*, with about four species, is confined to South Africa north only to Transvaal, and whether or not it is a leptodactylid (see under Leptodactylidae in list of families), it is an isolated and presumably relict genus. And *Breviceps*, with a dozen or more species from South Africa northeast to Tanganyika, is the principal genus of African Brevicipitidae and almost defines the present range of the family in Africa; it is apparently retreating into South Africa as *Heleophryne* may have done.

The frog fauna of Madagascar (Noble 1931, Grandidier and Petit 1932, etc.) is very different from that of Africa. The island lacks the pipids, bufonids including *Bufo*, phrynomerids, and *Heleophryne* of the continent. Ranids are few. Rhacophorids are relatively numerous, with two genera shared with Africa (*Megalixalus* and *Hyperolius*), one shared with the Orient but not Africa (*Rhacophorus*), and about four endemic. Brevicipitids are still more numerous (Parker 1934), with two subfamilies represented: Cophylinae, with eight genera, are confined to Madagascar; Dyscophinae have one endemic genus (*Dyscophus*) in Madagascar but are otherwise confined to the Orient. There are also in Madagascar a few other frogs, of which the relationships are still doubtful.

The frogs of the Seychelles Islands number only five species (Parker 1936). One (*Rana mascareniensis*) may be introduced, and another (*Megalixalus seychellensis*) is a distinct endemic species of an African and Madagascan genus. The remaining three species in two genera form an endemic subfamily, Sooglossinae, of the family Pelobatidae.

The frogs of tropical Asia, from India (Boulenger 1890) to the continental islands of Sumatra, Java, and Borneo (van Kampen 1923) include at least 6 families and 37 or more—perhaps many more—genera, some shared with Africa, others not, and some widely distributed in the Orient, others localized. Those that reach the north-temperate zone in China have already been mentioned. About a dozen of the genera reach the Philippines (Inger 1954) and include most of the frogs of those islands, which have otherwise only the relict discoglossid *Barbourula* and a couple of genera shared with the Australian Region via Celebes or the Moluccas. All the tropical Asiatic families (except the Hylidae) and at least nine of the genera range beyond Java and Borneo toward or into the Australian Region (van Kampen 1923, Mertens 1930, Parker 1934, etc.), but they follow no single pattern and reach no common limit. Details are given in Chapter 7 (p. 464).

The frogs of New Guinea (Parker 1934, Loveridge 1948) are dominated by *Hyla* (with *Nyctimystes*), with something like 40 species on this one great island, and by brevicipitids, of which 2 subfamilies with 9 genera and more than 50 species are almost confined to New Guinea and neighboring islands. Also on New Guinea are a few Australian leptodactylids, one true *Rana*, and a few other ranids, of the subfamily Cornuferinae. The frogs of the Moluccas are like those of New Guinea but less diverse: a few *Hyla*, 2 genera of breviciptids, and a few *Rana* and Cornuferinae. The Cornuferinae, although few on New Guinea and the Moluccas, are relatively numerous on the Solomon Islands, where there is otherwise only *Hyla*, and are the only frogs on the Fiji Islands, where they may not be native (p. 507).

The frogs of Australia and Tasmania (Loveridge 1935, Parker 1940) belong to only 2 main groups: *Hyla*, with 24 or more Australian and Tasmanian species, and Australian Leptodactylidae (Noble's Criniinae). The latter form one or 2 endemic subfamilies with 14 genera and 56 full species in Australia and Tasmania, 3 genera (none endemic) and 5 species (only 2 endemic) in New Guinea, and none beyond—but the whole assemblage is considered rather closely related to the American leptodactylids. Besides these, a very few New Guinean frogs reach tropical northern Australia: 2 closely related species of *Hylarana* (or *Rana*) do so, and also 2 New Guinean genera of brevicipitids, which are represented in North Queensland by 4 species, of which 2 are endemic.

New Zealand has a single, very primitive genus of frogs, *Leiopelma*, already mentioned.

The frog fauna of tropical and south-temperate America (Noble 1931) is very rich, with about 7 full families and 60 or more genera, the exact count depending on a given author's ideas at a given time—but this is true of all large frog faunas. A fair part of this fauna reaches Central America (see below), and a much smaller part reaches the West Indies (Chapter 8). The few tropical genera that reach the north-temperate zone in North America have already been mentioned. Different subfaunas of frogs can be distinguished in South America in different forested areas, in the grasslands, and in the mountains. Toward the southern end of South America the fauna diminishes, and in the wet south-temperate forests of southern Chile there is a very distinct subfauna characterized by endemic genera of leptodactylids and atelopodids, by *absence of all hylids,* and by absence also of brevicipitids, pipids, and *Rana* (Dunn, in letter of October 6, 1946).

Above are summarized a number of more or less distinct frog faunas of tropical and south-temperate lands around the world. Now the geographical relationships that exist among these faunas are to be reviewed briefly.

Africa, Madagascar, the Seychelles, and tropical Asia show a mixed pattern of frog relationships. Africa and tropical Asia share some groups of frogs, but each continent has several families that the other lacks (Fig. 20). Two African genera (*Megalixalus* and *Hyperolius*) that reach Madagascar and in one case (*Megalixalus*) the Seychelles too are absent in the Orient. Two other genera (*Bufo, Hylarana*) occur in Africa and the Orient but are absent on Madagascar and the Seychelles. The brevicipitid subfamily Melanobatrachinae is known only in East Africa and South India. Bufonidae other than *Bufo* occur in Africa and the Malay region but are absent in Madagascar, the Seychelles, and the main part of the Orient. *Rhacophorus,* absent in Africa, has many species on Madagascar, none on the Seychelles, and many in the Orient from Ceylon and India eastward. The brevicipitid subfamily Dyscophinae, absent in Africa, has one genus confined to Madagascar, none on the Seychelles, and none in Ceylon or peninsular India, but two farther east, in southwestern China, Burma, the Malay Peninsula, Sumatra, and Borneo. Pelobatidae are absent in the main part of Africa and absent in Madagascar, have one small endemic subfamily on the Seychelles, none in Ceylon or peninsular India, but one subfamily widely distributed in the eastern part of the Orient, and of course another in the north-temperate zone. Some of these distributions will probably be changed by improvements in classification, discovery of new species, and recognition of species introduced by man; but frog relationships in this part of the world are certainly mixed. There is here an example of the importance of knowing a situation as a whole. The whole pattern of distribution of frogs around and across the Indian Ocean is full of discontinuities and mixed relationships which indicate that the fauna is partly a relict one and that existing relationships are partly the result of extinction and survival rather than of simple dispersal. Under these circumstances, the relationships of some Madagascan frogs to Oriental ones do not necessarily indicate direct dispersal or a direct land bridge. On the whole, the frog faunas of tropical Africa, Madagascar, and tropical Asia are surprisingly different.

From the Orient to New Guinea and Australia there is so great a change of frog faunas that, excepting *Hyla*, of which the distribution is widely discontinuous, no family of tropical Asia reaches the temperate part of Australia. Much of the change occurs between the Oriental Region and New Guinea; however, there is no single boundary there for frogs, but rather a zone of transition including all the Lesser Sunda Islands, Celebes, and the southern Philippines (p. 464). Another great change occurs between New Guinea and the temperate part of Australia, but again the change is not at the water barrier and not at any single boundary. Some nine frogs (two *Rana* or *Hylarana*, three leptodactylids, two *Hyla*, and two brevicipitids) occur in both New Guinea and adjacent tropical Australia. I have collected seven of them. They are small frogs, most living in obscure habitats, and are not likely to have been carried by man. They show that New Guinea and Australia have been able to exchange some frogs. Yet the frog fauna of New Guinea is primarily a hylid-brevicipitid fauna, and that of Australia is primarily a hylid-leptodactylid fauna, and *only* hylids and leptodactylids occur in the temperate part of Australia.

The frog faunas of tropical Asia and tropical America are very different. The pelobatids, true bufonids, ranids, and rhacophorids of the Orient are represented in tropical America by only the ubiquitous genera *Rana* and *Bufo;* and the pipids, leptodactylids, and hylids of tropical America are absent (or in the case of *Hyla* nearly so) in the Orient. The only frogs that occur in tropical Asia and tropical America and nowhere else (except the adjacent parts of the north-temperate zone) are brevicipitids of the subfamily Microhylinae, including *Microhyla* itself. Atelopodids, once considered exclusively tropical American, may occur in the Orient, but they may occur in Africa too; their distribution is not properly known (see list of families).

The frog faunas of Africa and South America are very different. The African fauna is dominated by bufonids, ranids, and rhacophorids; of these only *Bufo* and *Rana* are in South America. The South American fauna is dominated by leptodactylids, atelopodids, and hylids; of these only one genus of doubtful leptodactylids and a few (doubtful?) atelopodids are known in Africa. The only frogs that occur in Africa and South America and nowhere else are the pipids, of which one subfamily is confined to Africa and the other to the northern half of South America. Pipids are primitive frogs according to Noble and are aquatic. Their distribution parallels that of some fresh-water fishes.

There are apparently no direct relationships between Madagascan and American frogs.

The frog faunas of Australia and South America are much alike, and this is the more surprising because most frog faunas are so different. Temperate Australia has only *Hyla* and leptodactylids, and both are numerous also in South America. But the distributions of these groups are neither exclusive nor simple, and there are many other groups of frogs in South America that are not in Australia, and some in the tropical part of the Australian Region that are not in South America. And it should be noted once more that the one group of frogs in New Zealand is not related to anything now known in either Australia or South America.

On the whole the frog faunas of different parts of the world are astonishingly different. The wide distributions of *Rana* and *Bufo* only emphasize how little most frogs have spread in recent times. This is, I think, consistent with the hypothesis that spread of dominant frogs is balanced by localization and retreat of other, less dominant ones.

Amphibians of Central America

Central America, with Mexico, is a significant transition area for amphibians, as it is for fishes.

From the north, several groups of North American amphibians reach the northern edge (in some cases just the northwestern corner) of Mexico. Ambystomid salamanders and spadefoot toads extend south to within the tropics on the Mexican plateau but do not reach tropical lowlands. Plethodontid salamanders and *Rana* extend through Central America to the northern half of South America. And *Bufo* extends still farther south.

From the south, Central America receives more amphibians. Of caecilians, according to Dunn (1942), 2 out of 6 South American genera reach Panama, and one of them continues through Central

America to tropical Mexico. Of frogs, according to Noble (1922, p. 67), 21 out of 60 South American genera reach Central America, but they undergo progressive subtraction there. Dunn (1931) says that of at least 25 South American genera of leptodactylids, only 5 enter Central America, only 3 extend to Mexico, and only one reaches the Mexican plateau (where, however, it has given rise to 2 endemic genera); of 10 South American genera of atelopodids, only 4 enter Central America, and none extends north of Nicaragua (but there is one on Cuba); of at least 13 South American genera of hylids, not more than 6 reach Central America (but 2 more are endemic there), and only *Hyla* reaches the main part of North America (but 2 more genera are endemic there); and of about 8 South American genera of brevicipitids, only 3 reach Central America, but 2 of them extend through Mexico into the United States.

This summary is superficial, but it is enough to show in a general way how North and South American amphibian faunas overlap, with progressive subtractions from each. Most of the transition occurs in Mexico and Central America, but the overlapping is actually much wider than this, for a few North American groups extend well into South America, and a few South American ones extend into North America.

This transition has been described as it is now, in terms of present distributions, without regard to direction of movements. Movements have evidently occurred in both directions, but there may have been retreats too; the whole process has probably been very complex.

Since Noble's and Dunn's summaries (above), much work has been done on Mexican and Central American amphibians. Smith and Taylor (1948) have summarized the mass of new information in a very useful way for Mexico, but no one has done it for Central America, and I cannot do it. Part of the difficulty is that there has been much splitting of genera of Central American amphibians. Whether or not this is good taxonomically, it is bad zoogeographically, for it complicates things beyond the understanding of anyone except specialists. One solution might be to retain the old, broad genera so far as they are natural, and to reduce most of the new ones to subgenera. This would allow as fine splitting as taxonomists wish, without confusing zoogeographers.

One Central American frog has turned out to be more isolated than Noble and Dunn realized. It is *Rhinophrynus dorsalis*, a specialized, burrowing frog formerly included in the Leptodactylidae but now put in a monotypic family, confined to parts of tropical Mexico and Guatemala (Smith and Taylor, p. 34, footnote 24, and references given there). This frog, in spite of its specialization, is apparently much more primitive than any other in Central America—perhaps near the evolutionary level of the discoglossids and pipids. It is presumably a relict, the only striking amphibian relict in Mexico or Central America. Unfortunately its ancestry and geographical source (whether North, Central, or South America) are unknown.

Transitions and barriers in distribution of amphibians

Where different amphibian fauns meet there is usually no sharp line between them but an area of transition or overlapping with progressive subtractions from both faunas, in opposite directions.

In eastern Asia, where tropical and north-temperate amphibian faunas meet without major barriers other than climate, caecilians and a number of tropical frogs reach the eastern Himalayas or South China, five genera of the frogs extend north to central China, and one of them reaches Manchuria, and another, Japan; and a few northern salamandrids and *Hyla* extend south into limited areas of tropical southeastern Asia.

In Central America the transition of northern and tropical faunas is much more extensive among amphibians than among strictly freshwater fishes: relatively more amphibians are involved and some of them extend farther in each direction than any of the fishes do. Amphibians have evidently moved across Central America more often and farther than fresh-water fishes have done.

Between Asia and Australia, across Wallace's Line, where strictly fresh-water fishes hardly enter, Oriental and Australian frogs overlap in a broad and complex way. Frogs have evidently crossed the narrow gaps of salt water often.

It is hardly necessary to give further details. Zonal climate is evidently a barrier to amphibians as it is to fishes, but some of them cross it, as some fishes do. Other barriers are much less important to amphibians than to fishes. If deserts and mountains are important, they are so only temporarily. Nothing—neither climates, nor deserts, nor mountains, nor even narrow gaps of salt water—has prevented the spread of dominant frogs over the world in the course of time.

Dominance and competition in relation to distribution of amphibians

The predominant existing amphibians are frogs of the genus *Rana*. This genus is represented by many species throughout the main tropical and temperate regions of the Old World and North America, although not more than two species (actually species of *Hylarana*) reach northern Australia and only one reaches (northern) South America. Individuals of some of the species are numerous too. The genus as a whole, considered in a fairly broad sense, is rather diverse in structure and habits, and includes amphibious, aquatic, terrestrial, and burrowing forms. The genus is dominant in both main climatic zones, tropical and north-temperate, and it occurs in a great variety of habitats including cold and arid ones: for example, it includes the northernmost of all amphibians, and it is one of the two frog genera that are widely distributed in the Sahara. Its distribution is therefore nearly continuous over a large part of the world, with less regard for climates and barriers than any other amphibians show. All these things go together to make a picture of a dominant group with a radial distribution, comparable in general, though not in detail, to the dominant fresh-water fish family Cyprinidae.

The fossil record is not good enough to show whether the radiation of *Rana* has been correlated with the retreat of other frogs, but it may have been. The distribution of leptodactylids, which are now numerous only in Australia and the warm parts of America, is complementary to the area of abundance of *Rana*. The rise of ranids may account for the disappearance of leptodactylids from the main, accessible part of the world, while the leptodactylids have survived in the barrier-protected continents where *Rana* is just arriving. Brevicipitids may be retreating before ranids in another way. Brevicipitids are mostly small and seldom found in numbers, and they tend to live in hidden places (Parker 1934, pp. 1 and 14). They may be retreating ecologically as well as geographically before the dominant ranids.

I know of no obvious explanation of the success of ranids and *Rana*. They evidently have superior qualities of some sort which make them better than other frogs in many different situations.

Bufo has a pattern of geographical radiation and dominance second only to that of *Rana* among amphibians. It is less numerous in species and less diverse in structure and habits than *Rana*, but it can exist in a variety of places including both warm and cold and both humid and arid ones.

Some other frogs are dominant only in smaller areas or special habitats. The distribution and complementarity of dominant frogs over the world is diagrammed in Figure 26.

Competition among frogs is repeatedly assumed in the preceding pages. It is inherent in the concept of dominance, in the idea of replacement of some frogs by others, and in the idea of complementarity: these are of course all related matters. It seems to me that it is not possible to understand or explain frog distributions satisfactorily without taking competition as a premise. Different groups of

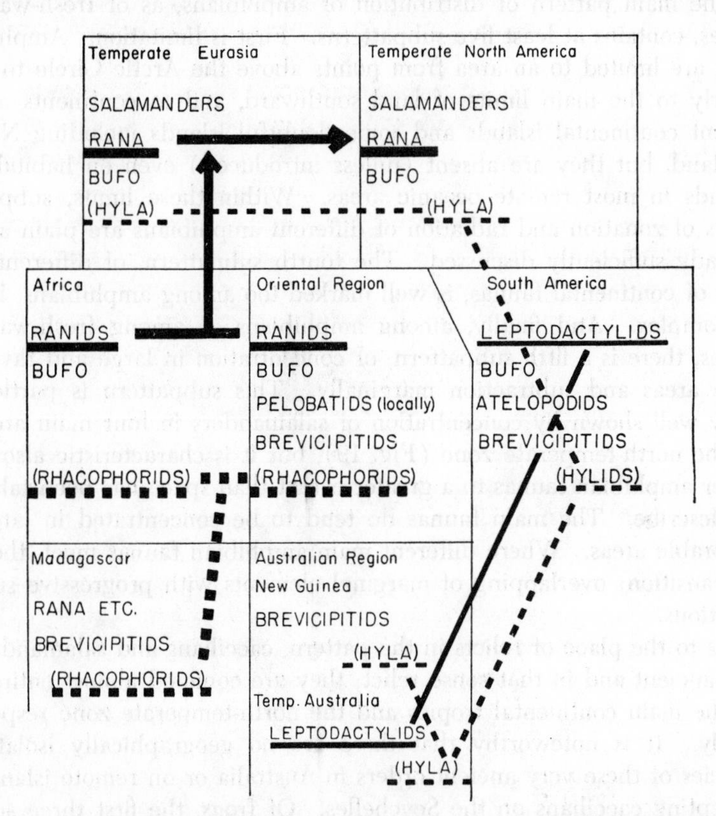

Fig. 26. Dominant amphibians of seven principal areas. Diagram stresses complementarity of, among terrestrial groups, ranids (heavy solid lines) and leptodactylids (lighter solid lines), and, among arboreal groups, rhacophorids (heavy broken lines) and hylids (lighter broken lines).

frogs apparently often compete in wide areas rather than on narrow fronts, and the outcome may be decided during long periods of time, perhaps more by adaptability than by initial adaptations. The competition must be complex, and the details of it are almost all beyond present knowledge. Competition may occur also between frogs and salamanders. That the former jump and the latter do not does not change the fact that the requirements of typical forms of both orders are about the same, and that, typically, both have aquatic, probably sometimes competing, larvae.

Summary: the pattern of distribution of amphibians

The main pattern of distribution of amphibians, as of fresh-water fishes, contains at least five subpatterns. First is limitation. Amphibians are limited to an area from points above the Arctic Circle to or nearly to the main limits of land southward, and to continents and recent continental islands and some doubtful islands including New Zealand, but they are absent (unless introduced) even on habitable islands in most remote oceanic areas. Within these limits, subpatterns of zonation and radiation of different amphibians are plain and already sufficiently discussed. The fourth subpattern, of differentiation of continental faunas, is well marked too among amphibians, but is complex. And finally, among amphibians as among fresh-water fishes, there is a fifth subpattern, of concentration in large and favorable areas and subtraction marginally. This subpattern is particularly well shown by concentration of salamanders in four main areas in the north-temperate zone (Fig. 19), but it is characteristic also of other amphibian faunas to a greater extent than space has been taken to describe. The main faunas do tend to be concentrated in large, favorable areas. Where different main amphibian faunas meet, there is transition: overlapping of marginal elements with progressive subtractions.

As to the place of relicts in the pattern, caecilians and salamanders are ancient and in that sense relict; they are confined almost entirely to the main continental tropics and the north-temperate zone respectively. It is noteworthy that there are no geographically isolated species of these very ancient orders in Australia or on remote islands, excepting caecilians on the Seychelles. Of frogs, the first three suborders are phylogenetic relicts, but only one genus of them, *Leiopelma* on New Zealand, is geographically isolated, beyond the reach of higher frogs. It is comparable to the lungfish in Australia. (But Australia has no notably primitive, relict frogs!) A small group of pelobatids is confined to the Seychelles. Other phylogenetically relict frogs (the first three suborders) occur on the main continents and are divided between the tropics and the north-temperate zone. The tropical Orient has a discoglossid and pelobatids; Africa and South America, pipids; tropical Mexico and Guatemala, the only known rhinophrynid; and the north-temperate zone, a leiopelmid, discoglossids, and pelobatids. In proportion to its whole frog fauna, the north-temperate zone has more groups and more species of relict frogs than the tropics have.

History of dispersal of amphibians

Rather than discuss separately how and why the pattern of distribution of amphibians has evolved and what it tells of ancient lands

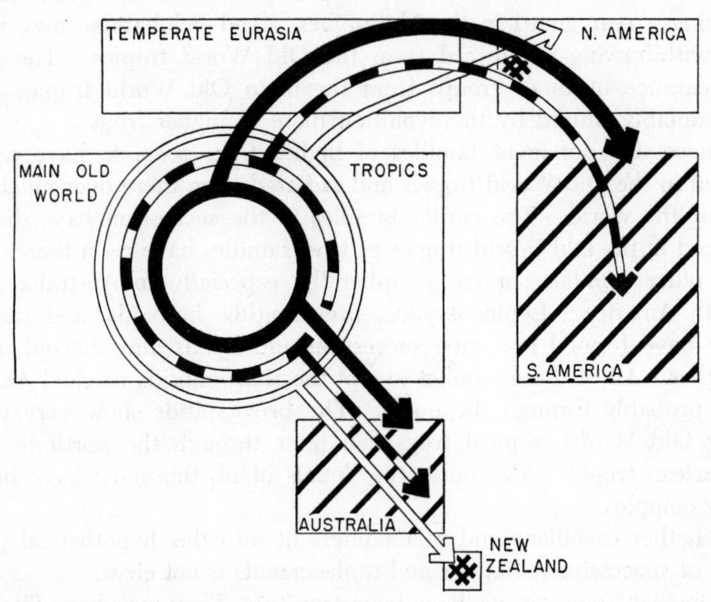

Fig. 27. Diagram of (hypothetical) successive dispersals of three families of frogs: Leiopelmidae (white circle and arrows), now surviving only in New Zealand and an isolated habitat in western North America (double crosses); Leptodactylidae (broken circle and arrows), surviving principally in Australia and South America (diagonal hatching); and Ranidae (black circle and arrows), which has apparently replaced the preceding families in the main part of the world but is just reaching Australia and South America.

and climates, I shall outline a hypothetical history of amphibian dispersal which gives or implies the answers (Fig. 27).

I shall begin in the Cretaceous, with frogs. At that time, the ancestors of most existing frog families were probably in the Old World tropics. There may have been other, now extinct families in other parts of the world, but there is no clear record of them.

Perhaps during the Cretaceous, early leiopelmids, once present in at least the Old World tropics, became extinct in the tropics but survived in the north- and south-temperate zones; and later they were reduced to relicts in northwestern North America and New Zealand. Probably late in the Cretaceous, aquatic pipids got from the Old World tropics to South America. If fresh-water fishes filtered through the north-temperate zone then, so probably did the pipids. Later, pipids withdrew from the north and from tropical Asia, surviving in Africa and South America. Perhaps still later, in the Tertiary, discoglossids withdrew from the Old World tropics into temperate Eurasia, leaving a rear guard in the Philippines. And Pelobatidae may now be withdrawing northward from the Old World tropics. The disappearance of these groups from the main Old World tropics was presumably caused by the evolution there of higher frogs.

Successive dominant families of higher frogs seem to have originated in the Old World tropics and radiated from there over much or all of the world. The earlier families in the succession have disappeared in the Old World tropics as later families have risen there, but the older families survive peripherally, especially in Australia and South America. Leptodactylids, brevicipitids, bufonids, and ranids may have formed one such succession and hylids and rhacophorids another. All of these families except the rhacophorids reached America, probably through the north. The brevicipitids show very well how Old World tropical frogs can filter through the north to the American tropics. Of course the details of all this must have been very complex.

Whether caecilians and salamanders fit into this hypothetical pattern of successive radiations and replacements is not clear.

Caecilians may always have been tropical. They may have filtered through the north from the Old to the New World tropics as some frogs have done. They may never have reached Australia; or they may have reached it and become extinct there later, as leiopelmid frogs probably did.

Salamanders are more of a problem. Have they been confined to the north-temperate zone since their origin, probably before the Mesozoic, until one group of them recently became able to push deep into the American tropics? Or were salamanders once widely distributed in the Old World tropics, and have they disappeared there without leaving any relicts on isolated mountains or islands (the few salamanders in the edge of the tropics in Asia are not relicts but outlying species of existing temperate-Asiatic groups)? Neither of these alternatives is fully satisfactory, but I prefer the second one. The present northern zonation of most salamanders could be a stage in a slow shift from the Old World to the New. Competition with dominant frogs might account for the disappearance of salamanders from the main part of the Old World tropics. And the distribution of existing salamanders in the north-temperate zone (Table 3), with many discontinuities, and with the greatest concentration now in eastern North America, suggests a withdrawal from the Old World toward the New. If it is occurring, the withdrawal is complicated by radiation of hynobiids in eastern Asia, and in other ways, but in complex groups during long times movements should not be expected to be simple.

This hypothetical history accounts for the distribution of existing amphibians without startling land bridges or drifting continents. However, the present distribution of these animals may have little to do with very ancient times. Existing orders of amphibians are probably old, pre-Mesozoic in origin, but their present distributions are probably much more recent, perhaps products of late Cretaceous and Tertiary events, and may tell nothing of the world before that.

LIST OF FAMILIES OF AMPHIBIANS

This list follows the general plan described in Chapter 1 (p. 13). It includes only existing groups of amphibians and fossils that concern existing groups. The classification used and the relation of existing groups to ancient, extinct ones are indicated at the beginning of the present chapter. A dagger (†) before a name indicates that the group is extinct.

Order Apoda

Caeciliidae, caecilians: the tropics of Africa, Asia, and America, and the Seychelles Islands in the Indian Ocean. In Africa there are 6 genera, about 17 known species; they are widely distributed in both western and eastern tropical Africa but have not yet been found in central Africa; the eastern and western African caecilians mostly belong to different genera. Madagascar has no caecilians. The Seychelles Islands were formerly thought to have 2 or 3 genera, but Parker (1941) thinks there is only one, endemic genus there, *Hypogeophis*, with 6 species, all presumably derived from one ancester. In Asia there are 4 genera, 6 recognized species, distributed from Ceylon and India north into the edge of the north-temperate zone to Sikkim (27° or 28° N.) in the eastern Himalayas (Boulenger 1890), possibly to the southern edge of China (Pope and Boring 1940, p. 14), and south and east to Sumatra, Java, and Borneo (van Kampen 1923), with one species reaching Palawan and Mindanao in the southern Philippines (Inger 1954). In America, Dunn (1942) recognizes 6 genera, 44 species, ranging from latitude 20° N. (Veracruz and Guerrero in

Mexico) to latitude 35° S. (Buenos Aires, Argentina, well within the south-temperate zone), and from sea level to 4500 feet in Costa Rica and 6200 feet in Ecuador. All the American genera occur in South America, 2 reach Panama, and one of the 2 continues through Central America to tropical Mexico. Caecilians reach Trinidad but not the West Indies proper. There are no wide-ranging genera of caecilians; all are confined to single regions: southeast Asia, the Seychelles Islands, the African tropics, or the American tropics. Dunn (1942, p. 452) says that in America caecilians inhabit tropical rain forest, tropical deciduous forest, tropical savanna, "temperate" (see p. 5) forest and savanna, and mountain cloud forest; but in savanna areas they probably occur only in strips of forest along rivers. The life histories and habits of most caecilians are not well known; those of a few are summarized by Gadow (1909), Noble (1931), and Dunn (1942, pp. 451–453). Some are amphibious, with an aquatic larval stage and terrestrial (burrowing) adults; some are aquatic throughout their lives; and others, including at least some of the Seychelles species (Gadow), have suppressed the aquatic larval stage and are entirely terrestrial.

No fossil caecilians are known.

Order Urodela, salamanders

Hynobiidae, Asiatic salamanders: temperate Asia, from the Urals to the Pacific, and from the edge of the arctic south to the mountains of southwestern Asia (at least to the Pamirs of Samarkand), southern China, and Formosa, and on Japan and Sakhalin Island; 5 genera, 26 or more species (Dunn 1923, Liu 1950 and earlier papers). Most of these salamanders are in Japan and eastern Asia; only *Hynobius kerserlingii* occurs across Siberia, reaching 60° N. in the Urals, across the Arctic Circle to Verkhoyansk, and to Kamchatka. The hynobiids are sometimes called "Asiatic land salamanders," but this is a misnomer. Their larvae are probably always aquatic, and the adults, though typically terrestrial, are aquatic too in some specialized genera. Fertilization is external, in water, in this family, and that may be why the aquatic larval stage has not been suppressed as it has been in many plethodontids.

This family, though only doubtfully known fossil (Miocene of Europe), is apparently older than (ancestral to) the cryptobranchids, which go back at least to the Upper Oligocene, and which have spread around the world in the northern hemisphere.

Cryptobranchidae, giant salamanders, hell-benders: eastern Asia and eastern North America; in Asia, only *Megalobatrachus,* with one species in the interior of China, from parts of Shansi and Shensi south and southwest to parts of Kwangsi, Kweichow, Szechwan, and southeastern Sikang; and another species or subspecies in Japan, apparently confined to the southern and western half of the main island, Honshu; in North America, only *Cryptobranchus,* with one species in western New York and the Susquehanna River and from the Ohio system west at least to central Missouri and south to Georgia and Louisiana, and a second closely related species or subspecies in southeastern Missouri, in rivers draining into the south-flowing Black River of Missouri and Arkansas. This family is entirely aquatic. The Asiatic forms, though gigantic among salamanders, live in mountain streams; the American ones, which are smaller but still large, live in larger rivers.

Megalobatrachus is fossil in Europe in the Miocene, and other fossils of extinct genera of this family have been found in Europe (Upper Oligocene and Miocene) and Nebraska (Lower Pliocene).

Ambystomidae, various salamanders, axolotls: North America; from the southern corner of Alaska, James Bay, and southern Labrador (north to about 53°), south to the southern edge of the Mexican plateau (into the edge of the tropics but not at low altitudes there) and to the northern half of Florida; but apparently absent in parts of northern Mexico and southwestern United States. The supposed occurrence of a species in Siam has been shown by Noble (1926) to be an error. (The supposed Siamese species is still listed with a "?" by Bourret, 1942, who says no specimens of it have been taken since the original find.) The principal genus is *Ambystoma* which, even in a restricted sense, occurs throughout the range of the family and includes about 22 species, of which 10 are Mexican. One of the more northern species (*A. tigrinum*) is the only salamander that ranges more or less continuously between the main eastern and western areas of salamander distribution in North America (Bishop 1943, Map 20). *A. macrodactylum,* a western species which occurs inland to western Montana, is recorded also from Iowa (Bishop, Map 15), but it may not be native there. Two additional, monotypic, genera occur along the Pacific coast, and one of them (*Dicamptodon ensatus*) occurs also in apparently isolated localities in the northern Rockies, in Idaho and Montana (Bishop, Map 21). Three more genera with together 7 species, including the permanently larval axolotl (*Siredon*), are recognized in Mexico by Smith and Taylor (1948) but go in *Ambystoma* in the broad sense of Noble (1931).

Fossils assigned to this family are all North American and are of Upper Cretaceous, Pliocene, and Pleistocene age. I do not know whether the earlier ones are surely ambystomids.

Salamandridae, typical salamanders, newts: all four main areas of distribution of salamanders in Eurasia and North America, probably not tropical West Africa, but certainly into the edge of the tropics in eastern Asia. In Europe there are 6 genera, 16 species, most in southern Europe but *Triturus* extends north to about 63° N., east to the Urals, and into southwestern Asia to northern Iran and Palestine, and *Salamandra* too extends to Asia Minor, Syria, and Palestine. Three of the European genera, including *Triturus,* occur also on the temperate northwestern corner of Africa, where they are represented by 4 species (2 endemic). One species, *Pleurodeles waltl* of Spain and Morocco, has been recorded also from West Africa, but this is now admitted to be an error (Chabanaud 1954). Beyond the limits of *Triturus* (above) the family is absent in western and northern Asia. In eastern Asia, in much of China and on Japan and the Riu Kiu Islands, are again *Triturus* and 4 or more endemic genera and a total of about 15 or more species. Several salamandrids occur in the edge of the tropics in Asia: *Triturus chinensis* is on Hong Kong Island (Pope and Boring 1940, p. 23); *Paramesotriton deloustali* is endemic in Tonkin; and *Tylototriton* is found on the Shan Plateau in Burma at about the same latitude as Tonkin (Myers in letter of July 14, 1948) and extends south even to northern Siam (Bourret 1942). In America is only *Triturus* in a broad sense (or 2 related genera if one wishes to divide them): one group of 4 species is in the west, from the southern corner of Alaska to southwestern California and the adjacent corner of Lower California (this group is supposedly related to eastern Asiatic forms); and another group of 3 species is in the east, from Gaspé and Minnesota to Florida and northeastern Mexico; the family is absent in a wide area across the Great Basin and Rocky Mountains. The members of this family, including *Triturus,* are primarily amphibious, but they vary in the amount of time spent on land.

Fossil salamandrids occur back to the Eocene, but only in Europe. They suggest a European or Eurasian origin of the family, as does the distribution of living forms. *Tylototriton,* living in eastern Asia, is listed fossil in Europe from the Eocene to the Miocene.

Amphiumidae, aquatic "lamper eels": southeastern United States; 2 forms, which are subspecies or closely related species (Baker 1947), one from the lower Mississippi west to northeastern Texas and north to the near corners of Missouri and Kentucky, and the other east along the Gulf from Louisiana through Florida and north on the Atlantic coastal plain to Virginia.

This family is listed from Europe in the Eocene and from North America in the Miocene, but the European fossil, at least, is doubtful.

Plethodontidae, lungless salamanders: North, Central, and part of South America, and one species isolated in Europe.

In Europe there is only *Hydromantes genei* (= *fuscus*), with 2 subspecies, localized partly in mountains and sometimes in caves in the southeastern corner of France, in parts of Italy, and in Sardinia. Two other species of this genus (one described by Gorman and Camp in 1953) are in California, localized at high altitudes. That the European and Californian species are congeneric is apparently agreed by herpetologists, and studies of the bone marrow, etc. (Barrett 1947), apparently confirm that the European species is a plethodontid.

In America north of Mexico, Bishop (1943) recognizes 17 genera of plethodontids (including *Hydromantes*) and 53 full species: 6 genera, 14 species in the west, from the lower corner of Alaska to the western upper corner of Mexico; and 13 genera, 37 species in the east, from James Bay and the north shore of the Gulf of St. Lawrence to the northern half of Florida, west to Wisconsin, the Ozarks, eastern Kansas, eastern Oklahoma, and central Texas. Only *Plethodon* and *Aneides* occur in both of these areas, and between the main areas are 2 additional species of *Plethodon* (one described since Bishop's book) isolated at single localities in Idaho and New Mexico, and one of *Aneides* in New Mexico (Lowe 1950; Stebbins and Riemer 1950); they are probably the only plethodontids in west-central North America.

Plethodontids are apparently absent across most of northern Mexico and most of the adjacent edge of the United States. Southward in Mexico, however, beginning even in Nayarit, Zacatecas, and southern Nuevo Leon, they reappear. They occur through central and southern Mexico and Central America (a diversity of species including both mountain and lowland forms; see Taylor 1944; Schmidt 1936, Figs. 15 and 16) and part of South America (comparatively few species) south to the eastern slope of the Andes in northern Bolivia, east to western Venezuela, and to the mouth of the Amazon. The existence of a plethodontid (*Oedipus paraensis*) at the last locality seems now well established (Myers and Carvalho 1945; Schmidt and Inger 1951). The genus has been recorded at low altitudes also at Nauta (about 150 meters altitude) on the Marañón, at Benjamin Constant (about 100 meters) on the Javary, and at about 200 meters in northern Bolivia (Dunn in letter of October 6, 1946). However, old records of a species of the western North American genus *Ensatina* from the La Plata region in southern South America are now discredited (Myers and Carvalho), and so is a supposed *Oedipus* from Haiti. The number of species and subspecies of *Oedipus* now known from Mexico and Central and South America must be near 100. [Taylor (1944) has broken up *Oedipus,* which he calls *Bolitoglossa,* into at least 7 smaller genera, but almost all of them apparently form one natural group (Taylor, p. 207). As in other similar cases, I have no opinion on the correctness of this splitting but wish that a way could be found to retain the old arrangement for zoogeographical purposes.]

In habits, plethodontids vary. Many are terrestrial and many are aquatic, especially in rapid streams, a few live in caves or subterranean water, and a few north-temperate and some tropical plethodontids are arboreal. The fertilization of these salamanders is internal, and this has permitted the suppression of the aquatic larval stage in many of the terrestrial forms.

No fossil plethodontids are known. The *Hydromantes* in Europe is presumbaly a relict, surviving from a time when plethodontids were widely distributed in Eurasia. The family evidently entered South America recently, after a land connection was made in the Pliocene.

Proteidae, the Olm, mud puppies: Europe and eastern North America. In Europe there is only *Proteus anguinus,* the Olm, with 2 subspecies, in underground water in a small area near the Adriatic coast (Trieste, Istria, coastal Yugoslavia). In North America there is only *Necturus,* with 7 recognized species, occurring west of the Appalachians from southern Canada to Louisiana and northwestern Florida, and east of the Appalachians from North Carolina to Georgia. Above North Carolina, north to the lower Connecticut River, *N. maculosus* occurs in several Atlantic coastal streams but may not be native. See Bishop (1943, Map 1) for distribution of the different species.

Fossils indicate that this family has been in Europe since the Eocene and perhaps since the Cretaceous.

Sirenidae, mud eels: southeastern North America; 2 genera, 3 full species, distributed coastwise from Virgina to Florida and around the Gulf to the northeastern corner of Mexico, and north along the Mississippi to northern Illinois and Indiana (Bishop 1943, Maps 54–56); in ponds, ditches, swamps, etc.

Not known fossil before the Pleistocene.

Order Anura, frogs (including toads)

(*Suborder Amphicoela*)

Leiopelmidae (Ascaphidae): northwestern North America, and New Zealand. In North America there is only *Ascaphus truei*, with 3 subspecies, occurring from southern British Columbia (to just above 50° N.) south to Humboldt County in northern California and east to western Montana, in cold mountain streams. Mittleman and Myers (1949) give the history of discovery of this frog, which was not described until 1899, divide it into subspecies, and list some of the literature on it. In New Zealand there is only *Leiopelma*, with 2 or 3 species. *L. hochstetteri* and *L. archeyi* (which may be forms of one species) are localized on the north side of North Ireland (Turbott 1942; Stephenson and Thomas 1945), where they occur on open, fog-dampened ridges as well as in brooks, and where one or both lay eggs on the ground under stones and logs. *L. hamiltoni* is known only from small Stephen Island in Cook Strait, an island which is without surface water at some seasons (Archey 1922). Leiopelmids are not yet known on South Island, although apparently suitable habitats occur there.

Fossils tentatively assigned to the suborder Amphicoela are known from the Upper Jurassic.

(*Suborder Opisthocoela*)

Discoglossidae, fire-bellied toads etc.: Europe etc., eastern Asia, and an island in the Philippines; 4 genera. *Bombina* has 2 species in Europe north to southern Sweden and 57° N. in Russia, and south to Italy, the Balkans, and the northern Caucasus, and 2 more species in eastern Asia, from Manchuria etc. to southwestern China and Tonkin. *Discoglossus* (one species) is in southern France and Spain, some Mediterranean islands, and North Africa. *Alytes* (2 species) is in western Europe, from Switzerland and western Germany to Spain. And *Barbourula* (one species) is known only from Busuanga Island in the Calamian group north of Palawan in the Philippines; it has been collected twice on this island (Myers 1943); it is the largest species of the family and is completely aquatic, in hill brooks. Other discoglossids vary from amphibious to aquatic.

Discoglossids are fossil in Europe back to the Miocene, but the family is probably much older than that.

Pipidae, clawed frogs etc.: Africa and South America. In Africa there are 3 genera: *Xenopus*, with about 5 species, is widely distributed in Africa south of the Sahara; the other 2 African genera, with together 4 or 5 species, are more localized. In America there is only *Pipa*, with 5 species, which are rather diverse in structure but which Dunn (1948) thinks are best considered congeneric; together they cover about the northern half of South America east of the Andes (Dunn). The family does not reach Central America. It does reach Trinidad, but not the West Indies proper. It is the most aquatic family of frogs.

Fossils assigned to this family are too doubtful or too recent to be very significant zoogeographically.

Rhinophrynidae: tropical lowlands of Mexico (from Tamaulipas on the Atlantic side and the Isthmus of Tehuantepec on the Pacific side) southward to Guatemala, and including Yucatan; one species only, *Rhinophrynus dorsalis.* Concerning the primitive nature of this frog, see references given by Smith and Taylor (1948, p. 34, footnote 24). The adult is a specialized burrowing frog, without ribs but apparently primitive in other ways; the tadpole thought to be of this species is rather pipid-like—but Dunn (1948, p. 7) doubts that the tadpole in question belongs to this species.

This family is unknown fossil.

(*Suborder Anomocoela*)

Pelobatidae: eastern tropical Asia etc., temperate Europe etc., and North America (but not most of temperate Asia), and the Seychelles Islands; 3 subfamilies. (1) Subfamily Megophryinae: eastern part of the Oriental Region (not Ceylon or peninsular India) north to central China (Kansu and southwestern Shansi), west to southwestern China (Liu 1950) and the eastern Himalayas, and south and east to Java and Bali, Borneo, and part of the Philippines; 7 genera, about 50 or more species. *Megophrys,* with something like 30 species, occurs throughout the range of the subfamily or nearly so; the other genera are localized in southwestern China, Borneo, etc. The frogs of this subfamily appear to be normally amphibious, their larvae often living in mountain streams. (2) Subfamily Pelobatinae, spadefoot toads etc.; discontinuously north-temperate; 2 genera in Europe etc. and one in North America. In Europe there are *Pelobates,* with 3 species, from Spain to the Urals, Caucasus, northern Iran, Asia Minor, Syria, and Palestine, and the north-temperate corner of Africa; and *Pelodytes,* with 2 geographically separated species, one in *western* Europe and the other in the Caucasus. In America there is only *Scaphiopus,* with 6 or 8 species, from southern British Columbia to Massachusetts, south to Florida and Mexico, reaching the southern edge of the Mexican plateau in Oaxaca, within the tropics but not at low altitudes there. These frogs usually live in rather dry or sandy places but have aquatic larvae. (3) Subfamily Sooglossinae: confined to the Seychelles Islands; 2 genera, 3 species; some or all lay their eggs on land and carry their tadpoles, so that they are independent of surface water. I am not sure that it has been settled whether these frogs are really pelobatids.

Of the 3 subfamilies, only the Pelobatinae is reported fossil, in Europe back to the Miocene, in North America back to the early Oligocene (Zweifel 1956), and in Mongolia (far outside the present range of the subfamily) in the Oligocene (Noble 1924*a*). But the relationships and distribution of existing forms suggest that pelobatids are older than this and that they have had a complex history. Zweifel (pp. 13–16) suggests some details of it.

(*Suborder Procoela*)

†Palaeobatrachidae: European fossils only. The taxonomic limits and geological range (Upper Jurassic to Miocene, or only Miocene) of this

family are not yet settled. The family may be composite, and the relationships of its components to living frogs are unknown.

Leptodactylidae (Fig. 24): South and Central America etc., the Australian Region, and perhaps South Africa; probably fossil in India but not existing there (an Indian tadpole resembling *Heleophryne* is now known not to be related to it—Ramaswami 1944). In America there are 25 or 30 genera even by old standards and more if they are finely split, and several hundred species. They are numerous through South and Central America; southward several genera reach or are endemic to the forests of southern Chile, and one or 2 approach and may reach the Straits of Magellan or even Tierra del Fuego. The northern limits of the family are set by 3 genera which reach Texas and in one case Arizona too. The genus *Eleutherodactylus* (in which the larval stage is suppressed), with more than 200 recognized species (perhaps some should be considered subspecies), is widely distributed within the limits given and is dominant in Central America and especially the West Indies, forming the main part of the frog fauna of the latter, and at least one West Indian species has been introduced into Florida. *Leptodactylus* too reaches some of the West Indies. In the Australian Region, Parker (1940) recognizes 14 genera, 58 full species; they are concentrated in Australia itself, in both the temperate and the tropical parts of the continent; 3 of the genera reach Tasmania; 3 of the genera reach New Guinea and are represented there by a total of 5 known species of which, however, only 2 are endemic. In South Africa there is only *Heleophryne,* with 4 species, discontinuously distributed (on mountain ranges) from Table Mountain (near the Cape of Good Hope) north only to Natal and Transvaal, and thus probably entirely within the south-temperate zone; most authors since Noble (1931) treat this genus as some sort of leptodactylid, but a few including du Toit (1934) doubt that it is one. However, all authors seem agreed that the South American and Australian leptodactylids are related (Parker 1940, p. 2). In habits, American leptodactylids are diverse and include many aquatic, terrestrial, and arboreal forms; Australian ones are mostly terrestrial or amphibious; *Heleophryne* is associated with mountain brooks, in which its highly specialized larva lives.

A fossil (Eocene) frog (†*Indobatrachus*) apparently referable to the Leptodactylidae is known from near Bombay, in the tropical part of India. Noble (1930) examined a series of about 20 fragmentary specimens of it—but the skeleton is still not completely known—and considered it closely allied to *Crinia* of Australia. Parker (1940, pp. 10–11) apparently accepted it as a leptodactylid but was not sure of its exact relationship to living forms. This is perhaps the only known fossil frog of much zoogeographic significance. But an apparent fact of frog phylogeny may be equally significant: true bufonids have apparently been derived from leptodactylids in the main part of the Old World tropics, which indicates the former presence of leptodactylids there. I give this idea on its own obvious merits but should add that Noble (1922, p. 71) had the idea long ago.

Bufonidae, typical toads etc. (Fig. 22): nearly cosmopolitan on the principal land masses but absent in Madagascar and the Australian Region. *Bufo* occurs throughout the range of the family; various species of it are common and widely distributed in every main region where the genus occurs: through all Africa south to the Cape and including isolated oases in the Sahara, north to 65° N. in Scandinavia, across Eurasia to Japan, and south and east through the Oriental Region to Java, Bali, Lombok, Celebes, and Palawan and Mindanao in the Philippines; and in America from at least 60° N. through North, Central, and South America nearly to the Straits of Magellan and perhaps to Tierra del Fuego, and on Cuba, Hispaniola, and Puerto Rico in the West Indies. The 4 other genera of the family are confined to 2 widely separated parts of the Old World tropics: tropical Africa (*Werneria* and *Nectophrynoides*) and the Malay region (*Pseudobufo* and *Pedostibes*). That these genera and *Bufo* form a natural group is indicated by the occurrence of "Bidder's organ" in all of them except possibly *Werneria,* which has not been examined (Davis 1936). The typical toads (*Bufo*) have aquatic larvae and usually terrestrial adults, but the other genera include aquatic and arboreal forms.

Fossils indicate that *Bufo* has been in Europe since the Oligocene and in North America since the Pliocene, but these are minor details in the history of the genus, which has apparently radiated from the Old World tropics. Bufonids probably arose in the main part of the Old World tropics from leptodactylids, although the latter no longer exist there.

Atelopodidae (Brachycephalidae): South America etc. and a few in West Africa and the Malay region. In South America there are about 10 genera, of which 4 enter Central America; on the mainland none occurs beyond Nicaragua (so far as Dunn knew in 1931), but there is an endemic genus of the family on Cuba; southward, at least one genus (*Rhinaderma*) occurs in southern Chile. In West Africa are *Nectophryne* and *Didynamipus,* which Parker (1934, p. 5) says closely resemble American atelopodids and may be related, and absence of Bidder's organ in *Nectophryne* (Davis 1936) tends to confirm this relationship. In the Malay region, only *Cacophryne borbonica* is now recognized as an atelopodid (Davis 1935, 1936)—but Griffiths (1954) thinks it is separately derived (from a Malayan bufonid) and not directly related to other "atelopodids." This family is badly in need of study. Until the relationships of the genera assigned to it are better understood, it cannot be of much significance zoogeographically. American atelopodids are mostly small and terrestrial.

Hylidae, tree frogs etc. (Fig. 25): the genus *Hyla* is nearly world-wide except for a wide gap in most of Africa and tropical Asia. Species of *Hyla* are very numerous in South and Central America and there are a few on Jamaica, Hispaniola, and Cuba. The genus extends north without interruption, but in decreasing numbers, in temperate North America (about 12 species), reaching Great Slave Lake and central Labrador (the hylid *Pseudacris* goes farther north, almost to the Arctic Circle in western North America). A few *Hyla* (just one variable species?), but no other hylids, occur across temperate Eurasia north to about 57° in Europe and above 52° east of Lake Baikal (Berg 1950, p. 310), and reach the temperate northwestern corner of Africa, Japan, the Riu Kius, and Formosa, and south on the mainland of Asia to limits given below. Hylids are absent or nearly so in most of Africa and the Oriental tropics, but *Hyla* reappears in both the tropical and temperate parts of the Australian Region. There are about 35 species of *Hyla* on New Guinea (Loveridge 1948) and a few on the Solomons, the Moluccas, and the eastern Lesser Sundas (2 on Timor, one extending to Ombai, Sawoe, and Soemba—van Kampen 1923), and about 24 (including 2 shared with New Guinea) in Australia and Tasmania (Loveridge 1935). Except for *Hyla,* most hylids are tropical American. Noble (1931) and Dunn (1931) recognized more than a dozen endemic genera in the American tropics, many of them confined to small areas, and many more can be distinguished if generic lines are finely drawn; for example, in Mexico alone Smith and Taylor (1948) recognize about a dozen genera in addition to *Hyla,* and they still leave 37 Mexican species (some perhaps subspecies) in *Hyla* itself. Outside the American tropics, besides *Hyla,* there are only *Acris* and *Pseudacris* in North America and *Nyctimystes* in New Guinea. In spite of the numbers and diversity of hylids in tropical South America, none of them reaches the wet south-temperate forest region of southern Chile—this is a fact which has long been known and which Dunn (letter of October 16, 1946) has confirmed for me.

Hyla is certainly absent or nearly so in the main part of the Old World tropics but there is still doubt about details. The temperate Eurasian group of the genus does enter the edge of the tropics in eastern Asia, reaching Tonkin, Hainan, and Annam (Bourret 1942). *Hyla* does reach Syria, Palestine, and even southern Arabia (Schmidt 1953*a*) but apparently not Egypt (Flower 1933, p. 844). In Africa, a *Hyla* has been described as a new species from Abyssinia, but in error; it proved to be a mislabeled specimen of a common South American species (Noble 1926). Another *Hyla* has been described by Ahl as a new species from Togo, on the Gulf of Guinea in tropical West Africa; this too was at least partly an error, for Noble (1926) has shown that it is not a good species but the same as *Hyla arborea meridionalis* of southwestern Europe, North Africa, etc.; the supposed Togo specimen was apparently not actually labeled from there but was simply included with a lot of specimens shipped from there by a German army officer in World War I, who might have collected it *en route* from Germany, and who died before he could be questioned about it; the locality is therefore doubtful. In the Oriental Region, at least one *Hyla* has been described from Java, by Ahl (1926); so far as I know, it has not been restudied, and the record is too surprising to accept without confirmation. A recent record of *Hyla* from Negros in the Philippines is presented as doubtful by Inger (1954, p. 247). In these cases, and in that of the "Togo" specimen, the burden of proof should be on those who think that *Hyla* really occurs in these unexpected places. I say this in spite of the fact that a *Hyla* in Java or the Philippines would fit very well with the idea of an original radiation of the genus from the Old World tropics.

Although most hylids are tree frogs, a few are aquatic or terrestrial or more or less fossorial. Even within the genus *Hyla,* although most are arboreal, some species are terrestrial, and some of those adapted to live in trees can and do live among rocks instead. Some even live in deserts. *Hyla rubella* has a very wide range in Australia, from humid North Queensland to the arid western part of the continent (Loveridge 1935, p. 42); in the deserts of central Australia this frog survives not by any obvious structural adaptations and not by burrowing but by very rapid reproduction and dispersal from one water hole to another when rains occur (Buxton 1923, p. 96). *Hyla arenicolor* may have the same habits in the deserts of southwestern North America (Bogert and Oliver 1945, p. 335). These examples show that "tree frogs" can cross deserts, and this must improve their chances of dispersal over the world.

Fossil hylids are few and doubtful (see text, p. 144) and are not much help in working out the history of the family. The wealth and diversity both of hylids and of *Hyla* itself in the American tropics may be the result of either a primary or a secondary radiation. If the latter, the primary radiation may have been in the Old World tropics, as suggested in the text (p. 141), and if so, hylids may have arisen from an early bufonid or from the same, now extinct, Old World tropical leptodactylids that produced the Bufonidae (*cf.* Noble 1922, p. 71); but whereas this origin is probable for the bufonids, it is only one possibility for the hylids.[1]

(*Suborder Diplasiocoela*)

Ranidae, typical frogs etc. (Fig. 21): nearly world-wide, within the limits of frogs; but absent in most of Australia and the southern half of South America. The great genus *Rana,* with something like 250 species, occurs through most of the range of the family: through all Africa including habitable parts of the Sahara, on Madagascar, through

[1] Metcalf (1923) tried to trace the history of hylids by their opalinid parasites. He thought that hylids originated in tropical America; that *Hyla* spread from there to North America and changed its opalinids or at least acquired a new one there and carried it to Eurasia; and that *Hyla* crossed a South Pacific (not antarctic) land bridge from South America to Australia and again changed opalinids or at least acquired in Australia a primitive opalinid which it got from a now-extinct, hypothetical, Australian leiopelmid frog. There are two comments to be made on this. First, the primitive "*Protoopalina* of subgenus II," which *Hyla* is supposed to have got in Australia, is said to occur also on discoglossids, pelobatids, and some other frogs, so *Hyla* might have got it almost anywhere on its way to Australia, perhaps in Eurasia or the Oriental Region where discoglossids and most pelobatids now occur. The second comment is that if frogs change their opalinids or take up new ones so easily, the opalinids are not much use in tracing the history of frogs, as Noble (1925) and Dunn (1925) have said. Metcalf's ideas have already been noted (p. 144, footnote). I mention them again here because they have recently been revived in connection with *Hyla* by Beaufort (1951, pp. 152 and 166).

Eurasia from the arctic to and through tropical Asia, on Japan, the Riu Kius, and Formosa, through the Indo-Australian Archipelago to the Philippines and New Guinea, and in America from the arctic to part of tropical South America. Toward Australia, *Rana* decreases in numbers, but related forms are numerous and reach the tropical northern edge of Australia and also the Solomons. The occurrence of ranids on the Palau and Fiji Islands is discussed in Chapter 8. In America, *Rana* is well represented in North and part of Central America, but probably only one species reaches South America (Dunn, in letter); it is *R. palmipes,* which extends from southern Mexico through Central America and northern South America at least to the Amazon region. Except for *Rana,* the family Ranidae is confined to the Old World and is almost entirely confined to the tropics there. Noble (1931) recognized 6 subfamilies with more than 30 genera. Most of them are confined to Africa, but the Cornuferinae, with about 10 genera, are well represented in the Oriental-Australian area. The classification of some subfamilies, including the Cornuferinae, is still unsettled, and their distributions are not worth giving in detail. In habits most ranids, including most *Rana,* are amphibious, but some have become aquatic, terrestrial, burrowing, or arboreal.

Fossils show the presence of *Rana* in Eurasia since the Miocene and in North America since the Pliocene, but these are minor details. The distribution of living forms shows clearly that ranids have evolved in the main part of the Old World tropics and that *Rana* has spread from there through the northern parts of the world and thence, very recently, to South America.

Rhacophoridae (Polypedatidae), Old World tropical tree frogs (Fig. 25): main Old World tropics etc.: 13 genera (Noble 1931). Six genera are African and 2 of them reach Madagascar and one (*Megalixalus*) the Seychelles too; 4 are confined to Madagascar; 2 are confined to the Orient; and the final genus, *Rhacophorus* (*Polypedates*), though absent in Africa, has many species on Madagascar, none on the Seychelles, and many in the Orient. This genus extends north into about the southern half of China and to Japan, and it occurs on the Riu Kius and Formosa, and east and south on the Indo-Australian Archipelago to and through the Philippines and to Celebes (4 species) and Timor (2 species), but probably not beyond. Beaufort (1951, p. 70) refers to rhacophorids in New Guinea, but this is probably an error; van Kampen (1923) and Loveridge (1948) knew none there. In habits, most rhacophorids are tree frogs, although a few have "returned to the sod" (Noble).

Rhacophorids are unknown fossil. Their diversity in Africa and Madagascar and their failure to reach the Australian Region suggest an African origin. These tree frogs, and especially *Rhacophorus* itself, may have displaced *Hyla* in the main area of the Old World tropics (see text, p. 142); and *Rhacophorus* may have been displaced in turn in Africa by the related (derived?) *Chiromantis* and other genera (see Noble 1931, p. 525).

Brevicipitidae (Microhylidae), narrow-mouthed toads etc. (Fig. 23): the tropics (excepting West Africa) and some adjacent warm-temperate areas; 7 subfamilies (Parker 1934). (1) Dyscophinae: 3 genera: one, with 6 species, confined to Madagascar; and 2, with together 6 species, in the *eastern* part of the Orient, scattered from southwestern China (Liu 1950) to Sumatra and Borneo. (2) Cophylinae: Madagascar; 8 genera, 21 species, diverse in form and structure, "the product of an isolated evolutionary center" (Parker). (3) Asterophryinae: 4 genera, 24 or more species, most in New Guinea but a few also on Halmahera, Amboina, etc. Myers (letter of July 14, 1948) writes that he has traced the record of a *Metopostira* from the "Sulu" Islands and that it is an error, the frog actually being from Sanana, Sula (Soela) Islands in the Moluccas. (4) Sphenophryninae: 5 genera, 36 or more species, most on New Guinea etc., but 4 species (2 endemic) are on the adjacent corner of Australia, and *Oreophryne* occurs not only on New Guinea but also on the Moluccas, Celebes, Mindanao and Biliran Islands in the southern Philippines, and along the Lesser Sunda Islands from Flores to Bali. (5) Microhylinae: the Orient and America. *Microhyla* (often called *Gastrophryne* in America) occurs in both areas, with 15 species in the Orient from Ceylon and India to Java, Bali, Borneo, Formosa, and the Riu Kiu Islands (not Japan), and on the mainland north at least to central China; and with 12 or more species in America from Brazil and Ecuador north to Kansas, southeastern Iowa (Klimstra 1950), Indiana, and Maryland. There are 9 more genera with at least 30 species in the Orient, mostly within the range of *Microhyla,* but *Kaloula* goes farther north, at least to central Manchuria, and reaches Celebes and the Lesser Sundas to Flores; and several genera reach the Philippines. Six more genera with at least 30 species (including recently described ones listed by Dunn, 1949) are in tropical America, one of them (*Hypopachus*) ranging north to southern Texas, and one ranging south to Buenos Aires, but none reaching the south-temperate forests of southern South America. (6) Brevicipitinae: South and East Africa: *Breviceps,* with 13 species, is mostly South African but occurs north to Tanganyika; the other 3 genera, with 4 species, are localized in Tanganyika. (7) Melanobatrachinae: East Africa and South India; in Africa occur 2 genera, 3 species, all localized in the Usambara and Uluguru Mountains in Tanganyika; in India there is only *Hoplophryne,* with one species, localized in the Travancore and Cochin Hills.

Brevicipitids are mostly small and retiring in habits. Many burrow and probably feed on ants or termites, but some are more active and are probably more or less arboreal, and some live in grass.

No fossil brevicipitids are known. Parker (1934, p. 3) thinks that the family is a monophyletic assemblage which has differentiated geographically. He thinks (p. 9) that it may be an early offshoot of the same stock which produced the ranids and rhacophorids, and that its place of origin may have been southeastern Asia. However this may be, it is clear that brevicipitids have evolved primarily in the Old World tropics and that the Microhylinae have reached the American tropics through the north. It is likely that they did so more than once. *Microhyla* may have reached America relatively recently, for the Oriental and American species are still very similar and the genus is better developed in North than in South America (Parker, p. 14). The other American genera may be derived from one or more earlier immigrants.

Phrynomeridae: Africa south of the Sahara; one genus, 5 or more species. They are tree frogs apparently derived from brevicipitids, as hylids have been from leptodactylids or primitive bufonids, and as rhacophorids have been from ranids (Parker 1934, p. 3).

REFERENCES

Ahl, E. 1926. Neue Eidechsen und Amphibien. *Zoologischer Anzeiger* (Leipzig), **67,** 186–192.

Angel, F. 1947. *Vie et moeurs des amphibiens.* Paris, Payot (Bibliothèque Scientifique).

Angel, F., and H. Lhote. 1938. Reptiles et amphibiens du Sahara central et du Soudan. *Bull. Comité d'Études Historiques et Scientifiques de l'Afrique Occidentale Française,* **21,** 345–384.

Archey, G. 1922. The habitat and life history of *Liopelma hochstetteri. Rec. Canterbury Mus.,* **2,** 59–71.

Baker, C. L. 1947. The species of Amphiumae. *J. Tennessee Acad. Sci.,* **22,** 9–21.

Barbour, T. 1926. *Reptiles and amphibians.* Boston, Houghton Mifflin.

Barrett, W. C., Jr. 1947. Hematopoiesis in the European plethodontid . . . . *Anatomical Record* (Philadelphia), **98,** 127–136.

Beaufort, L. F. de. 1951. *Zoogeography of the land and inland waters.* London, Sidgwick and Jackson; New York, Macmillan.

Berg, L. S. 1950. *Natural regions of the U.S.S.R.* New York, Macmillan.

Bishop, S. C. 1943. *Handbook of salamanders.* Ithaca, New York, Comstock.

Bogert, C. M., and J. A. Oliver. 1945. . . . herpetofauna of Sonora. *Bull. American Mus. Nat. Hist.,* **83,** 297–425.

Boulenger, E. G. 1936. Batrachia and Reptilia. (In) Regan (ed.), *Natural History,* pp. 297–392. London, Ward, Lock.

Boulenger, G. A. 1890. Reptilia and Batrachia. (In) *Fauna of British India including Ceylon and Burma.* London, Taylor and Francis.

———. 1903. Report on the batrachians and reptiles. (In) Annandale and Robinson, *Fasciculi Malayensis, Zool.,* **1,** 131–176.

Bourret, R. 1942. *Les batraciens de l'Indochine.* Institut Océanographique de l'Indochine, Mem. 6.

Bridges, E. L. 1948. *Uttermost part of the earth* [Tierra del Fuego]. London, Hodder and Stoughton.

Buxton, P. A. 1923. *Animal life in deserts* . . . . London, Edward Arnold.

Chabanaud, P. 1954. [On the supposed salamander larva from French Guinea.] *Bull. Inst. française d'Afrique noire,* **16,** Ser. A, 1293–1294.

Darlington, P. J., Jr. 1948. The geographical distribution of cold-blooded vertebrates. *Quarterly Review Biol.,* **23,** 1–26, 105–123.

Darwin, C. 1859, 1950. *On the origin of species* . . . . London, Murray; reprinted 1950, London, Watts and Co.

Davis, D. D. 1935. A new generic and family position for *Bufo borbonica*. *Zool. Ser. Field Mus. Nat. Hist.*, **20,** 87–92.

———. 1936. The distribution of Bidder's organ in the Bufonidae. *Zool. Ser. Field Mus. Nat. Hist.*, **20,** 115–125.

Dunn, E. R. 1923. The salamanders of the family Hynobiidae. *Proc. American Acad. Arts Sci.*, **58,** 443–523.

———. 1923*a*. The geographical distribution of amphibians. *American Naturalist,* **57,** 129–136.

———. 1925. The host-parasite method and the distribution of frogs. *American Naturalist,* **59,** 370–375.

———. 1931. The herpetological fauna of the Americas. *Copeia,* **1931,** 106–119.

———. 1942. The American caecilians. *Bull. Mus. Comparative Zool.*, **91,** 439–540.

———. 1948. American frogs of the family Pipidae. *American Mus. Novitates,* No. 1384.

———. 1949. Notes on South American frogs of the family Microhylidae. *American Mus. Novitates,* No. 1419.

du Toit, C. A. 1934. A revision of the genus *Heleophryne. Ann. U. Stellenbosch,* **12,** Section A, No. 2.

Ekman, S. 1922. *Djurvärldens Utbrednings-historia pa Skandinaviska Halvön.* Stockholm, Alb. Bonniers Boktryckeri.

Flower, S. S. 1933. . . . reptiles and amphibians of Egypt . . . . *Proc. Zool. Soc. London,* **1933,** 735–851.

Gadow, H. 1909. *Amphibia and reptiles.* The Cambridge Natural History, Vol. 8. London etc., Macmillan. The volume I have used is dated 1909. It is not clear whether or not it is an exact reprint of the "first edition," which is dated 1901.

Gorman, J., and C. L. Camp. 1953. A new . . . *Hydromantes* from California . . . . *Copeia,* **1953,** 39–43.

Grandidier, G., and G. Petit. 1932. *Zoologie de Madagascar.* Paris, Soc. d'Editions Géographiques, Maritimes et Coloniales.

Griffiths, I. 1954. On the "otic element" in Amphibia Salientia. *Proc. Zool. Soc. London,* **124,** 35–50.

Guibé, J. 1950. Batraciens. (In) Contribution à l'étude de l'Aïr. *Mem. l'Inst. français d'Afrique noire,* **10,** 329–332.

Günther, A. 1858. On the geographical distribution of batrachians. *Proc. Zool. Soc. London,* **26,** 390–398.

Harper, F. 1956. Amphibians and reptiles of the Ungava Peninsula. *Proc. Biol. Soc. Washington,* **69,** 93–104.

Hoffstetter, R. 1945. [Concerning fossil *Amphignathodon.*] *Bull. Soc. Geologique France* (5), **15,** 167–169.

Inger, R. F. 1950. . . . amphibians of the Riu Kiu Islands. *American Naturalist,* **84,** 95–115.

———. 1954. Systematics and zoogeography of Philippine Amphibia. *Fieldiana* (Chicago Nat. Hist. Mus.), *Zool.*, **33,** 181–531.

Klimstra, W. D. 1950. Narrow-mouthed toad taken in Iowa. *Copeia,* **1950,** 60.

Liu, C. 1950. Amphibians of western China. *Fieldiana* (Chicago Nat. Hist. Mus.), *Zool.*, **2.**

Logier, E. B. S. 1952. *The frogs, toads and salamanders of eastern Canada.* Canada, Clarke, Irwin.

Logier, E. B. S., and G. C. Toner. 1955. Check-list of the amphibians and reptiles of Canada and Alaska. *Contrib. R. Ontario Mus. Zool. and Palaeontology,* No. 41.

Loveridge, A. 1935. Australian Amphibia in the Museum of Comparative Zoology. *Bull. Mus. Comparative Zool.,* 78, 1–60.

——. 1945. *Reptiles [and amphibians] of the Pacific World.* New York, Macmillan.

——. 1948. New Guinean reptiles and amphibians . . . . *Bull. Mus. Comparative Zool.,* **101**, 305–430.

Lowe, C. H., Jr. 1950. . . . biogeographical problems in *Aneides. Copeia,* **1950**, 92–99.

Mertens, R. 1930. Die Amphibien und Reptilien der Inseln Bali, Lombok, Sumbawa und Flores. *Abhandlungen Senckenbergischen Naturforschenden Gesellschaft,* **42**, 117–344.

Mertens, R., and L. Müller. 1940. Die Amphibien und Reptilien europas (Zweite Liste). *Abhandlungen Senckenbergischen Naturforschenden Gesellschaft,* Abh. 451.

Metcalf, M. M. 1923. The opalinid ciliate infusorians. *Bull. United States National Mus.,* No. 120.

——. 1940. Further studies on the opalinid ciliate infusorians and their hosts. *Proc. United States National Mus.,* **87**, 465–634.

Mills, R. C. 1948. A check list of the reptiles and amphibians of Canada. *Herpetologica,* **4**, Supplement 2.

Mittleman, M. B., and G. S. Myers. 1949. Geographic variation in the ribbed frog, *Ascaphus truei. Proc. Biol. Soc. Washington,* **62**, 57–68.

Moore, J. A. 1954. Geographic and genetic isolation in Australian Amphibia. *American Naturalist,* **88**, 65–74.

Myers, G. S. 1943. Rediscovery of the Philippine discoglossid frog, *Barbourula busuangensis. Copeia,* **1943**, 148–150.

——. 1953. Ability of amphibians to cross sea barriers, with especial reference to Pacific zoogeography. *Proc. Seventh Pacific Sci. Congress,* **4**, 19–27.

Myers, G. S., and A. L. de Carvalho. 1945. Notes on . . . Brazilian amphibians . . . Plata salamander. *Boletim Mus. Nacional* [Brazil], *Zool.,* No. 35.

Noble, G. K. 1922. The phylogeny of the Salientia . . . . *Bull. American Mus. Nat. Hist.,* **46**, 1–87.

——. 1924. . . . herpetology of the Belgian Congo . . . . Part 3, Amphibia [with check list of the Amphibia of Africa]. *Bull. American Mus. Nat. Hist.,* **49**, 147–347.

——. 1924*a*. A new spadefoot toad from the Oligocene of Mongolia with a summary of the evolution of the Pelobatidae. *American Mus. Novitates,* No. 132.

——. 1925. The evolution and dispersal of the frogs. *American Naturalist,* **59**, 265–271.

——. 1926. . . . remarkable cases of distribution among the Amphibia . . . . *American Mus. Novitates,* No. 212.

——. 1928. Two new fossil Amphibia of zoogeographic importance from the Miocene of Europe. *American Mus. Novitates,* No. 303.

——. 1930. The fossil frogs of the Intertrappean beds of Bombay, India. *American Mus. Novitates,* No. 401.

Noble, G. K. 1931, 1954. *The biology of the Amphibia.* New York, McGraw-Hill; reprinted 1954 by Dover Publications.

Oliver, J. A., and C. E. Shaw. 1953. The amphibians and reptiles of the Hawaiian Islands. *Zoologica* (New York), **38,** 65–95.

Parker, H. W. 1929. Two fossil frogs from the Lower Miocene of Europe. *Ann. and Mag. Nat. Hist.* (10), **4,** 270–281.

———. 1930. The status of the extinct frog, *Lithobatrachus. Ann. and Mag. Nat. Hist* (10), **6,** 201–205.

———. 1932. Parallel modifications in the skeleton of the Amphibia Salientia. *Archivio Zool. Italiano,* **16,** 1239–1248.

———. 1934. *A monograph of the frogs of the family Microhylidae.* London, British Mus.

———. 1936. Revised list of reptiles (excluding chelonians) and amphibians collected in the Seychelles. *Tr. Linnaean Soc. London, Zool.* (2), **19,** 444–446.

———. 1940. The Australasian frogs of the family Leptodactylidae. *Novitates Zoologicae* (London), **42,** 1–106.

———. 1941. The caecilians of the Seychelles. *Ann. and Mag. Nat. Hist.* (11), **7,** 1–17.

Pearse, A. S. 1911. Concerning the development of frog tadpoles in sea water. *Philippine J. Sci. D* (General Biol. etc.), **6,** 219–220.

Pope, C. H., and A. M. Boring. 1940. A survey of Chinese Amphibia. *Peking Nat. Hist. Bull.,* **15,** 13–86.

Ramaswami, L. S. 1944. The chondrocranium of two torrent-dwelling anuran tadpoles. *J. Morphology,* **74,** 347–374.

Romer, A. S. 1945. *Vertebrate paleontology,* 2nd ed. Chicago, U. of Chicago Press.

———. 1947. Review of the Labyrinthodontia. *Bull. Mus. Comparative Zool.,* **99,** 1–368.

Schaeffer, B. 1949. Anurans from the early Tertiary of Patagonia. *Bull. American Mus. Nat. Hist.,* **93,** 41–68.

Schmidt, K. P. 1936. Guatemalan salamanders of the genus *Oedipus. Zool. Ser. Field Mus. Nat. Hist.,* **20,** 135–166.

———. 1951. Annotated bibliography of marine ecological relations of living amphibians. *Marine Life Occasional Papers,* **1,** 43–46.

———. 1953. *A checklist of North American amphibians and reptiles.* American Soc. Ichthyologists & Herpetologists (printed by U. of Chicago Press). Additions and corrections in *Copeia,* **1954,** 304–306.

———. 1953*a*. Amphibians and reptiles of Yemen. *Fieldiana* (Chicago Nat. Hist. Mus.), *Zool.,* **34,** 253–261.

Schmidt, K. P., and R. F. Inger. 1951. Amphibians and reptiles of the Hopkins-Branner Expedition to Brazil. *Fieldiana* (Chicago Nat. Hist. Mus.), *Zool.,* **31,** 439–465.

Smith, H. M., and E. H. Taylor. 1948. An annotated checklist and key to the Amphibia of Mexico. *Bull. United States National Mus.,* 194.

Smith, M. (A.). 1927. Contributions to the herpetology of the Indo-Australian region. *Proc. Zool. Soc. London,* **1927,** 199–225.

———. 1951. *The British amphibians and reptiles.* London, Collins.

Stebbins, R. C. 1954. *Amphibians and reptiles of western North America.* New York, McGraw-Hill.

Stebbins, R. C., and W. J. Riemer. 1950. A new species of plethodontid salamander from the Jemez Mountains of New Mexico. *Copeia,* **1950,** 73–80.

Stephenson, N. G., and E. M. Thomas. 1945. A note concerning the occurrence and life history of *Leiopelma* Fitzinger. *Tr. and Proc. R. Soc. New Zealand,* **75,** 319–320.

Taylor, E. H. 1944. The genera of plethodontid salamanders in Mexico. Part 1. *U. of Kansas Sci. Bull.,* **30,** 189–232.

Thorson, T., and A. Svihla. 1943. Correlations of the habitats of amphibians with their ability to survive the loss of body water. *Ecology,* **24,** 374–381.

Turbott, E. G. 1942. The distribution of the genus *Leiopelma* in New Zealand . . . . *Tr. and Proc. R. Soc. New Zealand,* **71,** 247–253.

van Kampen, P. N. 1923. *The Amphibia of the Indo-Australian Archipelago.* Leiden, E. J. Brill.

Wright, A. H., and A. A. Wright. 1949. *Handbook of frogs and toads of the United States and Canada,* 3rd ed. Ithaca, New York, Comstock.

Wynne-Edwards, V. C. 1952. Freshwater vertebrates of the arctic and subarctic. *Bull. Fisheries Research Board Canada,* No. 94.

Zweifel, R. G. 1956. Two pelobatid frogs from the tertiary of North America and their relationships to fossil and Recent forms. *American Mus. Novitates,* No. 1762.

chapter 4

Reptiles

In my work on the geography of reptiles I have been aided by the same persons and in much the same ways as acknowledged at the beginning of the preceding chapter, and in addition Charles M. Bogert, Samuel B. McDowell, Jr., Garth Underwood, and especially Ernest E. Williams have very generously answered questions and given me special, often unpublished, information. Romer (1945) presents a good, general account of the evolution and fossil record of reptiles, which should be read by everyone unfamiliar with these animals who wishes to understand them and their distribution. The main system of classification used here follows Romer's (1956) new *Osteology of the Reptiles,* which he kindly allowed me to see in manuscript. However, I shall be concerned only with existing reptiles and their immediate fossil relatives, and with a few ancient groups (especially dinosaurs) which may be geographically significant.

The age of reptiles was the Mesozoic. Existing reptiles represent only a few groups of all that there were then, but they belong to three different subclasses which have been separate since the Triassic. These subclasses are represented by turtles, crocodiles, and lizards and snakes with *Sphenodon.* Turtles have existed since the Triassic and have changed very little since then. Crocodiles too are very old, dating from the late Triassic, and they too have undergone little visible change during their long history. Within the third subclass, the order Rhynchocephalia appeared in the Triassic and was probably cosmopolitan in the early and middle Mesozoic. Rhynchocephalians are not known fossil later than the Lower Cretaceous, but nevertheless one of them (*Sphenodon*) still exists, on remote New Zealand. Lizards, which may be derived from the same stock as rhynchocephalians, appeared in the Jurassic; and snakes, which are derived from lizards, appeared in the Cretaceous. Some existing families of turtles and also crocodiles and their allies and rhynchocephalians have left revealing fossil records. Lizards and snakes have left a record sufficient to show their origin and evolution in a general way, but not sufficient to tell much about their geographical histories.

Among existing reptiles, as among amphibians, there is a correlation between numbers, taxonomic and ecological diversity, and extent and diversity of area occupied.

The ancient Rhynchocephalia, with one existing species, are now strictly limited in all ways.

Existing crocodilians form one family of 9 genera, 25 species; all are amphibious or aquatic (except that their eggs are laid on land); and they are confined to the tropics and some warm edges of the temperate zones.

Existing non-marine turtles form 7 families, 50 or more genera, and about 200 or more species; they are mostly amphibious or aquatic (except that their eggs are laid on land), but several groups of Testudinidae have become terrestrial; and turtles occur throughout the tropics and widely but irregularly in the temperate zones.

Lizards and snakes are much the most numerous and most widely distributed existing reptiles. The following careful estimate of their numbers was given in my *Quarterly Review* article (1948, pp. 23 and 24): of lizards, about 21 (now reduced to 18) families, more than 300 genera, and nearly 3000 recognized species; and of snakes, about 11 (now reduced to 8) families, 300 genera, and 2600 species. Both lizards and snakes are nearly cosmopolitan and occur in a wide variety of climates and habitats. Together they are the dominant existing reptiles. Of all of them one family of snakes, Colubridae, is predominant, as will be described.

Although both lizards and snakes are very widely distributed, they vary in numbers in different places. Lizards outnumber snakes (in number of species) in Europe and probably Africa and in New Guinea and Australia. Snakes outnumber lizards (in species) in southern and eastern Asia. In North America as a whole, the two groups are about equal in number of species, but there are more lizards in western North America and more snakes in the east. In South America, species of snakes may be more numerous in the main tropical part of the continent; of lizards, in the south-temperate part. On islands, except some recent continental ones, lizards are usually much more numerous than snakes. However, regardless of how they compare in numbers of species, there are many more individual lizards than snakes in most parts of the world.

Ecologically, reptiles are true land animals, basically independent of water as an environment. Some of them are secondarily amphibious or aquatic, but almost all of them return to the land to lay their eggs or bear their young. Of existing reptiles, only some sea snakes give birth in the water. Reptiles are much more numerous in a greater variety of land habitats than amphibians are, and they can disperse over land more rapidly. They are also more able to disperse across salt water. Their skins are relatively impervious (but not completely so—Bogert and Cowles 1947), and some of them enter the sea or even live in it without injury. They drink little fresh water, and some of them can exist for considerable periods without any water at all. Their eggs are much more resistant than amphibian eggs and more likely to be dispersed by various means. All this is reflected in the relatively wide and complex distributions of reptiles and in their occurrence on islands. Only in cold places do amphibians have the advantage over reptiles. Reptiles are the only true land animals that are cold-blooded. They have greater heat requirements than amphibians and less cold tolerance than birds and mammals. This too is reflected in their distributions.

Although reptiles are primarily land animals, existing crocodilians are amphibious and so are most turtles. *Sphenodon* lives on land but usually near water. And lizards and snakes are primarily terrestrial but diverse in their adaptations. Both lizards and snakes include a variety of ground-living, burrowing, and arboreal forms. There is an obvious tendency for lizards to occur in relatively dry, open places; snakes, in damper, better-covered ones. But this is a tendency rather than a rule, and there are many exceptions to it. Both lizards and snakes occur in deserts, and both are abundant in tropical rain forest. Only a few lizards are amphibious and none is fully aquatic now (some were in the Cretaceous), while several groups of snakes are aquatic; but lizards probably have more adaptations to special niches. Lizards and snakes tend to differ in their food, and this must affect their distributions. Most lizards are wholly or partly insectivorous, and insects occur nearly everywhere. Some snakes too eat insects, but many require larger prey, which may not be available in some places. It is probably because of their food requirements that individual snakes are usually relatively few; and, aside from the food itself, sparseness of populations may affect both chances of dispersal across barriers and ability to maintain populations in small, isolated areas or localized niches.

Lizards are disproportionately numerous on islands, and this is proof that they have unusual powers of crossing salt water (Chapter 8), but other factors are probably involved too. For reasons just given, lizards are probably better able than snakes to find food and to maintain populations in small, ecologically limited areas. Absence or scarcity of snakes on remote islands may in itself favor survival and multiplication of lizards there.

Some reptiles, even more than amphibians, have been carried about by man, by accident or design. Some lizards are so regularly associated with man that they are called house lizards, and some of them have been carried great distances in the tropics and to many remote islands (see under Gekkonidae and Scincidae in list of families). A small worm snake (*Typhlops*) has been so commonly carried about the Pacific and elsewhere in soil with plants that it is sometimes called the Flowerpot Snake. And turtles and less often other reptiles are commonly carried alive for food or as pets. On the other hand, man has caused the extinction of some reptiles in some places. He has destroyed giant tortoises on some islands within historic times, and he may have destroyed other reptiles elsewhere, because they were good food, like turtles or large snakes, or because they were dangerous, like poisonous snakes.

Limits of distribution of reptiles

The limits of distribution of reptiles are indicated in Figure 28.

The northernmost reptile is a lizard, *Lacerta vivipara,* which extends above the Arctic Circle, to at least 70° N. and perhaps to the limit of land in northern Europe, and nearly reaches the Circle in eastern Siberia; it is the only viviparous (ovoviviparous) species of the family Lacertidae. (The approximate limits of distribution of this species and those of all the other European species here listed have been mapped by Hecht, 1928.) Next is a snake, *Vipera berus,* the European Viper, which too crosses the Arctic Circle, to above 67° N., in Europe; it and its representatives come within a degree or two of the

Circle in part of Siberia; it too is viviparous. Both the *Lacerta* and the *Vipera* occur throughout the northern forest zone of Eurasia (Berg 1950, p. 48); I do not know whether they occur to any extent north of the forest. Third in order of northern reptiles is an oviparous snake, *Natrix natrix,* which reaches about 65° N. in Scandinavia. Fourth is another viviparous lizard, *Anguis fragilis,* which reaches about 64½° N. in Sweden (Professor C. H. Lindroth in letter—the

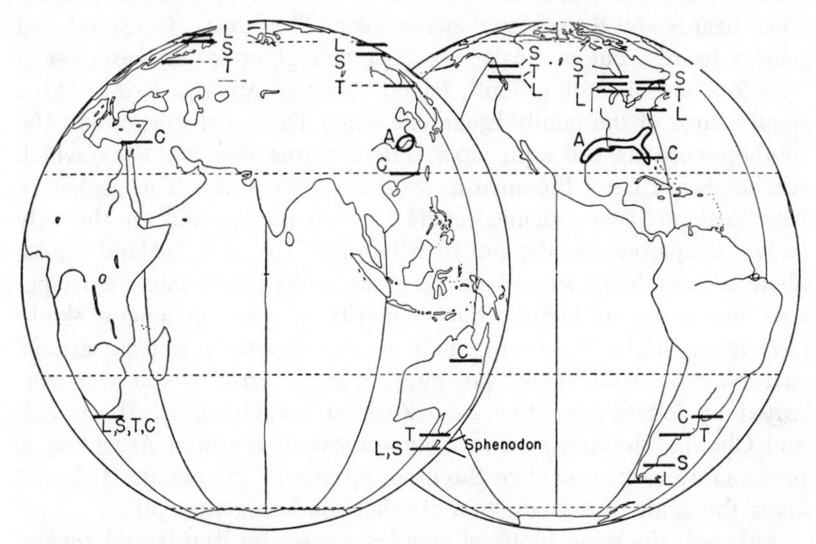

Fig. 28. Approximate northern and southern limits of reptiles. L, Lizards; S, Snakes; T, Turtles; C, Crocodilians other than *Alligator;* A, approximate ranges of *Alligator* in eastern Asia and eastern North America.

record of this species on the Arctic Circle was an error). Fifth and sixth are probably a viviparous snake, *Coronella austriaca,* which occurs above 63° in Sweden, and a lizard, *Lacerta agilis,* which reaches about 62½° N. in Russia, and which belongs to the same genus as the northernmost lizard but is oviparous.

In America, viviparous snakes of the genus *Thamnophis* (garter snakes—close relatives of *Natrix*) are the northernmost reptiles by a wide margin, reaching at least 56° N. in British Columbia, 60° at Fort Smith in the Northwest Territories, and 49° in the east (Logier and Toner 1955). Other snakes, including both viviparous and oviparous ones, reach only 51° or 52° in the west and 47° or 48° in the east. Three lizards, of three different families, reach 50° or 51° N. in British Columbia; two of them are viviparous and one oviparous. The northernmost eastern lizard is oviparous *Eumeces fasciatus*, but it reaches only about 45° in southern Ontario, and on the Atlantic coastal plain it stops slightly above 42°. There are no lizards north of this along the Atlantic coast of North America, although about ten species of snakes and three families of turtles extend farther north there. (This situation gives New England naturalists the erroneous idea that lizards are inherently unable to tolerate cold.)

Southward, many lizards and snakes reach the tip of South Africa. Some lizards and three elapid snakes reach Tasmania. Gekkonid and scincid lizards, but no snakes (except straggling sea snakes), reach New Zealand, including South Island. And in America two or three small lizards of the family Iguanidae reach Tierra del Fuego; and the southernmost snake is a pit viper, *Trimeresurus ammodytoides*, which reaches Santa Cruz Province in southern Argentina. The geckos of New Zealand (two endemic genera) are viviparous, and are the only known viviparous members of their family. The New Zealand skinks, about nine endemic species of the great genus *Lygosoma*, are viviparous too so far as known, but viviparity is common among skinks elsewhere. (The statement made in my *Quarterly Review* article, that New Zealand skinks lay eggs, was an error which was corrected by letters from two New Zealand herpetologists, H. B. Fell and Charles McCann.) The *Trimeresurus* in southern Argentina is presumably viviparous, like the other species of the genus. I do not know the mode of reproduction of other southernmost reptiles.

Although the main limits of reptiles are set by lizards and snakes, the limits of the other groups are to be noted. *Sphenodon*, on New Zealand, is near the southern limit of distribution of reptiles; that this ancient reptile survives in a climate which is marginal for modern reptiles is especially noteworthy. Crocodilians occur (or did so within recent historic times) north of the tropics to Syria in western Asia (*Crocodylus*), along the lower Yangtze in China (*Alligator*), and in the Carolinas in eastern North America (*Alligator*); they extend a little south of the tropics in Africa and South America. Of turtles, *Emys orbicularis* (Emydinae) in Europe reaches at least 57½° N., but no other European turtle reaches even 50°. A *Trionyx* (Trionychidae) reaches about 48° in eastern Asia, but no other eastern Asiatic turtle occurs north of about 40°, and a great part of interior and northern Asia is without turtles. In America, the most northern turtle, a *Chrysemys* (Emydinae), reaches at least 51° in British Columbia and Manitoba, 49° or 50° above Lake Superior, and perhaps 48° along the east coast of North America. Southward, turtles reach the tip of Africa, southern Australia but not Tasmania (except for straggling sea turtles), and northern Argentina in South America.

Although a few species reach the arctic and also the southern tips of the continents, reptiles begin to decrease in numbers far short of these final limits. In Europe, they decrease northward much more rapidly than amphibians do: in southern Europe there are 37 genera, 89 species of reptiles, and 14 genera, 39 species of amphibians; but in northern Europe, only 5 genera, 6 species of reptiles, and 6 genera, 11 species of amphibians (Schreiber 1912). In South America, reptiles decrease from the tropics southward. This is illustrated by the snakes of Argentina, which decrease progressively from 55 known species in near-tropical Misiones to one species in Santa Cruz (Table 4).

TABLE 4. PROGRESSIVE SUBTRACTION IN THE SNAKE FAUNA OF ARGENTINA

The eastern column of Provinces and Territories of Argentina listed from north to south, with the number of species of snakes known from each (data from Serié 1936, pp. 57–59).

Misiones	55
Corrientes	51
Entre Rios	32
Buenos Aires	22
La Pampa	15
Rio Negro	5
Chubut	5
Santa Cruz	1
Tierra del Fuego	0

As to the limits of reptiles on islands, there are almost no limits for lizards: they are numerous on every climatically suitable island within reasonable reach of a continent, and they are fewer but surely native on many remote islands, including the Mascarenes and Seychelles in the Indian Ocean, the islands of the western Pacific out at least to the Tongas (beyond the Fijis) and to New Caledonia and New Zealand, the Galapagos in the eastern Pacific, and not only the West Indies but also Bermuda, the Canaries, Madeira, and the Cape Verde Islands in the Atlantic. Beyond these limits, on still more remote islands especially in the tropical Pacific, there are a few geckos and skinks, but they have probably been carried by man. Further details of the occurrence of lizards and other reptiles on islands are given in Chapter 8.

A few families and genera of reptiles are very widely distributed.

Of non-marine turtles, the family Testudinidae reaches all habitable continents except Australia. The genus *Testudo* (land tortoises) occurs in much of Africa, the warm parts of Eurasia etc., and South America, and on various islands. The most widely distributed genus of fresh-water turtles is *Trionyx* (soft-shelled turtles of the family Trionychidae), in Africa, southern and eastern Asia, and eastern North America. The principal genus of crocodiles (*Crocodylus*) is almost tropicopolitan. Of lizards, the families Gekkonidae and Scincidae are tropicopolitan and extend into some warm-temperate areas, but no other lizard family is so widely distributed. The most widely distributed lizard genus is probably *Lygosoma,* which occurs in most of the warm part of the Old World, eastern North America, and Central America. Of snakes, the family Colubridae (common snakes) is cosmopolitan except that it does not reach the more remote parts of Australia; the family Viperidae (vipers etc.) is almost equally widely distributed, except that it does not reach even part of the Australian Region; and the families Typhlopidae (burrowing worm snakes), Boidae (constricting snakes), and Elapidae (cobras etc. and coral snakes) are tropicopolitan and enter some warm-temperate areas. The genus *Typhlops* is tropicopolitan. *Natrix* (mostly semi-aquatic water snakes, family Colubridae) occurs over most of the Old World including part of Australia and in much of North America. But these cases are exceptional. In spite of the great powers of dispersal of some reptiles and the wide distributions of a few of them, many families, most genera, and almost all species of reptiles are rather limited in distribution.

Zonation of reptiles

Crocodilians are primarily tropical. Except for the genus *Alligator,* the few that occur in the edges of the temperate zones are tropical species which extend into limited temperate areas. *Alligator,* however, is confined to warm-temperate areas in China and the southeastern United States.

Turtles too are mostly tropical. The only genus that occurs around the north-temperate zone is *Clemmys,* with two species in the Mediterranean area (one of them reaching the tropics in West Africa), five in eastern Asia (south into the edge of the tropics), one on the west coast of North America, and three in eastern North America. One other genus in Europe etc. (*Emys*) and four or five in eastern North America are confined to limited temperate areas but do not have zonal distributions. Within the north-temperate zone, only eastern North America has many turtles, and the majority of them are emydines, which are even more numerous in the tropics of Asia and which enter the tropics in America. All the families and many of the genera of northern turtles occur also in the tropics. Details are given in the list of families.

Most lizards too are tropical, and the north-temperate ones are not very different from tropical ones. All widely distributed northern lizards belong to families that are also tropical, and many belong to tropical genera. *Lacerta,* for example, which includes the northernmost reptile, ranges from arctic Eurasia to tropical Africa. Some genera of lizards and one small family (Anniellidae) are localized in parts of temperate Eurasia or North America, but every genus that occurs in both these areas occurs also somewhere in the tropics. The nearest to a truly Holarctic lizard genus is probably the anguid *Ophisaurus,* with one or two species in southeastern Europe, temperate North Africa, etc.; two or three in the eastern Himalayas, Burma, southern China, and Formosa; one in eastern North America; and one in the mountains of Borneo, almost on the equator. There are some endemic genera of lizards in some south-temperate areas too, but they do not form a zonal fauna; that is, the genera are not common to the south-temperate parts of the different southern continents.

Snakes, too, are primarily tropical, with no well-defined north-temperate fauna. All northern snakes belong to tropical families, many to tropical genera. *Vipera,* which includes the northernmost snake, occurs also through the tropics of Africa and Asia. There is probably no truly Holarctic snake genus. Some genera are confined to parts of temperate Eurasia or North America, but every well-defined genus that occurs in both places occurs also somewhere in the tropics. Schmidt (1946) cites *Opheodrys* as confined to eastern Asia and eastern North America, as it is for the most part, but it is well represented in the tropics in Indochina and there is a species in Yucatan in the edge of the American tropics. There is no zonal fauna of snakes in the south-temperate zone.

In summary, reptiles have a zonal pattern of distribution only in that they are primarily tropical. There is no distinct, generally distributed, zonal reptile fauna in the north-temperate zone. North-temperate reptiles are much fewer than tropical ones, and they all belong to tropical families, many to tropical genera. It looks as if tropical reptiles often invade the north-temperate zone but rarely becomes specially adapted to it and usually do not persist there long, not long enough to become very different from their tropical relatives. The south-temperate zone too lacks a distinct, zonal reptile fauna.

Another sort of zonal pattern, which might be called longitudinal zonation, is characteristic of certain families of lizards. These families are distributed north and south along continuities of land, across the climatic zones. In the Old World, the family Lacertidae ranges from the arctic to South Africa. It makes up the main part of the lizard fauna of temperate Eurasia but does not reach America. In America, the family Iguanidae ranges from the northern limit of lizards in western Canada to the southern tip of South America. It makes up the main part of the lizard fauna of temperate North America but does not now occur in the continental part of the Old World. The lizard families Agamidae in the Old World and Teiidae in the New have somewhat similar distributions but are less important in the north. Of course longitudinal distributions occur to some extent among other animals and merge with other sorts of distributions; *i.e.*, transitional situations occur. The significant thing is the large proportion of north-temperate lizards that belong to strongly longitudinal families, so that temperate Eurasia and North America differ more in their lizards than in most other groups of animals, except migratory land birds.

Radial distribution of reptiles

Radial distributions are less obvious among reptiles than among fishes and amphibians. Dominant terrestrial reptiles probably disperse rapidly and pass through the radial stage of dispersal quickly, and clues to their histories are probably lost quickly. However, a few families and genera of reptiles are radially distributed.

The largest existing family of reptiles is the Colubridae, the family of common snakes. It is also the most widely distributed family: colubrids are numerous in all the main tropical and temperate parts of the world except Australia, where they are few and are apparently newcomers. Colubrids are also ecologically diverse: they occur in all ordinary habitats, hot and cold, wet and dry, etc., so that the distribution of the family is nearly continuous across places that are inhospitable for many reptiles. The family Colubridae has certainly radiated geographically as well as in an evolutionary sense. However, the family's present distribution is complex rather than simply radial. The best that can be said of the family as a whole is that the dominant, derivative forms seem to occur mostly in Africa and tropical Asia and, in smaller numbers and less diversity, in temperate Eurasia and North America; that primitive forms tend to occur in Madagascar and South America; and that only a few, recent arrivals occur in Australia (see list of families for further details). This seems to be a pattern of (very complex) radiation from the main part of the Old World tropics. As dominant snakes, colubrids are replaced in Australia by a very distinct, probably rather old, fauna of elapids. This is an unsolved anomaly. Elapids are apparently derived from colubrids, were therefore presumably later in origin, and ought not to have reached Australia first. Why did elapids reach Australia so long ago that all the Australian genera of them are now endemic, colubrids so recently that no genus is endemic?

The most widely distributed genus of Colubridae is *Natrix,* with about 80 species, in Africa and Madagascar, tropical and temperate Eurasia north to about 65° in Scandinavia, south through the islands to northern Australia, and in eastern (not now western) North Americe south to Guatemala, and on Cuba; and *Thamnophis* (garter snakes), derived from *Natrix,* occurs through the whole of North America north to the limit of snakes and south to Costa Rica. More species of *Natrix* occur in the Orient than anywhere else. This pattern of distribution is radial, centering on the main area of the Old World tropics, or, more precisely, on tropical Asia. The species of *Natrix* are usually associated with water but are not fully aquatic. They probably have the powers of dispersal of terrestrial snakes plus some of the ability of aquatic ones to cross water barriers. But *Natrix* probably has other, unknown, qualities which make it superior to most other snakes and which have impelled its radiation. Old World *Natrix* are oviparous; American ones, viviparous; and (since viviparity in reptiles is sometimes an adaptation to cold climates) this may be a hint that the American forms are descended from an ancestor which came from Eurasia through the north. (I have used *Natrix* in a fairly broad sense. Some groups of the species, including for example those on Madagascar, are sometimes put in separate small genera.)

A few other large groups of reptiles have somewhat radial distributions. The turtle family Testudinidae centers in tropical Asia: emydines are most diverse and numerous there and extend in smaller numbers to temperate Eurasia and North and South America etc.; testudinines, probably derived from emydines in tropical Asia, extend even farther in some directions. Trionychid turtles are most diverse in the main part of the Old World tropics; one genus extends northward in eastern Asia and to North America; and one has reached New Guinea. Scincids are very diverse in the Old World tropics; a few extend northward; and three Old World genera extend to America. Viperids are most diverse in the main Old World tropics, especially tropical Asia; a few extend northward; and two Asiatic genera extend to America; but the rattlesnakes in America complicate this radial pattern.

Regional distribution of reptiles

The two northern regions of the world (Eurasia and North America above the tropics) have reptile faunas characterized by what they lack more than by what they have. No family of reptiles is exclusively Holarctic. The only family confined to any north-temperate area is the Anniellidae, a minor family localized in southwestern North America. The only exclusively Holarctic genus of reptiles may be *Alligator*. (A few other nearly Holarctic genera are noted under zonation of reptiles, above.) Most dominant northern reptiles belong to families that are primarily tropical, especially emydine turtles, agamid lizards in Eurasia and iguanids in North America, and colubrid and viperid snakes. Besides these, each northern region has other reptiles derived from or shared with the tropics (mostly the Old World tropics in the case of Eurasia, partly the Old World and partly the American tropics in the case of North America) and often occupying rather small areas in the temperate zone. They include, of turtles, chelydrids in eastern North America, *Testudo* in southern Eurasia and *Gopherus* in southern North America, and *Trionyx* in eastern Asia and eastern North America; of lizards, a chameleon in the southern corners of Europe, geckos especially in central and eastern Asia, *Xantusia* in southwestern North America, a few skinks in large parts of Eurasia and North America, lacertids far north in Eurasia, one genus of teiids in North America, anguids widely scattered in Europe and North America, *Shinisaurus* in China, *Heloderma* in southwestern North America, *Varanus* in the southern edge of Eurasia, and different amphisbaenids in southern Europe etc. and the southern corners of North America; and of snakes, *Typhlops* in southern Eurasia and *Leptotyphlops* in southwestern North America, different small boids in Asia and western North America, and cobras in the southern edge of Asia and coral snakes in southern North America. In some of these cases tropical genera or species extend northward locally; in others, species or genera are endemic in the temperate zone but belong to tropical families. But all these groups together in the north-temperate zone, those that are dominant plus those that are localized or marginal, make up faunas of much less than tropical richness.

Not only are north-temperate reptiles relatively few and not very different from tropical ones; also they are very unevenly distributed. Temperate Eurasia and North America differ especially in their dominant lizards, most of which belong to different families (agamids and lacertids in Eurasia, iguanids and teiids in North America), which are shared with the Old and New World tropics respectively. The Eurasian and North American reptile faunas differ in many other details, some of which are indicated in the preceding paragraph. Of course, temperate Eurasia and North America have some families in common and even genera: *Clemmys* and *Trionyx; Eumeces, Lygosoma,* and *Ophisaurus;* several genera of colubrid and viperid snakes; and *Alligator.* But the differences outweigh the similarities. Temperate Eurasia and North America have largely different reptile faunas, evidently partly derived from different tropical regions, from the Old and New World tropics respectively.

The distribution of reptiles within each northern region is uneven. The number of reptiles falls off rapidly northward, and there is east-west differentiation. For example, western North America has few turtles and many lizards (and many snakes); eastern North America, many turtles and relatively few lizards (and many snakes). This sort of unevenness is an effect of unevenness of climate, dry or wet, but the effect is increased because northern reptiles, especially lizards, are in a marginal zone where they can exist only under conditions particularly favorable for them. In the tropics, lizards are not so concentrated in dry, open, sunny places; many occur in wet forests, and many others are nocturnal and avoid the sun. It is particularly in cool climates that many reptiles depend on the sun to maintain their temperatures during their daily activities (Bogert 1949) and also to provide the heat necessary for reproduction (see p. 558).

In short, north-temperate reptiles are relatively few and not much different from tropical ones but are so unevenly distributed that the temperate Eurasian and North American reptile faunas differ in many ways, especially in their lizards; and the regional faunas are uneven in distribution too.

In the tropical and south-temperate parts of the world, reptiles are so numerous and their distributions and relationships are so complex that they can be described only briefly. There are just too many details to give in a small space. Additional details will be found in the list of families at the end of this chapter.

In Africa below the Sahara are trionychid and pelomedusid turtles and testudinine land tortoises but only one emydine, apparently a newcomer, which has entered only the northwestern corner of the African tropics; a few agamid lizards, many true chameleons, many geckos, many skinks, many lacertids, cordylids, *Varanus*, and amphisbaenids; and worm snakes of both families (Typhlopidae and Leptotyphlopidae), pythons, a sand boa (confined, in tropical Africa, to the east), many colubrids, elapids, and typical viperids; and crocodiles. Most of these groups are widely distributed in tropical Africa and the majority, but not all, extend to South Africa too. Cordylid lizards are chiefly South African but extend north well into the tropics. There is some differentiation of reptile faunas between the wet forest and the drier, opener country, but this cannot be gone into here; it and some other features of the distribution of reptiles in Africa are discussed by Schmidt (1919–1923).

Madagascar has pelomedusid but not trionychid turtles, and testudinine tortoises; relict iguanids but no agamids, many chameleons, many geckos, skinks, a few cordylids (only Gerrhosaurinae, not Cordylinae), but no lacertids, no *Varanus*, and no amphisbaenids; and typhlopid but not leptotyphlopid worm snakes, an endemic subfamily of boids but not pythons, and colubrids including supposedly primitive genera, but no elapids and no viperids; and the Nile Crocodile.

This Madagascan reptile fauna is chiefly African in its relationships but lacks many African families and includes many distinct endemic genera. Very few Madagascan reptiles are directly related to Oriental ones. *Sibynophis* is the only Madagascan-Oriental (but not African) genus that occurs to me (see under Colubridae in list of families), although there may be a few others. The Indian chameleon is related to a Mediterranean, not Madagascan, species. *Phelsuma*, a Mascarene genus of geckos which reaches the Andaman Islands in the Oriental Region, is an insular genus rather than a Madagascan-Oriental one (see list of families).

Tropical Asia etc. (the Oriental Region) has the one existing platysternine and many emydine turtles, testudinine land tortoises, and trionychids; many agamids, a chameleon, many geckos, many skinks, a few lacertids, a dibamid, a few *Ophisaurus* and *Varanus*, and *Lanthanotus;* representatives of every family and almost every subfamily of snakes, although some of them are only marginal; and crocodiles and gavials. Some of these groups are widely distributed in the Oriental Region but others are localized: *e.g.*, the platysternine and most of the emydine turtles are concentrated in the eastern part of the region; the chameleon is confined to southern India and Ceylon; the dibamid occurs only from southern Indochina southeastward on the islands; most of the lacertids are on the mainland, only one species reaching the Sunda Islands; *Lanthanotus* is confined to Borneo; and the whole subfamily Uropeltinae of burrowing snakes, with some 43 species, is confined to part of peninsular India and Ceylon. The agamid genus *Cophotis* is known from one species in Ceylon and one in Sumatra and Java; the agamid *Draco* has one species in southern India (not Ceylon), isolated by a gap of at least a thousand miles from the main range of the genus in the eastern Orient; in each of the scincid genera *Lygosoma, Dasia,* and *Riopa* there are pairs of species divided between southern India and the eastern Orient, with no closely related species in northern India; and the aniliid genus *Cylindrophis* has a species isolated on Ceylon but is otherwise confined to the eastern Orient etc.; and the species *Varanus salvator* occurs on Ceylon and in the eastern Orient but is absent from most of peninsular India. These groups show a pattern of discontinuity like that of some fresh-water fishes (p. 63). Smaller discontinuities perhaps proceeding in the same direction, of isolation of populations in southern India, occur among emydine turtles and viperid snakes (see list of families). At least five species and two additional genera of snakes show a different discontinuity: they occur in Indochina and also in the Greater Sunda Islands but skip the Malay Peninsula and peninsular Siam (M. A. Smith 1943, pp. 25–26). *Ophisaurus* shows a wider discontinuity in this direction, skipping from Burma etc. to Borneo.

As to relationships between the tropical African and Oriental reptile faunas, they share a number of important families, most of which occur in other parts of the world too, and they share also a number of genera: *Testudo** (land tortoises) and *Trionyx* (aquatic turtles); *Agama, Mabuya**, *Lygosoma, Lacerta,* and *Varanus; Typhlops**, *Python, Boiga* and *Natrix** and some other genera of colubrids, *Naja* (cobras), and *Vipera; Crocodylus**; and others. Most of these genera occur also outside the limits of Africa and the Oriental Region. Some of them, starred above, occur on Madagascar, but the others do not (or are probably not native there), and the latter are so numerous and varied as to show that Africa and tropical Asia have exchanged a variety of reptiles by mainland routes, while there is little to suggest that they have exchanged reptiles by way of Madagascar. But the differences between the African and Oriental reptile faunas show that the exchange has been limited. Some of the more striking differences are that Africa has pelomedusid while the Orient has emydine turtles; that Africa has more chameleons and lacertids while the Orient has more agamids; that Africa alone (of these two regions) has amphisbaenids and cordylids, as well as leptotyphlopids, which barely reach the Orient, while the Orient has more snakes than Africa, including aniliids, several subfamilies of colubrids lacking in Africa, pit vipers, and most sea snakes.

The Australian Region has chelyid turtles and (in New Guinea) *Carettochelys* and a trionychid; many agamids and geckos, the endemic family Pygopodidae, many skinks, a *Dibamus* (in New Guinea), and *Varanus;* many *Typhlops,* a number of python genera, small tree boas (in New Guinea, etc.), a few colubrids, and many elapids (and sea snakes); and crocodiles (in the tropics). Except for the restrictions indicated above, and except that pygopodids are few in New Guinea and colubrids absent in much of southern and western Australia, the groups listed are fairly evenly distributed over the main parts of New Guinea and Australia. However, many of them fail to reach Tasmania, which has no turtles (except straggling sea turtles), of course no crocodiles, no *Varanus,* no *Typhlops,* no pythonids, and no colubrids.

The relationships of Australian reptiles are mostly toward the Orient. The trionychid turtle in New Guinea (if it is really native there) is an Oriental species, which sometimes enters the sea. One of the crocodiles is an Oriental species, which habitually enters the sea. Of lizards, agamids are numerous in Australia as well as in the Orient; *Goniocephalus* and *Physignathus* occur in the Oriental and Australian Regions and nowhere else, but their ranges are discontinuous, for both genera are absent from many of the intervening islands; but, on the other hand, the aquatic agamid *Lophura* (*Hydrosaurus*) occurs only from the Philippines and Celebes to New Guinea. The gekkonid *Gymnodactylus* is numerous from the Orient to Australia but occurs elsewhere too, and two or three other widely distributed gekkonid genera include both the Orient and Australia in their ranges. Skinks, especially the genus *Lygosoma* in a broad sense, swarm from the Orient to New Guinea and Australia; *Tropidophorus, Ophioscincus,* and *Rhodona* are each divided between the Oriental and Australian Regions but, like some agamids named above, are absent from parts of the intervening archipelago; *Emoia* and *Otosaurus,* on the other hand, are almost entirely or completely confined to the islands, from the Greater Sundas to New Guinea. *Dibamus* ranges from southern Indochina to New Guinea. Monitor lizards (*Varanus*) range continuously (so far as the land allows) to Australia, and so do worm snakes (*Typhlops*). The aniliid snake genus *Cylindrophis* extends from the Orient to New Guinea. The widely distributed Old World tropical genus *Python* in a strict sense extends to Timor; this genus in a broader sense reaches New Guinea and Australia. Of colubrid snakes, the following genera extend from the Orient to Australia: four fully aquatic genera (not counting sea snakes), semi-aquatic *Natrix*, and arboreal *Boiga* and *Ahaetulla;* and terrestrial *Steganophis* extends from the Philippines to northeastern Australia. Against this list of Oriental ties, Australian reptiles show only one clear case of relationship toward any other part of the world: chelyid turtles occur only in the Australian Region and South America. Some Australian geckos have been thought to be directly related to South African or American ones, but these cases are doubtful.

The transition of the Oriental and Australian reptile faunas is too complex to describe in more detail here (many details are given by Rooij 1915–1917 and Mertens 1930). Examples are given above of families, genera, and species that range continuously from Asia to Australia, of groups that occur in both regions but are absent on intervening islands, and of groups that are confined to the islands. There are also many Oriental reptiles that stop at Wallace's Line, and many others that extend for varying distances toward the Australian Region but do not reach it. In short, much of the Oriental reptile fauna extends toward Australia, with progressive subtractions increasing with distance, but with many groups actually reaching New Guinea, somewhat fewer, Australia, and still fewer, Tasmania; but the pattern is complicated by discontinuities in some groups and by endemic groups localized in the zone of subtraction. There seems to have been a great but very complex flow of reptiles from the Orient toward and into the Australian Region. But the movement seems to have been almost all in this one direction. There seems to have been little opposite movement, of Australian groups such as chelyid turtles, pygopodid lizards, and Australian elapid snakes toward the Orient: of these groups, only a few of the snakes extend west even to the Moluccas.

The South American reptile fauna consists of *Chelydra* (in the northwestern corner of the continent) and *Kinosternon,* two non-endemic genera of emydine turtles, *Testudo,* and both pelomedusids and chelyids; many iguanids, many geckos, a single non-endemic genus of skinks, many teiids, many amphisbaenids, and some anguids; and both families of worm snakes, an aniliid, boines, many colubrids (especially primitive ones of the subfamily Xenodontinae, if recognizable), coral snakes, and pit vipers; and crocodiles and caimans. This fauna is, of course, not evenly distributed (but the details of this cannot be gone into here), and southward it undergoes progressive subtractions, with only a few iguanid lizards reaching Tierra del Fuego.

The South American reptile fauna is divisible into two main parts (Table 21, p. 563). The older part has probably been in South America at least since the beginning of the Tertiary, when the continent became isolated. The more recent part has probably come since the late Pliocene, when South America became connected with North America. But land tortoises apparently reached South America between times, across the Tertiary water barrier (Simpson 1943), and some other reptiles may have done so too. That is the sort of thing reptiles do.

Most of the more recent members of the South American reptile fauna, including *Chelydra, Kinosternon,* the emydines, coral snakes, pit vipers, and some non-endemic genera in other families, have obviously reached South America from or through North America. Most of them still occur in North America, but a few have already disappeared or are disappearing there. One example is *Geoemyda,* now in tropical Asia etc. and in Mexico and Central and South America but apparently fossil in Europe and temperate North America (see list of families). Another example is *Testudo,* now in the warm parts of the Old World and South America north to Panama (and on islands); but *Testudo* was in North America from the Eocene to the Pleistocene. *Sibynophis* in Madagascar, the Orient, and tropical America and *Trimeresurus* (including *Bothrops*) in southeastern Asia and tropical and subtropical America probably passed through North America too. *Mabuya,* the only genus of skinks in South America, is in the Old World as well as the New World tropics. It too may have dispersed through the north, or it may have reached America across the Atlantic; skinks have great powers of crossing water. The occurrence of *Mabuya* over the West Indies and the existence of an endemic species of it on Hispaniola suggest that the West Indies may have been the point of arrival of the genus in America, and that it may have come from Africa on drift. Certain genera of geckos, which are even better water crossers than skinks, may be common to the warmer parts of America and the Old World, and they too may have crossed the Atlantic (see especially *Tarentola* in list of families).

The older groups of South American reptiles, marked by generic endemism, have diverse geographical relationships. Aniliid snakes and also the colubrid subfamilies Xenoderminae and Pareinae plus

Dipsadinae link South (and in some cases Central) America with the Orient. Pelomedusid turtles are now shared with Africa and Madagascar (*Podocnemis,* only with Madagascar) but are fossil elsewhere; and amphisbaenid lizards and leptotyphlopid worm snakes are now confined to the warmer parts of America and to Africa and adjacent areas, but certain amphisbaenids are fossil farther north. Chelyid turtles are shared only with the Australian Region. Of course some of the older groups of South American reptiles, for example iguanids, are still shared with North America too. Also, there has been some recent redispersal of genera of some old families: *e.g.*, one South American genus of teiids has extended far into North America, and South and North America may have exchanged a few colubrids recently. It will be seen that, in terms of existing distribution, different groups of South American reptiles show relationships toward almost every other part of the world, and this situation can have come about only by complex dispersals followed by complex extinctions. Extinctions in North America have already begun to isolate some more recent reptiles in South America (preceding paragraph), and some of the older groups have evidently been isolated by broader extinctions occurring complexly, on various geographical patterns, over longer periods of time. Complexity of dispersals and extinctions is suggested especially by the present distributions of iguanids and boids (see list of families).

Reptiles of Central America

Central America has the only surviving dermatemydid (one species) and staurotypine (two genera) turtles, which belong to groups known fossil in the north, and *Chelydra* (snapping turtles), *Kinosternon* (musk turtles), and *Pseudemys* and *Geoemyda* (emydine turtles), all of which extend to parts of South America but evidently came from the north. Central America lacks *Testudo* (except in Panama), lacks soft-shelled turtles (*Trionyx*), which have been in North America since the Upper Cretaceous, and lacks also side-necked turtles (pelomedusids and chelyids), which have been in South America since the Cretaceous. Of lizards, Central America has many. Some Central American iguanids and geckos seem to have come from South America, but others have probably come from the north (some of these are mentioned below), and at least one genus, the iguanid *Anolis,* has radiated in Central America and may have dispersed from there. The small family Xantusiidae is almost confined to southwestern North America and Central America. Central American teiids seem to have come recently from South America. Central American scincids include *Mabuya,* which may have come across the Atlantic, and two other genera, received from the north. A few anguids occur in Central as well as North and South America. One genus of xenosaurids occurs from Guatemala to southern Mexico; a probable relative is in China; and doubtfully assigned fossils are in North America, etc. in the early Tertiary. Amphisbaenids seem to be absent in Central America (except Panama), although different genera occur in South America, southern North America, and the West Indies. Of snakes, Central America has representatives of both families of worm snakes; a localized aniliid (*Loxocemus*); boids, at least some of which have probably been derived from South America; many colubrids, some of which have probably reached Central America from South and others from North America; coral snakes; and pit vipers, derived from the north. Of crocodilians, Central America has *Crocodylus,* probably derived through the north, and caimans, probably derived (the recent ones) from South America. Further details of the Central American distributions of some of these groups are given below as well as in the list of families.

For purposes of analysis of the reptile fauna, Mexico and the adjacent southwestern corner of the United States should be added to Central America, for they form part of the same transition-relict area. *Anelytropsis,* anniellids, helodermatids (Gila monsters), and an isolated genus of amphisbaenids (*Bipes*) are confined to this additional area, and the testudinine *Gopherus* enters it.

This Central American-Mexican reptile fauna is notable for three things: it is diverse; it is transitional; it includes many relicts.

Central America (without Panama) and Mexico have (by present classification) representatives of 21 families plus 8 additional subfamilies of reptiles (not counting the marginal Trionychidae and Erycinae), against only 18 families plus 6 subfamilies (not counting the marginal Chelydrinae) in South America. It is particularly in lizards that the Central American-Mexican fauna exceeds that of South America.

That the Central American-Mexican reptile fauna is transitional, with many North American and many South American groups overlapping for varying distances, in some cases widely, is clear from the general composition of the fauna listed above. The transition is too complex to describe in full detail here. However, certain groups that have spread from the north through all or most of Central America but have not entered South America, or have done so very little, are worth itemizing. *Chelydra* (snapping turtles) extends from eastern North America through Central America to just the northwestern corner of South America. Iguanid lizards of the large genus *Sceloporus* extend from southern British Columbia and southern New York to Panama. Gekkonid lizards of the genus *Coleonyx,* which probably represents *Eublepharis* of Asia, occur from the southern edge of the United States to Panama. Of skinks, *Eumeces,* which occurs also in parts of the Old World, extends from the southern edge of Canada to Nicaragua; and the great, Old World genus *Lygosoma* is represented in America by one group of a few species, which extends from eastern United States to Panama. The anguid *Gerrhonotus* occurs from southern British Columbia etc. to Panama. Of colubrid snakes, *Thamnophis* (garter snakes) ranges from the northern limit of reptiles in North America to Costa Rica; the widely distributed genus *Elaphe* reaches Costa Rica; and *Lampropeltis* extends from most of the United States through Central America to the northwestern corner of South America. Finally, *Agkistrodon,* a genus of pit vipers which occurs also in Asia, extends from parts of Iowa, Massachusetts, etc., to Nicaragua. These are not temperate-zone reptiles which enter the tropics on mountain tops. Most of them occur in the tropical lowlands in Central America.

The obvious or probable relicts in the Central American-Mexican reptile fauna are the dermatemydid and staurotypine turtles and *Gopherus* (which does not represent the same stock as *Testudo*); xantusiids, *Anelytropsis,* anniellids, *Xenosaurus, Heloderma,* and *Bipes;* and *Loxocemus.* Existing relatives or fossils (see list of families) show that some of these relicts represent groups that reached the Central American-Mexican area from the north; some are of unknown history; none is clearly a relict of a primarily South American group.

The transition and the occurrence of relicts in the Central American-Mexican reptile fauna are partly correlated. Many different groups of reptiles extend into the transition area or have moved through it for varying distances in both directions. Some (not all) of the reptiles that have come northward from South America stop at or near the edge of the tropics in southern Mexico. This is a common pattern, of tropical groups which reach but do not enter the north-temperate zone. But of the many reptiles that have come southward from or through North America, although many have crossed the line of the tropics and spread through much or all of Mexico and Central America, comparatively few have entered South America; and some of the relicts seem to represent additional, older groups that reached Mexico and Central America from the north but did not enter South America. This is not a common pattern among other vertebrates. It has a special significance, which will be discussed in Chapter 9.

Transitions and barriers in the distribution of reptiles

Where different faunas of reptiles meet there are not sharp boundaries but broad areas of transition—overlapping with progressive subtractions.

In eastern Asia, where there is complex transition between tropical and north-temperate fishes and amphibians, reptiles simply decrease northward. Many tropical Asiatic reptiles stop short of the north-temperate zone; others extend northward for varying distances; but there is no primarily northern fauna of reptiles to make the opposite half of a transition pattern. For the same reason, because reptiles are few in the north and absent in much of the far north, the temperate Eurasian and temperate North American reptile faunas are now separated by an empty gap rather than an area of transition. However, Eurasia and North America exchanged a number of reptiles (various turtles, lizards, snakes, and probably crocodilians) during the Tertiary.

North and South American reptile faunas overlap, with progressive subtractions, in Central America and Mexico, and the farthest elements of the overlap go much farther: *e.g.*, primarily North American emydines reach part of Argentina, and primarily South American teiids reach the edge of British Columbia.

Oriental and Australian reptiles overlap in the eastern part of the Indo-Australian Archipelago. The great extensions into and across the transition area are all made by Oriental reptiles, some of which range (with progressive subtractions) across all the islands and into Australia, while primarily Australian reptiles range only to the Moluccas.

Further details are unnecessary. In general, cold is apparently an effective barrier for reptiles. Even moderate-temperate climate is a partial barrier for tropical reptiles. But no part of the world has been cold enough long enough to stop the dispersal of all reptiles permanently. Other kinds of barriers affect different reptiles differently. Dry country is a barrier for water-loving reptiles, wet country for desert ones. Moderate gaps of salt water stop some reptiles but not others. Barriers of all these sorts retard the dispersal of reptiles and stop some of them. But nothing has stopped the dispersal of appropriate dominant groups, such as the Colubridae, over the world.

Dominance and competition in relation to distribution of reptiles

The most dominant existing group of reptiles is the family Colubridae, of common snakes. This single family contains about two-thirds of all existing snakes and a still higher proportion of those in ordinary habitats, excluding the burrowing worm snakes and the sea snakes. Colubrids are very numerous in genera and species in all the principal regions of the world except Australia, and individuals are often numerous too. Colubrids are only moderately diverse in basic structure, but they are diverse ecologically, with many terrestrial, arboreal, burrowing, and aquatic forms. They are dominant in both main climatic zones (tropical and north-temperate) and in a variety of habitats; even in the north-temperate zone they are numerous in woods and wet places as well as in open and arid country; and some of them come near the northern limit of reptiles in Eurasia and set it in North America. The distribution of Colubridae is therefore nearly continuous over a large part of the world, with less regard for climates and barriers than any other reptiles show. All these things together make the colubrids an outstandingly dominant group, comparable in a general way to the family Cyprinidae among fresh-water fishes or the genus *Rana* among frogs.

No other reptiles compare with colubrids in dominance over the world as a whole, although others are dominant in limited areas. In certain areas, for example, emydine turtles are dominant, among turtles. They are numerous in genera and species and also in individuals primarily in part of tropical Asia and secondarily in temperate eastern North America, but their occurrence is irregular even within their limits of distribution. They are diverse ecologically: different ones are aquatic, amphibious, and terrestrial.

The rise and spread of the dominant colubrids presumably had a profound effect, through competition, on other snakes and perhaps still more on lizards: the localization and discontinuous distribution of so many lizard families may well be partly a result of competition with snakes and especially with colubrids. However, there is little actual evidence of this. We do not know how snakes and lizards were distributed before the appearance of colubrids, and we do not know where the colubrids themselves first appeared or how they spread; the fossil record is not much help. One significant detail is that colubrids and elapids are complementary in their areas of dominance (not in their total ranges), the colubrids in most of the world, the elapids in Australia. If colubrids have not replaced most elapids, they have probably at least limited their radiation outside of Australia. Within the Colubridae, the apparent tendency for derivative groups to occur in the main part of the world and for more primitive ones to be concentrated in South America and perhaps Madagascar suggests complex competition and replacement in the main part of the world. Clearer signs of replacement in most of the Old World are shown by the complementary distributions of iguanid and agamid lizards, the

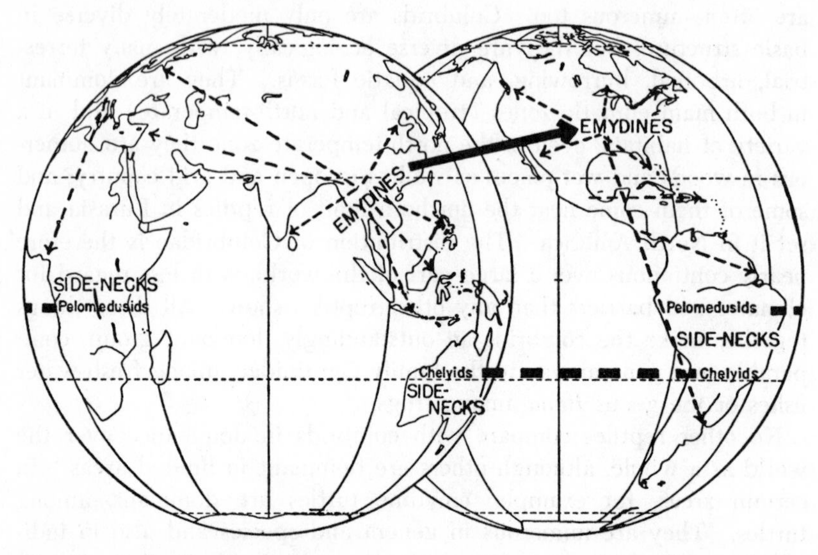

Fig. 29. Radiation of emydine turtles (diagrammatic) and distribution of side-necks.

former in America and relict on Madagascar and Fiji, the latter on the continents of the Old World. And still clearer, because shown by fossils as well as by existing distributions, is the replacement of formerly cosmopolitan side-necked turtles by the dominant emydines in the Oriental Region and the north-temperate zone (Fig. 29).

In the absence of direct evidence, the best evidence of competition between different groups of animals comes from the complementary distributions of dominant, ecologically equivalent groups. There is some evidence of this sort among reptiles. If the reptiles show less complementarity than fresh-water fishes and amphibians do, this may mean not that there is less competition among reptiles, but that the reptiles disperse more rapidly and cross barriers more easily, so that complementary patterns produced by competition and replacement are soon lost.

Summary: the pattern of distribution of reptiles

The main pattern of distribution of reptiles contains the same five subpatterns found among fresh-water fishes and amphibians, but some of the subpatterns are modified or weak. First is limitation: reptiles are more limited than fresh-water fishes or amphibians toward the cold north, but less limited on islands. The subpattern of zonation is modified: reptiles have a zonal pattern of distribution only in that they are primarily tropical; fringes of the tropical fauna extend northward, but there is no separate, zonal fauna of reptiles in the north-temperate zone. The subpattern of radiation is less conspicuous or more complex than among fresh-water fishes or amphibians. If there is a main center of geographical radiation of reptiles, it is probably the main part of the Old World tropics, which is the apparent main center of distribution of testudinid turtles (emydines plus testudines), trionychid turtles, scincid lizards, the most dominant colubrids, and viperid snakes. The fourth subpattern, of differentiation of different faunas on different continents, is very complex among reptiles. There is no single, main pattern of differentiation of reptile faunas but rather a number of separate patterns made by different groups, each covering much or all of the world. Turtles make a complex, world-wide pattern of their own in which (setting aside the complexities) the main differentiation is between an emydine fauna in Eurasia and North America and a side-neck fauna on the southern continents. Among agamid-iguanid lizards and also boid snakes the main differentiation is between the Old and New Worlds. Among colubrid-elapid snakes, the main differentiation is between Australia and the rest of the world. Other, smaller groups of reptiles, with their own patterns, multiply the complexity. The fifth subpattern, of concentration in large and favorable areas and subtraction marginally, is weaker among reptiles than among fresh-water fishes and amphibians. Reptiles are primarily land animals, and they occur in numbers almost everywhere on land except where it is too cold for them. However, reptiles are concentrated in warm places, and special groups are further concentrated locally. For example, in cool climates lizards tend to be concentrated in open, sunny places. Turtles tend to be concentrated in large areas which are favorable for them: *e.g.*, emydines in eastern tropical Asia and eastern North America. However, most reptiles are not so concentrated. Where different faunas of them come together, there is usually complex transition of many groups in a wide area.

The place of relicts in the main pattern of distribution of reptiles is this. Turtles as a group are relict. They have existed with little visible change since the Triassic. They are most diverse in the tropics, but some occur far into the north-temperate zone. Crocodilians too are relict as a group; they date from the Triassic and are the only surviving archosaurians. They are now all in or near the tropics. *Sphenodon,* the only surviving rhynchocephalian, isolated on New Zealand, is a phylogenetic and geographical relict, but it stands alone. No other ancient reptiles survive in geographical isolation. Australia is not now a refuge for any especially archaic reptiles, although *Meiolania,* the last of an ancient lineage of turtles (see list of families), survived there until the Pleistocene.

History of dispersal of reptiles

The present pattern of distribution of reptiles has evolved in a very complex way, even more complex than among fresh-water fishes and amphibians, by evolution, radiation and dispersal, adaptation, competition, limitation, retreat, etc., of many different groups. The actual pattern formed has been determined by the nature of reptiles (especially by the facts that they are primarily land animals, that they prefer warm places but can enter cool ones temporarily, and that some of them disperse rapidly and cross barriers easily) and by zonation of climate and by other controlling factors of the world. The movements of the main groups are worth tracing, so far as they can be traced. Here, only existing groups and related fossil ones will be considered, and only their major movements during the Tertiary, or not long before. The Mesozoic dispersal of dinosaurs is treated briefly in Chapter 10. The histories here outlined are partly hypothetical. See elsewhere in the text and the list of families for details and clues on which the histories are based.

Of turtles during the Tertiary, common turtles (emydines) have radiated in tropical Asia and spread from there through the north-temperate zone and to South America etc. Land tortoises (*Testudo* etc.), presumably derived from emydines in Asia, have spread over most of the warm parts of the world except Australia. *Testudo* apparently reached South America in the Miocene, emydines perhaps in the Pliocene. Soft-shelled turtles (*Trionyx*) apparently spread from the Old World tropics to eastern Asia and North America, but this was only a late movement in a complex, pre-Tertiary dispersal of trionychids. And musk and snapping turtles (chelydrids) have extended from North to South America recently. On the other hand, side-necked turtles (at least pelomedusids) have withdrawn from Eurasia and North America; dermatemydids have withdrawn from north-temperate regions and have been reduced to one species in Mexico and Guatemala; and chelydrids have withdrawn from temperate Eurasia; and the last meiolaniids, persisting in Australia etc. until the Pleistocene, have disappeared.

Of lizards, there seem to have been few recent exchanges between the Old and New Worlds. Skinks have apparently extended from the Old World to America: *Eumeces* and *Lygosoma* through the north to North and Central America, and *Mabuya* perhaps across the Atlantic from Africa to tropical America. The gekkonid genus *Tarentola* may have crossed the Atlantic to the West Indies. Only a few other genera of lizards are common to parts of the Old and New Worlds, and their histories are doubtful. Agamids have radiated on the continents of the Old World and have apparently replaced iguanids there, but have not reached America. Other movements of lizards within the Old World and also between North and South America have occurred but cannot be considered here. As for withdrawals, iguanids have withdrawn from the Old World continents; some anguids and amphisbaenids have withdrawn from some northern areas; and varanids withdrew from North America early in the Tertiary. This is about all of importance that can be said of the geographical movements of special families, but the present distribution of lizards in general, with many discontinuities and localizations, suggests that they have retreated more than advanced in recent times. Their retreat may be correlated with the rise of colubrid snakes.

Of snakes, the older dispersals (of worm snakes, aniliids, boids, etc.) are unknown. (The supposed replacement of boas by pythons on Old World continents was, apparently, a fiction suggested by faulty classification.) Common snakes (colubrids) have radiated complexly over most of the world. Their place of origin is unknown, but their most recent radiation seems to have been from the main part of the Old World tropics, from which many different genera have spread northward and to America, and a few have spread to Australia. The direction of movement of some of the genera is not clear, but *Natrix*, *Elaphe*, *Coluber*, and *Opheodrys* seem to be among those that have spread from the Old World to North America. Elapids arose from early colubrids probably somewhere in the Old World. Somehow they reached Australia before colubrids did, and one reached America and became our coral snakes. Viperids too were probably derived from colubrids in the main part of the Old World; pit vipers probably arose (from true vipers) in southern Eurasia, and two or three stocks reached America, one producing the rattlesnakes. As to withdrawals of snakes, little can be said in detail. Some groups have evidently withdrawn from northern areas, and others (aniliids, erycine boids, some subfamilies of colubrids) have disjunct ranges which suggest more extensive withdrawals.

True crocodiles (*Crocodylus*) may have dispersed early in the Tertiary, probably from the Old World to America, but this was only the last part of a complex dispersal of crocodilians. *Alligator* and both existing genera of gavials have evidently retreated.

The history of aquatic snakes is worth separate consideration. Various snakes in many parts of the world are more or less aquatic, but three groups of them are more highly adapted than most others to life in water, with valvular nostrils (M. A. Smith 1943, pp. 17–19) and sometimes flattened tails. The groups are the Acrochordinae and Homalopsinae, separately derived from colubrids, and the Hydrophiidae or sea snakes, derived from elapids. Details of the distributions of these groups are given in the list of families. All three of them, the three principal existing groups of fully aquatic snakes, seem to have arisen in the same place, tropical Asia and the Indo-Australian area; probably (judging from their ancestry—there are no pertinent fossils) they arose within the Tertiary or not long before it.

A number of dominant reptiles seem to have moved from the Old World to America. They include several emydines, *Testudo,* and *Trionyx;* three genera of skinks; several genera of colubrids, an elapid, and two or three pit vipers; and *Crocodylus.* On the other hand, it is not certain that any reptiles have moved from America to the Old World, although a few may have done so. There are also several groups of reptiles that seem to have withdrawn from parts of the Old World and survived in America: dermatemydid and chelydrid turtles, iguanid lizards, and perhaps primitive colubrids ("Xenodontinae"); but varanids have withdrawn in the other direction, from North America. Withdrawal is not the same as spreading but may be correlated with it: withdrawal of some groups may be caused by spreading of others and may proceed in the same direction. All these cases together do not amount to proof, but they do suggest a tendency for reptiles to move from the Old World to America more than the reverse. Moreover, reptiles are primarily tropical animals which often invade the north-temperate zone but usually do not do well there. And reptiles seem to have spread from the main area of the Old World toward Australia (see p. 193). This leaves the main part of the Old World tropics, Africa plus tropical Asia, as the apparent main center of geographical radiation of reptiles, so far as there has been a main center. There are at least indications that successive groups of dominant reptiles have risen and spread from there, as successive dominant groups of fresh-water fishes and frogs seem to have done.

Existing reptiles probably do not tell much about ancient lands and climates, and what they do tell is best carried over to Chapter 10. However, a few special cases are collected here for reference.

Pelomedusid turtles and amphisbaenid lizards are completely confined, or nearly so, to Africa and warm parts of America, but fossils suggest that they dispersed through the north. Leptotyphlopid worm snakes, also mostly African and American, may have done so too. The lizard genera *Mabuya* and *Tarentola* may have reached South America or the West Indies directly from Africa, but not necessarily by land; they may have crossed the Atlantic on drift; they belong to families that are notorious for water crossing.

Chelyid turtles occur only in South America and the Australian Region and are unknown elsewhere even as fossils. In South America they are almost entirely confined to the tropics and do not reach the cool southern extension of the continent. In the Australian Region they occur in both tropical and temperate areas, but not on Tasmania, and not on New Zealand. They are aquatic, in fresh water, but their distribution is not matched by fresh-water fishes or any other aquatic vertebrates. They have probably dispersed in some unusual way, perhaps through the sea from South America to Australia.

The distribution of chelyids is paralleled by extinct meiolaniid turtles. These turtles occurred in southern South America about the beginning of the Tertiary, and on eastern Australia, Lord Howe Island, and Walpole Island southeast of New Caledonia in the Pleistocene, but their record is blank for the intervening 50 million years or so, and they are unknown elsewhere at any time. They were giant land turtles, but several facts suggest that they could disperse across the sea. First, in South America they occurred with early placental mammals, which did not reach Australia, as if the turtles crossed a barrier which the mammals could not cross. Second, Lord Howe and Walpole Islands (and also New Caledonia) are biologically oceanic islands, which lack strictly fresh-water fishes, amphibians, and terrestrial mammals. That the turtles occurred there in the Pleistocene is almost proof that they crossed hundreds of miles of ocean. Finally, existing land tortoises, the ecological counterparts of meiolaniids, have crossed ocean gaps hundreds of miles wide. Matthew (1915, p. 283) long ago suggested that meiolaniids crossed the ocean barriers between South America and Australia, and this is probably still the best guess to account for their distribution.

The common factor of all these histories is that they account for the distributions of different reptiles without resorting to special land bridges or continental drift.

In conclusion, I see that I have been able to do less with reptiles than with any other vertebrates. The distributions of fresh-water fishes and amphibians, though complex enough in detail, form comparatively simple, very significant main patterns. The distribution of birds is very complex, but birds are the best-known animals and their distribution has special points of interest. The distribution of mammals, backed by their fossil record, is most significant of all. But existing reptiles have a complex distribution which is particularly difficult to reduce to main patterns, movements, and significances. The turtles are an exception to this rule.

LIST OF FAMILIES OF REPTILES

This list includes all existing groups of reptiles but only a few fossil ones, chiefly those that are directly connected with existing groups and those that seem likely to be geographically significant. The classification is abridged accordingly, but otherwise follows Romer (1956). A dagger (†) before a name indicates that the group is extinct.

(*Subclass Anapsida*)

Order Chelonia (*Testudinata*), turtles. Romer's (and therefore my) classification of turtles follows Williams (1950, p. 554), with slight modification. Carr (1952) gives good distribution maps of North American turtles.

(†*Suborder Amphichelydia*): an extinct order of 10 families including the following one.

†Meiolaniidae: extinct, gigantic, horned land tortoises known only in southern South America (2 genera, late Cretaceous and Eocene) and on eastern Australia, Lord Howe Island, and Walpole Island southeast of New Caledonia (another genus, Pleistocene). The dispersal of this group is discussed in the text (p. 205). Some pertinent references to it are Woodward (1901), Matthew (1915, p. 283; 1939, pp. 113–114), Anderson (1925), and Simpson (1938; 1943, pp. 422–423). The meiolaniids were the last survivors of a suborder which appeared in the Triassic and was dominant and probably cosmopolitan in the

Mesozoic (Romer), but which, except for the Pleistocene fossils in Australia etc., is unknown later than the Eocene.

(*Suborder Cryptodira*)

(*Superfamily Testudinoidea*)

Dermatemydidae: southern Mexico and northern Central America (rivers on the Atlantic side, from central Veracruz to Guatemala—H. M. Smith and Taylor 1950, p. 19); one species. Fossil genera provisionally placed in this family are in North America and eastern Asia in the later Cretaceous and early Tertiary.

Chelydridae, snapping and musk turtles etc.: North, Central, and part of South America; the often repeated record of a supposed endemic genus of snapping turtles in New Guinea was an error (Loveridge and Shreve 1947); 3 subfamilies, all of them aquatic, in fresh water. (1) Chelydrinae, snapping turtles: North and Central America etc.; 2 genera: *Chelydra* (one or 2 species) in United States east of the Rockies north to the southern edge of Canada, and in parts of Mexico, Central America, and the northwestern corner of South America (Ecuador); and *Macroclemys* (one species) in southeastern United States from northern Florida to central Texas, north in the Mississippi Valley to central Illinois etc. (2) Staurotypinae: southern Mexico and northern Central America (Veracruz to Guatemala and British Honduras—Smith and Taylor 1950, pp. 26–27); 2 genera, 3 species. (3) Subfamily Kinosterninae, musk turtles etc.: eastern North America north to southern Ontario, southwest to the southern interior corner of California, and south in Mexico, Central America, and about the northern half of South America, south to the Chaco (Williams); one or 2 genera, 12 or more species.

Fossil chelydrines are in the mid-Tertiary of North America and (doubtfully) Europe; staurotypines, in the Oligocene of North Americe (Williams 1952*a*); and kinosternines, in the Pliocene of North America. This fossil record suggests a North American radiation of the family, but the record is not complete. Williams thinks the family has existed "perhaps since the Cretaceous, certainly since the earliest Tertiary."

Testudinidae, common turtles, land tortoises, etc.: land tortoises (Testudininae) occur in all the main, warm areas of the world except the Australian Region; the other, primarily aquatic, subfamilies are restricted as stated below. There are 3 subfamilies.

(1) Platysterninae: southern Burma, Siam, French Indochina, extreme southern China, and Hainan; one species, in mountain streams.

(2) Emydinae, common turtles (Fig. 29): tropical and temperate Eurasia etc. and the Americas: most in *tropical* eastern Asia and *temperate* eastern North America; a few in other parts of the north-temperate zone; one reaching the tropics in northwestern Africa; and a few in Central and South America and the West Indies. In eastern Asia etc. (M. A. Smith 1931, Pope 1935, Bourret 1941, Rooij 1915) there are some 17 or more genera, 38 species, most in an area from northern India and extreme southern China to Sumatra, Java, and Borneo. Only 6 of the genera reach northern India, and only one species

(*Geoemyda trijuga*) reaches southern India and Ceylon, and its population there is isolated (M. A. Smith 1931, p. 97, map). In China north of the southern provinces of Kwangsi, Kwantung, and Fukien, only 4 species are definitely known (Pope 1935). The 2 that go farthest north are *Clemmys mutica* and *Geoclemys* (or *Chinemys*) *reevesii,* which reach to Kiangsu and to southern Shantung and Hopeh respectively—but the natural limits of *reevesii* are uncertain, for it is carried and sold in markets. In Japan occur *G. reevesii* and an endemic *Clemmys,* which occurs north at least to Tokyo; 2 additional emydids occur in the Riu Kiu Islands and several in Formosa. In the Philippines (Taylor 1920) there is a supposedly endemic species in southern Leyte, but otherwise there are only 2 widely distributed Oriental species, of which one, *Cyclemys* (or *Cuora*) *amboinensis,* is a common household pet; other emydines have been recorded from the Philippines but are doubtful. On Celebes and islands farther east is only the same *C. amboinensis,* widely but erratically distributed (Mertens 1930, p. 335) and perhaps carried by man. In southwestern Asia, parts of Europe, and the northwestern coast of Africa there are 2 genera, 3 species; one (*Clemmys leprosa*) apparently extends to western tropical Africa, for it is recorded from the Gambia River by Gray (1852—as *Emys laticeps*), from Porto Novo, Dahomey, by Chabanaud (1917), and from Air in the south-central Sahara by Angel and Lhote (1938, p. 376). In America north of Mexico east of the Rockies and north into the southern edge of Canada there are some 8 genera, 18 species, and numerous subspecies (Carr); additional species of some of the genera are in Mexico. A single eastern form, *Chrysemys picta bellii,* extends northwestward across the Rockies to southern British Columbia etc., to the Cascade Mountains and perhaps to the Pacific coast—Evenden (1948) says it is fairly common in the lower Willamette Valley of Oregon—but its natural western limits are perhaps doubtful; and one species, *Clemmys marmorata,* is endemic to the Pacific coast from the southern corner of British Columbia to northern Lower California, but its northern limit is doubtful (Evenden). Southward, 2 eastern North American genera extend into the tropics: terrestrial *Terrapene* (box turtles) to southern Mexico, to Nayarit and northern Yucatan (H. M. Smith and Taylor); aquatic *Pseudemys* (which includes *Trachemys,* referred to by Simpson 1943, p. 419) through Central America, part of the West Indies, and much of South America, to northern Argentina. The range in South America is apparently discontinuous. Williams (1956) says there are no records of the genus between Venezuela and southern Brazil. The only other emydines in the American tropics are several species of *Geoemyda* in South and Central America north to parts of Mexico (to southern Sonora and central Veracruz) but not farther north; this genus occurs also in tropical Asia, north to the southern edge of China and the Riu Kiu Islands. *Clemmys* and *Geoemyda* are the only genera of emydines now considered common to the Old and New Worlds. European and eastern North American species formerly put together in *Emys* are apparently not really related (Williams). Emydines are mostly aquatic but some are partly or completely terrestrial. The genus *Geoemyda* includes both aquatic and terrestrial species and is mainly vegetarian (M. A. Smith 1931, pp. 89–90); other emydines eat a variety of food. According to Williams (1950, p. 554) the earliest fossil surely assignable to this subfamily is in the Paleocene of North America, but emydines are fossil in Eurasia back to the Eocene, and the distribution of living forms suggests that the primary radiation of the subfamily was in southeastern Asia. Fossils probably representing *Geoemyda,* which is now divided between tropical Asia and tropical America, are known from the Eocene of Europe and Oligocene of North America (Williams).

(3) Testudininae, land tortoises: warm parts of all continents except Australia; many islands. Williams (1952) recognizes 6 genera. Four of them, with few species, are confined to tropical or South Africa or Madagascar. Another, *Gopherus,* with 3 existing species, is discontinuously distributed across southern North America, from South Carolina and Florida along the coast to Texas and northeastern Mexico, and in a limited area in southwestern United States and northwestern Mexico; this genus is now considered to represent a group (distinct from *Testudo*) which has been in North America at least since the Oligocene and which was in Eurasia too in part of the Tertiary (Williams). The sixth genus is *Testudo,* with some 50 Recent species, distributed through Africa, Madagascar, etc., the warm-temperate and tropical parts of Eurasia etc., and South America and part of Panama, but not now existing in North America or Central America above Panama. In Africa, especially South Africa, there are many species of *Testudo.* Madagascar has a few. Within historic times gigantic forms inhabited more than 30 islands in the Indian Ocean, including the Aldabras, Réunion, Mauritius, Rodriguez, and the Seychelles; but of these giants (no one knows just how many forms there were) only the one native to South Aldabra Island still exists. In Mediterranean Europe, the Balkans, and southwestern Asia north to Mariupol (Berg 1950, p. 108) and the Caspian and Aral Seas there are 6 species. In tropical Asia are 7 species, one of which reaches Ceylon; one, extreme southern China; and one, Sumatra, Borneo, and perhaps Java; and an eighth species is endemic to Celebes and Halmahera. In South America there are 2 species: one widely distributed in the tropics east of the Andes north to Panama and one in the warmer part of south-temperate South America. Finally, the Galapagos Islands possess a number of gigantic forms (some now extinct or rare) considered by Mertens (1934) to be about 13 subspecies of one species. Fossil testudinines are not known definitely until the Eocene (Williams). *Testudo* has been widely distributed since the Eocene, although apparently it never reached Australia and perhaps did not reach South America until the Miocene (Simpson 1943, p. 420). Although *Testudo* is now absent in North America, it was present there from the Eocene to the Pleistocene; and although it is now absent in Central America (except Panama) and the West Indies, species were on Cuba, Mona, and Sombrero Islands (Williams 1950*a*, map on p. 6;

1952, p. 552), and presumably other islands in the Pleistocene. Testudinines, though now most diverse in Africa and Madagascar, probably originated from an emydine in tropical Asia. Williams' (1952) arrangement of the fossil and Recent species of *Testudo,* in 14 subgenera, suggests that the dispersal of the genus has been complex.

(*Superfamily Chelonioidea*)

Cheloniidae, sea turtles: warm seas; 4 existing genera, few species.

(*Superfamily Dermochelyoidea*)

Dermochelyidae, leatherback turtles: warm seas; one existing genus, one or 2 species.

(*Superfamily Carettochelyoidea*)

Carettochelyidae: confined to New Guinea; one species, thoroughly aquatic, "in rivers, down to the sea" (Rooij). *Carettochelys* has been in New Guinea since the Miocene; other, rather distantly related genera were in Eurasia and North America in the early Tertiary.

(*Superfamily Trionychoidea*)

Trionychidae, soft-shelled turtles: tropical Africa, southern and eastern Asia etc., and eastern North America; 7 genera. Only *Trionyx* (often called *Amyda*) is widely distributed. In tropical Africa there are one species of *Trionyx* and 2 endemic genera with about 5 species; the *Trionyx* descends the Nile to lower Egypt and occurs accidentally on the Palestine coast (Flower 1933, pp. 753–754). Another *Trionyx* is in the Euphrates-Tigris system of southwestern Asia. In southeastern Asia there are some 5 genera (including *Trionyx*) and 12 species, most in an area from northern India and extreme southern China to Sumatra, Java, and Borneo; one species reaches Ceylon; *Pelochelys bibroni* (*cantorii*), which is "often caught on the sea coast" (Rooij), reaches Luzon in the Philippines and has a population isolated on New Guinea, but other records of trionychids beyond Java and Borneo are dubious; one *Trionyx* follows the lowlands of eastern Asia north to the Amur or at least to Lake Khanka and the Ussuri (Berg 1950, p. 66); and one, perhaps the same species, is in Japan. In America there is only *Trionyx,* 2 full species and several subspecies, widely distributed in east-central and southern United States and reaching southern Ontario and northeastern Mexico, and introduced westward in the Gila-Colorado system. Trionychids are thoroughly aquatic turtles. Fossils show that they have had a rather complex history on the main continents of the Old World and North America since the Cretaceous. They have apparently never reached South America. Certain fossil fragments in the Pleistocene of northeastern Australia (De Vis 1894) may or may not be trionychids (Williams); if they are, they may be *Pelochelys bibroni* (above), which enters the sea and which still survives in New Guinea, and which probably reached the Australian Region rather recently.

(*Suborder Pleurodira*), side-necked turtles

Pelomedusidae: tropical and southern Africa, Madagascar, etc., and South America; 3 genera, all more or less aquatic. The African forms have been revised by Loveridge (1941). The single species of *Pelomedusa* occurs in Africa and Madagascar. *Pelusios* has 5 species in Africa, and one of them extends to the Cape Verde Islands and to Madagascar, Mauritius, and the Seychelles. Apparently neither this species nor the *Pelomedusa* is differentiated on Madagascar or the other islands (Loveridge); perhaps they have been dispersed in part by man. Finally, *Podocnemis* has one species on Madagascar and 7 in South America (Williams 1954) in the Amazon, Orinoco, Magdalena, etc., but apparently none in the Paraguay-Paraná system, and none in Central America. *Podocnemis* was in Africa as late as the Pleistocene and on other continents earlier, but some of the fossils assigned to it probably do not belong there (Zangerl 1947, Williams). However, pelomedusids of some sort were on all the main continents except Australia in the Cretaceous or early Tertiary, and they persisted in Europe and Asia until the Miocene and Pliocene. There is also an old record for New Zealand in the Upper Cretaceous (Lydekker 1889), but it is based on a doubtful fragment (Williams); on geographical grounds it should be a chelyid rather than a pelomedusid; but on the other hand fossil *Podocnemis* have been found in marine deposits, which implies that they entered the sea.

Chelyidae: the Australian Region and South America. In Australia (but not Tasmania) and New Guinea etc. there are 4 genera and some 14 species. In South America are 6 other genera, some 13 species, chiefly in the tropics (south to northern Argentina) but not in the Magdalena or western drainages (Simpson 1943, p. 422) and not in Central America. The whole family is primarily aquatic. Fossils from Eurasia etc. formerly assigned to this family are at best doubtful (Williams 1953, 1954*a*); there is no good evidence that chelyids ever occurred outside South America and the Australian Region. The supposed *Hydraspis* in the Eocene of India is apparently really a pelomedusid (Williams 1953).

(*Subclass Lepidosauria*)

Order Rhynchocephalia

Sphenodontidae: New Zealand; one existing species, *Sphenodon*, the Tuatara. Formerly it lived in numbers on both main islands of New Zealand, but now it occurs chiefly on a few islets in the bays on the northeast side of North Island and in Cook Strait. It is a lizard-like reptile up to 2 feet or more long. Formerly it frequented sandy river banks as well as (as now) rocky and sandy islets in the sea. It is more or less amphibious, likes to lie in water, and swims well. It is active mainly at night, even on cool or cold nights, and it has a body temperature far lower than that of modern reptiles and a low metabolic rate (Bogert). Adults can live for months without food or fresh water. The eggs are laid in sand and develop very slowly, taking about a year to hatch. [Most of this information is from Hutton and Drummond (1923), Newman (1877), Gadow (1909, pp. 298–300), Conrad (1940), and Bogert (1953).] Fossils show that rhynchocephalians were on most continents in the early and middle Mesozoic, but they are unknown fossil later than the Lower Cretaceous, perhaps 135,000,000 years ago. That one should still exist anywhere is astounding; but that such an animal as *Sphenodon* has reached New Zealand is less surprising than that a frog has done so.

Order Squamata

(*Suborder Lacertilia*), lizards

(*Infraorder Iguania*)

Iguanidae, iguanas etc.: the Americas, and 2 relict genera in Madagascar and one on Fiji and the Tonga Islands in the western Pacific. In America there are about 50 genera, hundreds of species, most in the tropics of South and Central America and the West Indies, but some in the temperate zones. Of several genera in temperate North America, chiefly in the West, the most northern reach southern British Columbia, southern Alberta, southwestern North Dakota, Indiana, Pennsylvania, and southern New York. Southward, at least one genus, *Liolaemus,* reaches Tierra del Fuego. In habits, American iguanids are diverse: they live in deserts, grasslands, woods, and jungles; many are terrestrial; many, arboreal; a few, semi-aquatic. Large arboreal iguanas occur on the mainland of tropical America; large terrestrial forms of different genera occur on the mainland and on the West Indies and the Galapagos. Of many genera of smaller forms, the chiefly arboreal genus *Anolis* has 250 or more species and subspecies in northern South America, Central America, and the West Indies, etc., and also one endemic species (the "American Chameleon") in the southeastern United States north to the Carolinas. *Sceloporus,* with more than 100 species and subspecies (H. M. Smith 1946, p. 179; Smith and Taylor 1950, p. 105) terrestrial or semi-arboreal, occurs from southern British Columbia and southern New York to western Panama (H. M. Smith 1939); the greatest concentration of species of this genus is in Mexico. A number of ground-living genera are confined to arid western or southwestern United States and northern Mexico. Other American genera are too numerous to mention. On Madagascar are *Chalarodon,* with one species, localized in dry sandy areas in southwestern Madagascar, and *Oplurus* (*Hoplurus*), with about 6 species; they are small forms resembling American ground-living genera; that *Chalarodon* at least is really an iguanid is vouched for by Camp (1923, p. 308). On the Fiji and Tonga Islands occurs an endemic genus and species, *Brachylophus fasciatus,* a fine, iguana-like, arboreal lizard (Loveridge 1945, p. 92).

Fossils, mostly unsatisfactory fragments, suggest the presence of Iguanidae in North America at or before the beginning of the Tertiary but do not really throw much light on the history of the family. Fossil iguanids have been recorded from Europe too but, so far as I can judge from the literature (especially Camp 1923, p. 309, and Hoffstetter 1942), they are open to question, although the family's present distribution leaves little doubt that iguanids did occur on the Old World continents long ago.

Agamidae: main tropical and warm-temperate parts of the Old World, excepting Madagascar; most numerous in the Orient; more than 30 genera and some 280 species (M. A. Smith 1935, p. 132), some terrestrial and some arboreal, the latter including the "flying dragons"

(*Draco*). Northward, *Agama* reaches southern Greece and regions east of the Caspian, and there are several genera in central Asia, the most northern reaching Sinkiang and central Mongolia. *Agama* extends also over the whole of Africa, but only 2 other, localized, agamid genera occur in Africa. Of the many Oriental genera, 2 occur also in the Australian Region, where there are also some 6 endemic genera, one of which reaches Tasmania. The following genera have noteworthy ranges (all checked against M. A. Smith 1935). *Goniocephalus* has about a dozen species in the eastern Orient from South China to Java, Borneo, and some of the Philippines, apparently none on Celebes, Ceram, Halmahera, or the Lesser Sundas, but about a dozen in New Guinea etc. and northeastern Australia. *Physignathus* has one species in South China, eastern Siam, and French Indochina, and several in Australia, New Guinea, the Kei Islands, and Timor Laut, but apparently none otherwise in the great stretch of islands between Asia and New Guinea. However, some other genera fill in this gap. *Lophura* (*Hydrosaurus*), with 3 large, aquatic species, ranges from the Philippines and Celebes to New Guinea. *Calotes,* with about 27 species, ranges from tropical Asia through the islands to the Philippines, Celebes, Halmahera, Ceram, Timor Laut, etc., and to Bali, but not beyond. *Draco* has about 40 species from the southern edge of China to the Philippines, Celebes, Ceram, the Lesser Sundas to Timor etc., but apparently not beyond; and one species of this genus is in South India (not Ceylon), isolated from the others by a gap of at least 1000 miles. *Cophotis* is known only from one species in Sumatra and Java and one in Ceylon.

Agamids are related to iguanids, probably derived from them, and have probably replaced them on the Old World continents. Fossil agamids include (supposedly) *Agama* itself in the Eocene of Europe.

Chamaeleontidae, true chameleons: Madagascar, Africa, etc., to southern India and Ceylon; 4 or more genera, more than 80 species; specialized arboreal lizards derived from agamids. The main part of the family is about equally divided between Madagascar and the whole of Africa except parts of the Sahara. The common *Chamaeleo chamaeleon* of North Africa extends to southern Spain, eastern Mediterranean islands, etc.; and closely related forms occur in southern Arabia, on Socotra Island, and in Ceylon and India, in wooded districts south of the plain of the Ganges. That the Indian chameleon is close to a common African (not Madagascan!) one has long been known (*e.g.*, Gadow 1909, p. 579; M. A. Smith 1935, p. 253). There is an endemic species of *Chamaeleo* on the Seychelles; I do not know its geographical relationships.

"The fossil record shows no undoubted chameleons; all those so described are jaw fragments and can equally well be placed with the Agamidae" (Camp 1923, p. 312).

(*Infraorder Nyctisauria*)

Gekkonidae (including Eublepharidae, Uroplatidae), geckos: all tropical and some warm-temperate regions; more than 70 genera (M. A. Smith 1935, p. 22) and nearly 700 species (Loveridge 1945, p. 67). The northern limits of the family are set by several genera that reach parts of the Mediterranean coast of Europe, the Balkans north to Bulgaria, or the Crimea; by desert-living genera that range from the Kirgiz Steppes or Transcaspia across central Asia to southern Mongolia; by house lizards of the genus *Gekko* that reach northeastern China, Korea, and the main island of Japan; and in America by desert-living *Coleonyx*, which crosses the Mexican border into the United States and reaches southern Nevada and southwestern Utah (Klauber 1945), and by 3 genera *introduced* into southern Florida and the Florida Keys (H. M. Smith 1946). Southern limits are harder to fix, but geckos certainly reach South Africa, southern Australia (but they are probably not native on Tasmania), both islands of New Zealand, and central Chile and Argentina. Geckos are also widely distributed on islands within the tropics. Different geckos live in deserts, other open country, and (in the tropics) in wooded country and wet jungles, and certain species of half a dozen genera are house lizards, closely associated with human habitations and very widely distributed in the tropics by commerce. For example, *Hemidactylus brooki,* the commonest house lizard of India (M. A. Smith 1935), occurs also in parts of the Indo-Australian Archipelago, much of Africa, and the West Indies and parts of South America.

Space cannot be taken to give the distributions of most genera of this great family, which swarms in the tropics but is unfamiliar to most of us in the north-temperate zone, but a few genera need mention. M. A. Smith (1935) and Loveridge (1947) give much information about the geckos of India etc. and of Africa respectively. *Phyllodactylus,* with about 50 or perhaps more species, has perhaps the widest natural distribution of any gekkonid genus, but its distribution is discontinuous: there are several species on Madagascar, a few localized in tropical Africa, one on islands in the Mediterranean, 2 in Iran (none in India), one in Thailand etc. (none in the Indo-Australian Archipelago), a number in Australia, many in western and northern South America and the Galapagos, and a few scattered in Central America, Mexico, and the West Indies; some species of this group are sometimes placed in a separate genus, *Diplodactylus,* said to be confined to Africa, Madagascar, and Australia (Schmidt 1923, p. 14, Map 6), but M. A. Smith (1935, p. 80) doubts the distinctness of this subsidiary genus. *Gymnodactylus* (if it is a natural group) occurs from the eastern Mediterranean and tropical Asia to Australia and many Pacific islands, and in South and Central America. *Tarentola* is confined to countries bordering the Mediterranean (several species) and to Cuba and the Bahamas in the West Indies (one distinct endemic species); this uniquely distributed genus has been carefully studied recently (Loveridge 1944*b*), but some persons still doubt whether the West Indian species really belongs with the others. Other cases of supposed generic identity of geckos in the Old and New Worlds are doubtful: *Hemidactylus* may not be native in America; *Gonatodes* has been divided (M. A. Smith 1933); and *Thecadactylus* has been divided (Brongersma 1934). *Eublepharis,* with 3 species in arid southwestern

Asia and northern India and one in Hainan and the Riu Kiu Islands, and *Coleonyx,* with about 5 species plus a number of subspecies (Klauber 1945) irregularly distributed from arid southwestern United States to the jungles of Panama, may be related.

To turn to some exclusively Old World groups, *Diplodactylus* is mentioned above. *Oedura* is now restricted to numerous Australian species; the related or at least similar *Afroedura* is confined to southern Africa (Loveridge 1944*a*). *Phelsuma* (Loveridge 1942*a*), with some 18 full species, is confined to Madagascar, the Comoros, the Seychelles, and the Mascarene Islands in the Indian Ocean, except that one Madagascan species, which enters houses, is represented on the Andaman Islands off Burma by a supposedly endemic color form, and that another Madagascan species occurs around East African seaport towns. *Uroplates,* with about 5 species, is confined to Madagascar and was formerly placed in a family of its own. *Hoplodactylus* and *Naultinus* are confined to New Zealand, both North and South Islands; they are the only geckos on New Zealand proper and are the only known viviparous geckos (M. A. Smith 1935, p. 27). Most of the many remaining genera of geckos are localized within one or another of the following areas: Africa, Madagascar, North Africa plus southwestern Asia, tropical Asia etc., the Australian Region, and tropical America.

Underwood (1954) has upset some of the distributions given above. His paper is important and may lead to a basic reclassification of geckos, but it cannot be accepted zoogeographically until other specialists have passed on it. In the meantime Underwood's suggestions emphasize how little is surely known of the relationships of different geckos.

Although they are a large, widely distributed, probably old family, geckos are unknown fossil excepting a few Pleistocene ones and 3 genera recently described from the Tertiary of France.

Pygopodidae: Australia etc.; some 8 genera, 14 species (Kinghorn 1926); terrestrial or burrowing lizards with reduced limbs. All the genera are Australian, but *Pygopus* reaches Tasmania, and *Lialis* reaches New Guinea and the Aru Islands. This family may be related to the geckos. It may be primitive in a general way, as the geckos probably are, but it does not seem to be an ancient relict group. It is unknown fossil.

(*Infraorder Leptoglossa*)

†Ardeosauridae: an excellently described and figured fossil lizard from the Upper Jurassic of Europe; perhaps ancestral to the skinks (Romer) or the geckos (Camp).

Xantusiidae: southwestern North America, Central America, Cuba; 4 small genera. *Xantusia* (H. M. Smith 1946, p. 499; Stebbins 1954; Slevin 1949), with some 5 species, is in arid and semi-arid places in Arizona, the southwestern corner of Utah, southern Nevada, parts of the southern half of California, and parts of lower California and Sonora, and on several coastal islands. *Lepidophyma* and *Gaigeia* (H. M. Smith 1939*a*, p. 24; 1942), with several species, are in southern Mexico and Central America, in moist tropical lowlands and the high-mountain coniferous zone respectively; *Lepidophyma* goes south to Panama.

Cricosaura, with one species, is known only from eastern Cuba, in a few square miles of arid country between Belig and Cabo Cruz.

A fossil, †*Euposaurus,* from the Upper Jurassic of Europe, is placed in this family provisionally by Romer (1956), but it would be unwise to make too much of this zoogeographically.

Teiidae: the Americas; about 37 genera, 200 species. All the genera occur in South America, a few extending south to central Chile and Argentina. At least 6 of the genera, but especially *Cnemidophorus* (Burt 1931) and *Ameiva,* occur also in Central America. Only *Cnemidophorus* extends to the United States, but it reaches parts of Oregon, Idaho, South Dakota, Wisconsin, and Maryland; southward, this genus reaches southern Brazil. Only *Ameiva* extends widely into the West Indies, but 2 or 3 other genera reach the Lesser Antilles. This is a terrestrial family, related to the Old World lacertids etc.

Convincing fossils close to the living South American genus *Tupinambis* are known from the Oligocene and Pliocene of Argentina (Camp 1923, p. 316).

Scincidae (including Anelytropsidae, Feyliniidae), skinks: all tropical and warm-temperate areas including many islands, but more numerous in the Old World than in the New; some 40 genera, more than 600 species (M. A. Smith 1935, p. 254). The northern limits of the family are set by a few species that reach southern France, Austria, Rumania, or the Caspian in Europe; by several genera in the mountains of south-central Asia; and by *Eumeces,* which reaches Hopei in North China, the North Island of Japan, and in America the southern edges of British Columbia, Manitoba, and Ontario, and southern Massachusetts. Southward, skinks reach South Africa, Tasmania, both islands of New Zealand, and central Chile and Argentina. There are several geographically notable genera and species of skinks. *Lygosoma* in a broad sense (M. A. Smith 1937) is widely distributed in the Old World and occurs also in eastern North America and in Central America. This genus has 175 or more species in the Old World, most of them in the region from southern Asia to Australia etc., with a few in Africa but perhaps none in Madagascar. In New Zealand are some 9 endemic species supposed to represent 2 different sections of the genus; they are the only native New Zealand skinks, and they all are ovoviviparous so far as known (H. B. Fell and Charles McCann, in letters). In America this genus is represented by only some 9 species and subspecies, all in section *Leiolopisma* (which is also in the Orient etc.), and occurring only from eastern United States through Mexico and Central America to Panama (Taylor 1937, Stuart 1940—Mittleman, 1950, has suggested different generic names but they do not affect the relationships). The single species in the United States is confined to the southeast, north to southern New Jersey and west to eastern Kansas and Texas; there is none in western United States; but the eastern species is closely related to Chinese forms (Schmidt 1946, p. 149). *Eumeces* (Taylor 1935), with about 50 full species, occupies an area in North Africa and southwestern Asia north to the Caspian Sea and east to northwestern India; a separate area in eastern Asia from the

Indochinese peninsula to northern China and Japan; North and Central America from the southern edge of Canada to Nicaragua; and Bermuda, where there is a distinct endemic species. *Mabuya,* with more than 80 species (M. A. Smith 1935, p. 258), is in Africa and Madagascar, southern Asia to New Guinea (a single widely distributed species reaches the latter island), and South and Central America, southern Mexico (north to Colima and Veracruz), and the West Indies; it is the only genus of the family on the continent of South America; there are 9 species and a number of additional subspecies of this genus in America (Dunn 1935). The only other native American skinks (besides the 3 genera named above) are *Neoseps,* an endemic monotypic genus confined to central and southern Florida (H. M. Smith 1946, pp. 395 and 505), and *Anelytropsis,* a limbless burrowing genus of one species confined to southern Mexico. *Ablepharus* (or *Cryptoblepharus* or *Emoia*) *boutonii,* with many described subspecies, some of which may not be recognizable (Burt and Burt 1932, p. 514), occurs in tropical Africa, Madagascar, islands of the Indian Ocean (but not India etc.), the Indo-Australian Archipelago (especially the eastern part), most or all of Australia (not Tasmania), and across the tropical Pacific to islands off the west coast of Ecuador and Peru (Burt and Burt 1932, p. 481); it was long ago recorded also from the mainland of tropical America, but there is no recent record of its occurrence there (M. A. Smith 1935, p. 309); this species has probably been spread partly by man. Some other skinks of the genera *Emoia* and *Lygosoma* have been very widely spread across the islands of the tropical Pacific (Burt and Burt 1932). Several genera of skinks are common to southeastern Asia and Australia but are localized or absent on intervening islands: *Tropidophorus* (M. A. Smith 1935, p. 322) has 9 species in and around Indochina, apparently none in the Malay Peninsula or Sumatra or Java, 5 confined to Borneo, 5 on the Philippines (Taylor 1922*a*), and one Philippine species on Celebes, apparently none otherwise between Borneo and New Guinea, one on New Guinea (Loveridge 1945*a*), and one in northeastern Australia; *Ophioscincus* (M. A. Smith 1935, p. 333) has 3 species in the vicinity of Indochina and one in Australia; and *Rhodona* (M. A. Smith 1937, p. 230) has one species in Thailand and 20 in Australia. On the other hand, some genera of skinks are almost confined to the Indo-Australian Archipelago and islands to the east: *Emoia* in a restricted sense (M. A. Smith 1937, p. 226), with more than 30 species, occurs from Java, Borneo, and the Philippines to New Guinea and the Pacific islands, with one wide-ranging species reaching Australia; and *Otosaurus* (M. A. Smith 1937, p. 218), with 18 species, occurs from Sumatra, Borneo, and the Philippines to New Guinea and the Solomons. Finally, in each of the genera *Lygosoma, Dasia,* and *Riopa,* there are certain pairs of species divided between southern India and the eastern Orient, with no closely related species in northern India (M. A. Smith 1935, p. 15). Most skinks are terrestrial and many are burrowing, and some of the latter have reduced limbs or have lost their limbs entirely; a few have arboreal tendencies, but none is highly adapted to an arboreal life (M. A. Smith 1935, p. 254). As for reproduction, many skinks are oviparous and many are viviparous (M. A. Smith 1935, p. 256); the New Zealand ones are viviparous; most of the few American ones, including all the more northern ones, are oviparous, but viviparous species of *Eumeces* occur in Mexico (H. M. Smith 1946, p. 340).

Fossil skinks (Camp 1923, p. 315, Hoffstetter 1944–1945) are few and not very satisfactory and are of little use in unraveling the geographical history of the family, but the distribution of living forms leaves not much doubt that the group has radiated in the Old World tropics, especially in the Oriental-Australian area, and that 3 widely distributed genera have reached America independently, at least 2 of them by northern routes. How *Mabuya* may have reached tropical America is discussed in the text (p. 194). *Neoseps* and *Anelytropsis* may be relicts of earlier invasions.

Lacertidae: Africa (not Madagascar), Eurasia, etc.; 22 genera, 145 species (Boulenger 1920–1921); primarily terrestrial. The family is most numerous in Africa; several genera are common to the dry parts of northern Africa and southwestern Asia; several genera are in the Oriental Region, but only one, wide-ranging species reaches Sumatra, Java, and Borneo. Northward several genera are widely distributed in temperate Europe and Asia; the northernmost species, *Lacerta vivipara,* occurs across northern Eurasia from the extreme north of Scotland and Lapland to the Amur and Sakhalin, reaching 70° N. in Scandinavia; it is the northernmost reptile and the only known viviparous lacertid.

Fossils indicate that *Lacerta* has been in Europe since the Eocene (Camp 1923, p. 316). However, the record of an African genus, *Nucras,* in Baltic amber may be false; the supposed fossil may really be an existing African species embedded in East African copal (Loveridge 1942*b*, p. 338).

Cordylidae (Zonuridae, including Gerrhosauridae): Africa and Madagascar; 2 subfamilies; terrestrial lizards occurring chiefly in dry country. (1) Gerrhosaurinae: Africa and Madagascar. In Africa (Loveridge 1942) there are 4 genera, 13 species, plus additional subspecies, most in southern Africa, but *Gerrhosaurus* reaches to 13° N. in the Anglo-Egyptian Sudan. In Madagascar there are 2 other genera with about 10 species. (2) Cordylinae: Africa; 4 genera, 23 species, plus subspecies (Loveridge 1944), chiefly in southern Africa, but a few reach the Belgian Congo and East Africa, and one, Ethiopia. Records of this subfamily in Madagascar are errors (Mocquard 1909, p. 4, footnote).

Dibamidae: southern Indochina and the Nicobar Islands to the Philippines and New Guinea; one genus, 3 species; worm-like burrowing lizards of doubtful relationships.

(*Infraorder Diploglossa*)

(*Superfamily Anguoidea*)

Anguidae: the Americas, Eurasia, etc.; 3 existing subfamilies plus a fossil one; the existing forms are mostly terrestrial or burrowing. (1) Diploglossinae: tropical America; 4 genera, several of them localized especially on the main islands of the West Indies; one occurring south to part of Argentina. A doubtfully related fossil genus is in the Eocene-Oligocene of North America. (2) †Placosaurinae: Eurasia and western North America in the late Cretaceous and early Tertiary; rather large, armored lizards. (3) Gerrhonotinae: Central and western North America, Eurasia, etc.; 2 existing genera. *Gerrhonotus* ranges from Panama through Central America and western North America to southern British Columbia, northern Idaho, and western Montana (see Tihen 1949). *Ophisaurus* has one species in western Eurasia from the Balkans to Turkestan and in North Africa, possibly another in North Africa, 2 or 3 in the eastern Himalayas, South China, Burma, and Formosa, one in Borneo (confirmed by a fresh specimen in the Museum of Comparative Zoology), and one in eastern North America, from Wisconsin and Virginia to Texas and Florida. Three fossil genera assigned to this subfamily are in Europe (Eocene) and North America (Upper Cretaceous and Oligocene-Miocene). (4) Anguinae: one existing species, *Anguis fragilis,* the legless Blindworm of Europe; occurs from about 64½° N. in Sweden (see text, p. 181) to the northern corner of Africa, northern Asia Minor, and the Caucasus. The same genus is in the Miocene of Europe.

Anniellidae: southern half of California and northern Lower California; one genus, 2 species, plus one additional subspecies; worm-like, burrowing, ovoviviparous lizards which have probably "degenerated" from anguids.

Xenosauridae: now localized in China and in southern Mexico etc.; one fossil and 2 existing subfamilies, doubtfully associated. (1) †Melanosaurinae: North America (Eocene) and ?Europe (Paleocene-Eocene). (2) Shinisaurinae: one species, *Shinisaurus crocodilurus,* known only from the Yaoshan region, Kwangsi, China. McDowell and Bogert (1954) discuss its relationships. (3) Xenosaurinae: southern Mexico (north to San Luis Potosí) to Guatemala; one genus, one or 2 species.

(*Superfamily Varanoidea*)

Helodermatidae, Gila monsters: only *Heloderma,* with 2 living species, one in southern and western Arizona and adjacent corners of Utah and Nevada and in northwestern Mexico, and one in western and southern Mexico, chiefly coastal, from southern Sonora to Chiapas (Bogert and del Campo 1956, p. 15). They are large, ornate lizards, the only really poisonous lizards known. The same genus is fossil in Colorado in the Oligocene.

Varanidae, monitor lizards: warm parts of the Old World; one existing genus, *Varanus,* with about 24 full species plus additional subspecies (Mertens 1942). Monitors occur over the whole of Africa except its extreme northwestern corner, but not now in Europe and not in Madagascar; in southwestern Asia north to the Caspian Sea; through the Oriental Region north to South China, and south and east through the Indo-Australian Archipelago to the Philippines, the Solomons, all Australia (not Tasmania), and islands of the western Pacific to the Marianas, Carolines, and Marshalls. Monitors are large, carnivorous lizards which occur (different species) in both arid country and wet jungle, and which are terrestrial, arboreal, or semi-aquatic in habits. One of the semi-aquatic ones, *Varanus salvator,* is discontinuously but very widely distributed: on Ceylon but not most of peninsular India, from the eastern Orient to Celebes and Wetar, and perhaps to Australia, though Mertens doubts the latter. *Varanus komodoensis,* the Komodo Dragon, is known only from small islands between Sumbawa and Flores and from a small part of western Flores, in the Lesser Sunda Islands; it is the largest living lizard, reaching a length of at least 10 feet and being, for a lizard, bulky.

Fossil *Varanus* are known from Europe, India, and Australia. Europe had 2 or 3 species, the earliest in or before the Lower Miocene, and the latest persisting apparently into Neolithic times. Giant species occurred in India in the Pliocene and in Australia probably in the Pleistocene—the Australian †*Megalania* was hardly more than a heavily built *Varanus* perhaps 15 feet long (Dunn 1927). Other, very different fossil genera (placed by Romer in a different subfamily) are known from the Upper Cretaceous and Eocene of the northern hemisphere, including North America.

Lanthanotidae: Borneo; one species; not poisonous, and not so closely related to *Heloderma* as formerly thought (McDowell and Bogert 1954).

†Aigialosauridae, †Mosasauridae, †Dolichosauridae: 3 families of aquatic, *marine* lizards known only from the Cretaceous.

†Palaeophidae: poorly known Cretaceous and Eocene forms which Romer (1956) thinks may have been snake-like, aquatic, varanoid lizards. They are listed from Europe, North America, North Africa, and Madagascar.

(*Infraorder Annulata*)

Amphisbaenidae: Africa etc. (not Madagascar), South America north to Panama (not most of Central America), and isolated areas scattered in and near the northern edge of the American tropics; about 24 existing genera, 150 species; worm-like, usually limbless burrowers; 4 subfamilies (Vanzolini 1951). (1) Amphisbaeninae: Africa etc. and America; 16 genera. *Amphisbaena* is in Africa (one species, in Mozambique), South America north to Panama, and on Cuba, Hispaniola, and Puerto Rico. Other genera are localized in parts of Africa (8 genera besides *Amphisbaena*); the Mediterranean region north to Spain and Portugal, Rhodes, and the southwestern corner of Asia (*Blanus*); South America (3 genera); northwestern Mexico (Lower California) (*Bipes*); Cuba etc. (one genus); and the Virgin Islands (one genus). This subfamily is unknown fossil. (2) Rhineurinae: 2 genera in Africa, 2 in South America, and *Rhineura* in Florida. *Rhineura* is fossil in western United States in the Oligocene, and other genera are listed in western United States in the early Eocene and Oligocene. (3) Trogonophinae: parts of northern Africa, Arabia, Iran, and Socotra Island; 3 genera; no fossils. (4) †Crythiosaurinae: Oligocene of Inner Mongolia; one genus.

(*Suborder Ophidia*), snakes

(*Superfamily Typhlopoidea*)

Typhlopidae, worm snakes: the tropics and some warm-temperate areas; 4 genera, close to 200 species. *Typhlops* occurs throughout the range of the family, with many species in Africa and Madagascar; many in southern Asia, one extending northwest to Greece and the Caspian Sea, several in the Himalayas, and one in South China north to southern Kiangsi and Fukien; many in the Indo-Australian Archipelago; and many in all Australia, but not Tasmania; and relatively few in America, from southern Mexico (north to central Veracruz–Smith and Taylor 1945, p. 19) to northern Argentina, and in the West Indies. The other 3 genera, commonly put in a separate subfamily, are confined to America, within the range of *Typhlops*. *Typhlops braminus,* the Flowerpot Snake (Schmidt and Davis 1941, p. 91), is native in the Orient and ranges through the islands almost to New Guinea, and occurs also, probably introduced, in southern Africa, Madagascar, islands of the Indian and Pacific Oceans including the Hawaiian Islands, and southern Mexico (Smith and Taylor, p. 19). Typhlopids are small, worm-like, burrowing snakes. The family is apparently primitive and probably old, but the only known fossils of it are one or two in the Tertiary of Europe.

Leptotyphlopidae, worm snakes: Africa etc. and warm parts of America; one genus, 50 or more species; tiny, worm-like, burrowing snakes. In more detail their range is Africa south of the Sahara (not Madagascar) with northeastern Africa and southwestern Asia to northwestern India; and America from southern Kansas and the deserts of Arizona and adjacent parts of California, Nevada, and Utah to central Argentina, and the Lesser Antilles and (perhaps introduced) the Bahamas. The family is almost unknown fossil.

(*Superfamily Booidea*)

†Dinilysiidae: Upper Cretaceous of South America; one genus; probably the oldest known sure snake.

Aniliidae (Ilysiidae): localized in tropical Asia etc. and tropical America; 4 subfamilies, provisionally associated. (1) Loxoceminae: Central America and southern Mexico (Costa Rica to Colima and Nayarit); one genus, *Loxocemus,* 2 species; formerly thought to be related to the Old World pythons. (2) Xenopeltinae: eastern tropical Asia etc., from Upper Burma and Kwangtung, China, to the Greater Sundas, Palawan, southern Sulu Islands, and Celebes; one genus, *Xenopeltis,* one species; terrestrial-burrowing. (3) Aniliinae: tropical Asia etc. and South America; 3 genera (M. A. Smith 1943, p. 94): *Cylindrophis,* one species isolated on Ceylon and about 6 in the eastern Orient and islands from Burma and Indochina to the Aru Islands off New Guinea; *Anomochilus,* 2 species, on the Malay Peninsula, Sumatra, and Borneo; and *Anilius* (*Ilysia*), one species, widely distributed in northern South America east of the Andes. Aniliids are rather small, terrestrial-burrowing snakes. (4) Uropeltinae, rough-tailed snakes: peninsular India and Ceylon; 7 genera, 43 species, all in this limited area (M. A. Smith 1943, pp. 61–94). They are small, burrowing snakes, living in mountainous country and forest.

?†Archaeophidae: marine Eocene of Europe; one fossil, perhaps an eel rather than a snake (McDowell and Bogert 1954, pp. 66–67).

Boidae, constricting snakes: all tropical and some temperate regions. It has been customary to recognize two main groups of boids, the boas proper with their allies and the pythons with their allies, but M. A. Smith (1943, p. 103), Guibé (1949), and especially McDowell (MS.) say or imply that the family is more complex than this simple division shows. Romer (1956), following McDowell, recognizes 4 main groups or subfamilies, and this (together with the removal of *Loxocemus*) makes the distribution of the family more orderly and eliminates most of the striking discontinuities. It will be a blow to zoogeographers to learn that the boids of Madagascar are probably not closely related to American boas! I made too much of this case in my *Quarterly Review* article (1948). I made the additional mistake then of saying that the pythons are derived from boines. In fact, the pythons seem to be structurally the more primitive (Underwood 1951, McDowell). This family includes the huge constricting snakes and also a number of smaller, less conspicuous ones. The Old World pythons, incidentally, are oviparous; the New World boas, ovoviviparous (Schmidt 1950, p. 82). There is a useful check list of the species of this family (Stull 1935), although it is out of date in some ways. The 4 subfamilies recognized by Romer are:

(1) Pythoninae, pythons etc.: the Old World, excepting Madagascar; (the Central American *Loxocemus* has been transferred to the Aniliidae); 7 genera, some 18 full species. *Calabaria,* with one small terrestrial species, is confined to the rain forest of West Africa. *Python* has 3 species in Africa south of the Sahara; 2 in southern Asia from northern India and the southern tier of China south to Ceylon and the Indo-Australian Archipelago, to the Philippines and almost to New Guinea (and McDowell thinks that some Australian forms should be included in *Python*); one additional species on the Malay Peninsula, Sumatra, and Borneo; and one on Timor and Flores. The other 5 genera, 10 species, of Stull's check list occur in Australia and/or New Guinea etc., with one species or another reaching Timor, Wetar, southern Mindanao, and the Solomons; in Australia, some pythonids occur in nearly every part of the continent, in dry, open country as well as wet forest, and in the south-temperate zone as well as in the tropics, but there is none on Tasmania.

(2) Boinae, boas, anacondas, and related smaller forms: tropical America south to central Argentina and including the West Indies; about 7 genera, 20 or so species. One, a form of the Boa Constrictor, extends north to Sonora and may reach southern Arizona (Bogert and Oliver 1945, p. 350).

(3) Sanziniinae: Madagascar etc.; 4 genera. Boa-like *Sanzinia,* with one species, and *Acrantophis,* with 2 species, are on Madagascar, and one of the species of *Acrantophis* is recorded from Réunion, too (Guibé 1949). *Casarea* and *Bolyeria,* each with one species, also superficially boa-like, are known only from Round Island just off Mauritius, but one or both are subfossil on Mauritius itself (Hoffstetter 1946). McDowell (in letter) notes that the Madagascan boids (and also at least some of the Madagascan colubrids) have hypapophyses on

all the thoracic vertebrae, that this is unusual among snakes and may be a swimming adaptation, and that it suggests that these snakes are derived from aquatic ancestors which reached Madagascar etc. by swimming. (It has recently been suggested that the Mauritius "boids" are really primitive colubrids.)

(4) Erycinae: northeastern Africa and part of Asia, New Guinea etc., and western North America; 4 genera; mostly small, retiring snakes. *Eryx,* terrestrial sand boas, with about 7 species, occurs in north and east Africa south to the northern edge of Tanganyika (Loveridge), through southwestern Asia to India and Ceylon, north to the Balkan Peninsula, the Caucasus, and Uralsk (above 50° N.), and across central Asia to southern Mongolia. *Enygrus* (*Candoia*), very small tree boas, have 3 full species in the New Guinea-Bismarcks-Solomons area; one of the 3 extends west to Celebes; and another of the 3 is widely spread across the islands of the western Pacific to Samoa and Tonga and perhaps beyond—but it may have been spread partly by man. The 2 other genera, both terrestrial, are in western North America: *Charina* (one species) occurs from Utah and central California to southern British Columbia; *Lichanura* (2 species), in southern California, Lower California, and western Arizona.

Fossil boids, mostly poorly known, occur in various Tertiary deposits (Romer). They tell little of importance about the geographical history of the family, except that representatives of it occurred in Europe and North America during some times in the Tertiary.

(*Superfamily Colubroidea*)

Colubridae, common snakes etc.: all main parts of the world except part of Australia. The northern limits of the family in Europe are set by oviparous *Natrix natrix,* which reaches about 65° N., with viviparous *Coronella* a close second; I do not know the northern limits in Asia; in America, viviparous colubrids of the genus *Thamnophis* (common garter snakes) are much the northernmost reptiles, reaching about 60° N. in western Canada. Southward, colubrids reach South Africa and Chubut Province in southern Argentina. The family is a huge one, comprising nearly two-thirds of all existing snakes, and a still higher proportion of those in ordinary habitats (excluding primarily burrowing families and sea snakes). There are several subfamilies, but authorities do not agree on the number or composition of them. Generally speaking, subfamily Colubrinae is dominant in tropical and temperate Eurasia, Africa, and North America; Xenodontinae (Ophiinae), if recognizable, is dominant in tropical America; the other subfamilies are smaller and more localized. Some colubrid genera are endemic on Madagascar, and some of the Madagascan forms are said to be primitive (Schmidt 1923, p. 35). Several diverse genera of Oriental colubrids reach northern Australia (see under subfamilies), and a few extend southward in eastern Australia, but colubrids are absent in a large area in southern, central, and southwestern Australia, and they are absent in Tasmania. The following subfamilies of Colubridae are those listed by Romer (1956) plus the Xenodontinae (Dunn's Ophiinae). Most of these subfamilies were recognized by

M. A. Smith (1943), but Romer's arrangement of them is different. Dunn (1928), Pope (1935), and Bogert (1940) have divided the family somewhat differently.

(1) Colubrinae (including Boiginae, Natricinae, Coronellinae, Sibynophiinae): all main parts of the world except part of Australia; more than 100 genera. There are many in Africa (Bogert 1940), Eurasia, the Orient (M. A. Smith 1943), and North America (Dunn 1928); fewer in South America (Dunn 1928); and only the following 4 genera (of this subfamily) reach Australia: *Boiga*, widely distributed in tropical Africa and the Orient, and *Ahaetulla* (*Dendrophis*), of the Orient, are arboreal genera which extend through the islands to Australia—*Boiga* has one and *Ahaetulla* 2 species in northern and eastern Australia (Kinghorn 1929, Loveridge 1934, Glauert 1950); *Steganotus*, terrestrial, is known only from a few species scattered in the Philippines, Moluccas, and New Guinea, and 2 in Queensland; and *Natrix*, in part aquatic and widely distributed over the world, includes 3 species that reach northern Australia, and one of them extends south in eastern Australia to northern New South Wales. The classification of genera in this subfamily is in flux. Many old genera have been broken up, but the relationships of the resulting fractions have not always been made clear. Below are outlined the ranges of a few geographically interesting groups, but they should be treated with caution, and it should be remembered that most colubrine genera occupy more or less continuous, geographically restricted, logical areas, and that those with discontinuous ranges are exceptions. The most widely distributed genus of colubrids is *Natrix* (*Tropidonotus*) (see p. 187); and *Thamnophis* is derived from it and increases its American range. *Elaphe* and *Coluber* (in a broad sense) are primarily terrestrial genera which occur around the world in the northern hemisphere, and they occur also in the tropics in the Orient and elsewhere; *Coluber* has been partitioned by Inger and Clark (1943), and in a restricted sense is confined to North America and eastern Asia, but Bogert (in letter) doubts that the Asiatic species is really closely related to the American ones. *Helicops* in a broad sense has a few species scattered and isolated in tropical Africa and in Ceylon, India, and Yunnan, and a number in America, from Mexico (perhaps) to Argentina. The genus has been divided (Bogert 1940, pp. 37–38; M. A. Smith 1943, p. 319), but the resultant smaller genera "may be regarded as direct derivatives from the ancestral stock of *Helicops* surviving as relict forms now separated from the main [American] range of the parent stock" (Bogert). *Sibynophis* (*Polyodontophis*, *Scaphiodontophis*), a terrestrial and often mountain-dwelling genus, is on the Comoro Islands and Madagascar (2 species); in the Oriental Region (7 species) from Ceylon and India to Lombok, Borneo, and Palawan and Busuanga in the Philippines, and north to Shensi and Kiangsu in China; and in Central America etc. (9 species and subspecies) from the Atlantic slope of Mexico to Panama and Colombia. Bogert (1940, pp. 9–10) apparently thinks that the genus should be divided, but that the products of division will form a natural group of genera of perhaps

subfamily rank, as suggested by Dunn (1928); and Taylor and Smith (1943) have already separated the American from the Old World forms. *Opheodrys* (Schmidt and Necker 1936, Grobman 1941) (called *Ablabes* by Rooij 1917, *Eurypholis* by Pope 1935, and *Liopeltis* by Bourret 1936) occurs in tropical and temperate eastern Asia including Formosa and the Riu Kiu Islands, and in eastern North America and Yucatan. In eastern Asia etc. there are perhaps 6 or more species of the genus, occurring rather more in the tropics than in the north-temperate zone, and additional species, formerly placed in the genus but now usually separated from it, occur on the Greater Sunda Islands; on the mainland of Asia, the most northern species reaches Shanghai, part of Hupeh, etc. In eastern North America there are 2 species of the genus, one or the other reaching at least 50° N., west to Wyoming, Utah, and New Mexico (but not to the west coast), and south to the northeastern corner of Mexico; and an additional species is isolated in northern Yucatan within the tropics. This genus, *Opheodrys,* in the difficulty of deciding what name to use for it and what species to put in it, illustrates the taxonomic confusion of many colubrids.

(2) Xenodontinae (Ophiinae): America and perhaps parts of the Old World; 60 or 70 genera or more (Dunn 1928; 1931, p. 116, paragraph 7). These snakes are dominant in South America and are the only colubrids in the West Indies except for an aquatic *Natrix* that has reached Cuba; a few occur in North America; and it is likely that some Old World genera would be included, if the subfamily were recognized. Actually, recent ophiologists do not recognize it. It is listed here to emphasize what seems to be a fact, that there is in South America a great assemblage of relatively primitive colubrids which represent the diverse, surviving stock from which most of the Colubrinae of other parts of the world seem to have been derived by a number of separate evolutionary routes (Bogert 1940, p. 9). This situation parallels that found among perching birds, of which a great aggregation of relatively primitive forms (Suboscines) has survived or evolved in South America while the higher Oscines have evolved elsewhere, mainly in the Old World (pp. 274ff.).

(3) Dasypeltinae (Dasypeltidae), egg-eating snakes: Africa etc. and India; 2 genera. *Dasypeltis,* with one species, 2 or more subspecies (Bogert 1940, p. 85), is widely distributed in tropical and South Africa and reaches southern Arabia (Gans, in conversation); and *Elachistodon,* with one rare species, is found only in northeastern India. It has been doubted whether these 2 genera are related, but M. A. Smith (1943, p. 403) evidently thinks they are.

(4) Acrochordinae: tropical Asia to northern Australia; 2 species, both now placed in *Acrochordus* and both aquatic, one in fresh water, the other in estuaries and coastal seas; both species range from the mainland of tropical Asia to northern Australia, and the salt-water species reaches the Philippines and Solomons.

(5) Xenoderminae: the eastern Orient and tropical America. In the Orient there are 4 genera, 7 or 8 species, from southern and eastern China (north to Kiangsu) and the main island of Japan to Java

and Borneo. In Central and South America there are one or 2 other genera, few species.

(6, 7) Pareinae, Dipsadinae (Amblycephalidae), snail-eating snakes: the eastern Orient and tropical America respectively. In the Orient there are 2 genera, about 16 species, occurring from the eastern Himalayas, southern China (north to Szechwan and Fukien), and Formosa (but not most of India) to Java, Borneo, and the southern Philippines. America has several genera, about 40 or more species, mostly in tropical South America, but reaching northern Argentina and (*Sibynomorphus* or *Dipsas* only) Central America and southern Mexico, to Colima and central Veracruz. Several authors have questioned the relationship of the Asiatic and American "Amblycephalidae," and Romer puts them in separate subfamilies, but M. A. Smith (1943, p. 115) thinks nevertheless that the American and Asiatic genera are closely allied and probably had a common origin.

(8) Homalopsinae: India, southern China, Formosa, and the Philippines to northern Australia etc.: 10 genera, more than 25 species, aquatic, at home in both fresh and salt water (M. A. Smith 1943, p. 379). Eight of the genera occur on the mainland of southern Asia; several of the same genera range into the Indo-Australian Archipelago, and 3 of them reach northern Australia; the other 2 genera are confined to New Guinea and northern Australia respectively.

A few Tertiary and Pleistocene fossil colubrids are known, but most are fragmentary, and they tell almost nothing of the geographical history of the family.

Elapidae, cobras, coral snakes, Australian snakes, etc.: the tropical and warm-temperate regions of the world; 2 subfamilies.

(1) Elapinae: range of the family. Of some 30 genera, about half are confined to the Australian Region. A few of the Australian forms are large and dangerous snakes, but most, though poisonous, are small and inoffensive. Different Australian elapids therefore take the place not only of cobras and viperids but also of the usually non-venomous colubrids of the rest of the world. No elapid genus is common to the Australian Region and other parts of the world; *Acanthophis*, the Australian Death Adder, reaches the Moluccas and Tanimbars, but an old record for Borneo is probably wrong. *Ogmodon* is confined to Fiji; it is considered closely related to a group of burrowing genera in New Guinea (Bogert and Matalas 1945). In Africa and southern Asia, north at least to Transcaspia, the Kopet-Dagh Mountains on the Iran-Turkestan border (Berg 1950, p. 191), and northern India and parts of Hunan, Kiangsi, and Chekiang, and the Riu Kiu Islands, are various genera, including the cobras (mapped by Bogert 1943, p. 312). There are no elapids in Madagascar. In America there are only coral snakes of 3 genera: *Micrurus*, with some 65 species and subspecies (H. M. Smith and Taylor 1945, p. 169), occurs from North Carolina and southwestern Indiana to the Territory of Rio Negro in east-central Argentina (Serié 1936; Pope 1945, p. 18), but not in western North America or the West Indies; *Micruroides*, with one

species, is localized in southwestern North America; and *Leptomicrurus,* with 2 species, in northern South America.

(2) Dendroaspinae: tropical and South Africa; one genus, *Dendroaspis,* several species, the mambas or tree cobras.

The only known elapid fossil before the Pleistocene is †*Palaeonaja* in the Miocene and Pliocene of France; Bogert (1943, p. 294) reduces this to a synonym of *Naja,* the genus of existing cobra ; there are no cobras and in fact no elapids in Europe today, but cobras still reach the Mediterranean coast of Africa and come close to the southeastern corner of Europe in Asia.

Hydrophiidae, sea snakes (M. A. Smith 1926, Mertens 1934*a*): warmer parts of the Indian and Pacific Oceans; 15 genera, about 50 species, most in the tropical seas from southern Asia to northern Australia and the western Pacific. They straggle as far north as Japan and southern Siberia, and south to Tasmania and tc New Zealand (Phillips 1941). Several species occur west to the Persian Gulf; 2, to Madagascar (Mocquard); and one, *Pelamis platurus,* to the east coast of Africa. The same species, which is the only pelagic sea snake, is also the only one that reaches the west coast of America, where it occurs from Ecuador north to the Gulf of California; it is also around the Hawaiian and Galapagos Islands etc. There is no sea snake in the Atlantic. Although most of the species are marine, fresh-water forms closely allied to and evidently derived from marine ones occur in Lake Bombon (Taal) on Luzon in the Philippines and in Lake Tungano on Rennell Island in the Solomons (Loveridge 1945, p. 170). The sea snakes are allied to and derived from elapids and are poisonous. Mertens thinks they may have had a double origin, both times in the Indo-Australian area, late in the Cretaceous or early in the Tertiary. Romer (1956) recognizes 2 subfamilies, which need not be distinguished here.

Viperidae, vipers: all continents except Australia, but few islands; 2 subfamilies.

(1) Viperinae, typical vipers: Africa and Eurasia; 10 genera (M. A. Smith 1943, p. 478). *Vipera,* with about 10 full species (M. A. Smith 1943, p. 482), occurs through tropical and northern Africa, Europe, and Asia, and is local on the Indo-Australian Archipelago, where a single species (*V. russelli*) is definitely known from Lombok and an island off Flores (Mertens 1930) and doubtfully recorded from Sumatra and Java (Rooij 1917, p. 279); the range of this species seems to be discontinuous on the mainland of South Asia too (Pope 1935, pp. 384–385); it occurs also on Formosa. *Vipera berus* sets the northern limit of the family, and of all snakes, across northern Eurasia; in Scandinavia it reaches 67° N., just above the Arctic Circle, and it may go still farther north in the forests of the Kola Peninsula (Berg 1950, p. 48); it is viviparous. Six additional genera occur in Africa, some ranging over almost the whole of the continent, others localized; there are no vipers in Madagascar; 3 of the African genera extend into adjacent parts of Asia, and 3 more genera are confined to parts of southern Asia. No surely viperine fossils are known be-

fore the Miocene. *Vipera* was in Europe in the Miocene, as now. *Bitis,* now confined to Africa and Arabia, is recorded from Europe in the Pliocene.

(2) Crotalinae, pit vipers including rattlesnakes: Asia etc. and America; 6 genera. *Agkistrodon* occurs in Asia etc. and America: in southern Asia there are about 9 full species; one, the viviparous *A. halys,* ranges north in eastern Asia to at least 53° N. and to Lake Baikal, and west to the steppes of southern Russia east of the lower Volga, in the edge of Europe; one is in Japan; southward, the genus occurs in peninsular India below 16° N. and in Ceylon, but not in northern peninsular India, and one species reaches Sumatra and Java; in North America (Gloyd and Conant 1943) there are 2 species, the Copperhead and Water Moccasin, occurring in southeastern United States, the Copperhead extending north to southeastern Iowa, eastern Massachusetts, etc.; and there is an additional species of the genus in Mexico and Central America to Nicaragua. *Trimeresurus,* with about 22 species (M. A. Smith 1943, p. 502), occurs in Asia and islands, from Ceylon and India north to the Himalayas and central China (Hupeh and Chekiang) and the Riu Kiu Islands, and east and south to the Philippines, Celebes, and Timor and Sangir Island; the range of this genus too is apparently interrupted in northern peninsular India, above about 22° N., except that *T. albolabris,* which occurs in northern India, has an apparently isolated population in central India (M. A. Smith 1943, p. 524, map); this same species ranges through much of the Indochinese region; it is absent in Siam and the Malay Peninsula south of 13° N., but it occurs again in Sumatra, Java, etc., to Timor etc. In America, the representative of *Trimeresurus* is *Bothrops,* which cannot yet be separated morphologically (M. A. Smith 1943, p. 502); the many species occur from east-central Mexico (Tamaulipas–H. M. Smith and Taylor) to Santa Cruz Province in southern Argentina (Serié 1936). *Lachesis,* with one species, the Bushmaster, ranges from tropical Brazil to Panama. The rattlesnakes are strictly American. *Crotalus* (Gloyd 1940), the principal genus of them, with 22 full species plus additional subspecies, is best represented in southern United States and in Mexico; but species range north to parts of Maine, southern Ontario, Wisconsin, the Great Plains region of southern Canada, and southern British Columbia; and one species with 2 or more subspecies occurs in Central America and the more arid parts (not wet forests) of South America, south to northern Argentina. *Sistrurus* (pigmy rattlesnakes–Gloyd, 1940, p. 76, map), has 2 species in eastern North America north to southern Ontario, south to Florida and probably the northern edge of Mexico, and west to Arizona; and a third species is isolated on the southern end of the Mexican plateau. Crotalines are unknown fossil before the Pliocene (when rattlesnakes were in North America). Other clues suggest that the subfamily originated in Asia and that both *Agkistrodon* and *Trimeresurus* (*Bothrops*) moved from Asia to America, one of these general (or an earlier stock) presumably giving rise to the rattlesnakes.

Many, but not all, vipers and pit vipers feed chiefly or partly on warm-blooded animals. In the Old World, although most vipers are viviparous, some, including even a few *Agkistrodon,* are oviparous (Pope 1935, p. 387); in America probably all vipers are viviparous, except the Bushmaster.

(*Subclass Archosauria*)

Order Crocodilia: 5 suborders, 13 families of which only the following family still exists.

Crocodylidae, crocodiles etc.: the warm parts of the world; about 8 genera, 25 existing species (Schmidt 1944, slightly modified by Inger 1948). All are primarily aquatic as adults, but all lay their eggs on land. There are 3 subfamilies.

(1) Crocodylinae, true crocodiles: 2 genera, of which one, with 2 species, is confined to West Africa and the Upper Congo (Inger 1948), and the other, *Crocodylus,* with 12 species, occurs through the Old World tropics and in the northern part of the American tropics, from southern Florida and the lowlands of Mexico to Ecuador, Colombia, and the Orinoco, but not the Amazon etc. In each of the principal tropical regions (Africa, the Orient, tropical Australia etc., and tropical America) there is one wide-ranging species of *Crocodylus* which sometimes enters salt water, and one to 3 additional, localized species which are, I think, confined to fresh water. In Africa the wide-ranging species is *C. niloticus,* which occurs in suitable localities south to the Cape of Good Hope, north into parts of the Sahara (Angel and Lhote 1938), down the Nile, sporadically to the coast of Palestine (Flower 1933, p. 755), and now or formerly to Syria, and which occurs also on Madagascar, (formerly?) on the Seychelles, and probably formerly on Sicily in the Mediterranean (Flower); this species often enters brackish and even salt water (Schmidt 1919, p. 419). In the Oriental and Australian Regions the wide-ranging species is *C. porosus,* the Salt-water Crocodile, which occurs from tropical Asia to northern Australia and east through the Philippines and Solomons and beyond—Loveridge (1945, pp. 41 and 244) says to the Palau Islands and Fijis, "etc."; Adamson (1939) says the eastern limit of crocodiles in the Pacific is undetermined and notes the skeleton of one from Makatea in the western edge of the Tuamotus, beyond Samoa and the Society Islands; this species commonly enters the sea. In America the wide-ranging species is *C. acutus,* which occurs from southern Florida, northern Mexico (Tamaulipas and Sinaloa), etc., to Ecuador and Colombia, and on Cuba, Hispaniola, and Jamaica, and which frequents coastal swamps and estuaries.

(2) Alligatorinae, alligators etc.: the true alligators (*Alligator*), 2 existing species, now occur only in southeastern United States from the lowlands of the Carolinas and Florida to the Rio Grande, and in the lower Yangtze Valley in China—Pope (1935, p. 66) says study of local records would probably show that the Chinese alligator ranged widely in eastern China not long ago. The more-or-less-related caimans, 3 genera and 9 species, are in Central and South America, north at least to Oaxaca in Mexico and south to the Paraná system, but not in the West Indies.

(3) Gavialinae: the 2 monotypic genera of fish-eating gavials are localized in India and the Malay region respectively. (The Malayan genus may really be a crocodyline.)

The extensive fossil record of the crocodylids is summarized by Romer (1945, p. 222) thus: "All three [sub]families were present in the Eocene in northern continental areas, and both crocodile and alligator types were abundant and varied in these regions in the early Tertiary; finds from the southern continents, too, are not uncommon." The genus *Alligator,* which is apparently the only well-defined genus of reptiles strictly confined to temperate areas in *both* halves of the north-temperate zone, is known back to the Oligocene in North America but is unknown fossil in Eurasia, although related genera were in Europe early in the Tertiary; Mook (1925) says a fossil *Alligator* in the Lower Miocene of Nebraska is so much like the existing Chinese species that it could be its direct ancestor. The 2 existing genera of gavials, now confined to different parts of the Oriental Region, are both fossil in Europe.

(†Dinosaurs): The following 2 orders and 6 suborders are the dinosaurs. Their distribution is summarized in Chapter 10 (pp. 598ff.).

†*Order Saurischia*

(†*Suborder Theropoda*): 3 infraorders, 13 families

(†*Suborder Sauropoda*): Brachiosauridae, Titanosauridae

†*Order Ornithischia*

(†*Suborder Orthopoda*): 5 families

(†*Suborder Stegosauria*): 1 family

(†*Suborder Ankylosauria*): 2 families

(†*Suborder Ceratopsia*): 3 families

(Subclass uncertain)

†*Order Mesosauria,* mesosaurs

†Mesosauridae: South Africa and South America, probably Lower Permian; one genus; see text (p. 612).

[It should be repeated that this classification of reptiles, although complete for existing families, omits many ancient groups.]

REFERENCES

Adamson, A. M. 1939. Review of the fauna of the Marquesas Islands . . . . *Bull. Bernice P. Bishop Mus.,* No. 159.

Anderson, C. 1925. . . . *Meiolania* . . . . *Rec. Australian Mus.,* **14,** 223–242.

Angel, F., and H. Lhote. 1938. Reptiles et amphibiens du Sahara central et du Soudan. *Bull. Comité d'Études Historiques et Scientifiques de l'Afrique Occidentale Française,* **21,** 345–384.

Berg, L. S. 1950. *Natural regions of the U.S.S.R.* New York, Macmillan.

Bogert, C. M. 1940. . . . arrangement of African Colubridae. *Bull. American Mus. Nat. Hist.,* **77,** 1–107, pls.

———. 1943. Dentitional phenomena in cobras . . . . *Bull. American Mus. Nat. Hist.,* **81,** 285–360.

———. 1949. Thermoregulation in reptiles . . . . *Evolution,* **3,** 195–211.

Bogert, C. M. 1953. The Tuatara: why is it a lone survivor? *Scientific Monthly,* **76,** 163–170.

Bogert, C. M., and R. M. del Campo. 1956. The Gila Monster and its allies. *Bull. American Mus. Nat. Hist.,* **109,** 1–238.

Bogert, C. M., and R. B. Cowles. 1947. Moisture loss . . . in some Floridian reptiles. *American Mus. Novitates,* No. 1358.

Bogert, C. M., and B. L. Matalas. 1945. A review of the elapid genus *Ultrocalamus* of New Guinea. *American Mus. Novitates,* No. 1284.

Bogert, C. M., and J. A. Oliver. 1945. . . . herpetofauna of Sonora. *Bull. American Mus. Nat. Hist.,* **83,** 297–425.

Boulenger, G. A. 1920–1921. *Monograph of the Lacertidae.* London, British Mus., 2 vols.

Bourret, R. 1936. *Les serpents de l'Indochine.* Toulouse, Henri Basuyau, 2 vols.

———. 1941. *Les tortues de l'Indochine.* Institut Océanographique de l'Indochine, Note 38e.

Brongersma, L. D. 1934. Contributions to Indo-Australian herpetology. *Zoologische Mededeelingen* (Leiden), **17,** 161–251.

Burt, C. E. 1931. . . . *Cnemidophorus* . . . . *Bull. United States National Mus.,* No. 154.

Burt, C. E., and M. D. Burt. 1932. Pacific Island amphibians and reptiles . . . . *Bull. American Mus. Nat. Hist.,* **63,** 461–597.

Camp, C. L. 1923. Classification of the lizards. *Bull. American Mus. Nat. Hist.,* **48,** 289–481.

Carl, G. C. 1944. *The reptiles of British Columbia.* British Columbia Provincial Mus. (Victoria), Handbook No. 3.

Carr, A. 1952. *Handbook of turtles.* Ithaca, New York, Comstock (Cornell U. Press).

Chabanaud, P. 1917. . . . reptiles . . . Afrique occidentale . . . . *Bull. Mus. National d'Hist. Nat.,* **23,** 83–105.

Conrad, G. M. 1940. By boat to the Age of Reptiles [*Sphenodon*]. *Natural History* (New York), **45,** 224–231.

Darlington, P. J., Jr. 1948. The geographical distribution of cold-blooded vertebrates. *Quarterly Review Biol.,* **23,** 1–26, 105–123.

De Vis, C. W. 1894. The lesser chelonians of the Nototherian drifts. *Proc. R. Soc. Queensland,* **10,** 123–127.

Dunn, E. R. 1927. Notes on *Varanus komodoensis. American Mus. Novitates,* No. 286.

———. 1928. . . . arrangement . . . American . . . Colubridae. *Bull. Antivenin Inst. America,* **2,** 18–24.

———. 1931. The herpetological fauna of the Americas. *Copeia,* **1931,** 106–119.

———. 1935. Notes on American Mabuyas. *Proc. Acad. Nat. Sci. Philadelphia,* **87,** 533–557.

Evenden, F. G., Jr. 1948. Distribution of the turtles of western Oregon. *Herpetologica,* **4,** 201–204.

Flower, S. S. 1933. . . . reptiles and amphibians of Egypt . . . . *Proc. Zool. Soc. London,* **1933,** 735–851.

Gadow, H. 1909. *Amphibia and reptiles.* The Cambridge Natural History, Vol. 8. London, Macmillan.

Gilmore, C. W. 1928. Fossil lizards of North America. *Mem. National Acad. Sci.,* **22,** No. 3.

Glauert, L. 1950. *A handbook of the snakes of Western Australia.* Perth, Western Australian Naturalists' Club.

Gloyd, H. K. 1940. *The rattlesnakes* . . . . Chicago Acad. Sci. Special Pub. No. 4.

Gloyd, H. K., and R. Conant. 1943. . . . American . . . *Agkistrodon. Bull. Chicago Acad. Sci.,* **7,** 147–170.

Gray, J. E. 1852. Description of . . . new . . . tortoises. *Proc. Zool. Soc. London,* **20,** 133–135.

Grobman, A. B. 1941. . . . *Opheodrys vernalis* . . . . *Misc. Pub. Mus. Zool. U. Michigan,* No. 50.

Guibé, J. 1949. Revision . . . boidés de Madagascar. *Mem. Inst. Sci. Madagascar,* Ser. A, **3,** 95–105.

Günther, A. 1858. On the geographical distribution of reptiles [snakes]. *Proc. Zool. Soc. London,* **26,** 373–389.

Hecht, G. 1928. Zur Kenntnis der Nordgrenzen der mitteleuropäischen Reptilien. *Mitteilungen Zool. Mus. Berlin,* **14,** 501–595.

Hoffstetter, R. 1942. Sur les restes de . . . Iguanidae. *Bull. Mus. National d'Hist. Nat.* (Paris) (2), **14,** 233–240.

———. 1944–1945. Sur les Scincidae fossiles. *Bull. Mus. National d'Hist. Nat.* (Paris) (2), **16,** 547–553; **17,** 80–86.

———. 1946. . . . Boidae des Mascareignes . . . . *Bull. Mus. National d'Hist. Nat.* (Paris) (2), **18,** 132–135.

Hutton, F. W., and J. Drummond. 1923. *The animals of New Zealand,* 4th ed. Auckland etc., Whitcombe and Tombs.

Inger, R. F. 1948. The systematic status of the crocodile *Osteoblepharon osborni. Copeia,* **1948,** 15–19.

Inger, R. F., and P. J. Clark. 1943. Partition of the genus *Coluber. Copeia,* **1943,** 141–145.

Kinghorn, J. R. 1926. A brief review of the family Pygopodidae. *Rec. Australian Mus.,* **15,** 40–64.

———. 1929. *Snakes of Australia.* Sydney, Angus and Robertson.

Klauber, L. M. 1945. . . . *Coleonyx* . . . . *Tr. San Diego Soc. Nat. Hist.,* **10,** 135–213.

Logier, E. B. S., and G. C. Toner. 1955. Check-list of the amphibians and reptiles of Canada and Alaska. *Contrib. R. Ontario Mus. Zool. and Palaeontology,* No. 41.

Loveridge, A. 1934. Australian Reptiles in the Museum of Comparative Zoology. *Bull. Mus. Comparative Zool.,* **77,** 243–383.

———. 1941. . . . Pelomedusidae. *Bull. Mus. Comparative Zool.,* **88,** 467–524.

———. 1942. . . . Gerrhosauridae. *Bull. Mus. Comparative Zool.,* **89,** 485–543.

———. 1942*a*. . . . *Phelsuma. Bull. Mus. Comparative Zool.,* **89,** 439–482.

———. 1942*b*. [Reptiles of forested areas in East and Central Africa.] *Bull. Mus. Comparative Zool.,* **91,** 237–373.

———. 1944. . . . Cordylidae. *Bull. Mus. Comparative Zool.,* **95,** 1–118.

———. 1944*a*. . . . *Afroedura* . . . . *American Mus. Novitates,* No. 1254.

———. 1944*b*. . . . American . . . *Tarentola. Copeia,* **1944,** 18–20.

———. 1945. *Reptiles of the Pacific World.* New York, Macmillan.

Loveridge, A. 1945*a*. New scincid lizards of the genera *Tropidophorus* and *Lygosoma* from New Guinea. *Proc. Biol. Soc. Washington,* **58,** 47–52.
———. 1947. . . . African . . . Gekkonidae. *Bull. Mus. Comparative Zool.,* **98,** 1–469.
———. 1948. New Guinean reptiles . . . . *Bull. Mus. Comparative Zool.,* **101,** 305–430.
Loveridge, A., and B. Shreve. 1947. The "New Guinea" snapping turtle. *Copeia,* **1947,** 120–123.
Lydekker, R. 1889. *Catalogue of the fossil Reptilia and Amphibia in the British Museum.* Part 3. *Chelonia.* London, British Mus.
McDowell, S. B., and C. M. Bogert. 1954. . . . *Lanthanotus* . . . . *Bull. American Mus. Nat. Hist.,* **105,** 1–142.
Matthew, W. D. 1915, 1939. Climate and evolution. *Ann. New York Acad. Sci.,* **24,** 171–318; reprinted (1939) as *Special pub. New York Acad. Sci.,* **1.**
Mertens, R. 1930. Die Amphibien und Reptilien der Inseln Bali, Lombok, Sumbawa und Flores. *Abhandlungen Senckenbergischen Naturforschenden Gesellschaft,* **42,** 117–344.
———. 1934. Die Insel-Reptilien . . . . *Zoologica* (Stuttgart), **32,** 6. Lieferung, 1–209.
———. 1934*a*. . . . Verbreitung . . . der Seeschlangen . . . . *Zoogeographica* (Jena), **2,** 305–319.
———. 1942. Die Familie der Warane (Varanidae). *Abhandlungen Senckenbergischen Naturforschenden Gesellschaft,* Abh. 462, 1–115; 465, 117–233; 466, 235–391.
Mertens, R., and L. Müller. 1940. Die Amphibien und Reptilien Europas (zweite Liste). *Abhandlungen Senckenbergischen Naturforschenden Gesellschaft,* Abh. 451.
Miller, R. C. 1948. A checklist of the reptiles and amphibians of Canada. *Herpetologica,* **4,** Supplement 2, 1–15.
Mittleman, M. B. 1950. The generic status of *Scincus lateralis* Say, 1823. *Herpetologica,* **6,** 17–20.
Mocquard, F. 1909. . . . reptiles . . . batraciens de Madagascar. *Nouvelles Archives Mus. d'Hist. Nat.* (Paris), Ser. 5, **1,** 1–110.
Mook, C. C. 1925. Ancestry of the alligators. *Natural History* (New York), **25,** 407–408.
Newman, A. K. 1877. Notes of the physiology and anatomy of the Tuatara. *Tr. New Zealand Inst.,* **10,** 222–239.
Oliver, J. A., and C. E. Shaw. 1953. The amphibians and reptiles of the Hawaiian Islands. *Zoologica* (New York), **38,** 65–95.
Phillips, W. J. 1941. . . . *Pelamis* . . . in New Zealand. *Tr. and Proc. R. Soc. New Zealand,* **71,** 23–24.
Pope, C. H. 1935. The reptiles of China. *Nat. Hist. Central Asia,* **10.** New York, American Mus. Nat. Hist.
———. 1944–1945. The poisonous snakes of the New World. *Bull. New York Zool. Soc.* ("Animal Kingdom"), **47,** 83–90, 111–120, 143–152; **48,** 17–23, 44–47.
Romer, A. S. 1945. *Vertebrate paleontology,* 2nd ed. Chicago, U. of Chicago Press.
———. 1956. *Osteology of the reptiles.* Chicago, U. of Chicago Press.
Rooij, N. de. 1915–1917. *The reptiles of the Indo-Australian Archipelago:*

Vol. 1 (1915) Lacertilia, Chelonia, Emydosauria; Vol. 2 (1917) Ophidia. Leiden, E. J. Brill.

Schmidt, K. P. 1919–1923. Contributions to the herpetology of the Belgian Congo . . . . *Bull. American Mus. Nat. Hist.,* **39,** 385–624; **49,** 1–146.

———. 1944. Crocodiles. *Fauna* (Zool. Soc. Philadelphia), **6,** 67–72.

———. 1946. On the zoogeography of the Holarctic Region. *Copeia,* **1946,** 144–152; slightly modified from *Lingnan Sci. J.* (Canton), **10,** 1931, 441–449.

———. 1950. Modes of evolution discernible in the taxonomy of snakes. *Evolution,* **4,** 79–86.

———. 1953. *A check list of North American amphibians and reptiles.* Chicago, U. of Chicago Press.

Schmidt, K. P., and D. D. Davis. 1941. *Field book of snakes of the United States and Canada.* New York, Putnams.

Schmidt, K. P., and W. L. Necker. 1936. The scientific name of the American Smooth Green Snake. *Herpetologica,* **1,** 63–64.

Schreiber, E. 1912. *Herpetologia europaea,* 2nd ed. [with Nachtrag in 1913]. Jena, Fischer.

Serié, P. 1936. . . . distribución geográfica de los ofidios argentinos. *Obra Cincuentenario Museo de la Plata,* **2,** 33–61.

Simpson, G. G. 1938. *Crossochelys,* Eocene horned turtle from Patagonia. *Bull. American Mus. Nat. Hist.,* **74,** 221–254.

———. 1943. Turtles and the origin of the fauna of Latin America. *American J. Sci.,* **241,** 413–429.

Slevin, J. R. 1949. Range extension of *Xantusia vigilis. Herpetologica,* **5,** 148.

Smith, H. M. 1939. The Mexican and Central American lizards of the genus *Sceloporus. Zool. Ser. Field Mus. Nat. Hist.,* **26,** 1–397.

———. 1939*a.* Notes on Mexican reptiles and amphibians. *Zool. Ser. Field Mus. Nat. Hist.,* **24,** 15–35.

———. 1942. Mexican herpetological miscellany. *Proc. United States National Mus.,* **92,** 349–395.

———. 1946. *Handbook of lizards. Lizards of the United States and Canada.* Ithaca, New York, Comstock Pub. Co.

Smith, H. M., and E. H. Taylor. 1945. An annotated checklist and key to the snakes of Mexico. *Bull. United States National Mus.,* No. 187.

———. 1950. An annotated checklist and key to the reptiles of Mexico exclusive of the snakes. *Bull. United States National Mus.,* No. 199.

Smith, M. (A.). 1926. *Monograph of the sea-snakes.* London, British Mus.

———. 1931–1943. Reptilia and Amphibia. (*In*) *Fauna of British India including Ceylon and Burma.* Vol. 1 (1931) Loricata, Testudines; Vol. 2 (1935) Sauria; Vol. 3 (1943) Serpentes. London, Taylor and Francis.

———. 1933. Remarks on some Old World geckoes. *Rec. Indian Mus.* (Calcutta), **35,** 9–19.

———. 1937. A review of the genus *Lygosoma* . . . . *Rec. Indian Mus.,* **39,** 213–234.

———. 1951. *The British amphibians and reptiles.* London, Collins.

Stebbins, R. C. 1954. *Amphibians and reptiles of western North America.* New York, McGraw-Hill.

Stuart, L. C. 1940. Notes on the "*Lampropholis*" group of Middle American *Lygosoma* . . . . *Occasional Papers Mus. Zool. U. Michigan,* No. 421.

Stull, O. G. 1935. A check list of the family Boidae. *Proc. Boston Soc. Nat. Hist.*, **40**, 387–408.

Taylor, E. H. 1920. Philippine turtles. *Philippine J. Sci.*, **16**, 111–144.

———. 1922. The snakes of the Philippine Islands. *Philippine Bureau Sci. Pub.*, No. 16.

———. 1922*a*. The lizards of the Philippine Islands. *Philippine Bureau Sci. Pub.*, No. 21.

———. 1935. . . . *Eumeces* . . . . *U. of Kansas Sci. Bull.*, **23**, 1–643.

———. 1937. . . . *Leiolopisma* from Mexico . . . . *Copeia*, **1937**, 5–11.

Taylor, E. H., and H. M. Smith. 1943. A review of American sibynophine snakes . . . . *U. of Kansas Sci. Bull.*, **29**, 301–338.

Tihen, J. A. 1949. The genera of gerrhonotine lizards. *American Midland Naturalist*, **41**, 580–601.

Underwood, G. 1951. On the distribution of snakes of the family Boidae. *Proc. Zool. Soc. London*, **120**, 713–714.

———. 1954. On the classification and evolution of geckos. *Proc. Zool. Soc. London*, **124**, 469–492.

Vanzolini, P. E. 1951. A systematic arrangement of the family Amphisbaenidae. *Herpetologica*, **7**, 113–123.

Williams, E. 1950. . . . cervical central articulations of living turtles [with a summary of classification of turtles]. *Bull. American Mus. Nat. Hist.*, **94**, 505–562.

———. 1950*a*. *Testudo cubensis* and the evolution of Western Hemisphere tortoises. *Bull. American Mus. Nat. Hist.*, **95**, 1–36.

———. 1952. . . . tentative arrangement of the tortoises of the world. *Bull. American Mus. Nat. Hist.*, **99**, 541–560.

———. 1952*a*. A staurotypine skull from the Oligocene of South Dakota. *Breviora* (Mus. Comparative Zool.), No. 2.

———. 1953. Fossils and the distribution of chelyid turtles. *Breviora* (Mus. Comparative Zool.), No. 13.

———. 1954. . . . the living species of . . . *Podocnemis*. *Bull. Mus. Comparative Zool.*, **111**, 277–295.

———. 1954*a*. Fossils and the distribution of chelyid turtles. 2. Additional reputed chelyid turtles on northern continents. *Breviora* (Mus. Comparative Zool.), No. 32.

———. 1956. *Pseudemys scripta callirostris* from Venezuela with a general survey of the *scripta* series. *Bull. Mus. Comparative Zool.*, **115**, 145–160.

Woodward, A. S. 1901. . . . *Miolania* . . . . *Proc. Zool. Soc. London*, **1901**, 169–184.

Zangerl, R. 1947. Redescription of *Taphrosphys olssoni* . . . . *Fieldiana* (Chicago Nat. Hist. Mus.), *Geology*, **10**, 29–40.

chapter 5

Birds

In some ways, birds are the best-known animals. Almost all existing species of them are probably known, some 8600 full species (Mayr 1946*a*, Mayr and Amadon 1951) plus thousands of geographical subspecies, and the distributions of many of the species are known in detail. Of all vertebrates, birds are the ones I know best myself. I have watched them almost all my life and have collected them in a small way in northern South America and Australia. I have had the benefit of many conversations about them with the late James L. Peters and with Ludlow Griscom and James C. Greenway, Jr., of the museum staff. And Dr. Josselyn Van Tyne and Professor Ernst Mayr have read stages of the manuscript of this chapter and made useful criticisms of it; Professor Mayr has allowed me to use his carded references on bird geography. I have therefore had unusual opportunities. Nevertheless, I still find the distribution of birds very hard to understand. The present pattern is clear enough, though complex. But the processes that have produced the pattern—the evolution and dispersal of birds—are very difficult to trace and understand.

Perhaps because the subject is so difficult, most American ornithologists (with some outstanding exceptions) seem very little interested in the geographical relationships and sources of our birds. The geography of migration is widely known, but not much else, and there is much else worth knowing in spite of the difficulties. Our bird fauna is not a stable thing. It is a temporary assemblage formed in the course of time, some of it recently, by complex movements of many birds in many directions, and it is destined to be partly scattered or destroyed and added to and reformed continually. To me the "chank" of a Scarlet Tanager and the screech of a Crested Flycatcher are authentic sounds of the tropical rain forest of South and Central America, from where these birds were derived not long ago. The rattle of our Kingfisher is, I suppose, a sound from the Old World tropics: our bird belongs to a genus of Africa and southern Asia; it is one of the more specialized members of a mainly Old World family, many members of which eat insects, lizards, and the like instead of fishes. Our eagles, falcons, and some of our other hawks have evidently come from the Old World by way of Eurasia rather recently; our vultures have probably come from tropical America; but *Buteo* (including the fine Red-tailed Hawk, and the Red-shouldered, Broad-winged, and others) has been ours for 30 million years or so. Europe and Asia have evidently given us the Barn Swallow (although it now winters in South America etc.), our Robin (which is really a thrush, not related to the English Robin), our chickadees and titmice, goldfinches and crossbills, jays and crows, and many other songbirds; tropical America has given us mockingbirds and thrashers, wrens (one of which has spread across Asia and Europe), vireos, wood warblers, orioles (which are not the same as Old World orioles), blackbirds (which have nothing to do with the English Blackbird, which is a thrush), many sparrows, and others. Details like these have an importance beyond their individual interest. Some knowledge of the birds of the world and of the geographical history of birds is necessary to understand our own bird fauna.

Ornithologists are still far from agreement on many points of classification of birds, on what families to recognize, what genera go in what family, and how the families are related to each other. There are in fact two recent classifications, one by Wetmore (1951) and the other by Mayr and Amadon (1951). The two do not differ much in families recognized, except that Mayr and Amadon reduce a number of conventional families to subfamilies, but the arrangements are different. Mayr and Amadon go further than Wetmore in arranging the songbirds in groups according to probable or possible lines of evolution, and the groups suggest dispersals better than Wetmore's unbroken column of songbird families does. For this reason Mayr and Amadon's classification is used here.

One order of birds, the Passeres or perching birds, including the songbirds, is overwhelmingly dominant on land. It contains more than half of all existing birds. This one order will repeatedly be opposed to all other birds, the non-passerines, in the following pages.

Fossil birds are few and often fragmentary and doubtful in their relationships. Birds are a good example of the fact that, in zoogeography, a poor fossil record interpreted too literally is almost worse than no record at all. In no case do fossils clearly show the place of origin of a widely distributed family of birds. In some families the earliest known fossils are in Europe or North America, but this probably reflects the distribution of paleontologists more than that of birds. Parrots, for example, have been found in the Upper Oligocene or Miocene in Europe and the Miocene in North America, and nowhere else until the Pleistocene. Probably no one would say that parrots originated in the north and did not spread southward until the Pleistocene, but efforts have been made to fix the places of origin of, for example, ducks and trogons in Europe and cracids and limpkins in North America on the same sort of geographically one-sided evidence.

In special cases fossils do suggest that certain groups of birds have always been confined to certain places: elephant birds to Madagascar, moas to New Zealand, penguins to southern and auks to northern parts of the world, etc. In some other cases fossils show that certain birds once occurred outside of (usually north of) their present limits: ostriches, across southern Europe and Asia; tropic birds, pelicans, secretary birds, pheasants, jungle fowls, painted snipes, parrots, trogons, hornbills, and edible-nest swifts in France, Germany, or England; flamingoes, in South Dakota and Oregon; cracids, limpkins, and parrots, in Nebraska etc.; turkeys, in California; the California Condor, in Florida; and Old World vultures, in North America. But the changes of distribution shown by these fossils are withdrawals which probably have little to do with places of origin and usually not much more to do even with dispersal routes. The exceptions are those few cases, for example parrots and trogons, in which the fossils show a northward extension in the past of groups that are now more or less confined to tropical or southern parts of both Old and New Worlds. In these cases the fossils suggest dispersal through the north.

In the list of bird families at the end of this chapter, fossil families are interpolated according to Wetmore (1951), and known fossils of existing families are noted briefly, but they should be interpreted with caution.

Although fossils tell so little about the geographical history of birds, they do give some idea of the general course of bird evolution and of the time available for dispersals. *Archaeopteryx* and *Archaeornis* of the Upper Jurassic probably carry birds back nearly to their reptilian ancestors. Cretaceous birds were much more like existing ones. So far as is known they were all toothed, but only two are adequately known and they were both sea birds. Toothless birds may well have appeared before the end of the Cretaceous. Known Eocene (including Paleocene) birds are all toothless. A number of them are currently assigned to existing families, but this is partly a matter of convenience. Many of the fossils are too fragmentary to show degrees of relationship. Wetmore (1951*a*, pp. 53–54) thinks that most Eocene birds were probably really different enough from existing ones to belong in different, extinct families, and that most existing families therefore date from later-than-Eocene times, but this is not much more than a well-informed guess. Mayr (1946, p. 3) thinks that many existing bird families are older, that they date from or before the beginning of the Tertiary. It may be that some existing families of non-passerine birds do go back to the Eocene or even to the late Cretaceous, but it does not necessarily follow that their distributions date from then. The older groups may have redispersed repeatedly during the Tertiary. Passerines too are listed fossil back to the Eocene, but their early record is doubtful. It seems likely that most of them are more recent than most non-passerines, and that many passerine families may date from the mid-Tertiary (see pp. 276–277), but this too is not much better than a guess. Wetmore (1951*a*, p. 63) thinks that living genera and species of birds evolved mostly in the Miocene and Pliocene. He thinks that birds in general had their maximum abundance in the Tertiary and have declined since then. This may be true of most non-passerines on land, but the passerines seem to be in the midst of their radiation and may be at a maximum now.

Birds are reptiles which have evolved feathers, warm-bloodedness, and the power of flight. Feathers and warm-bloodedness enable birds to withstand cold better than reptiles can, and this has probably facilitated their dispersal over the world and affected their pattern of distribution in other ways. But obviously the most important thing about birds, zoogeographically, is that they can fly.

Flight affects the distribution of birds in several ways: by enabling them to cross barriers, by increasing dispersal rates regardless of barriers, and by making possible great annual migrations which not only complicate the distributions of the migratory forms but affect the whole main pattern of bird distribution.

That birds can fly across barriers is one of those apparently simple facts that are not simple. Not all birds do it. Not only flightless birds but also many flying ones are often stopped by barriers such as moderate water gaps. For example several families of birds that are well represented and probably old in South and Central America do not cross the moderate water gaps to the West Indies. These families include the tinamous, cracids, puffbirds etc., toucans, the whole superfamily Furnarioidea (500 species), and manakins. That none of these birds reaches the West Indies proper is a remarkable fact of bird geography. Some other families that are well represented in South and Central America, for example quails, trogons, and cotingas, reach only one or two islands of the West Indies. But still other families and subfamilies, including such strictly New World ones as hummingbirds, tyrant flycatchers, and wood warblers, and also such widely distributed ones as pigeons and parrots, have spread widely over the West Indies.

New Zealand supplies an informative series of barrier-crossing birds. New Zealand is separated from Australia by about 1000 miles of ocean. Many Australian bird families, some old and some recent from an evolutionary point of view, have failed to reach New Zealand. They include (besides the flightless emus and cassowaries) megapodes, button quails, goatsuckers and their allies, lyrebirds, scrubbirds, flowerpeckers, sunbirds, weaver finches, etc. Some other Australian groups, including parrots and "crow-like" birds, are represented in New Zealand by much-differentiated forms which probably arrived long ago. Others have apparently reached New Zealand more recently. And others are still reaching it. Among the last are an Australian roller which occasionally reaches New Zealand, sometimes in small flocks; a swallow which appears occasionally in small flocks and is said to have nested once; an Australian cuckoo shrike which has been recorded from New Zealand at least four times; an Australian *Zosterops* which appeared suddenly in New Zealand about 1856 and became established there; the White-faced Heron, first reported from New Zealand before 1865, which began nesting there about 1940 and is now established; the Royal Spoonbill and Spur-winged Plover, first found nesting in 1950; and the Australian Coot, which is suspected of nesting (Oliver 1930 and 1955, Falla 1953).

The general rule derived from situations like these is that, where birds come against barriers, some of the birds are stopped while others cross the barriers at different times and for different distances.

There is no final limit to the width of barriers that some birds cross. Most of the remote islands of the world, if they are habitable, have been reached and colonized by land birds. For example, the Hawaiian Islands have received a few land birds across at least 2000 miles of ocean from America as well as from the Old World; and Tristan da Cunha etc. in the South Atlantic, about 2000 miles from the nearest mainland (Africa) and perhaps the most isolated habitable islands in the world, have five endemic genera of land birds representing five separate arrivals (two rails, a thrush, two finches). It is a reasonable possibility that some land birds may have crossed the Atlantic Ocean between Africa and South America, a distance of less than 2000 miles. This is within the powers even of some small land birds, like the thrush and finches that have reached Tristan da Cunha. This possibility should be kept in mind. However, the dispersal of most land birds evidently usually occurs along continuity of land rather than across wide ocean gaps.

Although there is no final limit to the width of water or the number of water gaps that some land birds may cross, water gaps often profoundly affect bird faunas by limiting and selecting the birds which compose them and by isolating them to a greater or lesser degree.

Why some land birds do and others do not habitually cross water barriers is not easy to say. Large birds probably have an advantage; other things being equal, they can fly farther than small birds can. Strong fliers (*e.g.*, swallows) evidently have an advantage too. But this is only part of the matter and not enough to explain most actual cases. As a rule it is not possible to tell from a bird's physical characteristics or from watching it fly whether or not it is a water crosser. The best example of this is the rails. When we flush a rail and see how reluctant it is to fly and how it struggles to get into the air and fly even a few yards, we might doubt that it could fly across even a good-sized ditch. But rails more than any other non-marine birds have reached and populated the most remote oceanic islands. Of course there is more to this than just the ability or habit of flying across water. The birds which successfully disperse across water barriers must be able to establish themselves in the new lands they reach, as rails can often do even on small islands, and this brings in a whole new set of complications, including habitat limitations, competition, and adaptability.

The dispersal rates of birds must depend on many things: on dominance and adaptability, for example, as well as on power of movement. Birds may be stationary or may retreat regardless of their ability to fly. But when other things are favorable, flight allows some birds to disperse so rapidly that they reach all the principal parts of the world within the evolution times of single species. The existence of a number of more or less cosmopolitan species is proof of this.

The rapidity with which some dominant birds have spread over the world is probably one of the reasons why their histories are so hard to trace. If dispersal is very rapid, there will be little time for evolution during dispersal and few clues as to how dispersal has occurred. Suppose, for example, that the ancestors of a rising, dominant group of birds spread over the world in less than a million years; then the occurrence of species and genera of the group 10 million years later will depend mostly on the evolutionary possibilities in different places —on available area, continuity of area, number of ecological niches, amount of competition, etc.—and not on the length of time the group has been in different places. To find the place of origin and lines of dispersal of such a group by the usual clues may be very difficult.

The difficulty is increased by the fact that the dispersal of dominant groups of birds is often multiple. Several genera or species of one group may disperse over the world at the same time. For example, at least four genera of existing hawks have become nearly cosmopolitan, *Circus, Accipiter, Falco,* and *Pandion; Falco* breaks into subgenera of which in turn three are cosmopolitan or nearly so; and one species of *Falco* and the one of *Pandion* are nearly cosmopolitan. Of herons, at least half a dozen genera and two species are cosmopolitan or tropicopolitan. Four genera of Rallidae are cosmopolitan or nearly so. And three genera or subgenera of swallows are cosmopolitan or at least very widely distributed. Moreover there has probably been successive as well as multiple dispersal in some of these groups. Herons, rails, and one family of hawks apparently go back at least to the Eocene. It seems likely that existing cosmopolitan genera of these groups have dispersed recently and that they were preceded by earlier widely distributed genera.

In short, the dispersal of dominant birds has probably often been so rapid and also so complex as to be beyond analysis. From the point of view of zoogeography this may be the most important effect of flight, and it is one of the main reasons (the other being the fewness of fossils) why it is so difficult to trace and understand the geographical history of birds. I shall interrupt discussion of these matters now, but in the proper place (pp. 269ff.) I shall try to show what the clues do suggest about the recent dispersals of dominant groups of birds.

Geographical effects of bird migration

The migration of birds is still mysterious in some ways. Why certain stimuli cause certain birds to migrate at certain times and in certain directions, and how the birds find their way over great distances of land and water are still essentially mysteries, although we have some clues. But these matters are not important here. Birds do migrate, and they do find their way at least most of the time. What is important here is the general nature and pattern of migration and the main geographical effects of it. I have recently refreshed my knowledge of this subject from Thomson (1926, 1949), Griscom (1945), Moreau (1951), Phillips (1951), Mayr (1953), and other sources including Mayr and Meise (1930), who are credited by Moreau with formulating the basic theory of the origin of migration.

Bird migration is usually thought of in terms of movement, but it may be thought of also in terms of range. The range of an animal is the particular area it inhabits. On maps, ranges seem simple, stable things, but one should not be deceived (as I am afraid some zoogeographers are) by that. Ranges are inherently complex and unstable. One of the complexities is that the ranges of most free-living animals change slightly with the animals' activities. The gross range of a species usually consists of two or more slightly different ranges which are occupied alternately or successively. For example, both the exact limits and internal range patterns of many species of vertebrates which are not considered migratory are slightly different when the animals are successively feeding, breeding, dispersing after breeding, and wintering. Each of the alternate ranges which make up the gross range of a species can change to some extent *independently* of the others, the amount of change being limited by the powers of movement of the animals concerned. In the case of animals that can move far and fast, alternate ranges can and sometimes do move far apart. This is what has happened in the case of strongly migratory birds.

If the duality of range of migratory birds is thought of not as unique but as a special case of the less conspicuous alternation of ranges characteristic of free-living animals, then it becomes, if still marvelous and still mysterious in some ways, perhaps somewhat easier to understand in other ways. There is no need any more for the old idea, almost discredited anyway, that bird migration is a special thing imposed on northern birds by the ice age.[1] There is no need to think

[1] It has been suggested by Wolfson (1948) that the more highly developed forms of migration were imposed on birds by continental drift. This idea was

of migratory species as having originated either where they now breed or where they now winter; their breeding and wintering ranges may both have changed within the widest limits allowed by the birds' powers of movement. For example it is not necessary to suppose that the Arctic Tern, which breeds in the northern part of the northern hemisphere north into the arctic and winters in the southern part of the southern hemisphere south to the antarctic (Fig. 30), was originally either an arctic or an antarctic bird, or that it began to migrate across the tropics suddenly. It may have begun as a tropical species (several other species of the same genus are tropical) which became first moderately migratory, breeding in not very cold parts of the north-temperate zone and wintering in the tropics (as the Common Tern does now), and it may then have moved its breeding range northward and its wintering range southward. Double movements like this may explain known cases in which the northernmost breeding form of a species is the southernmost wintering form, migrating across the ranges of less migratory forms. Perhaps the individual birds which moved their breeding ranges northward at first returned to winter with the resident birds but could not compete with them, perhaps because the resident ones were more familiar with the local food supply, so that the migratory individuals eventually moved their wintering ranges southward to entirely new grounds.

If this way of looking at migration is correct, all birds are potentially migratory. All of them have or can have the beginnings of alternate, seasonal ranges, and the seasonal ranges can be moved apart in any directions that are favorable and to any distances that the powers of movement of the birds allow.

It follows that migratory habits ought to have originated many different times, in many different birds, in many different places, and on many different patterns, and this has been the case. It is probable too that migratory habits have been lost many times, that many different birds belonging to migratory groups have stopped migrating and become resident in favorable places. This has probably happened, for example, among some wood warblers (see p. 282).

The theoretical complexity is matched by the actual complexity of

sensational enough to be mentioned in a national news magazine and answered in *Science* (Amadon 1948), but it hardly fits the facts. If the continents drifted apart, they did so much too long ago to affect existing species of birds. At best, even if the drift explanation were satisfactory otherwise, it would explain only certain special cases which are only a fraction of the whole phenomenon of migration, leaving the rest unaccounted for.

bird migration. Multitudes of different birds make small movements which grade almost imperceptibly from local changes of feeding grounds in some cases, to irregular or regular movements of individuals or subpopulations or whole populations for a few miles or a few hun-

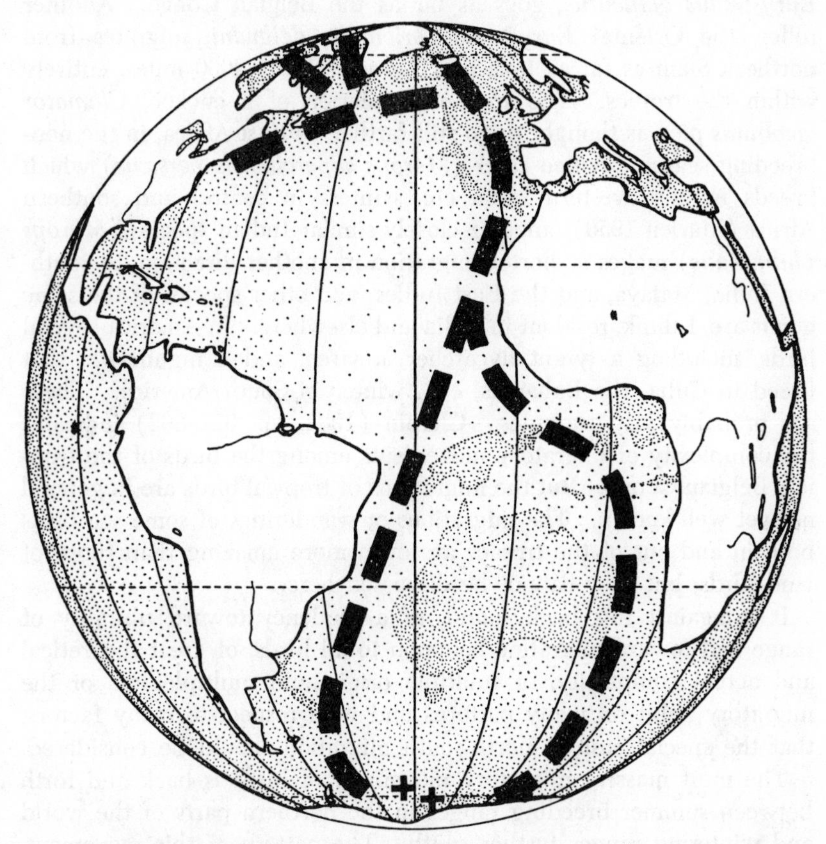

Fig. 30. Migration routes of the Arctic Tern in the Atlantic (diagrammatic). Crosses show actual localities where wintering individuals have been taken (Murphy 1936, pp. 1099–1105).

dred miles in other cases, to great migrations of a thousand or several thousand miles in other cases. These movements occur in all parts of the world. Even the great migrations are not all made by northern birds escaping winter. Moreau (1951) lists some of the others. Several species of birds migrate from North Africa or Egypt southward for a thousand miles or more; others that breed in warm-temperate South Africa migrate equal distances northward; and still others move as far or farther entirely within the tropics in Africa. These movements are correlated with wet and dry seasons rather than with warm and cold ones. Rand (1936, p. 300) lists five birds which nest in Madagascar and "winter" in Africa; at least one of them, a roller, *Eurystomus glaucurus,* goes as far as the Belgian Congo. Another roller, the Oriental *Eurystomus orientalis deignani,* migrates from northern Siam as far as Java and Borneo, nearly 2000 miles, entirely within the tropics. The Indian population of a cuckoo, *Clamator jacobinus pica,* is thought to go still farther, to East Africa, in the non-breeding season. A bee eater (*Merops superciliosus persicus*) which breeds in northwestern India etc. winters in central and southern Africa (Marien 1950); another closely similar Indian species (*Merops philippinus*) makes a shorter migration in another direction, to southern India, Malaya, and the East Indies; and other species of the same genus are, I think, resident in India and elsewhere. In America several birds, including a tyrant flycatcher, a vireo, and a nighthawk, that breed in Cuba, the Bahamas, etc., winter in South America. These are probably just examples. Chapin (1932, pp. 322–362) indicates the complexity of migratory movements among the birds of the tropical Belgian Congo. But the migrations of tropical birds are in general not yet well known. The migrations or wanderings of some sea birds both in and out of the tropics are even more amazing than those of land birds, but they cannot be discussed here.

It is against this background, of a tendency toward plurality of range and toward migration common to all birds, of great theoretical and actual complexity, of multiple origin and multiple loss of the migratory habit on many patterns and with respect to many factors, that the special case of north-south migration should be considered.

The most massive migratory movement of birds is back and forth between summer breeding ranges in the northern parts of the world and wintering ranges farther south. The pattern of this movement, the reasons for it, the origin of it, and the effect of it on the main pattern of distribution of birds are all to be considered.

The main pattern of north-south migration is well known, but the details are very complex. Many different kinds of birds that breed in the north go south by many different routes; different individuals of one species often follow different routes; and the main populations of some species go south by one route and return north by another. The movement is not always directly north and south. Some birds go very indirectly, moving east or west before they go south. Some Asiatic birds that have extended their breeding ranges to Alaska return to the Old World to winter, and vice versa. The Wheatear, which breeds across northern Eurasia, has extended its breeding area into North America from two directions, from Europe across Greenland into eastern North America south to northern Quebec, and from Asia into Alaska, but American individuals still winter in the Old World: the eastern ones go through Europe to West Africa; most of the Alaskan ones, through the length of Asia to East Africa (Meinertzhagen 1954, p. 240, map). The East Siberian form of the Willow Warbler (*Phylloscopus trochilus*) winters in East Africa, and another species of the same genus (*P. borealis*), which breeds from Asia to northern Norway, winters in Siam etc., crossing the migration route of the first bird at right angles (Ticehurst 1938). There are also very great differences in the distance that different north-south migrants go, some scarcely leaving their northern breeding grounds and others going to the tropics or even beyond, and these differences in some cases occur between closely related species or even between individuals of one species. Thomson (1949, pp. 107–122) and Mayr (1953) describe the complexity of north-south migration in more detail.

In the south-temperate zone there is a south-north migration of birds, but fewer species are involved and most of them do not go far. In Chile, for example (Hellmayr 1932, pp. 25–26), some birds that nest high in the southern Andes winter in adjacent valleys or move for moderate distances northward; various land birds that nest at the southern tip of South America winter in central or northern Chile etc.; and there is some coastal migration of sea birds, including penguins. In Australia, too, some south-breeding birds migrate northward, more in the cooler eastern part than in the warmer western part of the continent. In Western Australia (Serventy and Whittell 1948, pp. 5–6) there is only a little "classical" migration of land birds but a number of species make regular, local, partial migrations or are nomadic. In New Zealand there is some local movement, and two New Zealand land birds make migrations which are among the most remarkable in the world. Both are cuckoos. Both leave New Zealand in winter. One of them, *Chalcites lucidus lucidus,* winters in the Solomon Islands etc. (Fell 1947); the other, *Eudynamis taitensis,* throughout Polynesia etc. in the central Pacific (Bogert 1937). The second of these birds is a very distinct species and its pattern of migration may be long established, but the first is only a New Zealand form of a species which is represented in Tasmania and southern Australia by another migratory form (which winters in the Lesser Sunda Islands and New Guinea) and on certain tropical islands by resident forms, so that the migratory pattern of the New Zealand bird may be of recent origin. Both the New Zealand cuckoos apparently migrate across the ocean, not via Australia, but Australia may nevertheless have been concerned in the evolution of their migrations.

The main reason for the north-south and south-north migrations of birds is evidently seasonal variation in food, although other factors are probably involved in some cases. The extent of migration of different birds varies with their food. Most northern insectivorous birds are migratory and so are a good many fruit eaters, while the seed eaters are on the whole less migratory; but this is a general rule to which there are many exceptions. Some northern birds of prey are migratory but others are not or move only sporadically. Many northern fresh-water birds are strongly migratory. They are very numerous in the far north in summer, but after breeding they come south for varying distances. Many that breed by fresh water take to the seashore on migration, and some of them, especially waders, go enormous distances, to the south-temperate zone of South America, Australia, or New Zealand, to find food during the northern winter.

The origin of the mass north-south migration of birds has probably been as complex as the pattern of it. Very many different birds have evolved north-south migrations, and some of the migratory forms have probably further complicated the situation by later setting off local resident forms. It has long been argued whether the north-south migratory groups are in general southern ones which have moved their breeding ranges northward or northern ones which have moved their wintering ranges southward, and if the latter, whether north-south migration occurred before the Pleistocene or whether it was forced on northern birds by Pleistocene climate. Probably all these possibilities have been realized in different cases.

If birds, like other animals, move to secure advantages, many north-south migrations probably began before the Pleistocene by northward movements of the breeding ranges of tropical or warm-temperate birds to take advantage of the abundant food in the north in summer—of course the northern parts of the world had some alternation of seasons even before the Pleistocene. The migrations of most northern insectivorous birds probably began in this way. These birds in Eurasia and in North America mostly belong to different families which seem to have come independently from the Old and New World tropics. But northern water birds may have evolved migration in the other direction. They usually breed farther north, often within the arctic.

They are adapted to primarily northern habitats, which the insectivorous birds are not, and their migrations may have originated as southward movements of northern birds in search of food in the nonbreeding season. This too is suggested by the birds' relationships. Many of the northern Eurasian and North American migratory water birds are closely related, as if they are descended from originally northern, circumpolar stocks. However, whatever the mode of origin, the Pleistocene probably increased both the number and length of migrations of northern birds. If, for example, northern insectivorous birds and water birds were only moderately migratory before the Pleistocene, afterward (and during each interglacial period) they presumably moved their breeding areas northward to take advantage of new habitats and reduced competition, and then, because of the pressure of increased populations, they may have moved their wintering areas southward, widening the discontinuities between breeding and wintering ranges, as I have suggested that the Arctic Tern may have done.

Although migration extends and complicates the distributions of many different groups of birds, the total effect of it is orderly. It has not scattered birds at random over the world. Migrating birds do sometimes get lost and turn up in unexpected places, but they do not often establish themselves there.

The principal effect of migration on the main pattern of distribution of birds has probably been to increase relationships between tropical and north-temperate bird faunas in each half of the world, and to decrease relationships between the Old and New Worlds, especially among insectivorous birds. In each half of the world, the Old World half and the New World half, different groups of insectivorous and some other small land birds have apparently extended northward from the tropics as migratory birds and become intrenched in the north in summer, while most of them have been kept within their half of the world by the habit of returning to specific wintering areas. In the Old World, Muscicapinae, Sylviinae, and Motacillidae, all of which are numerous in the Old World tropics, extend northward in numbers as migratory birds, but no Muscicapinae and only a few Sylviinae and Motacillidae have reached North America. In the New World, Tyrannidae, Vireonidae, and Parulinae, all of which are well represented in the tropics, extend far northward in numbers as migratory birds but do not reach Eurasia, except that two species of Parulinae have extended their breeding ranges to include small areas in eastern Siberia. These birds not only have been limited themselves by their pattern of migration but also have probably tended to block exchange of other insectivorous birds between the Old and New Worlds. Probably no other animals that are well represented in the north are so different in Eurasia and North America as these birds are. This effect does not occur, or at least not to the same extent, among large birds (which have greater powers of dispersal) or among water birds.

Among shore birds, migration has had another special effect. It has allowed many species that breed in the far north to winter in remote south-temperate areas, and this has probably prevented the evolution of any considerable number of shore birds breeding in the south-temperate zone.

Limits of distribution of birds

Some of the limits and other features of bird distribution are given in Figure 31.

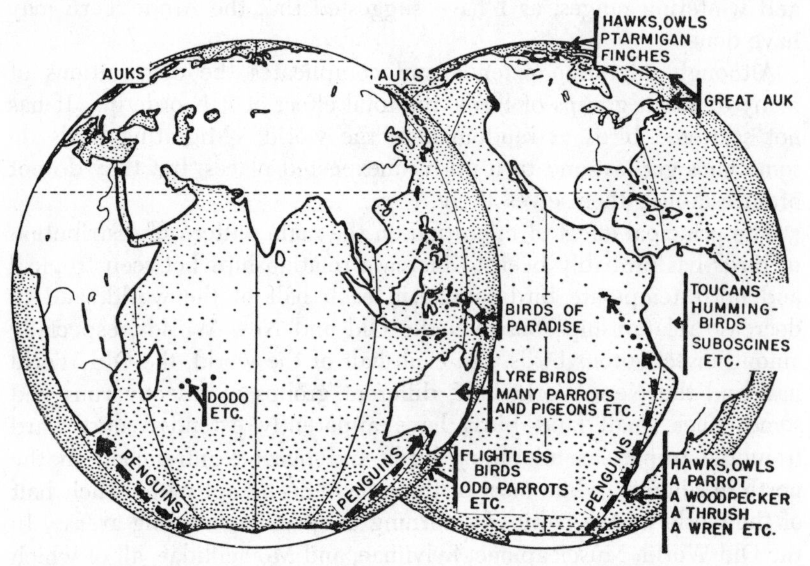

Fig. 31. Northernmost and southernmost land birds, and some other details of the distribution of birds.

Northward, birds reach no absolute limit except the end of land, and I suppose sea birds feed or wander beyond that. But birds decrease in variety northward. A good many different sorts of birds breed north to the limit of trees. Many water birds (geese, ducks, shore birds, etc.) but few land birds (some birds of prey, ptarmigan, a few finches, etc.) breed farther north, although a few of them extend well above 80° N. on the arctic islands above North America. Most of them migrate southward to winter, but a few including ptarmigan and the Snowy Owl often winter in the far north. Even some redpolls (seed-eating finches of the genus *Acanthis*) are apparently resident as far north as northern Greenland.

Southward, many birds reach the southern tips of the continents; some breed on antarctic islands still farther south; some penguins breed on ice on the edge of the Antarctic Continent; and the Snow Petrel and several other Tubinares nest in rocky places or on bare ground even somewhat inland on the Antarctic Continent. The most southern birds that can reasonably be considered land birds are the sheath bills (Chionididae, see list of families), which breed on antarctic islands and perhaps also on the edge of the Antarctic Continent, and which wander somewhat but are not strongly migratory.

There is no final limit to the occurrence of birds on islands. A few of them, even a few small land birds, have reached most of the remotest habitable islands in the world. The effect of distance or of a succession of water gaps is not to set a definite boundary for land birds but to decrease their variety progressively, fewer and fewer kinds of them occurring on more and more remote islands. On islands of the western and central Pacific, for example, the variety of land birds, especially of songbirds, decreases with increasing distance from the continents and large islands of the Indo-Australian area (Fig. 32), and the decrease occurs not only on successive archipelagos but in some cases on successive islands within the archipelagos (Mayr 1933, p. 315, map). This is another example of the fact that, where birds come against barriers, some of the birds are stopped while others cross the barriers for varying distances.

Some species and genera of birds are almost without limits in their distribution. A falcon, *Falco peregrinus* (the European Peregrine Falcon and American Duck Hawk), breeds in parts of all the continents and on many islands. The Fish Hawk, *Pandion haliaetus,* breeds or winters over most of the world, but in some regions it is mainly coastal. A large heron, *Casmerodius albus* (one subspecies is the American Egret), is almost cosmopolitan except in cold places. A small heron, *Butorides striatus,* is almost tropicopolitan except that it is replaced by a related species in North and Central America and the West Indies. An ibis, *Plegadis falcinellus,* is nearly tropicopolitan except that the South American representative of it may be a separate species. The Snowy Plover, *Charadrius alexandrinus,* is represented on at least part of every continent. The Black-necked Stilt,

Himantopus himantopus, occurs on all the continents, but its breeding areas are localized. And the Barn Owl, *Tyto alba,* is almost cosmopolitan except that it is absent in much of temperate Asia and in cold

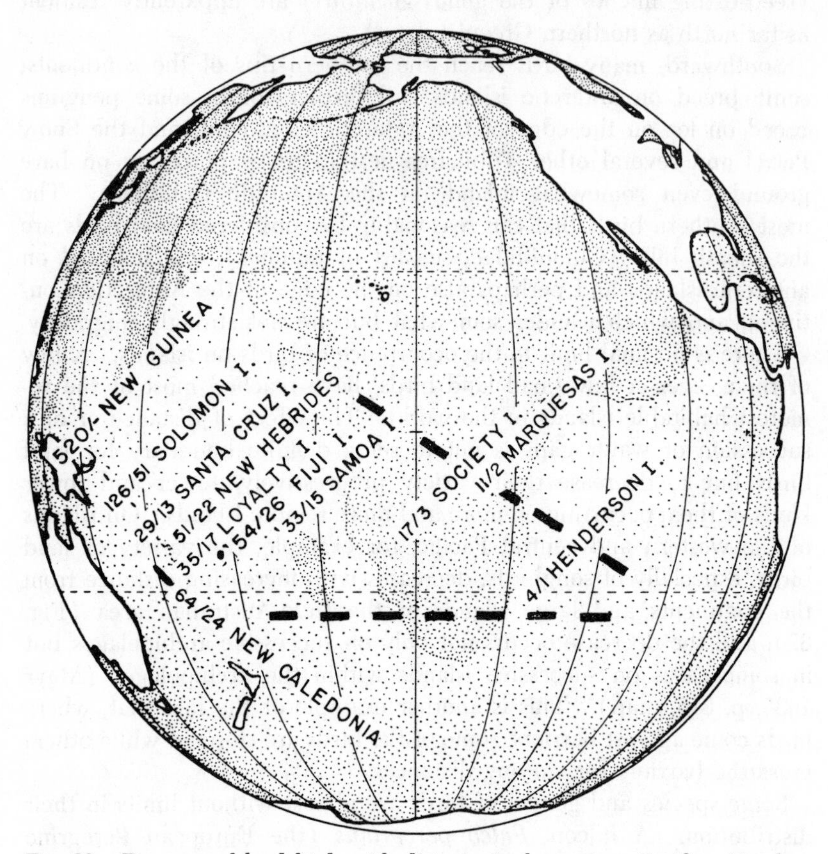

Fig. 32. Decrease of land birds with distance in the western Pacific. Numbers (*e.g.* 126/51) are total numbers (species) of land birds over numbers of song-birds (after Mayr 1933). The numbers are placed approximately on the positions of the islands named.

places. The distributions of most of these species are given in more detail in the list of families. All of them break up into subspecies in different parts of the world. All are non-passerines; all except the *Charadrius* are medium-sized or large; and all are either water birds or predators.

Genera of non-passerine land and fresh-water birds, besides those named above, that are cosmopolitan or tropicopolitan or nearly so include two genera of grebes; *Phalacrocorax* and *Pelecanus* (cormorants and pelicans—but these are partly sea birds); *Circus* and *Accipiter* among the hawks; about five genera of herons; *Dendrocygna* (tree ducks) and *Anas* (common ducks); *Rallus* and *Porzana* among the rails, *Gallinula* among the gallinules, and *Fulica* of coots; *Burhinus* (thick-knees); *Recurvirostra* (avocets), which is very widely but discontinuously distributed (besides this genus and *Charadrius,* some other genera and even species of wading birds cover much of the world in their migrations and winterings); *Columba* (common pigeons); and *Caprimulgus* among goatsuckers.

Eight species and 21 additional genera of non-passerine birds that are more or less cosmopolitan or tropicopolitan are listed above, and there may be more.

Passerines, in general, are not so widely distributed. No species of them is quite cosmopolitan or tropicopolitan. The Barn Swallow (*Hirundo rustica*) probably comes nearest to being so: with its several subspecies, and including its wintering as well as breeding areas, this species occurs over most of the world except Australia, and it occasionally reaches the north coast of the latter; but it only winters and does not breed in most of Africa, tropical Asia etc., and South America etc. The most widely distributed genus of perching birds may be *Hirundo* (barn swallows etc.), which breeds on every continent except South America and winters there. *Petrochelidon* (cliff swallows etc.), which may be only a subgenus of *Hirundo,* breeds on some part of every continent, but its different breeding areas may be more restricted and discontinuous than those of *Hirundo* proper. A third genus of swallows, *Riparia,* is widely distributed too in both Old and New Worlds, but I do not know its exact limits in the former, and in the latter its breeding range does not include South America, although it winters there. *Anthus* (pipits or titlarks, of the family Motacillidae) breeds in part of every continent but is discontinuously distributed, as if in its dispersal it had skipped wide areas and settled only in places especially suitable for it. The birds of this genus are lark-like (but not true larks) and are good fliers, but no better than some other passerines and not so good as swallows.

After the four genera named above, the most widely distributed genera of passerines are probably *Turdus* and *Corvus. Turdus* (typical thrushes including the American Robin, but not including most other American thrushes), although not quite cosmopolitan, is within its limits probably the most ubiquitous genus of passerine birds. It is well represented by breeding species on every continent except

Australia and on many islands, and different species range ecologically from open heath in temperate and subarctic regions to the deepest tropical jungle (Ripley 1952). The species of *Turdus* are not exceptionally strong fliers and not very large passerines, but they seem often to be dominant or successful or adaptable birds, and this probably has something to do with the wide distribution of the genus. Our Robin, for example, is at home both on suburban lawns and in the spruce forests of the North Woods. *Corvus* (crows etc.) is on every continent except South America, and it has reached Madagascar, New Zealand (where it is now extinct), and some West Indies, and even the Hawaiian Islands, where there is an endemic species. Crows do not have exceptional powers of flight for their size, but they are very large passerines and are supposedly unusually intelligent birds.

That most species and genera (and families) of passerines are relatively limited in distribution is apparently not (or not entirely) owing to any peculiarities of classification or rate of evolution or geological age. Passerines are probably relatively recent birds, but many of them have certainly been in existence long enough to spread over the world, if they had the power to do so rapidly. Apparently most of them do not have the power to disperse rapidly, perhaps partly because they are mostly small birds. The swallows, with three nearly cosmopolitan genera, show what passerines can do when their powers of dispersal are increased.

Zonation of birds

Although birds are ubiquitous, and although many tropical and north-temperate birds are closely related, there are striking differences between tropical and north-temperate bird faunas. The differences involve not so much different families as differences in the nature and composition of the faunas as wholes.

The bird fauna of the tropics is immensely richer in families, genera, and species than that of the north-temperate zone. Griscom (1945, p. 114) estimates that 85 per cent of all species and subspecies of birds are tropical, or perhaps he means that only 15 per cent occur north of the tropics. Pigeons and parrots swarm in the tropics but are relatively few or absent in most of the north-temperate zone, and some smaller but widely distributed families of birds are much better represented in the tropics than northward, for example ibises, cuckoos, goatsuckers, and kingfishers. Small families which are widely distributed in the tropics but do not or scarcely enter the north-temperate zone are sun grebes, jacanas, trogons, and barbets. Many other families of birds are characteristic of the tropics in one half of the world or the other but do not enter the north-temperate zone or do so only slightly. In the Old World, they include pheasants and jungle fowls (mostly Oriental, but see details in list of families) and guinea fowls (mostly African), touracos (African), hornbills, mousebirds (African), honey guides, broadbills, pittas, bulbuls (which extend north to Japan etc.), cuckoo shrikes, babblers (of which a few occur also in temperate Eurasia and one is localized in western North America), helmet shrikes (mostly African), wood swallows (Oriental etc.), flowerpeckers (Oriental etc.), sunbirds, white-eyes (which extend north to Japan etc.), and drongos, not to mention weaverbirds etc. and starlings, which are much more numerous in the tropics than northward. In the New World, the tropical families include tinamous, guans, potoos, motmots, puffbirds, jacamars, toucans, ant birds, ovenbirds (which do not include our North American Ovenbird), wood hewers, manakins, cotingas, and honey creepers, not to mention hummingbirds and tanagers, which are very numerous in tropical America and comparatively few northward. Besides these a number of small families of birds are further localized in different places in the tropics and unrepresented northward. The unfamiliarity of many of these names to northern ornithologists is some measure of the wealth and diversity of bird life in the tropics.

Compared with the tropics, the north-temperate land-bird fauna is essentially a fauna of fewer birds rather than different ones. The north-temperate zone has very few families or subfamilies of birds peculiar to it or even nearly so, about the only ones among land birds being the grouse (Tetraoninae), hedge sparrows (Prunellidae, in the Old World north-temperate only), waxwings (Bombycillinae), and typical creepers (Certhiidae), and these 4 groups total only 39 species. Of course, a number of genera and many species of land birds of other families are confined to north-temperate areas, and some of them have zonal distributions, being widely distributed around the northern part of the world but absent in the tropics, but this is zonal endemism at a lower-than-family level.

Of fresh-water and sea birds there is a better list of families or subfamilies that are entirely or chiefly northern at least as breeding birds, but they are more characteristic of the far north than of the north-temperate zone as a whole. The loons (four species) breed well northward and winter within the north-temperate zone. Some groups of ducks and especially geese and swans breed in numbers in the north, but they are represented in some other parts of the world too.

Most sandpipers and their immediate allies (Scolopacinae) breed in the north, but they cover the world in their migrations, and a few breed in scattered places over the world. Phalaropes (three species) breed far northward but winter on tropical seas. Skuas (one species) breed in both far northern and far southern parts of the world. Jaegers (three species) breed only in the north. Gulls are numerous in the north but occur in other parts of the world too. And auks breed only in the north (a few south to Lower California) and winter within the north-temperate zone.

The principal zones of bird life from the tropics northward, then, are (1) the tropics themselves, in which birds are very numerous and diverse; (2) the north-temperate zone, in which birds are fewer and much less diverse than in the tropics but otherwise not much different; and (3) the subarctic and arctic zone, in which land birds are very few and not much different, but in which there is a unique concentration of breeding (but mostly migratory) water and shore birds.

The boundaries between these zones are not sharp. In Africa, many tropical groups of birds stop at the Sahara, but a few cross it. In Asia, especially in eastern Asia, some (but of course not all) tropical bird families extend north through much or all of China or to Japan; the line between tropical and north-temperate bird faunas is probably less distinct here than anywhere else; in several cases, noted in the list of families, the eastern Asiatic representatives of tropical families are migratory. In America, many tropical bird families stop at the northern edge of rain forest in the tropical lowlands of Mexico (Mayr 1946, p. 32), but other families extend farther north for varying distances. Of course I am speaking of the limits of primarily tropical groups, which give the tropics their peculiar richness; some other families of birds are distributed over most of the world with little regard for the line of the tropics, and they further blur the boundary. Farther north, the sharpest boundary is the limit of trees (Mayr 1946, pp. 30–31) beyond which a good many sorts of land birds cannot live, but even this limit is passed by a few land birds as well as many water birds.

The comparative poverty of bird life in the north-temperate zone, measured in families, genera, and species, may not extend to individuals. I know of no accurate, comparable counts, but my impression is that, although there are probably more individual birds in a given area in the tropics than in the north-temperate zone, the difference may not be very great. If so, the average number of indi viduals per species is greater in the north-temperate zone than in the tropics. This is a subject very much worth detailed study.

From the tropics southward, too, the greatest change in bird faunas is a disappearance of tropical groups. In South America, for example, many tropical birds reach southern limits in northern Argentina (a few degrees below the actual line of the tropics) but others extend farther south for varying distances, a few reaching Tierra del Fuego. Some large tropical families that do not enter the north-temperate zone or do so only a little extend far southward. For example, parrots are numerous in southern Australia and Tasmania, and several are endemic in New Zealand, and in America one is common as far south as Tierra del Fuego; and ovenbirds and tapaculos are well represented in south-temperate South America south to Tierra del Fuego, although the whole great tropical American superfamily (Furnarioidea) to which they belong stops abruptly, northward, in southern Mexico.

It is surprising and noteworthy that there is no zonal fauna of land birds in the south-temperate zone; there are no special groups common to the southern tips of different continents. In fact there are probably *no* direct relationships between the land birds of South America and of Australia or New Zealand. Supposed relationships between the South American ant pipits (Conopophagidae) and the New Zealand wrens (Xenicidae) and between the South American tapaculos (Rhinocryptidae) and the Australian scrubbirds (Atricornithidae) are very doubtful (Mayr and Amadon 1951, p. 12). The rheas of South America, emus of Australia, and moas etc. of New Zealand, which were once thought to be related, are now usually considered independent, convergent groups.

The breeding fresh-water birds of the south-temperate zone are scattered genera and species of widely distributed families and do not form a distinct zonal fauna. The northern shore birds that winter in southern South America, Australia, and New Zealand are in some cases closely related. They form a southern, zonal fauna of non-breeding birds. Of sea birds, there is an antarctic fauna of breeding species (supplemented by some non-breeders) characterized by penguins and diving petrels and including various gulls etc. This fauna is primarily zonal, but parts of it, including penguins, follow the cold Antarctic and Peruvian Currents up the west coast of South America far into the tropics.

The principal zones of bird life southward, then, are like those to the north but differ in important details. The zones are (1) the rich tropics; (2) the south-temperate zone, in which birds are fewer and less diverse than in the tropics but otherwise not much different (except for a few groups localized mostly in Australia and New Zealand), and in which there are no strongly zoned elements except wintering shore birds; and (3) the antarctic zone, in which there is a primarily zonal fauna of sea birds.

A longitudinal zonation like that of some lizards is shown by some groups of birds, especially by small, migratory, insectivorous land birds already discussed: in the Old World, especially by Muscicapinae, Sylviinae, and Motacillidae, and also Sternidae and Ploceidae, which are less insectivorous and less migratory; and in America especially by Tyrannidae, Vireonidae, and Parulinae. The tyrannids (American flycatchers) range from Alaska to Tierra del Fuego. Icterids too occur from Alaska to Tierra del Fuego and are less insectivorous and less migratory. None of these American birds occurs in the Old World, except for minor invasions of eastern Siberia by two Parulinae. Some genera of other families have north-south distributions, examples in America being *Coccyzus* (cuckoos) and *Chordeiles* (nighthawks), both of which are migratory in the north. Even a few species have north-south distributions. An example in America is the Sparrow Hawk (*Falco sparverius*) which, in numerous subspecies, ranges from Alaska to Tierra del Fuego, and which is only moderately migratory in the north.

Radial distributions of birds

Well-defined radial patterns of distribution, reflecting centers of origin and outward dispersals, are rare among birds and usually involve single genera or small groups rather than whole large families. The latter probably usually disperse so complexly and also sometimes so rapidly that the first radial stages of their dispersal, soon completed, are soon obliterated by redispersals and by evolution, multiplication of species, and extinction in different places.

The most important radial pattern of distribution of a dominant group of birds is that of the songbirds (the suborder Oscines of the Passeres). Songbirds far outnumber other land birds on all continents except South America. Many groups of them extend from the Old World tropics northward, from the Old World to North America, and (fewer groups) from North to South America. There are complexities (*e.g.*, some groups extend northward again from the American tropics), but the pattern as a whole is one of a very complex radiation from the warmer part of the Old World. Other, less important, radial patterns will be noted in discussing the history of dispersal of birds.

Regional distribution of birds

The broadest patterns of distribution of land birds are made by the cosmopolitan, transversely and longitudinally zoned, and radial groups, which make a very complex, partly crisscross network of relationships over the world. Within this network and part of it, but accentuated by localization of other groups, are six regional bird faunas. They occupy the six conventional faunal regions, which were in fact first proposed for birds. The regional faunas are large and complex. In order to avoid repetition, I shall describe them very briefly here, partly in general terms, with cross references to details in other chapters. The composition of the regional bird faunas is given in a little more detail in Chapter 7, but even there details are limited. I hope that some readers will go on to some of the good regional and local bird books referred to below or in the chapter bibliography. There is no good, modern book on the bird faunas of the whole world. One would be useful. It would, among other things, help local ornithologists everywhere orient themselves.

The two north-temperate regions, Eurasia and North America above the tropics, have bird faunas characterized by poverty, by fewness of families, genera, and species, although individual birds may be relatively more numerous. The poverty of birds in the north-temperate zone has already been described, and many of the tropical groups that are lacking there have been listed under zonation of birds (pp. 254–255). Besides being alike in what they lack, the two north-temperate bird faunas are alike in some positive ways. They share most of the few families and subfamilies of birds that are confined to the northern parts of the world (loons, auks, grouse, waxwings, and creepers), and they share many genera of land birds of other families, especially of hawks, owls, woodpeckers, swallows, thrushes, kinglets, titmice etc., finches, and corvids (see list of families), and finally they share a large proportion not only of genera but even of species of far-northern water and shore birds. On the other hand these two regions differ in some ways. Temperate Eurasia has a small endemic family (hedge sparrows) and North America has one almost endemic family (turkeys), but these are minor groups. Each region has many genera of land birds confined to it. And the two regions have different sets of birds shared with or derived from the Old and New World tropics respectively. They are primarily tropical groups which extend into adjacent north-temperate regions; most but not all are migratory in the north; and many but not all are insectivorous (for the nature and probable history of these groups see under longitudinal zonation of birds and under north-south migration, pp. 249 and 258). In temperate Eurasia these birds include chiefly larks, Old World flycatchers, Old World warblers, motacillids, weaver finches, and starlings, and also (in smaller numbers or only into the southern part of the north-temperate zone or only into eastern Asia) one or more Old World vultures, storks, pheasants (E. Asia), bustards, rollers, bee eaters, hoopoes, cuckoo shrikes (E. Asia), sunbirds (to central China), white-eyes (E. Asia), Old World orioles, and drongos (E. Asia). In America the tropical birds that extend northward are chiefly American flycatchers, wrens and mimids, vireos, wood warblers, cardinal grosbeaks, American sparrows, and American orioles and blackbirds and their allies, and also (in smaller numbers) New World vultures, hummingbirds, and tanagers. (Of course, these are not complete lists of the tropical birds that enter the edge of the north-temperate zone.) It is these birds, the longitudinally zoned ones and the tropical derivatives, which best differentiate the temperate Eurasian and North American bird faunas.

As to boundaries, Mayr (1946) shows that the bird fauna of the whole of North America is essentially transitional, with Eurasian elements numerous in the north and decreasing southward, and with tropical American elements numerous in the south and decreasing northward. In terms of existing distributions (disregarding places of origin and directions of movement), the temperate Eurasian bird fauna is transitional too, with North American elements numerous in the north and decreasing southward, and with Old World tropical elements numerous in the south and decreasing northward. Under these conditions faunal boundaries are not sharp. The boundary of the full-scale tropical bird fauna is about at the Tropic of Cancer, near which many tropical groups of birds stop in both the Old and New Worlds, but many other groups cross it. The boundary at Bering Strait is even less well defined, as Mayr (1946) shows. The Eurasian and North American bird faunas form a complex transition in the north, although it is limited by the comparative fewness of northern land birds.

The two main tropical regions of the Old World, Africa below the Sahara and tropical Asia with its recent continental islands (the Oriental Region), have bird faunas of which the principal components are itemized in Chapter 7. These faunas are alike in many ways.

Both are rich in genera and species of many cosmopolitan, zonal (tropical), and Old World families, but less rich in endemic families. Of endemic or almost endemic families and subfamilies that do not occur in the Orient, Africa has ostriches (1 species), secretary birds (1), hammerhead storks (1), guinea fowls (7), touracos (19), tree hoopoes (6), mousebirds (6), river martins (1), bush shrikes (42), helmet shrikes (13), buffalo weavers (3), widow birds (9), and tick-birds (2), but these groups make up a rather small part of the whole African bird fauna. Africa has also most honey guides, most true shrikes, and most sunbirds. Of birds not or poorly represented in Africa, the Orient has most pheasants, frogmouths (also Australian), most pittas, crested swifts (one genus, which extends to New Guinea etc.), fairy bluebirds, most timalines, wood swallows (one genus, primarily Australian), and some flowerpeckers (most are Australian); but only the fairy bluebirds (Irenidae) are strictly endemic to the Oriental Region. The African and Oriental bird faunas differ further in having unequal representations of other families, and each fauna has many endemic genera and species. But the similarities outweigh the differences. Both faunas are dominated by the widely distributed or shared families. The amount and level of endemism in these two faunas are greater (more in Africa than the Orient) than in the north-temperate bird faunas but less than in the Australian and South American ones.

I cannot take space to discuss the distribution of birds within the main part of Africa or the tropical Orient, but one point must be noted for later reference. A number of birds of South India and Ceylon are geographically isolated from their nearest relatives, which in different cases are in the Himalayas, Assam, Burma, or Malaya; five birds confined to Ceylon have their nearest relatives in Malaya (Ripley 1949).

There is no sharp boundary between the African and Oriental bird faunas. Southern Arabia is a subtraction-transition area between them. A good many tropical African birds (but only small parts of the African bird fauna as a whole) extend to southwestern Arabia, to the area "visible in the distance from the southern Red Sea" (Bates 1937). In southeastern Arabia, however, in Muscat and Oman, the few resident tropical land birds are mostly Indian, not African (Ripley 1954, pp. 245–246). In the main part of Arabia the characteristic breeding birds (various larks, wheatears, vultures, bustards, sand grouse, the Ostrich, etc.) are primarily desert birds rather than African or Eurasian ones. Their proper home is the dry country from the

Sahara to (in some cases) Iran and Central Asia, with outlying representatives both north into Eurasia and south into Africa (Bates). These desert birds form a sort of transition between the tropical African and temperate Eurasian bird faunas. The situation is, of course, more complex than this. Meinertzhagen (1954) describes it in more detail in his recent, long-needed volume on the birds of Arabia.

The bird fauna of the Oriental Region is itself to some extent transitional between three surrounding faunas. It is primarily African in its general nature, and additional African elements reach India: Ripley (in a manuscript which he has very kindly allowed me to see) says that 11.5 per cent of tropical Indian birds seem to be directly related to African rather than to eastern Oriental ones. Temperate Eurasian elements in the Oriental bird fauna include certain waders (*Scolopax*), owls (*Strix*), woodpeckers (*Dryocopus*), nuthatches, and titmice, etc., and in winter many northern migratory birds. Australian elements are a few megapodes, fruit doves and imperial pigeons, frogmouths, a wood swallow, flowerpeckers, and, if Bali and the Philippines are Oriental, lories and cockatoos. See list of families for further details.

The bird fauna of the Australian Region (Australia, New Guinea, and closely associated islands) lacks some important Oriental and some widely distributed families, including pheasants, trogons, barbets, woodpeckers, broadbills, bulbuls (which reach the Moluccas), fairy bluebirds, and true finches, but it has a good share of representatives of the usual cosmopolitan groups and more than its share of pigeons, parrots, and kingfishers, as well as many cuckoo shrikes, Old World flycatchers, flowerpeckers, and estrildine weaver finches, etc., and Australian birds include also a number of groups which are endemic or nearly so: emus and cassowaries, megapodes, several more or less distinct groups of pigeons and parrots, frogmouths, lyrebirds, scrubbirds, Australian warblers, honey eaters, and several families of crow-like birds including the birds of paradise (see also Chapter 7, pp. 451–452). The main part of this fauna is almost equally divided between Australia and New Guinea. These two very different pieces of land have generally similar bird faunas, although many details differ (Mayr and Serventy 1944). Almost all the ascertainable relationships of Australian land birds are toward Asia. The few supposed African relationships, for example that of the honey eaters with *Promerops* of South Africa, are doubtful, and there is no sure, direct relationship between any Australian and South American land birds.

There is no sharp boundary between the Oriental and Australian bird faunas, but a broad zone of gradual transition covering all the islands between Java and Borneo on one side and New Guinea and Australia on the other. The transition actually involves a still wider area, for some Oriental birds (for example hornbills and crested swifts) reach New Guinea but not Australia, and some others (for example pittas and drongos) reach northern and eastern Australia but not the rest of the continent, and in the other direction a few Australian birds extend across Wallace's Line into the Oriental Region, as already noted. See Chapter 7, pages 465–466, for a few more details of this transition and for references.

South America (the "Bird Continent") has the richest of all bird faunas, with many more species, especially of small land birds, than any other equal area. This richness is due partly, but only partly, to multiplication of species at different altitudes on different mountain ranges (for examples see Chapman 1917, 1926). Part of the fauna, but not all of it, reaches Central America, and much less, the West Indies. This fauna has its share of herons, ibises, storks, ducks, rails, hawks, owls, quails, cuckoos, plovers, pigeons, parrots, goatsuckers, trogons, swifts, barbets, woodpeckers, swallows, thrushes, and some other, smaller, widely distributed groups. It lacks cranes, bustards, turnicids, hornbills, broadbills, and many other Old World families, and it has few kingfishers, but its greatest lack is in songbirds (Oscines); it is the only continental bird fauna in which songbirds are in a minority. But all the lacks are more than made up for by endemic or at least New World groups, which are many, often taxonomically isolated, and in some cases very rich in species: rheas, tinamous (33 species), New World vultures, screamers, cracids (38 species), the Hoatzin, seriamas, trumpeters, the Limpkin, the Sun Bittern, seed snipes, potoos, the Oilbird, motmots, hummingbirds (319 species, mostly in South and Central America), puffbirds (32 species), jacamars (14 species), toucans (37 species), the superfamily Furnarioidea (about 500 full species, all in South and Central America), American flycatchers (365 species, including the North American and West Indian ones), manakins (59 species), cotingas (90 species), and plant cutters. These groups, most of them confined to South or South and Central America, a few reaching North America and the West Indies, but none reaching the Old World, total more than 1500 species, more than one-sixth of all the known species of birds. Moreover, South American songbirds, though a minority of the total fauna, are fairly numerous and are mostly in groups which have radiated in the American tropics: thrushes of endemic groups, wrens, mimids, and members of the chiefly American vireo-warbler-honey creeper-tanager-grosbeak-finch-icterid complex. Except for these, the only songbirds that reach the continent of South America are a lark, swallows, a few aberrant sylviines, recent thrushes, and a few dippers, pipits, siskins, and jays. A few others reach parts of Central but not South America: kinglets, silky flycatchers, waxwings (in winter), creepers, titmice, evening grosbeaks, crossbills, and crows.

The geographical relationships of many South American birds are obscure. Some of the more distinct endemic groups are probably not closely related to any other existing birds. Some of the tropicopolitan groups, including parrots, trogons, and barbets, are apparently relict or at least retreating, and there may be no *direct* relationships between those now in South America and those in any other particular part of the world; this is likely to be true also of the cuckoos, most of the suboscine passerines (which are now dominant in South America), and some other widely distributed groups. Where the geographical relationships of South American birds are discernible, they are mostly toward North America, or through North America to Eurasia. There are a few other relationships: *Spizaëtus,* a genus of eagles, occurs only from eastern Asia to New Guinea and in tropical America; *Dendrocygna viduata,* a tree duck, occurs in Africa and Madagascar and South America; *Porphyrula,* a genus of gallinules, occurs in Africa and Madagascar and tropical and warm-temperate America; *Ciccaba,* a genus of owls, is confined to Africa and tropical America, and owls of the tropical American genus *Lophostrix* and African genus *Jubula* are related; *Picumnus,* a genus of piculets or small woodpeckers, occurs in the Oriental Region and tropical America; some swallows of South America and other southern continents may be related; and the aberrant tropical American sylviines *Ramphocaenus* and *Microbates* are apparently related to an African genus; but these are rare exceptions. There are few direct relationships between South American and African land birds and probably none between South American and Australian ones. The South American bird fauna as a whole is fundamentally different from that of Africa or Australia.

Central America has a bird fauna which, though more South than North American, is in many ways transitional. It lacks some South American groups and others extend only part way through it, and it is reached by some North American birds, including some songbirds mentioned above, that do not reach South America. Central America has also some distinct endemic birds of its own. The area of transition includes not just Central America but much of North and South America, for many tropical American birds extend into North America for different distances (see discussion of the North American bird fauna, pp. 259–260), and some North American birds extend into South America for different distances, a few of them reaching the southern tip of South America, *e.g.*, pipits, siskins (*Spinus*), and some woodpeckers, not to mention some migratory birds, especially shore birds. The situation in Central America is really much more complex than this; Griscom (1942; 1950, especially pp. 375ff.) describes it in more detail. The nature and history of the Central American bird fauna are further discussed in connection with "tropical North America" at the end of this chapter.

The following is a summary of the nature and principal relationships of the regional bird faunas. The land bird faunas of temperate Eurasia and North America are little more than depauperate fringes of adjacent tropical faunas, lacking many tropical families, and with a low level of endemism, but there is a more distinct assemblage of northern water birds. These two north-temperate bird faunas are alike in many of their birds but different in others, especially different in families of small, mostly insectivorous, mostly migratory land birds shared with, respectively, the Old and New World tropics. The bird faunas of the two main regions of the Old World tropics, Africa and the Oriental Region, are rich especially in representatives of more or less widely distributed families. Their level of endemism is moderate: Africa has some endemic families; the Oriental Region, only one. These two faunas are alike in general but different in many details. The Australian bird fauna is less rich, lacks some important families, but is more endemic. Its ascertainable relationships are almost all with the Oriental fauna. The bird fauna of South America is richest and most endemic of all. So far as it has ascertainable relationships, they are mostly with North American birds or through North America to the Old World.

Transitions and barriers in distribution of birds

The transitions between bird faunas, formed by overlapping and progressive subtraction of faunal elements, are broad and complex. Mayr (1946) is correct in stressing that the bird fauna of the whole of North America is transitional between the faunas of Eurasia and tropical America, but transition is even broader than this. The temperate Eurasian bird fauna is largely transitional and the tropical Oriental one partly transitional too. In fact, the bird faunas of the Old World tropics and of the American tropics meet in a two-way transition which involves at least the whole of Eurasia, North America, and Central and much of South America, if not the whole world. Nevertheless the six conventional regional faunas of birds are more or less distinct, as already described. This anomaly will be further considered in Chapter 7.

No barriers completely stop the dispersal of birds. Some birds always cross them for different distances or at different times. But some barriers, especially those of salt water, profoundly affect some bird faunas by selection of immigrants and by partial isolation. Examples have been given in preceding pages.

Dominance and competition in relation to distribution of birds

The pattern of distribution of dominant groups is more complex among birds than among other vertebrates. Every main ecological group of birds has its own dominant family or families, which are often cosmopolitan. Some of them are named in the discussion of the history of bird dispersal (pp. 269ff.).

The most obviously dominant existing birds are the passerines, which far outnumber all other land birds on every continent: oscine passerines (songbirds) are dominant on all continents except South America; suboscine passerines (supposedly more primitive), in South America. Passerines are dominant in most land habitats from the tropics to the arctic and from wet jungles to deserts. Nevertheless there are ecological limits to their dominance. Few are predaceous; they do not threaten the place of hawks and owls. And few occur on fresh water and none is marine; they do not compete much with water birds.

Competition is suggested by or is consistent with many details of the distribution of birds, including the geographical complementarity of Oscines and Suboscines, but the evidence is almost limited to deduction from existing distributions. The fossil record is not good enough to show whether retreat of some birds has been closely correlated with spread of others.

Summary: The pattern of distribution of birds

Although the relationships of birds make a very complex network over the world, there is a main pattern in their distribution, with indications of the usual five subpatterns.

The broadest subpattern, of limitation and complementarity, is poorly marked. Birds reach no absolute limits of distribution but do decrease very much in numbers and diversity in the far north and far south and on remote islands. Complementarity is limited by the existence of numerous cosmopolitan groups of birds and usually involves minor groups more than it does major faunal elements. The general complementarity of Oscines and Suboscines is an exception.

The zonation of land birds is mainly a matter of decreasing diversity. Temperate-zone birds are fewer than tropical ones but for the most part not much different otherwise. And arctic birds, at least land birds, are still fewer but not much different otherwise. Arctic and antarctic water birds are more distinct. Longitudinal zonation is characteristic of some small land birds, especially migratory insectivorous ones.

The subpattern of radiation is not obvious among most birds, probably because many dominant groups have dispersed so rapidly and so complexly that their main radiations have been rapidly completed and the traces of them obliterated. The songbirds (Oscines) form a radial pattern of a sort, around the warm part of the Old World, but it is a very complicated, not very obvious, one.

The fourth subpattern, of differentiation of faunas in different continental areas, is well marked among birds but is very complex in detail. The continental (regional) bird faunas are made up of different combinations of cosmopolitan, zonal, and radial groups of birds plus different proportions of endemic groups. In general, temperate Eurasia and North America have poor combination faunas, with low endemism; Africa and the Oriental Region, richer combination faunas, with moderate endemism; Australia, a moderately rich fauna, with more endemism; and South America, a very rich, highly endemic fauna.

As to the fifth subpattern, of concentration in favorable areas and subtraction marginally, each main fauna of birds is concentrated in favorable places and many elements are subtracted in less favorable ones, for example in the arctic and in deserts. Nevertheless there are fair numbers of birds almost everywhere, even in unfavorable transition zones. Probably partly for this reason, as well as because many birds cross barriers and disperse rapidly, the transition (overlapping with progressive subtraction in both directions) of bird faunas is very complex and very broad, covering much of the world.

Of the place of relicts in the existing pattern of distribution of birds, little can be said. No living bird is a relict comparable to the lungfish in Australia or *Sphenodon* on New Zealand. Just which existing birds are most primitive is unknown. The "ratites" (Fig. 33), in Africa etc., Australia, and South America and formerly on Madagascar and New Zealand (kiwis are still on the latter), may not be primitive relicts.

That is, they may not be remnants of an old cosmopolitan group of birds but may be independent, convergent groups which originated where they now are. Tinamous *may* be survivors of a primitive assemblage of flying birds from which some ratites *may* be derived. Tinamous are confined to South and Central America but are still

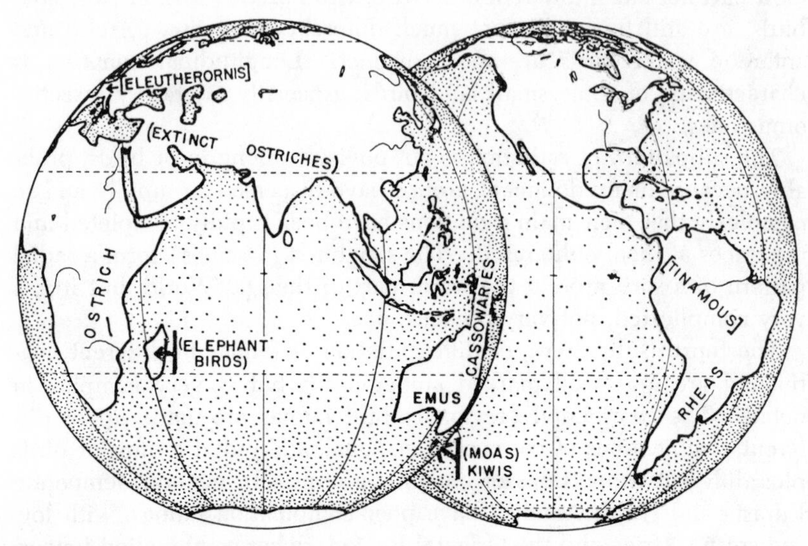

Fig. 33. Distribution of flightless "ratite" birds and possible fossil ancestor (*Eleutherornis*) and existing flying relatives (tinamous).

numerous and successful there. My impression is that there are more primitive birds in South America than anywhere else.

History of dispersal of birds

The existing pattern of distribution of birds has been formed by a very complex process of evolution, radiation, complex (multiple and successive) and probably often rapid dispersal, adaptation, presumably competition, and limitation, retreat, and extinction, according to the nature of birds and to the distribution of climates and the position of more or less effective barriers, especially oceans. This process has apparently been going on partly independently in each major ecological group of birds. In each such group, dispersal has probably involved the rise, diversification, and complex and more or less rapid spread over much or all of the world of a succession of dominant groups, and the complex withdrawal and disappearance of older groups as new ones spread. I shall try to trace some of the details of this, but it is not easy to do. I go much further than Mayr (1946) in considering "unanalyzable" not only families of birds in which clues are lacking but also many families in which clues seem to be present but are, I think, untrustworthy. An example of untrustworthy clues is discussed in connection with the parrots (pp. 271–272).

What I am going to do is to take the birds by broad ecological groups, pick out the dominant families in each group (the dominant families are most likely to show significant clues), and if possible trace their histories. I shall do this briefly, without much detailed explanation. The reasons for my conclusions can usually be deduced from details given in the list of families. This method of analysis of birds' geographical history is an oversimplification in several ways, but it is the best I can do.

Among sea birds, gulls and terns are dominant over the world as a whole, but their origin and radiation are probably beyond analysis. Auks have apparently always been northern and penguins southern in distribution.

Among birds of prey, hawks (in a broad sense) are dominant by day and owls by night, but these are old and complex groups and their original patterns of dispersal are lost. However, a few details of recent movement are clearly indicated. A *Circus* (the Marsh Hawk), two subgenera of *Falco*, the eagles *Aquila* and *Haliaeetus*, and a *Tyto* (the Barn Owl) have apparently reached North America from the Old World rather recently. On the other hand *Buteo* seems to have radiated from North America. This genus of soaring hawks (including the Red-tailed, Red-shouldered, and Broad-winged Hawks, etc.) has most of its species in North America (and the genus is fossil there back to the Oligocene), half a dozen species in temperate Eurasia, fewer in Africa (one on Madagascar) and South America, none breeding in the Oriental Region, and none at all in the Australian Region; and the occurrence of endemic species of the genus on the West Indies, Galapagos, and Hawaiian Islands is at least consistent with an American origin. Analysis of other genera of hawks and owls might show other details of movement in other parts of the world, but there is space here to consider only movements between the Old and New Worlds.

Among fresh-water swimming birds, Anatidae (ducks etc.) are dominant. Their distribution is complex now, and the family is probably old, and its original pattern of dispersal has probably long been lost. I have not attempted to unravel recent directional movements in this family.

Among large, long-legged wading and ground birds, Ardeidae (herons) are dominant. They are an old group which has probably reradiated recently. Half a dozen genera and at least a couple of species of them are cosmopolitan or nearly so now. Herons may be replacing other birds in this ecological group, including perhaps Grues (cranes, limpkins, bustards, rails, etc.), which Mayr and Amadon (1951, p. 8) consider a declining order, and also ibises and storks. Most herons are water birds, and some of the other birds just named live away from water, but this may be a result rather than a disproof of competition. Even the reradiation of herons is complex, and I have not attempted to analyze it in detail. The recent spread of the Cattle Egret from Africa to South America and then to North America is traced by Sprunt (1955).

Among smaller wading and shore birds, plovers (Charadriinae) and sandpipers (Scolopacinae) are dominant, but their histories are doubtful. Plovers are widely and nearly evenly spread over the world. Sandpipers breed mostly in the north, but whether they originated there (as Mayr 1946, p. 14, thinks probable) or became concentrated there perhaps during the Pleistocene is a question.

Rails (Rallidae) dominate an ecological group of birds characteristic of thick vegetation beside or over water, and some of them live on the ground away from water, especially in tropical forests. They overlap on one side the wading birds and on the other the fowl-like ones. The family is old, and its distribution is complex, and I can see no trace of its original pattern of radiation. I have not tried to work out its recent minor movements.

Among fowl-like birds, the family Phasianidae is dominant over much (but not all) of the world. It is the first large family of this survey which shows an apparent pattern of radiation from one main center, the Oriental Region. The true pheasants, jungle fowls, and peacocks are almost all in the Orient, and from there the ancestors of *Afropavo* (the Congo Peacock) and of the guinea fowls may have reached Africa; the ancestors of the grouse (Tetraoninae) may have gone northward and around the northern part of the world, eventually differentiating most in North America; and the quails may have radiated over the whole world, with a secondary evolution center in southern North America. The fowl-like birds probably have relatively low powers of dispersal, so their present distribution may reflect something of their early history. Nevertheless the history even of the Phasianidae has probably been more complex than suggested, and there are other families which further complicate the pattern, the megapodes in Australia etc. and the cracids in South and Central America, which may be survivors of one or more earlier radiations of fowl-like birds; and the primitive but superficially fowl-like tinamous are still dominant in South and Central America.

Parrots (Fig. 34) are almost equally numerous in species in the Old and the New Worlds but are much more diverse in the former,

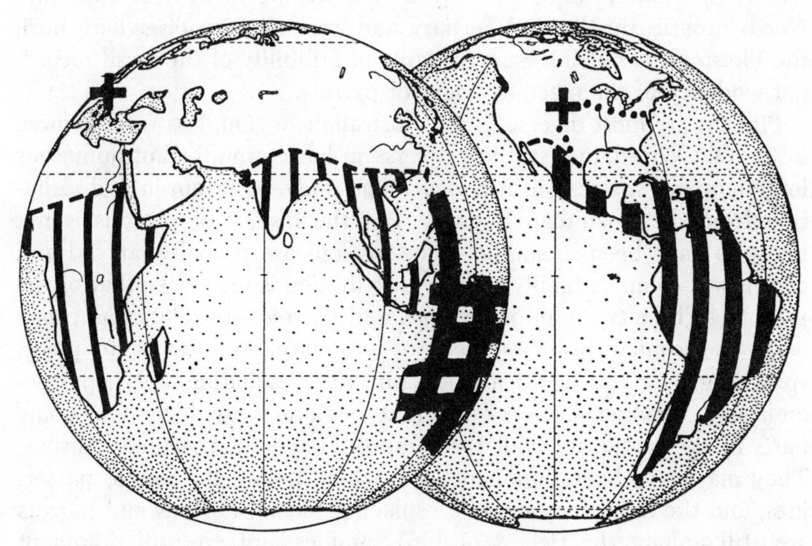

Fig. 34. Distribution of parrots. Heavy crossed bars indicate many diverse parrots in Australia etc.; heavy longitudinal bars, many but less diverse ones in South America etc.; and lighter bars, fewer and less diverse ones in Africa and tropical Asia etc. Broken line is approximate present northern limit of parrots; dotted line, approximate former limit of the Carolina Paroquet; crosses, approximate localities of fossil parrots.

all the New World forms belonging to one of the several Old World subfamilies; this suggests an Old World origin of parrots. But this does not carry the analysis far enough. It is only in the Australian Region and adjacent island areas that parrots occur in great diversity. Elsewhere in the Old World parrots are relatively few and all belong to one subfamily, the same one that is in America. Does the unique diversity of parrots in the Australian Region reflect place of origin, or withdrawal of some groups of parrots from other parts of the world, or local radiation? I cannot answer this question, but it seems to me that parrots need not have originated in the Australian Region. Outside the Australian Region, parrots are much more numerous in tropical America than anywhere else, but I think that they need not have originated there either. To doubt first an Australian and then an American origin, and to suppose that parrots may have originated in some part of the Old World where they are now neither particularly diverse nor particularly numerous, is to deny the value of existing clues in this family. In fact it is to deny, with good reason, that apparent clues are in general very trustworthy among old, complex, widely distributed families of birds. Parrots are fossil in Europe and North America in the mid-Tertiary and are unknown elsewhere until the Pleistocene, but this is an example of fallibility of the fossil record, not evidence of the place of origin of parrots.

Pigeons are most diverse in the Australian Region, less so elsewhere in the Old World, and still less diverse in basic groups (but numerous in genera and species) in America. This suggests origin in and radiation from the Australian Region. But the history of pigeons is not likely to have been so simple. The pigeons are probably an old and certainly a complex family. Their distribution is much like that of the parrots. These two groups, pigeons and parrots, are related and may have had similar histories, perhaps originating in the same place, spreading over the world at the same time, radiating (perhaps secondarily) together in Australia, and beginning to decline in some parts of the world at about the same time and for the same reasons. They may have been dominant everywhere before the rise of passerines, and the latter may now be replacing them. Pigeons and parrots are still among the richest of bird families and are still dominant over much of the world, but there are two indications that they are declining. One is the retreat at least of parrots from the north-temperate zone. The other is that no genus of parrots and only one of pigeons are common to the Old and New Worlds, which suggests a cessation of active dispersal some time ago. Perhaps passerines are more successful migrants than pigeons or parrots (although these birds can migrate too), and that may be why the passerines are replacing the pigeons and parrots in the north-temperate zone, if they are doing so. This may be an example of how migratory birds may keep other, competing, birds out of northern areas and prevent them from passing between the Old and New Worlds. It is not hard to see how a declining group of birds could pass from the state of the parrots to that of the trogons, and then to the end of the cycle, to extinction.

The one genus of pigeons common to the Old and New Worlds is *Columba* (from which domestic pigeons are derived). This genus is an example of ambiguity of numbers clues. It is nearly cosmopolitan. There are about 32 species of it native in the Old World and about

20 in the New, and the Old World species are more diversified, which suggest an Old World origin. But all 20 New World species occur in South and Central America and the West Indies. One of the Central American species extends into western North America north to southwestern Canada, but the genus is otherwise absent from the main part of North America, above southern Florida. There are about 14 species in temperate Eurasia and associated islands; 11 in the main part of Africa and closely associated islands, but none in Madagascar; 5 in the tropical Oriental Region etc.; and 2 in the Australian Region, but only one of them reaches Australia proper, and only the eastern part of the continent; and none reaches New Zealand. Thus detailed, the numbers suggest a tropical American origin of *Columba,* dispersal to the Old World through the north (not by the existing western North American species but perhaps by an earlier one), and spread through the Old World from the north. The absence of the genus in Madagascar and the more remote part of Australia is consistent with this history. Alternatively the genus may have originated in temperate Eurasia and radiated from there and then radiated secondarily in tropical America. Or (and I think this is most likely) it may have had a still more complex history.

Cuckoos are cosmopolitan but more numerous in the tropics than in the north-temperate zone, and no genus of them is common to the Old and New Worlds. The family is probably old and perhaps declining, and it has probably had a complex, undecipherable history.

The Caprimulgi (goatsuckers etc.) are the principal insectivorous birds of the night. Their radiation has probably been at least threefold: an early radiation which left small families isolated in the Oriental-Australian area and tropical America, then one of the family Caprimulgidae, and then one of the genus *Caprimulgus.* From what centers these radiations occurred I do not know.

Swallows and swifts dominate an ecological group of powerfully flying, diurnal, insectivorous birds. The swifts have the older pattern of distribution, with more discontinuity and more localization of genera. The swallows are more dominant. Three genera of them are very widely distributed and have clearly reached North America from the Old World rather recently. But the early history of swallows is not so clear. The present main concentration of them is in Africa, and there is a second concentration in South America, but swallows have plainly undergone successive as well as multiple dispersal, and I hesitate to deduce any simple history from their present distribution.

Kingfishers are numerous and diverse in the Old World, and there is not much doubt that the two fish-eating genera in America are of Old World derivation.

Woodpeckers are dominant within their ecological limits. They are numerous in both America and the Oriental Region, less numerous in temperate Eurasia and Africa, and absent in Madagascar and the Australian Region. This pattern suggests radiation from America, but the first center could also have been the Orient, with a secondary

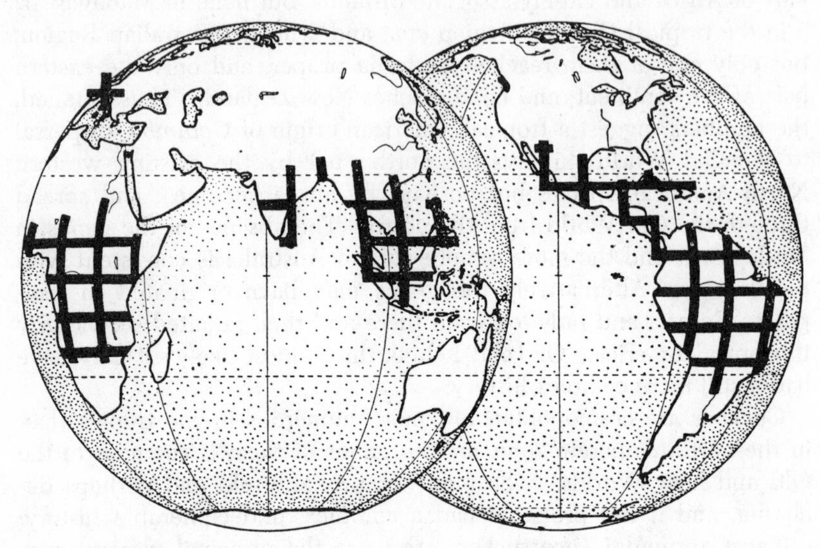

Fig. 35. Distribution of trogons. Isolated cross shows place of fossil trogon, in Europe.

center in America. As in so many other groups of birds, the dispersal of woodpeckers has been multiple (shown by the wide distribution of several genera), and this is a hint against trusting existing clues too far.

Trogons (Fig. 35) and barbets are not dominant but are few in number, discontinuous in distribution, and evidently retreating, and these are good reasons to doubt that their present numbers and distribution reflect their early history. That trogons are fossil in Europe proves that they have retreated and suggests that they may have dispersed by northern routes but, in the absence of an adequate fossil record in other parts of the world, does not show their place of origin.

Of dominant, widely distributed groups of birds there remain to be discussed only the perching birds, excepting the swallows. The order Passeres, the passerines or perching birds, contains more than half of all existing birds, about 5100 out of a total of about 8600 species, and it is even more dominant in number of individuals, within its ecological limits (see p. 266). Most of the individual birds we see about us everywhere on land are passerines. The order is divided by Mayr and Amadon (1951) into several suborders, and its actual phylogeny may have been even more complex than the classification shows, but it is best to begin by dividing the order into only two groups, Suboscines and Oscines.

The Suboscines, or presong perchers, are supposedly the more primitive. Most of them (there are about 1100 species of them in all) are concentrated in South and Central America. One of the tropical American families (Tyrannidae, the American flycatchers) has spread through North America too (its pattern of distribution plainly reflects extension northward of at least four different stocks—Mayr 1946, p. 24) and over the West Indies, and another family (Cotingidae) reaches southern Arizona and Jamaica. There are no other Suboscines in North America north of southern Mexico or in the West Indies, but several distinct families of them are in the Old World: broadbills and pittas in the tropics of Africa and the Oriental Region with pittas extending to northeastern Australia, philapittas in Madagascar, and xenicids in New Zealand.

The Oscines, or songbirds, number about 4000 species and are considered the most advanced, successful, and perhaps the latest to evolve of all the main groups of birds. Their basic structure is very uniform, although they are superficially diverse. Most families of them are weakly characterized, and many of the families are connected by intermediate or doubtful genera. There has been much parallelism of different groups of them, often on different continents. For example, the American orioles and blackbirds (Icteridae) parallel the Old World starlings (Sturnidae), although they are probably not related to them. Because of all this we are still in "abysmal ignorance" of how the Oscines have evolved and how the different groups of them are related to each other (Mayr and Amadon 1951, p. 13).

Oscines are dominant over all the continents except South America, where they are outnumbered by Suboscines in a ratio of about three to two. This general complementarity plus the fact that no family of Suboscines is common to the Old and New Worlds and that some of the Old World ones are localized and may be relict, while some Oscine families are very widely distributed, suggest that over the world as a whole Suboscines are being replaced by Oscines. However, there are indications that this is not the whole story. Existing

Old World Suboscines may not be relicts. The broadbills and pittas are not geographically isolated but occur in the main regions of the Old World tropics. Madagascar and New Zealand do not have remnants of an old, diverse Suboscine fauna, but in each case just one small family which may be autochthonous rather than relict. And in America certain Suboscines (American flycatchers) have recently advanced, not retreated. These details are not necessarily inconsistent with a complex, gradual replacement of Suboscines by Oscines, but they do suggest modifications of the pattern. Existing South American Suboscines may not be part of an old, world-wide fauna but may have evolved in South America from a few ancestors. In the Old World, Suboscines may never have been numerous and may soon have become mixed in faunas with primitive Oscines (of which the Australian lyrebirds and scrubbirds may be survivors) and then with evolving higher Oscines, and the mixed faunas may have passed through many complex changes in the Old World while Suboscines were evolving independently in South America. It is even possible that Suboscines originated in South America and that a few of them spread north to Alaska (as the American flycatchers have done) and thence to the Old World. In other words Suboscines and Oscines may have evolved mostly in different parts of the world, and each may have sent invaders into the territory of the other, the Oscines doing so more often, more recently, and more successfully, but still not yet forcing a general retreat of Suboscines on their home grounds in South America.

When did the dominant Oscines originate and radiate? Fossils (listed by Lambrecht 1933; Romer 1945, pp. 609–610) seem to answer at least part of this question. Two existing families of Oscines (Paridae and Laniidae) are listed in the Upper Eocene of Europe, and several modern Oscine genera (*Lanius, Motacilla, Sylvia, Passer*) are listed in the Oligocene of Europe. This seems to prove at least the existence of modern families and genera of Oscines fairly early in the Tertiary. But it may be doubted whether the fossils are what they are supposed to be. Some of them are skeletal impressions (which are not the same as complete skeletons), and some are fragments, but all seem to have been identified only by superficial comparison with existing birds. Their phylogenetic characters and real relationships have not been determined, so they are not very good evidence of the age of songbirds. These fossils need critical reexamination.

The appearance of Oscines in North America might help to date their radiation, but the North American record of the group is worse than the European one: in approximate order of time, a doubtful passerine from the Eocene of Wyoming; then a broken leg bone assigned by Wetmore (1943) to the wood warblers, from the Lower Miocene of Florida; then *Palaeospiza* (see below) and a feather fragment (originally described as a moss) from the Miocene of Colorado; then *Palaeostruthus,* an upper mandible apparently of a finch near *Pipilo* from the Miocene or Pliocene of Kansas; and then a few Pliocene and a good many Pleistocene remains mostly assigned to existing North American genera or species. The Miocene *Palaeospiza* is an unusually good fossil bird, a skeletal impression nearly complete and "handsome to look upon" but not showing all the characters that could be wished. Wetmore (1925) thinks it is an Oscine but a primitive one, in some ways approaching the Mesomyodi (Suboscines), and he puts it in a separate, extinct family. There is thus no fossil evidence of the occurrence of Oscines in North America before the Miocene, and very little even then.

Since the fossil record is inconclusive, it is necessary to look for other clues to date the Oscines, and clues can be found in their patterns of evolution and distribution. The Oscines have evolved explosively, and the fact that most families of them are still not much differentiated suggests that the explosion and the main dispersal of the group have been recent. And if Oscines existed and were dominant in the Eocene and Oligocene, when the northern parts of the world were warmer than now, many families of them ought to have reached America. That many of the dominant Old World families have not reached America or have done so only recently strongly suggests a more recent origin. That some Oscines (but not many separate stocks) were evidently in South America before the late Pliocene connection with North America does not date them, for Oscines sometimes cross wide water gaps. Many have reached Australia. All this evidence suggests that, whenever they originated, Oscines radiated and dispersed no longer ago than the mid-Tertiary, perhaps primarily in the Miocene. The radiation of these birds is probably still going on and may even be at a climax now.

It remains to summarize the recent movements of different Oscines between the Old and New Worlds. The following have apparently dispersed from the Old World to North America: a lark, which has reached northern South America too; three swallows, all of which have reached South America too as migrants and one of which has established breeding populations there; a timaline; two or more sylviines, one of which, the ancestor of the gnatcatchers etc., reached

South America too some time ago; four or more thrushes etc., which came at different times, and of which three have reached and radiated in South America (see Mayr 1946, p. 19); a pipit, which has extended through the whole length of North and South America; one or two shrikes; perhaps six members of the creeper-nuthatch-titmouse group, some of which have reached Central but none South America; perhaps six genera of fringiline finches, of which *Spinus* has extended through South America too; and perhaps four jays, crows, etc., which came at different times—of these the jays have reached South America; and if the Oscines are originally Old World birds, the ultimate ancestors of the American mimid-wren and vireo-warbler-honey creeper-tanager-grosbeak-finch-icterid assemblages should be added to the list of Oscines that have come from the Old World and radiated in America, including South America. The only Oscine that has clearly dispersed in the other direction, from North America to the Old World, is the wren *Troglodytes,* of which one American species (the American Winter Wren) has extended across temperate Eurasia; but some thrushes, titmice, finches, and corvids *may* have dispersed in this direction recently too; and the ancestors of the Old World emberizine and fringiline finches *may* have done so some time ago.

I have not included in these lists certain Eurasian birds (the Willow Warbler, Wheatear, Bluethroat, and Yellow Wagtail) which have extended their breeding ranges into Alaska (and the Wheatear into northeastern North America too) but which return to the Old World to winter, or certain American birds (the Gray-cheeked Thrush, Myrtle Warbler, and Northern Water Thrush) which have extended their breeding ranges into the corner of Siberia but which return to America to winter. This kind of range extension is probably unimportant in dispersal unless the birds shift their wintering as well as breeding areas from one half of the world to the other.

Songbirds (Oscine Passerines), then, probably arose in the Old World, radiated rapidly and complexly in the mid-Tertiary, and are still radiating. Many different stocks of them have reached North America from time to time, and a good many have reached South America too and some have radiated secondarily there, and a few have probably returned northward and/or to the Old World in minor ways.

I have made a formal list of all the cases mentioned in the preceding pages in which there seems to be evidence of direction of movement of birds, non-passerines as well as passerines, between the Old World and America, but the list is best not published. Too many of the cases are too doubtful, and the total is too small a fraction of all the movements that must have occurred. Nevertheless something can be said in summary of the movements. They have not been all in one direction. There has surely been an exchange, America not only receiving much but giving something back, for example probably *Buteo,* perhaps *Columba,* perhaps woodpeckers, almost certainly *Troglodytes,* and perhaps finches. But it is almost equally sure that the exchange has been unequal, that there has been much more movement of birds from the Old World to America than in the opposite direction, and this seems to have been true both of relatively recent movements and of older ones. Moreover, as Mayr (1946, p. 38) points out, a number of Old World birds have reached South as well as North America, and some of them have radiated secondarily there, while no (or very few) South American birds have reached the Old World. There seems to have been a continual flow of birds more from the Old World to North and South America than the reverse. This is what Simpson has found among mammals and what I have found among cold-blooded vertebrates. I suggest it here for the birds as a reasonable probability, consistent with what is known and supported by some evidence, but as yet not a proved fact.

Birds probably tell little about ancient lands and climates. The present distribution of birds on the continents is consistent with a complex, fairly recent dispersal mainly over existing land, with some withdrawals and extinctions and possibly some crossings of wide oceans in exceptional cases. The relationships between a few land birds of Madagascar and tropical Asia, or of tropical Asia and South America, or of Africa and South America are rare exceptions which can hardly have anything to do with ancient land bridges. There seem to be no direct relationships between the land birds of South America and those of Australia or New Zealand. Birds tell something about the history of some islands, but this will be carried over to Chapter 8. However, it has been supposed that birds have something more important to tell about the history of North America.

It is supposed by some ornithologists (Lönnberg 1927, Mayr 1946, Bond 1948) that in the Tertiary much of southern North America was tropical, and that it had a great, tropical bird fauna which was effectively isolated from South America and which was the place of origin of whole families of American birds and the source of most existing West Indian birds. If the birds really showed all this, they would tell important things about the climate and degree of isolation of North America in the Tertiary. But do they show it?

This subject has been confused by inexact terms. In discussing Tertiary zoogeography a clear distinction should be made between North America proper and Central America, which may have been an island or archipelago during the Tertiary (Fig. 36) and which may have had a separate fauna. Ornithologists discussing "tropical North

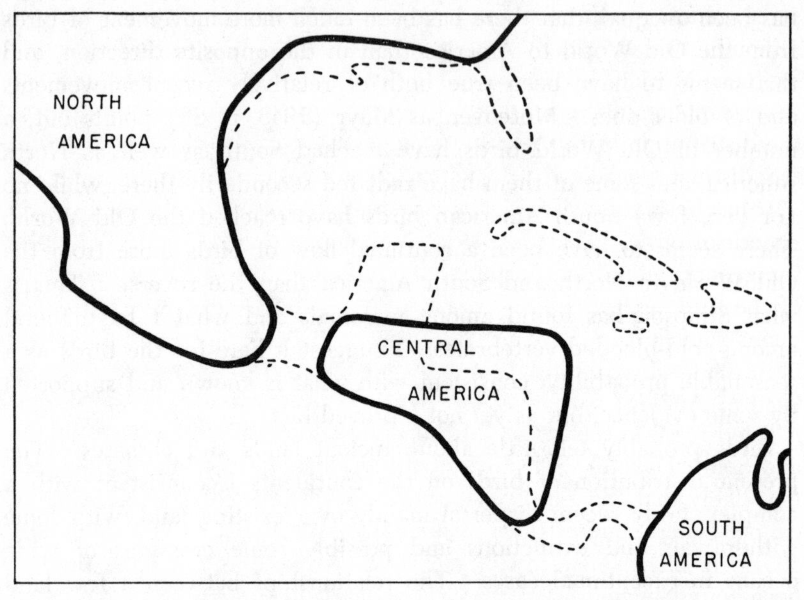

Fig. 36. Diagram of (hypothetical) Tertiary Central America.

America" have sometimes not made this distinction. For example, motmots have been called tropical North American birds, but they are really Central American (see below). The commonly employed term "Middle America," which is useful in some other connections, makes for imprecision here, for it lumps several different pieces of land which have probably had very different histories: southern North America, Central America, and sometimes the West Indies.

The evidence of particular groups of birds supposed to have originated in tropical North America seems to me to have been misinterpreted.

Motmots, supposedly tropical North American birds, are actually distributed as follows. There are six genera, eight species of them. Two of the genera are widely distributed in South and Central America and reach Mexico; each of these genera consists of two species, of which one is widely distributed in South and Central America and the other confined to Central America or Mexico. This might be the result either of origin in or of double invasions of Central America. It is one of these widely distributed genera (*Momotus*) rather than a localized relict which sets the northern limit of motmots, in northern Mexico. Another genus, with one species, is mainly South American and enters Central America only to Nicaragua. And the remaining three genera, each with one species, are mainly Central American, although two of them extend across the Isthmus of Tehuantepec in southern Mexico. No genus of the family occurs chiefly on the North American side of the isthmus. It seems to me that there is nothing in the distribution of this family to suggest that southern North America (as distinct from Central America) was its place of origin.

It has been supposed that cracids too were originally tropical North American birds. There are eleven existing genera of them. Three of the genera range from northern Argentina to parts of Mexico, and one of the three (*Ortalis,* the chachalacas) reaches southern Texas. One genus ranges from Peru to Costa Rica. Five genera are confined to parts of South America. And two genera are confined to parts of Central America and southern Mexico. The existing genus *Ortalis* is listed fossil in the Lower Miocene and Lower Pliocene of Nebraska and South Dakota, well north of the present limits of the family, but it is likely that the Lower Miocene forms were really not *Ortalis* but extinct genera, one of them smaller than any living cracid (Wetmore 1951*a,* p. 59). That fossil cracids of comparable age have not yet been found in South America probably indicates no more than that the fossil record of the family is very incomplete. The present distribution of genera makes it almost certain that cracids were in South America long before the late Pliocene connection with Central and North America. The distribution of cracids seems to reflect multiple dispersal and differentiation during a long period in the Americas as a whole (excepting the West Indies), not radiation from southern North America.

Wood warblers (Parulinae) are another group of birds that supposedly originated in tropical North America. Their distribution is more complicated. Many of them are resident in South America, Central America, the West Indies, and southern North America (Mexico etc.). Many others are migratory, wintering usually in or near the tropics (often in Central America, northern South America, or the West Indies) but breeding far north, some of them to the northern limit of trees, in North America. Of the genera, some contain only species resident in the tropics or subtropics; others, only northern migratory species; and still others, both resident and migratory species, in the same genus. Statistically, the majority of existing genera and species are either northern and migratory or resident in rather small areas in Mexico or Central America or on islands in the West Indies. No genus is confined to South America, but *Myioborus* and *Basileuterus* occur chiefly there; if numbers are to be used as clues to the history of this family, these genera must be considered originally South American ones which have spread into Central America and Mexico. In addition South America has widely distributed resident species of *Parula* (this genus contains also the northern Parula Warbler) and *Geothlypis* (which contains also the northern yellow throats) and shares *Granatellus* with Central America and Mexico. In the genus *Dendroica*, the relation between migratory and resident forms is complex. This genus contains many of the most northern, most migratory wood warblers. The northern, migratory Yellow Warbler (*Dendroica aestiva*) has resident subspecies or closely related species in Mexico etc. and the West Indies. The migratory Yellow-throated Warbler (*D. dominica*) has a resident form on some of the Bahamas. The northern, migratory Pine Warbler (*D. pinus*) has resident forms on the northern Bahamas and the mountains of Hispaniola. The northern, migratory Prairie Warbler (*D. discolor*) has a representative (*D. vitellina*) resident on a few small islands in the West Indies (the Caymans and Swan Island). And additional species of *Dendroica* are resident or slightly migratory in parts of Mexico and the West Indies. In the genus *Vermivora*, a form of the migratory Orange-crowned Warbler (*V. celata*) is resident on Todos Santos Island off northwestern Lower California, and apparent representatives of the migratory Virginia Warbler (*V. virginiae*) are resident or slightly migratory in parts of Mexico. Many of these details are from Bond (1936) and Blake (1953).

Mayr and Bond have counted the numbers of genera and species of Parulinae in different places and have concluded that these birds originated in "tropical North America," but this method of analysis would be good only if the birds had spread from one place in a fairly simple and orderly way. They have not. They have had a complex, multiple dispersal in the Americas as a whole, and they have originated (and lost) migratory habits many times. I think that no single place can be fixed as their place of origin. They were probably widely distributed in North and South America almost from the first. The number of genera localized in Central America and the West Indies probably reflects not place of origin but past and present discontinuity of land. The number of northern, migratory species was probably increased by the Pleistocene. Many Parulinae (but perhaps not all) are good water crossers; they migrate across water gaps to the West Indies; they were probably not blocked by Tertiary water gaps. It is likely, too, that these birds have inherent qualities which make them unusually good migrants, more ready than most birds to move their breeding ranges northward when opportunities occur. In short, the distribution of Parulinae seems to be the product of very complex dispersal in the Americas as a whole, of past and present conditions in many places, and of special qualities of the birds themselves, rather than of simple spread from one place. They should, I think, be struck off the list of birds that are supposed to have originated in "tropical North America."

So far as I can see, there is no good evidence that any family of birds (except perhaps turkeys) has really originated in or spread from southern North America, although some groups of less than family rank have probably done so. The clearest case is that of the American quails, considered a subfamily by some authors but not by Mayr and Amadon. Of the ten genera of American quails listed by Peters, nine are confined to southern North America or may have spread from there. The tenth is mainly Central American. Three of the genera reach South America, but none is endemic there. Mexico has been the apparent center of radiation of these birds. But turkeys and quails do not make a whole bird fauna, and few other important groups of birds have distributions like theirs.

Fossil birds have been overstressed in connection with "tropical North America." Cracids and limpkins, now mainly tropical, are fossil in mid-Tertiary North America, but they need not have originated there. The fossil record of these birds and of birds in general is too poor and too one-sided to be safely interpreted that way. The cracids have already been discussed above. What the fossils show of limpkins is not that they are North American in origin but that they are a declining group (now reduced to one species) not suitable for analysis by numbers clues.

There is another, apparently untried source of evidence about "tropical North American" birds. If whole families of birds evolved in southern North America without crossing the water barriers to South America, they were not good water crossers. If such birds spread across the West Indies from North America (as the theory supposes), they presumably did so with difficulty, and they should be distributed across the islands in a sort of cline (Fig. 37). The greatest number of them should be on Cuba, and their numbers should decrease with distance from North America, somewhat as land birds decrease with distance from New Guinea across the western Pacific (p. 252, Fig. 32). So far as I can see, there is no such cline in the distribution of most West Indian birds. There are more birds on the Greater

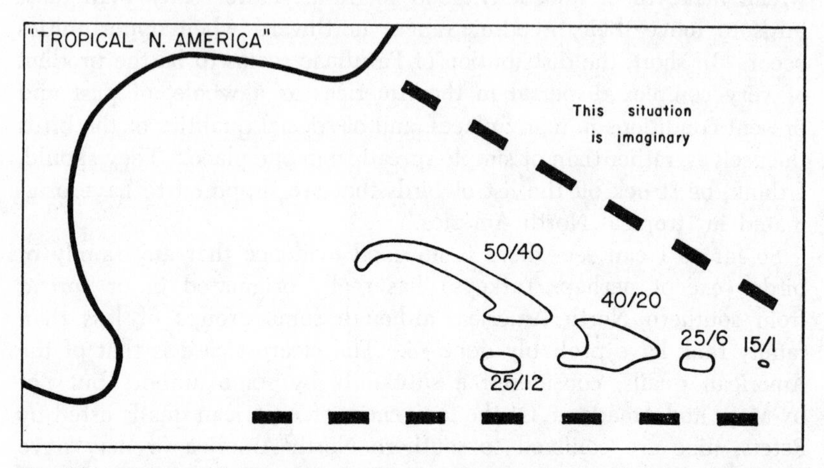

Fig. 37. Diagram of cline that should (but does not) exist if certain families of birds reached the West Indies from "tropical North America." Numbers (*e.g.* 50/40) are (hypothetical) total numbers (species) of land birds over numbers of "tropical North American" ones.

Antilles than on the smaller islands, but not proportionately more of the supposedly tropical North American ones such as wrens, mockingbirds, vireos, and wood warblers. This matter is complicated by doubt about the Tertiary history of some of the islands, but the best that can be said of it is that what might have been decisive evidence of a great tropical bird fauna spreading from southern North America is negative. The quails, incidentally, do show the sort of distribution on the West Indies to be expected of other birds, if they really spread from southern North America: quails have reached Cuba and the Isle of Pines (one species) but not beyond.

That much of North America was warmer in the Tertiary than now is certain. That birds were more numerous in southern North America and somewhat more isolated from South America than now is probable, although the birds themselves seem to give little actual evidence of it. But that southern North America had a great tropical bird fauna, so isolated from South America that it was a separate center of evolution of important families of birds is, I think, doubtful and against the evidence. I think that the idea of tropical North America is not so much wrong as exaggerated by ornithologists. I do not mean just to condemn the idea. Ideas like this ought to be proposed and tested, and this particular idea was a very interesting one. It almost seems as if it ought to have been correct. If there is not evidence that it is correct, perhaps there is not yet sufficient evidence that it is wrong. But that is the point I am trying to make: that birds tell little about ancient lands and climates, and little about Tertiary North America.

Since I have gone so far in criticism, I had better say what I think probably was the situation among the birds of southern North America and Central America in the Tertiary. My conclusions are drawn partly from what the birds show, partly from deduction from other sources: from what is known of the Tertiary histories of North, Central, and South America, and from the nature of the existing vertebrate faunas of Central America and the West Indies.

North and South America were certainly separated during most of the Tertiary. North America was certainly warmer than now but was probably more subtropical than tropical. Central America was, I think, probably completely cut off by water from North as well as South America through much of the Tertiary. My reason for thinking this is that many of the older West Indian vertebrates seem (judging from their patterns of distribution on the islands) to have come from Central America, and they include animals, especially hystricomorph rodents, of sorts which would probably not have been in Central America if it had been connected to North America. I think that southern North America, Central America, and the West Indies probably all had bird faunas which (compared to the South American bird fauna) were depauperate, and I think that (in spite of the water gaps) all the depauperate faunas were made up largely of the same more or less widely distributed families of birds, except that the West Indies evidently lacked some of the families. *I think that this whole area plus appropriate parts of South America was the area of evolution of such families as the wrens, mimids, and wood warblers,* all of which probably had complex geographical histories, with complex origins and sometimes losses of migratory habits and much water crossing. Birds probably reached the West Indies from several directions (but most of all from Central America) and spread over the islands so freely and complexly that they formed no simple cline of numbers with distance. Cracids and some other birds that did not reach the West Indies were nevertheless (I think) widely distributed in North, Central, and South America at this time. There was probably a good deal of endemism of genera as well as of species in isolated places, and some of the endemic genera are still localized in southern North America and elsewhere, but they are evidence of past isolations, not of places of origin of families. Few families seem to have been localized. Southern North America does not seem to have been the place of origin of any families of birds except perhaps of turkeys, although quails and some other groups probably had secondary radiations there. Central America seems to have been the place of origin of motmots, but they are a lone case; no other family of birds is distributed like them. If other families were confined to Central America, they have redispersed beyond recognition or have become extinct. But perhaps motmots were the only ones; their close relatives, the todies, are the only family of birds confined to the West Indies now.

What I have just said demotes "tropical North America" to an area which was not very large or at least not very tropical and, as a separate area, not very important in evolution and dispersal of birds. The water barriers that separated North, Central, and South America were apparently only partly effective. There must have been a very rich tropical bird fauna in South America during the Tertiary. Part of it, but only part, probably spread northward across the water barriers to Central America and southern North America, and a still smaller part reached the West Indies. Other birds probably crossed the water barriers in the other direction, from North to Central and South America, and some of these birds too reached the West Indies, directly or indirectly. Actual movements of some birds (*e.g.*, wood warblers) across the water barriers were probably very complex. This is all in accordance with the general rule derived from existing situations elsewhere in the world, that where birds come against barriers, some of the birds are stopped while others cross the barriers at different times and for different distances.

Wolfson (1955) has tried to fit the history of North American birds to the theory of continental drift. It is worth while to try this kind of thing now and then, but in this case the results are not convincing. The present distribution of birds is clearly the result of complex dispersal. Many different birds have spread over the world from different centers, for different distances, and at different times; and the size, relationships, and relative endemism of different continental bird faunas are correlated with the present distribution of land, climates, and barriers of the world. In other words, the present distribution of birds is plainly the result of movement of birds over something like the existing pattern of land, not of movement of continents. Continental drift, if it occurred, was evidently too long ago to affect bird distribution now.

LIST OF FAMILIES OF BIRDS

Class Aves, birds. This list follows Mayr and Amadon's (1951) classification, for reasons given in the text (p. 237), but the details of distributions are largely from Peters' (1931–1951) check list, so far as it goes. Additional fossil families are interpolated from Wetmore (1951). The fossil records of existing families are noted briefly, from Lambrecht (1933) and Wetmore (1951*a* etc.), usually checked against Romer (1945, pp. 602–610). The presence or absence of all African families in Madagascar, of all Australian families in New Zealand, and of all American families in the West Indies is noted for later reference. A dagger (†) before a name indicates that the group is extinct.

The great power of movement of birds makes special difficulty in describing their ranges. Some birds occasionally straggle far beyond their normal limits, but records of stragglers cannot be included here. Many sea birds, which breed only very locally, move about over great areas of ocean during parts of the year, but this cannot be gone into here. The migrations of land birds are mentioned when they affect the limits or patterns of distribution of families in important ways, but very many details of migration even of land birds have had to be omitted.

(†*Subclass Archaeornithes*), ancestral birds

†Archaeopterygidae and †Archaeornithidae: Europe (Bavaria) in the Upper Jurassic; each family based on one good skeleton. These famous, reptile-like birds are of great evolutionary but little geographical importance.

(*Subclass Neornithes*), true birds

†*Orders Ichthyornithes and Hesperornithes,* Cretaceous toothed birds

†Ichthyornithidae, †Apatornithidae, †Hesperornithidae, †Enaliornithidae, †Baptornithidae: Cretaceous of North America, Europe, and southern South America. Only the North American †Ichthyornithidae and †Hesperornithidae are adequately known. Both were toothed and both were sea birds. The ichthyornithids were strongly winged. The hesperornithids were flightless swimming birds with vestigial wings. The other Cretaceous fossil birds listed above are inadequately known and are placed here only tentatively. The South American one is †*Neogaeornis* (Lambrecht 1933, pp. 258 and 654) of southern Chile,

tentatively assigned to the †Enaliornithidae. All the Cretaceous birds are of evolutionary rather than geographical importance.

†*Order Caenagnathi*

†Caenagnathidae: Upper Cretaceous of North America (Alberta); a single mandible, *without* teeth—but it may not be a bird.

[*The following 5 orders are "ratites," 5 probably unrelated groups of birds which have become flightless secondarily and attained large size* (*Mayr and Amadon 1951, pp. 3 and 4*) (*Fig. 33*).]

†*Order Aepyornithes*

†Aepyornithidae, elephant birds: Madagascar; extinct, Pleistocene and post-Pleistocene; 2 genera, several species, including good material. Supposed records of this family from Africa are based on unsatisfactory fragments and should not be accepted (Mayr and Amadon, p. 4).

Order Apteryges

†Dinornithidae and †Anomalopterigidae, moas: New Zealand (both main islands); known in the Pliocene (and perhaps late Miocene), numerous in the Pleistocene, and surviving until the arrival of the earlier Polynesians on New Zealand; 7 genera, about 20 or more species (Archey 1941, Oliver 1949); known from a number of more or less complete skeletons as well as thousands of separate bones. The supposed occurrence of moas in Australia is based on doubtful fragments probably misidentified (Mayr and Amadon, p. 4).

Apterygidae, kiwis: New Zealand, including both main islands and Stewart Island, and from subtropical forest at sea level to the snow line (Falla 1953); one genus, 3 species; and another genus and species subfossil. The supposed records of kiwi bones in Australia are erroneous (Mayr and Amadon, p. 4).

Order Struthiones

†Eleutherornithidae, †*Eleutherornis:* Eocene of Europe (Switzerland); a fairly well-preserved pelvis; apparently representing a group ancestral to the living ostriches and connecting them with flying ancestors (Wetmore).

Struthionidae, ostriches: almost all the drier parts (not heavy forests) of Africa from the Atlas Mountains to the Cape of Good Hope (not Madagascar) and (until recently) deserts of Syria and Arabia; one existing species; and fossil bones, eggs, and egg shells across southern Eurasia to China, back to the Lower Pliocene (Lambrecht 1933, p. 109, map). The end of the Ostrich in Syria and Arabia, where it became extinct in 1941, is described by Meinertzhagen (1954, p. 574).

Order Casuarii

Casuariidae, cassowaries: New Guinea and neighboring islands (Aru Islands, Ceram, Salawati, Jobi, New Britain) and the northeastern corner of Australia (not New Zealand) [the poet who put a cassowary on the banks of the Timbuctoo, in Africa, was not a zoogeographer]; one genus, 6 species; and fossil (end of a tibiotarsus) in the Pleistocene of New South Wales, south of the present range in Australia. Unlike the other big existing "ratites," these birds live in heavy forest.

Dromaeidae, emus: Australia (not New Zeland); extinct on Tasmania and King and Kangaroo Islands south of Australia; one genus, 2 species; and fossil in Australia in the Pleistocene.

†Dromornithidae, †*Dromornis*, †*Genyornis:* Pleistocene of Australia (not New Zealand); fairly good material; giant, emu-like birds.

Order Rheae

Rheidae, rheas: eastern and southern South America (not West Indies), from northern and eastern Brazil and the southeastern corner of Peru to the Straits of Magellan; 2 living genera and species; and fossil fragments.

Order Crypturi

Tinamidae, tinamous: South and Central America, north to parts of Mexico (2 genera) and south nearly to the tip of South America, and including Trinidad (but not West Indies proper); about 9 genera, 33 or more species, some in forest, others in grassland. They are primitive, superficially grouse- or quail-like birds, which fly powerfully for short distances, and which may be distantly related to the rheas (McDowell 1948; Mayr and Amadon 1951, p. 4). A few fossils of the family are known within its present range, back only to the Pliocene.

Order Sphenisci

Spheniscidae, penguins: antarctic seas north to the coasts of southern and eastern Australia (north to southern Queensland), New Zealand, South Africa, and southern and western South America (north to Peru), with an endemic species on the Galapagos Islands, on the equator; 6 genera, 16 species; and fossil in southern South America (Patagonia), Australia, New Zealand, and Seymour Island in Antarctica, all probably in the Miocene. (The somewhat earlier †Cladornithidae are now thought not to be related to the penguins—Simpson 1946.) Simpson (p. 71) stresses the absence of fossil penguins outside the present area of distribution of the birds and concludes that the limits of distribution of the penguin family in the Miocene were probably about the same as now.

Order Tubinares

Diomedeidae, albatrosses: chiefly southern tropical and subtropical seas, and a few in the North Pacific; 2 genera, 13 species; none now regularly in the North Atlantic, but fossil (a tarsometatarsus) in England in the Lower Pliocene.

Procellariidae, petrels, shearwaters, fulmars: most seas; 2 subfamilies, both widely distributed: (1) Hydrobatinae, storm petrels, several genera, 20 species; and (2) Procellariinae, shearwaters etc., several genera, 53 species; some geographically unimportant fossils.

Pelecanoididae, diving petrels: southern seas, north to the coasts of southeastern Australia, New Zealand, and southern and western South America north to Peru; one genus, 4 species.

Order Podicipides

Podicipitidae, grebes: cosmopolitan, including Madagascar, New Zealand, and West Indies; 5 genera (of which *Poliocephalus* and *Podiceps* are more or less cosmopolitan), 20 species; primarily fresh-water birds; listed fossil back to the Upper Eocene (of Europe).

Order Gaviae

Gaviidae, loons: breed in the colder parts of the northern hemisphere and winter more or less southward within the north-temperate zone; one genus, 4 species; primarily fresh-water birds; a few fossil fragments back to the Eocene, all in the northern hemisphere.

Order Steganopodes

Phaëthontidae, tropic birds: warm seas; one genus, 3 species; and a fossil genus represented by fair material in the Lower Eocene of England.

Fregatidae, frigate birds: warm seas; one genus, 5 species.

Phalacrocoracidae, cormorants etc.: 2 subfamilies. (1) Phalacrocoracinae, cormorants: almost cosmopolitan, including Madagascar, New Zealand, and some West Indies; 3 genera, 30 species; fossil back to the Eocene of North America and Oligocene of Europe. (2) Anhinginae, anhingas: warmer regions of the world including Madagascar and some West Indies but not (except accidentally) New Zealand; one or 2 species, several subspecies; geographically unimportant fossils.

†Pelagornithidae, †*Pelagornis:* Miocene of France; a humerus.

Sulidae, boobies, gannets: tropical and temperate seas; 2 genera, 9 species; some fossils of little geographical importance.

†Elopterigidae, †*Elopteryx,* †*Eostega,* †*Actiornis:* Upper Cretaceous to Upper Eocene of Europe; fragments.

Pelecanidae, pelicans: widely scattered over the world in tropical and some temperate regions, north to southeastern Europe and to Great Slave Lake in Canada, including West Indies but not (except doubtfully or accidentally) Madagascar or New Zealand; one genus, 6 species; some fair fossil material back to the Upper Eocene (of France).

†Cyphornithidae, †*Cyphornis,* †*Palaeochenoides:* North America, in the Tertiary (probably Miocene) of British Columbia and Miocene of South Carolina; fragmentary leg bones.

†Odontopterygidae and †Pseudodontopterygidae, false-toothed birds: the former is based on a skull and fragments from the Lower Eocene of England; the latter, on a skull said to be from Brazil and of doubtful age. These birds are "of highly doubtful status" (Wetmore 1951, pp. 4 and 15).

†Cladornithidae, †*Cladornis,* †*Cruschedula:* Upper Cretaceous of southern South America; fragmentary metatarsi. The affinities of these forms and even their relationship to each other are very doubtful (Simpson 1946; Wetmore 1951, pp. 2 and 15).

Order Falcones or Accipitres: a heterogeneous and perhaps polyphyletic group

Accipitridae, hawks in part, eagles, Old World vultures, etc.: cosmopolitan, including Madagascar, New Zealand, West Indies, and many more remote islands; many genera, 205 species. Many genera are localized but some are widely distributed. *Circus* (harriers or marsh hawks) is cosmopolitan, including Madagascar, New Zealand, West Indies (in part), and other islands, but it only winters and does not breed in most of the Oriental Region. *Accipiter* (goshawks etc.) is almost completely cosmopolitan except that it is not on New Zealand; it does occur on Madagascar, the West Indies, and some more remote islands. *Aquila* occurs around the northern parts of the world and south through Africa (not Madagascar), to India etc. in Asia (and in Australia, if *Uroaëtus* is included), and south to central Mexico in America; the genus consists of several Old World species of which one (*A. chrysaëtos,* the Golden Eagle) reaches North America. *Haliaeetus* occurs through most of the Old World including Madagascar and Australia (not New Zealand) and in North America south to northern Mexico; this genus too consists of several Old World species and one (*H. leucocephalus,* the Bald Eagle) in North America. *Buteo* (including many common soaring hawks) occurs through Africa, Madagascar, the cooler parts of Eurasia, and North, Central, and South America, and the West Indies, and on many remote islands including Hawaii, the Galapagos, Juan Fernandez, and several Atlantic islands; but it does not breed in the Oriental Region and winters in only part of it, and it does not reach the Australian Region at all (see also text, p. 269). *Spizaëtus,* a genus of eagles, with few species, occurs from southern and eastern Asia and Japan to New Guinea (not Australia), and in tropical America from southern Mexico to northern Argentina (not West Indies). Fossil accipitrids are listed from widely scattered places back to the Eocene; *Buteo* was apparently present in North America as long ago as the Oligocene; Old World vultures ("Aegypiinae"), which are not now represented in the New World, were common in North America from the Lower Miocene to the Pleistocene.

Falconidae, hawks in part, falcons, etc.: cosmopolitan, including Madagascar, New Zealand, and West Indies; various genera, 58 species. Most of the genera are more or less restricted in range, but *Falco* (falcons) is fully cosmopolitan, including the islands named. This genus is an example of multiple dispersal of a dominant group. It consists of 10 subgenera (Peters), of which 3 are nearly cosmopolitan. One species (*F. peregrinus,* with a number of subspecies including the Duck Hawk) is nearly cosmopolitan, breeding on parts of all the continents and on many islands including Madagascar and Fiji but not New Zealand or the West Indies, although it reaches the latter in winter. A few fossils of the family are known, back to the Miocene (of North America); *Falco* itself is fossil in the Miocene of North America.

Pandionidae, ospreys: one species (several subspecies) of which the breeding and wintering areas together cover most of the world. It breeds around the northern parts of the world above the tropics, and from Java, Celebes, and the Lesser Sunda Islands to Australia, Tasmania, and New Caledonia (not New Zealand), and locally in the Bahamas (not main West Indies) and Yucatan and British Honduras in Central America. It winters in parts of the breeding range and through most of Africa (not Madagascar), part of the Oriental Region, and Central and South America and the West Indies. The same species is in the Pleistocene of Florida.

†Neocathartidae, †*Neocathartes:* Upper Eocene of North America (Wyoming); "a small-winged, strong-legged vulture that evidently was terrestrial with limited powers of flight" (Wetmore 1951, p. 5); a nearly complete skeleton.

Cathartidae, New World vultures: the New World, except the colder parts; in the West Indies, native only on Cuba and Jamaica; 5 genera, 6 species; and various fossils in the New World back to the Oligocene of

North America, including bones of the California Condor east to Florida in the Pleistocene. It has been supposed that †*Eocathartes* in the Middle Eocene of Germany and †*Plesiocathartes* in the Upper Eocene or Lower Oligocene of France show the former occurrence of this family in the Old World, but both are fragmentary and their assignment here requires confirmation (Wetmore, in Mayr and Amadon 1951, p. 6).

†Teratornithidae, †*Teratornis,* †*Cathartornis,* giant fossil vultures: Pleistocene of California and Florida; good material.

Sagittariidae, secretary birds: Africa (not Madagascar), from Senegambia and the Egyptian Sudan to Cape Province; one living species; and a fossil genus (fragments) in the Oligocene (and Upper Eocene?) of France.

Order Gressores

Ardeidae, herons: cosmopolitan, including Madagascar, New Zealand, and West Indies; 32 genera, 59 species; the most northern ones are migratory. Six of the genera (*Ardea, Butorides, Casmerodius, Nycticorax, Ixobrychus,* and *Botaurus,* including respectively the Great Blue Heron, Green Heron, American Egret, Black-crowned Night Heron, Least Bittern, and American Bittern) are more or less cosmopolitan or tropicopolitan; the many other genera of the family are confined to parts of either the Old or the New World, and some are still more localized. The species *Casmerodius albus* (5 subspecies including the American Egret) is almost cosmopolitan except in cold places and reaches Madagascar, New Zealand, and the larger islands of the West Indies. The species *Butorides striatus* (numerous subspecies) too is almost cosmopolitan except in cold places, occurring in tropical Africa, Madagascar, remote islands of the Indian Ocean, southern and eastern Asia north to the Amur River and Japan, south through the islands to northern and eastern Australia (not New Zealand), on Tahiti, probably other remote islands of the tropical Pacific, the Galapagos, and South America from Panama to northern Argentina; but the species is not in the main part of Central America, not on the West Indies, and not in North America—another species of the genus (*B. virescens,* the Green Heron) occurs in these places. The genus *Cochlearius* (boat-billed herons), with one species (3 subspecies), is sometimes placed in a separate family; it occurs in tropical America from Mexico to southern Brazil (not West Indies). Fossil herons are listed back to the Eocene (of Europe and North America).

Threskiornithidae, ibises, spoonbills: the warmer parts of the world, including Madagascar, New Zealand (occasional), and some West Indies; numerous genera, 28 species. Peters recognizes 20 genera in this small family, only one of them occurring in both Old and New Worlds. It is *Plegadis,* of which the principal species (3 subspecies, including the Glossy Ibis) is almost tropicopolitan. The other 19 genera are each confined to either the Old or the New World and in many cases to part of a single continent. Fossil ibises include fair material from the Upper Oligocene of Europe and scanty material from the Upper Eocene of England.

Ciconiidae, storks: Old World (including Madagascar) to northern and eastern Australia (not New Zealand), and New World from southeastern United States to southern Argentina, and some West Indies; the northern Eurasian species make tremendous migrations; numerous genera, 17 species. Of the 12 genera recognized by Peters, 9 are confined to parts of the Old World, 3 are confined to the New World, and none occurs in both (but the Old World *Ciconia* is listed fossil in North America in the Pliocene and Pleistocene). The genus *Balaeniceps* (whale-headed storks), with one species, confined to a limited area in eastern Africa (the White Nile etc.), is sometimes placed in a family of its own. Fossil storks are known back to the Oligocene (of Europe etc.).

Scopidae, hammerheads: all Africa below the Sahara, Madagascar, and southwestern Arabia; one species.

Order Phoenicopteri

Phoenicopteridae, flamingos: now only in widely scattered localities in Africa, Madagascar, southern Europe, *western* Asia to India, and the warmer parts of America from southern Mexico, the West Indies, and the Florida Keys to Tierra del Fuego; 3 genera, 6 species. Fossil flamingos include a nearly complete skeleton from the Upper Oligocene of France, a humerus from the Upper Eocene of England, an extinct genus from the Lower Miocene of North America (South Dakota), and fragments of the existing genus *Phoenicopterus* (the only genus of the family which occurs in both Old and New Worlds) in the Pliocene and Pleistocene of North America, north to Oregon.

†Agnopteridae, †*Agnopterus:* Upper Eocene of France and England; fragments; and now recorded also from central Asia.

†Scaniornithidae, †*Scaniornis,* †*Parascaniornis:* Upper Cretaceous of Europe; odd bones.

Order Anseres

†Paranyrocidae, †*Paranyroca:* Lower Miocene of North America (South Dakota).

Anatidae, ducks, geese, swans: cosmopolitan, including Madagascar, New Zealand, West Indies; many northern forms are migratory; many genera, 145 species. The family has been revised by Delacour and Mayr (1945). Swans (all of which Delacour and Mayr put in *Cygnus*) occur around the northern parts of the world (5 species) and isolated in southern South America (one species), from Paraguay etc. southward, and in Australia and Tasmania (one species, the Black Swan); a species (now extinct) once occurred on New Zealand. Typical geese (2 genera, 14 species) are entirely northern except for an endemic species in Hawaii, but unrelated "goose-like" genera (large, long-legged, often terrestrial and grazing forms) occur in many other parts of the world. Whistling or tree ducks (*Dendrocygna,* 8 species) occur in all principal tropical and some warm-temperate areas, including Madagascar, Australia and Tasmania, and islands of the western Pacific to New Caledonia and Fiji (not New Zealand), and some West Indies; some species of this genus are very widely and discontinuously distributed (see Peters' *Check-list,* Vol. 1, pp. 152–154).

Other ducks are numerous, complex in their interrelationships, and complexly and often widely distributed. Mayr (1946, p. 14) mentions some of the species and species groups that have notably wide or discontinuous ranges. Fossil anatids occur back to the Eocene (of Europe and North America) and very doubtfully to the Upper Cretaceous (of Europe); to what extent the rather numerous Tertiary forms are geographically significant I am unable to judge.

Anhimidae, screamers: most of continental South America, but not Central America and not West Indies; 2 genera, 3 species.

Order Galli, fowl-like birds

Megapodiidae, mound builders: Australia (not Tasmania, not New Zealand), New Guinea etc., Moluccas, Celebes, with the widely distributed genus *Megapodius* extending to the Solomons and some (perhaps formerly more) islands of the western Pacific to the Marianas and central Polynesia, and to the Lesser Sunda Islands (to Lombok), all the Philippines, small islands off the eastern and northern coasts of Borneo and between Borneo and Java, and the Nicobar Islands *west* of the Malay Peninsula (but not Borneo, Java, or Sumatra, and not any part of the mainland of Asia); 7 genera, 10 species. In Australia different megapodes inhabit dry brushy country as well as tropical jungle. No significant fossils of this family are known.

†Gallinuloididae, †*Gallinuloides:* Eocene of Wyoming; a nearly complete skeleton.

Cracidae, guans etc.: tropical America (not West Indies) north to southern Texas and south to northern Argentina; 11 genera, 38 species; see text (p. 281) for further details.

Phasianidae, pheasants etc.: 3 subfamilies. (1) Phasianinae, pheasants, peafowls, fowls, quails: most diverse in the Orient, but one pheasant is localized in Africa, and quails are nearly cosmopolitan; many genera, 165 species. Pheasants (Delacour 1951) are chiefly Oriental (well represented in both tropical lowland and cool mountain habitats); forms of the Ring-necked Pheasant (*Phasianus colchicus*) extend west to the Caucasus etc. and north to southeastern Siberia below latitude 44° N.; and one or more pheasants reach or occur on Japan, Formosa, the Greater Sunda Islands (several genera), Palawan (but not the other Philippines), and the Lesser Sundas to Sumba (jungle fowls only), but there are no pheasants on Celebes etc. (It is supposed that jungle fowls on the Philippines, Celebes, etc., are introduced.) Peafowls (peacocks—2 species), which are pheasants, occur through most of the continental part of the Oriental Region and on Ceylon and Java (but not Sumatra or Borneo etc.), and the single African pheasant, the very distinct monotypic genus *Afropavo,* confined to a small area in the virgin rain forests of the east central Congo basin, may be related to them. Jungle fowls (*Gallus,* 4 species), which too are pheasants and from which domestic chickens are derived, are native to most of the tropical Oriental Region from Ceylon and India east to Java (not Borneo etc.) and the Lesser Sundas. Quails occur through the Old World including Madagascar, Australia, and (formerly) New Zealand, and in North, Central, and tropical South America (but not south-temperate South America), and on Cuba etc. (but not most of the West Indies); the American ones are sometimes considered to form a separate subfamily, Odontophorinae (see text p. 283). (2) Numidinae, guinea fowls: most Africa north to western Morocco, with undifferentiated populations of single species (perhaps introduced) in southwestern Arabia and Madagascar; 5 genera, 7 species. That the Guinea Fowl in Arabia is introduced is suggested by Meinertzhagen (1954, p. 572). (3) Tetraoninae, grouse etc.: north-temperate and arctic regions of Old and New Worlds; 11 genera, 18 species: one genus (*Lagopus,* the ptarmigans) is circumpolar in the far north, reaching the arctic islands above Asia and North America and also Greenland and Iceland; 3 genera are spread across Eurasia and another is Asiatic; and 6 genera are confined to North America, one being characteristic of northern spruce forest, 2 now or formerly distributed across the more temperate part of North America, and 3 being western; this pattern seems to be the result of a complex dispersal and ecological radiation rather than of spread from any single center. Fossils appear to show that both pheasants and jungle fowls once inhabited Europe, where now only domestic and introduced species occur, but supposed pheasants (fossil fragments) in North America are doubtful; fossil quails are listed back to the Lower Miocene of North America; and fossil Tetraoninae, back to the Eocene of North America.

Meleagridae, turkeys: eastern and southern North America etc. (not West Indies); 2 genera and species. The common Wild Turkey formerly occurred from southern Maine, southern Ontario, South Dakota, etc. to the tableland of Mexico, but its range is now more restricted. The very distinct Ocellated Turkey is confined to the Yucatan Peninsula and adjacent parts of Guatemala and British Honduras. Fossil turkeys go back to the Oligocene (of North America) and include good material of an extinct genus from the Pleistocene of Texas and California.

Opisthocomidae, hoatzins: Amazon forests etc.; one species; an odd, arboreal bird of doubtful relationships.

Order Cuculi

Musophagidae, touracos: Africa below the Sahara (not Madagascar); 6 genera, 19 species.

Cuculidae, cuckoos etc.: cosmopolitan, including Madagascar, New Zealand, and West Indies; many genera, 128 species; none of the genera is common to the Old and New Worlds. Most cuckoos are tropical; those that breed in the cooler parts of the north- and south-temperate zones are migratory; the 2 that breed on New Zealand migrate long distances northward across the ocean (p. 247). Fossil cuckoos are listed back to the Lower Oligocene or Upper Eocene (of France).

Order Grues: "The large number of relict families of a genus or two each in the Grues mark it as a declining and ancient order, as does its fossil history" (Mayr and Amadon 1951, p. 8)

Cariamidae, seriamas: part of South America (not West Indies), from the tableland of Brazil to northwestern Argentina; 2 genera and species.

†Bathornithidae, †*Bathornis:* Oligocene of North America.

†Hermosiornithidae, †*Hermosiornis,* †*Procariama:* Lower Pliocene of Argentina; fair material.

Psophiidae, trumpeters: the main part of tropical South America (not West Indies); one genus, 3 species.

†Geranoididae, †*Geranoides:* Lower Eocene of North America (Wyoming); parts of leg bones.

†Eogruidae, †*Eogrus:* Upper Eocene of Asia (Inner Mongolia); leg bones and a wing fragment.

Gruidae, cranes: the Old World and North America (the more northern forms are migratory); somewhat discontinuously distributed: in Africa but not Madagascar; in Eurasia etc. including tropical Asia, but not Sumatra, Java, Borneo, Celebes, etc. (but *Grus grus* occurred on Java in the ice age, perhaps as a migrant—Wetmore 1951*a,* p. 61); locally resident on Luzon but not on other Philippines; in Australia but not New Zealand; in North America, south to Mexico only on migration, but resident on western Cuba and the Isle of Pines (not other West Indies); the 2 American species belong to the widely distributed Old World genus *Grus;* 4 genera, 14 species; and fossil back to the Eocene (of Europe and North America).

Aramidae, limpkins: the warmer part of the Americas from southern Georgia, Florida, some West Indies, and southern Mexico to eastern Argentina; one existing species (5 subspecies); and fossil genera in the Oligocene and Miocene of North America (South Dakota and Nebraska).

Eurypygidae, sun bitterns: much (but not all) tropical South and Central America from central Brazil to the Caribbean slope of Guatemala (not West Indies); one species (3 subspecies).

Heliornithidae, sun grebes: tropical and part of South Africa (not Madagascar), the Oriental Region from northeastern India to Sumatra, and tropical America from southern Mexico to northeastern Argentina etc. (not West Indies); each of these 3 regions has one endemic genus and species.

Rhynochetidae, kagus: New Caledonia, one species; rather large, with well-developed wings, but flightless.

Otididae, bustards: continents of the Old World: Eurasia (somewhat migratory at the northern edge of the range), Africa (not Madagascar), and Australia (not New Zealand), but not the islands between Asia and Australia; 11 genera, 23 species; fossil back to the Eocene (of Europe).

Rallidae, rails, etc.: cosmopolitan, including Madagascar, New Zealand, and the West Indies, and many remote islands; migratory in the colder part of the north-temperate zone; 52 genera, 132 species. The genera *Rallus* and *Porzana* among the rails, *Gallinula* among the gallinules, and *Fulica* of coots are cosmopolitan and occur on the islands named above, except that *Gallinula* and *Fulica* do not reach New Zealand (where, however, there are many other rails). *Coturnicops* is known from widely scattered and perhaps discontinuous areas in eastern and southeastern Africa (not Madagascar), eastern Asia, North America, Mexico, British Guiana, and southern South America (not West Indies). *Porphyrula* has one species in Africa below the Sahara and in Madagascar, and 2 in America from Texas, South Carolina, and the West Indies to northern and eastern Argentina. Other genera of the family are confined to either the Old or the New World or (many of them) to single islands or island groups; many of the island genera are flightless. Fossil rails have been found in various parts of the world back to the Eocene (of Europe, Asia, and North America).

†Orthocnemidae, †*Orthocnemis*, †*Elaphrocnemis:* Lower Oligocene of France; leg bones only; position provisional.

Mesoenatidae, roatelos: Madagascar; 2 genera, 3 species; relationships doubtful; medium-sized birds with well-developed wings but flightless.

Turnicidae, button quails etc.: 2 subfamilies. (1) Turnicinae, button quails: Africa (north to Spain), Madagascar, and southern Asia to Australia (not New Zealand); 2 genera, 15 species. (2) Pedionominae, collared hemipodes: arid plains of the interior of Australia (not New Zealand); one species.

(*Additional fossil Grues*), not placed in relation to existing families

†Phororhacidae, †*Phororhacos* etc.: Tertiary (Oligocene to Pliocene) of southern South America; much material; giant, flightless, predaceous birds; for notes on this and the following 2 related families see Patterson (1941).

†Psilopteridae, †*Psilopterus*, †*Smiliornis:* Tertiary (Oligocene and Miocene) of southern South America.

†Brontornithidae, †*Brontornis* etc.: Tertiary (Oligocene and Miocene) of southern South America; much material.

†Opisthodactylidae, †*Opisthodactylus:* Miocene of southern South America; fragments.

†Cunampaiidae, †*Cunampaia:* Oligocene of western Argentina; position doubtful.

†*Order Diatrymae:* possibly related to (derived from) Grues (Mayr and Amadon 1951, p. 8)

†Diatrymidae, †*Diatryma* etc.: Eocene of North America (Wyoming, New Mexico, New Jersey) and perhaps also of Europe; giant, predaceous, flightless.

†Gastornithidae, †*Gastornis*, †*Remiornis:* Lower Eocene of Europe; various bones.

Order Laro-Limicolae

Jacanidae, jacanas: most continental tropics and some islands, including Madagascar and Australia but not New Zealand, north in America to southern Texas and some West Indies; 6 genera, 7 species; freshwater birds.

Thinocoridae, seed snipe: western and southern South America, from Ecuador to Tierra del Fuego etc.; 2 genera, 4 species; the relationships of the family are doubtful.

Chionididae, sheathbills: breed on antarctic islands and perhaps also on the edge of the Antarctic Continent, and one species reaches but probably does not breed on the Falkland Islands and the coast of Tierra del Fuego and extreme southern Argentina; resident or sporadically migratory; one genus, 2 species. They are the most southern non-marine birds (if they can be called non-marine) and are the only birds without webbed feet that reach the antarctic. They are said to be pigeon-like in appearance and movements, crow-like in beak and voice, with the habits of waders but afraid of the water, pugnacious, and practically omnivorous scavengers (Murphy 1936, pp. 999–1006).

Dromadidae, crab plovers: coasts and some islands of the Indian Ocean etc. from eastern Africa to India etc., and Madagascar; one species.

Burhinidae, thick-knees: Africa (not Madagascar); southern Europe and Asia to Australia, Tasmania, New Caledonia, etc. (not New Zealand); and tropical America from southern Mexico to northern and western South America south to southern Peru, with one subspecies isolated on Hispaniola in the West Indies; 3 genera, 9 species; rather large, ground-living birds which inhabit open places. *Burhinus,* with 7 species, has the range of the family except that, though present in Australia, it is absent on the Indo-Australian Archipelago. The other 2 genera are monotypic and are confined to the Oriental Region and the Indo-Australian area respectively. A burhinid is listed from the Miocene of Europe.

†Rhegminornithidae, †*Rhegminornis:* Lower Miocene of Florida; only the foot of this bird is known (Wetmore 1951, p. 6).

†Presbyornithidae, †*Presbyornis:* North America (Utah) in the Eocene; several bones and fragments.

Haematopodidae, oyster catchers: widely but discontinuously distributed: sea coasts and some inland waters chiefly of the cooler zones both north and south of the tropics in the Old World, reaching New Zealand but not Madagascar; and coasts of most of the New World including the tropics and including the West Indies; one genus, 6 species; and another genus listed fossil (a tarsometatarsus) in the Lower Miocene of Nebraska.

Charadriidae, plovers, sandpipers, etc.: cosmopolitan; 5 subfamilies. (1) Charadriinae, plovers: cosmopolitan, including Madagascar, New Zealand, and West Indies; 32 genera, 63 species. *Charadrius* (typical plovers) breeds in the whole main range of the family including the islands named, but some northern species of the genus migrate great distances. *Squatarola* (the Black-bellied Plover) and *Pluvialis* (the Golden Plover etc.) breed only in northern regions but cover much of the world in their migrations. Other genera are either northern migrating southward, far southern (in either Old or New Worlds but not both) in some cases migrating northward, or localized; different localized genera occur in different places in the tropics as well as in cold places and on islands. Fossil plovers are listed back to the Oligocene of Europe and Miocene of North America. (2) Scolopacinae, sandpipers, etc.: cosmopolitan in their migrations, but most breed only in the cold parts of the northern hemisphere; 29 genera, 77 species. Of the 29 genera, 22 breed only in arctic or north-temperate regions. The 7 exceptions are *Scolopax,* which breeds from the northern parts of the Old World south to the mountains of the Indo-

Australian Archipelago (to New Guinea); *Catoptrophorus* (the Willet) which breeds in the northern part of North America and down the Atlantic coast to Texas and on the Bahamas and probably the Greater Antilles in the West Indies; *Capella* (snipe), which breeds in northern regions and in South and East Africa, Madagascar, Mauritius in the Indian Ocean, and parts of South America (not West Indies); *Chubbia,* confined to mountains of South America and the Falkland Islands; one genus confined to small islands off New Zealand; and 2 genera confined to remote islands in the tropical Pacific. Large parts of Africa, parts of tropical Asia etc., Australia, and parts of tropical America have no breeding members of this subfamily. Fossil scolopacines are listed back to the Eocene (of Europe and North America). (3) Phalaropinae, phalaropes: breed in northern regions and winter at sea in certain tropical and far-southern areas; 3 genera and species. (4) Recurvirostrinae, stilts, avocets: cosmopolitan (but breeding areas often localized), including Madagascar, New Zealand, and West Indies; 4 genera, 7 species. The single species (several subspecies) of *Himantopus* (stilts) has the range of the family. *Recurvirostra* (4 species) breeds locally across Eurasia, in parts of tropical and southern Africa and possibly Madagascar, in Australia and Tasmania (not New Zealand), North America, and a small area in the mountains of southern Peru, northern Chile, etc., in South America (not West Indies). The other 2 genera are confined to the high plateaus of central Asia and to Australia respectively. Recurvirostrines are unknown fossil before the Pleistocene (of North America). (5) Rostratulinae, painted snipe: Africa below the Sahara, Egypt, Madagascar, southern Asia to Wallace's Line (not Celebes to New Guinea), and Australia and Tasmania (not New Zealand)—all one species; and southern South America (not West Indies)—another genus and species; and a third genus fossil (a nearly complete skeletal impression) in the Eocene of Germany.

Glareolidae, swallow plovers: most of the Old World: Africa and Madagascar, southern Europe and Asia, and Australia; some (both Asiatic and Australian species) winter in the Indo-Australian Archipelago but none breeds there; accidental in New Zealand; 6 genera, 16 species.

Laridae, gulls etc.: 4 subfamilies. (1) Stercorariinae, skuas, jaegers: skuas (one species, several subspecies) breed in and near both arctic and antarctic regions; jaegers (one genus, 3 species), only in the far north; both winter widely over warmer seas; a jaeger is fossil in the Pleistocene of Oregon. (2) Larinae, gulls: cosmopolitan, including Madagascar, New Zealand, and West Indies; several genera, 43 species; listed fossil back to the Eocene (of Europe and North America). (3) Sterninae, terns: cosmopolitan, including Madagascar, New Zealand, West Indies; many breed in tropical areas, many on remote islands, but comparatively few in far northern parts of the world; some northern ones are strongly migratory (p. 244); several genera, 39 species. (4) Rynchopinae, skimmers: coasts and rivers of tropical Africa (not Madagascar); large rivers of India, Burma, and Indochina; east coast of North America; and coasts and rivers of South America south to the

Straits of Magellan, straggling to the West Indies; one genus, 3 closely related species.

Alcidae, auks etc.: breed in the cold parts of the northern hemisphere (a few south to Lower California) and winter chiefly on cool north-temperate seas; none in the tropics except by accident and none in the southern hemisphere; 13 genera, 22 species. The large, flightless, extinct Great Auk (†*Pinguinus impennis*) bred on Funk Island off Newfoundland, on Iceland, the Faroes, St. Kilda, and the Orkney Islands, and wandered or migrated southward on both sides of the North Atlantic, to Massachusetts, the British Isles and Denmark, and casually to Florida and northern Spain (Peters). It lived from Pleistocene to historic times. Other alcids are listed fossil (all in the northern hemisphere) back to the Eocene (of western North America). One of them, †*Mancalla,* in the Pliocene of California, was flightless, comparable to the Great Auk in large size and in reduction of wings (Miller and Howard 1949).

Order Columbae

Pteroclidae, sand grouse: Africa and Madagascar, southern Europe, and central and southwestern Asia east to India, mostly in deserts and plains; 2 genera, 16 species; and listed fossil back to the Oligocene (of Europe).

Columbidae, pigeons etc.: cosmopolitan, including Madagascar, New Zealand, and West Indies; many genera, 289 species; most in tropical or warm places; some northern species are migratory. Mayr and Amadon do not recognize subfamilies but Peters distinguishes the following which, although they may not be sharply defined, help to show the pattern of distribution of the family: Treroninae, confined to the Old World, in Africa and Madagascar, very numerous in the Oriental and Australian Regions and on islands of the western Pacific, with one endemic genus on New Zealand; Columbinae, nearly cosmopolitan including Madagascar and the West Indies (but not New Zealand)—it is the only subfamily in the New World, where it is numerous in genera and species; Gourinae, confined to New Guinea and neighboring islands; one genus, 6 species; and Didunculinae, confined to Samoa, one species. The genus *Columba,* with many species, from one of which the domestic pigeon is derived, occurs naturally over most of the world (see text, pp. 272–273). No other genus of pigeons is common to the Old and New Worlds. Fossil pigeons are few and not remarkable and are known back only to the Miocene (of Europe).

†Raphidae, dodos etc.: islands of the Indian Ocean: Réunion, Mauritius, and Rodriguez; 2 genera, 3 species (one species on each island), all extinct more than 200 years. These birds were, phylogenetically, very large, flightless pigeons. Hutchinson (1954) tells what is known of them and reproduces the best of the old pictures of them.

Order Psittaci

Psittacidae, parrots etc. (Fig. 34): the tropical and south-temperate parts of the world including Madagascar, New Zealand, and the West Indies; absent in Europe (but fossil there); north in Asia to the Himalayas etc. and southern China; north in America now only to northern

Mexico (several genera) with one species wandering sporadically to the mountains of southern Arizona, but the extinct Carolina Paroquet formerly occurred north to southern Wisconsin, Indiana, Ohio, and western New York (Peters); southward psittacids do not quite reach the southern tip of Africa west of Knysna; a variety of parrots etc. reach Tasmania; several stocks have inhabited New Zealand apparently for a long time; and several occur in Chile etc. and at least one reaches Tierra del Fuego; many genera, 316 species. Mayr and Amadon do not recognize subfamilies, but Peters recognizes 6: Strigopinae, New Zealand, one species; Nestorinae, New Zealand and Norfolk Island (extinct on latter), one genus, 3 species; Loriinae (lories), the Australian Region (not New Zealand) to all the Lesser Sundas including Bali, Celebes, the Philippines, and islands of the western Pacific, 15 genera, many species, especially numerous on New Guinea etc.; Micropsittinae, New Guinea etc. and the Solomons, one genus, various species; Kakatoeinae (cockatoos), Australia (not New Zealand) and New Guinea etc. to the Lesser Sundas including Bali (but perhaps introduced there), Celebes, and the Philippines, 5 genera, 17 species; and Psittacinae (typical parrots etc. including macaws), the whole range of the family (except perhaps some minor islands), many genera and species, including all the psittacids of all parts of the world except the Australian Region and adjacent island areas named above. It should be added that no genus of psittacids is common to the Old and New Worlds, that almost all the Australian genera are endemic or extend only to the Lesser Sundas, Celebes, the Philippines, etc., and that many genera of the family are confined to small areas in other parts of the world. In Australia, where parrots are diverse, their habits are diverse too. Different ones live in heavy forest, scattered trees, and scrub; some live in grassland and feed on the seeds of grass and herbs; and a few inhabit semi-deserts, nesting among rocks or on the ground. Fossil parrots are few. They include wing and leg bones, from the Upper Oligocene or Miocene of France, which are said to resemble those of an African parrot, and a humerus from the Middle Miocene of North America (Nebraska) from an apparent relative of the Carolina Paroquet. But parrots are probably much older than this, and the few fossils tell nothing about their main dispersal.

Order Striges

†Protostrigidae, †*Protostrix:* Eocene of North America; fragments apparently representing 4 species; probably more closely related to the typical owls than to the barn owls (Wetmore).

Strigidae, owls: cosmopolitan, including Madagascar, New Zealand, and West Indies; 2 subfamilies. (1) Striginae, typical owls: cosmopolitan, including the islands named; 27 genera, 134 species. No single genus of strigines is cosmopolitan, but *Otus* (screech owls), *Bubo* (horned owls), and *Glaucidium* (pygmy owls) occur in all the principal parts of the world except the Australian Region. These genera do not reach quite the same limits: *Otus* reaches Madagascar and the West Indies and crosses Wallace's Line, going as far east as the Moluccas and Biak Island off New Guinea; *Bubo* and *Glaucidium* do not reach Madagascar or the West Indies (except for a *Glaucidium* on Cuba) and do not cross Wallace's Line, although both reach Java, Bali, Borneo, etc.; and *Glaucidium* is absent in eastern North America. *Strix* (barred owls) occurs in North Africa and temperate and tropical Eurasia to Sumatra, Java, Borneo, and Palawan; North America south to Honduras; and southern South America from southern Brazil etc. to the Straits of Magellan (not West Indies). *Asio* (long-eared and short-eared owls) occurs in Africa and Madagascar, across temperate but not (except rarely in winter) tropical Eurasia, and in North, Central, and South America and some West Indies. The species *Asio flammeus* (the Short-eared Owl), with several subspecies, occurs around the northern parts of the world (more or less migratory), at high altitudes in the mountains of northern South America, in southern South America from southern Brazil etc. southward, and on some of the West Indies, the Falkland Islands, the Galapagos, the Hawaiian Islands, and Ponape in the Caroline Islands. *Aegolius* (Saw-whet Owl etc.) occurs around the northern parts of the world and in parts of Central and South America (not the West Indies). *Nyctea* (the Snowy Owl) and *Sturnia* (the Hawk Owl), each with one species, occur around the far-northern parts of the world. *Ninox* has a number of species from eastern and southern Asia to Australia, Tasmania, and New Zealand, and one species in western Madagascar. *Ciccaba* has one species (several subspecies) in Africa below the Sahara (not Madagascar) and 4 species in tropical America from Mexico to northeastern Argentina (not West Indies). It should not be forgotten that there are 17 other genera of owls, not named here, most of which have more limited, more or less continuous ranges. Fossil strigine owls (including the existing *Bubo*) are listed back at least to the Oligocene (of Europe), and *Strix* occurs back to the Miocene of North America (South Dakota). (2) Tytoninae, barn owls: almost cosmopolitan, including Madagascar, New Zealand, and the West Indies, but absent in most of temperate Asia and absent in other cold places where typical owls occur; 2 genera. *Tyto*, with 10 species, has the range of the subfamily, but all except one of the species are confined to the Old World, where they are widely scattered in the tropics and south-temperate areas; the second genus, with one species, occurs in the Oriental Region. The tenth species of *Tyto* (*alba*, the Barn Owl), with numerous subspecies, is almost cosmopolitan, occupying almost the whole range of the genus (except New Zealand and perhaps some other islands), including Madagascar, many islands in the Atlantic and western Pacific Oceans (to Samoa and the Society Islands), the West Indies, and the Galapagos. It is the only existing tytonid in America. The distribution of existing forms would seem to show that tytonids are an Old World group of which a single species has recently extended to America—but 3 giant fossil species of *Tyto* are known from the Pleistocene of the West Indies (Hispaniola, Cuba, and the Bahamas—Wetmore 1951*a*, p. 62). The subfamily is unknown fossil before the Pleistocene.

Order Caprimulgi

Aegothelidae, owlet frogmouths: Australian Region to Halmahera and New Caledonia (not New Zealand); one genus, 8 species.

Podargidae, frogmouths: discontinuously Oriental-Australian: one genus, 9 species in the Oriental Region to Java, Borneo, Philippines; and another genus, 3 species in Australia, Tasmania, New Guinea, Solomons, etc. (but none on the Lesser Sundas, Celebes, or Moluccas, and not on New Zealand).

Caprimulgidae, goatsuckers: almost cosmopolitan, including Madagascar, Australia (not New Zealand), and the West Indies; migratory in the north; 19 genera, 67 species, most in warm places. *Caprimulgus* (including the Whippoorwill etc.) is almost cosmopolitan; the other genera are confined to parts of either the Old or the New World. *Caprimulgus* is fossil back to the Pliocene (of Europe).

Nyctibiidae, potoos: tropical America from southern Mexico to Paraguay etc., and Jamaica and Hispaniola in the West Indies; one genus, 5 species.

Steatornithidae, oilbirds: widely but locally distributed in northern South America including Trinidad, but not the West Indies proper; one species.

Order Trogones

Trogonidae, trogons (Fig. 35) in 3 separate tropical areas: tropical Africa (not Madagascar) (2 genera, 3 species); the Oriental Region (one genus, 11 species) from Ceylon and India etc. to Java, Borneo, and the Philippines; and tropical America (5 genera, 20 species) north to southern Texas and Arizona, south to northern Argentina, and on Cuba and Hispaniola in the West Indies. No genus of trogons now occurs in more than one of the 3 main areas. Fossil wing and leg bones of trogons are in the Oligocene (and Upper Eocene?) of France. *Trogon*, now confined to America, is recorded from Europe in the Miocene, but cautious persons might question the generic identity of the fossil.

Order Coraciae

Coraciidae, rollers: warmer parts of the Old World; 3 subfamilies. (1) Leptosomatinae, cuckoo rollers: Madagascar and the Comoro Islands; one species. (2) Brachypteraciinae, ground rollers: Madagascar; 3 genera, 5 species. (3) Coraciinae, typical rollers: most of the Old World including Madagascar and part of Australia—the Australian species occasionally reaches New Zealand, sometimes in small flocks; the north-temperate (and also the Australian) species are migratory; 2 genera, 11 species; and a fossil humerus in the Upper Eocene or Lower Oligocene of France.

Alcedinidae, kingfishers: cosmopolitan including Madagascar, New Zealand, and the West Indies; migratory in the north; 14 genera, 87 species, most in the warmer parts of the Old World. Only 2 genera occur in America: *Ceryle*, which occurs in Africa (not Madagascar) and southern Asia (not reaching the Sunda Islands), occurs also in North, Central, and South America and some West Indies; and *Chloroceryle* is confined to the warmer parts of America from Texas to northern

Argentina (not West Indies). Some Old World alcedinids feed on insects, lizards, etc. instead of on fishes; both American genera (and some other, Old World ones) are specialized fisheaters. The family is unknown fossil before the Pleistocene.

Meropidae, bee eaters: tropical and warm-temperate parts of the Old World including Madagascar and Australia (but not New Zealand); the northern and some tropical species are migratory (p. 246); 7 genera, 25 species.

Momotidae, motmots: tropical and subtropical America from Mexico to northern Argentina (not West Indies); 6 genera, 8 species; further details are given in the text (pp. 280–281).

Todidae, todies: West Indies (Greater Antilles); one genus, 5 closely related species.

Upupidae, hoopoes: most of the Old World except the Australian Region; 2 subfamilies. (1) Upupinae, typical hoopoes: Africa, Madagascar, and temperate and tropical Eurasia to Sumatra; migratory in the north; one species, various subspecies. (2) Phoeniculinae, tree hoopoes; Africa below the Sahara (not Madagascar); 2 genera, 6 species.

Bucerotidae, hornbills: much, but not quite all, of the Old World tropics: Africa below the Sahara (not Madagascar), one species reaching southwestern Arabia; and the Oriental Region and islands to the Philippines, New Guinea, and Solomons (not Australia or New Zealand); 12 genera, 45 species; and fossil in the Eocene of Germany.

Order Colii

Coliidae, mousebirds: Africa below the Sahara (not Madagascar); 2 genera, 6 species.

Order Macrochires

†Aegialornithidae, †*Aegialornis:* Upper Eocene or Lower Oligocene of France; leg and wing bones; position provisional (Wetmore 1951, p. 19).

Apodidae (Micropodidae), swifts: 2 subfamilies. (1) Apodinae, typical swifts: nearly cosmopolitan including Madagascar, the West Indies, and (accidental) New Zealand; migratory in the north; 16 genera, 76 species. There are no swifts breeding in the main part of Australia below the tropics, but 2 northeastern Asiatic species winter there. *Chaetura* (chimney swifts etc.) occurs in tropical Africa (not Madagascar), parts of the Oriental Region, and North, Central, and South America and some West Indies. *Apus* occurs in Africa and Madagascar, temperate and tropical Eurasia to Java, Borneo, etc., and the central part of the Andes in South America, in Peru, northern Chile, etc. The other genera of the subfamily are confined to parts of either the Old or the New World. Fossil apodines are listed back to the Lower Oligocene or Upper Eocene (of Europe); *Collocalia,* which makes the edible birds' nests of commerce and which is now widely distributed from the Orient to the northeastern corner of Australia and islands of the western Pacific and Indian Oceans (to Mauritius, Réunion, and the Seychelles), is recorded from France in the Miocene (a tibiotarsus). (2) Hemiprocninae, crested swifts: India etc. to the Philippines and New Guinea etc. (not Australia, not New Zealand); one genus, 3 species.

Trochilidae, hummingbirds: the New World, including the West Indies; many genera, 319 species, and many additional subspecies. Most are in the tropics, but many of the tropical forms are in mountains and some live at very high altitudes, even in the páramo zone above the forest line; a few species occur in temperate North America, chiefly in the West (one reaches southeastern Alaska and one breeds through most of eastern North America north to southern Canada—the northern forms are migratory); southward, a number reach parts of Argentina and Chile, and at least one reaches the Straits of Magellan.

Order Pici

Bucconidae, puffbirds: tropical America from southern Mexico to northern Argentina (not West Indies); 10 genera, 32 species.

Galbulidae, jacamars: tropical America from southern Mexico to northeastern Argentina (not West Indies); 5 genera, 14 species.

Capitonidae, barbets: 3 separate, chiefly tropical areas: Africa from the southern Sahara and lower Nile southward (not Madagascar); southern Asia to Java, Bali, Borneo, and the Philippines; and tropical America from Costa Rica to Brazil etc. (not West Indies); various genera, 76 species, more in the Old World than the New. No genus of the family occurs in more than one of the 3 main areas of distribution. Ripley (1945) has recently revised the members of this family.

Picidae, woodpeckers etc.: most of the temperate and tropical parts of the world except Madagascar and the Australian Region (to Celebes and Flores but not farther east) but including the West Indies; 3 subfamilies. (1) Jynginae, wrynecks: breed in parts of tropical and southern Africa and across temperate Eurasia etc.; and some of the northern forms migrate southward into the tropics in Africa, India, etc.; one genus, 2 species. (2) Picumninae, piculets: 4 genera; one genus and species in West Africa; another genus with 2 species confined to part of the Oriental Region from northern India and the Himalayas to western Java and Borneo; *Picumnus* with one species (5 subspecies) in southeastern Asia from India etc. to Sumatra and Borneo and north to central China, and about 25 species in tropical America from Honduras to northern Argentina (not West Indies); and one genus and species confined to Hispaniola etc. in the West Indies. (3) Picinae, woodpeckers: the whole range of the family; 33 genera, 179 species. No single genus covers the whole range. *Dendrocopos* (*Dryobates,* including the Downy and Hairy Woodpeckers etc.) is the most widely distributed: in much of Africa (but not the southern part and of course not Madagascar); across the whole of Eurasia and the Oriental Region to the Lesser Sunda Islands (to Flores etc.), Celebes, and the Philippines; in North and Central America south to western Panama and the Bahamas (not main West Indies); and in a separate area in southern South America from southeastern Brazil etc. well southward; Voous (1947) thinks this genus has spread from eastern Asia. *Dryocopus* (including the Pileated Woodpecker) occurs across temperate Eurasia and through the Oriental Region to Bali, Borneo, and the Philippines, and in North, Central, and South America to northern Argentina (not West Indies). *Picoïdes* (three-toed woodpeckers) is circumpolar. *Campephilus* (ivory-billed woodpeckers) is localized in the southeastern United States, Cuba, the Sierra Madre of Mexico, and the Magellanic forest region of far-southern South America. Except for the first 3 named above, all genera of the subfamily are confined to parts of either the Old or the New World, and most of them are still further localized; for example, almost all the genera of Africa and of the Oriental Region are different. Fossil woodpeckers (leg bones) are known back to the Upper Oligocene (of France).

Ramphastidae, toucans: tropical America from southern Mexico to northern Argentina (not West Indies); 5 genera, 37 species.

Indicatoridae, honey guides: Africa below the Sahara (not Madagascar) and parts of the Oriental Region; 4 genera, 12 species. The family is entirely African except that *Indicator*, chiefly an African genus, has 2 species isolated respectively in India and in Siam, the Malay Peninsula, Sumatra, and Borneo.

Order Passeres, perching birds: see text, pp. 274ff. Although a number of fossil Passeres are listed, they tell little about the history of the order. None is older than the Eocene, and most of the older ones are doubtful in their relationships (text p. 276); very few even of the later fossils are of much geographical importance. For these reasons fossils will usually not be mentioned under the individual families of Passeres

[*Suboscines*, presong perchers]

(*Suborder Eurylaimi*)

Eurylaimidae, broadbills: tropical Africa (not Madagascar) (2 genera, 4 species) and the Oriental Region (6 other genera, 10 species) from the Himalayas (not peninsular India) to Java, Borneo, and some Philippines.

(*Suborder Tyranni*)

(*Superfamily Furnarioidea*): a group of about 500 full species entirely confined to South and Central America, north only to southern Mexico, and entirely unrepresented in the West Indies

Rhinocryptidae, tapaculos: South and part of Central America (not West Indies), north to Costa Rica and south to Tierra del Fuego; 11 genera, 26 species, most on mountains in the tropics or in south-temperate South America.

Conopophagidae, ant pipits: tropical South America (not Central America or the West Indies) south to northeastern Argentina; 2 genera, 10 species.

Formicariidae, ant birds etc.: South and Central America; several genera reach southern Mexico (not West Indies); various genera reach northern Argentina but none goes far south there; many genera, 221 species.

Furnariidae: 2 subfamilies. (1) Furnariinae, ovenbirds: the whole of South and Central America; several genera reach southern Mexico (not West Indies) and several reach Tierra del Fuego etc.; a few of the far-southern forms make short migrations northward (Hellmayr 1932, p. 25) or descend from nesting grounds in the Andes to neighboring valleys to winter; many genera, 212 species. (The North American Ovenbird is a wood warbler, not a member of this family.) (2) Dendrocolaptinae, wood hewers: South and Central America (not West Indies), several genera reaching southern Mexico, but none south of central Argentina; 13 genera, 47 species.

[This brings the present list to the end of the uncompleted *Check-list of Birds of the World* by the late J. L. Peters. Lack of this consistent, dependable source necessitates a somewhat less exact and less-detailed treatment of the distributions of the remaining bird families. Other good modern lists exist for the birds of most parts of the world (see references at end of chapter), but the different lists are not all easily comparable with each other.]

(*Superfamily Tyrannoidea*): chiefly an American group, but with a few well-differentiated representatives widely scattered in the Old World. The Old World families may not be directly related to any particular American ones. Mayr and Amadon (1951, p. 10) in fact think that the 3 Old World families assigned to this group, though not really closely related among themselves, are very likely more nearly allied to one another than to American forms.

Pittidae, pittas, jewel thrushes: tropical parts of the Old World: 2 closely related species in Africa (not Madagascar); the rest from tropical Asia (north to southern China) south through the Indo-Australian Archipelago to New Guinea, the Solomons, and northeastern Australia (not New Zealand); one genus, 23 species.

Philepittidae, philepittas: Madagascar; 2 genera, 4 species.

Xenicidae (Acanthisittidae), New Zealand wrens: New Zealand; 3 genera, 4 species; superficially similar to certain Conopophagidae of tropical South America, but Mayr and Amadon (1951, p. 12) think this is parallelism. This family includes *Traversia lyalli,* the only known apparently flightless perching bird, now extinct.

Tyrannidae, American flycatchers etc.: 2 subfamilies. (1) Tyranninae, American flycatchers: entire New World, several occurring north (in summer) to central Alaska etc. and others south to Tierra del Fuego (the northern and some far-southern forms are migratory), and many reaching the West Indies; none in the Old World; many genera, 365 species. (2) Oxyruncinae, sharp-bills: tropical America from Brazil to Costa Rica (not West Indies); one species.

Pipridae, manakins: South and Central America, several genera occurring north to southern Mexico (not West Indies) and several south to northeastern Argentina; various genera, 59 species.

Cotingidae, cotingas: South and Central America; several genera range well up into Mexico and one is recorded from the southern edge of the United States (southern Arizona); an endemic species on Jamaica is the only cotinga in the West Indies; southward several genera reach northern Argentina; many genera, 90 species.

Phytotomidae, plant cutters: southern South America from parts of Peru, Paraguay, etc., to south-central Chile and Argentina; one genus, 3 species.

[*Oscines,* songbirds], see text, pp. 275ff.

(*Suborder Menurae*), primitive songbirds

Menuridae, lyrebirds: southeastern Australia, from southern Queensland to Victoria (not New Zealand); one genus, 2 species.

Atrichornithidae, scrubbirds: Australia (not New Zealand); one genus, 2 species, one in northern New South Wales and southern Queensland, the other (which may be extinct) in southwestern Australia.

(*Suborder Oscines s. s.*), true songbirds

Alaudidae, larks: most in Africa (one on Madagascar) and north-temperate Eurasia, a few from the Orient to Australia (not native on New Zealand), and one species reaching North America etc.; this species, the Horned Lark (*Otocoris alpestris*), with numerous subspecies, occurs around the north-temperate parts of the Old and New Worlds, and south in North America to southern Mexico, with a subspecies isolated at high altitudes in Colombia in northern South America; there are no other larks in the New World, and there is none at all in Central America, most of South America, or the West Indies; some northern forms are more or less migratory; numerous genera, 75 species.

†Palaeospizidae, †*Palaeospiza:* Upper Miocene of Colorado; an exceptionally good fossil passerine (see text, p. 277).

Hirundinidae, swallows: cosmopolitan, including Madagascar and West Indies, and one species (*Hylochelidon nigricans*) occurs on New Zealand, where it appears occasionally, from Australia, in small flocks and is said to have nested once; 2 subfamilies. (1) Hirundininae: range of the family; the northern ones are migratory and so, to some extent, are some far-southern ones; a number of genera, 74 species. In their tentative classification of swallows, Mayr and Bond (1943) recognize about 8 genera of swallows in the Old World and 6 "natural groups" which may be equivalent to genera in the New World; the 3 genera or subgenera that are common to both Old and New Worlds are "obvious" immigrants to America from the Old World. *Hirundo* and *Riparia* (barn and bank swallows) have most of their species and their largest areas of distribution in the Old World, but in each genus one Eurasian species extends to North America or is represented there, and the North American form winters in South America; some individual Barn Swallows go as far south as parts of Chile. *Petrochelidon* (cliff swallows) (the genus is not well separated from *Hirundo*) too is apparently best represented in the Old World and has a North American species which winters in South America, but additional species and subspecies of this genus are resident in scattered areas in the American tropics and in southern South America. In the Old World, swallows are most numerous in Africa, where all except one of the Old World genera (the exception is a monotypic genus in Australia) and even most of the species are represented. In America swallows are most numerous in South America, but different ones occur north to Alaska etc. (in summer) and south to Tierra del Fuego. Certain swallows of different southern continents are similar and may (or may not) be directly related. It seems at least possible that, if they are related, their ancestors crossed the oceans from, say, Africa to South America, as one species sometimes does now from Australia to New Zealand. (2) Pseudochelidoninae, African river martins: West Africa (not Madagascar); one species.

(*Bulbuls and allies*): an exclusively Old World assemblage

Pycnonotidae, bulbuls: all Africa, Madagascar etc., southern Asia north to Japan and Korea etc. (migratory), and islands east to the Philippines and Moluccas (not New Guinea or Australia); 13 genera, 109 species. Delacour (1943, p. 18) summarizes the distribution of genera of the family.

Irenidae, fairy bluebirds etc.: confined to the Oriental Region, but well distributed and common there, particularly in Malaysia (Delacour 1947); 4 genera, 14 species.

Campephagidae, cuckoo shrikes etc.: tropical and southern parts of the Old World (one migratory species breeds north to the Amur in eastern Asia), including Madagascar, Australia, Tasmania, and islands of the Indian and Western Pacific Oceans; not New Zealand except that an Australian species straggles there (at least 4 records); numerous genera, 72 species.

(*"Primitive insect eaters"*)

Muscicapidae: in the sense of Mayr and Amadon (1951, pp. 17–20 and 36–37) this is a great, cosmopolitan family which includes a number of our most familiar birds and totals 1460 species, more than a third of all existing songbirds—but some ornithologists doubt whether the family is natural. The assemblage as a whole is most numerous in the Old World and is especially rich and diverse in the Oriental-Australian area, but the mockingbirds etc. are entirely American and the wrens chiefly so, and some other groups are represented in America. Mayr and Amadon divide the family into 8 subfamilies. (1) Muscicapinae, typical (Old World) flycatchers: entire Old World including Madagascar, New Zealand, and islands even to Hawaii; none in the New World (unless Hawaii is New World); migratory in the north; many genera, 378 species. (2) Timaliinae, babblers etc.: warm parts of the Old World including Madagascar and Australia (not New Zealand), with a few in parts of temperate Eurasia (one north to the Ussuri region), and one in western North America; many genera, 282 species, most numerous and diverse in the Oriental Region. The American form is the Wren Tit, *Chamaea,* which occurs along the Pacific coast from Oregon to northwestern Lower California; Delacour (1946) and Mayr and Amadon (p. 18) think this bird is closely related to a Chinese genus, but Wetmore (1951, p. 11) doubts it. Delacour (1946) has reviewed and discussed the whole subfamily. (3) Sylviinae, typical (Old World) warblers etc.: nearly cosmopolitan, but most in the Old World; reaching Madagascar and many western Pacific islands (but I think not New Zealand—see following subfamily), and in America reaching (*Polioptila* only) the Bahamas and Cuba, etc. (not most West Indies); many genera, 313 species; some northern forms are migratory. The principal tribe, Sylviini, is confined to the Old World except that the kinglets (*Regulus* etc.) (Regulidae of Wetmore) extend from temperate Eurasia to North America and are widely distributed there (south to Guatemala) (not West Indies), and that a willow warbler (*Phylloscopus*) extends its breeding range to western Alaska (but winters in southeastern Asia). A second, small tribe of sylviines (if they really are sylviines), the Polioptilini, is confined to America; it contains the gnatcatchers (*Polioptila*) which are few in species but widely distributed in temperate and tropical America (migratory in the north) including Cuba and the Bahamas, and perhaps also the aberrant tropical America genera *Ramphocaenus* and *Microbates*. However, Rand and Traylor (1953) think the last two genera represent a separate stock, with African relationships. (4) Malurinae (Acanthisidae), Australian warblers: Australia, New Guinea, Polynesia, and New Zealand (Mayr and Amadon, p. 19); about 25 genera, 85 species. (5) Turdinae, thrushes, etc.: nearly cosmopolitan (more or less migratory in the north) including Madagascar, the West Indies, many islands of the western Pacific including the Hawaiian Islands, and Tristan da Cunha in the South Atlantic, but probably not New Zealand (the New Zealand "thrush," *Turnagra*, is probably an aberrant muscicapine); many genera, 304 species. The genus *Turdus*, which includes the American (not the English) Robin, is spread over almost the whole range of the family except a few remote islands and except the actual continent of Australia (a *Turdus* does reach New Guinea and western Pacific islands and even New Caledonia, Norfolk, and Lord Howe Islands). Ripley (1952) has recently reviewed the subfamily. (6) Miminae, mockingbirds, thrashers: most of the New World (migratory in the north) including the West Indies; various genera, 30 species. (7) Troglodytinae, wrens: most in the New World especially in the tropics and including some West Indies; some into the cooler parts of North America, the northernmost being migratory; a typical wren of the widely distributed American genus *Troglodytes* occurs across temperate Eurasia etc. and is now considered to be the only wren in the Old World (Mayr 1946, p. 7); various genera, 63 species. The Eurasian wren is probably the same species as the North American Winter Wren; it has spread west to Iceland and south to North Africa and (at high altitudes) Formosa; it is migratory only in the most northern parts of its range. (8) Cinclinae, dippers: confined to mountainous regions, in Eurasia, North Africa, and Formosa, western (not eastern) North America, and Central and South America (in America, from Alaska to northwestern Argentina) (not West Indies); one genus, 5 species.

Prunellidae, hedge sparrows: temperate Eurasia, North Africa, Formosa (in mountains), and one species has an isolated colony in southwestern Arabia, within the tropics but at considerable altitudes, above 6000 feet (Meinertzhagen 1954, pp. 271–272); only weakly migratory; one genus, 12 species. Ripley (1952, p. 4) considers these birds "chat-like thrushes" which have become secondarily bunting-like, and he thinks they should be included in the Turdinae.

Motacillidae, wagtails, pipits: nearly cosmopolitan including Madagascar and New Zealand, but few in America and none in the West Indies; various genera, 48 species; the northern forms are migratory. The ones that reach America are only an Asiatic *Motacilla* which extends its breeding range to western Alaska but winters in the Asiatic tropics (a second species of *Motacilla* may extend to Alaska too—Bailey 1948,

p. 284), and *Anthus* (pipits) which occurs in America scattered (not continuous) from the arctic to Tierra del Fuego, as well as in much of the Old World.

(*Shrikes and allies*)

Laniidae, shrikes: 2 subfamilies. (1) Laniinae, typical shrikes: Africa (not Madagascar), temperate and tropical Eurasia and islands (irregularly) to the Philippines, Timor, and New Guinea (not Australia), and North America south to the Isthmus of Tehuantepec; 3 genera, 25 species; migratory in the north. Two of the genera are confined to Africa; the third, *Lanius,* occurs throughout the range of the subfamily and is of course the only genus in America, where there are only 2 species, probably derived from one Eurasian stock (Mayr). (2) Malaconotinae, bush shrikes: almost all Africa north to Morocco and Tunisia, one species reaching southwestern Arabia (not Madagascar); 6 genera, 42 species; it is not certain that these birds are really related to the true shrikes.

Prionopidae, helmet shrikes etc.: 2 subfamilies. (1) Prionopinae, helmet shrikes: Africa below the Sahara (not Madagascar); 3 genera, 13 species. (2) Pityriasidinae, bristleheads: Borneo; one species. It is not certain that this bird is really related to the African prionopids.

Vangidae, vangas: Madagascar; about 9 genera, 11 species; diverse in isolation on Madagascar (Mayr and Amadon, p. 22).

(*Waxwings and wood swallows*)

Artamidae, wood swallows: Australia (and Tasmania) to the Oriental Region and New Caledonia, Fiji, etc. (not New Zealand); one genus, 10 species. (The West African *Pseudochelidon* is placed with the true swallows by Mayr and Amadon.)

Bombycillidae, waxwings etc.: 4 subfamilies (but Wetmore 1951, p. 11, doubts if they all belong together). (1) Hypocoliinae: part of southwestern Asia including Arabia, and perhaps the adjacent corner of Africa (not Madagascar); one species. (2) Dulinae, palm chats: the island of Hispaniola in the West Indies; one species. (3) Ptilogonatinae, silky flycatchers: southwestern United States to western Panama (not West Indies); 3 genera, 4 species. (4) Bombycillinae, waxwings: breed in the northern parts of Eurasia and North America; winter more or less southward, in America to Central America and rarely the Greater Antilles, etc.; one genus, 3 species.

(*Creepers, nuthatches, and titmice*)

Certhiidae, typical creepers: temperate Eurasia and North America, south in the Old World to the Himalayas, and in America to the highlands of Nicaragua (not West Indies); one genus, 6 species, of which only one species is in America.

Sittidae, nuthatches etc.: about 8 genera and 25 species are very widely scattered over the Old World including Madagascar and Australia (not New Zealand), and one of the genera, *Sitta,* extends to North America and has 4 additional species there; the full range of *Sitta* (typical nuthatches) covers the north-temperate part of the Old World, the Oriental Region (and east to Timor), North America south through the highlands of Mexico, and Grand Bahama Island (but not the West Indies otherwise). Mayr and Amadon (pp. 23–24) discuss the genera and divide the family into 3 subfamilies, but since the family as a whole is "something of a scrap basket" and since the relationships of several of the genera are in doubt, the subfamilies are of doubtful geographical significance.

Paridae, titmice, chickadees, etc.: widely distributed in Africa (not Madagascar), temperate and tropical Eurasia and associated islands to the Philippines and Lesser Sundas, and North America south to southern Mexico and Guatemala (not West Indies); several diverse genera, 64 species. Mayr and Amadon (pp. 24–25 and 37) recognize 3 subfamilies. (1) The Parinae (titmice and chickadees, 2 or 3 genera, 46 species) and (2) the Aegithalinae (bush tits etc., 3 genera, few species) are widely distributed within the range of the family, although their limits are not identical. (3) The Remizinae (penduline titmice and verdins, 3 genera, few species) have an apparently relict distribution in separate areas in the southern edge of the north-temperate zone: southwestern Asia (*Remiz*), perhaps the Himalayas (*Cephalopyrus,* if it belongs here), and a limited area across southwestern United States and northern Mexico (*Auriparus,* the verdins). Mayr and Amadon say that this family may be polyphyletic, and that the Aegithalinae may be independently derived from timalines and the Remizinae perhaps from Old World Dicaeidae.

(*"Old World nectar eaters"*): a group of about 400 species entirely confined to the Old World and scarcely entering the north-temperate zone even there

Dicaeidae, flowerpeckers: the Oriental and Australian Regions (not New Zealand): most numerous in New Guinea and the Philippines, occurring east and south to the Solomons, Australia, and Tasmania, and north and west to South China, India, and Ceylon (Mayr and Amadon 1947); various genera, 54 species.

Nectariniidae, sunbirds: Africa below the Sahara, Madagascar, southern Asia north to southern Syria and central China (somewhat migratory in latter), and islands to northeastern Australia (not New Zealand), but only a few reach the Australian Region; 5 genera, 104 species. Delacour (1944) has reviewed this family. Mayr and Amadon (1951, p. 26) note that this and the preceding family are almost complementary in distribution within the Old World.

Meliphagidae (Melithrepidae), honey eaters: 2 subfamilies. (1) Meliphaginae: most in the Australian Region including New Guinea, but reaching Bali, Celebes, many islands in the western Pacific to Hawaii, and New Zealand; many genera, 159 species. (2) Promeropinae: South Africa; one species; whether or not this bird is really related to the Australian meliphagids is an open question (Mayr and Amadon, pp. 26–27).

Zosteropidae, white-eyes: Africa below the Sahara, Madagascar, and the Oriental Region, north in eastern Asia to North China (migratory) and Japan, and islands to Australia and Tasmania; represented in New Zealand by an Australian species which appeared suddenly and became established about 1856; few genera, 80 species, most in the genus *Zosterops* itself.

(*Vireos, finches, and allies*)

Vireonidae, vireos, etc.: the New World, from well into Canada (migratory in the north) to northern Argentina, including the West Indies; several genera including the aberrant *Vireolanius* and *Cyclarhis,* 41 species.

Drepaniidae, Hawaiian honey creepers: Hawaiian Islands to Laysan; various genera, 22 species; probably derived from an American honey creeper or some other thraupid; a classic example of radiation in isolation. Amadon (1950) has made a thorough study of the family.

Thraupidae, wood warblers, tanagers, etc.: the New World; 5 subfamilies. (1) Parulinae, wood warblers or American warblers: the New World including the West Indies, north in North America to the limit of trees (2 species, the Myrtle Warbler and Northern Water Thrush, have extended their breeding ranges to include small areas in northeastern Siberia, but they return to America to winter) and south to northern Argentina; the northern forms are migratory; numerous genera, 109 species. See text (pp. 281–283) for further details. (2) Coerebinae, honey creepers: South and Central America, north to southern Mexico and south to northern Argentina, including the West Indies and Bahamas and accidental in Florida; several genera, 36 species. (3) Catamblyrhynchinae, plush-capped finches: mountains of South America from Colombia etc. to Bolivia; one species. (4) Thraupinae, tanagers: the New World including the West Indies; most are tropical or subtropical, but 3 strongly migratory species of *Piranga* (a genus which occurs in the tropics too) breed widely in the United States and the southern edge of Canada, 2 in the east and one in the west; southward, various tanagers reach Argentina, but none goes far south there; many genera, 196 species. (5) Pyrrhuloxiinae, cardinal grosbeaks etc.: the New World, most in the tropics but some widely distributed in temperate areas, north to southern Canada (migratory) or south into Argentina; numerous genera, 132 species.

Tersinidae, swallow tanagers: South America from Colombia etc. to northeastern Argentina (not West Indies); one species.

Fringillidae, typical finches etc.: a large but possibly artificial (polyphyletic) family, widely distributed over much of the world but absent in Madagascar and absent in the Australian Region; 2 subfamilies. (1) Emberizinae, buntings and American sparrows: most in the New World, from the arctic to Tierra del Fuego, and numerous in both temperate and tropical areas (most of the far-northern species are more or less migratory); but *Calcarius* (longspurs) and *Plectrophenax* (snow buntings) occur across far-northern Eurasia as well as North America; and *Emberiza* (with its close relatives) is confined to the Old World and is widely distributed in Eurasia, Africa (not Madagascar), and India (Hartert); many genera, 171 species. The Galapagos finches (Geospizinae of some authors) are not set apart by Mayr and Amadon. (2) Fringillinae, chaffinches, linnets, etc.: most in the Old World, but none in Madagascar and none in the Australian Region; fewer in the New World, from the arctic to Tierra del Fuego; many genera,

122 species. In the Old World, comparatively few fringillines reach the Oriental Region; only one species breeds on Sumatra and Java (on high mountains), 2 on Luzon (one of them is a subspecies of the Red Crossbill, confined to pine forests on the mountains of northern Luzon), and one on Mindanao; none occurs (except by accident) on the other Philippines, in Borneo, Celebes, or beyond. There is no agreement about just what New World genera belong in this subfamily. Hellmayr, in his *Catalogue of Birds of the Americas* (Part XI, 1938), lists 28 genera of fringillines and carduelines, but Mayr and Amadon (p. 28) combine these subfamilies and rule out all except 13 of the genera, and 5 of the 13 are only accidental or introduced in America. Of the remaining 8 genera which Mayr and Amadon accept as true fringillines and which really belong to the American fauna, all occur also in Eurasia except *Hesperiphona* (which is related to a Eurasian genus) and *Loximitris* (confined to Hispaniola). Of these 8 genera in America, *Pinicola* (pine grosbeaks), *Leucosticte* (rosy finches), and *Acanthis* (redpolls) do not range south even in winter as far as the southern edge of the United States; *Carpodacus* (purple and house finches) reaches the southern end of the Mexican tableland; *Hesperiphona* (evening grosbeaks) reaches Guatemala; and *Spinus* (siskins and American goldfinches) ranges in America from Alaska etc. to Tierra del Fuego, with forms scattered over the high mountains in the tropics, and also some in tropical lowlands, especially in arid places. *Loxia* (crossbills) ranges in America from the northern limit of trees to Guatemala, with a form isolated in the mountains of Hispaniola in the West Indies (*cf.* the Philippine crossbill, above); surprisingly, the Hispaniolan Crossbill is a subspecies of the more northern of the 2 American species, the White-winged Crossbill, which does not breed on the mainland south of the mountains of New York and New England and which does not usually occur south of North Carolina even in winter. The eighth genus, *Loximitris* (the Hispaniolan Siskin), is monotypic and confined to the mountains of Hispaniola but is presumably derived from *Spinus* (above) by an earlier wandering comparable to that of the Hispaniolan Crossbill. It will be seen that the pattern of distribution of these American fringillines is strikingly consistent with an Old World origin of the subfamily, movement of seven genera to America by northern routes, and spreading or straggling of several of them for various distances southward; but this comparatively simple history depends on the removal of the many tropical American genera included in the Fringillinae (plus the Carduelinae) by Hellmayr.

Icteridae, American orioles, American blackbirds, etc.: the New World, from Alaska etc. to Tierra del Fuego, including the West Indies; numerous genera, 88 species; well represented in both tropical and temperate zones but more or less migratory in the north and in some cases slightly migratory in the far south.

(*Weaverbirds, starlings, etc.*)

Ploceidae, weaverbirds, weaver finches, etc.: the Old World including Madagascar and Australia (not New Zealand); none native in the

New World but several introduced there (Hellmayr 1938, Part XI, pp. 1–4), including the English Sparrow, which is now widely distributed in North America and also in the southern part of South America, and which is established on Bermuda, some West Indies, etc.; many genera, 263 species. Mayr and Amadon recognize 5 subfamilies, of which the distributions are in general as follows, although I am not sure of the limits in all cases. (1) Bubalornithinae, buffalo weavers: Africa; 2 genera, 3 species. (2) Passerinae, sparrow weavers: Africa, temperate Eurasia, and the Oriental Region; various genera, 35 species. (3) Ploceinae, typical weavers, Africa, Madagascar, and the Oriental Region; various genera, 109 species. (4) Estrildinae, waxbills etc.: Africa, Madagascar, southern Asia, to Australia etc.; various genera, 107 species. (5) Viduinae, widow birds: Africa; one genus, 9 species.

Sturnidae, starlings etc.: the Old World; 2 subfamilies. (1) Sturninae, starlings: the Old World including Madagascar and Australia but not native on New Zealand (the supposed starlings of New Zealand are now placed in the Callaeidae, below); not native in the New World, but the European Starling was introduced in New York City in 1890 and is now widely distributed in North America, and the Crested Mynal of southern China is introduced and established in a small area in northwestern North America, around Vancouver; about 24 genera (Amadon 1943), 101 species; the northern species are migratory or partly so. (2) Buphaginae, tickbirds: Africa below the Sahara (not Madagascar); one genus, 2 species.

Oriolidae, Old World orioles: tropical and temperate parts of the Old World north through much of Europe and parts of temperate Asia (migratory), and including Madagascar (*Tylas* only, if it is an oriolid) and Australia (not New Zealand); 2 or more genera, 20 species.

Dicruridae, drongos: the warmer parts of the Old World: Africa below the Sahara, Madagascar, southern Asia and north in eastern Asia to the Amur (migratory), south and east through the Indo-Australian Archipelago, to northern and eastern Australia (migratory); 2 genera (one in the whole range of the family, the other monotypic and confined to New Guinea), 20 species, many subspecies (Vaurie 1949). The evolution and dispersal of the drongos is discussed by Mayr and Vaurie (1948).

(*Crows and Australian crow-like birds*); it is not certain that all the families placed in this group really belong together

Corvidae, crows, jays, magpies, etc.: almost cosmopolitan including Madagascar, New Zealand (subfossil only), and some West Indies; 19 genera, 100 species. The genus *Corvus* (crows etc.) in a rather broad sense is nearly cosmopolitan and is the only genus of the family in much of Africa and in the Australian Region and on the islands named above (and an endemic species is on Hawaii), but the genus does not reach South America, where the only corvids are jays. The southern limit of *Corvus* in America is Nicaragua; of jays, northern Argentina. The genera and relationships of corvids are discussed by Amadon (1944).

Cracticidae, Australian magpies etc.: Australia and New Guinea and some smaller islands including Tasmania and Lord Howe Island (not New Zealand); 3 genera, 11 species; see Amadon (1951).

Grallinidae, magpie larks etc.: Australia and New Guinea (not New Zealand); 3 genera, 4 species; see Amadon (1950*a*).

Callaeidae, wattlebirds: New Zealand; 3 genera, 3 species.

Ptilonorhynchidae, bowerbirds: Australia and New Guinea and a few adjacent small islands (not New Zealand); various genera, 17 species; numerous in New Guinea, less so in eastern Australia, very few in Western Australia.

Paradisaeidae, birds of paradise: most in New Guinea and adjacent small islands, a few in the Moluccas, and a few in northeastern Australia south only to northern New South Wales (not New Zealand); many genera, 43 species. Iredale (1950) has covered this whole family (and the bowerbirds) with text and colored plates, but (as readers of his book will gather) there is disagreement about some of the species he recognizes.

REFERENCES

Amadon, D. 1943. The genera of starlings and their relationships. *American Mus. Novitates,* No. 1247.

———. 1944. The genera of Corvidae and their relationships. *American Mus. Novitates,* No. 1251.

———. 1948. Continental drift and bird migration. *Science,* **108,** 705–707.

———. 1950. The Hawaiian honeycreepers . . . . *Bull. American Mus. Nat. Hist.,* **95,** 151–262.

———. 1950*a*. Australian mud nest builders. *Emu,* **50,** 123–127.

———. 1951. . . . Cracticidae. *American Mus. Novitates,* No. 1504.

American Ornithologists' Union. 1931. *Check-list of North American birds,* 4th ed. Lancaster, Pa., American Ornithologists' Union. Various supplements in the *Auk.*

Archey, G. 1941. The moa . . . . *Bull. Aukland Inst. and Mus.,* No. 1.

Bailey, A. M. 1948. *Birds of arctic Alaska.* Colorado Mus. Nat. Hist., Popular Ser. No. 8.

Bates, G. L. 1937. Birds of 'Asir . . . [part of Arabia]. *Ibis* (14), **1,** 786–830.

Blake, E. R. 1953. *Birds of Mexico.* Chicago, U. of Chicago Press.

Bogert, C. 1937. The distribution and the migration of the Long-tailed Cuckoo . . . . *American Mus. Novitates,* No. 933.

Bond, J. 1934. The distribution and origin of the West Indian avifauna. *Proc. American Philosophical Soc.,* **73,** 341–349.

———. 1936. *Birds of the West Indies.* Philadelphia, Acad. Nat. Sci.

———. 1948. Origin of the bird fauna of the West Indies. *Wilson Bull.,* **60,** 207–229.

Chapin, J. P. 1932. The birds of the Belgian Congo. Part I. *Bull. American Mus. Nat. Hist.,* **65.**

Chapman, F. M. 1917. The distribution of bird-life in Colombia . . . . *Bull. American Mus. Nat. Hist.,* **36.**

Chapman, F. M. 1926. The distribution of bird-life in Ecuador. *Bull. American Mus. Nat. Hist.*, **55.**

Delacour, J. 1943. A revision of the genera and species of the family Pycnonotidae (bulbuls). *Zoologica* (New York), **28,** 17–28.

———. 1943*a*. A revision of the subfamily Estrildinae of the family Ploceidae. *Zoologica* (New York), **28,** 69–86.

———. 1944. A revision of the family Nectariniidae (sunbirds). *Zoologica* (New York), **29,** 17–38.

———. 1946. Les timaliinés. *L'Oiseau,* **16,** 7–36.

———. 1947. *Birds of Malaysia.* New York, Macmillan.

———. 1951. *The pheasants of the world.* London, Country Life; New York, Scribner's.

Delacour, J., and E. Mayr. 1945. The family Anatidae. *Wilson Bull.*, **57,** 3–55.

———. 1946. *Birds of the Philippines.* New York, Macmillan.

Falla, R. A. 1953. The Australian element in the avifauna of New Zealand. *Emu,* **53,** 36–46.

Fell, H. B. 1947. The migration of the New Zealand Bronze Cuckoo . . . . *Tr. and Proc. R. Soc. New Zealand,* **76,** 504–515.

Griscom, L. 1942. Origin and relationships of the faunal areas of Central America. *Proc. Eighth American Sci. Congress,* **3** (Biol. Sci.), 425–430.

———. 1945. *Modern bird study.* Cambridge, Mass., Harvard U. Press.

———. 1950. Distribution and origin of the birds of Mexico. *Bull. Mus. Comp. Zool.*, **103,** 341–382.

Hartert, E. 1910–1922. *Die Vögel der paläarktischen Fauna.* Berlin, Friedlander, 3 vols. Supplements 1932–1938.

Hellmayr (different parts actually by C. B. Cory; C. E. Hellmayr; Hellmayr and B. Conover). 1918–1949. Catalogue of birds of the Americas. *Zool. Ser. Field Mus. Nat. Hist.*, **13,** in 11 parts, some of the parts in 2 or 3 numbers—actually 15 book-sized items.

Hellmayr, C. E. 1932. The birds of Chile. *Zool. Ser. Field Mus. Nat. Hist.*, **19.**

Henry, G. M. 1955. *A guide to the birds of Ceylon.* London, Oxford U. Press.

Hutchinson, G. E. 1954. The Dodo and the Solitaire [etc.]. *American Scientist,* **42,** 300–305.

Iredale, T. 1950. *Birds of paradise and bower birds.* Melbourne, Georgian House.

Lack, D. 1947. *Darwin's finches.* London, Cambridge U. Press.

Lambrecht, K. 1933. *Handbuch der Palaeornithologie.* Berlin, Gebrüder Borntraeger.

Lönnberg, E. 1927. Some speculations on the origin of the North American ornithic fauna. *Kungl. Svenska Vetenskapsakademiens Handlingar* (3), **4,** No. 6.

McDowell, S. 1948. The bony palate of birds. Part I. The Palaeognathae. *Auk,* **65,** 520–549.

Marien, D. 1950. Notes on some Asiatic Meropidae. *J. Bombay Nat. Hist. Soc.*, **49,** 151–164.

Mayr, E. 1933. Die Vogelwelt Polynesiens. *Mitteilungen Zool. Mus. Berlin,* **19,** 306–323.

———. 1940. The origin and history of the bird fauna of Polynesia. *Proc. Sixth Pacific Sci. Congress,* **4,** 197–216.

Mayr, E. 1944. Timor and the colonization of Australia by birds. *Emu,* **44,** 113–130.

———. 1945. *Birds of the Southwest Pacific.* New York, Macmillan.

———. 1946. History of the North American bird fauna. *Wilson Bull.,* **58,** 1–41.

———. 1946*a.* The number of species of birds. *Auk,* **63,** 64–69.

———. 1953. On the origin of bird migration in the Pacific. *Proc. Seventh Pacific Sci. Congress,* **4,** 387–394.

Mayr, E., and D. Amadon. 1947. A review of the Dicaeidae. *American Mus. Novitates,* No. 1360.

———. 1951. A classification of Recent birds. *American Mus. Novitates,* No. 1496.

Mayr, E., and J. Bond. 1943. Notes on the generic classification of the swallows, Hirundinidae. *Ibis,* **85,** 334–341.

Mayr, E., and W. Meise. 1930. Theoretisches zur Geschichte des Vogelzuges. *Der Vogelzug* (Berlin), **1,** 149–172.

Mayr, E., and D. L. Serventy. 1944. The number of Australian bird species. *Emu,* **44,** 33–40.

Mayr, E., and C. Vaurie. 1948. Evolution in the family Dicruridae. *Evolution,* **2,** 238–265.

Meinertzhagen, R. 1954. *Birds of Arabia.* Edinburgh and London, Oliver and Boyd.

Miller, L., and H. Howard. 1949. The flightless Pliocene bird *Mancalla. Carnegie Inst. Washington Pub.,* No. 584, 201–228.

Moreau, R. E. 1951. The migration system in perspective. *Proc. Tenth International Ornithological Congress,* pp. 245–248.

Murphy, R. C. 1936. *Oceanic birds of South America.* New York, American Mus. Nat. Hist., 2 vols.

Oliver, W. R. B. 1930. *New Zealand birds.* Wellington, Fine Arts.

———. 1949. The moas . . . . *Dominion* (*New Zealand*) *Mus. Bull.,* No. 15.

———. 1955. *New Zealand birds,* 2nd ed. Wellington, A. H. and A. W. Reed.

Patterson, B. 1941. A new phororhacoid bird . . . . *Geological Ser. Field Mus. Nat. Hist.,* **8,** 49–54.

Peters, J. L. 1931–1951. *Check-list of birds of the world.* Cambridge, Mass., Harvard U. Press; 7 vols. published, others to be prepared by various authors.

Phillips, A. R. 1951. Complexities of migration: a review . . . . *Wilson Bull.,* **63,** 129–136.

Rand, A. L. 1936. The distribution and habits of Madagascar birds. *Bull. American Mus. Nat. Hist.,* **72,** 143–499.

Rand, A. L., and M. A. Traylor, Jr. 1953. The systematic position of the genera *Ramphocaenus* and *Microbates. Auk,* **70,** 334–337.

Ripley, S. D. 1945. The barbets. *Auk,* **62,** 542–563.

———. 1949. Avian relicts and double invasions in peninsular India and Ceylon. *Evolution,* **3,** 150–159.

———. 1952. The thrushes. *Postilla* (Yale Peabody Mus. Nat. Hist.), No. 13.

———. 1954. Comments on the biogeography of Arabia with particular reference to birds. *J. Bombay Nat. Hist. Soc.,* **52,** 241–248.

Roberts, A. 1940. *The birds of South Africa.* London, Witherby; Johannesburg, Central News Agency.

Romer, A. S. 1945. *Vertebrate paleontology,* 2nd ed. Chicago, U. of Chicago Press.

Sclater, P. L. 1858. On the general geographical distribution of the members of the class Aves. *J. of Proc. Linnean Soc.* (*Zool.*), **2,** 130–145.

Serventy, D. L., and H. M. Whittell. 1948. . . . *birds of Western Australia* . . . . Perth, Patersons Press.

Simpson, G. G. 1946. Fossil penguins. *Bull. American Mus. Nat. Hist.,* **87,** 1–100.

Sprunt, A., Jr. 1955. The spread of the Cattle Egret. *Smithsonian Report,* **1954,** 259–276.

Thomson, A. L. 1926. *Problems of bird-migration.* London, Witherby.

———. 1949. *Bird migration. A short account.* London, Witherby.

Ticehurst, C. B. 1938. *A systematic review of the genus* Phylloscopus. London, British Museum.

Vaurie, C. 1949. A revision of the bird family Dicruridae. *Bull. American Mus. Nat. Hist.,* **93,** 199–342.

Voous, K. H., Jr. 1947. . . . history of . . . *Dendrocopos. Limosa* (Orgaan der Club Nederlandsche Vogelkundigen), **20,** 1–142.

Wetmore, A. 1925. . . . *Palaeospiza* . . . . *Bull. Mus. Comparative Zool.,* **67,** 183–193.

———. 1943. Fossil birds from the Tertiary deposits of Florida. *Proc. New England Zool. Club,* **22,** 59–68.

———. 1951. A revised classification for the birds of the world. *Smithsonian Misc. Coll.,* **117,** No. 4.

———. 1951*a*. Recent additions to our knowledge of prehistoric birds 1933–1949. *Proc. Tenth International Ornithological Congress,* pp. 51–74.

Wolfson, A. 1948. Bird migration and the concept of continental drift. *Science,* **108,** 23–30.

———. 1955. Origin of the North American bird fauna: critique and reinterpretation from the standpoint of continental drift. *American Midland Naturalist,* **53,** 353–380.

chapter 6

Mammals

Simpson (1945) has made a complete classification of fossil and existing orders and families of mammals, with nearly complete lists of genera. This classification is, of course, followed here, and I have in general followed Simpson's ideas of the relationships and histories of mammals. Dr. Ernest Williams and Dr. Karl Koopman have been kind enough to read what I have written about mammals and have made many useful suggestions and corrections, but of course (as is usual in such cases) they are not responsible either for any errors that may remain or for my conclusions. Bourlière (1955) presents a good series of photographs of living mammals, including many species of zoogeographic importance.

Mammals are in some ways uniquely important in zoogeography. Their fossil record is unequaled and allows an almost magical view into the past although, geographically, there are still wide gaps in the record (pp. 342–343). But the species of mammals are not so well known as those of birds.

The number of existing species of mammals is estimated by Mayr (1946) as 3500. An actual count is not possible. In many places, especially in the tropics, the real species have not been worked out, and there are far too many names on some faunal lists. Fortunately Ellerman and Morrison-Scott (1951) have brought reasonable order to the mammals of almost all of temperate and tropical Eurasia, with North Africa. They list 809 species in this area. This may be guessed to be something like a quarter of all the species of mammals of the world, the other three-quarters being in the main part of Africa plus Madagascar, Australia plus the Indo-Australian Archipelago, and America. Four times 809 is 3236. This is a very rough approximation, but it is not too far from Mayr's estimate. Of valid genera of mammals, Simpson (1945, p. 35) thinks there are about 1000 existing and 2000 additional known fossil ones.

Table 5 shows the composition of the north-temperate and tropical mammal fauna covered by Ellerman and Morrison-Scott.

TABLE 5. FULL SPECIES OF MAMMALS OF THE PALEARCTIC REGION AND MOST OF TROPICAL ASIA (FROM ELLERMAN AND MORRISON-SCOTT 1951)

Terrestrial, other than lagomorphs and rodents	273
Lagomorphs (rabbits etc.)	27
Rodents	281
Bats	160
Aquatic (most marine)	68
	809

This is probably a fair sample of the composition of the mammal fauna of the world, although details vary considerably in different places. There are relatively more bats in wholly tropical areas. There are relatively fewer rodents in Australia and Madagascar, but probably more rodents in some other regions and certainly more on some other islands. Of course the first category of the table, of terrestrial mammals other than lagomorphs and rodents, is made up of a great variety of mammals, very different in different parts of the world. The place of rodents in relation to other mammals is further discussed on page 340.

The main course of evolution of mammals is described by Simpson (1945, especially p. 165). Mammals originated in the Jurassic, but until the end of the Mesozoic they were insignificant animals, overshadowed by the ruling reptiles. Their fossil record during this time is very poor. It was only about the beginning of the Tertiary that mammals became dominant, radiated, and began to leave a better record of themselves. Their geographical history during the Tertiary is outlined on pages 342ff. In discussing the history of mammals, the Paleocene is customarily distinguished from the Eocene, at least by

American paleontologists, and I shall distinguish it here, although it is usually not distinguished in preceding chapters.

It used to be thought that the earliest mammals were monotremes, that marsupials were derived from monotremes, and that placentals were derived from marsupials, but it is now known that mammalian phylogeny was not so simple. Mammals split into two or three subclasses very early in their history. Whether the monotremes were one of the earliest groups is doubtful. They are not known ever to have existed outside of Australia, and for this and other reasons it has been suggested that they may not be very ancient but may be degenerate (though superficially specialized) descendants of primitive marsupials. Marsupials and placentals belong to one ancient subclass, but, instead of being successive groups, they were parallel ones which existed together well before the end of the Cretaceous. There apparently never was a time when only marsupials, and not placentals, were widely distributed over the world.

Phylogenetically, mammals are reptiles which have evolved hair, warm-bloodedness, improved reproduction, and eventually increased intelligence. Warm-bloodedness has enabled mammals to live in and pass through cold places. However, the dominance of mammals is more important than their warm-bloodedness. It is their general dominance, arising presumably from a combination of superior qualities, which accounts for the wide and sometimes rapid dispersal of so many mammals over the available parts of the world.

The dispersal of mammals is limited especially by salt-water barriers. The extent to which land mammals can cross salt water has been argued bitterly, too often simply in terms of "They can!" and "They can't!" I shall not repeat the arugments but shall simply say what it seems to me the mammals themselves show. The actual distribution of different land mammals seems to show that salt water is a formidable obstacle to them, but that any mammal may sometimes get across a narrow water gap, and that some mammals have crossed fairly wide ones. Rodents seem to be relatively good water crossers. They have reached Australia several times, and they occur on a number of islands beyond the limits of other flightless land mammals. No limit can be set to the width of water an exceptional rodent might cross. The ancestor of the native cricetids of the Galapagos apparently crossed more than 600 miles of ocean. It is possible—it has been suggested before—that the ancestor of the South American hystricomorph rodents came across the Atlantic from Africa, perhaps on a raft like the one described in Chapter 1 (p. 15). I do not say that this happened, but just that it may have happened, if the distribution of hystricomorphs requires it. Bats can and do cross water barriers more easily. Many of them, but only a few stocks of rodents and probably no other placentals, have reached Australia without the aid of man, and a few bats have reached very remote islands.

Limits of distribution of mammals

The northern limits of terrestrial mammals (Fig. 38) are the limits of land, and marine mammals (whales and seals etc.) live beyond

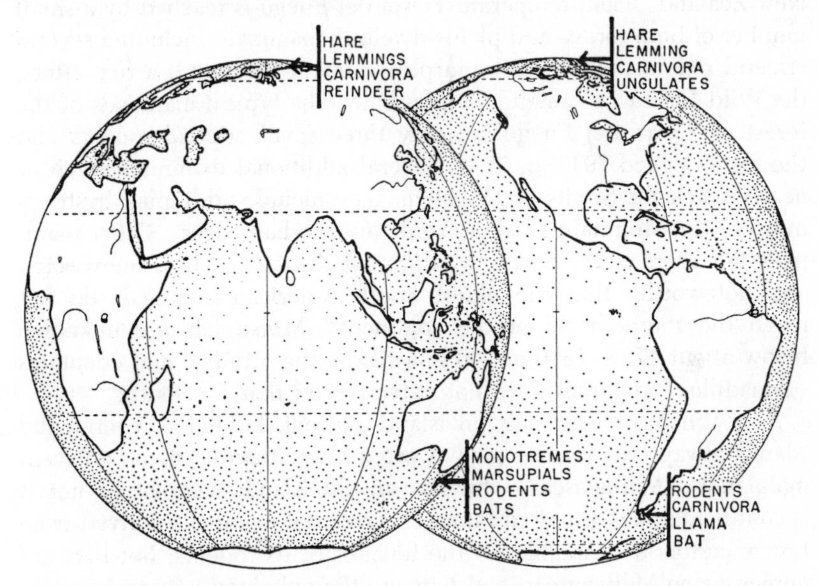

Fig. 38. Northernmost and southernmost land mammals.

that. In America, the northernmost arctic lands, including the Arctic Archipelago and the north coast of Greenland, are reached by the Arctic Hare, Collared Lemming, Wolf, Arctic Fox, Polar Bear (mainly along the sea coast), Short-tailed Weasel, Caribou, and Musk Ox. All of these animals except the Musk Ox are represented in the extreme north of Eurasia too, by the same or related species—the Short-tailed Weasel is the Eurasian Ermine, and the Caribou is in the same genus and perhaps the same species as the Eurasian Reindeer. In fact the mammals of the northern Taimyr (the northernmost point of continental Asia) listed by Berg (1950, p. 19) correspond exactly to the northernmost American mammals except that, on the Taimyr, the Musk Ox is absent and an additional lemming is present. Several other mammals, including shrews, ground squirrels (*Citellus*), additional cricetid rodents (besides the arctic lemmings), and additional mustelids (besides the Short-tailed Weasel), extend onto parts of the arctic tundra without reaching the northern limit of land. And still more mammals, including bats (*Myotis*), reach the northern limit of forest without extending onto the tundra to any significant extent.

Southward, mammals swarm to the warm-temperate tip of Africa. Cool-temperate Tasmania is reached by the Echidna, Platypus, many marsupials, five native rodents, and several bats; two bats are on New Zealand. Cold-temperate Tierra del Fuego is reached by a small number of both forest- and plains-dwelling mammals, including several cricetid rodents, one hystricomorph rodent (*Ctenomys*), a fox, otters, the Wild Llama (a camelid), and a bat. The typical mammals of the forest of Tierra del Fuego are only three small cricetid rodents and the fox (Osgood 1943, p. 31). Several additional mammals reach or nearly reach the Straits of Magellan; they include additional hystricomorphs, the Mountain Lion, a skunk, and perhaps deer. Other mammals fall short of the Straits by different distances. It is unexpected and noteworthy that the oldest South American mammals do not reach the southern tip of the continent. Marsupials are unknown below about 47° S. (300 or 400 miles from the Straits), and edentates (armadillos) apparently do not extend even that far south.

As to limits of mammals on islands, recent, habitable, continental islands always have mammals but often fewer than occur on adjacent mainlands. Madagascar, which is an old island, whether or not it is continental, has a limited fauna of terrestrial mammals derived from few ancestors and notable for the fewness of its rodents; but bats are numerous on Madagascar, and bats are the only land mammals native on the Mascarene Islands and the Seychelles. Australia, which is an old island continent, has been reached by perhaps one monotreme, at least one marsupial, several different murid rodents, and many bats, but by no placental mammals except rodents and bats. Eastward from Australia and New Guinea, one genus of arboreal marsupials and several rodents reach the Solomons; but beyond that are only rats and mice probably carried by man, and bats. The latter are numerous in the Solomons but decrease rapidly eastward, the last of the Old World ones reaching Samoa. An American bat (a *Lasiurus*) is on the Hawaiian Islands. The Galapagos have been reached by one group of small, American cricetid rodents and by a *Lasiurus* bat. The West Indies have or have had many rodents, two groups of insectivores, ground sloths, a few other scattered terrestrial mammals of doubtful history, and many bats. Some of the smaller West Indies, including some of the Bahamas, the Caymans, and Swan Island, may never have had any native land mammals except rodents and bats. The more remote islands of the Atlantic have no land mammals except bats: a small European bat has reached even the Azores; bats are not resident on Bermuda but appear there occasionally. It will be seen that most flightless land mammals have narrow limits on islands; rodents have somewhat wider limits; and bats are almost without final limits, although very few occur on very remote islands.

Most families of mammals are less than world-wide in distribution. The only cosmopolitan or tropicopolitan families of land mammals (excepting man) are of bats: Vespertilionidae, Emballonuridae, and Molossidae. The family Muridae (mice and rats) would be world-wide if the cricetids were included, as they sometimes are. The following terrestrial families are native and widely distributed on all continents except Australia: Leporidae (rabbits), Sciuridae (squirrels), Canidae (dogs etc.), Mustelidae (weasels), and Felidae (cats); and one or two families of elephants and also Equidae (horses) may have had this distribution in the past. All other existing families of flightless land mammals are even more limited in distribution.

The most widely distributed genera of land mammals, excepting *Myotis* and a few other genera of vespertilionid and molossid bats (see list of families), and excepting man and forms carried by man, are *Lutra* (otters, which are semi-aquatic) and *Felis* (cats), both of which are widely distributed on all continents except Australia. One or two genera of elephants and *Equus* (horses) may have had this distribution in the past. The genera *Lepus* (hares), *Mustela* (weasels), and *Canis* (wolves etc.) have very wide climatic ranges, from the highest arctic (northern Greenland) to parts of the hot tropics.

Some species of mammals have wide ranges, both geographically and climatically. The Wolf (*Canis lupus*) occurs around the northern hemisphere north to the highest arctic (northern Greenland) and extends south in Asia (in peninsular India) to several degrees within the tropics. There is a question whether the wolves that reach these opposite limits really are one species, but Ellerman and Morrison-Scott think so. The Leopard (*Panthera pardus*) ranges through the whole of Africa and southern Asia to southeastern Siberia. Several species of *Lepus* (hares) have enormous ranges: *L. capensis,* the length of Africa and the width of temperate Eurasia, from Cape Town to Spain to China; *L. europaeus,* from South Africa to Germany and Denmark, and now spreading into Asia beyond the Urals; and *L.*

timidus, the width of temperate and arctic Eurasia and probably also across North America (Ellerman and Morrison-Scott). Some other species of mammals are more or less circumpolar. And in America, the Mountain Lion (*Felis concolor*) ranges from cold northern British Columbia through the tropics of Central and South America to the Straits of Magellan, and it occurs at high as well as low altitudes and in dry as well as wet country.

Some other mammals with notable climatic ranges are *Tachyglossus* and *Ornithorhynchus* (monotremes), from cool-temperate Tasmania to tropical Australia; *Didelphis* (opossums), from cool-temperate North America through the American tropics; *Crocidura* (a genus of shrews), from Germany and eastern Siberia south through Africa and the tropical Oriental Region; *Myotis* (common bats), from the northern limit of trees south through the tropics of both hemispheres. And *Sciurus* (squirrels), *Rattus* (rats), *Equus* (horses), *Sus* (pigs), *Cervus* and *Odocoileus* (deer), and certain genera of elephants etc., rhinoceroses, and wild cattle all range or formerly ranged from some cold to some tropical areas.

Zonation of mammals

Although some genera and species of mammals are distributed without much regard to the climatic zones, and although very few large groups of mammals are strictly zoned, climate does profoundly influence the distribution of mammals. In the tropics, there is a diversity of mammals; in the temperate zones, less diversity; and in the arctic, still less, only about eight species of land mammals reaching any given northernmost point of land (p. 323).

Tropical mammal faunas are not only very rich but also very diverse in composition. No family of mammals, excepting one or two of bats, is pan-tropical: none occurs in all tropical regions but no temperate ones. Most of the primarily tropical families of terrestrial mammals are confined to either the Old or the New World. The family Tapiridae, now in part of the Oriental Region and tropical America, is an unimportant exception. Primates as a whole (excepting man) are almost pan-tropical, but no single family of primates is. That there are no pan-tropical families of flightless land mammals obviously reflects the Tertiary isolation of South America, the rapidity of mammalian evolution during the Tertiary, and the fact that the Old World mammals that have reached South America recently have passed through the north-temperate zone and usually still exist there. However, something else, not so obvious, may be implied: that no large fauna of tropical mammals accumulated in southern North America during the Tertiary. If such a fauna had accumulated, it might have included Old World tropical families which had filtered through the north in the earlier, warmer part of the Tertiary and which, in the Pliocene, would have spread through the American tropics and become pan-tropical. Whether or not this reasoning is correct, there seems to be very little evidence of a distinct, separate, tropical mammal fauna in southern North America during the Tertiary (see p. 368).

North-temperate mammal faunas are not only less rich than tropical faunas but have relatively fewer important endemic groups. The primarily north-temperate families and subfamilies of land mammals are the *Soricinae* (red-toothed shrews), but one genus of them extends to northern South America, and the subfamily may not be natural; *Talpidae* (moles), but they extend into the Oriental tropics, to the Malay Peninsula; *Ochotonidae* (pikas), but they are absent from most of Europe and from eastern North America, and the family consists of only one existing genus; *Castoridae* (beavers) with one existing genus, two species; *Microtinae* (common field mice etc.), which comes closer than any other large group of mammals to being fully and exclusively north-temperate; *Zapodidae* (jumping mice), a small family confined to but not reaching all the limits of the north-temperate regions; *Aplodontidae* (sewellels, one species), in western North America; *Spalacidae* (mole rats), in southeastern Europe and adjacent parts of Asia and North Africa; *Seleviniidae* (related to dormice, one species), in central Asia; and *Antilocapridae* (pronghorns, one existing species), in western North America. This is not an impressive list of endemic groups for a part of the world that has been supposed (by Matthew and others) to be the main theater of evolution of mammals. Most north-temperate mammals belong to families that are also widely distributed in the tropics, usually the Old World tropics.

I have tried to analyze and compare a tropical and a north-temperate mammal fauna so as to show the effect of climate more exactly. The faunas chosen are those of southern and eastern Asia, for in eastern Asia tropical and temperate faunas come together on a broad front, without major barriers, so that climate is the principal differentiating factor. First I counted and tabulated separately, by families, the full species listed by Ellerman and Morrison-Scott (1951) from tropical Asia and from temperate Asia east of about 90° E., using my best judgment in assigning species that extend across the boundary of the tropics: if a species is mostly on one side of the boundary, I assigned it accordingly; if it is widely distributed on both sides, I counted it with both faunas. Then I added to the tropical fauna additional species listed from Malaysia by Chasen (1940), reducing Chasen's species to the standards of Ellerman and Morrison-Scott when the latters' comments made it possible. My lists then covered all the tropical Oriental Region (except the Philippines) and a roughly equal temperate-to-arctic area in eastern Asia. Finally, as insurance against serious error, I tabulated the mammals of Ceylon (Phillips 1935) and of the British Isles (Matthews 1952) as samples of the tropical Oriental and temperate Eurasian faunas and found that the samples had about the same general characteristics as the whole faunas. The results of all this are given in Table 6. For simplicity, I have combined the lagomorphs with the rodents, as "gnawers."

TABLE 6. COMPARISON OF TROPICAL ORIENTAL AND TEMPERATE EASTERN ASIATIC MAMMAL FAUNAS

	Land Mammals Except Gnawers and Bats (in widely distributed families [a])	Gnawers—Lagomorphs and Rodents (in widely distributed families [b])	Bats (Vespertilionidae)
Temperate eastern Asia: temperate China (except Sinkiang and Tibet), Mongolia, Siberia east of 90°, etc., and Japan	100 (83: 83%)	109 (85: 78%)	43 (32: 74%)
British Isles	19[c] (18: 95%)	10[c] (8: 80%)	12 (10: 83%)
Oriental Region (except Philippines)	180 (85: 47%)	135 (123: 91%)	154 (59: 38%)
Ceylon	32 (18: 56%)	23 (21: 91%)	28 (12: 43%)

Note: These figures are from actual counts of full species in Ellerman and Morrison-Scott (1951) and Chasen (1940), as described in the text, but the totals should be considered approximations.

[a] The "widely distributed" (*i.e.*, widely distributed in both tropical and temperate areas) families of land mammals except gnawers and bats are Erinaceidae, Soricidae, Canidae, Ursidae, Mustelidae, Felidae, Equidae, Suidae, Cervidae, Bovidae.

[b] The widely distributed families of gnawers are Leporidae, Sciuridae, Cricetidae, Muridae.

[c] The figures for the British Isles include the Wolf, Bear, Wild Pig, and Beaver, which have become extinct in Britain within historic times.

This table is probably affected by factors that are unknown or beyond analysis now. Temperate eastern Asia is continuous with an enormous additional temperate area, but the tropical Orient is isolated from other tropical areas, and this may increase the relative number of species in temperate eastern Asia. This effect does not extend to the bats, for which the Oriental Region is not so isolated, and this is probably one of the reasons why there are so many bats there. On the other hand, Chasen splits species more finely than Ellerman and Morrison-Scott do, and this may in some cases have increased the tropical totals unduly. I have tried to correct this but have not been able to do so in some groups, particularly in some groups of bats. This factor does not affect the figures given for Ceylon and the British Isles. The figures for these islands are consistent with the main totals and suggest that the latter are approximately correct.

There are two main differences between the tropical and temperate faunas tabulated above. The tropical fauna is larger, containing nearly twice as many species as the temperate one. And the tropical fauna is more diverse, consisting of widely distributed families plus many additional families, while the temperate fauna consists of the same widely distributed families plus few additional ones. In the first column of the table, which includes the main variety of terrestrial mammals, ten widely distributed families make up only about half of the tropical fauna but more than four-fifths of the temperate one; and among bats, one widely distributed family, Vespertilionidae, makes up less than two-fifths of the tropical fauna but about three-fourths of the temperate one.

The rodents etc. do not fall into the same pattern as the other mammals. They seem to be relatively more numerous than the others in temperate eastern Asia. This may be an effect of area, of multiplication of different species in different parts of the great area of temperate Eurasia. Or rodents may really be disproportionately numerous in parts of the north-temperate zone. They may be especially able to survive northern winters because they can eat the overwintering seeds of northern plants. And most of the rodents of the tropical Orient and temperate Eurasia belong to the same widely distributed families. This may mean either that rodents are less affected by climate than other mammals are, or that they have evolved and dispersed more recently. The dominant, Old World murids do seem to be recent; they are apparently relatively late (late Miocene?) derivatives of cricetids (Simpson 1945, pp. 205–206). But there are many strictly tropical groups of rodents in other tropical regions.

The boundary between tropical and temperate mammal faunas is not sharp. Not only the widely distributed families but also some primarily tropical ones (*e.g.*, monkeys and viverrids) as well as many genera and some species (*e.g.*, the Tiger and Leopard) extend from the Oriental tropics for varying distances into temperate eastern Asia. The line between the tropical African and temperate European mammal faunas is sharper, but that is probably owing partly to the desert and water barriers. A few families and even genera and species (*e.g.*, species of *Lepus* already mentioned) extend across the barriers, from tropical Africa to Europe. In America, most tropical mammals (*e.g.*, most marsupials, monkeys, most edentates, and most hystricomorph rodents) do not extend north of southern Mexico, but the Virginia Opossum and Canadian Porcupine (which represents a tropical American family) extend much farther north, and an armadillo is spreading into the southeastern United States, and many families, genera, and species of widely distributed families cross the climatic boundary in America.

Southward, there is no real zonation of land mammals but a diminution of the tropical faunas; this is most marked in South America. A few small families are confined to south-temperate areas, but there is no special south-temperate fauna common to the different southern continents.

Radial distribution of mammals

Clear patterns of recent geographical radiation of dominant families are rare among mammals, perhaps partly because many mammals have declined rather than radiated recently. Rats and mice of the family Muridae have a pattern of complex radiation from the main part of the Old World tropics into temperate Eurasia and to Australia, but not to America (except as carried by man). Bovidae (cattle etc.) have a pattern of complex radiation from the main part of the Old World to North America. Some other, less important geographical radiations are indicated by the present distributions of other families, and genera. However, the best evidence of the geographical radiation of most mammals is not their present distribution but their fossil record.

Regional distribution of mammals

The two northern regions, Eurasia and North America above the tropics, have mammal faunas which are much alike in both what they have and what they do not have, and which differ from each other mostly in small or restricted families, except that temperate Eurasia has many murid rodents. North America has the marsupial genus *Didelphis* (opossums), which extends into eastern North America from the American tropics, and which of course is lacking in Europe and Asia. Of insectivores, both northern regions have shrews (mostly of a primarily northern subfamily) and moles (an almost wholly northern family); and temperate Eurasia also has hedgehogs (shared with Africa and the Oriental Region). One genus of monkeys extends from the Old World tropics into a small part of temperate eastern Asia. Both northern regions have pikas (exclusively northern, but restricted in distribution within the northern regions) and rabbits etc. (widely distributed elsewhere). Of rodents, both regions have many squirrels etc. and cricetids (both families widely distributed elsewhere, but the subfamily Microtinae is almost confined to these regions), a few zapodids (north-temperate), and beavers (a species in each northern region); temperate Eurasia also has murids (dominant throughout the Old World) and mole rats, dormice, and jerboas (all more or less localized or ecologically restricted); and temperate North America has also the Sewellel (localized in the West), pocket gophers and pocket mice (mostly in the West, and extending into the edge of the American tropics), and the Canadian Porcupine (which represents a tropical American family). Of Carnivora, both regions have representatives of widely distributed groups: wolves and foxes, bears, weasels etc. (including skunks in America), and cats; Eurasia has pandas (localized); North America, raccoons etc. (also in the American tropics); and certain warm corners of temperate Eurasia are reached by viverrids and a hyaena (from the Old World tropics). An African hyrax reaches the southwestern corner of Asia. Of higher ungulates, *Equus* (horses etc.) is in temperate Asia (and Africa); wild pigs, temperate Eurasia (and the Old World tropics); peccaries, in part of southern North America (and the American tropics); camels, localized in temperate Asia (and South America); deer, widely distributed in both northern regions (and in the Oriental and American tropics); pronghorns (one species), in western North America; and bovids, in both northern regions but especially in Eurasia (also Africa and tropical Eurasia). Finally, bats occur in both regions: many are vespertilionids, a few, molossids, and a few belong to primarily tropical families which are confined to either the Old or the New World and which extend into limited temperate areas in Eurasia or North America.

Both the two main regions of the Old World tropics, Africa below the Sahara and the Oriental Region, have many shrews (mostly Crocidurinae) and some hedgehogs (but "hairy hedgehogs" or gymnures are confined to the eastern Orient); Africa alone has *Potamogale* (in West Africa), golden moles (most in South Africa), and elephant shrews; and the Oriental Region has a mole, which extends locally into the region, to the Malay Peninsula. Flying lemurs are confined to part of the eastern Oriental Region. Tree shrews are Oriental; tarsiers, eastern Oriental; and lorises, Old World monkeys, and apes are common to Africa and the Oriental Region. Manids are African and Oriental. Rabbits etc. are numerous and widely distributed in Africa, less so in the tropical Orient. Of rodents, squirrels etc. and murids are numerous in both regions; cricetids occur in both but are relatively few; dormice (of different families), rhizomyids, and Old World porcupines occur in both; and Africa has also anomalurids, pedetids, and three families of supposed hystricomorphs. Of Carnivora, both regions have numerous canids (including jackals), mustelids, viverrids, and cats; Africa has hyaenas, one of which reaches India; and the Orient has bears. Only Africa has the Aardvark. Each region has its own genus and species of elephant, its own two genera of rhinoceroses, wild pigs, chevrotains (more or less localized), and many bovids (most in Africa); but (of these regions) only Africa has hyraxes, *Equus*, hippopotamuses, and giraffes; and only the Orient has tapirs (localized) and deer. Both suborders of bats are numerous and diverse in both regions. It will be seen that these two regions have mammal faunas which are rich, alike in many ways, but different in other ways.

Madagascar has a limited, highly endemic mammal fauna described in Chapter 8.

South and Central America have a mammal fauna made up of surviving fractions of an old (Tertiary), endemic fauna plus a number of new (late Tertiary and Pleistocene) arrivals. Table 7 gives the approximate composition of this fauna now.

This fauna is tabulated by genera rather than species because the real species have not yet been worked out. Rodents appear to outnumber all other flightless land mammals in this fauna, but the figures may not be exact. The genera of cricetids may be more finely split than those of other South American mammals (*cf.* Simpson 1945, pp. 206–207). On the other hand, the species of cricetids may be even more numerous than the number of genera suggests.

Table 7. Existing Land Mammals of South and Central America

	Approximate Numbers of Existing Genera in Region	
Marsupials (2 families) (old)	14	
Shrews (new)	2	
Monkeys (2 families) (old)	14	
Edentates (3 families) (old)	14	
Canids (new)	4	
A bear (new)	1	
Procyonids ([a]new)	5	
Mustelids (new)	10	
Felids (new)	2	
Tapirs (new)	1	
Peccaries (new)	1	
Camels (new)	1	
Deer (new)	6	
Terrestrial, other than lagomorphs and rodents		75 (42 old)
Rabbits (new)	1	
Rabbits		1
Squirrels (new)	4	
Pocket gophers (new)	2	
Pocket mice (new)	1	
Cricetids (new)	40+	
Hystricomorphs (11 families) (old)	39	
Rodents		86+ (39 old)
Bats (9 families)		53 (not classified by age)

[a] Reached South America in the Miocene, before the other "new" families.

All of the families marked "new" in this table, except the tapirs and camels, are also represented in North America, and most of them occur in parts of the Old World too.

Central American mammals are included in this table, but they require some further comment. The Central American mammal fauna is transitional in many details. A number of the "old" South American families extend to Central America. Several (but not all) South American genera of marsupials do so, and four of them reach parts of Mexico, and one, eastern United States. All three existing genera of anteaters extend from South into Central America, two of them reaching southern Mexico. Both existing genera of tree sloths extend from South into Central America, reaching Nicaragua and

Honduras respectively. Of nine South American genera of armadillos, two extend into Central America, one reaching Honduras, the other, Texas etc. Of South American hystricomorph rodents, seven of the families are confined to parts of South America; one (of three) South American genera of porcupines extends through Central America to parts of Mexico (and the Canadian Porcupine, a fourth genus of the family, ranges from about the Mexican border of the United States to the northern limit of forest); a capybara reaches Panama; two (of four) South American genera of agoutis extend through Central America to southern Mexico; and three (of fourteen) South American genera of echimyids extend to Panama or Nicaragua. However, there are no endemic genera of these old families in Central America; it is especially noteworthy that there are no endemic genera of hystricomorphs there, for some do occur in the West Indies. Of the "new" families of Table 7, all except bears and camels are still represented in Central as well as South America. Of shrews, *Sorex* extends from the north into the tropics only to Guatemala, and *Cryptotis,* to northern South America. Of canids, *Canis* extends south to Costa Rica; *Urocyon,* to northern South America; and there are additional, endemic, genera in South America. Procyonids do not form a pattern of recent southward spread, but one of (somewhat older?) differentiation of genera in Central and South America. Of mustelids, *Mustela* extends south through Central and part of South America to Peru; the three genera of skunks extend south to Honduras, Costa Rica, and the Straits of Magellan respectively; *Lutra* (otters) reaches Tierra del Fuego; and there are additional, endemic genera in South and Central America. The two genera of felids are very widely distributed. Tapirs, peccaries, camels, and deer all originally reached South America from the north, but the distributions of genera of these groups no longer reflect the southward movements. A single genus of rabbits extends through Central and South America to northern Argentina. Of "new" rodents, *Sciurus* extends south through Central and South America to northern Argentina; pocket gophers reach only to Panama; and pocket mice, to northern South America. I have not tried to analyze the distributions of cricetids. This is far from a full list of details, but it is enough to show that Central America (with southern Mexico) is the main transition area of the tropical and north-temperate American mammal faunas, although a few South American groups extend farther into North America and one South American family (porcupines) reaches the northern limit of trees, and although some North American groups extend far in the other direction, through part or all of South America. The transition is complicated by the existence of endemic genera in South America, even in some "new" families, and by genera of doubtful history. Some genera even of the "new" families may have made northward extensions recently.

The Australian Region (Australia, New Guinea, and closely associated small islands) has, of native land mammals, only monotremes, marsupials, murid rodents, and bats, in proportions given in Table 8.

TABLE 8. SUMMARY OF THE NATIVE LAND MAMMALS OF AUSTRALIA AND NEW GUINEA

The figures are chiefly from various papers by Tate (see list of references). The species of rodents may be a little more finely split than those of marsupials; the species of bats are admittedly lumped by Tate. Tate's genera are in general more finely split than Simpson's.

	Australia (and Tasmania etc.)			New Guinea (and close islands)		
	Families	Genera	Species	Families	Genera	Species
Monotremes	2	2	2	1	2	3
Marsupials	6	52	119	4	24	47
Rodents (murids)	1	13	67	1	20	56
Bats	7	21	41+	6	21	45+

The "wild" dogs (Dingo etc.) and pigs of Australia and New Guinea were probably brought by prehistoric man, and the European Rabbit, Hare, and Fox have been introduced, as have the Norway and Black Rats and the House Mouse, the last two now being widely distributed in Australia even in natural habitats.

Table 8 shows that Australia and New Guinea have equally limited mammal faunas which are similar in general composition, except that Australia has more and more diverse marsupials and New Guinea more genera of rodents. These differences are probably due to the different histories of the marsupials and rodents. The marsupials have been in Australia a long time and have apparently differentiated principally there, although their recent movements into and out of New Guinea have been numerous. The rodents have come more recently, from Asia, and most of them probably reached New Guinea first and have had more time to evolve there than in Australia, although their movements between New Guinea and Australia have been numerous.

Australian marsupials are the best existing example of evolutionary or adaptive radiation in an isolated place. Different ones have become adapted to most of the different ways of life of placental mammals elsewhere. The small dasyurid marsupials are shrew-like and are insectivorous or fiercely carnivorous. Most of them run, but *Antechinomys* hops like a jumping mouse. Larger dasyurids are weasel- or wolverine- or wolf-like superficially and carnivorous in habits. An aberrant dasyurid, *Myrmecobius,* is an "anteater" (actually a termite eater), with long snout and degenerate teeth. *Notoryctes,* placed in a family of its own, is extraordinarily mole-like and lives like a mole in the ground in semi-deserts. The rat- or rabbit-like bandicoots (*Peramelidae*) live on the ground and are more or less omnivorous. Different ones occur in all sorts of country from rain forest to desert. Most phalangerids are mouse- or opossum-like. Most are arboreal, but a few have come down to the ground and live in dry, rocky places. Most of the smaller ones are probably at least partly insectivorous; most of the larger ones, herbivorous; but some are specialized. Small *Tarsipes,* with a long snout and weak teeth, feeds on flower nectar and soft food. *Dactylopsila,* the Striped Opossum, has chisel-like incisors, with which it breaks into dead wood, and a long fourth finger, with which it pulls wood-boring insect larvae out of their galleries. Three different phalangerids of different sizes (size of a small mouse to size of a small cat) have evolved gliding membranes and long feathery tails and can glide like flying squirrels. This is a particularly useful adaptation in the Australian eucalyptus woods, where the trees are often spaced apart. The Koala or Native Bear (*Phascolarctos*) is a specialized phalangerid. It is bear-like in appearance but more sloth-like in habits, although it lives right side up. It is arboreal and eats the leaves of only a few species of eucalypts. Wombats (*Phascolomidae*) are large terrestrial marsupials, powerful diggers, with rodent-like gnawing teeth. They feed on roots and vegetation. Macropodids include a variety of small rat kangaroos, larger wallabies, and still larger kangaroos. Most are terrestrial and all are herbivorous. Different ones occur in numbers in every habitable part of Australia, although comparatively few reach New Guinea. The big kangaroos are grazing animals which take the place of the hoofed grazers of other parts of the world. Superficially, in profile, the skull of a kangaroo looks something like that of a horse. Both are adapted for cropping grass. Finally, tree kangaroos (*Dendrolagus*) have become secondarily arboreal in the rain forest of northeastern Australia and New Guinea. They are rather clumsy climbers, but they do spend most of their time in trees, and their hind legs are shortened, their hands and claws modified for climbing, and their tails (stiff in ordinary kangaroos) changed to slender, flexible counterweights. Besides all these, in the Pleistocene or post-Pleistocene there were also in Australia kangaroos larger than any existing ones, giant herbivorous diprotodonts the size of rhinoceroses, and *Thylacoleo* the size of a lion, which may or may not have been carnivorous.

There are limits to the adaptive radiation of Australian marsupials, however. None is aquatic, and this is surprising, for parts of eastern Australia as well as New Guinea are well watered; the strictly aquatic Platypus has been able to survive in Australia for unknown ages; and a marsupial (*Chironectes*) is at least amphibious in South America. There are no small, dominant, gnawing marsupials in Australia or anywhere else. The striped opossums and the wombats have gnawing teeth, but these animals have special habits and do not take the place of dominant rodents. Perhaps there were rodent-like, gnawing marsupials in Australia in the past, and perhaps they have been replaced by the real rodents that have reached Australia. There are no marsupial bats, but wings are not easy to evolve, and there are placental bats in Australia to pre-empt the air.

The distribution of marsupials within the Australian Region is complex. Dasyuridae (Tate 1947, pp. 115 and 150) are mostly in Australia, but about six stocks of them have reached New Guinea, some evidently earlier than others. *Peramelidae* (bandicoots) (Tate 1948*a*, p. 322) are about equally numerous in Australia and New Guinea. Besides older movements of them, there have evidently been recent movements in both directions: a species of *Isoodon* has extended from Australia to southern New Guinea, and one of *Echymipera*, from New Guinea to northeastern Australia. *Phalangeridae* (Australian opossums) are more diverse in Australia than in New Guinea, but their movements have been complex. Besides older movements, there have evidently been recent extensions of *Petaurus* (flying-squirrel-like gliders) and *Eudromecia* (dormouse opossums) from Australia to New Guinea, and of two species of *Phalanger* (cuscuses) and one of *Dactylopsila* (striped opossums) from New Guinea to northeastern Australia, and there have probably been both older and rather recent movements in *Pseudocheirus* (ring-tailed opossums) (Tate 1945*a*, p. 2) too. *Macropodidae* (kangaroos etc.) (Tate 1948, pp. 255–258) are mostly Australian, but small wallabies reached New Guinea long enough ago to have become endemic genera; the larger Agile Wallaby (*Protemnodon agilis*) has extended to southern New Guinea recently, and tree kangaroos (*Dendrolagus*) have probably made two movements, one older and one recent, between Australia and New Guinea in one direction or the other.

The latest movements in all these families probably occurred across late Pleistocene land connections between Australia and New Guinea. The earlier movements may be no older than the late Tertiary. Australia has a number of phylogenetically isolated or specialized marsupials, including two peculiar families (marsupial moles and wombats) and the Marsupial Wolf, Marsupial Anteater, *Tarsipes,* Koala, three groups of gliding phalangerids, etc. New Guinea has no such old, endemic, separately specialized marsupials, none which may not have come from Australia or evolved from an Australian ancestor within the last few million years. This suggests that existing marsupials may not have reached New Guinea until a few million years ago.

The rodents of the Australian Region are all classified as Muridae, and this has misled some zoogeographers, who have dismissed them as "mice" or simply ignored them. Actually Australian murids are not only rather numerous (Table 8) but also rather diverse. The subfamily Hydromyinae is almost confined to the Australian Region. There are eight genera of it in New Guinea, one of them widely distributed in Australia too, and a ninth genus is localized in northeastern Australia. Only two other, monotypic, genera are assigned to the subfamily, and they are on a mountain in northern Luzon in the Philippines. Most Hydromyinae in the Australian Region are water rats, but a few are terrestrial, and so are the Philippine ones. The other Australian murids are all Murinae, but even they are rather diverse. There are various endemic genera of them in all parts of New Guinea and Australia. Some are terrestrial and mouse- or rat-like, others jeboa-like (jumping), and others arboreal. Some of the latter are very large (for rats) and have more or less prehensile tails. Besides the endemic genera, there are many endemic species and species groups of the dominant Old World genus *Rattus* in New Guinea and Australia.

Apparently at least five different murids have reached the Australian Region at different times (see Tate 1951, pp. 217–221). An ancestral hydromyine (probably terrestrial) probably came early in murid history, perhaps in the Miocene. One or more murines probably came rather early too. The ancestor of the *Uromys* group of genera probably came later. And *Rattus* probably came still later, and more than once. Apparently one or more groups of this genus came via New Guinea, and another group came via the Sunda Islands and Timor. After their arrival, different stocks of murids tended to evolve separately in New Guinea and Australia, but there were successive exchanges too, in both directions, the latest ones listed by Tate (1951, p. 220).

The bats of the Australian Region, so far as I can judge from Tate (1946), are an accumulation of many different stocks which have come from Asia during a long period. Some of the older ones have evolved to some extent in Australia and New Guinea (but much less than the marsupials and rodents have done), and some of them may have returned westward for short distances, but the main flow of bats has apparently been into, not out of, the Australian Region.

The geographical history of monotremes is unknown. It may be guessed that, wherever they originated, monotremes radiated in Australia before marsupials did, and that the old Australian monotreme fauna may have been at least as diverse as the later marsupial fauna. That the Platypus and the echidnas are the last survivors of an ancient fauna is consistent with their specialization. Although they are basically primitive (or possibly degenerate), they are individually more specialized than any marsupials, as if they are at the ends of longer lines of adaptation. The Platypus is unique, and the monotreme anteaters (the echidnas) are much more specialized than the marsupial anteater. As for more recent history, if Australian marsupials were confined to Australia proper until late in the Tertiary (see above), monotremes probably were too. In that case, the echidnas are primarily Australian animals which have reached New Guinea twice, perhaps once before and once during the Pleistocene.

Transitions and barriers in distribution of mammals

Transition always occurs between adjacent regional mammal faunas, but it is often limited by barriers. Transition of mammal faunas between tropical Africa and temperate Europe is very limited; between tropical and temperate eastern Asia, broad and complex; and between tropical and north-temperate America, also broad and complex but more one-sided—many northern groups extend into South America, but few South American groups extend far into North America. There is also transition of the temperate Eurasian and North American faunas through the north; both faunas diminish northward, progressive subtractions beginning above the tropics and continuing northward until both faunas are reduced to similar northern fractions. Between the mammal faunas of tropical Africa and tropical Asia there is only a little transition now. These faunas have much in common, but they are now separated by a wide barrier of dry, not-tropical country, and few tropical mammals meet across it, although many must have crossed it in the past. Finally there is some transition of mammal faunas between the tropical Orient and Australia, even across the water barriers. Of Oriental mammals, many reach Wallace's Line; shrews, tarsiers, monkeys, squirrels, viverrids, wild pigs, deer, wild cattle, and a fossil pygmy elephant occur on Celebes; and certain shrews, monkeys, viverrids, pigs, and deer extend to some of the Moluccas, Lesser Sundas, and Timor, although there is some doubt about their natural limits; and murid rodents extend across the islands to Australia and Tasmania. In the other direction, most marsupials stop on New Guinea and its coastal islands, but a few reach the Moluccas, and *Phalanger* (cuscuses) extends to Celebes. On Celebes, endemic species of *Phalanger* occur with endemic monkeys and other placentals named above. Of these flightless land mammals, more extend from the Orient toward Australia than from Australia toward the Orient. This is true of the bats too. Many Oriental bats extend toward or to Australia, but only a few Australian bats extend toward the Orient.

No barriers except those of salt water have prevented the spread of dominant groups of mammals over the world. Some flightless land mammals have crossed even salt water to some extent, and bats have done so often. Nevertheless barriers of different sorts have had various effects on the distribution of mammals, and moderate salt-water barriers have almost completely (but not quite completely) isolated mammal faunas in South America and Australia through much or all of the Tertiary.

Dominance and competition in relation to the distribution of mammals

The most dominant order of mammals is the Rodentia. Wood (1947, p. 154) suggests that rodents are about twice as numerous in species as all other mammals now and that they may have been so throughout most of the Tertiary, but this is probably an overestimate (see Tables 5, 7, and 8). However, rodents probably at least equal all other flightless land mammals in species and outnumber them in individuals. The Rodentia is (excepting man) the only order of flightless land mammals which is cosmopolitan. Different rodents are dominant in many habitats, from the wettest forests to the driest deserts where mammals exist. Many different ones are terrestrial, arboreal, or more or less aquatic, and several different groups have become subterranean. However, there are ecological limits to the dominance of rodents which recall the limits of the perching birds. Rodents are rarely predaceous; they do not compete seriously with carnivorous mammals. And, although rodents compete to some extent with some grazing and browsing animals, they hardly threaten the dominance of hoofed mammals in the latters' sphere. Some other groups of mammals are obviously dominant (*e.g.*, canids, felids, bovids, and bats), but only within still narrower geographical or ecological limits.

Competition—any interaction among animals that has an adverse effect on any of them—is obvious among mammals. Introduction of rats, rabbits, mongooses, cats, etc., into new parts of the world has produced many examples of the effect of it. The effect of competition is plain in the history of mammals in South America where, in a few million years since the late Pliocene, new mammals have replaced much of the old South American mammal fauna. In fact the fundamental role of competition is indicated by the successions of mammal faunas which have occurred in many parts of the world, as fossils show. The radiation of marsupials in Australia emphasizes competition by showing what happens when it is limited. All this seems so obvious that I am almost ashamed to put it down in detail. I do so in preparation for further discussion (p. 552) and because some good biologists still doubt the fundamental importance of competition in animal distribution.

Summary: the pattern of distribution of mammals

Broad patterns of limitation and complementarity are plain in the present distribution of mammals. Mammals reach no final limits northward and southward except the limits of land; but most flightless land mammals do not extend beyond the limits of continents and continental islands or islands very near continents, although rodents extend somewhat farther, and a few bats much farther. There is striking complementarity in the occurrence of a primarily placental fauna of terrestrial mammals in the main part of the world, a mixed fauna in South America, and a primarily marsupial fauna in the Australian Region.

Zonation of land mammals is well marked but is mostly a matter of subtraction. North-temperate mammal faunas are essentially tropical faunas from which much has been subtracted. And the arctic mammal fauna is essentially a temperate fauna from which much more has been subtracted. Southward, too, mammal faunas are reduced by subtraction, especially toward the cold southern end of South America, but there is no zonal fauna of south-temperate mammals common to different southern continents.

The subpattern of radiation is not conspicuous in the present distribution of mammals, although fossils show that in the past many groups of them have radiated over most of the world.

Differentiation of six regional mammal faunas is a result of the broad complementarity and zonation of some groups of mammals and the further localization of others. The mammal faunas of the two main regions of the Old World tropics are rich in widely distributed as well as localized groups. These two faunas share many groups but differ in others. The mammal faunas of the two north-temperate regions are characterized primarily by poverty, by lack of many tropical groups. These two faunas are alike in general, especially toward the north, but different in detail. The mammal fauna of South and Central America is a unique mixture of old, endemic groups and new ones shared with other parts of the world. The Australian mammal fauna consists primarily of old, endemic families of marsupials and monotremes, plus murid rodents and many bats.

The fifth subpattern, of concentration in favorable areas and subtraction marginally, is moderately marked among mammals. Mammals are most numerous in favorable places and fewer in less favorable ones such as the arctic and deserts. But there are some mammals almost everywhere, even across the most unfavorable transition zones of land.

The oldest relicts among existing mammals are probably the monotremes in Australia, which are probably the last survivors of an ancient subclass. The oldest existing family of mammals is the marsupial family Didelphidae (American opossums); didelphids have survived substantially unchanged since the middle of the Upper Cretaceous (Scott 1937, p. 746). But existing Australian and South American marsupials are not relicts in quite the sense that was once supposed. They are apparently not so much survivors of a group that once dominated the world as products of evolution in isolation. Of course they are relicts in a sense, but so are some placentals. Existing insectivores are more or less specialized survivors of an ancient order which was ancestral to the placentals at the beginning of the Tertiary and which may be as old as the first marsupials. What should be stressed about the marsupials in Australia and South America is not that they are older than placentals but that they have evolved separately, in separate parts of the world, for a very long time.

History of dispersal of mammals

Although the fossil record of mammals is wonderful, there are still enormous gaps in it, geographically. Mammals originated in the

Triassic, but so few are known during the Mesozoic that they are not worth discussing geographically. The rise to dominance and the main "deployment" of mammals occurred about the beginning of the Tertiary, but even the Tertiary record is fragmentary except in Europe, parts of temperate Asia, western North America, and southern South America. The whole of Africa through about the first third of the Tertiary is a blank so far as land mammals are concerned, and this was probably one of the most important parts of the world and the most important span of time in the history of mammals (Simpson 1945, p. 250); later Tertiary African deposits are also few (Hopwood 1954). In tropical Asia, a few fossil mammals are known from Burma in the Upper Eocene. Otherwise most of the Tertiary mammals of southern Asia are in the Siwalik Hills of northern India. The Siwaliks actually lie about 6 degrees north of the tropics, and their record does not begin until the Miocene. In Australia the situation is still worse. Australia is a complete blank for mammals throughout the Tertiary, except for one geographically insignificant marsupial in the Pliocene of Tasmania. In the whole of the Old World tropics there is no record of mammals in the earlier part of the Tertiary and only an infinitesimal record later. The central importance of the Old World tropics in the present distribution of mammals emphasizes the importance of this great gap in the record. There is another gap in the American tropics. The Tertiary mammals of southern South America are remarkably well known, but comparatively little is known of those of tropical South America, although this gap is beginning to be filled (Couto 1952 etc., Stirton 1953). These gaps and others, especially the relatively poor record of small mammals almost everywhere and at all times, make some of the following histories somewhat hypothetical.

The following summaries of the geographical histories of the better known and more significant groups of land mammals are derived partly from the fossil record, partly by deduction from present distributions (see list of families). The histories have been checked against Simpson (1945 etc.), Romer (1945), and Scott (1937). The histories are arranged in the order of the list of families (Simpson), except that rodents and bats have been transferred to the end. I have tried to state the facts and probabilities simply but also to indicate the complexity of dispersals and to emphasize what is still unknown. Directions of movements are usually not discussed in detail. They will be summarized later.

Of monotremes, the ancestry and time and place of origin are all unknown. Their ancestor was probably an early Mesozoic mammal or mammal-like reptile but may possibly have been a late Mesozoic, primitive, degenerate marsupial. In the former case, monotremes may have originated almost anywhere; in the latter, they may always have been confined to the Australian Region. In any case they have probably been in Australia a very long time.

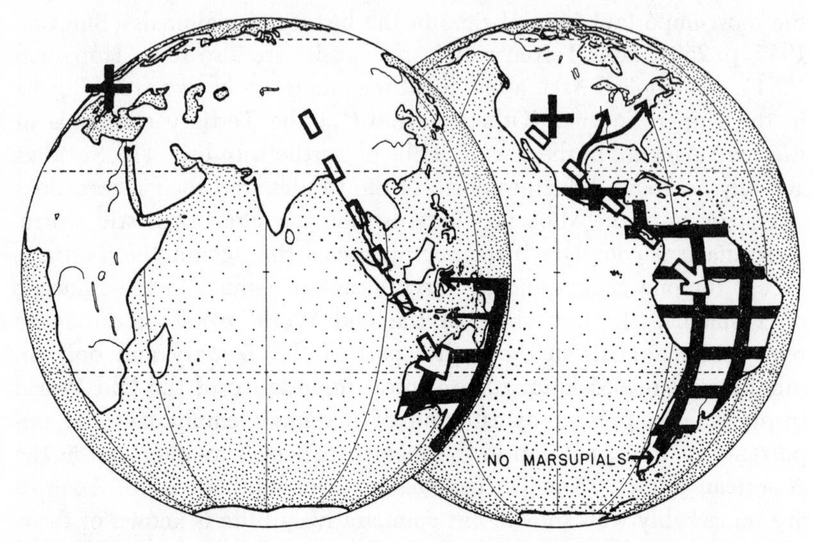

Fig. 39. Distribution of marsupials. Heavy crossed bars show main areas occupied by marsupials; arrows, (probably recent) extensions of *Phalanger* west of New Guinea and of *Didelphis* north of the American tropics; crosses, places where marsupials are fossil in the late Cretaceous and/or early Tertiary. Large broken arrows suggest directions of dispersal of early marsupials.

The distribution and apparent dispersal of marsupials is indicated in Figure 39. Didelphoids occurred in North America in the Upper Cretaceous and in Europe too in the early Tertiary, and were presumably more or less world-wide then. In the main part of the world they occurred with many placentals and apparently did not radiate extensively. They reached South America and did radiate there, although some placentals were present. The South American radiation produced a variety of didelphoids (opossums etc.), a number of which still exist; a variety of predaceous marsupials, all extinct, including a marsupial saber-toothed tiger (*Thylacosmilus*); and a variety of caenolestoids, of which a few obscure relicts (opossum-rats) still exist in the Andes and Chile. All three of these main groups are fossil in southern South America back to the Paleocene. Their earlier beginnings are unknown. They may all be derived from one didelphoid which reached South America about the beginning of the Tertiary, or their origins may have been more complex. In North America, didelphids persisted until the Lower Miocene, then apparently were absent until *Didelphis* returned in the Pleistocene. [Scott (1937, pp. 702–703) thinks didelphids survived the whole Tertiary in North as well as South America and that *Didelphis* originated in North America and reached South America across an early Pliocene water gap.] Of course marsupials reached Australia too and radiated there free of competition with placentals, except bats and eventually murid rodents. As to how many marsupials reached Australia, and when, the fossil record is mute. Australian marsupials may all be derived from one didelphoid or from several different ancestors, and so far as the actual evidence goes, the ancestor(s) may have reached Australia at any time in the late Cretaceous or early Tertiary. The radiation and recent movements of marsupials within the Australian Region have already been described.

Insectivores are fossil in the Upper Cretaceous of Asia and North America, but the full extent of their distribution then is unknown. They apparently never reached Australia, and they may not have reached South America (unless *Necrolestes* was an insectivore) until a modern shrew did so, very recently. The radiation of their descendants in the rest of the world was without parallel—unspecialized proto-insectivores were apparently the ancestors of all placental mammals. The insectivores themselves are a complex group and have had a complex dispersal in the main part of the world. They had some sort of dispersal before the end of the Cretaceous. Tenrecoids probably dispersed to the geographical limits of insectivores early in the Tertiary, then declined, surviving on Madagascar and the West Indies and (*Potamogale*) in West Africa. Erinaceoids too probably dispersed to the geographical limits of insectivores early in their history. They are still widely distributed in Africa and Eurasia (hedgehogs etc.), and through much of the Tertiary they were in North America, although they have disappeared there. Soricoids (shrews etc.), which date from the early Tertiary, are still widely distributed. Each of these widely distributed groups has probably had a more complex history than the actual record shows. There is no evidence whether chrysochloroids (which are probably derived from tenrecoids) and macroscelidoids are endemic or relict in Africa, or whether moles (which are related to soricoids and which have existed since the Upper Eocene) have ever extended much beyond the limits of the north-temperate zone, but the existence of these and other localized and specialized groups suggests further complexities in insectivore dispersal.

Dermoptera, now reduced to one specialized genus (the flying lemurs) in the eastern Orient, occurred in North America early in the Tertiary. Their history is otherwise unknown.

Primates are probably derived from something like tree shrews (which are now confined to the Oriental Region), and they, from something like macroscelidoid insectivores (which are now confined to Africa). Primitive primates (lemurs and tarsiers) occurred in Eurasia and North America (and presumably in some other parts of the world) early in the Tertiary and apparently had a complex dispersal then. Whether one of them reached South America and evolved the New World monkeys, paralleling the evolution of monkeys in the Old World, or whether American monkeys were derived from a more monkey-like ancestor which came from the Old World later, without leaving traces in North America, is unknown. Monkeys are not known fossil in South America until the Miocene, but perhaps they were there earlier, in the unfossiliferous tropical forest. In the Old World, lemurs, or perhaps just one lemur, reached Madagascar and radiated there. Otherwise the primitive primates declined. In North America they may not have survived beyond the Eocene. In the Old World, some still survive, confined to more or less restricted areas in the Old World tropics. But higher monkeys, apes, and finally man dispersed and redispersed very complexly in Africa and Eurasia during the middle and late Tertiary and Pleistocene. Their movements were probably mostly within the warmer parts of the Old World, for none of them is known to have reached America until man did so, probably by way of eastern Asia and Alaska, late in the Pleistocene.

Edentates were probably derived from a proto-insectivore. No unspecialized insectivore is known from South America; rather primitive edentates were in North America early in the Tertiary, although they were too late and too specialized to be ancestral; and the structure of all the South American forms implies that their immediate ancestor was already somewhat specialized in teeth, vertebrae, and limbs. All this suggests that edentates originated in North America (or even in the Old World) and had a short history of specialization there, and that one or more somewhat specialized (perhaps armadillo-like but unarmored, and perhaps arboreal) forms reached South America, probably in the Paleocene—armadillos are fossil in South America back to the Paleocene. In any case, edentates diversified in South America during the Tertiary, and some of them were dominant. They included a variety of small-to-gigantic ground sloths, armadillos, and huge armored glyptodonts, and fewer anteaters and tree sloths. At the end of the Tertiary, some edentates spread northward into North America, where the older ones had long been extinct. One of the ground sloths apparently reached North America in the mid-Pliocene, before the Central American land bridge was quite complete. This, plus the fact that ground sloths reached some of the West Indies, suggests that they crossed narrow water gaps. By the late Pliocene or Pleistocene, various ground sloths, glyptodonts, and armadillos were more or less widespread in North America, but they disappeared again (ground sloths apparently persisted in southwestern North America until the arrival of man) and ground sloths and glyptodonts became extinct everywhere, so that edentates are now much reduced in numbers and variety and are again almost confined to South (and Central) America, with only an armadillo extending north to Texas etc.

Edentate-like Pholidota (scaly anteaters) have had an unknown and probably unimportant history in the main part of the Old World.

The Carnivora have had a long series of multiple and successive dispersals in the main part of the world including North America, but they have not reached the Australian Region (except as carried by man) and probably did not reach South America until the Upper Miocene (procyonids) or late Pliocene and Pleistocene (other families). The earliest Carnivora, the extinct creodonts, appeared in the Paleocene and were dominant in the Eocene. They had a complex evolution and dispersal. Simpson's (1945) classification suggests that at least ten of them moved, one way or the other, between the Old World and North America. Fissiped carnivora had their main, complex dispersal later, mostly during and after the mid-Tertiary. The mostly Eocene miacids were at least common to Eurasia and North America. Canids (wolves, dogs, etc.) have dispersed successively and complexly. The relationships and distributions of living and fossil forms indicate that at least a dozen different groups of them have moved from Eurasia to North America or the reverse at different times, and at least three or four have reached South America. Bears, which (phylogenetically) are big dogs with flat feet, apparently originated in Eurasia well along in the Miocene but never reached the main part of Africa. They have moved between Eurasia and North America at least three or four times and have reached South America at least once. Procyonids have been in Eurasia (pandas etc.) and North America (raccoons etc.) since the Miocene; they reached South America by the Upper Miocene, before the Central American land bridge was complete. Mustelids have had a long and complex history in the main part of the world since the Lower Oligocene, with apparent Miocene and Pliocene radiations preceding the radiation of existing forms, and with at least nine movements indicated between the Old World and North America (not necessarily all in one direction) and at least four movements to South America. The South American immigrants were the ancestor(s) of the endemic musteline genera, one or more otters, a skunk, and the Long-tailed Weasel. Viverrids have evidently had a complex dispersal, extending through more than half the Tertiary, in the main, warmer part of the Old World. They have reached Madagascar and also Celebes and perhaps the Moluccas etc., but there is no good evidence that any viverrid ever reached North America. Hyaenas, which are derived from viverrids, have had a minor radiation in Eurasia and Africa. Cats (felids) have radiated complexly and successively, the extinct false saber-tooths and saber-tooths, from the mid-Tertiary to the Pliocene and Pleistocene respectively, and modern felines, mostly in the Pliocene and Pleistocene. Several felids have moved between the Old and New Worlds at different times; and several, including a saber-tooth as well as *Felis* and *Panthera,* have reached South America. All these details probably no more than suggest the actual complexity of dispersal of the Carnivora. The felids suggest the possible rapidity and completeness of replacement of successive dominant groups. The family Felidae has had a long and complex history, but existing felids almost all belong to two closely related, dominant genera which have dispersed comparatively recently and replaced all other felids (except the Cheetah) everywhere.

Early ungulates had a complex, important, but poorly known geographical history from about the beginning of the Tertiary. They are unknown in the Cretaceous but were diverse in the Paleocene. The most generalized ones, the condylarths, were not primarily hoofed mammals but were probably ancestral to them—preungulates, so to speak. They were close to the common ancestors of carnivores and ungulates, and presumably only a step from proto-insectivores. They were diverse in size and adaptations. Some were small. Some had claws instead of hoofs and may have been arboreal. Condylarths may have reached all continents except Australia at the beginning of the Tertiary. If they were the ancestral ungulates (which is perhaps hypothetical), they presumably reached Africa and fathered the proboscideans there. Four or five families of them were in North America and/or Europe in the Paleocene and Eocene, and one of the North American families was apparently represented in South America. A sixth family, Didolodontidae, is known only in South America, from the Upper Paleocene to the Miocene. The didolodontids are rather poorly known. They may have been conservative survivors of the common ancestor (if there was a common ancestor) of all Tertiary South American ungulates (Simpson 1945, p. 234).

The ungulates in South America during the Tertiary had an extraordinary evolutionary radiation. Simpson classifies most of them as protungulates and calls them archaic, but what should be stressed here is their separateness. They did not precede modern ungulates but paralleled them in a separate part of the world. South American ungulates comprised four or five orders besides the didolodontid condylarths. Litopterna, with two families and about forty listed genera, were confined to South America, from the Upper Paleocene to the Pliocene and Pleistocene. Notoungulata are known from thirteen families and about a hundred listed genera in South America, from the Paleocene to (in some cases) the Pliocene and Pleistocene, and are represented outside of South America by a single additional family, Arctostylopidae, with two genera, in Mongolia in the Upper Paleocene and North America in the Lower Eocene. Whether the northern notoungulates were survivors of a northern, ancestral stock or whether they came from South America is unknown. Astrapotheria, with two families, nine listed genera, were confined to South America, from the Eocene to the Miocene. Pyrotheria were confined to South America, with four genera in the Eocene and Oligocene. They were very distinctive animals, elephant-like in some ways (this is thought to have been parallelism), and not obviously related to anything else in South America. Two additional, Paleocene, genera doubtfully assigned to this order by Simpson have been put in a new order, Xenungulata, by Couto (1952). They are probably the oldest known South American mammals. That such early forms seem to belong to a peculiar order and are not broadly ancestral protungulates emphasizes how little is known of early ungulate history.

The beginnings of the South American ungulates are unknown. If the present classification of early ungulates surely reflected their phylogeny, it would seem that at least four or five of them reached

South America independently: one or two condylarths, a notoungulate, and the ancestors of the Xenungulata and Pyrotheria—unless notoungulates spread out of instead of into South America. But the classification is still partly hypothetical. All the endemic ungulate groups of South America may possibly be derived from one ancestor, presumably a condylarth.

The South American ungulates, whatever their origin, evolved in almost complete isolation from the rest of the world. I shall not try to describe them, except to say that they were extraordinarily diverse, varying in size, adaptations, and probable habits from rodent-like to horse-, camel-, rhinoceros-, elephant- and chalicothere-like. But these similes are inexact. Some of the South American ungulates were not much like anything now alive. As Romer (1945, p. 392) says, our understanding of these animals is hampered by the fact that we have no mental pictures of them and no popular names for them. Both Romer (1945) and Scott (1937) describe what is known of them in some detail. They must have existed in almost indescribable variety in South America through much of the Tertiary. Their fossil record is mostly confined to the southern part of South America, below the tropics. There were probably many additional, specialized, ones in the tropical forests. They reached their climax about the Oligocene, then declined, and by the end of the Pleistocene, when modern ungulates and carnivores had occupied South America, the entire assemblage of old South American ungulates became completely extinct. They apparently did not invade North America even temporarily, as some other South American mammals did, and they left no specialized relicts, as the South American edentates etc. have done.

Outside of South America, several different main lines of ungulates have evolved and dispersed in multiple and successive radiations. Pantodonts (amblypods) and uintatheres, which are separate orders, were temporarily dominant in North America in parts of the Paleocene and Eocene. They occurred also in Eurasia, where pantodonts survived until the Middle Oligocene. These orders may or may not have been directly related to early African proboscideans and South American pyrotheres. Two superfamilies of odd-toed ungulates, the titanotheres (brontotheres) and chalicotheres, were temporarily dominant in northern regions in parts of the Eocene and Oligocene. Chalicotheres survived later, until the Pleistocene in parts of Asia, and reached Africa; the last of them may have been in the Old World tropics. Most other major groups of ungulates which have been dominant in the main part of the world still survive at least as relicts.

The Aardvark, in Africa, is the only existing representative of the order Tubulidentata, which has a scanty and doubtful fossil record in Eurasia and North America and which may be derived from an old protungulate condylarth—but this is just an hypothesis.

Proboscideans—elephants etc.—are the "nucleus" of Simpson's superorder Paenungulata, which includes also the Embrythopoda (*Arsinoi-*

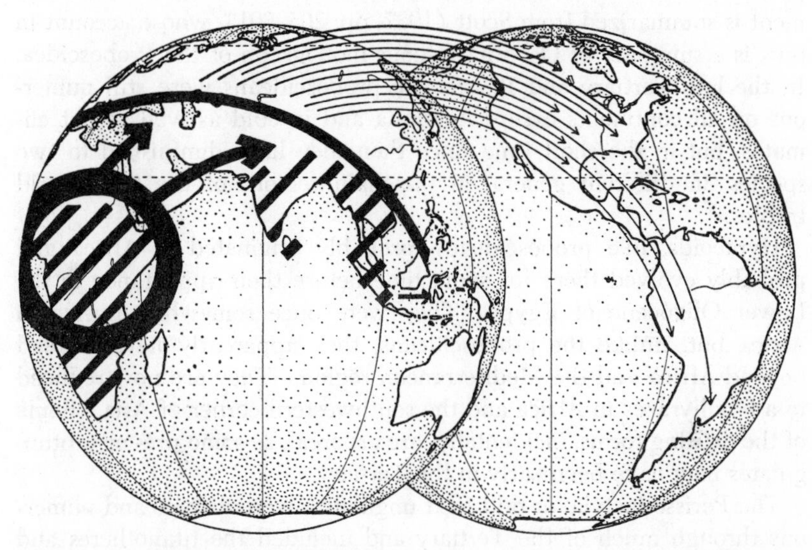

Fig. 40. Distribution of proboscideans. Heavy circle and arrows show origin in and spread of group from Africa; heavy arc, extension of the area of evolution to include southern Eurasia; arrows, movement of at least twelve stocks to North America, three reaching South America. Diagonal bars show approximate distribution of the two surviving elephants.

therium) of the Lower Oligocene of Egypt, the aquatic Sirenia, probably the African hyracoids (below), and hypothetically the pantodonts and uintatheres (above) of North America and Eurasia and the pyrotheres of South America. Something of the probable geographic history of proboscideans is shown in Figure 40. They are known first in Egypt, in the Upper Eocene and Lower Oligocene, but had probably been evolving and diversifying in Africa for some time before that. Africa may have been separated from Eurasia by the Tethys Sea during this time. The later history of proboscideans was one of spectacular radiation of successive groups from Africa or at least from the main part of the Old World. As new groups arose, they spread with surprising rapidity, and most of them sooner or later (from the

Miocene to the Pleistocene) spread to North America, although a few stocks did not. The number of different proboscideans that reached North America was probably at least a dozen and may have been more. There may have been some return movement, from North America to Eurasia, but this is not clear. And finally, in the late Pliocene and Pleistocene, at least three groups of mastodons and probably also a mammoth reached South America. This brief statement is summarized from Scott (1937, pp. 265–301), whose account in turn is a summary of Osborn's famous monograph of the Proboscidea. In the late Tertiary and Pleistocene proboscideans were still numerous on all continents except Australia and in cold as well as hot climates, but in the short time since then they have diminished to two species, in different genera, in the main regions of the Old World tropics.

Hyracoids, like proboscideans, probably originated in Africa and probably evolved there for some time before their appearance in the Lower Oligocene of Egypt. They were once somewhat diverse in Africa but, unlike the proboscideans, they apparently never spread beyond Africa and the Mediterranean region. They are now reduced to a few hyraxes in Africa and the southwestern corner of Asia. Some of the existing forms, incidentally, are arboreal, as some pre- or protungulates may have been early in the Tertiary.

The Perissodactyla, or odd-toed ungulates, were diverse and numerous through much of the Tertiary and included the titanotheres and chalicotheres (above); but now they are reduced to a few species of horses etc., tapirs, and rhinoceroses.

Horses, with their ancestors and relatives (the family Equidae), have the best fossil record, or at least the best described one, of any animals. The first, primitive equids, probably of Eurasian origin, are in the Lower Eocene of Europe and North America. It is probable (but not absolutely certain) that the whole main line of further evolution that culminated in *Equus* was in North America, although limited secondary lines evolved and became extinct in the Old World. Even in the Eocene and Oligocene but especially in the Miocene and Pliocene successive radiations of genera occurred in North America; at least three of these genera spread to Eurasia and at least one of them to Africa too, but all that reached the Old World became extinct. Then most of the five or more Pliocene lines became extinct even in North America; but one survived, radiated again, and produced *Equus* in North America (still in the Pliocene); and *Equus* evolved numerous species in North America and spread widely over the world. *Equus* (in a rather broad sense) occurred on all continents except Australia in the Pleistocene and was the only genus of equids existing then outside of South America—in South America it co-existed with three endemic genera representing one or more earlier (late Pliocene?) stocks. Finally, *Equus* and all its relatives suddenly became extinct in North and South America, leaving only the few existing *Equus* in Eurasia and Africa. Even this summary does not do full justice to the complexity of equid dispersal.

Tapiroids had a probably moderately complex but not very well-known geographical history in Europe and North America in the Eocene and later. They are not known to have reached Africa. True tapirs date from the Oligocene and still existed in Europe, Asia, and North America in the Pleistocene, when they extended into cool climates, but since then they have become restricted to a part of the Malay region and tropical America.

Rhinocerotoids were diverse in North America and Eurasia in the Eocene and had a multiple dispersal then. Rhinocerotids in a stricter sense had a complex evolution—much more complex than that of horses—and presumably a complex dispersal in the main area of the Old World, and it is probable that several different Old World stocks reached North America, but the details and in some cases the directions of movement are not certain. Rhinocerotids are fossil in Eurasia from the Middle Eocene through the Pleistocene, in Africa in the Miocene (but they may have been there earlier), and in North America from the Upper Eocene to part of the Pliocene. They are unknown at any time in South America. In Pleistocene Eurasia they occurred in cold as well as warm climates. Now, the few that remain have withdrawn into the Old World tropics.

The Artiodactyla, or even-toed ungulates, include several extinct Tertiary groups, some of which were at least common to Eurasia and North America but most of which cannot be discussed in detail here (see list of families), several groups which have declined, and the dominant pigs, deer, and antelope-cattle assemblage. Generally speaking, the Artiodactyla are now the dominant ungulates and as such have replaced most Perissodactyla.

True pigs (Suidae) have existed in variety in the main part of the Old World since the Oligocene and Miocene and have reached Celebes, etc., but not America. Peccaries etc. (Tayassuidae) were apparently represented by a separate subfamily in Eurasia from the Lower Oligocene to the Lower Pliocene. They have been in North

America since the Lower Oligocene and appeared in South America in the Pleistocene.

Existing hippopotamuses, in Africa, are the last survivors of a superfamily which included the anoplotheres in Europe (Middle Eocene to Middle Oligocene) and the anthracotheres in Eurasia (Middle Eocene to Pleistocene) and Africa and North America (Oligocene and Lower Miocene). Hippopotamuses themselves are fossil in Eurasia as well as Africa in the late Pliocene and Pleistocene, and in Madagascar in the Pleistocene.

Camels and their immediate ancestors were apparently confined to North America during almost the whole of their complex evolutionary history, from the late Eocene through the Tertiary. They were numerous in North America in the later Tertiary. Not until the Pleistocene do camels appear in Eurasia, North Africa, and South America. But they now survive as native animals only in parts of Asia and South America and have disappeared in their old home in North America.

Traguloids were numerous in Eurasia (and perhaps also in Africa, although they have not been found fossil there) in the late Eocene and Oligocene, and hypertraguloids had a parallel radiation in North America at about the same time. Both groups declined later. The North American forms apparently came to an end in the Miocene and early Pliocene. Most of the Old World forms have become extinct too, but *Hyemoschus* still survives in West Africa and *Tragulus* in the Oriental tropics. This is a simple summary of what was probably a complex geographical history.

The deer family, Cervidae, probably arose from early traguloids in Eurasia in the Oligocene. The primitive stock (subfamily Palaeomerycinae) extended to North America in the Miocene but disappeared there during the Pliocene; it may still be represented in Asia by the Musk Deer (*Moschus*), widely distributed in mountains from the Himalayas to parts of Siberia and Sakhalin Island. The main line of evolution of deer continued in Eurasia, but a peculiar subfamily (Dromomerycinae) radiated in North America in the Miocene and Pliocene, then became extinct. The main line passed through a stage (subfamily Muntiacinae) confined to Eurasia and still represented by two existing genera in Asia, and culminated in the Cervinae and Odocoileinae, the former dominant in and confined to Eurasia etc. except that *Cervus* has extended to North America (as the American Elk), and the Odocoileinae represented in Eurasia but especially characteristic of the Americas, including South America. The Odocoileinae probably originated in Eurasia and invaded North America in the Pliocene and South America in or just before the Pleistocene. *Alce* and *Rangifer* (called Moose and Caribou in America but Elk and Reindeer in Europe) are northern Odocoileinae which Scott (1937, p. 322) thinks extended from Eurasia to North America in the Pleistocene. Deer apparently never reached the main part of Africa, below the Sahara. This history is mainly stated or implied by Simpson (1945). Whether or not the history is entirely correct in detail, it is plain that the evolution and dispersal of cervids have been complex.

Giraffids are related to deer and to some extent have paralleled them in the Old World tropics. They were numerous in Asia (and probably also in Africa?) from the Miocene to the Pleistocene, and several reached eastern Europe in the Lower Pliocene, but now they are reduced to two species: the Giraffe in the more open parts of Africa below the Sahara, and the Okapi in a limited area of heavy forest in West Africa.

The Pronghorn (*Antilocapra*) in western North America is the last survivor of a family which radiated in North America (mostly in the West, but at least one reached Florida) in the later Miocene, Pliocene, and Pleistocene. The ancestor of the family presumably came from Eurasia.

The family Bovidae, which includes antelopes, sheep, goats, etc., as well as cattle, was one of the last families of mammals to become dominant and to disperse, but its dispersal has been complex nevertheless, as the following summary will show (Fig. 41). The classification and fossil record are, of course, from Simpson. In the subfamily Bovinae, the tribe Strepsicerotini (six listed fossil, two existing genera, including the Kudu and Eland) was in Eurasia from the Upper Miocene to the Pleistocene and is recorded from North Africa in the Lower Pliocene and from the rest of Africa not until the Pleistocene, but survives only in Africa below the Sahara; the tribe Boselaphini (sixteen listed fossil, two existing genera, including the Nilgai and Four-horned Antelope) was in Europe in the Miocene and Lower Pliocene and in Asia since then, but survives only in peninsular India; and the tribe Bovini (eight fossil, six existing genera, including cattle, buffaloes, etc.) is known fossil in Eurasia since the Middle Pliocene and in Africa and North America only in the Pleistocene, and is still widely distributed within the geographical limits of the family. Two genera of bovines reached North America: *Bos* extended to Alaska in the Pleistocene but disappeared there; and several species of *Bison* apparently reached North America successively. The subfamily Cephalophinae contains only one tribe (three existing genera, the duikers) confined to Africa (Pleistocene and Recent). In the subfamily Hippotraginae, the tribe Reduncini (eight fossil, five existing

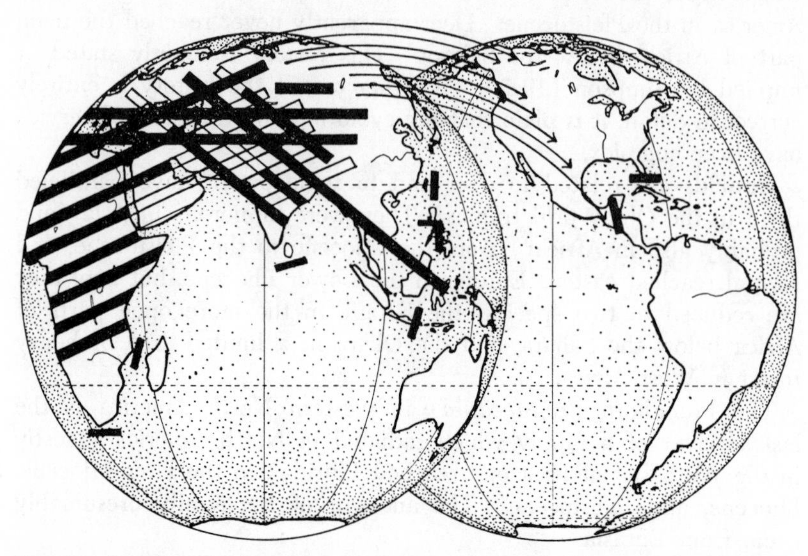

Fig. 41. Distribution of bovids. Each heavy bar represents one tribe in the Old World; the solid part of the bar indicates (roughly) present occurrence (whether in Eurasia or Africa), and the open part, past occurrence indicated by fossils. Smaller bars show extreme limits of the family. Arrows indicate movement of eight stocks to North America. Horizontal bars (from top) represent Ovibovini, Rupicaprini, Caprini; diagonal bars in Asia, Saigini, Bovini, Boselaphini, Antilopini; diagonal bars in Africa, Bovini (repeated), Antilopini (repeated), Hippotragini, Strepsicerotini, Reduncini, Alcelaphini, Cephalophini, Neotragini. Arrows to North America represent *Ovibos, Bos* (Pleistocene), *Saiga* (Pleistocene), Rocky Mountain Goat, Rocky Mountain Sheep, and three *Bison*. Diagram is intended to emphasize complexity of distribution.

genera, including various African antelopes) is in the Pliocene and Pleistocene of Asia and Pleistocene of Africa but survives only in Africa below the Sahara; the tribe Hippotragini (fourteen fossil, three existing genera, including various antelopes) is fossil in Europe in the Lower Pliocene, in Asia in the Pliocene and Pleistocene, and in Africa in the Pleistocene, but now occurs only in Africa and the southwestern corner of Asia; and the tribe Alcelaphini (three fossil, five existing genera, including the Hartebeest, Gnu, etc.) is listed fossil in the Lower Pliocene of eastern Europe and the Pleistocene of Asia and Africa, but now occurs only in Africa below the Sahara. In the subfamily Antilopinae, the tribe Neotragini (one fossil, eight existing genera, including various African antelopes) is known only from Africa below the Sahara (Pleistocene and Recent); and the tribe Antilopini (eight fossil, seven existing genera, antelopes etc.) is known in the Lower Pliocene of Europe, the Pliocene and later of Asia, and the Pleistocene of Africa, and survives widely distributed in Asia and Africa. And in the subfamily Caprinae, the tribe Saigini (two existing genera) occurs from the eastern edge of Europe to central Asia, with both existing genera fossil in the Pleistocene of Eurasia and one of them (*Saiga*) in Alaska too in the Pleistocene; the tribe Rupicaprini (three fossil, four existing genera, including the Chamois etc.) has been in Eurasia since the Pliocene and is still widely distributed there especially in mountains (from parts of Spain, the Alps, etc., to the Caucasus, and from the Himalayas to Manchuria, Japan, Formosa, and Sumatra) and is represented in western North America by the Pleistocene and existing Rocky Mountain Goat (*Oreamnos*), which is not a true goat; the somewhat doubtful tribe Ovibovini (nine fossil, two existing genera) is fossil in Eurasia in the Lower Pliocene and again in the Pleistocene and in North America in the Pleistocene, and now survives in east-central Asia (*Budorcas,* the Takin, placed in this tribe "tentatively") and part of arctic North America (*Ovibos,* the Musk Ox); and the tribe Caprini (five fossil, five existing genera, including goats and sheep) is fossil in Europe and Asia in the Pliocene and Pleistocene and in northern Africa and western North America in the Pleistocene, and still occurs, mostly in mountains, scattered through this range, but only one genus of the tribe (*Ovis*, sheep) has reached (western) North America. Some of the details of classification and distribution of this long summary may prove to be wrong and many other details of the history of bovids are unknown (their fossil record is probably very incomplete in Africa), but the main pattern, of complex evolution and dispersal in the main part of the Old World, with about six stocks reaching North America and with some groups withdrawing from wide areas, is certainly correct. The Musk Ox may have returned from North America to northern Eurasia and then become extinct there during the Pleistocene. Bovids of course did not reach Australia and apparently did not reach South America, although *Bison* did reach northern Central America. Bovids are probably not exceptional in the complexity of their dispersal. Many other dominant families of mammals have probably been as complex in their geographical movements but have left less-detailed fossil records.

Of dominant gnawing mammals, the first of the two orders, Lagomorpha, has had the less complex history. The oldest "rabbit" is *Eurymylus* of the Upper Paleocene of Asia, but I do not know how it is related to other, later, lagomorphs. Pikas have evolved in Eurasia and apparently formerly also in Africa at least since the mid-Tertiary, and have declined since then. They apparently did not reach (western) North America until the Pleistocene. Leporids (rabbits etc.) have had successive dispersals in the main part of the world including North America: probably one or more of Palaeolaginae (if this is a natural group), which left relict genera (not directly related to each other) in southern Africa, the Riu Kiu Islands, and Central Mexico; one or more of late Tertiary Leporinae; and one of *Lepus,* which is still radiating. *Sylvilagus,* the only leporid to reach Central and South America, is not known there until the Pleistocene.

Rodents are very numerous. Their interrelationships are not well understood. And their fossil record is inadequate. They may have had the most complex dispersal of any order of terrestrial mammals, but the details of it are largely unknown, and even the main features of it are partly doubtful. The following summary is therefore extremely oversimplified and partly hypothetical.

Ancestral rodents (Aplodontoidea) appear in the Upper Paleocene (of North America). They deployed in the early Tertiary. They have been found fossil only in North America and Eurasia, but they presumably occurred in Africa too. The last of the aplodontoids, or rather their least-modified descendants, are the aplodontids, which have been in North America since the Upper Eocene and were in Asia too at least in the Lower Pliocene, but which are now reduced to one species (the Sewellel, *Aplodontia*) in western North America.

Squirrels etc. (Sciuroidea) have had an evidently complex dispersal in the main part of the world including North America, and true squirrels have reached South America, probably not very long ago.

Pocket gophers and pocket mice etc. (Geomyoidea) have existed in North America, probably mostly in the West, since the mid-Tertiary. They extended to Panama (pocket gophers) and northern South America (pocket mice) probably not long ago.

Beavers and their ancestors and relatives (Castoroidea) were numerous in North America and Eurasia from the mid-Tertiary to the Pleistocene but are now reduced to one genus with one Eurasian and one North American species.

Anomaluroids are now confined to Africa. Ones that are supposed to have existed in the Tertiary of Eurasia and Africa are doubtful. Pedetids have been in southern Africa at least since the Lower Miocene.

The first muroids were the cricetids, which appear in the Oligocene of Eurasia and North America and were probably once dominant throughout the main part of the world including Africa—otherwise they would not have been likely to reach Madagascar, which at least one of them did. Cricetids have been much reduced in Africa and in tropical Asia (if they were numerous there), but they are still dominant in temperate Eurasia and especially in North America, and they have reached South America (apparently in the Upper Pliocene) and become dominant there. Three genera are common to North and South America, and additional stocks presumably reached South America to produce the endemic groups: about forty genera of cricetids are now endemic in Central and South America, but I do not know how many primary immigrants they represent. The second great complex of muroids is the family Muridae, which was apparently derived from cricetids somewhere in the Old World tropics, perhaps in tropical Asia, for murids have reached Australia but probably not Madagascar. Murids apparently did not rise until late in mammalian history, perhaps toward the end of the Miocene (Simpson), but they have had time to radiate overwhelmingly in the warmer part of the Old World. They have replaced most cricetids in tropical Asia and in Africa (if cricetids were numerous there), and several (five?) of them have reached the Australian Region at different times and radiated there, but they have not reached America, except as brought by man. This is at best the barest outline of a history that was probably full of multiple and successive dispersals, and of specialization and localization of minor groups, such as the spalacids and rhizomyids, in different parts of the Old World.

Dormice and their allies (Gliroidea) have had an unknown history in the main part of the Old World. Many of them are localized or discontinuous in distribution, which suggests that the group is declining.

Jumping mice etc. (Dipodoidea) are derived from non-jumping forms, of which one genus still exists in temperate Eurasia. Jumping Zapodinae may have originated in Eurasia (one is still in western China) and extended to North America, perhaps in the Miocene. Jerboas probably had a separate origin (as jumpers) in the drier parts of Eurasia and Africa.

Hystricomorphs are derived from one or more ancestors which lived somewhere in the main part of the world early in the Tertiary. They have had a minor radiation in the Old World, in Eurasia as well as Africa—Africa has been somewhat overstressed in this connection. Old World porcupines are fossil in Eurasia back to the Oligocene and still occur in the Orient as well as Africa, and an apparent thryonomyid is fossil in India. That hystricomorphs do not occur in Madagascar indicates that they may never have been numerous in Africa. Hystricomorphs are unknown in North America before the late Pliocene. Nevertheless, somehow, they got to South America a long time ago. They are fossil there back to the Lower Oligocene, and the earlier ones converge as if they were all derived from one ancestor which may have reached South America in the late Eocene (Wood 1950, p. 88). From this hypothetical ancestor radiated a variety of rat- and rabbit-like forms including guinea pigs, chinchillas, agoutis, jutias, etc.; several semi-aquatic forms including capybaras; arboreal porcupines; and many extinct forms, some of which were very large, with skulls approaching 2 feet in length. This radiation reached its maximum in the Pliocene. Whether or not the area of radiation originally included Central America and the West Indies is unknown because pertinent fossils are lacking, but sooner or later some hystricomorphs reached these places. In the late Pliocene and Pleistocene a few South American hystricomorphs extended to the main part of North America. One, and only one, of them still survives there: our Porcupine. How many people who meet the Porcupine in our northern forests know that it is a recent immigrant from the South American tropics? Capybaras (two genera) reached southeastern United States, at least to South Carolina, in the Pleistocene but have now withdrawn from North America and also from most of Central America. This may have been part of a recent, extensive withdrawal of all but the most dominant hystricomorphs from Central and northern South America. Such a withdrawal would account for the present concentration of these animals in southern South America (see list of families), for the absence of old endemic genera in Central America, and for the taxonomic and geographic isolation of the West Indian genera. The recent entry of non-hystricomorph rodents and other mammals into Central and South America from the north accounts well enough for the withdrawal of many hystricomorphs southward, and also for the extinction of the giant Pliocene forms.

How hystricomorphs reached South America is a problem, but not a unique one. Monkeys too appear suddenly in South America, in the Miocene, but are not known fossil in North America. Nor is the problem insoluble. There are too many solutions. Both hystricomorphs and monkeys may have reached South America through North America in spite of the absence of fossils there. This is, I think, the most probable solution. If both groups were then forest living and arboreal, they would have been the less likely to leave fossils in North America and the more likely to reach South America across narrow water gaps. Or South American hystricomorphs and monkeys may both be products of parallel evolution, and not directly related to their Old World counterparts (see Wood 1950). Or hystricomorphs, at least, may have rafted across the Atlantic from Africa to South America. A final choice cannot be made among these alternatives without more evidence.

Bats have dispersed complexly. Megachiroptera (fruit bats) date from the Oligocene or before and have evidently had a complex geographical history in the warmer part of the Old World. They have considerable powers of dispersal and have reached many islands in the Indian and western Pacific Oceans. But they have not reached America. Microchiroptera ("insectivorous" bats, really diverse in their food habits: most do feed on insects, but some take fruit, fish, other bats etc., or blood) have evidently had multiple and successive dispersals over the whole world, but there is almost no fossil record of their movements. Their earliest movements are therefore unknown. Emballonurids have probably had a fairly old radiation: they may go back to the Eocene; they have a discontinuous, tropicopolitan distribution; and the Old and New World genera of them are all different. Existing molossids may have dispersed somewhat later: they are less strictly tropical, and one genus of them is still common to the Old and New Worlds. Existing vespertilionids may have dispersed still more recently: they are cosmopolitan, and several genera of them are common to the Old and New Worlds. The vespertilionid genus *Myotis*, which is dominant and cosmopolitan, may have dispersed most recently of all. It is noteworthy that the two bats that have reached New Zealand have evidently come at different times, one long ago, the other recently.

This completes the summaries of geographical histories of the principal groups of land mammals. It might be added that all the major groups except the rodents and perhaps the bats and higher Artiodactyla (pigs, deer, cattle, etc.) seem to have declined since the Miocene or Pliocene (Simpson 1945, p. 34), and that many large mammals in many parts of the world retreated or disappeared during the Pleistocene.

In nearly every one of the preceding histories I have mentioned complexity, but the whole process must have been far more complex than I have been able to say. The main dispersals of orders and families have been complicated enough, but they are just the comparatively simple sums of the movements of multitudes of genera and species. The dispersal of mammals must have been almost endlessly,

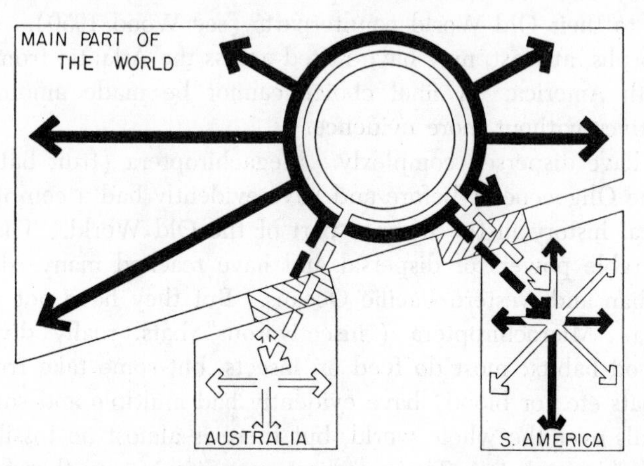

Fig. 42. Main pattern of radiation of mammals at the beginning of the Tertiary. Black circle and arrows represent placentals; white, marsupials; diagonally hatched bars, early Tertiary barriers.

inconceivably complex—a kaleidoscope of mingled and changing patterns.

Nevertheless there has been a rather simple main pattern in mammalian dispersal (Fig. 42). At the end of the Cretaceous, when reptiles ceased to rule, there were two groups of mammals in the main part of the world ready to radiate: proto-insectivores (placentals) and didelphoids (marsupials). All existing mammals, probably excepting the Australian monotremes, are derived from these two groups. During the Tertiary they evolved, in different combinations, more or less separately, in three principal theaters. The proto-insectivores produced a diversity of placentals which dispersed very complexly in the main part of the world. Marsupials reached Australia and radiated separately there. And marsupials and a few selected placentals reached and radiated together in South America.

Two misconceptions exist about the endemic mammal faunas of

Australia and South America. One is that they are more ancient than the mammal fauna of the rest of the world. They are not. Placentals are apparently as old in their origins as marsupials. Both existed well back into the Cretaceous. The great evolutionary radiation of mammals began in South America at about the same time as in the main part of the world, and there is no reason to think that the radiation of Australian marsupials was earlier. These radiations occurred in separate places, not at separate times. The other misconception is that early mammals must have reached Australia and South America across (different) land connections. This is at least doubtful. The Australian and South American mammal faunas are not derived from the whole of any known world-wide fauna but from fractions which may have reached the isolated continents at different times, across water gaps.

In Australia, the two existing monotremes are more specialized than any marsupials (p. 339), and it may be guessed that they are the last relicts of a monotreme fauna which was in Australia long before marsupials arrived, and which (if it existed) probably dated from the Mesozoic. Marsupials probably came later, but the exact time of their arrival is unknown. So far as any actual evidence goes (there is no record of when or whether marsupials were in Asia, and no record of them in the Australian Region until the Pliocene), the ancestor(s) of Australian marsupials may have come at any time from the late Cretaceous through perhaps the first half of the Tertiary. Murid rodents came still later, different ones probably in the late Miocene or Pliocene and Pleistocene. Bats, of course, have probably reached Australia from time to time during much of the Tertiary. The Australian mammal fauna seems thus to be derived from a very few ancestors which apparently came at different times, and the presumption is that they all came across water barriers. There is no direct evidence of the direction from which monotremes and marsupials reached Australia. The rodents and bats evidently came from Asia.

In South America, the story is more complicated but better documented. Marsupials, edentates, and ungulates were in South America in at least part of the Paleocene, and the South American forms were diversified or specialized enough to suggest that their ancestors reached South America early in the Paleocene or even before. But there is no certainty of just how many ancestral forms there were or that they all came at exactly the same time, and there is some evidence that they were selected fractions of a more diverse fauna which existed in the main part of the world: the edentates and ungulates were apparently already edentates and ungulates (or protungulates) when they reached South America (this is indicated by certain related fossils especially in North America), and some important Paleocene mammals apparently failed to reach South America, *e.g.*, insectivores and early placental carnivores. The best explanation of all this seems to be that one or more marsupials, one or more edentates, and one or two or three or more ungulates reached South America about the beginning of the Tertiary, but not necessarily all at one time, and that they came across a water barrier which blocked other mammals. This was, of course, not the end of the story. A hystricomorph rodent reached South America probably in the late Eocene. Monkeys appeared in South America in the Miocene but probably got there earlier. Procyonids (the raccoon family) reached South America before the end of the Miocene. Bats evidently came at various times, some of them (*e.g.*, phyllostomatids) probably in the early or middle Tertiary. Whether or not some other mammals arrived before a complete connection was made with North America is unimportant. The whole of the earlier mammal fauna of South America seems to have been derived from fractions of the fauna of the main part of the world, and all the fractions may have been selected by passage across a water barrier, although the barrier may have been less formidable at the beginning of the Tertiary than later.

As to direction, marsupials, edentates, and one or two ungulates seem to have reached South America from North America. This is suggested by fossils in North America. Another group of early South American ungulates, the Pyrotheria, is curiously like the (primarily African) Proboscidea, although this is supposed to be parallelism. The ancestor of the South American monkeys and hystricomorphs may have come directly from Africa, or through or from North America. Procyonids and all later mammals that have reached South America have come from North America. It seems fair to say that all the mammals known to have reached South America at any time *may* have come from or through North America. Where there is doubt, the doubt is based less on contrary evidence than on lack of evidence. That all the mammals that reached South America, at intervals widely spaced through the Tertiary, did so from one direction, from North America, across a changing archipelago which became a land bridge in the late Pliocene, is an hypothesis consistent within itself and, I think, consistent enough with the evidence, so far as it goes.

In the main part of the world, from the beginning of the Tertiary, radiated the great placental fauna from which a few chips now and then reached South America and, fewer and later, Australia. The radiation of this fauna involved dispersal and repeated redispersal of successive stocks of many different evolutionary lines: insectivores, primates, various carnivores, many ungulates, lagomorphs, many rodents, and bats, to name only the main groups. The details of all this need not be repeated, but three special aspects of it are important. They are the role of Africa especially early in the Tertiary, the exchange between the Old World and North America, and the exchange between North and South America when a land connection was finally established.

The earliest fossil mammals known in Africa are in the Upper Eocene and Lower Oligocene of El Faiyûm, in northern Egypt, a few degrees north of the tropics, strictly speaking. They are a mixture of contemporaneous European groups and distinct, apparently autochthonous African groups. This virtually proves that Europe and Africa were connected then. The African groups in El Faiyûm are so very distinct, however, including the first known proboscideans and hyracoids as well as *Arsinoitherium,* that they suggest that Africa had been isolated for some time just before. But, if so, there had presumably been a still earlier connection with Eurasia or at least an opportunity for Africa to receive ancestral placental stocks. All this suggests that Africa was included in the main theater of radiation of placental mammals at the beginning of the Tertiary, that it was then isolated and that some mammals evolved there independently until the late Eocene, and that it was then reunited with Eurasia. This hypothetical history is consistent with the geological history of the Tethys Sea, which once separated Africa from Eurasia and across which land connections may have been made and broken many times (p. 99).

The history of exchange of mammals between the Old World and North America during the Tertiary and Pleistocene is established in detail by Simpson's admirable analysis (1947, summarized 1947*a*) of the successive fossil mammal faunas of Europe, Asia, and North America. There was intense exchange of land mammals between Eurasia and North America in the early Eocene, late Eocene, early Oligocene, late Miocene, middle to late Pliocene, and Pleistocene. There was probably some, but not much, exchange in the early and middle Miocene and early Pliocene. There was little or no exchange in the middle Eocene and middle and late Oligocene, and of course there has not been much in Recent times. It seems likely, therefore, that the connection between Eurasia and North America which evidently existed through much of the Tertiary was interrupted (presumably submerged) perhaps during part of the Paleocene,[1] probably in the middle Eocene, again in the middle to late Oligocene, perhaps more briefly in the first half of the Pliocene, and possibly at other times, as it is now. The mammals do not indicate any Tertiary connection across the Atlantic between Europe and North America. Up to the early Eocene, the evidence is indecisive, but it does not require an Atlantic bridge and is consistent with a North Pacific (Bering) bridge. After the early Eocene, the evidence clearly favors a single, North Pacific bridge and does so increasingly in later epochs. In the early Eocene, exchange of mammals involved many major groups and large parts of the continental faunas, but subsequent exchanges were increasingly selective and tended to involve smaller fractions of faunas, progressively less distinctive types of animals, lower taxonomic categories, and ecological types less novel in the invaded regions. The major selective factor was apparently climate. The groups exchanged were, as a rule, those probably belonging to relatively cool, but not alpine, environments. Selection of such groups became increasingly clear and increasingly strong from early to late Tertiary, which suggests a gradual cooling of the northern climate through much of the Tertiary, ending in Pleistocene glaciation. Each successive exchange certainly involved movements in both directions, but there seems always to have been more movement from Eurasia to North America than in the opposite direction, although the direction of movement of many groups is doubtful.

The groups that were not exchanged between the Old World and North America are significant. There were always some such groups, and some of them were important and long-lasting. Viverrids (civet cats, mongooses, etc.) and Megachiroptera (fruit bats) have had great radiations in the main, warmer part of the Old World since the Eocene or Oligocene, but none has ever reached North America, so far as is known. Higher Old World primates too have radiated diversely in the main, warmer part of the Old World since at least the Lower Oligocene, but none is known to have reached North America until man did so. Representatives of all of these groups extend into limited warm-temperate areas now, and fossils show that the groups extended north of the strict tropics in the past. Murid ro-

[1] Although Simpson does not say so, the nature of the exchange in the early Eocene seems to imply a preceding interruption sufficient to allow different, distinctive groups of mammals to evolve in Eurasia and North America, after an exchange perhaps early in the Paleocene.

dents have risen to extreme dominance in the Old World without reaching America (until man brought them), but their rise may date only from the Pliocene. Less important groups that seem to have been confined to the main part of the Old World through their whole history include gliroid rodents, true pigs, and giraffes. In North America, geomyoid rodents have evolved since at least the Lower Miocene without reaching the Old World. Procyonids and peccaries have had independent lines in North America since the Lower Miocene and Lower Oligocene respectively. Camelids were apparently confined to North America from the late Eocene until the Pleistocene. And pronghorns have radiated in North America since the late Miocene without reaching the Old World. But the strictly North American groups are fewer, less important, in some cases more restricted ecologically (geomyoids and pronghorns probably radiated mostly in the drier part of western North America), and usually less long-lived than the Old World groups. All this suggests that, although the northern regions were certainly warmer than now, there was probably enough zonation of climate back at least to the Oligocene and perhaps before to prevent some mammals which were dominant in the Old World tropics from reaching North America, and to prevent some (but fewer) mammals which inhabited the warmer part of North America from reaching the Old World.

The exchange of mammals between North and South America after a land connection was established, rather late in the Pliocene, is dramatically summarized by Simpson (1940, p. 158). Before the connection, North America had about 27 and South America 29 families of terrestrial mammals, and only one or 2 families (procyonids and perhaps opossums) occurred on both continents. After the connection, there was a rush in both directions. In the Pleistocene North and South America had 22 families in common, and there were still further movements. But there were even more withdrawals and extinctions. After all the shuffling, North America was left with only 23 families, and South America had 29, the same number as before. Except that it was a little reduced, the North American fauna was not much changed as a result of all this. But the South American fauna was profoundly changed. Many South American mammals became extinct, and many North American ones survived in South America and became dominant there.

The South American mammals that are known to have extended for significant distances into North America were opossums (unless they survived the Tertiary in the north); three families (at least four different stocks) of ground sloths, two stocks of armadillos, and two stocks of glyptodonts; and, of hystricomorph rodents, at least porcupines and two capybaras. (One of the ground sloths apparently reached North America a little before the land connection was made.) Of all these, only an opossum and a porcupine are still widely distributed in North America, and an armadillo reaches Texas etc. The once-dominant marsupial carnivores and South American ungulates apparently did not enter North America even temporarily, and caenolestid marsupials, monkeys, anteaters, tree sloths, and many groups of hystricomorph rodents either failed to reach North America or reached only southern Mexico, so far as is known.

The North American mammals that have extended to (made separate invasions of) South America are a shrew; a genus of rabbits; a typical squirrel, a genus of pocket mice, and an undetermined number of stocks of cricetid rodents; at least three canids, a bear, one or two procyonids (one of them before the land connection was made), four mustelids, and three or more cats (including an extinct saber-tooth); three or four elephants (in a broad sense); two horses; tapirs; peccaries; camels; and deer. There is no animal on this list, incidentally, that seems to have come from a separate, previously hidden, tropical North American mammal fauna. The principal Pleistocene and Recent North American groups that have not reached South America are moles, pikas, sciurids other than typical squirrels, beavers, jumping mice (zapodids), pronghorns, and bovids. All the North American groups except elephants, horses, and saber-tooths that are known to have reached South America have survived there, although tapirs and camels as well as elephants, horses, and saber-tooths have disappeared in North America. Some of the northern groups are now dominant in South America. Northern Carnivora have entirely replaced the carnivorous marsupials. Northern ungulates have entirely replaced the old South American ungulates. And northern rodents are probably replacing hystricomorphs: there are already nearly as many genera of cricetids as of hystricomorphs in South America, and the hystricomorphs are apparently retreating southward.

The evidence that mammals have dispersed in certain definite directions is worth recapitulating (Fig. 43). Apparently at least five groups of flightless land mammals (different murid rodents) have moved from Asia to the Australian Region comparatively recently, but only one group (*Phalanger*) has extended from the Australian Region toward Asia even as far as Celebes; and bats have moved into the Australian Region more often than out of it. There has apparently been continual movement of mammals from the Old World to North America, more than the reverse, as far back as the evidence goes. The recent movement of mammals from North into South America has been overwhelming. And the recent movement from Old World to South America (through North America) has been entirely one way: a number of originally Old World groups have reached South America, but no South American ones have reached the Old World.

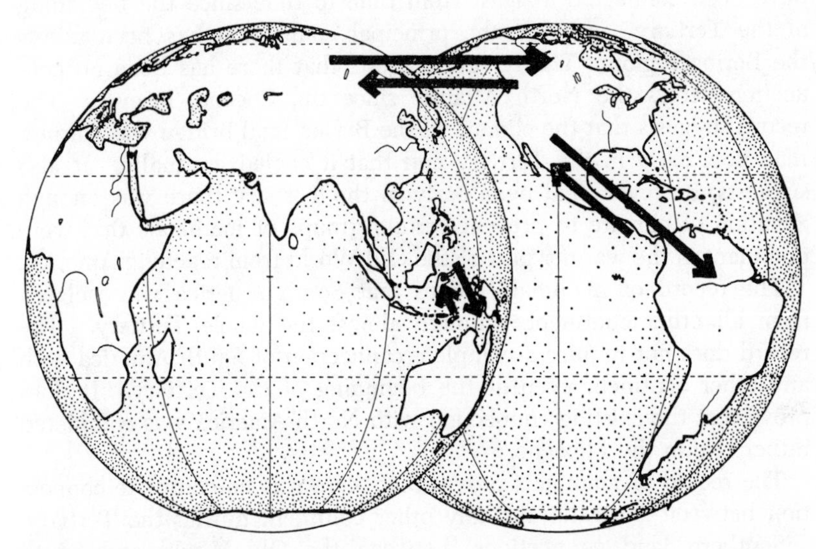

Fig. 43. Known directions of movement of mammals in the Tertiary and Pleistocene. Length of arrows is roughly proportional to amount of movement in directions shown.

The nature of these directional movements is sometimes misunderstood. They are not simple movements like flowing of water or marching of armies, but primarily extensions, and the direction is a complex statistical effect. Complex exchange is usually involved, with many movements (extensions) in both directions, and it is only in the course of time that unequal exchanges and selective extinctions give direction to the process as a whole. In this sense there apparently has been direction in the dispersal of mammals, from the main part of the Old World toward Australia, from the Old World to North America, and from North to South America. Since the greatest mammal faunas are tropical, the greatest single center of geographical radiation of mammals would seem to have been the main tropical regions of the Old World.

Mammals, through their unique fossil record, suggest or prove definite things about past lands and climates.

The record of mammals suggests that Africa may have been cut off from Eurasia for a while before the end of the Eocene. The record proves that Africa and Eurasia have been connected at least from time to time since then.

The record of mammals proves that Eurasia and North America have been connected at least from time to time since the beginning of the Tertiary, and that the principal connection has been across the Bering region. The record suggests that there has been no connection across the North Atlantic since the Eocene, if ever. The record suggests that the climate of the Bering land bridge was warmer than now early in the Tertiary and that it cooled gradually. It also suggests, however, that even early in the Tertiary there was enough zonation of climate to prevent certain groups of mammals that were dominant in the warmer part of the Old World from reaching America.

The record of mammals proves that South America was isolated from all other continents through almost the whole Tertiary. The record does not require a complete connection of South America with any other continent even at the beginning of the Tertiary. It does prove that the existing connection with North America was completed rather late in the Pliocene.

The record of mammals proves that there was no complete connection between Australia and any other continent during the Tertiary.

Southern land connections, between the Old World and South America, are scarcely compatible with what is now known of the distribution of existing and fossil mammals, at least in the Tertiary. An Eocene connection between Africa and South America, suggested (but not required) by the distribution of monkeys and hystricomorph rodents, is virtually impossible in the light of other evidence. A very early Tertiary connection between South America and Australia, suggested by the distribution of existing marsupials and especially by a striking resemblance between extinct South American carnivorous marsupials (borhyaenids) and existing Australian dasyurids, is improbable too. The supposed relationship of the borhyaenids to the dasyurids has become increasingly doubtful. Didelphid marsupials occurred in Europe and North America, and presumably in Asia too, early in the Tertiary, and Simpson thinks that the didelphids are the central group from which all other marsupials may be derived. It therefore seems likely that different didelphids reached South America and Australia independently about the beginning of the Tertiary, from

North America and Asia respectively, and radiated independently, producing (among other groups) borhyaenids and dasyurids by parallel evolution (*cf.* Simpson 1945, p. 171). Evidence against a direct land connection within their time is given by the diversity of edentates and peculiar ungulates which existed in South America not only through the Tertiary but also into the Pleistocene or Recent, but of which there is no trace in the Pleistocene and Recent faunas of Australia. What connections there may or may not have been between continents before the Tertiary, the record of mammals cannot show.

LIST OF FAMILIES OF MAMMALS

Class Mammalia, mammals. This is a complete list of known families of Recent and fossil mammals. Taxonomically the list follows Simpson (1945), but his classification is simplified by omission of some higher categories and some subfamilies which do not seem to be geographically significant. The details of existing distributions are derived from many sources, of which the more important are listed in the chapter bibliography. The fossil record of each family is summarized from Simpson and checked against Romer (1945) and Scott (1937) and in some cases other authors. The numbers of genera are usually given for both existing and fossil families and the numbers of species for most existing ones, but of course the numbers are not final. Simpson notes that his lists of genera are not quite complete, and many ancient extinct genera must still be unknown—even the numbers of existing species are still unsettled in many cases. Pre-Pleistocene records of existing genera should be treated with caution; generic names are often used more broadly among fossil than among existing mammals. Many fossil mammals, even many important ones, are still inadequately known or doubtfully placed, and there are probably still many outright errors in the accepted fossil record. The Paleocene is distinguished from the Eocene throughout this list. A dagger (†) before a name in the following list indicates that the group is extinct.

(*Subclass Prototheria*)

Order Monotremata, egg-laying mammals: not known ever to have occurred outside the Australian Region

Tachyglossidae, spiny anteaters: Australia (including Tasmania) and New Guinea; 2 genera, 3 or 4 species. *Tachyglossus aculeatus,* the Echidna (several subspecies), is widely distributed in Australia and Tasmania and extends to southeastern New Guinea. It inhabits a variety of country including rain forest as well as drier open forest and brushy and rocky places, but probably not open desert. In the colder parts of its range, it hibernates. Although it is a very primitive mammal with very specialized food habits, it is still common and must be considered a still successful and in some ways adaptable animal. The closely related genus *Zaglossus* (*Proechidna*), with 2 or 3 species (several subspecies), is confined to New Guinea (and Salawati Island just to the west) but is widely distributed there especially in the mountains. *Tachyglossus* is fossil in Australia in the Pleistocene.

Ornithorhynchidae, duckbills, platypuses: widely distributed in eastern Australia from the base of the Cape York Peninsula (north to some 20 miles south of Cooktown–Tate 1952, p. 577) in tropical North Queensland, south to Victoria and Tasmania, and west to the Leichhardt River in North Queensland and into the edge of South Australia in the south; one genus (*Ornithorhynchus* or *Platypus*) and one existing species (several subspecies); still common in many places. "The waters inhabited range from clear and icy rapid alpine streams, at a height of 5000 feet on the Kosciusko tableland, to the warm and rather turbid rivers of the Queensland coast, and from large lakes to small waterholes" (Troughton 1951, p. 3). The existing genus is fossil in eastern Australia in the Pleistocene.

(†*Subclass Allotheria*)

†*Order Multituberculata:* an ancient order of 3 or 4 families. The doubtfully mammalian †Tritylodontidae are known from Europe (Upper Triassic to Middle Jurassic), China (Upper Triassic), and South Africa (Upper Triassic); the families †Plagiaulacidae, †Ptilodontidae, and †Taeniolabididae, with a total of 27 listed genera (mostly fragments), are from Europe, Asia, and North America at various times from the Upper Jurassic to the Lower Eocene.

(*Subclass and order doubtful*)

†Microcleptidae: Europe (Upper Triassic); 2 genera; isolated teeth.

(*Subclass doubtful*)

†*Order Triconodonta*

†Triconodontidae: Europe and North America (Jurassic, and doubtfully Upper Triassic); 8 listed genera.

(*Subclass Theria*)

†*Order Pantotheria:* 4 families (†Amphitheriidae, †Paurodontidae, †Dryolestidae, †Docodontidae) known from Europe or Europe and North America (†Paurodontidae from East Africa too) in the Jurassic; 22 listed genera in all.

†*Order Symmetrodonta:* 2 families (†Spalacotheriidae, †Amphidontidae), from Europe and/or North America in the Upper Jurassic; 5 listed genera; inadequately known.

[*Infraclass Metatheria*]

Order Marsupialia, marsupials or pouched mammals: of the 6 superfamilies of this order recognized by Simpson, the first 3, the Didelphoidea (living in America and including all the old North American and European fossil marsupials), †Borhyaenoidea (South America), and Dasyuroidea (Australian Region) are apparently closely related. The other 3 superfamilies, the Perameloidea (Australian Region), Caenolestoidea (South America), and Phalangeroidea (Australian Region to Celebes etc.) are phylogenetically somewhat more isolated.

(*Superfamily Didelphoidea*)

Didelphidae, American opossums: now confined to South, Central, and eastern North America; 12 existing genera (including *Caluromysiops*

Sanborn 1951), perhaps 80 species (more than half in *Marmosa*). All the genera occur in South America; several of them extend into or through Central America, 4 of them reaching parts of Mexico; and one of the 4, *Didelphis,* extends farther, into eastern North America. The northernmost species, the Virginia Opossum (*D. virginiana,* which is a representative of the South and Central American *D. marsupialis*), extends through most of the eastern half of the United States; it has been found as far north as southern New Hampshire, southern Ontario, central Michigan, and southern Wisconsin. In spite of its primitiveness, it is a common and aggressive animal which has extended its range northward in recent years. It has been introduced into part of California and at scattered places in Oregon and Washington State. Southward, *Didelphis* reaches central Argentina; *Lutreolus,* perhaps a little farther south, to Chubut; and *Marmosa,* to south-central Chile (Valdivia). *Dromiciops* is known only from Chiloe Island and a small area of the adjacent mainland of Chile. And *Notodelphis* is known only from the vicinity of the Gulf of St. George in southern Argentina, at about 47° S. latitude, this being the most southern known locality for any marsupial. The southern tip of South America, the last 300 or 400 miles, is without marsupials so far as is known. Fossils indicate that didelphids were numerous (2 extinct subfamilies, 11 listed genera) in North America in the Upper Cretaceous, and that Didelphinae (4 extinct genera) existed there in the first half of the Tertiary (until the Lower Miocene); that one of the North American genera occurred also in Europe from the Eocene to the Miocene; and that didelphids were present and probably numerous (though their record is scanty) in South America throughout the Tertiary—Couto (1952*a*) thinks they may have reached South America in the late Cretaceous. In North America, marsupials are unknown after the Lower Miocene, until the appearance of *Didelphis* in the Pleistocene.

†Caroloameghiniidae: South America (Lower Eocene); one genus.

(†*Superfamily Borhyaenoidea*)

†Borhyaenidae, sparassodonts, marsupial carnivores: South America (Paleocene to Pliocene); 21 listed genera.

(*Superfamily Dasyuroidea*)

Dasyuridae, marsupial mice, marsupial cats, etc.: Australian Region. The 9 genera listed by Simpson all occur or have occurred in Australia; several reach Tasmania; and *Phascogale* and *Sminthopsis* (marsupial mice) and *Dasyurus* (marsupial cats) extend to New Guinea (and certain small, close-lying islands) and the Aru (not Kei) Islands. However, the actual pattern of geographical relationships within the family is more complex than this. Tate's more finely divided classification (20 genera, 45 species) includes several primitive genera confined to New Guinea etc. and suggests half a dozen or more exchanges of different stocks between Australia and New Guinea (Tate 1947, Table 4). *Sarcophilus* (the Tasmanian Devil) and *Thylacinus* (the Tasmanian Wolf), both now confined to Tasmania, are fossil in Australia (Pleistocene), and so are the more widely distributed genera *Phascogale* and *Dasyurus.* *Myrmecobius* (marsupial anteaters, 2 species or subspecies), which almost within memory occurred across most of the southern part of Australia, is now apparently restricted to a small area in northwestern South Australia etc. and the adjacent part of Western Australia.

Notoryctidae, marsupial moles: Australia; one genus, 2 species; widely distributed in South and Central Australia "wherever the typical sand-ridge country covered with mulga and stunted salt-bush, or red sand-flats covered with spinifex and acacias, provides a suitable habitat" (Troughton 1951, p. 57), and west (continuously?) to where the Great Sandy Desert reaches the coast of Western Australia.

(*Superfamily Perameloidea*)

Peramelidae, bandicoots: Australian Region to Ceram; Simpson (1945) recognizes 5 genera; Tate (1948*a*), 8, with 19 species. The family is well represented in Australia and New Guinea (Tate 1948*a*, Table 1); 2 genera extend to Tasmania; one (*Echymipera*), to New Britain and New Ireland, the Aru and Kei Islands, and Misol; and one monotypic genus is confined to Ceram and is apparently still known only from the types, from an altitude of 6000 feet on Mt. Manusela. Several existing genera are fossil in Australia (Pleistocene).

(*Superfamily Caenolestoidea*)

Caenolestidae, opossum rats: South America, 3 existing genera, scattered along the Andes from the Venezuelan-Colombian boundary to Ecuador (*Caenolestes*, 4 species) and southern Peru (*Orolestes*, one species), at altitudes up to 14,000 feet, and isolated in Llanquihue Province including Chiloe Island in south-central Chile (*Rhyncholestes*, one species) in heavy forest at relatively low altitudes. Fossils (14 listed extinct genera) show that this family has existed in South America since the Lower Eocene, and that formerly it was rather numerous and diverse.

†Polydolopidae: South America (Paleocene and Eocene); 4 listed genera.

(*Superfamily Phalangeroidea*)

Phalangeridae, Australian possums, etc.: Australian Region to Timor, Celebes, and the Solomons; 14 existing or recently existing genera, about 44 species. Most of the genera are Australian; 4 extend to Tasmania; several extend to New Guinea; and 4 are confined to or center on New Guinea; but the distributions of some of the genera are complex. *Phalanger* (cuscuses) sets both western and eastern limits of marsupials in the Indo-Australian Archipelago: the genus is primarily New Guinean (2 species of it extend to the upper part of the Cape York Peninsula of Australia), and it occurs west to Timor and Wetar and to Celebes, and east to the remotest Solomons. Celebes has 2 distinct, endemic species of the genus, perhaps derived independently from New Guinea (see Tate 1945). *Petaurus breviceps* extends from northern Australia and New Guinea to New Britain, the Aru and Kei Islands, and (in the Moluccas) Misol, Batjan, Halmahera, and Ternate Islands (Tate 1945*b*). No other phalangerids occur in this direction beyond New Guinea and closely associated coastal islands (Salawati etc.) and the Aru Islands. Several existing Australian genera have very restricted ranges; *e.g.*, *Tarsipes*, the "Honey Mouse," confined to the southern corner of Western Australia. *Phascolarctos*, the Koala, formerly ranged in timbered areas from southern Queensland (19° 30′ S. lat.) through New South Wales and Victoria to (within recent years) southeastern South Australia and (in the Pleistocene) southern Western Australia, but is now much reduced in range and numbers (Harper 1945, Troughton 1951). The Australian Possum (*Trichosurus*) has been introduced into New Zealand and is now widely distributed on both main islands and Stewart Island and has unexpectedly become a pest (Wodzicki 1950). Fossil phalangerids, representing several existing and extinct genera, are known from Australia in the Pleistocene; and †*Wynyardia* is from Tasmania in the Pliocene.

†Thylacoleonidae: Australia (Pleistocene); one genus; very large ("the size of a lion"–Romer), of unknown habits.

Phascolomidae, wombats: eastern and southern Australia and Tasmania; 2 existing genera, 2 or more species; formerly plentiful north to central Queensland and west to the eastern edge of Western Australia (and to the southwest corner in the Pleistocene) and on islands in Bass Strait as well as on Tasmania, but now much reduced in range. Both existing and 2 extinct genera are fossil in Australia in the Pleistocene (Tate 1951*a*).

Macropodidae, kangaroos, wallabies, rat kangaroos, etc.: the Australian Region; 14 (Simpson) or 18 (Tate 1948) existing genera, about 46 species (Tate), many subspecies. The kangaroos and larger wallabies (*Macropus, Protemnodon*, etc.) are confined to Australia (including Tasmania and some coastal islands) except that *Protemnodon agilis* of northeastern Australia extends to southern New Guinea and the Trobriand Islands. Other, mostly smaller, forms are numerous and diverse in Australia (several reach Tasmania), less so in New Guinea, although 8 species groups occur there (Tate), some shared with Australia and some peculiar, and some reaching 10,000 feet or higher in the mountains. The distribution of the family is shown in more detail by Tate (1948, Table 2). *Thylogale* extends from Australia and New Guinea to the Aru and Kei Islands, and to New Britain and New Ireland. *Dorcopsis* reaches Mysol, Waigeo, and Salawati Islands just west of New Guinea. Secondarily arboreal tree kangaroos, *Dendrolagus*, with at least 3 full species, occur in the rain forests of New Guinea and North Queensland south to the lower Herbert River, and this genus may extend to the Aru and Kei Islands, although the evidence of this is "slender and unsatisfactory" (Rothschild and Dollman 1936, p. 477). Fossil macropodids, representing various existing and 6 listed extinct genera and including kangaroos larger than any living ones, are in the Australian Pleistocene.

†Diprotodontidae: Australia, including Tasmania: 6 listed fossil genera (Pleistocene); terrestrial marsupials, at least one of which reached the size of a large rhinoceros.

[*Infraclass Eutheria*, placental mammals]

Order Insectivora, insectivores: 8 superfamilies (for Tupaioidea, tree shrews etc., see under Primates)

(†*Superfamily Deltatheridioidea*)

†Deltatheridiidae: Asia (Upper Cretaceous) and North America (Paleocene and Eocene); 7 listed genera.

(*Superfamily Tenrecoidea*)

†Palaeoryctidae: North America (Middle Paleocene); one genus.

Solenodontidae, solenodons: existing only on Cuba and Hispaniola in the West Indies; one genus (*Solenodon*), with one species now localized in eastern Cuba (the same or slightly different species formerly occurred also in central and western Cuba—Aguayo 1950) and one in Hispaniola; and 3 fossil genera in North America (Oligocene).

Tenrecidae, tenrecs etc.: Madagascar (*Tenrec* extends also to the Comoro Islands); 10 (Simpson) or 13 (Allen) genera, 30 species (G. M. Allen 1939); ecologically diverse, presumably by radiation on Madagascar; and fossil on Madagascar in the Pleistocene.

Potamogalidae, otter shrews: West Africa (Cameroons to Angola and the Belgian Congo); one species, aquatic.

(*Superfamily Chrysochloroidea*)

Chrysochloridae, golden moles: Africa, most in South Africa but extending north to the southern Cameroons, Uganda, etc.; 4 or more existing genera, 29 species (Allen), and an additional genus in the Pleistocene of South Africa.

(*Superfamily Erinaceoidea*)

†Zalambdalestidae: Asia (Upper Cretaceous); one genus.

†Leptictidae: North America (Upper Cretaceous to Oligocene) and Europe (Upper Paleocene and Eocene); 11 or more genera.

Erinaceidae, hedgehogs etc.: now confined to the main part of the Old World (not the Australian Region); 2 subfamilies. (1) Echinosoricinae, hairy hedgehogs, gymnures: eastern Oriental Region from southwestern China to Java, Borneo, and Mindanao; 4 genera, 4 species; and 12 listed fossil genera in Europe and North America from parts of the Eocene to the Upper Miocene (Europe) or Lower Pliocene (North America)—that erinaceids were present and varied in North America through much of the Tertiary is an unexpected fact shown by recent studies (Simpson). (2) Erinaceinae, true or spiny hedgehogs: Africa (not Madagascar); Eurasia north to Scotland, central Sweden, central Siberia, and the Amur Region, and south to India (discontinuous areas in North and South India, not Ceylon) and central China (not Japan, probably not Formosa although listed from there, not the Indochinese area etc.); 5 genera, about 15 species; and several genera fossil in Eurasia (beginning in the Oligocene) and a doubtful one in North America (Upper Miocene).

†Dimylidae: Europe (Upper Oligocene to Upper Miocene); 4 listed genera.

(*Superfamily Macroscelidoidea*)

Macroscelididae, elephant shrews: Africa, including North Africa (not Madagascar); 5 or more genera, about 40 listed species (G. M. Allen 1939).

(*Superfamily Soricoidea*)

†Nyctitheriidae: North America (Middle Paleocene to Middle Eocene) and probably Europe (Lower Eocene); 5 genera; not adequately known.

Soricidae, shrews: 3 subfamilies (but Ellerman and Morrison-Scott 1951, p. 41, doubt their distinctness). (1) Soricinae: Eurasia and North America, mostly north-temperate, north into the arctic, south to central Spain, southern Italy, Palestine, apparently Kashmir (not most of India), northern Burma, Tonkin, and Formosa, and in America south through Central America to the high mountains of Colombia, Ecuador, and western Venezuela; 8 existing genera, 18 Eurasian and an undetermined number of American species. *Sorex* is widely distributed in Eurasia and North America and occurs south to Guatemala. *Cryptotis* occurs through most of eastern United States, Mexico, and Central America, and south into northern South America to the limits given above; it is the only insectivore known ever to have reached South America, unless the dubious †*Necrolestes* is an insectivore (see p. 378). Fossils (several genera) assigned to this subfamily go back to the Upper Eocene of Europe and Lower Oligocene of North America. (2) Crocidurinae: the Old World except the Australian Region —Africa, Madagascar(?), much of Europe, Asia north to parts of Siberia, south to Ceylon, and to Celebes, the Moluccas(?), Kei Islands (?), Lesser Sundas to Timor(?), and the Philippines(?); 12 genera, many species. *Crocidura* and *Suncus* are supposed to be native on the Comoro Islands, and *Suncus,* on Madagascar, and one or both of these genera are supposed to be native on the other islands questioned above. However, both genera include species which are more or less associated with man, and their natural limits are doubtful. Two existing genera are listed fossil in Eurasia back to the Upper Miocene and Lower Pliocene. An extinct genus is listed in North America in the Upper Miocene, but its assignment to this subfamily might be questioned on geographical grounds. (3) Scutisoricinae: east-central Africa (adjacent parts of the eastern Belgian Congo and Uganda); one specialized genus, 2 species.

Talpidae, moles: temperate Eurasia (south into the tropics) and North America; north to southern Sweden, apparently across the Arctic Circle in Asia, and to southwestern British Columbia and southern Labrador; and south to Spain, Italy and Sicily, Asia Minor and the Caucasus, Nepal, the Indochinese Region and Malay Peninsula, and Formosa, and to Lower California, the northeastern corner of Mexico, and Florida; 17 (or fewer) existing genera, 20 species. No genus is common to Old and New Worlds. The oldest fossil moles are in the Upper Eocene in Europe and Upper Oligocene in North America.

†Nesophontidae: West Indies (Cuba, Hispaniola, Puerto Rico); Pleistocene and subfossil; one genus, 5 or 6 species; never found alive and probably extinct, but possibly still existing on mountains in Hispaniola.

(†*Superfamily Pantolestoidea*)

†Pantolestidae: North America (Middle Paleocene to Middle Eocene) and Europe (Lower Eocene); 2 subfamilies, 6 listed genera.

(†*Superfamily Mixodectoidea*)

†Mixodectidae: North America (Middle Paleocene to Upper Eocene); 8 listed genera.

(*Insectivora, families and genera incertae sedis*)

†Picrodontidae, †Apheliscidae, †Necrolestidae, and unplaced fossil genera. These are mostly early Tertiary forms from North America or Eurasia, but †*Necrolestes* is from South America (Lower Miocene). It is a peculiar animal which may or may not have been an insectivore and which annoys and baffles evolutionists and zoogeographers (Simpson 1945, p. 178).

Order Dermoptera

†Plagiomenidae: North America (Upper Paleocene and Lower Eocene); 2 genera. Simpson (1945, p. 179) confirms the probable relationship of these fossils to the following family.

Cynocephalidae (Galeopithecidae), flying lemurs: Tenasserim and Cochin China to Java, Borneo, and the southeastern Philippines; one genus, 2 species.

Order Chiroptera, bats. I am particularly indebted to Dr. Koopman for recent information about bats.

(*Suborder Megachiroptera*)

Pteropidae, fruit bats, including flying foxes: warmer parts of the Old World, north to Cyprus, Syria, India, Sikkim etc., the coast of South China, and the Riu Kiu Islands; south through all Africa (except the Sahara and the northwest) and to northern and eastern Australia (one species occasionally migrating as far south as the edge of Victoria); and on many islands of the Indian and Pacific Oceans, east to Samoa; 21 genera, about 200 (or fewer) species. *Pteropus*, "flying foxes," with numerous species, occurs from India to Australia and on many islands in the western Pacific Ocean and across the Indian Ocean. Several species are on Madagascar, and single species occur on Pemba and Mafia Islands, about 40 and 20 miles off the African coast (Swynnerton and Hayman 1951); these huge, powerful bats could easily reach the mainland of Africa, but they do not do so. The large fruit bats of Africa belong to other genera. *Megaloglossus*, in the forests of West Africa, belongs to a group of flower-eating forms (Macroglossinae) otherwise confined to an area from Burma etc. south and east in the Indo-Australian Archipelago. The only fossil of this family listed by Simpson is from Italy (Oligocene).

(*Suborder Microchiroptera*)

(*Superfamily Emballonuroidea*)

Rhinopomatidae, mouse-tailed bats: northeastern Africa (west to parts of the Sahara and south to Lake Rudolph) and southern Asia (Arabia, Palestine, Iran, peninsular India, to Thailand and Sumatra); one genus, about 4 species.

Emballonuridae, sheath-tailed bats etc.: tropics etc. of Old and New Worlds including much of Australia; north to Egypt, Iraq, India, and Yunnan, and in America north into Mexico; 8 genera, none common to Old and New Worlds; numerous species. The genus *Emballonura* has several species occurring from Thailand to New Guinea and the Solomons and west to Samoa, and one isolated on Madagascar (the related *Coleura* occurs in eastern Africa, southern Arabia, and on the Seychelles). The single fossil assigned to this family by Simpson is from Europe in the Upper Eocene or Lower Oligocene.

Noctilionidae, hare-lipped (including fish-eating) bats: tropical and subtropical America, north to southern Mexico and West Indies; one genus, few species.

(*Superfamily Rhinolophoidea*)

Nycteridae, hollow-faced bats: tropical Africa to Corfu (Greece), Palestine, and Arabia, and Madagascar; and the Malay Region from Tenasserim to Java, Borneo, and Timor; one genus, 20-odd (or fewer) African and one Malayan species.

Megadermatidae, false vampires etc.: parts of tropical Africa including Abyssinia; India and southern China to Java, Celebes, Ternate, and the Philippines; and the more tropical northern half of Australia (and fossil or subfossil in South Australia); 3 or more existing genera, all more or less localized; 5 species; and 2 fossil genera in Europe (Upper Eocene or Lower Oligocene, and Middle Miocene).

Rhinolophidae, horseshoe bats (part): warmer parts of the Old World; north to England and Germany, South Russia, North India etc., and Korea and Japan; and south through Africa and from Asia to eastern Australia (to southern New South Wales); one existing genus, many species. The existing genus is listed fossil in Europe in the Eocene; an extinct genus, in the Upper Oligocene.

Hipposideridae, horseshoe bats (part): warmer parts of the Old World; north to North Africa (not Europe), Iraq, Iran, India, and southern China; and south through most of Africa (present in Madagascar) and from Asia to much of Australia; 6 existing genera, numerous species; and a fossil genus in Europe (Upper Eocene or Lower Oligocene and perhaps Miocene).

(*Superfamily Phyllostomatoidea*)

Phyllostomatidae, leaf-nosed bats etc.: the warmer parts of America; 35 (Simpson) or 50 or more (G. M. Allen 1939*a*, p. 196) genera, many species. Of the genera, many reach Mexico, but only 4 extend into the southern edge of the United States; and many are in the West Indies and a few in the Bahamas, but none reaches Florida except an occasional straggler. The family is now known fossil in northern South America and also in Florida in the Miocene.

Desmodontidae, true vampire bats: tropical America; central Mexico to central Chile and parts of Argentina; on Trinidad but not West Indies proper; 2 or 3 genera, 3 species.

(*Superfamily Vespertilionoidea*)

Natalidae, long-legged bats: tropical America north to Mexico, West Indies; one genus, few species.

Furipteridae, smoky bats: much of tropical South America; one or 2 genera, few species.

Thryopteridae, disk-winged bats: part of tropical America from eastern Brazil and Ecuador north to Honduras; one genus, 2 or 3 species.

Myzopodidae, sucker-footed bats: Madagascar; one species. G. M. Allen (1939*a*, p. 123) suggests that this may be (distantly) related to the preceding family.

Vespertilionidae, common bats etc.: cosmopolitan, from the northern limit of trees to South Africa, Tasmania and New Zealand, and Tierra del Fuego, and on many islands; 25 or more genera, many species. *Myotis* is virtually cosmopolitan, from the northern limit of forests in Eurasia and North America to South Africa, most of Australia (but apparently not the extreme southeast and Tasmania), and "probably to the Straits of Magellan" (Osgood 1943) (but apparently not on New Guinea, New Zealand, or most of the West Indies); this is the most widely distributed genus of bats; it is listed fossil back to the Middle Oligocene (of Europe). Besides *Myotis, Eptesicus* and *Pipistrellus* and also *Nycticeius* and *Plecotus* in Simpson's (1945) sense occur in both Old and New Worlds; the remaining genera are confined to one or the other. Some genera reach remote islands: *e.g., Chalinolobus,* New Zealand (from Australia); *Lasiurus,* Hawaii and The Galapagos Islands (from America); and *Pipistrellus,* the Azores (from Europe or North Africa). Of the few known fossil genera of the family, the oldest are in the Upper Eocene or Lower Oligocene (of Europe).

Mystacinidae (Mystacopidae), short-tailed bats: New Zealand; one species. It extends far south in New Zealand, to Stewart Island. It is now rare but probably still exists at least on some coastal islands (Powell 1947).

Molossidae, free-tailed bats: warmer parts of the world, north to southern Europe and Asia (to Korea in East Asia) and southern and western United States (to the corner of British Columbia in the west); 6 or more genera, a moderate number of species. Of the genera, only *Tadarida* (in Simpson's sense) occurs in both Old and New Worlds; this genus is listed fossil back to the Oligocene (of Europe).

(*Superfamily uncertain*)

†Archaeonycteridae, †Palaeochiropterygidae, and 2 unplaced genera: Europe (Middle Eocene to Lower Oligocene).

Order Primates, including man and all his post-insectivore ancestors and collateral relatives.

(*Suborder Prosimii*), primitive primates

(*Superfamily Tupaioidea*)

†Anagalidae: Mongolia (Lower Oligocene); one genus, beautifully preserved.

Tupaiidae, tree shrews: the Oriental Region, from the southwestern corner of China and parts of peninsular India to Java, Bali, Borneo, and the southern Philippines; 6 genera, a moderate number of species.

(*Superfamily Lemuroidea*) (arranged after Simpson, 1945; Hill, 1953, p. 317, arranges them somewhat differently)

†Plesiadapidae: Europe and North America (Paleocene and Eocene); 5 listed genera.

†Adapidae: Europe and North America (Eocene); 9 listed genera.

Lemuridae, typical lemurs: Madagascar (2 species extend to some of the Comoro Islands); 6 existing genera, 15 species; some existing and 4 extinct genera are in the Pleistocene or post-Pleistocene of Madagascar. Simpson divides the existing and extinct forms into 5 subfamilies.

Indridae, woolly lemurs etc.: Madagascar; 3 genera, 4 species; the existing genera and 3 additional extinct ones are in the Pleistocene or post-Pleistocene of Madagascar.

(*Superfamily Daubentonioidea*)

Daubentoniidae, aye-ayes: Madagascar; one species.

(*Infraorder Lorisiformes*)

Lorisidae, pottos, lorises: tropical Africa and the Oriental Region; 2 subfamilies. (1) Lorisinae: *Arctocebus* and *Perodicticus* (pottos), each with one species, are in the forests of West Africa, and *Perodicticus* extends east in isolated forest tracts to the Mau Forest, Kenya; *Loris* (one species, the Slender Loris) is in southern India and Ceylon; and *Nycticebus* (one or 2 species, slow lorises) occurs from northeastern India and Assam to Java and Borneo but probably not on Mindanao (Hill 1953, map between pp. 169 and 170). A fossil genus is in the Pliocene of India. (2) Galaginae, bush babies: most of Africa below the Sahara, except the extreme southern tip (Hill 1953, maps between pp. 219 and 220, 237 and 238); 3 genera, 5 species.

(*Infraorder Tarsiiformes*)

†Anaptomorphidae: North America (Middle Paleocene to Middle Eocene), Europe (Eocene), and perhaps Asia (Oligocene); 5 subfamilies, 26 listed genera.

Tarsiidae, tarsiers: East Indies—Sumatra etc., Borneo etc., Celebes etc., and southeastern Philippines (Mindanao, Bohol, Leyte, Samar); one genus, 3 species. The distribution is mapped by Hill (1955, pp. 112–113).

(*Unplaced and doubtful primitive primates*)

†Apatemyidae, †Carpolestidae, and numerous doubtful genera: North America and Eurasia in the early Tertiary (Paleocene and Eocene).

(*Suborder Anthropoidea*), higher primates

(*Superfamily Ceboidea*), American monkeys

Cebidae: tropical America; 12 genera, a moderate number of species. All the genera occur in South America; at least 5 extend to part or all of Central America; and 2 also reach Veracruz in southern Mexico. Southward, at least 2 genera reach forested parts of northern Argentina. Several existing genera are fossil in the Pleistocene and extinct genera are in the Miocene of South America. No living native monkeys have been found anywhere in the West Indies, but an extinct genus was on Jamaica in Pleistocene or Recent times, and a monkey of some sort may have occurred on Hispaniola (Williams and Koopman 1952).

Callithricidae, marmosets: tropical America, from southern Brazil to Panama; 2 or more genera, several species.

(*Superfamily Cercopithecoidea*)

Cercopithecidae, Old World monkeys: warmer parts of the Old World, except the Australian Region; 16 existing genera, numerous species. The northern limit of monkeys is set by *Macaca* (macaques, including the Rhesus Monkey) in Morocco and Algeria (introduced on Gibraltar), northern India etc., southern China north to the Yangtze Valley (introduced farther north?), and the main island of Japan; and a couple of other genera reach southwestern China. Southward and eastward monkeys inhabit all Africa (not Madagascar), India and Ceylon, and islands to Java and Bali; other Lesser Sundas to Timor

(*Macaca irus* only), Celebes (*Macaca* and a peculiar genus, *Cynopithecus*), Formosa (*Macaca* only), and the Philippines to Luzon (*Macaca irus* only). *Macaca* (above) extends from North Africa to the limits of distribution of monkeys in the Orient; the Sacred Baboon (*Comopithecus*) extends from northeastern Africa to southern Arabia; otherwise no genus of monkeys is common to Africa and Asia; but *Papio* (baboons) and *Cercopithecus* (common African monkeys), both now confined to Africa, are fossil in India in the Lower Pleistocene and Middle Pliocene respectively; and *Colobus* of tropical Africa represents a subfamily, Colobinae (leaf-eating monkeys), now living otherwise only in the Orient but fossil in Europe etc. in the Pliocene. A few other fossil Old World monkeys are known, the oldest in the Lower Oligocene of Egypt. *Macaca irus,* the Crab-eating Monkey, is known to have been introduced on certain islands, and I am not sure it is really native on the outer Lesser Sundas, Timor, and the Philippines; endemic subspecies are listed from some of these islands, but that is not necessarily proof that the animals are native there (see p. 467).

(*Superfamily Hominoidea*)

†Parapithecidae: Egypt (Lower Oligocene); one genus.

Pongidae, apes: parts of Africa and the eastern Oriental Region; 2 existing subfamilies. (1) Hylobatinae, gibbons: southeastern Asia etc. from northeastern India and western Yunnan to Sumatra, Java, Borneo, and perhaps (introduced?) the Sulu Islands: one or more genera, 7 species; and 4 more fossil genera in Africa (Lower Oligocene and Lower Miocene) and central Europe (Middle Miocene and Lower Pliocene). (2) Ponginae, great apes: 3 genera, 3 or more species. *Pongo* (*Simia*) *pygmaeus,* the Orangutan, occurs only on Sumatra and Borneo (2 slightly different subspecies?), but the genus was in India in the Lower Pleistocene. *Pan,* the chimpanzees (one or 2 species), occurs in West and Central Africa, from Sierra Leone and the Cameroons to the Congo region, east in forested areas to Lakes Albert and Victoria. *Gorilla,* the gorillas (2 species or subspecies), has one form in West Africa from southeastern Nigeria nearly to (but not across) the Congo River, and another in Central Africa in mountain forests in a narrow strip along the border of the Belgian Congo and Uganda, from just below the equator to Baraka on northwestern Lake Tanganyika (G. M. Allen 1939, p. 177). A fourth, fossil genus, †*Gigantopithecus,* of China (Pleistocene) is placed in this subfamily, and Simpson recognizes the following 2 additional subfamilies of fossils. (3) †Dryopithecinae: Europe, Africa, and India (Miocene and Pliocene); 8 listed genera. This subfamily is probably a heterogeneous one which represents a stage in evolution preceding and leading to the existing Ponginae (Simpson). (4) †Australopithecinae: South Africa (Pleistocene); one genus. (Le Gros Clark assigns this subfamily to the Hominidae—see Chapter 11.) Besides these, Simpson lists "Possible pongids of uncertain affinities" fossil in Burma (Upper Eocene) and East Africa (Lower Miocene).

Hominidae, men: cosmopolitan; one existing genus and species; and fossil (an undetermined number of genera and species) in Africa, Eurasia, and Java (Pleistocene). For further details see Chapter 11.

†*Order Tillodontia*

†Tillotheriidae: North America (Upper Paleocene to Middle Eocene) and Europe (Lower Eocene); 4 listed genera.

†*Order Taeniodonta*

†Stylinodontidae: North America (Paleocene and Eocene); 7 listed genera.

Order Edentata, edentates: apparently always confined to the New World

†Metacheiromyidae: North America (Upper Paleocene to Middle Eocene); 2 genera. †*Metacheiromys* is known from nearly complete skeletons. Zoogeographically, these are among the most important of all fossil mammals. They show that, early in the Tertiary, edentates were in North America, from where they may have reached South America.

†Epoicotheriidae: North America (Lower Oligocene); 2 genera.

(†*Superfamily Megalonychoidea*), ground sloths

†Megalonychidae: South America (doubtfully Upper Eocene and certainly Oligocene to Pleistocene), North America (Middle Pliocene to Pleistocene), and West Indies (Cuba, Hispaniola, Puerto Rico) (Pleistocene); 26 listed genera. At least 2 different stocks of this family reached North America in the Pliocene and Pleistocene; †*Megalonyx* got north to the Northwest Territories of Canada (Stock and Richards 1949).

†Megatheriidae: South America (Lower Miocene to Pleistocene) and North America (Pleistocene); 8 genera.

†Mylodontidae: South America (Oligocene or Lower Miocene to Pleistocene) and North America (Pleistocene); 13 genera.

(*Superfamily Myrmecophagoidea*)

Myrmecophagidae, anteaters: tropical America; 3 existing genera, 3 or more species. All the genera occur in both South and Central America, 2 reaching northern limits in southern Mexico, and 2 reaching southern limits in northern Argentina; and 5 or more fossil genera occur in South America (Miocene and Pliocene).

(*Superfamily Bradypodoidea*)

Bradypodidae, tree sloths: tropical America; 2 genera, few species. *Bradypus* (three-toed sloths) ranges from northern Argentina to Honduras; *Choloepus* (two-toed sloths), across northern South America and into Central America to Nicaragua. No fossil tree sloths are known.

(*Superfamily Dasypodoidea*)

Dasypodidae, armadillos: South and Central America, north into southeastern United States; 9 existing genera, a moderate number of species. All the genera occur in South America, only 2 extending into Central America: *Cabassous* reaches Honduras, and *Dasypus* continues through Central America and Mexico to Texas and has recently extended its range to adjacent parts of Oklahoma and Louisiana and perhaps farther eastward around the Gulf of Mexico. Southward, several genera including *Dasypus* reach the pampas of central Argentina, and *Zaedyus* reaches Chubut. Fossil armadillos are numerous (21 listed genera) in the Tertiary and Pleistocene of South America, the oldest in the Paleocene. *Dasypus* reached southern North America in the Pleistocene (if not before), and the very large †*Holmesina* was in Florida then.

†Peltephilidae: South America (Oligocene to ?Lower Pliocene); 5 listed genera.

(†*Superfamily Glyptodontoidea*)

†Glyptodontidae, glyptodonts: South America (Upper Eocene to Pleistocene) and North America (Upper Pliocene and Pleistocene); 37 genera. At least 2 different stocks reached North America.

Order Pholidota

Manidae, scaly anteaters, pangolins: much of Africa below the Sahara (but not far south of the Orange River and Zululand), and the Oriental Region from peninsular India and Ceylon, Sikkim, and southern China (Yunnan to Fukien and north to Kiangsu) to Java, Bali, Borneo, and Palawan etc.; one genus, with 3 or 4 African and 3 Oriental species, and fossil in the Pleistocene in Asia and doubtfully in Europe.

Doubtful Pholidota: Europe (Oligocene and Miocene); 3 genera.

Order Lagomorpha, rabbits etc.

†Eurymylidae: Asia (Upper Paleocene); one genus.

Ochotonidae, pikas: temperate Eurasia and western North America; one genus, about 12 Eurasian and 2 North American species. In Eurasia, pikas occur from eastern Russia (not now the rest of Europe) north to the arctic coast, east to the Anadir etc. and northern Japan (Hokkaido), and south to the Kopet-Dag Mountains etc., northern India, and northern Burma; most species are in the mountains of south-central Asia. In North America pikas occur rather locally and discontinuously from south-central Alaska to central-interior California and northern New Mexico, mostly in mountains. The existing genus *Ochotona* is fossil in the Pliocene and Pleistocene of Eurasia (it once extended west to England) and in the Pleistocene of North America; other, extinct, genera are fossil in Eurasia (back to the Upper Oligocene) and Southwest Africa (Lower Miocene).

Leporidae, rabbits etc.: Africa (not Madagascar), Eurasia, and North, Central, and South America; 9 existing genera, numerous species. Northward, *Lepus* (hares) reaches the coast of the Arctic Ocean in Eurasia and North America, and also the Arctic Archipelago and the north coast of Greenland. Southward, leporids reach the southern tip of Africa; India and Ceylon; Siam and Indochina, with an endemic genus isolated on Sumatra (and a hare has apparently been introduced into West Java from Ceylon); and, in America, northern Argentina. The only genus of the family in Central and South America is *Sylvilagus* (cottontails), which extends from the southern edge of Canada to northern Argentina. *Pronolagus* (South Africa north to Kenya, usually in rocky places), *Pentalagus* (confined to the Riu Kiu Islands south of Japan), and *Romerolagus* (known only from the middle slopes of Mts. Popocatepetl and Iztaccihuatl in central Mexico) are supposedly relicts of a formerly numerous (10 listed fossil genera), primitive subfamily, Palaeolaginae, which dates from the Upper Eocene (of North America); whether or not *Nesolagus* (isolated on Sumatra) belongs to it I do not know—this genus was unknown to Dice (1929). Modern rabbits (fossils assigned to existing genera of the subfamily Leporinae) appear in the Pliocene (of Europe and doubtfully North America). The European rabbit (*Oryctolagus*) has been introduced into Australia, with disastrous results; and some other leporids have been introduced far outside their natural ranges.

Order Rodentia, rodents: every order of mammals is to some extent mysterious and involves taxonomic problems that cannot now be satisfactorily settled, but no other order can compare with the Rodentia in difficulty. Most rodents are small, many are obscure and rare, and their relationships are confused by an intricate web of convergence, divergence, and parallelism, 'and their fossil record is (among mammals) peculiarly unsatisfactory (paraphrased from Simpson 1945, p. 197). All this makes doubt and difficulty in working out their distributions and histories. Nevertheless rodents are important in zoogeography, especially because of their occurrence in Australia and on many islands.

(*Superfamily Aplodontoidea*)

†Ischyromyidae (including †Paramyidae), ancestral rodents: North America (Upper Paleocene to Lower Miocene), Europe (Eocene), and Asia (Oligocene); 23 listed genera.

Aplodontidae, sewellels: one existing genus and species, in western North America, from southern British Columbia (north to Hope and the Chilliwack-Sumas region) south at least to Mono County in the Sierra Nevada of California and scattered localities near the coast nearly to San Francisco (to Marin County); and 5 listed fossil genera in North America (back to the Upper Eocene) and one in Asia (Mongolia, Lower Pliocene).

†Mylagaulidae: North America (Lower Miocene to Middle Pliocene); 6 listed genera.

†Protoptychidae: North America (Upper Eocene); one genus.

†Eomyidae: North America (Upper Eocene to Middle Oligocene) and Europe (Upper Eocene to Lower Miocene); 6 listed genera.

(*Superfamily Sciuroidea*)

Sciuridae, squirrels etc.: numerous and diverse in Africa, temperate and tropical Eurasia, and North America; fewer and less diverse (only more or less typical squirrels) in most of Central America and in South America; and absent in Madagascar and the Australian Region; 2 subfamilies. (1) Sciurinae, ground and tree squirrels: range of the family; 30 or more genera, many species. Northward, most northern sciurids stop at the limit of trees, but *Citellus* (susliks and American ground squirrels) extends onto parts of the arctic tundra in both Asia (Berg 1950, p. 21) and North America, and *Marmota* (marmots and woodchucks) commonly extends above the tree line on mountains. Southward, sciurines reach all Africa, India and Ceylon, Java, Bali, Borneo, Celebes, and the southern and eastern Philippines (Palawan etc., and Mindanao, Leyte, and Samar); and in America, northern Argentina. The most widely distributed genus, *Sciurus,* occurs across temperate Eurasia and in North, Central, and South America. Three genera of northern ground squirrels etc. (*Citellus* and *Marmota,* mentioned above, and *Eutamias,* Siberian and western American chipmunks) occur in parts of both Eurasia and North America. All other genera of the family are confined to either the Old or the New World, and most of those of Eurasia and Africa are different. Simpson lists no extinct fossil genera of squirrels, but *Sciurus* in a very broad sense is fossil in Europe and North America back to the Miocene, and some other existing genera go back at least to the Pliocene. (2) Petauristinae, flying squirrels: Eurasia and North America, from wooded areas in the arctic south to India and Ceylon, Java and Borneo and some southern Philippines, and to discontinuous areas in Mexico, Guatemala, and Honduras; 11 genera in Eurasia, one in America, a moderate number of species; unknown fossil before the Pleistocene (of North America).

(*Superfamily Geomyoidea*)

Geomyidae, pocket gophers: North and Central America, from the plains of Alberta, Saskatchewan, etc. (but north only to Georgia in the east) to Panama; 9 or fewer existing genera, numerous species; and fossil (7 or more genera, in 2 subfamilies) in North America since the Lower Miocene or earlier.

Heteromyidae, pocket mice, kangaroo mice and rats: western North America (north to southern Canada and east at least to the Dakotas and Texas), Central America, and northern South America (parts of Colombia, Ecuador, and Venezuela, and Trinidad); 5 existing genera, numerous species; and fossil (9 listed genera) in North America back to the Lower Miocene or earlier.

(*Superfamily Castoroidea*)

Castoridae, beavers (with their ancestors and extinct relatives, most of which were probably not beaver-like in life): temperate Eurasia and North America; one existing genus, one Eurasian and one slightly different American species, various subspecies. The Eurasian beaver formerly extended across much of temperate Europe and Asia but is now restricted to a few localities in Scandinavia, France (the Rhone and tributaries), Germany (the middle Elbe and tributaries), eastern Poland, parts of Russia, across the northern Ural Mountains, on the Upper Yenisei (extinct?), and in northern Mongolia (see Harper, 1945, for further details). The American beaver formerly occurred through most of North America from Alaska and Labrador (the northern limit of forest) to the Rio Grande and northern Florida; it was at one time extirpated in much of this range but has been reintroduced in many places especially in the east, even within a short drive of New York City (see G. M. Allen, 1942, for details). The existing genus *Castor* is fossil back to the Pliocene; of 13 other recognized, extinct genera of the family, the oldest are in the Lower Oligocene of North America, the Upper Oligocene of Europe, and the Upper Miocene of Asia.

†Eutypomyidae: North America (Oligocene); one genus.

(*Superfamily Anomaluroidea*)

Anomaluridae (no acceptable English name—they are not "squirrels" of any kind): tropical Africa; 3 genera, few species. The relationships of this family are doubtful, and fossils sometimes assigned to it are doubtful.

(*Doubtful Anomaluroidea*)

†Pseudosciuridae: Europe (Upper Eocene to Middle Oligocene); 3 listed genera.

†Theridomyidae: Europe (Upper Eocene to Middle Oligocene); 6 listed genera; and 9 doubtfully assigned genera from Africa and Asia (Oligocene and Lower Miocene).

Pedetidae, spring haas: South Africa etc. north to Kenya; one existing genus, 2 species; and another genus fossil in Southwest Africa (Lower Miocene).

(*Superfamily Muroidea*)

Cricetidae, diverse mice and rats: Africa, Madagascar, most of Eurasia, and the Americas; north to the limit of land, islands in the Arctic Ocean, and northern Greenland (lemmings); and south to the tip of Africa, India and Ceylon (Gerbillinae only), northern Burma and southeastern China (not the Indochinese Subregion), Formosa, and to the southern tip of South America and Tierra del Fuego; 5 subfamilies. (1) Cricetinae, Old World hamsters, most New World mice and rats, etc. Hamsters etc. (tribe Cricetini) occur across temperate Eurasia (5 existing genera) and isolated in South Africa (one genus); they are fossil (several genera) back to the Lower Pliocene of Eurasia. Mole mice (Myospalacini) occur in central and eastern Siberia and northern and central China (one genus, 5 species), and fossil (an additional genus) back to the Upper Miocene of Asia. All other existing Cricetinae form a tribe, Hesperomyini, of about 50 genera and many species, confined to the New World (about 40 of the genera confined to Central and South America), and including the rice, cotton, wood, and pack rats, and the harvest, white-footed, and grasshopper mice, etc., and occurring from parts of Alaska and Labrador to Tierra del Fuego (but only a few reach these extremes), with outlying forms on the Galapagos, Jamaica, and the Lesser Antilles; and fossil (a few additional genera) back to the Lower Pliocene of North America. Three additional fossil tribes with 12 listed genera, and also various unplaced and doubtful fossil genera of this subfamily, occur in the Oligocene and later, in Eurasia and North America. (2) Nesomyinae: Madagascar; 7 existing genera, 12 listed species; several of the existing genera and one additional extinct one are fossil (Pleistocene) or subfossil in Madagascar. Ellerman (1940–1949) scatters the genera among several subfamilies, but Simpson (1945, pp. 86 and 207) brings them together again. I have followed Simpson, but I do not know the rights of the matter. Ellerman's arrangement would imply that the "Nesomyinae" consist of about 5 different stocks which reached Madagascar independently. (3) Lophiomyinae, maned rats: tropical Africa; one genus, about 11 species. (4) Microtinae, lemmings, voles, etc.; 3 tribes. Lemmings (tribe Lemmini) occur from the highest arctic of Eurasia and North America south to the Siberian Altai and to parts of Kansas and Virginia; 4 existing genera, few species; *Dicrostonyx* and *Lemmus* occur in both Eurasia and North America; this tribe is fossil back to the Upper Pliocene (of North America). Some northern lemmings make spectacular, periodic mass emigrations (Elton 1942). The tribe Microtini includes the dominant genus *Microtus* (common voles and field mice), with about 260 named species and subspecies, occurring across Eurasia and North America, from the arctic south to North Africa (Cyrenaica), Palestine, Iran, the

Himalayas, eastern China north of the Yangtze, perhaps Formosa, and southern Mexico and Guatemala; most of the 15 other existing genera of the tribe are localized within the range of *Microtus,* but some extend slightly beyond it in North India, North Burma, and southeastern China. Besides *Microtus,* the following genera of the tribe are common to Eurasia and North America: *Clethrionomys* (red-backed mice) in cool-temperate and arctic regions of both hemispheres, *Pitymys* (pine mice) in widely separated areas in continental Europe to Asia Minor and the Caucasus and in *eastern* United States and east-central Mexico, and *Lagurus* from South Russia to Mongolia and in the interior of western North America. The North American Muskrat (*Ondatra*) is a specialized member of this tribe; it has been introduced into Europe. Fossil Microtini include some existing and 11 additional listed fossil genera and go back to the Upper Miocene of North America and the Pliocene of Eurasia. Mole lemmings (Ellobiini) occur in Eurasia from southern Russia and Asia Minor to Mongolia; one genus, 4 species; and another genus is fossil in Europe (Upper Pliocene). (5) Gerbillinae, gerbils etc.: throughout Africa in suitable places (but not in the heavy forests of West Africa), and from southwestern Asia to the Caucasus, adjacent parts of Russia, central Asia to Mongolia etc., and peninsular India and Ceylon; 10 genera, numerous species; the existing genus *Gerbillus* is listed fossil in the Pliocene of Asia.

Spalacidae, mole rats: from southeastern Europe (Hungary etc.) to the Caspian, Syria, Palestine, Egypt, and Libya; one genus, 2 or 3 species; and 2 fossil genera in Europe (Pliocene).

Rhizomyidae, bamboo rats etc.: eastern tropical Africa north to Abyssinia (one genus, few species) and from the eastern Himalayas and southern China to Sumatra (2 genera, 4 species); and 5 listed fossil genera in Europe and Asia (Upper Oligocene etc.) and North Africa (Pleistocene).

Muridae, typical mice and rats: native throughout the Old World probably excepting Madagascar, introduced into the New World; 6 subfamilies. (1) Murinae: the Old World; 68 existing genera, hundreds of species; many in all Africa (but probably none native in Madagascar); many in temperate and tropical Eurasia and associated islands, a few reaching the arctic but none widely distributed or especially characteristic there; many through the Indo-Australian Archipelago to Australia and Tasmania, and east to the Philippines and Solomons; small rats of the *Rattus concolor-exulans* group, which are house rats to some extent, are widely distributed farther east in the Pacific (probably carried by early Polynesians) as far as Hawaii, the Marquesas, and New Zealand; and 2 rats and one mouse have been carried over the whole world by commerce. *Rattus rattus* (the Black Rat) is native apparently in tropical Asia etc.; *R. norvegicus* (the Norway Rat), apparently in cool-temperate Asia; both are now nearly world-wide, but the Black Rat prefers warmer and the Norway Rat cooler places, and the details of their distribution depend not only on climate but also partly on competition with each other and with native animals in different places. *Mus musculus* (the House Mouse) is apparently represented by wild forms in southern Europe, North Africa, and central Asia to Manchuria and Japan, but "commensal" forms are of course now world-wide. *Rattus,* with more than 550 recognized forms (Ellerman)—more than any other genus of mammals—occurs naturally throughout the range of the family; not only native species but endemic species groups of this genus occur in every part of the Old World within the limits of the family (see Ellerman 1941, pp. 149–150). Most of the other genera are restricted to one or another of the conventional faunal regions of the Old World or to part of one region or to smaller natural areas, many to single islands or island groups in the Oriental-Australian area. This great subfamily is known fossil (only 5 listed fossil genera) only in the Pliocene (of Eurasia) and Pleistocene. (2) Dendromurinae, African tree mice: Africa south of the Sahara; 6 genera, more than 50 listed species. (3) Otomyinae: Africa below the Sahara; one genus, 11 species. (4) Phloeomyinae, giant Philippine rats, etc.: Oriental-Australian area, from Assam, Burma, and Yunnan to Java, Borneo, Celebes, and some Philippines; and on New Guinea, New Britain, and associated small islands, and perhaps Flores; 6 existing genera, more than 20 listed species; and one fossil genus on Timor (Pleistocene?); but it is very doubtful whether all these genera really belong together. (5) Rhynchomyinae, shrew rats: known only from Mt. Data, northern Luzon, Philippine Islands; one species. (6) Hydromyinae, Australian water rats, etc.: Australia and Tasmania, New Guinea, New Britain, and associated small islands (the Aru and Kei Islands, etc.); 9 genera, most in New Guinea, including terrestrial as well as semi-aquatic and specialized aquatic forms, 13 species (Tate 1951). And northern Luzon in the Philippines; 2 non-aquatic genera, 2 species; somewhat doubtfully related to the Australian forms.

(*Superfamily Gliroidea*)

Gliridae (Muscardinidae), dormice etc.: 2 subfamilies. (1) Glirinae: temperate Eurasia etc. (discontinuous); in most of Europe, North Africa, and southwestern and central Asia (5 genera, 6 species); and Japan (one genus and species); and fossil (4 additional genera) in Europe since the Miocene or earlier. (2) Graphiurinae: Africa from the southern Sahara to the Cape; one genus, 39 listed species.

Platacanthomyidae, spiny dormice: localized in tropical Asia; one genus and species in southern peninsular India, and another genus and species in southeastern China and northern Indochina.

Seleviniidae: west-central Asia (west of Lake Balkash); one species.

(*Superfamily Dipodoidea*)

Zapodidae, jumping mice etc.: temperate Eurasia and North America; 2 subfamilies. (1) Sicistinae (not modified for jumping): parts of Scandinavia and central Europe through central Asia to western China, Manchuria, and Sakhalin Island; one existing genus, 6 species; and 2 fossil genera in Eurasia (Oligocene and Lower Pliocene). (2) Zapodinae (jumping mice): western China (Kansu and Szechwan, one very distinct species sometimes put in a genus of its own, sometimes included in an American genus) and North America from parts of Alaska and Labrador south to parts of California, New Mexico, and the southern Appalachians (2 closely related genera, several species); and a fossil genus in North America in the Pliocene.

Dipodidae, jerboas: drier parts of northern Africa and temperate Eurasia, from the Sahara etc., Arabia, and southern Russia through southwestern and central Asia to Mongolia etc.; 12 existing genera (in 3 subfamilies), a moderate number of species; the family is much more diverse in Asia than in Africa; and 4 fossil genera in Asia, back to the Upper Miocene.

(*Suborder Hystricomorpha*) Most of the remaining families of rodents are hystricomorphs or have been so considered. Most of them are South or Central American or West Indian, with only a porcupine in the main part of North America. Certain Old World (chiefly African) groups have been or still are associated with the American ones, but Simpson (1945, pp. 210ff.) considers them doubtful and Wood (1950) thinks their similarity may be due to parallelism rather than relationship. See text (pp. 360–361) for further discussion.

(*Superfamily Hystricoidea*)

Hystricidae, Old World porcupines: main warm parts of the Old World (not the Australian Region); north to Morocco, Italy (introduced?), Syria, Russian Turkestan, and South China (south of the Yangtze), and south through most of Africa (not Madagascar), peninsular India and Ceylon, Java, the Lesser Sundas to Sumbawa (and perhaps Flores), Borneo, and Palawan etc. in the southern Philippines; 5 genera, about 18 species. Old World porcupines are terrestrial (the New World ones are arboreal). Fossil hystricids include an extinct genus and go back to the Oligocene (of Europe).

(*Superfamily Erethizontoidea*)

Erethizontidae, New World porcupines: in South America there are 2 localized monotypic genera plus *Coendou* (several species), which occurs south to the northern corner of Argentina and north to Central America and part of Mexico; and in North America there is *Erethizon* (one or 2 species), from the northern limit of forest south about to the Mexican border in the west and at least to southern Pennsylvania in the east. Seven additional fossil genera are listed from South America in the Oligocene and later; *Erethizon* appeared in North America in the Upper Pliocene.

(*Superfamily Cavioidea*)

†Cephalomyidae: South America (Oligocene); 2 genera.

†Eocardiidae: South America (Upper Oligocene to Lower Miocene); 5 listed genera.

Caviidae, guinea pigs etc.: South America, especially the southern part from Patagonia north to 15° S. latitude, only *Cavia* (typical guinea pigs) extending to northern South America; 5 existing genera, a moderate number of species; and 9 additional listed fossil genera in South America (Pliocene).

Hydrochoeridae: tropical South America (south to the delta of the Paraná in Argentina) and Panama; one existing genus, 2 species (capybaras, the largest existing rodents, semi-aquatic); and 10 fossil genera (in 2 subfamilies) in South America in the Pliocene and Pleistocene. Two genera of the family extended to southeastern North America in the Pleistocene, both reaching South Carolina.

Dinomyidae, false pacas: western South America (central Colombia to central Peru etc.); one species.

†Heptaxodontidae, extinct giant rats: South America (Miocene and Pliocene, 19 listed genera, in 3 subfamilies) and West Indies (Pleistocene and sub-Recent, a fourth subfamily, 7 genera including †*Quemesia,* from Jamaica, Hispaniola, Puerto Rico, and Anguilla and St. Martin—but none on Cuba).

Dasyproctidae, agoutis, pacas: South and Central America, south to southern Brazil and northern Argentina and north (2 genera) into southern Mexico, and *Dasyprocta* also reaches some of the Lesser Antilles; 4 genera, a moderate number of species; apparently not known fossil.

(*Superfamily Chinchilloidea*)

Chinchillidae, vizcachas, chinchillas: western and southern South America, from Peru, Bolivia, and northern Argentina to or nearly to the Straits of Magellan; 3 existing genera, about 7 species; and 7 listed fossil genera from South America (Oligocene and later). True chinchillas (*Chinchilla*) occur from the coastal hills of northern Chile to high altitudes in the Andes of Peru, Bolivia, and northwestern Argentina but have been much reduced by trappers (G. M. Allen 1942, pp. 389ff.).

(*Superfamily Octodontoidea*)

Capromyidae, hutias etc.: the Greater Antilles etc. in the West Indies, and parts of South America; 4 or 5 existing genera. The West Indian forms (below), including the extinct ones, are arranged after Williams (1951 MS. list; later unpublished information; and 1953, with Reynolds and Koopman). *Capromys* (4 existing species) is on Cuba and the Isle of Pines, and a fossil species is on the Cayman Islands. *Geocapromys* (3 existing species) is on the Bahamas, Jamaica, and Swan Island, and is fossil on Cuba as well as Jamaica and the Bahamas. The extinct †*Hexolobodon* (one species) on Hispaniola is related to the 2 preceding genera. *Plagiodontia* (2 existing and 2 extinct species), on Hispaniola, represents another group of genera to which belong also †*Aphaetreus* (one species) on Hispaniola and †*Isolobodon* (†*Ithyodontia*) (one species) on Hispaniola, Mona, Puerto Rico, and the Virgin Islands (perhaps carried by prehistoric man). All the extinct West Indian forms are Pleistocene or post-Pleistocene. *Procapromys* (one species) is supposed to be from the north coast of South America, from the mountainous central coastal region of Venezuela, and supposedly represents the mainland stock from which the West Indian *Capromys* etc. came, but this has apparently not been confirmed. The genus was not seen by Ellerman (1940). It has been suggested that it may be just a young West Indian *Capromys* with a wrong locality. *Myocastor* (one species), the semi-aquatic Nutria or Coypu, is in southern South America from extreme southern Brazil etc. to or nearly to the Straits of Magellan; it has been introduced in south-central and (locally) northwestern United States and in France. *Myocastor* differs considerably from the living West Indian capromyids but is apparently connected with them by fossils (Simpson). Seven fossil genera of the family are listed from South America (Lower Miocene to Pliocene).

Octodontidae, degus: southern South America, from southern Peru, Bolivia, and southern Brazil far southward, but apparently not reaching the Straits of Magellan; 5 existing genera, 8 species; and 3 fossil genera in South America (Upper Oligocene and Pliocene) and one (†*Alterodon*) on Jamaica (Pleistocene).

Ctenomyidae, tucu tucus: southern South America, from southern Peru, Bolivia, and southern Brazil to Tierra del Fuego; one existing genus, numerous species and subspecies; and 4 fossil genera in the Pliocene of South America.

Abrocomidae, rat chinchillas: southwestern South America, from southern Bolivia etc. to central Chile; one genus, few species; and one fossil genus in South America (Pliocene).

Echimyidae, spiny rats: tropical South and part of Central America; 14 existing genera, a moderate number of species; all the genera occur in some part of tropical South America, a few south to southern Brazil etc.; 2 extend into Central America to Nicaragua, and at least one more reaches Panama; and an *Echimys* is recorded from Martinique in the Lesser Antilles. About 7 additional genera are fossil in South America (Upper Oligocene and Lower Miocene), and 4 on the Greater Antilles (Pleistocene or post-Pleistocene), as follows: †*Boromys* (2 species), Cuba; †*Brotomys* (2 species), Hispaniola; †*Homopsomys* (one species) and †*Heteropsomys* (one species), Puerto Rico. And *Proechimys,* which now occurs from southern Brazil to Nicaragua, is apparently represented by a fossil (probably Pleistocene) on Puerto Rico (Williams and Koopman 1951).

Thryonomyidae, cane rats: most of Africa below the Sahara; one genus, about 7 species; and one fossil genus (†*Paraulacodus* Hinton 1933) in northern India (Pliocene). Three other fossil genera, from Africa (Miocene and Pliocene), are doubtfully assigned to this family.

Petromyidae, rock rats: southwestern Africa (Little Namaqualand to southern Angola); one genus, one or 2 species, several subspecies.

(*Doubtful Hystricomorpha*)

(*Superfamily Bathyergoidea*)

Bathyergidae, mole rats: Africa below the Sahara; 5 genera, a moderate number of species; and one fossil genus in the Pleistocene of South Africa.

(*Superfamily Ctenodactyloidea*)

Ctenodactylidae, gundis: northern Africa, from Senegal and Morocco (including the Palearctic coastal area) to Somaliland (Ellerman); 4 genera, about 8 species; one of the genera (*Pectinator*) is doubtfully represented in the Pliocene of northern India (Hinton 1933).

Order Cetacea, whales, dolphins, porpoises: primarily marine (all oceans), but some Delphinidae (common dolphins etc.) enter or live in fresh water (*e.g.,* species of the marine genus *Sotalia* in the Amazon), and the following family is almost confined to it. The strictly marine families of this order are omitted here.

Platanistidae, river dolphins: 4 existing monotypic genera, in (1) India (in the Indus, Ganges, and Brahmaputra Rivers), (2) China (Tungting Lake, on Yangtze drainage), (3) the Amazon and Orinoco systems in South America, and (4) the La Plata (estuarine but mainly fresh) and adjacent coastal (salt) waters in South America. No existing member of the family is wholly marine, but the existing fresh-water genera are relicts of a marine family (Miocene and Pliocene) which has been replaced in the sea.

Order Carnivora, placental carnivores.

(†*Suborder Creodonta*)

†Arctocyonidae, †Mesonychidae, †Oxyaenidae, †Hyaenodontidae, †Creotarsidae; with a total of 68 listed genera: North America and Eurasia (numerous in the Paleocene and Eocene, fewer in the Oligocene etc.) and Africa (few known, Lower Oligocene and Lower Miocene).

(*Suborder Fissipeda*)

†Miacidae: North America (Middle Paleocene to Upper Eocene) and Eurasia (Upper Eocene and perhaps Lower Oligocene); 12 listed genera.

Canidae, foxes, wolves, dogs: cosmopolitan but not native on Madagascar or the West Indies etc. and probably not in the Australian Region; 12 existing genera, a moderate number of species. Northward, both foxes and wolves reach the limit of continental land, the Arctic Archipelago, the north coast of Greenland, etc.; southward, different canids reach the tip of Africa, Australia (but they probably are not native beyond Java in this direction), and Tierra del Fuego and the Falkland Islands off the tip of South America. Typical foxes (*Vulpes*) occur in parts of Africa, through the whole of temperate Eurasia and south to peninsular India and northern Indochina, and in most of North America south about to the United States–Mexican border. The foxes of Mexico and Central and South America belong to other genera. *Vulpes vulpes,* the Red Fox of Europe, occurs across the whole north-temperate part of Eurasia south to North Africa, Arabia, northwestern India, and northern Indochina, and north into parts of the arctic (but not into the interior of the tundra), and the Red Fox of North America is sometimes considered only a subspecies of it, although it is more often treated as a related species. The Arctic Fox (*Alopex lagopus*) occurs in the arctic in both hemispheres, from Norway across northern Eurasia and North America to Greenland, Iceland, etc. Wolves and their allies (*Canis,* including jackals, coyotes, true dogs, etc.) occur naturally almost throughout Africa, Eurasia south to Ceylon and Siam, and North and Central America to Costa Rica, but not in South America (and South American Pleistocene fossils assigned to *Canis* probably do not belong to it in a strict sense–Simpson). The Wolf (*Canis lupus*) formerly occurred all the way across Eurasia including the British Isles but is now extinct in western Europe except locally in parts of Portugal, Spain, Italy, Sicily, and Sweden and Norway; the western limit of eastern wolves fluctuates, for they wander, but they reach East Prussia, Poland, Czechoslovakia, and parts of the Balkans, and wolves still occur in much of northern and central Asia east to

Japan (if not extinct there) and south to Arabia, at least to Dharwar in India (several degrees within the tropics), and parts of China; and forms of the same species formerly occurred almost throughout North America from the high arctic to the tableland of Mexico and are still common in the far north, but elsewhere in North America they are now confined to a few isolated, wild places. Domestic dogs were derived from the Wolf, presumably in Eurasia, in prehistoric times (jackals too may have been concerned in the origin of dogs), and have been carried over the world by man; the ancestors of the Dingo on Australia and of "wild" dogs on New Guinea etc. were presumably brought from Asia by prehistoric man. *Lycaon* (the Hunting Dog), widely distributed in Africa, *Cuon* (the Dhole or Red Dog), widely distributed in southern and eastern Asia north to the Amur region and south to India and Java, and *Speothos* (the Bush Dog), of tropical South America from Ecuador and the Guianas to Paraguay etc., are the only surviving members of a subfamily (Simocyoninae) which was well represented in North America and Europe in the mid-Tertiary. Other fossils, representing more than 50 extinct genera and including several extinct subfamilies, show the complexity of evolution of the Canidae, beginning in the Upper Eocene of North America and Europe.

Ursidae, bears: Eurasia and North and South America; 6 genera (one arctic, one Eurasian and North American, 3 Asiatic, and one South American), probably only 7 or 8 distinct living species. The Polar Bear (*Thalarctos*) occurs on the arctic coasts, islands, and ice of Eurasia and North America. In the Old World bears occur southward, often very discontinuously, to northern Spain (reputedly formerly North Africa), parts of Italy and Greece, Syria etc., Iran, India and Ceylon, and Sumatra and Borneo. In North America, the Black Bear (*Ursus americanus*), of which the Cinnamon Bear etc. are color phases, occurs across the continent in the north, and south in suitable areas to Florida and northern Mexico; the Grizzly (*U. horribilis*) is confined to the West, from Alaska to the Rockies south to New Mexico, and it or related species formerly reached California and northern Mexico; and Big Brown Bears (*U. middendorffi*) are confined to the south coast of Alaska and adjacent islands. There are no bears in Central America. In South America, the Spectacled Bear (*Tremarctos*) inhabits wooded slopes of the Andes in western Venezuela, Colombia, Ecuador, Peru, and Bolivia. Fossil bears (8 listed genera) are known back only to the Middle Miocene of Europe and Lower Pliocene of North America; 2 or 3 of the fossil genera were common to Eurasia and North America.

Procyonidae, raccoons etc., pandas: the Americas and eastern Asia; 2 subfamilies. (1) Procyoninae, raccoons etc.: the Americas; 6 existing plus 8 listed fossil genera. *Procyon* (raccoons, 2 main species) extends from the southern edge of Canada to northern Argentina etc., the North American species reaching Panama, and the South American one, Costa Rica, and isolated populations are or have been (introduced?) on New Providence Island in the Bahamas and Guadeloupe and Barbados in the Lesser Antilles. The 5 other existing genera of the family occur in tropical South and/or Central America (not West Indies), 3 of them reaching Mexico, and 2 of the 3 extending into parts of southwestern United States. The subfamily is fossil in North America since the Lower Miocene and in South America since the Upper Miocene, before the formation of the present land connection. (2) Ailurinae, pandas: localized in eastern Asia, the Red Panda (*Ailurus*) in southwestern China (Yunnan and Szechwan) and northern Burma, Sikkim, and Nepal, and the Giant Panda (*Ailuropoda*) in central and western Szechwan at altitudes up to 12,000 feet, in bamboo forests; 2 other genera are fossil in the Upper Miocene and Lower Pliocene of Europe and Lower Pliocene of Asia.

Mustelidae, weasels etc.: all principal parts of the world except Madagascar and the Australian Region, north to the high arctic, and south to the Cape of Good Hope, southern India and (otters only) Ceylon, and Java, Borneo, and Palawan etc., and to southern South America and (otters only) Tierra del Fuego; 6 subfamilies. (1) Mustelinae, weasels etc.: almost the whole range of the family, but not Ceylon or Tierra del Fuego; 12 existing genera, a moderate number of species. *Mustela* (various weasels, mink, etc.) occurs across the north-temperate and arctic parts of Eurasia and North America and south to North Africa, northern India, Java and Borneo and perhaps Palawan, and part of South America (see below); *M. erminea,* the Ermine or Short-tailed Weasel, is common to the cool-temperate and arctic parts of Eurasia and North America, and other species pairs in the genus show close relationships between these areas; *M. frenata,* the Long-tailed Weasel, ranges from southern Canada to Venezuela, Colombia, Ecuador, Peru, and Bolivia; and another species of the genus is endemic in the Amazon region etc. The genus *Martes* (several species, martens, the Sable, the Fisher) and the species *Gulo gulo* (the Wolverine) are common to Eurasia and North America, including parts of the arctic. Other genera of the subfamily are confined to parts of Eurasia, or Africa, or South America in some cases north through Central America to southern Mexico. Fossil Mustelinae are numerous (21 listed genera, some of them divergent) and go back to the Lower Oligocene of Eurasia and North America. Three of the fossil genera are common to Eurasia and North America, and the distribution of existing forms indicates at least 6 more movements between these continents. (2) *Mellivorinae:* one existing species, *Mellivora capensis,* the Ratel or Honey Badger, which occurs through much of Africa (south to the Cape) and southwestern Asia to western and northern India, to Nepal. The same genus is listed fossil in the Middle Pliocene of Asia, and another genus, in the Lower Pliocene of Asia and North America. (3) Melinae, badgers: temperate Eurasia etc. (5 genera) south to Palestine, Iran (not India), and through Indochina etc. to Java, Borneo, and Palawan etc.; and western North America (another genus, one species) from southern Canada to northern Mexico; and fossil (5 listed genera) in Eurasia since the Miocene and North America since the Upper Pliocene. (4) Mephitinae, skunks: the Americas;

3 existing genera, a moderate number of species. *Mephitis* occurs from cold-temperate Canada to Honduras; *Spilogale,* from southern British Columbia and northern Virginia to northern Costa Rica; and *Conepatus,* from southwestern United States to the Straits of Magellan. Fossil Mephitinae (6 listed genera) are in Eurasia (Upper Miocene and Pliocene) as well as North America (Pliocene and later) and South America (Pleistocene). (5) †Leptarctinae: North America (Miocene and Lower Pliocene); 3 listed genera. (6) Lutrinae, otters: almost to the limits of distribution of the family excepting the coldest arctic; 7 existing genera. *Lutra* (typical otters, various species) has almost the distribution of the subfamily but does not reach the Philippines (another genus reaches Palawan). The other genera are confined to Africa, southern Asia etc., or South America, except that the Sea Otter (*Enhydra*) occurs coastwise in the North Pacific Ocean from the Bering Sea etc. south to southern California and (at least formerly) northern Japan (Hokkaido). Of 11 listed fossil genera of otters, the oldest is in the Upper Oligocene (of Europe).

Viverridae, civets, mongooses, etc.: much of the warm part of the Old World; 36 existing genera, many species. In more detail the family's range is Africa, Madagascar, the warmer part of Eurasia north to Spain, France etc., Palestine, Iran, Afghanistan, northern India, and southern China (and *Paguma* extends north to Peking), and south and east to Ceylon, the Lesser Sundas to Timor(?), Celebes, the Moluccas(?), the Aru and Kei Islands(?), and the Philippines (but perhaps not native on the outer ones). Southeastward, *Macrogalidia* is peculiar to Celebes; *Viverra tangalunga* (the Malay Civet) is listed east to the southern Moluccas; and *Paradoxurus hermaphroditus* (the Palm Civet), to the Moluccas, the Aru and Kei Islands, and the Lesser Sundas to Timor; but there is doubt about the natural limits of these species. On Madagascar there are 7 endemic genera of viverrids (including *Cryptoprocta,* sometimes placed in the Felidae) divided by Simpson (1945) among 3 subfamilies and 4 tribes; the arrangement implies that they are derived from more than one mainland stock, but this is not certain (Simpson, p. 229); *i.e.,* there is a possibility that all the Madagascan forms have a single ancestor. The Burmese form of the Small Indian Mongoose (*Herpestes auropunctatus birmanicus*) is introduced and widely established on the West Indies. Fossil viverrids (7 listed genera) go back to the Upper Eocene or Lower Oligocene (of Europe); there is no good evidence that any viverrid ever reached Australia or the Americas.

Hyaenidae, hyaenas etc.: Africa (including North Africa) and southwestern Asia, north to the west side of the Caspian and east to India (not Ceylon); 3 existing genera (including *Proteles,* the Aardwolf), 4 species; all are African except that *Hyaena hyaena* extends from North and East Africa into Asia to the limits given. Fossil hyaenids (3 extinct as well as 2 of the existing genera) are in Europe as well as Asia (Pliocene and Pleistocene), and the oldest known fossil of the family is in Asia (Upper Miocene, referred to an existing African genus); the family is apparently not known fossil in Africa.

Felidae, cats: all principal parts of the world except the Australian Region and Madagascar (*Cryptoprocta* is placed in the Viverridae) and certain other islands; 3 existing genera, many species. *Felis* in a broad sense includes all the smaller existing cats and the American Mountain Lion, and sets all the geographical limits of the family, occurring naturally through all Africa, Eurasia north at least to the wooded parts of the arctic and south and east to India and Ceylon, and to Java, Bali (but probably not Timor), Borneo, and some Philippines (not Luzon), and in America from parts of the arctic to the Straits of Magellan (but not the West Indies). The northernmost species is *Felis lynx,* the Lynx, which ranges across the forested parts of cold-temperate and arctic Eurasia and North America. *F. concolor,* the Mountain Lion, occurs from northern British Columbia east to southern Quebec and south to the Straits of Magellan, in a variety of country: arctic and temperate woods, tropical rain forest, mixed and broken country, and more or less open plains and semi-deserts, and from sea level to high mountains. The domestic cat may have a dual origin from *Felis silvestris* (the European Wild Cat) and *F. lybica* (the African Wild Cat); it has been carried to all parts of the world and has become wild (or "feral") again in some remote places. *Panthera* contains 6 "big cats." *Panthera leo,* the Lion, formerly occurred in all Africa, except the West African forests and full-scale deserts, and in southwestern Asia to part of India, but now in Africa it occurs only below the Sahara, south to Southwest Africa and Transvaal or perhaps a little farther, and in Asia it survives (under protection) probably only in the Gir Forest, Kathiawar, northwestern India. It occurred in Morocco until the 1920's and in Algeria and Tunis until about 1891, and in Cape Province, South Africa, until about 1858. In southwestern Asia a few lions may still exist in Iraq and Iran, but this is doubtful; they occurred within historic times also in Arabia, Palestine, Syria, and Asia Minor; and they apparently existed in Greece in classical times, until about A.D. 80–100. The known history of the Lion is therefore one of retreat before man during a period of nearly 2000 years. *Panthera tigris,* the Tiger, occurs in Asia, west through Iran nearly to the Caucasus, north to the Aral Sea in the west and to Manchuria and the Middle Amur in the east, and south to most of India (not Ceylon) and to southeastern Asia, Sumatra, Java, and Bali. *Panthera pardus,* the Leopard, occurs in Africa, both north and south of the Sahara, south to the Cape, and across southern Asia, north to the Caucasus etc. and to the Amur region in southeastern Siberia, and south to Ceylon and the Malay Peninsula and on Java. Also in Asia are *Panthera nebulosa,* the Clouded Leopard, from South China to Sumatra and Borneo, and *P. uncia,* the Ounce or Snow Leopard, in the high mountains of Central Asia. *Panthera onca,* the Jaguar, is in America, north (formerly) at least to the "mountainous parts of eastern Arizona north to the Grand Canyon, [and the] southern half of western New Mexico" and also (coastwise) to central Texas (G. M. Allen 1942, p. 253), and south formerly nearly to the southern tip of South America (at least to the upper Santa Cruz River in Patagonia) but now only to the Chaco, and Misiones in extreme northern Argentina

(Cabrera and Yepes). The final genus and species of existing felids is *Acinonyx jubatus,* the Cheetah or Hunting Leopard, which occurs in most of Africa including North Africa, formerly south to the Cape (but now extinct there), and into Asia north to southern Turkmenia and east (at least formerly) to northern and central India; a supposed second species of the genus, localized in Rhodesia, is doubtful; the genus is fossil in Europe in the Upper Pliocene. Fossil felids are diverse. Of the existing subfamily Felinae, all 3 existing genera and several additional fossil ones occur back to or in the Pliocene of Eurasia; the subfamily is unknown in Africa before the Pleistocene but presumably occurred there earlier; *Felis* and *Panthera,* the only genera of the subfamily known to have reached America, are not definitely known there until the Pleistocene. There are 3 additional, extinct subfamilies of felids, as follows. The †Proailurinae (4 or 5 genera) were in Eurasia in the mid-Tertiary. The †Nimravinae, false saber-tooths (9 listed genera), were in Eurasia, Africa, and North America at various times in the mid-Tertiary; †*Pseudaelurus* was in Africa in the Lower Miocene, in Europe through the Miocene, and in North America from the Upper Miocene to Middle Pliocene. The †Machairodontinae, saber-tooths (10 listed genera), were in Eurasia and North America from the Lower Oligocene to the Pleistocene, in Africa in the Lower Pliocene (and presumably at other times), and in South America in the Pleistocene; †*Eusmilus* was in Europe through the Oligocene and in North America in the later Oligocene; †*Machairodus,* in Europe in the Upper Miocene and in Europe, Asia, Africa, and North America in the Lower Pliocene; and †*Smilodon,* in North and South America in the Pleistocene.

(*Suborder Pinnipedia*), seals etc.: most seas, chiefly coastwise, especially numerous in cold northern and southern seas; 3 existing families (not listed here), 16 existing genera. The genus *Phoca,* which is primarily marine (northern seas), sometimes enters fresh water (as do a few other seals) and has populations, some considered endemic subspecies, in fresh-water lakes in Finland and northern Russia, Lake Baikal in Siberia, on Baffin Island, and in Seal Lake, Ungava, Canada (Wynne-Edwards 1952), as well as in the Caspian Sea, which is salt. The fossil record of seals begins in the Miocene.

(*Superorder Protungulata*)

†*Order Condylarthra,* archaic ungulates or preungulates

†Hyopsodontidae, †Phenacodontidae, †Periptychidae, †Meniscotheriidae, ?†Tricuspiodontidae (5 families, 33 listed genera): North America (most) and Europe (a few) (Paleocene and Eocene). Couto (1952) has suggested that †*Asmithwoodwardia* in the Paleocene of South America is a hyopsodontid rather than a didolodontid.

†Didolodontidae: South America (Paleocene and Eocene); 7 or 8 recognized genera. This family is placed among the preceding ones by Simpson (1945, pp. 124 and 235) but is separated here to emphasize its geographical isolation in South America.

†*Order Xenungulata,* extinct South American ungulates

†Carodniidae: South America (Paleocene); one or more genera (Couto 1952). These are, I think, the oldest known South American mammals.

†*Order Litopterna,* extinct South American ungulates

†Proterotheriidae: South America (Paleocene to Lower Pliocene); 23 listed genera.

†Macraucheniidae: South America (Paleocene to Pleistocene); 15 or more genera.

†*Order Notoungulata,* extinct, chiefly South American ungulates

†Arctostylopidae: Asia (Upper Paleocene) and North America (Lower Eocene); 2 genera. This family is here listed separately from the following ones to emphasize its occurrence outside of South America; it is not set apart by Simpson.

†Henricosborniidae, †Notostylopidae, †Oldfieldthomasiidae, †Archaeopithecidae, †Archaeohyracidae, †Isotemnidae, †Homalodotheriidae, †Leontiniidae, †Notohippidae, †Toxodontidae, †Interatheriidae, †Mesotheriidae, †Hegetotheriidae (13 families, various suborders and subfamilies, about 100 recognized genera): South America (Paleocene to Pleistocene, but each family existed through only part of this time).

†*Order Astrapotheria,* extinct South American ungulates

†Trigonostylopidae: South America (Paleocene and Eocene); one genus.

†Astrapotheriidae: South America (Eocene to Miocene); 7 listed genera, in 2 subfamilies.

Order Tubulidentata

Orycteropodidae, aardvarks: most of Africa below the Sahara; one genus, perhaps only one existing species, various subspecies; the same genus is fossil in the Pliocene of Eurasia (France, Greece, Iran, India); supposed orycteropodids have been described from the Oligocene and Miocene of Europe; and a doubtful tubulidentate is in the Lower Eocene of North America.

(*Superorder Paenungulata*)

†*Order Pantodonta*

†Coryphodontidae, †Barylambdidae, †Pantolambdodontidae: combined range, North America (Middle Paleocene to Lower Eocene) and Eurasia (Eocene to Middle Oligocene); 9 listed genera.

†*Order Dinocerata*

†Uintatheriidae: North America and Asia (Upper Paleocene to Upper Eocene); 8 listed genera.

†*Order Pyrotheria*

†Pyrotheriidae: South America (Eocene to Oligocene); 4 listed genera; an isolated order of unknown origin. The Paleocene †*Carodnia,* doubtfully assigned here by Simpson, is now placed in a different order, †Xenungulata (above).

Order Proboscidea

†Moeritheriidae: Africa (Egypt) (Upper Eocene and Lower Oligocene); one genus.

†Gomphotheriidae (†Bunomastodontidae), mastodons (part): Africa (Lower Oligocene to Miocene), Eurasia (Miocene to Pleistocene), North America (Upper Miocene to Pleistocene), and South America (Pleistocene); 14 or more genera, of which 5 were common to the Old World and North America (at various times in the Upper Miocene and Pliocene) and 2 were common to North and South America (Pleistocene).

†Mammutidae, mastodons (part): Africa (Lower Miocene), Eurasia (Miocene and Pliocene), and North America (Middle Miocene to Pleistocene); one genus.

Elephantidae, elephants: 2 existing genera and species. *Elephas maximus,* the Indian Elephant, occurs in suitable places from the base of the Himalayas in northern India to Ceylon, and through Assam, Burma, etc. to the Malay Peninsula and Sumatra (and introduced on Borneo). *Loxodonta africana,* the African Elephant, occurs in suitable places through much of Africa below the Sahara; formerly it reached the Cape, but it now survives in South Africa only in remote or protected places; still earlier, in classical times, it occurred in North Africa too; and in the Pleistocene the genus extended to Europe and Asia. Another genus, †*Mammuthus,* including the mammoths etc., occurred in Africa, Eurasia, and North and perhaps South America in the Pleistocene, and a pygmy elephant (an †*Archidiskodon,* included in †*Mammuthus* by Simpson) is known to have been on Celebes, across Wallace's Line, in the Pleistocene (Hooijer 1951). Two additional genera, placed by Simpson in a separate subfamily, were in Eurasia in the late Tertiary and Pleistocene.

†Deinotheriidae: Africa and Eurasia (Miocene to Pleistocene); one genus.

†Barytheriidae: Africa (Egypt) (Upper Eocene); one genus.

†*Order Embrithopoda*

†Arsinoitheriidae: Africa (Egypt) (Lower Oligocene); one genus.

Order Hyracoidea

Procaviidae, hyraxes: Africa and the adjacent corner of Asia; 3 existing genera, few species; all the genera and species are African; one African species, *Procavia capensis,* extends to Asia, ranging in suitable places from the Cape of Good Hope to North Africa, Arabia, Palestine, Sinai, and Syria. Four additional fossil genera are listed in Africa (Oligocene and Miocene) and Greece (Pliocene).

†Geniohyidae: Africa (Egypt) (Lower Oligocene); 4 genera.

†Myohyracidae: Africa (Lower Miocene); 2 genera.

Order Sirenia, sea cows etc.

†Prorastomidae: West Indies (Jamaica) (Eocene); one genus.

†Protosirenidae: North Africa and Europe (Eocene); one genus.

Dugongidae, dugongs etc.: the existing *Dugong dugon* occurs (except where now exterminated) in the sea coastwise from East Africa etc. east to the Riu Kiu Islands and south through the Indo-Australian Archipelago to northern Australia. Steller's Sea Cow, †*Hydrodamalis stelleri,* apparently existed only around Bering and Copper Islands on the Asiatic side of the Bering Sea when Bering and Steller discovered the animal in 1741, and it was hunted to extinction (notwithstanding Kipling's story to the contrary) within a hundred years; it probably once had a wider range. Fossil dugongids are numerous and diverse and occur back to the Eocene (of Africa and Europe).

Trichechidae, manatees: coastal seas and suitable rivers of much of the Caribbean region including Florida; the Amazon and Orinoco Rivers in South America; and the lower parts of the rivers of tropical West Africa; one genus, 3 species; not known fossil before the Pleistocene.

†Desmostylidae: North Pacific area (Upper Oligocene and Miocene); 2 genera.

(*Superorder Mesaxonia*)

Order Perissodactyla, odd-toed ungulates

(*Superfamily Equoidea*)

†Palaeotheriidae: Eurasia (Eocene and Lower Oligocene); 7 listed genera.

Equidae, horses etc.: now existing naturally only in parts of Asia and Africa; one genus, 7 species. Horses in a strict sense (one wild species) are considered to be truly wild and not seriously mixed with domestic horses now only (if anywhere) in parts of Mongolia and Chinese Turkestan in central Asia. The Onager or Asiatic Wild Ass (but it is not a true ass) occurs in central and southwestern Asia, from Mongolia and Tibet to Syria. The true Wild Ass occurs in northern and eastern Africa south to Somaliland. And zebras (4 species) occur in southern and eastern Africa north to Abyssinia and Somaliland. Most of the existing wild equids are retreating before man (see Harper 1945); one of the zebras is already extinct. Domestic asses (donkeys) and horses have been carried over the world by man, and horses at least have become feral in many places. The domestic ass was derived from the wild African species (*Equus asinus*) in North Africa or the eastern Mediterranean region probably between 4000 and 3000 B.C. or earlier. The domestic horse (*E. caballus*) has interbred with truly wild forms, even with the existing wild horse of central Asia (which is considered a separate species or subspecies, *przewalskii*) until no one knows just where in Eurasia unmixed, indigenous wild horses really have occurred in recent times. They did occur once over much of Europe as well as Asia, and they are supposed to have survived in parts of Europe into historic times. It is likely that the horse was first domesticated perhaps as early as 3000 B.C. somewhere in the main part of Asia or eastern Europe, although it first appears in history later, in southwestern Asia and North Africa. The fossil record of horses, including their ancestors and relatives (3 subfamilies, 18 listed fossil genera) is probably the best described of any animals. It is summarized in the text (pp. 352–353). Simpson (1951) has written a popular but authoritative book on horses and their history.

(†*Superfamily Brontotherioidea*)

†Brontotheriidae, titanotheres: North America (Lower Eocene to Lower Oligocene) and Eurasia (Upper Eocene to Middle Oligocene); 39 listed genera; most were North American, but at least 2 different groups reached Eurasia.

(†*Superfamily Chalicotherioidea*)

†Chalicotheriidae, clawed ungulates: Eurasia (Eocene to Pleistocene), probably also Africa (Miocene, Pleistocene?, presumably at other times), and North America (Upper Eocene to Middle Miocene); 13 listed genera.

(*Superfamily Tapiroidea*)

†Isectolophidae: North America (Lower and Middle Eocene) and Asia (Upper Eocene); 4 listed genera.

†Helaletidae: North America (Lower Eocene to Middle Oligocene) and Asia (Upper Eocene and Lower Oligocene); 10 or more genera.

†Lophiodontidae: Eurasia (Eocene); 5 or more genera.

Tapiridae, tapirs: existing in Sumatra and the Malay Peninsula etc. to 18° N. (one species), and in tropical America from southern Mexico (Veracruz) to northern Argentina (3 species: one Central American, one widely distributed in South America, and one in the Andes of Colombia, Ecuador, and Peru—see maps by Hershkovitz 1954, pp. 493 and 495); all these existing tapirs can be placed in one genus, although they are sometimes divided into 2 or 3. Fossil tapirs (including several extinct genera) occur in North America (Lower Eocene to Pleistocene) and Europe (Lower Oligocene to Pleistocene). The existing genus *Tapirus* (in a broad sense) was in Europe and Asia in the Pliocene and Pleistocene and in North and South America in the Pleistocene. Tapirs are not known ever to have reached Africa.

(*Superfamily Rhinocerotoidea*)

†Hyrachyidae: North America and perhaps Asia (Eocene); 4 listed genera.

†Hyracodontidae: North America (Middle Eocene to Upper Oligocene) and perhaps Asia (Eocene); 4 American and 3 doubtfully assigned Asiatic genera.

†Amynodontidae: North America and Eurasia (Upper Eocene to Middle Oligocene and Lower Miocene respectively); 6 listed genera.

Rhinocerotidae, rhinoceroses and allies: existing in Africa below the Sahara (2 genera and species) and tropical Asia etc. (2 other genera, 3 species). One of the Asiatic genera, *Dicerorhinus,* occurs from northeastern India etc. to Sumatra and Borneo (not Java); the other, *Rhinoceros,* formerly occurred from much of India (not Ceylon) to Sumatra and Java (not Borneo) but is now extinct except in a few localities. Fossil rhinoceroses and allies (several subfamilies, about 30 extinct genera) occurred in Europe from the Middle Eocene to the Pleistocene and in North America from the Upper Eocene to the Lower Pliocene; they are not known in Asia and Africa until the Upper Eocene and Miocene respectively but may have been there earlier. The existing Oriental *Dicerorhinus,* or at least its ancestral line, was in Europe as well as Asia from the Upper Oligocene to the Pleistocene.

(*Superorder Paraxonia*)

Order Artiodactyla, even-toed ungulates

†Dichobunidae: North America (Eocene) and Europe (Eocene to Middle Oligocene); 15 or more genera.

†Choeropotamidae: North America (Eocene) and Eurasia (Middle Eocene to Middle Oligocene); 7 listed genera.

†Cebochoeridae: Europe (Middle Eocene to Lower Oligocene); 3 listed genera.

†Leptochoeridae: North America (Oligocene); 2 genera.

†Entelodontidae: North America (Upper Eocene to Lower Miocene) and Eurasia (Oligocene, and perhaps Upper Eocene); 4 listed genera.

(*Superfamily Suoidea*)

Suidae, pigs: all habitable parts of Africa (and a supposedly endemic but doubtfully native species on Madagascar), and temperate and tropical Eurasia etc.; 5 existing genera, few species. True pigs (*Sus*) now occur north to about 58° N. in eastern Europe and to the Amur region etc. in eastern Asia, and south and east to India and Ceylon; Java and Borneo; the Lesser Sundas to Timor(?); Celebes; the Moluccas, New Guinea, the Aru and Kei Islands, Bismarcks and Solomons(?), and the Philippines(?); 'but toward the southeast it is doubtful where pigs are really native and where they have been carried by early man. They are native certainly on Celebes (which has the endemic *Babirussa* and also Pleistocene fossil *Sus*) and perhaps on Buru, but perhaps not farther east and perhaps not on the outer Lesser Sundas, Timor, or the outermost Philippines. Domestic pigs probably have had a dual origin, from the Wild Boar (*Sus scropha*) of Europe, North Africa, etc., and from another form of the same species in eastern Asia. Fossil suids (17 listed genera) are known only from the main part of the Old World, the oldest being in the Oligocene (of Europe).

Tayassuidae, peccaries: living only in America; one genus, 2 species, of which one extends from southern Texas, New Mexico, and Arizona (formerly north to Arkansas) to northern (and formerly central) Argentina, and the other from southern Mexico to northern Argentina. Probable tayassuids occur in Eurasia (a separate subfamily, 3 listed genera, Lower Oligocene to Lower Pliocene); more certain ones, in North America (6 fossil genera, Lower Oligocene and later) and South America (not until the Pleistocene, 2 extinct genera, one of them North American).

(*Superfamily Anthracotherioidea*)

†Anoplotheriidae: Europe (Middle Eocene to Middle Oligocene); 7 listed genera.

†Anthracotheriidae: Eurasia (Middle Eocene to Pleistocene); Africa (Oligocene and Lower Miocene); and North America (Oligocene and Lower Miocene); 23 listed genera.

Hippopotamidae, hippopotamuses: existing only in Africa; 2 genera and species. The big *Hippopotamus* formerly occurred through all Africa where there was suitable water, north to the delta of the Nile, but not in the Sahara etc.; and the genus occurred also in Europe and Asia in the Pleistocene. The Pygmy Hippopotamus (*Choeropsis*) is localized in Liberia etc. at the lower-outer corner of the western bulge of Africa; this genus is listed fossil (Pleistocene) or subfossil on Madagascar and islands in the Mediterranean, but Simpson (1945, p. 148, footnote 1) thinks some of the fossils may be merely convergently small forms. The Madagascan pygmy hippo was very common and was contemporaneous with the extinct elephant birds and giant lemurs (White 1930). The earliest known hippopotamuses are in the Middle Pliocene of Africa (*Hippopotamus*) and Eurasia (an extinct genus).

(†*Superfamily Cainotherioidea*)

†Cainotheriidae: Europe (Upper Eocene to Lower Miocene); 6 listed genera.

(†*Superfamily Merycoidodontoidea*)

†Agriochoeridae: North America (Upper Eocene to Lower Miocene); 4 listed genera.

†Merycoidodontidae: North America (Upper Eocene to Middle Pliocene); several subfamilies, 24 listed genera.

(*Suborder Tylopoda*)

†Xiphodontidae: Europe (Middle Eocene to Lower Oligocene); 3 genera.

Camelidae, camels etc.: now existing naturally only in Central Asia and western and southern South America; 2 genera. The Old World genus is *Camelus*, with 2 existing species. The Two-humped Camel, *C. bactrianus*, is still (probably) wild in deserts of Mongolia and Chinese Turkestan but domestic or feral elsewhere. The One-humped Camel or Dromedary, *C. dromedarius*, the usual domestic camel of North Africa etc., is unknown in the wild state and is of doubtful origin. The South American genus (*Lama*, in a broad sense) includes 2 wild species (the Huanaco and the Vicuña) and 2 domestic forms (the Llama and the Alpaca); the combined range of the wild species (which overlap but do not coincide in range) is from southern Ecuador and Peru south through the Andes to Patagonia etc., and at lower altitudes to Tierra del Fuego. Fossil camelids (several subfamilies, 24 or more fossil genera) are in North America from the Upper Eocene to the Pleistocene; they appear in Eurasia etc. and South America in the Pleistocene.

(†*Superfamily Amphimerycoidea*)

†Amphimerycidae: Europe (Upper Eocene to Lower Oligocene); 2 genera.

(†*Superfamily Hypertraguloidea*)

†Hypertragulidae: North America (Upper Eocene to Lower Miocene) and Asia (Upper Eocene); 11 or more genera.

†Protoceratidae: North America (Lower Oligocene to Lower Pliocene); 4 genera.

(*Superfamily Traguloidea*)

†Gelocidae: Eurasia (Upper Eocene to Lower Oligocene); 10 listed genera.

Tragulidae, chevrotains: parts of the Oriental Region and West Africa; 2 genera. In the Orient is *Tragulus*, with 2 species from southern Burma to Java, Borneo, and Balabac Island at the southwestern corner of the Philippines, and a third species in peninsular India and Ceylon. In West Africa is *Hyemoschus*, one species, in the forest from Gambia to the eastern Congo. Fossil tragulids (2 listed genera) occur in the Upper Miocene and part of the Pliocene in Europe and Asia.

(*Superfamily Cervoidea*)

Cervidae, deer: Eurasia etc. with the northwestern corner of Africa, and the Americas; 17 existing genera, a moderate number of species. Northward, *Rangifer* is widely distributed on arctic tundras (as well as in some wooded areas) of Eurasia and North America including the Arctic Archipelago and the north coast of Greenland. Southward, native cervids reach the northwestern corner of Africa (one species); Asia Minor, Iran, etc.; India and Ceylon; Java, Bali, and Borneo; and perhaps the Lesser Sundas, Timor, Celebes, the Moluccas, and the Philippines (but they may have been carried to some of these places by man); and nearly to the southern end of South America. *Rangifer* (above) is circumpolar, and all the forms of it (the Reindeer and

Caribou) may belong to one species. *Alce,* the Moose ("Elk" in Europe), occurs across the cold northern forests of Eurasia and North America; the Old and New World forms are sometimes considered one species. *Cervus* is common to the north-temperate parts of both hemispheres; the Red Deer (*C. elephus*), which occurs across temperate Eurasia (and in North Africa), and the American Elk (*C. canadensis*), which formerly occurred through most of temperate North America, are sometimes considered one species. The American *Odocoileus* (White-tailed and Mule Deer etc.) extends from Canada to parts of South America, to the Amazons and Peru. Of the 13 other existing genera of the family, Asia has 8 (2 extending to Europe) including both primitive and derived forms; North America has none (*i.e.,* no peculiar genera); and South America has 5 (one extending to Central America and southern Mexico) all in the American tribe *Odocoileini.* Fossil cervids (34 listed genera) occur in Eurasia back to the Lower Oligocene, in North America back to the Lower Miocene, and in South America not until the Pleistocene.

(*Superfamily Giraffoidea*)

†Lagomerycidae: Eurasia and Africa (Miocene); 3 genera.

Giraffidae, giraffes etc.: Africa; 2 existing genera and species, representing different subfamilies. *Giraffa camelopardalis,* the Giraffe, occurs in suitable places (not heavy forests) in most of Africa below the Sahara. *Okapia johnstoni,* the Okapi, is confined to the heavy rain forest of the Upper Congo region. Fossil giraffids (15 listed genera representing the 2 existing subfamilies and an additional fossil subfamily) are known mostly from Asia (Miocene to Pleistocene, several groups reaching eastern Europe in the Lower Pliocene); a few have been found in Africa; a supposed record of the family in the Pleistocene of North America is doubtful.

(*Superfamily Bovoidea*)

Antilocapridae, pronghorns: western North America; one existing species, the Pronghorn, *Antilocapra americana,* ranging formerly from the prairie region of southern Canada (Manitoba to Alberta) to parts of Texas, the Mexican tableland, and Lower California, but now much reduced in range. Fossils show that antilocaprids were numerous (12 listed extinct genera in 2 subfamilies) in western North America in the past, back to the Middle Miocene, and that they occurred east to Florida in the Pliocene (†*Hexameryx* White 1941), but they are not certainly known anywhere else.

Bovidae, cattle, antelopes, sheep, etc.: Africa, temperate and tropical Eurasia, and North America; 5 subfamilies, 13 tribes, 54 existing genera (but the classification of bovids is difficult and still partly doubtful) (see text, pp. 355–357, for additional details). In Eurasia now the northernmost bovid is *Ovis canadensis,* the Bighorn Sheep, which occurs across part of northern Siberia (west almost to the mouth of the Yenesei) as well as in western North America; in America, the northernmost bovid is *Ovibos,* the Musk Ox, a truly arctic animal, which occurs (or recently did occur) from the northernmost point of Alaska to the

Arctic Archipelago and the north coast of Greenland, and which occurred in northern Eurasia too in the Pleistocene. Southward, the natural limits of the family include all Africa (not Madagascar); India (but probably not Ceylon, although the Water Buffalo, *Bubalus,* is feral there); and Java (probably not native on Bali), Borneo, and Celebes, and an endemic species of buffalo (*Bubalus*) is isolated on Mindoro in the Philippines. These southeastern limits in the Indo-Australian Archipelago are all set by wild cattle (Bovini); the Serow (*Capricornis*) reaches Sumatra and also Formosa; no other bovids reach even the Indochinese Subregion of Asia, except perhaps its northern edge. In America bovids occur and so far as is known have occurred south only to Mexico and the northern edge of Central America. Fossil bovids (including 86 fossil genera) are mostly in the Old World, a few in the Miocene but most in the Pliocene and Pleistocene (see text, pp. 355–357). The genus *Bison* now has a notable, discontinuous, relict distribution. This genus included a number of species in Pleistocene and probably post-Pleistocene times, when it occurred across temperate Eurasia and through most of North America. One Eurasian species (*Bison bonasus,* the Wisent) still occurred within historic times in Europe east perhaps to central Asia and the Lena River in Siberia, but it has receded steadily; the last "wild" (but protected) herds, in the Caucasus and Lithuania, died out in the 1920's or 1930's, although the species still exists in captivity. The American Bison or Buffalo (*Bison bison*) formerly occurred from the plains and some woodlands of Canada to the Mexican plateau, and from the inner edge of Oregon etc. at least to western New York (and to Cape Cod in the Pleistocene) and Georgia. The last wild Buffalo was killed in western Pennsylvania just after 1800, and the last anywhere in the East about 1825. The destruction of the Buffalo in the West is part of American history. By 1900 only 2 wild herds were left, a small one protected in Yellowstone Park, and "Wood Buffalo" in the Athabasca region of Alberta, Canada. Now, under protection, the American Buffalo has increased again, and small herds are widely scattered. Domestic cattle, goats, and sheep are all derived from Eurasian species. The humpless domestic cattle of Europe, and now of most other cool parts of the world, come from the Aurochs (†*Bos primigenius*), which was once widely distributed in Europe but became extinct as a wild animal in the early 1600's. The humped cattle of the tropics probably come from a different, tropical Asiatic, ancestor. The domestic Water Buffalo is derived from the existing *Bubalus bubalus;* it was probably first domesticated in India, then taken elsewhere in the Orient and west to Egypt etc., and recently to other warm parts of the world. The domestic Goat is probably derived from the existing *Capra hircus* of southwestern Asia and islands in the eastern Mediterranean. Domestic sheep may be of mixed origin, derived in part from the existing Mouflon (*Ovis musimon*) of Sardinia and Corsica and the Argali (*O. ammon*) of Asia. Some of these animals (at least humpless cattle, sheep, and goats) were probably first domesticated something like 8000 years ago, when farming began in southwestern Asia.

REFERENCES

Aguayo, C. G. 1950. Observaciones sobre algunos mamiferos cubanos extinguidos. *Boletin Hist. Nat. Soc. Felipe Poey,* **1,** 121–134.

Allen, G. M. 1938–1940. *The mammals of China and Mongolia.* New York, American Mus. Nat. Hist., 2 vols.

———. 1939. A checklist of African mammals. *Bull. Mus. Comparative Zool.,* **83.**

———. 1939*a*. *Bats.* Cambridge, Mass., Harvard U. Press.

———. 1942. *Extinct and vanishing mammals of the Western Hemisphere with marine species of all oceans.* American Committee for International Wild Life Protection (New York), Special Pub. No. 11.

Allen, J. A. 1878. The geographical distribution of the Mammalia . . . . *Bull. United States Geological and Geographical Survey of the Territories,* **4,** 313–377.

Berg, L. S. 1950. *Natural regions of the U.S.S.R.* New York, Macmillan.

Bourlière, F. 1955. *Mammals of the world.* New York, Alfred A. Knopf.

Burt, W. H. 1949. Present distribution and affinities of Mexican mammals. *Ann. Association American Geographers,* **39,** 211–218.

Burt, W. H. (illustrated by R. P. Grossenheider). 1952. *A field guide to the mammals* [of America north of Mexico]. Boston, Houghton Mifflin. Although designed as a field guide, this is in fact the best modern summary of the North American mammal fauna.

Cabrera, A., and J. Yepes. 1940. *Mamiferos sud-americanos* . . . . Tucuman and Buenos Aires, Argentina, Compania Argentina de Editores.

Carter, T. D., J. E. Hill, and G. H. H. Tate. 1945. *Mammals of the Pacific world.* New York, Macmillan.

Chasen, F. N. 1940. *A handlist of Malaysian mammals.* Bull. Raffles Mus. (Singapore), No. 15.

Couto, C. de P. 1952. Fossil mammals from the beginning of the Cenozoic in Brazil. Condylarthra, Litopterna, Xenungulata, and Astrapotheria. *Bull. American Mus. Nat. Hist.,* **99,** 355–394.

———. 1952*a*. Fossil mammals from the beginning of the Cenozoic in Brazil. Marsupialia: Polydolopidae and Borhyaenidae. *American Mus. Novitates,* No. 1559.

———. 1952*b*. Fossil mammals from the beginning of the Cenozoic in Brazil. Marsupialia: Didelphidae. *American Mus. Novitates,* No. 1567.

Dice, L. R. 1929. The phylogeny of the Leporidae . . . . *J. Mammalogy,* **10,** 340–344.

Ellerman, J. R. 1940–1949. *The families and genera of living rodents.* London, British Mus., 3 vols.

Ellerman, J. R., and T. C. S. Morrison-Scott. 1951. *Checklist of Palaearctic and Indian mammals* . . . . London, British Mus.

Ellerman, J. R., T. C. S. Morrison-Scott, and R. W. Hayman. 1953. *South African mammals 1758–1951: a reclassification.* London, British Mus.

Elton, C. 1942. *Voles, mice, and lemmings: problems in population dynamics.* New York, Oxford U. Press.

Goodwin, G. G. 1946. Mammals of Costa Rica. *Bull. American Mus. Nat. Hist.*, **87**, 271–474.

Harper, F. 1945. *Extinct and vanishing mammals of the Old World.* American Committee for International Wild Life Protection (New York), Special Pub. No. 12.

Hershkovitz, P. 1954. . . . tapirs . . . American species. *Proc. United States National Mus.*, **103**, 465–496.

Hill, W. C. O. 1953–1955. *Primates. Comparative anatomy and taxonomy.* Vol. I—Strepsirhini [lemurs etc.] (1953). Vol. II—Haplorhini: Tarsioidea (1955). Edinburgh, Edinburgh U. Press.

Hinton, M. A. C. 1933. . . . rodents from Indian Tertiary deposits. *Ann. Mag. Nat. Hist.* (10), **12**, 620–622.

Hooijer, D. A. 1951. Pygmy elephant and giant tortoise. *Sci. Monthly*, **72**, 3–8.

Hopwood, A. T. 1954. Notes on the recent and fossil mammalian faunas of Africa. *Proc. Linnean Soc. London*, **165**, 46–49.

Laurie, E. M. O., and J. E. Hill. 1954. *List of land mammals of New Guinea, Celebes and adjacent islands 1758–1952.* London, British Mus.

Lydekker, R. 1896. *A geographical history of mammals.* Cambridge, Cambridge U. Press; New York, Macmillan.

Matthew, W. D. 1915, 1939. Climate and evolution. *Ann. New York Acad. Sci.*, **24**, 171–318; reprinted (1939) as *Special Pub. New York Acad. Sci.*, **1.**

Matthews, L. H. 1952. *British mammals.* London, Collins.

Mayr, E. 1946. The number of species of birds. *Auk*, **63**, 64–69.

Miller, G. S., Jr., and R. Kellogg. 1955. List of North American Recent mammals. *Bull. United States National Mus.*, No. 205.

Murray, A. 1866. *The geographical distribution of mammals.* London, Day and Son.

Osgood, W. H. 1943. The mammals of Chile. *Zool. Ser. Field Mus. Nat. Hist.*, **30.**

Phillips, W. W. A. 1935. . . . *mammals of Ceylon.* Ceylon, Colombo Mus.; London, Dulau.

Powell, A. W. B. 1947. *Native animals of New Zealand.* Auckland Mus. Handbook Zool.

Reynolds, T. E., K. F. Koopman, and E. E. Williams. 1953. A cave faunule from western Puerto Rico with a discussion of the genus *Isolobodon*. *Breviora* (Mus. Comparative Zool.), No. 12.

Roberts, A. 1951. *The mammals of South Africa.* South Africa, Central News Agency.

Romer, A. S. 1945. *Vertebrate paleontology*, 2nd ed. Chicago, U. of Chicago Press.

Rothschild, Lord, and G. Dollman. 1936. The genus *Dendrolagus*. *Trans. Zool. Soc. London*, **21**, 477–548.

Sanborn, C. C. 1951. Two new mammals from southern Peru. *Fieldiana* (Chicago Nat. Hist. Mus.), *Zool.*, **31**, 473–477.

Sclater, P. L. 1875. The geographical distribution of mammals. *Manchester Sci. Lectures*, Ser. 5–6, 202–219.

Sclater, W. L., and P. L. Sclater. 1899. *The geography of mammals.* London, Kegan Paul, Trench, Trübner and Co.

Scott, W. B. 1937. *A history of land mammals in the Western Hemisphere,* revised ed. New York, Macmillan.

Simpson, G. G. 1937. The beginning of the age of mammals. *Biological Reviews Cambridge Philosophical Soc.,* **12,** 1–47.

———. 1940. Mammals and land bridges. *J. Washington Acad. Sci.,* **30,** 137–163.

———. 1943. Mammals and the nature of continents. *American J. Sci.,* **241,** 1–31.

———. 1945. The principles of classification and a classification of mammals. *Bull. American Mus. Nat. Hist.,* **85.**

———. 1947. Holarctic mammalian faunas and continental relationships during the Cenozoic. *Bull. Geological Soc. America,* **58,** 613–688.

———. 1947*a.* Evolution, interchange, and resemblance of the North American and Eurasian Cenozoic mammalian fauna. *Evolution,* **1,** 218–220.

———. 1951. *Horses* . . . . New York, Oxford U. Press.

Stirton, R. A. 1953. Vertebrate paleontology and continental stratigraphy in Colombia. *Bull. Geological Soc. America,* **64,** 603–622.

Stock, C., and H. G. Richards. 1949. A *Megalonyx* tooth from the Northwest Territories, Canada. *Science,* **110,** 709–710.

Swynnerton, G. H., and R. W. Hayman. 1951. A check list of the land mammals of the Tanganyika Territory and the Zanzibar Protectorate. *J. East African Nat. Hist. Soc.,* **20,** 274–392.

Tate, G. H. H. 1945. The marsupial genus *Phalanger. American Mus. Novitates,* No. 1283.

———. 1945*a.* The marsupial genus *Pseudocheirus* . . . . *American Mus. Novitates,* No. 1287.

———. 1945*b.* Notes on the squirrel-like and mouse-like possums . . . . *American Mus. Novitates,* No. 1305.

———. 1946. Geographical distribution of the bats in the Australasian Archipelago. *American Mus. Novitates,* No. 1323.

———. 1947. . . . Dasyuridae . . . . *Bull. American Mus. Nat. Hist.,* **88,** 97–156.

———. 1947*a. Mammals of eastern Asia.* New York, Macmillan.

———. 1948. . . . Macropodidae . . . . *Bull. American Mus. Nat. Hist.,* **91,** 233–352.

———. 1948*a.* . . . Peramelidae . . . . *Bull. American Mus. Nat. Hist.,* **92,** 313–346.

———. 1951. The rodents of Australia and New Guinea. *Bull. American Mus. Nat. Hist.,* **97,** 183–430.

———. 1951*a.* The wombats . . . . *American Mus. Novitates,* No. 1525.

———. 1952. Mammals of Cape York Peninsula . . . . *Bull. American Mus. Nat. Hist.,* **98,** 563–616.

Taylor, E. H. 1934. *Philippine land mammals.* Philippine Bureau Sci. Monograph 30.

Troughton, E. 1951. *Furred animals of Australia,* 4th ed. Sydney, Angus and Robertson.

White, E. I. 1930. Fossil hunting in Madagascar. *Nat. Hist. Mag.* (British Museum), **2,** 209–235.

White, T. E. 1941. Additions to the fauna of the Florida Pliocene. *Proc. New England Zool. Club,* **18,** 67–70.

Williams, E. E., and K. F. Koopman. 1951. A new fossil rodent from Puerto Rico. *American Mus. Novitates,* No. 1515.

———. 1952. West Indian fossil monkeys. *American Mus. Novitates,* No. 1546.

Wodzicki, K. A. 1950. *Introduced mammals of New Zealand* . . . . Wellington, New Zealand, Department of Scientific and Industrial Research.

Wood, A. E. 1947. Rodents—a study in evolution. *Evolution,* 1, 154–162.

———. 1950. Porcupines, paleogeography and parallelism. *Evolution,* 4, 87–98.

Wynne-Edwards, V. C. 1952. Freshwater vertebrates of the arctic and subarctic. *Bull. Fisheries Research Board Canada,* No. 94.

chapter 7

Continental patterns and faunal regions

The purpose of this chapter is, first, to combine the patterns of distribution of the five great classes of land and fresh-water vertebrates, treated separately in the five preceding chapters, into a common pattern, as simple as possible (but it will not be very simple), and, second, to divide the world into faunal regions, as simple and natural as possible.

The main pattern of vertebrate distribution is made up of several subpatterns. Five have been distinguished here, but the exact number could be increased or decreased by dividing or combining different ones or perhaps by adding others. The subpatterns have been described as if they were separate, but they are not. They run into each other and modify each other until they are in fact inextricably combined in one very complex whole.

The first and broadest of the five main subpatterns of distribution of land and fresh-water vertebrates is *limitation*. Vertebrates of all the classes reach final limits or are limited in numbers northward (and to some extent southward) in cold latitudes, and outward from the continents on islands. Fresh-water fishes extend far northward into the arctic, but strictly fresh-water ones scarcely extend outward beyond the limits of the great continents and closely associated continental islands. Amphibians extend northward to parts of the arctic but not to the northern limit of fishes; but they extend farther onto islands than strictly fresh-water fishes do. Reptiles fall far short of the northern limit of amphibians in most places in the north, but they extend still farther onto islands. Birds extend both far northward and far onto islands; they reach no final limits in either direction; but even birds, like other vertebrates, are limited in numbers in the north and on remote islands. Mammals too extend far northward, but (excepting bats) they are about as limited as amphibians on islands. The sequence of northern limits of the different classes is suggestive: the evolution of successive cold-blooded groups (fishes, amphibians, and reptiles) seems to have been accompanied by decreasing tolerance for cold (or by increasing heat requirements) and by retreat from the cold north, until warm-bloodedness enabled birds and mammals to extend northward again. The limits of the different classes on islands suggest at once that vertebrates occur on islands according to their powers of dispersal rather than according to their geological ages.

Within their broadest limits, the different classes of vertebrates exhibit different amounts of non-zonal complementarity, caused by *further limitation* of different, dominant, orders or families to different parts of the world. Among strictly fresh-water fishes there is a profound complementarity, involving large parts of faunas, between the dominant cypriniforms of Eurasia and the dominant non-cypriniforms of South America, the North American and African faunas being mixed (and Australia being outside this pattern). Amphibians exhibit a complementarity which is less profound (involving smaller parts of faunas) and more complex, but which includes dominance of ranids in the main part of the world and of leptodactylids in South America and Australia, and of rhacophorid tree frogs in Africa and the Oriental Region and of hylids elsewhere. Reptiles show a still less profound, still more complex complementarity in which the dominance of emydid turtles in Eurasia and North America and of side-necked turtles elsewhere, and of colubrid snakes in most of the world and of elapids in Australia are important features. Birds probably show the least profound, most complex complementarity of distribution of any vertebrates, although Oscines (songbirds), dominant over most of the world, and Suboscines, dominant in South America, are incompletely complementary. Terrestrial mammals show a more profound complementarity between the almost wholly placental fauna of the main part of the world and the chiefly marsupial fauna of Australia, the South American fauna being mixed. Of course all this is greatly oversimplified and somewhat subjective—I have not measured the complementarities in different classes of vertebrates but have estimated them rather arbitrarily—but it does suggest that non-zonal complementarity is inversely proportional to power of dispersal. The vertebrates that have the most difficulty dispersing over land (fishes) or crossing salt-water barriers between continents (fresh-water fishes and terrestrial mammals) show the most profound localizations and geographical complementarities of dominant groups. This is perhaps an obvious and expected finding, but it is worth stating nevertheless. It is evidence that, in general, existing vertebrates have dispersed over the world according to their own powers and not as passive cargo on drifting continents.

The second subpattern, of *zonation,* which includes zonal complementarity, is strongest among fresh-water fishes: most non-ostariophysan primary-division families are confined either to the tropics or to the north-temperate zone, and some of them go around the world in one zone or the other; and some Ostariophysi and some secondary-division and peripheral fishes are zoned too. Of amphibians, most families are mainly tropical; some of the tropical groups extend into the north-temperate zone; but salamanders and a few small groups of frogs are mainly north-temperate. Of reptiles, most families are tropical; some of the tropical families extend northward for different distances; but there are no well-defined, zonal, north-temperate groups of reptiles. This—concentration in the tropics and extension northward of some tropical families—is also the main pattern of zonation of land birds and mammals, although these animals extend northward farther and in greater numbers than reptiles do. It will be seen that the zonation of fishes involves much zonal complementarity; of amphibians, less; and of reptiles, land birds, and mammals, still less, the north-temperate faunas of these classes being formed mostly by subtraction rather than by differentiation. Apparently many important groups of fishes and some, but fewer, of amphibians become inherently adapted to north-temperate climates; many groups of birds and mammals tolerate northern climates without becoming inherently adapted to them; but reptiles neither become adapted to nor tolerate northern climates very well.

Longitudinal *zonation* is not an important pattern in the distribution of most vertebrates. It is best developed among some lizards and some small, migratory, especially insectivorous land birds. In both cases the longitudinal patterns seem to have been formed by primarily tropical families which have spread northward in one half of the world or the other (either the Old World or the New World) without crossing to the opposite half. The reasons for not crossing are probably different. The lizards have probably not got far enough north. The birds have probably been held back by the habit of returning to wintering grounds in one half of the world or the other.

The third subpattern, of *geographical radiation,* is most profound (most deeply marked in the most important groups) among fresh-water fishes (*e.g.,* the order Ostariophysi and, within the order, the cypriniforms), less profound among amphibians (*e.g.,* ranids), and still less so among reptiles (*e.g.,* colubrids), birds (*e.g.,* songbirds), and mammals (*e.g.,* murids and bovids), in which the radial patterns either are less clear or occur in groups which, though important, are less so than the Ostariophysi and cypriniforms are among fresh-water fishes. Of course this matter too is somewhat subjective, but there does seem to be a plain and expected correlation: the most profound patterns of geographical radiation, reflecting spread of great, dominant groups from definite centers, are found among the animals which disperse most slowly over land, so that evolution has the most time to mark length of stay in different places. The best-developed radial patterns in each class center on the Old World and usually on the main Old World tropics: that of cypriniform fishes, on tropical Asia; of ranid frogs, on the Old World tropics; of advanced colubrid snakes, probably on tropical Asia; of songbirds, on the whole warmer part of the Old World; and of murid rodents and bovids, on the warmer part of Eurasia and Africa.

The fourth subpattern, of *differentiation of faunas* on different continents, is different in each class of vertebrates. Among fresh-water fishes, the main pattern is one of diversity, with survival of old tropical families, in Africa; predominance of more recent cypriniforms in Eurasia; diversity, with survival of old north-temperate families, in North America; evolutionary radiation in isolation in South America; and accumulation of salt-tolerant groups, with survival of one very ancient relict, in Australia. Among amphibians, the main pattern is more complex, involving restriction of caecilians to Africa, tropical Asia, and South America etc.; concentration of most salamanders on the northern continents; and complex complementarities of dominant frogs. Among reptiles, the pattern of differentiation of faunas on the continents is still more complex: turtles form one pattern complex enough in itself; crocodilians, another; lizards, others; large constricting snakes, another; and small, dominant snakes, still others. Among birds, the main pattern (excessively complex in detail) is one of diversity in the main Old World tropics, subtraction northward, and differentiation in Australia and South America, the South American fauna being richest and most differentiated of all. Among mammals, there is the same Old World tropical diversity and subtraction northward, with differentiation in Australia and South America, but for mammals the Australian fauna is more differentiated than the South American one.

The fifth and final subpattern, of *concentration in the largest, most favorable areas* of each continent and subtraction in less favorable, often marginal areas, is most marked among fresh-water fishes, less among amphibians, and still less among reptiles, birds, and mammals, which are true land animals, and which exist in some numbers almost everywhere on land, although even these animals are most numerous in the largest, most favorable areas. This pattern of concentration and limitation within single continents is connected with the broader pattern of limitation of vertebrates northward and on islands over the world as a whole. As I have said before (of fresh-water fishes, p. 101), the fact that there are the most animals where there is the most space for them (and the most favorable conditions) may seem too obvious to emphasize, but the effect of space (extent of favorable area) will turn out to be neither simple nor obvious but very important.

Summary: the pattern of distribution of vertebrates

In still simpler terms, land and fresh-water vertebrates are most diverse and numerous in the largest and most favorable areas of the world, the continental tropics, and are reduced and limited northward (and southward) and also outward from the large land masses; the exact limits vary with the nature of each class and with its power of dispersal; and, within its limits, each class exhibits non-zonal complementarity in something like inverse proportion to power of dispersal. All the classes are zoned: zonation of fresh-water fishes involves much zonal complementarity (differentiation of north-temperate as well as tropical groups); that of amphibians, less; and that of reptiles, birds, and mammals, still less, the zonation of these classes involving mostly subtraction north of the tropics. Radial distributions occur in all classes, in something like inverse proportion to rate of dispersal. Differentiation of faunas on different continents is different in each class. And on each continent there is in each class further concentration in large and favorable areas and subtraction marginally. Of course, as I have said, these subpatterns run into each other and modify each other and are inextricably combined in one very complex whole.

This complex pattern has evidently been produced by a very complex combination of inorganic and organic factors and processes. The most important inorganic factors seem to have been extent and favorableness of area, zonation of climate, and position of salt-water barriers between continents. The organic factors have included the different requirements, tolerances, and powers of dispersal of different vertebrates. The processes have included the evolutions of different dominant vertebrates at different times and in different places (but especially in the Old World tropics) and their dispersals in different ways over the world.

I have revised this summary many times in an effort to make it as simple, clear, and true as I can. If it is not simple, that is because animal distribution is not simple. Perhaps I have left out things that ought to be put in, or put in things that ought to be left out. But, whether or not it is entirely correct, the summary does serve some useful purposes. To make such a summary, and also to read it and criticize it, requires an effort of understanding worth making for its own sake. The summary suggests what seem to be the most important factors in animal distribution, and they are not exactly what is usually supposed: many zoogeographers underrate the effect of zonation of climate and, still more, the effect of area on animal distribution. And the summary suggests that the distribution of vertebrates conforms (in a very complex way) to the world as it is now, to the present pattern of land, climate, and barriers, and this in turn suggests that existing vertebrates have dispersed by their own movements over a world not very different from now, not by means of drifting continents or extraordinary land bridges. These are preliminary findings. They will be considered in more detail in later chapters.

Discontinuities

Geographic discontinuities, occurrences of the same or related animals in more or less widely separated places, are very common within the main pattern of distribution of vertebrates. Special cases of discontinuity have been overemphasized and misunderstood by some zoogeographers, especially by bridge builders and movers of continents. I want to treat discontinuities in a more general way and to try to put them in perspective.

Animals can become discontinuously distributed in three ways: by jumping gaps, as when land animals reach oceanic islands across water; by changes of land, as when parts of faunas are cut off on continental islands by flooding of connecting land (or when continents drift apart, if they drift); and by withdrawals (*i.e.*, partial extinctions) which divide originally continuous distributions.

Single cases are relatively unimportant. For example, the discontinuous distribution of leiopelmid frogs, in northwestern North America and New Zealand, is a unique case and is unimportant as a pattern

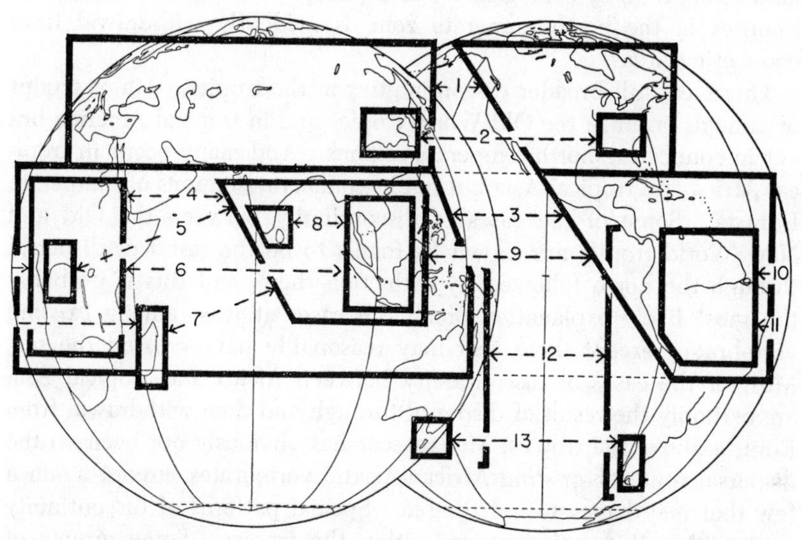

Fig. 44. Some patterns of discontinuity in animal distribution: 1, temperate Eurasia and temperate North America; 2, eastern Asia and eastern North America; 3, Old World and New World tropics; 4, tropical Africa and tropical Asia; 5, *West* Africa and tropical Asia; 6, Africa and *eastern* tropical Asia; 7, Madagascar and tropical Asia; 8, southern India or Ceylon and eastern tropical Asia; 9, tropical Asia and tropical America; 10, Africa and South America; 11, Madagascar and South America; 12, Australian Region and South America; and 13, southern Australia (New Zealand), and south-temperate South America.

of discontinuity, although the occurrence of *Leiopelma* on New Zealand is important for other reasons. However, discontinuities form a number of common patterns (Fig. 44). The distributions of most of the animals named below to exemplify the patterns are given in more detail in the family lists at the ends of Chapters 2 to 6.

There are broad discontinuities in the north-temperate zone. Many groups of animals occur in temperate Eurasia and temperate North America but not in connecting arctic areas. There are many special patterns within this main one. For example, many genera and species pairs of plants and animals occur in eastern Asia and eastern North America but not between. They include tulip trees, magnolias, sassafrases, witch hazels, skunk cabbages, and many other plants (Fernald 1929, Li 1952); and paddlefishes, cryptobranchid salamanders, alligators, certain small skinks, and *Opheodrys* snakes. Some other north-temperate patterns are suggested by salamanders (p. 145). Still other northern patterns have been described better by botanists (*e.g.*, Fernald 1929) than by zoologists. The explanation of most of the discontinuities in the north-temperate zone is evidently withdrawal from connecting areas.

There are still broader discontinuities in the tropics. Many groups of animals occur in the Old World tropics and in tropical America but not in connecting north-temperate regions. And many occur in tropical Africa and tropical Asia but not in connecting regions of temperate Eurasia. Some of the cases of discontinuity between the Old and New World tropics are shown by fossils to be the result of dispersal through the north followed by extinction there, and this is probably the most likely explanation of all the cases at least among *existing* vertebrates, except those that may reasonably have crossed the sea. Most of the cases of discontinuity between Africa and tropical Asia are certainly the result of dispersal through and then withdrawal from Eurasia above the tropics; Madagascar has obviously not been on the dispersal route of *existing* African-Asiatic vertebrates, except again a few that may have crossed the sea. Special patterns of discontinuity exist within this main pattern within the tropics. Some groups of animals (*e.g.*, microhyline frogs, aniliid and some colubrid snakes, a genus of eagles and one of woodpeckers, and tapirs) occur only or mainly in tropical Asia and tropical America (not in Africa). Others (*e.g.*, several groups of fishes, pipid frogs, pelomedusid turtles, amphisbaenid lizards, leptotyphlopid worm snakes, and a few birds) occur only or mainly in Africa and tropical America (not in tropical Asia). Still others (at least *Podocnemis* turtles and iguanid lizards) occur only in Madagascar and America (not in Africa—but there is an iguanid on Fiji). Some groups (*e.g.*, pheasants, chevrotains) are divided between West Africa and tropical Asia; others (*e.g.*, some fishes, reptiles, and birds—see pp. 63, 191, and 261), between southern India and the eastern Orient; others (*e.g.*, some snakes, p. 191), between Indochina and the Greater Sunda Islands. These patterns, and others like them, have evidently been formed mainly by withdrawals from connecting areas.

South of the tropics discontinuities are broader still, in distance over existing land. Leptodactylid frogs, chelyid turtles, and marsupials occur only or mostly in South America and Australia. However, although these animals extend into the south-temperate zone, they are all more tropical than south-temperate in total distribution, and they do not occur on New Zealand. Fossils and other evidence suggest that the frogs and marsupials have dispersed through the main part of the world; the chelyids may have reached Australia directly through the sea.

Some groups of salt-tolerant fishes (*e.g.*, galaxiids) and terrestrial invertebrates, and also certain plants, have zonal "antarctic" distributions: they are common to the cool southern extremities of South America and Australia, and many of them occur on New Zealand, and some are represented on the southern tip of Africa too. No terrestrial or strictly fresh-water vertebrates have such distributions. This is a significant fact referred to again in Chapter 10.

Most of the discontinuities listed above seem to be results of recessions or extinctions of many animals in large areas. This is as it should be if, as I think, recession and extinction are very common processes.

Faunal regions

I want now to divide the world into faunal regions and to describe the regional vertebrate faunas and their interrelationships. This is to be done first within the limits of the continents and closely associated continental islands. Later (Chapter 8), outlying islands and island faunas will be considered in relation to the continents.

The distinction between continents and islands is arbitrary. Australia is an island continent, and parts of its fauna show the effects of isolation more than, for example, the fauna of Madagascar does; South America was an island continent not long ago, and its fauna still shows many effects of isolation. However, it is customary and convenient to treat as continents Eurasia, Africa, North and South America, and Australia. They, with the islands that have recently been connected to them, form one nearly continuous system of land (Frontispiece), very irregular in shape, unsymmetrically placed on the round earth, and irregularly interrupted by mountains, deserts, and narrow water gaps. This is the land now to be divided into regions according to the distribution of animals on it.

Faunal regions are often misunderstood and undeservedly abused nowadays. There is, to begin with, a feeling among some zoogeographers that existing distribution is unimportant, or at least that present "static" patterns and faunal regions are not worth much consideration in themselves but only in relation to the past, and that it is only the past—the evolution and movement of animals—that really matters in zoogeography. This, as I have said in Chapter 1 (p. 22), seems to me to be a great mistake. It is only in the present that we can see exactly how animals really are distributed and how their distributions are related to space, climate, barriers, other inorganic factors, plant cover, and each other. Even a little way back in the past these things cannot be seen but for the most part can only be deduced by comparison with the present. Knowledge of the present is therefore necessary to understand the past. It is of course also true that knowledge of the past is necessary to understand the present. The present and past are equally important in zoogeography. Neither can be understood without the other. Nevertheless it is logically permissible and in fact necessary sometimes to consider animal distribution exclusively in the present, as if it were static, and to bring in past evolutions and movements later (Chapters 9 and 10).

Much of what I shall say in general about faunal regions has been said before, in different words, by Wallace (1876, Vol. 1, especially pp. 53–56). He described the nature and purpose of faunal regions; he admitted their imperfections and limitations but pointed out their reality and usefulness; and he answered with plain common sense most of the misunderstandings and criticisms of them that are still current now, 80 years later. He noted (p. 55), for example, that the faunal regions are designed to represent the main features of distribution of *existing* animals, and that, although the causes that have produced present distributions lie in the past, any attempt to combine the past with the present in one system of regions must lead to confusion. This is obviously true. Nevertheless some zoogeographers still try to combine geological history and Tertiary relationships with present distributions in defining faunal regions.

Some of the feeling against faunal regions has been put into words by Dunn (1922), who says, among other things, ". . . the zoogeographical realms are nothing save and except the great land masses with lines drawn to correspond to the physiographic barriers. There is a great philosophical difference between such terms as Holarctic Fauna and Holarctic Region. In the first case we speak of zoological matters in terms of zoology, in the second of geographical matters in terms of mythology." I have hesitated to cite this one passage for criticism, but it has been quoted and misquoted recently and the idea it expresses—that faunal regions are unreal, artificial things—needs to be counteracted. The faunal regions do *not* correspond exactly to the great land masses. It is true that the regions are, in general, separated by barriers but not by "the" barriers. There is no one set of barriers.

The world is full of different ones. That certain barriers rather than others have been found to separate major faunas and faunal regions is not an arbitrary decision but a very significant fact gradually discovered by zoogeographers. As to the philosophical difference between faunas and regions, faunal regions are areas occupied now by faunas. There is a philosophical difference, but if the faunas are real, the regions are real too, and not mythological.

The great difficulty with faunal regions is not philosophical but practical. It arises from the complexity of animal distribution. Scarcely any two groups of animals are distributed in exactly the same way, and different major groups (for example different classes of vertebrates) are in some cases very differently distributed. It is hard to reduce this diversity to a simple system. The difficulty is so great that it has been said that the search for faunal regions is a mare's-nest (Gadow); that there are no faunal regions, or too many of them, or different ones for each different group of animals, but no generally applicable and useful ones. This, I think, shows a lack of understanding of the nature and purpose of the regions.

It should be understood that the system of faunal regions represents an average pattern, not a common one. It is the best average of all the very different distributions of different animals that zoogeographers have been able to devise. It is not and is not intended to be a common pattern which many different groups of animals fit exactly. Wallace knew this. He exaggerated the sharpness of boundaries of some of the regions, but he understood the underlying diversity of distribution of different animals.

The nature of faunal regions can perhaps be clarified by analogy. Suppose a carpet were woven of many different sets of differently colored threads, representing different groups of animals, with the color differences proportional to relationships. Suppose the carpet were woven very irregularly, thick in some places and thin in others, with different threads not following a common pattern but variously scattered or concentrated in different places in the fabric. Such a carpet would have no regular pattern, but it might have a pattern of irregular, differently colored blotches formed by predominance of differently colored threads in different places. The pattern might be real and very obvious even though the different blotches might not be sharply limited and might run together at their edges. Plants and animals do form such a carpet over the world. It is thick in some places and thin in others, and different sorts of plants and animals are variously scattered or concentrated in different places in it. The regional faunas, which determine the faunal regions, tend to be great concentrations of animals in favorable places (corresponding to the thick places of the rug) and main faunas are often separated by unfavorable areas where animals are few (the rug thin) and where transitions occur. The pattern formed in this way by the distribution of animals is real and very obvious even though the regional faunas (like the blotches of the carpet) run together at the edges. This is the sort of pattern the faunal regions are designed to show.

The system of faunal regions, then, represents the average, gross pattern of distribution of many different animals with more or less different distributions. *There is such an average pattern of animal distribution.* It is real and very significant. It shows in the broadest way how animal distribution is fitted to the world and how climate and barriers affect it most deeply. A natural system of faunal regions would, I think, be worth making for this reason alone, but it has other uses.

The system of faunal regions is a standard pattern, a sort of meter stick, by which the distributions of different animals can be measured, described, and compared, their special features determined, and important things about the animals and their histories revealed. The standard pattern is useful for this purpose even if few groups of animals fit it exactly. There is no more reason to discard the standard pattern of animal distribution because few animals exactly fit it than there is to discard the meter because few things in nature are exactly a meter long. Deviations from the standard are expected and informative. If, for example, a particular group of animals reaches the Australian Region but shows less than standard differentiation there, this suggests recent dispersal and power of crossing water gaps. Or if, in a broader way, a group of animals tends to be unusually well represented in the climate-limited regions but more than usually differentiated or absent in the barrier-limited ones (as many fresh-water fishes are), this suggests fundamental things about the animals' relation to climate and salt-water barriers. To be useful as a standard, the system of faunal regions must be generally accepted. It must not be changed except for good reason and by general agreement. If a specialist finds that animals he works with do not fit the standard pattern of distribution, he should not damn the standard and make a new one to suit himself. That would be as if a surveyor changed the length of the meter to suit himself. The specialist should accept the standard, find how his animals differ from it, and try to find what the differences mean. Wallace (p. 58) knew this too. Nevertheless some zoogeographers still misunderstand it and still tamper needlessly with the standard system of faunal regions. Perhaps I should repeat that, although the standard system of regions should be arbitrarily accepted most of the time, the system is (unlike the meter) inherently not arbitrary but real. The standard system of faunal regions is the truest simple diagram of animal distribution that zoogeographers have been able to devise.

Faunal regions have other, less important uses. They help zoologists in different parts of the world set natural limits to regional studies. And the names of the regions are useful terms. For example, "Oriental Region" is simpler, more exact, and therefore more useful than "tropical Asia and certain closely associated continental islands."

It should be plain from what has been said in the last few pages that regional faunas are not homogeneous assemblages of animals uniformly distributed within common limits. Each regional fauna is more or less concentrated in favorable places, varies in composition in different places, and enters into complex transitions with adjacent faunas. Nevertheless, in spite of the differences, the animals in different parts of one faunal region are on the whole more related to those of other parts of the same region than to those of other regions.

I hope that all this is enough to make it clear what faunal regions and regional faunas are, and to persuade skeptics that the system of faunal regions is real and useful. If any doubts remain, the best reference I can give for further persuasion is Wallace (1876). Critics of faunal regions may be surprised to find how well Wallace understood what he was doing and how little their criticisms apply to him.

Six main faunal regions are now generally accepted. They were first proposed by Sclater for birds in a paper read before the Linnean Society in 1857 and published in 1858. Sclater's regions were confirmed by Wallace (1876) for vertebrates in general and for some invertebrates. Reasons for adopting these particular regions are given in detail by Wallace. Other regions have been proposed, but Sclater's regions seem best to combine reality and usefulness. They are listed in Table 9. Sclater's regions and his names for them still stand with minor changes of boundaries and some simplification of spelling, except that the name "Oriental Region," proposed by Wallace, has replaced Sclater's "Indian Region."

Although the separate regions are Sclater's, they have been arranged and rearranged in different ways by later authors. Sclater grouped the Old World regions against the New World ones in what

TABLE 9. SCLATER'S *Schema Avium Distributionis Geographicae*, OR SYSTEM OF FAUNAL REGIONS BASED ON THE DISTRIBUTION OF BIRDS

Sclater's names and essential definitions (excepting some islands) are given.

Creatio Palaeogeana

I. Palaearctic Region: temperate Eurasia and the northern corner of Africa, north of the Atlas Mountains
II. Aethiopian or Western Palaeotropical Region: Africa except the northern corner, and southern Arabia
III. Indian or Middle Palaeotropical Region: tropical Asia and closely associated continental islands
IV. Australian or Eastern Palaeotropical Region: New Guinea, Australia, and Tasmania

Creatio Neogeana

V. Nearctic or North American Region: North America south to central Mexico, and including Greenland
VI. Neotropical or South American Region: South and Central America and southern Mexico

he called (he was looking for centers of creation) "Creatio Palaeogeana" and "Creatio Neogeana," but this arrangement is not the most natural one. Huxley (1868) grouped together the four regions that cover Africa, Eurasia, and North America as "Arctogaea"; and he grouped together South America and Australia as "Notogaea." This was partly good and partly not. The faunas of South America and Australia are very different from those of the main part of the world, but they are also very different from each other. Wallace (1876) apparently was more interested in the individual regions than in grouping them; he combined Sclater's and Huxley's systems in one table (Vol. 1, p. 66). Finally Blanford (1890) suggested a grouping which better expresses the real interrelationships of the six regional faunas. He made three main divisions of the world: Arctogaea (the main part of the world, with four subdivisions), the "South-American Region," and the Australian Region. It remained for an anonymous author (1893) to propose that the South American and Australian divisions be called Neogaea and Notogaea, and for Lydekker (1896) to propose that the three divisions be called realms. One change of name is still desirable. "Arctogaea" is an unfortunate, misleading name for the great area which includes the main part of the Old World tropics and the north-temperate zone, with just an arctic fringe. I propose to call it Megagea, or the great part of the world. Thus amended, the generally accepted system of faunal regions (of the continents) is as given in Table 10. One additional term is in common

TABLE 10. THE ACCEPTED SYSTEM OF CONTINENTAL FAUNAL REGIONS

Realm Megagea (Arctogea): the main part of the world

1. Ethiopian Region: Africa (except the northern corner), with part of southern Arabia
2. Oriental Region: tropical Asia, with associated continental islands
3. Palearctic Region: Eurasia above the tropics, with the northern corner of Africa
4. Nearctic Region: North America, excepting the tropical part of Mexico

Realm Neogea

5. Neotropical Region: South and Central America with the tropical part of Mexico

Realm Notogea

6. Australian Region: Australia, with New Guinea, etc.

use. Heilprin (1887) combined the two northern regions, the Palearctic and Nearctic, into a Holarctic Region. They are not combined here, but they may be called the Holarctic Region*s*, and animals which occur in both are often called Holarctic.

This is, of course, a very incomplete history of ideas about faunal regions. For further details see Wallace (1876), Blanford (1890), Lydekker (1911), Gadow (1913), Scrivenor *et al.* (1943), Schmidt (1954), and others. The history of faunal regions is only one part of the history of zoogeography, which is treated in a little more detail in the introduction (pp. 20–23).

This system of faunal regions (Table 10) shows the relative distinctness of the different regions probably as well as any simple system can, but the regions can be grouped in another way to emphasize their natures (Table 11). The two northern regions (Palearctic and Nearctic) are separated from the tropics not by impassable bar-

TABLE 11. ANOTHER ARRANGEMENT OF THE FAUNAL REGIONS TO EMPHASIZE THEIR NATURES

Climate-limited regions

1. Palearctic Region
2. Nearctic Region

Main regions of the Old World tropics

3. Oriental Region
4. Ethiopian Region

Barrier-limited regions

5. Neotropical Region
6. Australian Region

riers but primarily by zonal climate, because of which their faunas are differentiated and limited. They are climate-limited regions. The Neotropical and Australian Regions owe their special natures to past and present salt-water barriers, which have isolated and in some ways limited their faunas. They are barrier-limited regions. The two remaining regions, the main regions of the Old World tropics (Ethiopian and Oriental), are less limited by climate and barriers. They are not unlimited; no regions are. But they are the least limited, most central ones, and the other regions form an unsymmetrical radial pattern of increased limitation around them (Fig. 45). This is the main,

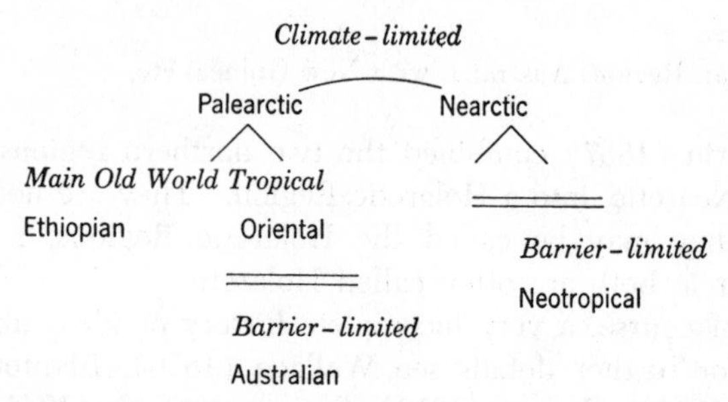

Fig. 45. Diagrammatic arrangement of the faunal regions.

average pattern of present distribution of animals, and something like it—radiation from the main Old World tropics—is also the apparent main pattern of animal dispersal (Chapter 9).

The six continental faunal regions are outlined on the accompanying map (Fig. 46) and are shown separately in more detail on later maps (Figs. 47–52). In treating each one, I shall first tell what continents or parts of continents compose it and what the general nature of its climate and plant cover is, then summarize the composition and relationships of its vertebrate fauna, then discuss its boundaries briefly, and then indicate the main pattern of distribution of the fauna within the region. Where the land is more or less continuous, exact boundaries are hard to define and are not very important; transitions, their nature and extent, are more important; and the faunas themselves, their natures and interrelationships, are most important of all.

It is impossible to describe the regional faunas adequately in a few pages. Wallace (1876) devoted nearly 500 pages to them, to "a review of the chief forms of animal life in the several regions and sub-regions." He called this "zoological geography" in contrast to "geographical zoology," which is the distribution of animals by taxonomic groups (my Chapters 2 to 6) rather than by regions. Wallace's old account of the regional faunas is still the best there is, although out of date in detail. A new, good, detailed and accurate regional or faunal zoogeography is much needed. It would help to relieve the tendency to provincialism of naturalists everywhere.

To analyze the regional faunas statistically in a few pages is also impossible. An adequate statistical treatment would have to deal

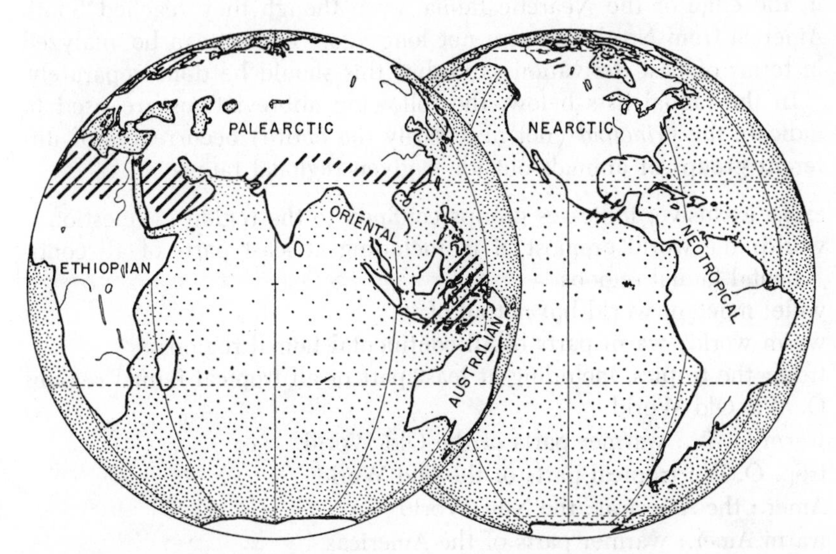

Fig. 46. The six continental faunal regions. Diagonal hatching shows approximate boundaries and transition areas.

not only with families but also with genera and perhaps with species. It would have to describe not only the composition and relationships of each fauna but also transitions between faunas. It might have to do all this for different parts of faunas separately, *e.g.*, the forest part, savanna part, desert part, fresh-water part, etc. And there are many other complications and many chances of error.

Since I can neither describe the regional faunas adequately nor treat them statistically, I have reduced them to little more than bare lists of families. The shortcomings of this method are obvious, but there is nothing else I can do here. More details can be assembled from the family lists at the ends of Chapters 2 to 6 by readers who care to take the trouble. If my method of presentation has any advantage, it is that it summarizes the regional faunas so briefly that readers can get some idea of the faunas as wholes.

One of the particular errors to be avoided both in describing the regional faunas and in treating them statistically is mixing the present with the past. In describing the relationships and transitions of existing faunas, animals should be classified strictly according to their present distributions, not according to their past derivations. Peccaries, for example, now mainly in South and Central America extending into southern North America, are at present a Neotropical element in the edge of the Nearctic fauna, even though they reached South America from North America not long ago. Faunas can be analyzed in terms of past derivations, too, but this should be done separately.

In the faunal lists below, the following abbreviations are used to indicate the *principal* (not necessarily the entire) occurrences of different groups of animals outside a given regional fauna.

excl.: exclusive; indicates groups confined to the region in question
world: the world or most of it, including at least parts of all continental faunal regions
wide: much of world but not all of it
warm world: warm parts of all continental faunal regions
trop.: the tropics, including at least part of all tropical faunal regions
O. W.: Old World
warm O. W.: warmer parts of the Old World
trop. O. W.: tropical parts of the Old World
Amer.: the Americas: the New World
warm Amer.: warmer parts of the Americas
trop. Amer.: tropical America
Afr.: Ethiopian Africa
O.: Oriental Region (some groups so designated extend into temperate Asia)
Hol.: Holarctic Regions
Pal.: Palearctic Region
N. A.: Nearctic North America
S. A.: South America (some groups so designated extend into Central America, etc.)
A.: Australian Region

The Ethiopian Region

The Ethiopian Region (Fig. 47) is Africa, less its northern corner, plus (if one wishes) part of southern Arabia. Madagascar is often included, but I prefer to consider it separately. In the following summaries, the presence or absence of groups in Madagascar is not mentioned except in a few special cases.

The Ethiopian Region is mainly tropical. There is a large block of rain forest in equatorial West Africa and many small patches of

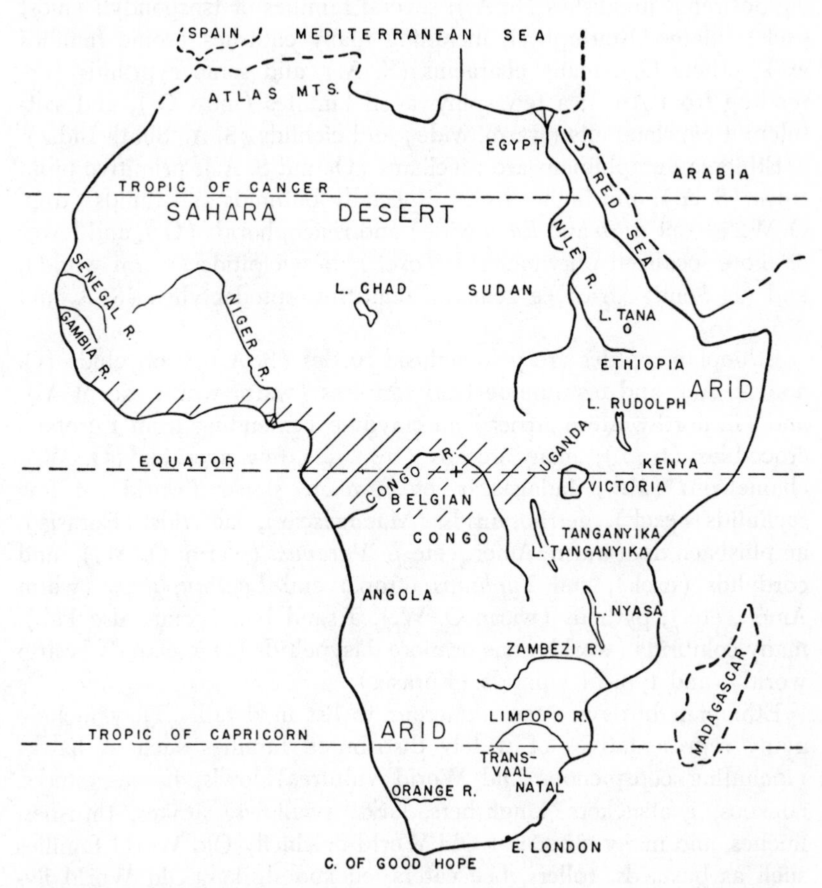

Fig. 47. The Ethiopian Region (solid lines) and adjacent land (broken lines). Orthographic projection. Diagonal hatching shows main areas of rain forest.

rain forest elsewhere, notably on isolated mountains in East Africa. Much of the rest of Africa is covered by dry or seasonal thorn scrub or grassland, grading into desert northward and southwestward. The southern tip of Africa is warm-temperate (reaching a latitude equivalent to that of South Carolina), with complexly mixed vegetations. (This, like the descriptions of the other regions below, is necessarily a very much simplified account of a huge, diverse piece of land. I have tried to describe it and the other regions as they were before man deforested large parts of them.)

Ethiopian fresh-water fishes are diverse: archaic bichirs (excl.) and lepidosirenid lungfishes (S. A.); several families of Isospondyli (most excl.); many Ostariophysi, including many catfishes (some families excl., others O.), many characins (S. A.), and some cyprinids (extending from Asia); a few spiny-rayed families (most O.); and salt-tolerant cyprinodonts (warm, wide) and cichlids (S. A., South India).

Ethiopian amphibians are caecilians (O. and S. A.); primitive pipid frogs (S. A.); and higher frogs, especially bufonids and ranids (trop. O. W., except *Bufo* and *Rana* wide) and rhacophorids (O.), and fewer or more localized phrynomerids (excl.), brevicipitids (warm world), and (in South Africa) a genus of doubtful leptodactylids (S. A. and A.).

Ethiopian reptiles are pelomedusid turtles (S. A.), trionychids (O. and N. A.), and testudinine land tortoises (warm world except A.), and (in northwestern Africa) an emydine (extending from Europe); crocodiles (trop.); many geckos (trop.), a few agamids (O. W.), chameleons (also Madagascar, etc.), many skinks (world), a few feyliniids (excl.), gerrhosaurids (Madagascar), lacertids (Eurasia), amphisbaenids (warm Amer., etc.), *Varanus* (warm O. W.), and cordylids (excl.); and *Typhlops* (trop.) and *Leptotyphlops* (warm Amer., etc.), pythons (warm O. W.), a sand boa (genus also Pal.), many colubrids (world), one or more dasypeltids (O.), elapids (warm world), and typical viperids (Eurasia).

Ethiopian birds are too numerous to list in detail. They include many representatives of widely distributed families such as hawks (including conspicuous Old World vultures), owls, herons, storks, cuckoos, goatsuckers, kingfishers, larks, swallows, shrikes, thrushes, finches, and many others; of Old World or chiefly Old World families such as bustards, rollers, bee eaters, cuckoo shrikes, Old World flycatchers and warblers, motacillids, weaverbirds, starlings, orioles, drongos, etc.; of tropical or Old World tropical families such as trogons (few), hornbills, barbets, most honey guides, pittas (few), bulbuls, and most sunbirds; moderate numbers of parrots and pigeons; and a few exclusively Ethiopian families named below. Of course this list is incomplete. Of 67 families of birds represented on land and fresh water in the Ethiopian Region, 53 occur over all or much of the world, or much of the Old World, or the warmer parts of it; 3 (honey guides, broadbills, and bulbuls) are shared only or mainly with the Oriental

Region; 5 (the Crab Plover, sand grouse, hoopoes, a doubtful bombycillid, and a doubtful honey eater) have special or doubtful relationships; and only 6 are exclusively Ethiopian. Of the 6 families that are exclusively Ethiopian or nearly so, 3 consist of single species (the Ostrich, Secretary Bird, and Hammerhead), and the others are the touracos (19 species), mousebirds (6 species), and helmet shrikes (13 species—and there is a doubtful member of the family in Borneo). The following additional subfamilies are exclusively Ethiopian or nearly so: guinea fowls (7 species), tree hoopoes (6 species), bush shrikes (42 species), buffalo weavers (3 species), widow birds (9 species), and tickbirds (2 species). This geographical classification of families is somewhat arbitrary, but it serves to show the diversity of Ethiopian birds and the small number of exclusively Ethiopian families and subfamilies, which total only 110 species. The relationships of Ethiopian birds are shown better by genera than by families. Many genera are endemic; many occur also in the Oriental Region or have relatives there; some extend into temperate Eurasia, but many of these are migratory; and few have other relationships.

The Ethiopian Region is notable for its big game, but some of the less conspicuous mammals are more important zoogeographically. Ethiopian mammals are the Otter Shrew (excl. fam.), golden moles (excl.), hedgehogs (Eurasia), elephant shrews (almost excl.), and shrews (wide); lemur-like lorisids (O.), Old World monkeys (O.), 2 great apes (O.); scaly anteaters (O.); canids (wide), mustelids (wide), viverrids (O. etc.), hyaenas (near part of Asia), and cats (wide); the Aardvark (excl. order); an elephant (O.); hyraxes (nearly excl.); equids (Pal.); 2 rhinoceroses (O.); pigs (Eurasia); 2 hippopotamuses (excl.); a tragulid (O.); 2 giraffids (excl.); many bovids (Eurasia and a few N. A.); rabbits (wide); many rodents: squirrels (wide), anomalurids (excl.), the Spring Haas (excl. fam.), a few cricetids (wide), bamboo rats (O.), many murids (O. W., but some subfamilies are excl. Afr.), dormice (Pal.), jerboas in the Sahara etc. (Pal.), Old World porcupines (O. etc.), and 4 additional families of probable or possible hystricomorphs (fams. excl., but some apparently related S. A.); and fruit bats (warm O. W.) and 8 families of insectivorous bats: rhinopomatids (northeastern Africa and southern Asia), emballonurids (trop.), nycterids (O., etc.), megadermatids (warm O. W.), 2 families of horseshoe bats (warm O. W.), vespertilionids (world), and molossids (warm world). This mammal fauna is a diverse mixture of more or less widely distributed families, families shared with the Oriental Region, and exclusive families, and a few families with other relationships. The Oriental relationships are relatively stronger than shown, for many Ethiopian genera and even some species of mammals occur also in the Oriental Region or have close relatives there.

In summary, the fresh-water fishes of the Ethiopian Region are a very distinctive assemblage, with some clear but distant relationships toward South America, some closer ones toward the Orient, and important exclusive groups. Ethiopian amphibians and reptiles are less distinctive, with again some South American relationships (pipid frogs, pelomedusid turtles, amphisbaenid lizards, leptotyphlopid worm snakes), more Oriental ones (more than the list of families shows, for many Ethiopian genera occur also in the Orient or have close relatives there), and some exclusive groups. Ethiopian birds are still less distinctive, strongly Oriental in their main relationships, with relatively few exclusive groups. And Ethiopian mammals are a mixed assemblage, rather strongly Oriental in their main relationships, but with a number of exclusive groups.

The Ethiopian vertebrate fauna as a whole (so far as so complex a thing can be treated as a unit) is most like the Oriental one. The two are roughly similar in size, roughly similar in composition (least so in fishes), and they share many families, very many genera, and some species. Parts of the Ethiopian fauna have also close relationships with the Palearctic, formed principally by northward extensions of tropical groups, including many migratory birds and certain families, genera, and species of other vertebrates. Some Ethiopian vertebrates are, much more distantly, related to South American ones; they include several important groups of fishes, certain amphibians and reptiles, but very few birds and mammals. There are hardly any direct relationships between Ethiopian and Australian vertebrates. The Ethiopian fauna also has important endemic groups, especially of fishes and mammals, which have no living relatives outside Africa.

The boundaries of the Ethiopian Region are the ocean, except in the north. Most Ethiopian vertebrates occur only south of the Sahara, or extend north only along the Nile. A very small northern corner of Africa, consisting of the northern parts of Morocco and Algeria north of the Atlas Mountains, has a depauperate fauna which is European in most of its relationships. In between, the great deserts of northern Africa have a very limited, specialized fauna with mixed relationships, part (most?) Ethiopian and part Eurasian. Under these circumstances it is not important whether the northern boundary of the Ethiopian Region is set at the Atlas Mountains (where Sclater put it) or across the middle of the Sahara at the northern boundary of the tropics (where Wallace and most later zoogeographers put it), or whether the Sahara is considered a transition area. Southwestern Arabia, across the southern Red Sea from Africa, has a vertebrate fauna which is in some ways Ethiopian, but depauperate. This part of Arabia can be included in the Ethiopian Region or it can be considered part of a complex transition area.

Within the Ethiopian Region, the fauna is richest and most diverse in the more favorable tropical areas. There is a strong, general reduction of the fauna northward into the Sahara and a less strong reduction and some differentiation southward into South Africa. There is also some differentiation of a West African wet-forest fauna and an East African open-forest and steppe fauna, although the two overlap and mix complexly. These general differences led Wallace to divide the continental Ethiopian Region into three subregions, corresponding to West Africa, East Africa, and South Africa, but I shall not make a formal division. Further details of distribution within the Ethiopian Region vary with different groups of animals and are too complex to discuss here.

The Oriental Region

The Oriental Region (Fig. 48) is essentially tropical Asia and closely associated continental islands, including Ceylon, Sumatra, Java, Borneo, Formosa, and (if one wishes) the Philippines.

The Oriental Region is mainly tropical but extends a few degrees north of the tropics, especially in northern India. In the eastern part of the region, from the northeastern corner of India, Burma, Indochina, etc., to the Greater Sunda Islands, there is much rain forest, although it is interrupted in places by more open, drier country. High mountains, with rich and diverse plant cover, extend northward from Burma etc. to southwestern China and the eastern Himalayas. Westward, the main part of peninsular India is relatively dry and open, but there is an isolated, wet, low-mountainous, forested strip along the southwest coast, and there is an interrupted range of low mountains (the Satpuras) across India from northeast to southwest which has apparently been a route of dispersal of wet-country animals in the past (Hora 1953). The Himalayas too become drier and less hospitable westward.

Oriental fresh-water fishes include no archaic groups; one osteoglossid (trop.) and a genus of notopterids (Afr.); many catfishes (some fams. Afr. and/or extending into Pal., others excl.) and an overwhelming diversity of cypriniforms, including cyprinids (Pal., N. A., Afr.), cobitids (Pal. etc.), and homalopterids (excl.); a few nandids (Afr., S. A.) and pristolepids (excl.); anabantids (most, but one genus extends to Afr.), channids (Afr.), and the only luciocephalid (excl.); mastacembelids (Afr.) and the only chaudhuriid (excl.); and a few secondary-division cyprinodonts (warm, wide) and, in peninsular India and Ceylon, a genus of cichlids (related Afr., Madagascar).

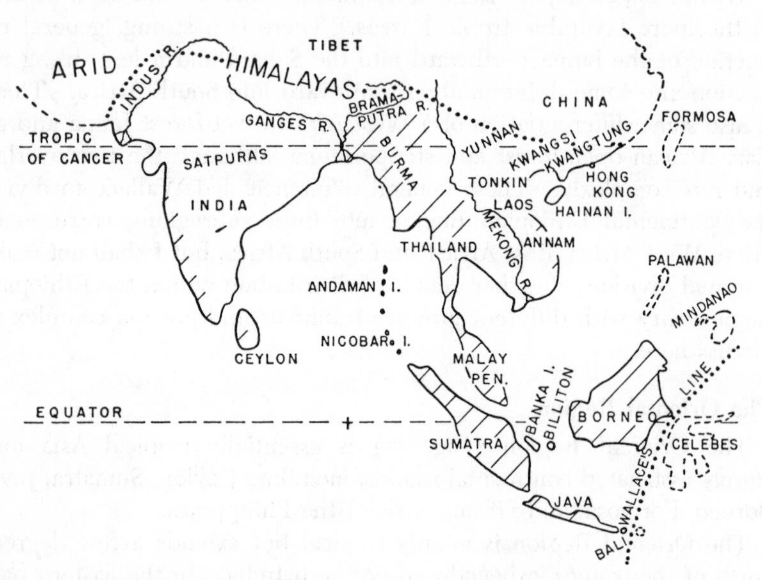

Fig. 48. The Oriental Region (solid lines) and adjacent land (broken lines). Orthographic projection. Dotted lines are approximate regional boundaries. Diagonal hatching shows main areas of rain forest.

Oriental amphibians are caecilians (Afr., S. A.); a few salamanders in northern Indochina etc. (extending from Pal.); a few localized discoglossids (Pal.); pelobatids (Hol.); buffonids (wide); many ranids (wide) and rhacophorids (Afr. etc.); *Hyla* in northern Indochina etc. (extending from Pal.); and brevicipitids (warm world).

Oriental reptiles are many emydine turtles (Pal. and Amer.), the only platysternine (excl.), trionychids (Afr., N. A.), and testudinine land tortoises (warm world except A.); crocodiles (trop.) and gavials (excl.); many geckos (warm world) and agamids (O. W.), a chameleon in southern India and Ceylon (related toward Mediterranean region), many skinks (world), a few dibamids (extending to New Guinea), a few lacertids (Afr., Pal.), *Varanus* (warm O. W.), localized anguids (Pal., Amer.), and *Lanthanotus* in Borneo (excl.); and many snakes: *Typhlops* (warm world) and *Leptotyphlops* (marginal, Afr. etc.); the localized family Uropeltidae (excl.); a few aniliids (trop. Amer.); the one xenopeltid (excl.); pythons (warm O. W.); sand boas (marginal, Pal. etc.); many colubrids [Dipsadinae (trop. Amer.), Xenoderminae (S. A.), Colubrinae (world), and aquatic Acrochordinae (A.) and Homalopsinae (A.)]; a dasypeltid (Afr.); elapids (warm world); sea snakes (Indo-Pacific); and true vipers (Afr. and Pal.) and pit vipers (Pal. and Amer.).

Oriental birds include almost all the widely distributed, Old World, tropical, and Old World tropical families that occur in the Ethiopian Region, but in somewhat different proportions. The Orient has fewer representatives of some groups (*e.g.*, of shrikes, hoopoes, honey guides, sunbirds, and finches), and more of others (*e.g.*, of woodpeckers, pittas, babblers, and corvids). The Orient also has many pheasants (only one in Africa), many pigeons, but (like Africa) only a moderate number of parrots. Of 66 families of birds represented on land and fresh water in the Oriental Region, 53 are more or less widely distributed elsewhere; 3 are shared only or mainly with Africa; 4 have special or doubtful geographical relationships; 5 are shared only or chiefly with the Australian Region (megapodes, frogmouths, wood swallows, flowerpeckers, and honey eaters—but the last does not extend into the Orient beyond Bali); and only one is exclusive. The exclusively Oriental family is the Irenidae, the fairy bluebirds and leafbirds, with 4 genera, 14 species. This partly arbitrary geographical classification shows the general nature of the Oriental bird fauna. It is most like the Ethiopian one, diverse, with even fewer exclusive groups, but with some Australian relationships too. The genera of Oriental birds include many endemic ones, many shared with or related toward Africa, many shared with Palearctic Eurasia (many of these are migratory), some shared with Australia, and very few with other relationships.

Oriental mammals are spiny and hairy hedgehogs (the former Afr. and Pal., the latter excl.), shrews (wide), moles (local, extending from Pal.); flying lemurs (excl.); tree shrews (excl.), lorisids (Afr.), tarsiers (excl.), Old World monkeys (Afr.), and apes (Afr.); scaly anteaters (Afr.); canids (wide), bears (Pal. and Amer.), mustelids (wide), viverrids (Afr. etc.), one hyaena (Afr.), and cats; an elephant (Afr.); a tapir (trop. Amer.); 2 rhinoceroses (Afr.); wild pigs (Afr., Pal.); chevrotains (Afr.); deer (Pal. and Amer.); bovids (Afr., Pal., a few N. A.); rabbits (wide); many rodents: squirrels (wide) and flying squirrels (Pal. and N. A.), a few cricetids (wide), bamboo rats (Afr.), many murids (O. W.), spiny dormice (excl.), and Old World porcupines (Afr.); and fruit bats and 8 families of insectivorous bats (same as Afr.). This Oriental mammal fauna is a mixture of more or less widely distributed families, families and genera shared with the Ethiopian Region, exclusive families, a few groups of murid rodents and many genera and species of bats shared with the Australian Region, and a very few groups with other relationships.

In summary, the fresh-water fishes of the Oriental Region are a rich and dominant assemblage, notable for lack of archaic groups and for dominance of cypriniforms. Oriental amphibians and reptiles are partly like and partly unlike the Ethiopian ones: some notable differences are that the Orient lacks pipids and has pelobatid frogs; lacks pelomedusids and has emydine turtles; has gavials; lacks or has only a few chameleons, gerrhosaurids, amphisbaenids, and cordylid lizards; has almost no leptotyphlopids, and has uropeltids, ilysiids, additional subfamilies of colubrids, and pit vipers; in general, Africa has a greater diversity of lizards, the Orient, of snakes. Oriental birds are strongly Ethiopian in their main relationships, with many Palearctic and some Australian relationships too, with few important exclusive groups. And Oriental mammals are rather strongly Ethiopian in general but include some groups with other relationships and some important exclusive groups.

The Oriental vertebrate fauna as a whole is most like the Ethiopian one, as indicated thereunder, in spite of the differences listed above. Parts of the Oriental fauna too have Palearctic relationships, especially through eastern Asia. A few Oriental vertebrates, chiefly certain reptiles, birds, and bats, are related to Australian ones. And a very few (exceptional frogs, snakes, and birds, as well as tapirs) have tropical American relatives. The Oriental fauna has few important endemic groups of vertebrates confined to it, fewer than any other tropical regional fauna. Either the Orient has been a center from which vertebrates have tended to spread into other regions, or it has been a main crossroads in dispersal, or both. (In fact, the Orient seems to have been a center of dispersal for some vertebrates, *e.g.*, cyprinid fishes, and a crossroads for others, *e.g.*, birds.)

The boundaries of the Oriental Region are not sharply defined, for the Oriental fauna enters into complex transitions in several directions. The dry country of northwestern India and beyond is "debatable land" (Wallace), with an impoverished fauna in which transitions occur both from east to west and from south to north (see under fresh-water fishes, pp. 65–66). Where regional boundaries are drawn here is perhaps not very important. A little farther east, the Himalayas form a natural boundary several degrees north of the tropics. This mountain barrier is not the main cause of differentiation of the Oriental and Palearctic regional faunas—the main cause is evidently the zonal climatic difference—but the mountains and the dry country west and north of them do form a barrier which permits parts of the Oriental fauna to extend somewhat north of the tropics without much mixture of northern elements. Still father east, in the mountains of Burma, southwestern China, etc., and along the coast of China, there is a very broad, complex transition, and the boundary of the Oriental Region is hard to fix; it is usually drawn across southern China not far above the Tropic of Cancer, where many tropical groups of animals reach northern limits. Southward and eastward, Ceylon and Formosa (which have somewhat depauperate continental faunas) as well as Sumatra, Java, and Borneo (which have almost full-scale Oriental faunas) are included in the Oriental Region, and the Philippines may be included, although their fauna actually forms a fringing pattern of strong, progressive subtraction with increasing distance from Borneo (Chapter 8). Beyond Java and Borneo, the Oriental fauna forms a broad, complex transition with the Australian fauna, as described in more detail at the end of the present chapter.

Within the Oriental Region, the fauna shows both east-west and north-south differentiation. It is richest and most varied in the wetter eastern part of the region (from South China etc. to Java and Borneo), where, for example, most Oriental pelobatid frogs, emydine turtles, and pheasants occur, and less rich and somewhat different westward, in India. And there is partial differentiation in the east of a northern subfauna (in South China, Indochina, Burma, Siam, etc.) and a southern one (in the Malay Peninsula, Sumatra, Java, Borneo, etc.), and in the west of a northern subfauna (in the main, drier part of India) and a southern one (in South India and Ceylon), the latter with some isolated Malayan and even a few Madagascan elements. Wallace divided the Oriental Region into four subregions along these lines, naming them (1876, Vol. 1, map facing p. 315) Indochinese, Indo-Malayan, Indian, and Ceylonese. Subregions are not formally recognized here, but the expressions "Indochinese Subregion" and "Malayan Subregion" (Wallace's "Indo-Malayan") are sometimes useful.

The Palearctic Region

The Palearctic Region (Fig. 49) is essentially Eurasia above the tropics, plus a small northern corner of Africa.

The Palearctic is north-temperate with an arctic fringe. In the wetter parts of eastern Asia, including parts of temperate China and Japan, is fine, mostly deciduous forest. Much of the interior of Asia is arid and open, and the arid area extends through southwestern Asia to North Africa. The arid zone grades into grassy steppes especially along its northern edge. In Europe there is again much deciduous or mixed forest, but it is less rich, with many fewer species of trees, than the forest of eastern Asia. Across northern Asia and Europe is a tremendous stretch of coniferous forest, and above it tundra.

Palearctic fresh-water fishes are a paddlefish in China (N. A.), 2 suckers in eastern Asia (N. A.), many cyprinids (O., Afr., N. A.), cobitids (O. etc.), a few localized catfishes (fams. O.), an umbrid in the Danube etc. (N. A.), *Dallia* in eastern Siberia (Alaska), the Pike (N. A.), some percids (N. A.), and a very few anabantids, channids, and a mastacembelid in eastern Asia (extending from O.); and a few secondary-division cyprinodonts southward (wide, warm). The Cyprinidae is the only family of this list that is numerous and dominant throughout the Palearctic Region, but shortage of other strictly fresh-water fishes is partly made up by peripheral ones, especially salmonids (N. A.) northward.

Palearctic amphibians are salamanders (N. A. etc.), discoglossids (localized in O.), pelobatids (N. A., O.), *Bufo* (wide), *Hyla* (wide), many *Rana* (wide), and a few *Rhacophorus* and brevicipitids in eastern Asia (extending from O.).

Reptiles that are more or less widely represented in the Palearctic Region are emydine turtles (O., N. A., etc.), skinks (wide), lacertids (Afr., O.), anguids (Amer., local in O.), many colubrids (wide), true vipers (Afr., O.), and in Asia pit vipers (O., Amer.). Additional reptiles which occur in parts of the southern edge of the Palearctic Region or eastern Asia are *Testudo* (warm, wide), *Trionyx* in eastern Asia (Afr., O., N. A.), an alligator in China (N. A.), geckos (warm world), agamids (warm O. W.), a chameleon (Afr. etc.), amphisbaenids (Afr., warm Amer.), *Varanus* (warm O. W.), *Shinisaurus* in China (excl. fam.), *Typhlops* (warm world), *Leptotyphlops* (Afr., warm Amer.), and sand boas (Afr., etc.).

Palearctic birds include representatives of 53 families on land and fresh water, not many fewer than occur in Africa or the Oriental Re-

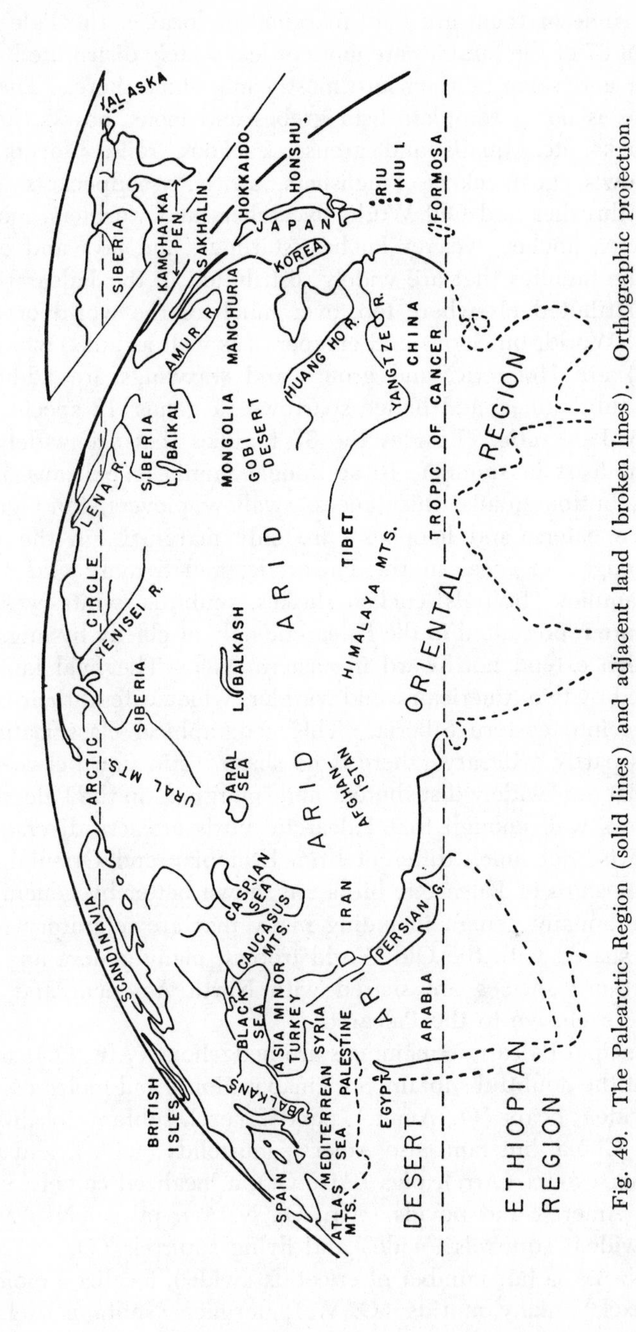

Fig. 49. The Palearctic Region (solid lines) and adjacent land (broken lines). Orthographic projection.

gion, but some of them are only marginal or local in the Palearctic. Only about 37 of the families are more or less widely distributed in the Palearctic, and some of them are mostly migratory there. They include (this is not a complete list) grebes and loons, hawks, herons, storks, ducks etc., quails and grouse, cuckoos, rails, plovers etc., pigeons, owls, goatsuckers, kingfishers, swifts, woodpeckers, larks, swallows, thrushes and Old World flycatchers and warblers, motacillids, tits etc., finches, weaver finches, starlings, and jays and crows. Most of the families that are widely distributed in the Palearctic are widely distributed elsewhere too, over much of the world or much of the Old World, but loons and creepers (as well as auks, which are sea birds) are Holarctic, and grouse and waxwings are additional Holarctic subfamilies, and hedge sparrows (1 genus, 12 species) are exclusively Palearctic. Besides the 37 families that are widely distributed at least in summer, 10 additional families (pelicans, ibises, flamingos, button quails, thick-knees, swallow plovers, sand grouse, rollers, bee eaters, and hoopoes) are only marginal (in the warm southern edge) or local in the Palearctic, and 5 more Old World tropical families (bulbuls, cuckoo shrikes, sunbirds, white-eyes, and drongos) are represented in the Palearctic only or chiefly by migratory forms which extend northward in eastern Asia. The final family is represented by two American wood warblers which extend their breeding ranges into eastern Siberia. This geographical classification of families is partly arbitrary (there is no sharp distinction between the families that are "widely distributed" and "marginal" in the Palearctic), but it shows well enough that Palearctic birds are less diverse than but otherwise not much different from Ethiopian and Oriental ones. The relationships of Palearctic birds are shown better by genera than by families: many genera including many that are migratory in the north are shared with the Old World tropics; many genera and even many northern species are shared with North America; and some genera are exclusive to the Palearctic.

The principal Palearctic mammals are hedgehogs (Afr., O.), shrews (wide, but the doubtful subfam. Soricinae is Hol.), and moles (N. A.); canids (wide), bears (O., Amer.), pandas (excl. subfam. localized in southwest China, but fam. also Amer.), mustelids (wide), and felids (wide); horses etc. (Afr.), pigs (Afr., O.), a localized camel (S. A.), deer (O., Amer.), and bovids (Afr., O., N. A.); pikas (N. A.) and rabbits (wide); squirrels (wide) and flying squirrels (O., N. A.), a beaver (N. A.), a fair number of cricetids (wide), localized mole rats (almost excl.), many murids (O. W.), dormice (subfam. excl., but fam. also Afr.), a localized seleviniid (excl.), jumping mice (N. A.), and jerboas (northern Afr.); and bats, mostly vespertilionids (world). Besides these, the following additional families reach small marginal areas in the Palearctic: elephant shrews in North Africa (Afr.), scaly anteaters in eastern China (O., Afr.), Old World monkeys in North Africa and parts of China and Japan (Afr., O.), viverrids (Afr., O.), a hyaena and a hyrax in parts of southwestern Asia (extending from Afr.), ctenodactylid rodents in North Africa, and fruit bats and several families of insectivorous bats (from tropics). The genera of Palearctic mammals are partly endemic, partly shared with the Old World (especially Oriental) tropics, and partly shared with North America. The North American relationships become very strong in the north.

In summary, Palearctic fresh-water fishes are cyprinids (extending from the Orient); a few, mostly localized species of other primary-division families (many with North American relationships); and peripheral fishes (many in the north, with North American relationships). Palearctic amphibians are salamanders (N. A.); frogs of 3 widely distributed dominant genera (*Rana, Bufo, Hyla*); and more or less localized representatives of 4 other families of frogs with Oriental and/or North American relationships. Palearctic reptiles are few; most (but not all) are related toward the African and/or Oriental tropics. Palearctic birds are of less-than-tropical diversity; most are related (at the generic or specific level) toward the Old World tropics or North America. Palearctic mammals are only moderately diverse; most are related (at the generic or specific level) toward the Old World (especially Oriental) tropics or North America.

The Palearctic vertebrate fauna as a whole is much less rich than the Ethiopian and Oriental ones. It is, in terms of existing relationships, a transitional fauna, a complex mixture of Old World tropical and North American groups, the former most numerous southward, the latter northward, with few important endemic groups of its own.

The Palearctic Region is bounded on the south by the Ethiopian and Oriental Regions, and on the north by the limit of land. Westward it includes the British Isles and, conventionally, Iceland, although the latter has a fragmentary fauna; and eastward it includes Japan and is bounded by Bering Strait.

Within the Palearctic Region the fauna is richest in the warmer areas, where they are wet enough, and diminishes northward, until, in most of the arctic, vertebrates are reduced to fresh-water fishes, mostly peripheral; one frog (*Rana*); no reptiles; a few land and more fresh-water birds, mostly migratory; and a few mammals, as few as 8 species in the highest continental arctic. The fauna is reduced and specialized in dry and mountainous places too. There is also some differentiation from east to west, more in the south than in the north: the vertebrate fauna of favorable parts of China is rich (the richest part of the Palearctic) and includes many Oriental elements; that of the Mediterranean Region is rather rich too, but with more Ethiopian elements. Wallace divided the Palearctic into four subregions: Siberian, European, Manchurian, and Mediterranean; but the first two are not very distinct, and the others are not worth formal recognition, although "Mediterranean" is a useful descriptive term for some distributions.

The Nearctic Region

The Nearctic Region (Fig. 50) is North America above the tropics.

The Nearctic too is north-temperate with an arctic fringe. The original cover of much of eastern North America in the mid-latitudes

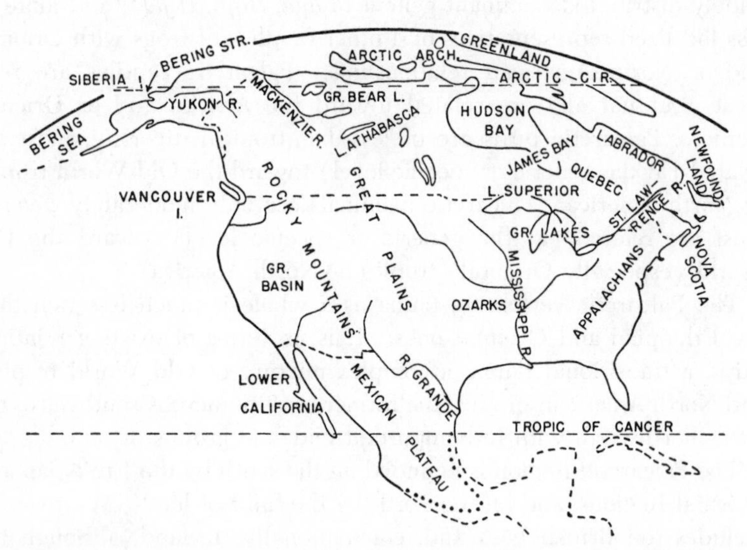

Fig. 50. The Nearctic Region (solid lines) and adjacent land (broken lines). Orthographic projection. Dotted line is approximate boundary with Neotropical Region.

is deciduous or mixed forest. Westward, in the middle part of the continent, are extensive grasslands, and, farther west, a complex of more or less arid country from which rise mountains with strips of mixed or coniferous forest. Northward, as in Eurasia, there is first a great stretch of coniferous forest and then tundra.

Nearctic primary-division fishes are one or 2 paddlefishes (China); the Bowfin (excl. order); a few mooneyes (excl.); suckers (E. Asia), many small cyprinids (O., Pal., Afr.), ameiurid catfishes (almost excl.); a few umbrids (eastern Europe), *Dallia* in Alaska (E. Siberia), and *Esox* (Pal.); a few trout perches and aphredoderids (both excl.); basses etc. (excl. fam.); and many percids (Pal., one subfam. excl. N. A.). Nearctic secondary-division fishes are gar pikes (also Central America) and Cyprinodontes (warm, wide). Besides these, a primary-division characin and a secondary-division cichlid reach the Rio Grande (extending from S. A.), and there are many peripheral fishes, especially northern salmonids (Pal.).

Nearctic amphibians are salamanders (Pal., etc.); a localized leiopelmid (New Zealand); pelobatids (subfam. also Europe); *Bufo* (wide), *Hyla* etc. (wide), and *Rana* (wide); and a few leptodactylids and brevicipitids (extending from S. A.).

Nearctic reptiles are snapping and musk turtles (extending to parts S. A.), emydines (Eurasia and trop. Amer.), an endemic genus of land tortoises in south (subfam. warm, wide), and *Trionyx* (Afr., O., E. Asia); an alligator in the southeast (China) and a crocodile in southern Florida (warm world); geckos in the southwest (warm world), many iguanids (trop. Amer. etc.), *Heloderma* (excl. fam.), xantusiids in southwest (Central Amer. etc.), skinks (wide), teiids (trop. Amer.), amphisbaenids in the south (S. A. etc., Afr.), anguids (trop. Amer., few Pal. and O.), and anniellids in southwest (excl.); and *Leptotyphlops* (trop. Amer., Afr.), a few small boines in west (subfam. also in parts Afr., Asia, etc.), many colubrids (world), coral snakes (trop. Amer., fam. warm world), and pit vipers (temp. and trop. Asia, trop. Amer.).

Nearctic birds include land and fresh-water representatives of about 49 families. About 39 of them are more or less widely distributed in the Nearctic or the warmer parts of it at least in summer. They are grebes, loons, pelicans (local), hawks (3 fams.) and New World vultures (trop. Amer.), herons, ducks etc., quails and grouse, turkeys (nearly excl.), cuckoos, cranes, rails, plovers and sandpipers, gulls, pigeons, owls (2 families), goatsuckers, kingfishers, swifts, hummingbirds (trop. Amer.), woodpeckers, tyrant flycatchers (trop. Amer.), larks (one sp.), swallows, thrushes and mockingbirds (subfam. trop. Amer.) and wrens (subfam. mostly trop. Amer.) etc., motacillids (pipits), shrikes, waxwings, creepers, nuthatches, titmice and chickadees, vireos (trop. Amer.), wood warblers and tanagers and cardinal grosbeaks (fam. trop. Amer.), finches, icterids (trop. Amer.), and crows and jays. This list shows a mixture of many widely distributed families, some dominant New World ones ("trop. Amer."), a few Holarctic ones, and only one (turkeys) exclusive or nearly so. Representatives of the following 10 additional families enter the southern edge of the Nearctic Region from the tropics, reaching to or nearly to the southern edge of the United States: ibises, a stork, a flamingo, a cracid, the Limpkin, a jacana, parrots, a trogon, a motmot (only to northern Mexico), and a cotinga. Of genera of Nearctic birds, some (including many migratory ones) are shared with Central and South America, some (especially northward), with the Palearctic Region, and some are endemic.

Nearctic mammals are an opossum (extending from S. A.); shrews (wide, subfam. also Pal.) and moles (Pal.); an armadillo in Texas etc. (extending from S. A.); canids (wide), bears (Eurasia, S. A.), raccoons etc. (trop. Amer., related pandas in Asia), mustelids (wide), and cats (wide); peccaries in south (extending from S. A.), deer (Eurasia, trop. Amer.), the Pronghorn in the west (excl.), and a few bovids (Pal., O., Afr.); pikas in the west (Pal.) and rabbits (wide); the Sewellel in the west (excl. fam.), squirrels (wide) and flying squirrels (Eurasia etc.), pocket gophers and pocket mice mostly in the west (extending to Central and northern South Amer. respectively), a beaver (Pal.), many cricetids (wide), jumping mice (Pal.), and a porcupine (trop. Amer.); and vespertilionids (wide) and southward a few leaf-nosed bats (trop. Amer.), long-legged bats (trop. Amer.), and molossids (warm world). This fauna is obviously a mixture of families shared with the Palearctic Region (and other parts of the Old World) and families shared with the American tropics, with few important exclusive groups. Nearctic genera of mammals show this same mixture, but with relatively more relationships toward the Palearctic, especially in the north, and with numerous endemic genera in some groups, especially rodents.

In summary, Nearctic fresh-water fishes are numerous and distinctive; most of the families are exclusive or are shared with the Palearctic. Dominant Nearctic amphibians belong to the same groups as Palearctic ones (salamanders, *Rana*, *Bufo*, and *Hyla*), and Nearctic pelobatids are related to Palearctic ones, but a few Nearctic frogs have South American or New Zealand relationships. Nearctic reptiles are a mixture of Palearctic (or Old World) and tropical American groups (the latter especially lizards), with few important exclusive groups.

Nearctic birds are a mixture of Palearctic (Old World) and tropical American groups, the former more numerous northward, the latter southward, although many that winter in tropical America breed far northward. Nearctic mammals too are a mixture of (relatively more) Palearctic or Old World groups and (relatively fewer) tropical American ones, with few important exclusive groups.

The Nearctic vertebrate fauna as a whole, like the Palearctic one, is much less rich than the tropical regional faunas and is mainly transitional, a mixture of Palearctic and tropical American groups, the former more numerous northward, the latter southward, with few important groups of its own except of fresh-water fishes.

Northward, the Nearctic Region includes Newfoundland, the Arctic Archipelago, and Greenland—Greenland mammals are all American, although some of the birds are European. Southward, the Nearctic fauna forms a complex transition with the Neotropical one as described later in this chapter; the boundary is usually set in southern Mexico, where the cooler Mexican highlands meet the tropical lowlands and where many tropical vertebrates reach northern limits.

Within the Nearctic Region, vertebrates are most numerous in the most favorable southern areas (which differ for the different classes), and the fauna diminishes progressively (but irregularly) northward until it is reduced to a small arctic fauna much like the Palearctic one. There is also east-west differentiation, stronger in some groups than in others: *e.g.*, most fresh-water fishes and turtles are concentrated east of the Rocky Mountains; salamanders are more numerous in the east than in the west but form an isolated western subfauna (Fig. 19); there are relatively more snakes in the east and more lizards in the west; and some families of mammals are represented only or chiefly in the west (the Pronghorn, pikas, the Sewellel, pocket gophers, and pocket mice); but birds as a whole, dominant mammals other than rodents, and some families and genera of other classes show much less east-west differentiation. Wallace divided the Nearctic Region into four subregions: a northern one extending across the continent and below it an eastern, a west-central (Rocky Mountain etc.), and Pacific coast ("Californian") one. These subregions obviously correspond to more or less different climatic and vegetation areas, and their faunas are to some degree different; but whether they should be formally distinguished is too unimportant and too complex a matter to be discussed here. Merriam (1892 etc.) divided North America into a more complex system of "life zones" based on temperature.

These zones are useful in some ways, but they have been overemphasized. They are not well-defined faunal subregions.

The Neotropical Region

The Neotropical Region (Fig. 51) is South and Central America and the tropical lowlands of Mexico with Trinidad and, if one wishes, the West Indies proper, but the West Indian fauna is limited and is in

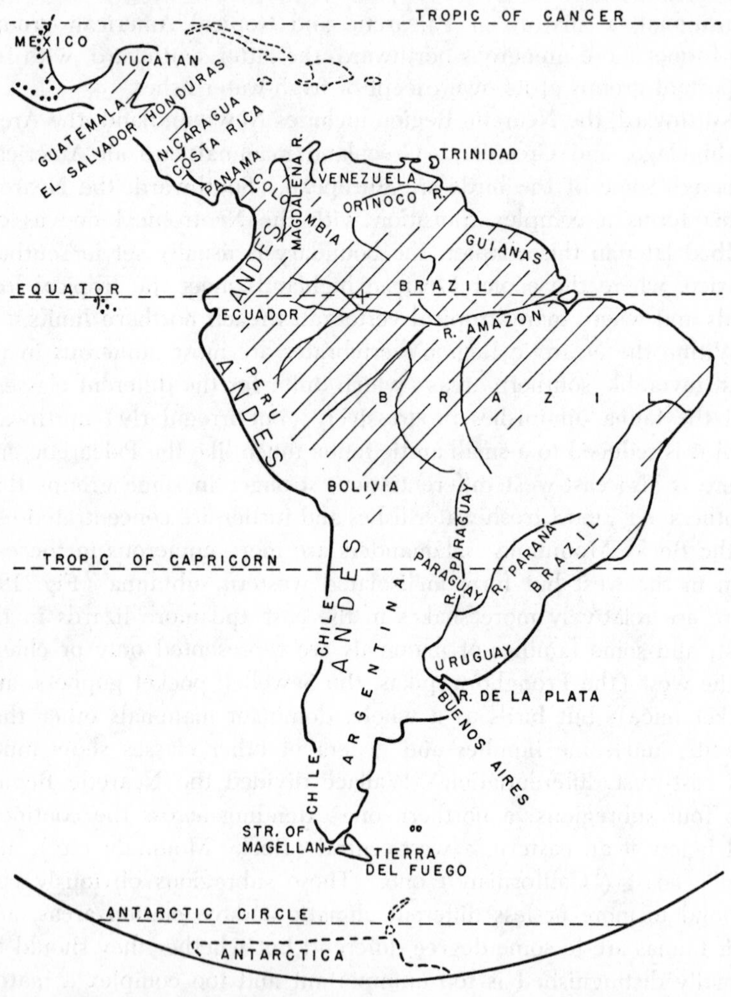

Fig. 51. The Neotropical Region (solid lines) and adjacent land (broken lines). Orthographic projection. Dotted line is approximate boundary with Nearctic Region. Diagonal hatching shows main areas of rain forest.

some ways transitional, and forms a complex fringing pattern rather than part of a continental one.

This region is mostly tropical, but southern South America extends into the south-temperate zone, reaching a latitude equivalent to that of Labrador or England, with temperatures between the two. There is a huge tract of rain forest in the Amazon Valley and separate, smaller tracts elsewhere in South and Central America, as well as tracts of drier forest. There are also extensive areas of savanna and grassland in the tropics, as well as grassland south of them in Argentina, and there are also desert and subdesert areas, especially in western South America but widely scattered elsewhere. Vegetation is in fact very complexly distributed in South America and perhaps still more so in Central America, where many sorts of wet forest, dry forest, savanna, subdesert, and mountain vegetations exist in comparatively small areas. The Andes support a succession of altitudinal forests and grasslands in northern South America, but the mountains become drier southward. Far south in western South America, in southern Chile and the western part of Tierra del Fuego, there is a special, wet, cool-temperate forest named for the southern beech, *Nothofagus,* which forms forests also in New Zealand and southern Australia and Tasmania.

Neotropical fresh-water fishes (pp. 69ff.) are very numerous but belong to few basic groups. They are dominated by certain Ostariophysi: characins (Afr.), gymnotid eels (excl.), and catfishes (excl. fams.). There are also a lungfish (fam. also Afr.), 2 osteoglossids (discontinuously trop.), 2 nandids (Afr., O.), a number of secondary-division Cyprinodontes (warm, wide) and cichlids (Afr.), and some peripheral fishes; and suckers, ameiurid catfishes, and secondary-division gar pikes extend to parts of Central America (from N. A.).

Neotropical amphibians are caecilians (Afr., O.), plethodontid salamanders (most N. A.), pipid frogs (Afr.), *Rhinophrynus* in southern Mexico etc. (excl. fam.), many leptodactylids (A. etc.), *Bufo* (wide), many atelopodids (few in trop. O. W.), many hylids (Hol., A.), a few *Rana* (wide), and brevicipitids (subfam. also O. and extends into N. A.).

Neotropical reptiles are a dermatemydid in northern Central America etc. (excl. fam.), a *Chelydra* in Central America etc. (N. A.), staurotypines in northern Central America etc. (excl.) and *Kinosternon* (N. A.), a few emydines (related N. A. and O.), *Testudo* (warm O. W.), and pelomedusid (Afr.) and chelyid (A.) side-necked turtles; crocodiles (trop.) and caimans (excl.); geckos (warm world), many iguanids (N. A., etc.), xantusiids in Central America etc. (also southern N. A.), a few skinks (world), an elytropsid in southern Mexico (excl. fam.), many teiids (N. A.), amphisbaenids (Afr. etc.), anguids (N. A., Eurasia, etc.), *Xenosaurus* in southern Mexico etc. (excl. fam.); and *Typhlops* (trop. etc.), *Leptotyphlops* (Afr., southern N. A.), one or 2 aniliids (O.), boas etc. (subfam. extending to southern N. A.), many colubrids (world), coral snakes (N. A., fam. warm world), and pit vipers (N. A., Asia).

The Neotropical bird fauna is very rich and very distinct. Of 67 bird families represented on land or fresh water in the region, about half are widely distributed elsewhere; a few have special or doubtful relationships; and nearly half are exclusive or are at least confined to the New World. See pp. 263ff. for further details.

Neotropical mammals are opossums (nearly exclusive, but one extends to N. A.) and caenolestid marsupials (excl.); a few shrews in Central America etc. (extending from N. A.); monkeys (2 excl. fams.); anteaters (excl.), sloths (excl.), and armadillos (nearly excl., one extending to southern N. A.); canids (wide), a bear (N. A., Eurasia), procyonids (N. A.), mustelids (wide), and cats (wide); tapirs (O.), peccaries (extending to N. A.), camels (Pal.), and deer (N. A., Eurasia); rabbits (wide); squirrels (wide), pocket gophers and pocket mice in Central America etc. (both extending from N. A.), many cricetids (N. A., parts of O. W.), New World porcupines (N. A.), and 10 additional families (all excl.) of hystricomorph rodents; and 9 families of insectivorous bats: emballonurids (trop. etc.), noctilionids (excl.), many phyllostomatids (extending to southern N. A.), true vampire bats (excl.), natalids (nearly excl.), furipterids (excl.), disk-winged bats (excl.), vespertilionids (world), and molossids (warm world). This mammal fauna is tabulated and discussed on pp. 332ff.

In summary, Neotropical fresh-water fishes are very numerous and highly endemic (most families exclusive) but with general African relationships. Neotropical amphibians and reptiles are mixed: some groups of them are primarily tropical and subtropical (*e.g.*, caecilians, *Testudo*, crocodiles, *Typhlops*); some, African in their relationships (*e.g.*, pipid frogs, pelomedusid turtles, amphisbaenid lizards, *Leptotyphlops*); some, Oriental (a subfam. of brevicipitid frogs, a genus of emydine turtles, and aniliid and some groups of colubrid snakes); some, Australian (leptodactylid and hylid frogs and chelyid turtles); and some, North American (plethodontid salamanders, iguanid and teiid lizards, and many genera of other families). Neotropical birds are very numerous, and very many of them belong to exclusive families, though others are wide-ranging. Neotropical mammals are a mixture of very distinct exclusive families and families shared with other parts of the world, usually including North America.

In short (to bring in the past for just a moment), the whole existing Neotropical vertebrate fauna is a mixture of surviving parts of an old, endemic, Tertiary fauna of which the geographical relationships are now diverse (presumably determined by extinctions and survivals more than by directions of dispersal) and a new fauna received via North America. The old fauna includes all the strictly fresh-water fishes (except a few recent arrivals in northern Central America), probably a majority of the amphibians, probably a majority of the reptiles in South America but relatively fewer of those in Central America, very many birds (but the old and new groups are not so clearly distinguishable among birds), and about half the mammals.

The northern boundary of the Neotropical Region is the southern boundary of the Nearctic (above).

The main pattern of distribution of vertebrates within the Neotropical Region is one of richness in the main tropical part of South America; subtraction, mixture with additional North American elements, and some differentiation in Central America; subtraction and some (but in most cases not much) differentiation southward, in southern South America; and less important differentiation of rain forest, grassland, desert, and mountain faunas. Wallace distinguished "Brazilian" (main South American), "Mexican" (Central American), and "Chilean" (south-temperate) subregions, which, of course, are not formally recognized here.

The Australian Region

The Australian Region (Fig. 52) is Australia and New Guinea, with Tasmania and certain smaller islands which have (depauperate) continental faunas.

The Australian Region is partly tropical and partly south-temperate. New Guinea is wholly within the tropics. Much of the island is covered with rain forest, but there are also areas of grassland as well as a series of altitudinal vegetations on the mountains, and on the south coast are areas of open eucalyptus woods like those of Australia. Northern Australia too is tropical, with diverse vegetations which include, in northern Queensland, a series of isolated patches of fine rain forest called "scrub" locally. Farther south, in the temperate zone, eastern and southeastern Australia is fairly well watered, with much open eucalyptus woods and some wetter, denser forest on the mountains. Most of the interior of Australia is arid. Southwestern Australia is moderately wet but is cut off by deserts from the east. There are some fine forests of big trees in the southwest. Tasmania is cool-temperate, equivalent in latitude to New York or Boston (but not so cold in winter), and is partly wet and forested.

Fig. 52. The Australian Region (solid lines) and adjacent land (broken lines). Orthographic projection. Dotted lines are approximate boundaries of the Australian and Oriental Regions. Diagonal hatching shows main areas of rain forest.

Australian fresh-water fishes are a ceratodontid lungfish (excl. fam.) and an osteoglossid (O. etc.), both restricted in distribution, and various peripheral fishes.

Australian amphibians are only frogs: leptodactylids (S. A. etc.), hylids (S. A. etc.), a few ranids (wide), and brevicipitids (warm world, but subfams. are nearly excl.).

Australian reptiles are *Carettochelys* (excl. fam.) and a trionychid (species also O.) in New Guinea, and chelyid turtles (S. A.); crocodiles (trop.); many geckos (warm world), agamids (warm O. W.), many skinks (world), a *Dibamus* in New Guinea (O.), *Varanus* (warm O. W.), and pygopodids (excl.); and *Typhlops* (trop. etc.), perhaps an aniliid in New Guinea (O. etc.), pythons etc. (warm O. W.), some colubrids in New Guinea and parts of Australia (world), and many elapids (warm world).

The birds of the Australian Region include representatives of about 58 families on land and fresh water. Of these, 44 families are more or less widely distributed elsewhere, over the world, or the tropics, or all or much of the Old World. Two small families (frogmouths and wood swallows) are divided between the Australian and Oriental Regions. The 12 remaining families are exclusive to the Australian Region or occur primarily there: cassowaries (6 species), emus (2), megapodes (to edge O.) (10), owlet frogmouths (also some islands) (8), lyrebirds (2), scrubbirds (2), flowerpeckers (into O.) (54), honey eaters (also many islands and doubtfully S. Afr.) (159), bell magpies (11), magpie larks (4), bowerbirds (17), and birds of paradise (43). These families total 318 species, including the few extra-Australian ones. The Australian Region also has more than its share of pigeons and parrots, including important exclusive groups, and many "Australian warblers" (Malurinae) (also Polynesia etc.) (85 species). The proportion of Australian birds in exclusive or nearly exclusive families is larger than in any other continental faunal region except the Neotropical.

The native land mammals of the Australian Region are only a few monotremes (excl. subclass), many marsupials (6 excl. fams.), rather diverse rodents of the family Muridae (O. W.), and fruit bats (warm O. W.) and 6 families of insectivorous bats (all occurring elsewhere). This mammal fauna is further analyzed and discussed on pp. 335ff.

In summary, the vertebrate fauna of the Australian Region consists of only one or 2 fishes of strictly fresh-water groups, one of them a striking relict, plus peripheral fishes; 4 families of frogs, none strikingly primitive, and none exclusive, although 2 subfamilies of brevicipitids are nearly so; a moderate variety of reptiles, none strikingly primitive and few in exclusive families; a fair variety of birds, a fair proportion in exclusive families; and a very limited mammal fauna, with strikingly primitive and exclusive elements.

To bring in the past again for just a moment, the Australian vertebrate fauna is made up of a few very old endemic groups which have survived or radiated in Australia for a very long time, plus an accumulation of additional groups of salt-tolerant (but perhaps no strictly fresh-water) fishes and of frogs, reptiles, birds, rodents, and bats. The most extreme endemics are the lungfish, monotremes, and peculiar superfamilies of marsupials. Australia has no comparable relict or endemic groups of amphibians, reptiles, or birds; if some of these animals have been in Australia as long as marsupials have, they have not radiated to the same extent. Some of the older Australian vertebrates (some frogs, turtles, and the marsupials) now find their living relatives principally in (*tropical*) South America; in most cases this is apparently a result of withdrawal or extinction in the main part of the world. Strictly fresh-water fishes, terrestrial reptiles, and birds seem to show no particular relationships between Australia and South America, and the Australian vertebrate fauna as a whole has little direct relationship with the African fauna. Many of the more recent Australian vertebrates (ranid frogs and many reptiles, birds, rodents, and bats) still have relatives in tropical Asia.

The northwestern boundary of the Australian Region is discussed under "Wallacea," below. Whether or not various islands to the east of Australia and New Guinea are included in the Australian Region is a matter of preference rather than of fact. I prefer to treat them separately (Chapter 8), as forming a fringing pattern.

Within the Australian Region, the faunas of Australia and New Guinea are somewhat differentiated. Among other differences, Australia alone has the lungfish; Australia has more leptodactylid frogs and New Guinea more brevicipitid frogs; Australia has most pygopodid lizards, New Guinea, relatively more colubrid snakes; Australia has emus, New Guinea, most cassowaries; Australia has lyrebirds and scrubbirds, New Guinea, most birds of paradise; Australia has the Platypus and more (and more specialized) marsupials, and New Guinea has relatively more rodents. But these details are not very important. The vertebrate faunas of Australia and New Guinea are much alike as wholes (see comparison of the mammals, p. 335, and of the birds, p. 262); they share many groups, even many genera and species; and there is extensive transition between them. Various Australian animals extend to the areas of open eucalyptus woods in southern New Guinea, and various New Guinean animals extend to the areas of rain forest in northeastern Australia. Within Australia, there is some further faunal differentiation from north to south, and between the wetter eastern zone, the great central and western arid area, and the moderately wet southwestern corner of the continent. However, as in other cases, I see no need to divide the region into formal subregions.

Transitions between regional faunas

Wherever regional faunas meet or are separated only by partial barriers, there is transition: overlapping of faunal elements, with progressive subtraction, in both directions.

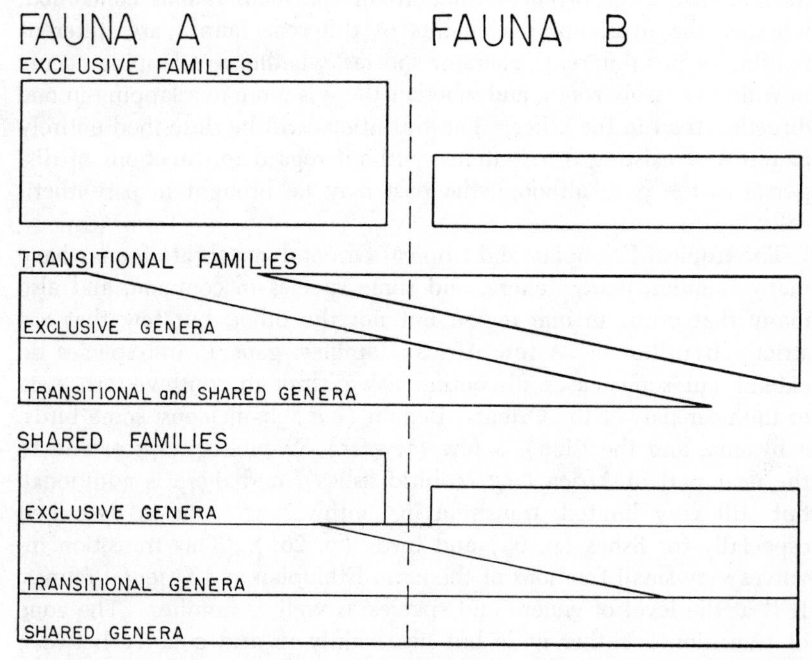

Fig. 53. Diagram of transition of two adjacent regional faunas (simplified). Each fauna consists of exclusive, transitional, and shared families; and the transitional and shared families include exclusive, transitional, and shared genera. The exclusive families and genera determine the boundary (broken line) between the regions. Fauna A is larger than B and contributes more to the transition; that is, transition is mainly toward B.

Faunal transition is a very complex thing (Fig. 53). Two adjacent regional faunas usually consist of some families, genera, and species which are shared, some which occur mostly in one region but extend into parts of the other (these are the strictly transitional elements), and some which occur in one region but not the other; but this is just the beginning of the complexity. A family shared by two regions may contain transitional genera; shared genera may contain transitional species; a family transitional in one direction may contain genera or species transitional in the other direction; different transitional groups extend for different distances; and situations are usually complicated in other ways and by special or doubtful cases. To analyze transitions in full detail would require careful planning and elaborate statistics. The most I can do here is to describe the principal transitions of regional faunas in a general way and tell whether they involve large or small fractions of the main faunas concerned, whether the overlapping elements of different faunas are different families or just different genera or species, whether overlapping occurs in wide or narrow zones, and whether there is more overlapping in one direction than in the other. The transitions will be described entirely in terms of existing distributions, without regard to directions of dispersal in the past, although the past may be brought in parenthetically.

The tropical Ethiopian and tropical Oriental vertebrate faunas have many families, many genera, and some species in common, and also many that occur in one region but not the other, but few that are strictly transitional. A few African families, genera, and species do extend (in some cases discontinuously) through southwestern Asia to the near part of the Oriental Region (*e.g.*, chameleons, some birds, a hyaena, and the Lion); a few (fewer?) Oriental groups extend to the near part of Africa (*e.g.*, cobitid fishes); and there is additional, but still very limited, transition in southwestern Asia as described especially for fishes (p. 65) and birds (p. 261). This transition involves very small fractions of the main Ethiopian and Oriental faunas. It is at the level of genera and species as well as families. The zone of transition is rather wide but lies mainly in arid country between more than within the limits of the main faunas. And the direction of transition is doubtful; it is doubtful whether there is more extension of African vertebrates toward the Orient or of Oriental ones toward Africa. There is also a mixture of (many) African and (fewer) Oriental vertebrates in Madagascar (Chapter 8) which makes a transition of a sort, but the African groups in Madagascar rarely extend to the Orient that way (salt-tolerant cichlid fishes are an exception), and the Oriental groups that reach Madagascar rarely extend to Africa. The Madagascan fauna therefore seems to be an accumulation of animals received from two directions (but most from Africa) rather than part of a fauna exchanged by Africa and India across Madagascar. (To turn to the past for a moment, the fact that the Ethiopian and Oriental faunas have much in common in spite of the small amount of actual transition now suggests that they have exchanged more vertebrates in the past than recently. That the exchange has been through southwestern Asia is indicated by the real if limited transition of African and Oriental faunas along that route now, while Madagascar does not seem to have been a route of exchange.)

The tropical Ethiopian and temperate Palearctic regional faunas share many groups of vertebrates, although there are many others (chiefly on the Ethiopian side) that are not shared. Actual transition, however, is limited, although some does occur in the Sahara and the Mediterranean area. Most of it is due to extension of Ethiopian (tropical) families, genera, and species into the southern edge of the Palearctic: *e.g.*, cyprinodontid fishes; *Testudo,* geckos, chameleons, amphisbaenids, etc.; birds, of which various Ethiopian (tropical) groups extend for different distances into southern Europe as residents or migrants; and elephant shrews and macaques (in Palearctic North Africa), a porcupine, a hyaena, viverrids, a hyrax, and bats. Extension of Palearctic groups into the edge of the Ethiopian Region occurs in comparatively few cases: *e.g.*, an emydid turtle probably extends into part of western tropical Africa, and *Microtus* (voles) extends to a small part of non-Palearctic or transitional North Africa. This transition between the Ethiopian and Palearctic vertebrate faunas involves rather small fractions of the main faunas. It is at the level of genera and species as well as families. The zone of transition is only moderately wide, and much of it is in arid country between the limits of the main faunas. And the direction of transition is mostly northward: many more Ethiopian (tropical) vertebrates extend into the edge of the Palearctic Region than vice versa.

Transition between the tropical Oriental and temperate Palearctic faunas is extensive in eastern Asia in all classes of vertebrates. A number of families, genera, and species of Oriental fresh-water fishes extend into temperate China and several reach the Amur. Several Oriental families and genera of frogs extend into temperate China, and the brevicipitid *Kaloula* reaches Manchuria, and *Rhacophorus,* northern Honshu in Japan; and, in the other direction, a few Palearctic salamanders and *Hyla* extend for comparatively short distances southward, into the edge of the Oriental Region. Many Oriental groups of reptiles extend for various distances into temperate Asia. So do many Oriental birds, both resident and migratory ones. And of mammals, many Oriental families and lesser groups extend for various distances into temperate Asia; and a few Palearctic ones extend into the Oriental tropics: *e.g.*, the Serow (representing a Palearctic tribe of bovids) extends to Sumatra; a mole (representing a Holarctic family), to the Malay Peninsula; and the Wolf (a Holarctic species), to part of tropical India. A substantial part especially of the Oriental vertebrate fauna (but still only a part of it) is involved in this transition. The transition is at the level of genera and species as well as families. The transition occurs in a broad zone of favorable country within the limits of the main faunas concerned. And the direction of transition is mainly northward; many more Oriental vertebrates extend into the Palearctic than vice versa.

That the transitions of both the Ethiopian and Oriental vertebrate faunas with the Palearctic fauna are formed mainly by northward extensions of tropical groups reflects the general reduction of vertebrate life north of the tropics and the fact that in many ways the Palearctic fauna is little more than a depauperate fringe of the tropical faunas.

Transition between the temperate Palearctic and temperate Nearctic faunas is extensive through the north in all classes of vertebrates. The northern parts of both faunas are alike in many ways, with even many species in common; and the common, northern, transitional subfauna grades into the more characteristic Palearctic and Nearctic faunas southward. This transition involves considerable parts of both faunas. Most of it is at the level of genera and species rather than families. It occurs in a broad zone, much of which lies in rather unfavorable, cold country. And the direction of the transition is doubtful: it is doubtful whether (in terms of existing distribution) there is more extension of Palearctic groups into the Nearctic than vice versa.

In a broader way, the Palearctic and Nearctic faunas are themselves in many ways transitional between the faunas of the Old and New World tropics, but this is beside the point here.

The transitions summarized above are within the limits of Megagea, the main part of the world. Transitions between the faunas of Megagea and those of the barrier-limited regions are more striking (because the faunas are more different) and are worth more detailed discussion.

Central American-Mexican transition area

The vertebrate faunas of the Nearctic and Neotropical Regions are very different in many ways, but they also share many families and genera and some species, and there is transition especially in Mexico and Central America (Fig. 54), and to some extent in a still wider area.

Among fresh-water fishes, some Nearctic primary-division families do not even reach the Rio Grande; centrarchids and percids reach the Tropic of Cancer in Mexico; cyprinids reach the Rio Balsas in southern Mexico; and suckers and ameiurid catfishes reach Guatemala. Of Neotropical primary-division families, some do not even reach Central America; others do not extend beyond Panama; the northern-

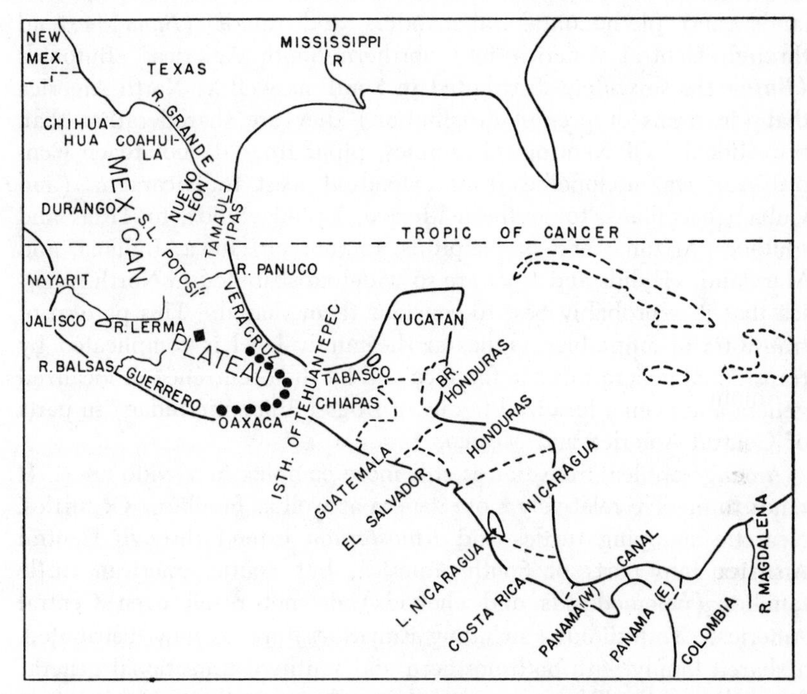

Fig. 54. The Mexican-Central American transition area. Heavy dotted line is approximate boundary between Nearctic and Neotropical Regions.

most gymnotid eel reaches Guatemala; the northernmost South American fresh-water catfish, the Rio Panuco on the east coast of Mexico; and the northernmost characin, the Rio Grande. Among strictly fresh-water fishes, then, transition involves only very small fractions of the Nearctic and Neotropical faunas; the fractions represent different families; and overlapping occurs only from the Rio Grande to Guatemala (see Fig. 15). Secondary-division families overlap more widely: Nearctic gar pikes extend south to Lake Nicaragua; Neotropical cichlids, north to the Rio Grande; and Cyprinodontes extend so far into North America that they might be considered shared rather than transitional.

Among amphibians, transition involves larger parts of the two faunas and a much wider area, but transition is still mostly at the family level. Of Nearctic families, some (especially some families of salamanders) extend south only to Texas or the northern corners of Mexico; ambystomid salamanders and pelobatid frogs cross the Tropic of Cancer on the plateau of Mexico but do not reach the tropical lowlands; and plethodontid salamanders and ranids (*Rana*) extend through Central America into northern South America. Bufonids (*Bufo*) are so widely distributed in South as well as North America that (in terms of present distribution) they are shared rather than transitional. Of Neotropical families, pipid frogs do not reach Central America; atelopodid frogs extend at least to Nicaragua (and Cuba); caecilians, to southern Mexico; leptodactylids, to Texas and southern Arizona; and brevicipitids, to southern Iowa, Indiana, and Maryland. Hylids and *Hyla* are so widely distributed in North America that it is probably best to consider them shared. This pattern of transition of amphibian faunas at the family level is complicated by transitional genera in some families and by the occurrence of localized genera and even a localized family of frogs (Rhinophrynidae) in parts of Central America and Mexico.

Among reptiles, transition is still more complex in a wide area. It tends to involve relatively more genera as well as families. Of turtles, Nearctic snapping turtles and *Kinosternon* extend through Central America into parts of South America; but South American turtle families (pelomedusids and chelyids) do not reach even Central America. Testudinids (including Emydines) are, as now distributed, a shared family with both northern and southern transitional genera: from the north, *Terrapene* extends south to southern Mexico, and *Pseudemys,* through Central and much of South America (perhaps it should be reckoned a shared rather than transitional genus); and from the south, *Testudo* extends north to Panama, and *Geoemyda,* through Central America to parts of Mexico. Of crocodilians, the Nearctic alligator does not extend south of the Rio Grande; but Neotropical caimans and a crocodile extend north to parts of Mexico and (the crocodile) southern Florida. Of lizards, the principal families in America are mostly tropical, extending for different distances northward, but the pattern of distribution of genera is very complex; some genera are mainly Nearctic and extend into Central America etc. as northern transitional elements, and others extend northward from South America. Of snakes, the northern transitional elements in Mexico and Central America are mostly genera of colubrids and pit vipers. Of Neotropical snakes, *Typhlops* reaches southern Mexico; *Leptotyphlops,* southernwestern United States; and many South American genera of colubrids extend for varying distances into Central and southern North America. Additional details of the complex transition of reptile faunas in Central America and Mexico are given in Chapter 4 (pp. 195–198). A surprising number of North American reptiles extend into Central America without reaching South America. And the transition is complicated by the presence of a number of localized, endemic or relict genera and even families of reptiles in Mexico and Central America (see again Chapter 4, pages indicated).

Among birds, transition of Nearctic and Neotropical faunas is very extensive and complex, involving many genera and species as well as families and covering not only Central America and Mexico but also much of North and South America. Some (but comparatively few) Nearctic groups extend into the Neotropics as breeding birds: *e.g.,* turkeys, to Guatemala; kinglets, creepers, titmice, evening grosbeaks, crossbills, and crows, to parts of Central but not South America; the Horned Lark, to northern South America; etc.; but the most conspicuous Nearctic elements in the transition are the many genera and species which (regardless of their geographical origin) now breed in the Nearctic but winter in the edge of the Neotropical Region. In the other direction, many (but far from all) families of the rich South American bird fauna extend through Central America; many important groups (*e.g.,* tinamous, puffbirds, jacamars, toucans, the superfamily Furnarioidea, manakins, honey creepers, and many smaller families as well as very many genera and species) reach a northern limit in southern Mexico; and many other Neotropical groups extend for different distances into temperate North America: *e.g.,* motmots reach northern Mexico; cracids, jacanas, parrots, trogons, and cotingas reach southern Texas and/or Arizona; ibises and limpkins reach southeastern United States; New World vultures extend farther north, and hummingbirds, still farther, until it becomes a question what groups are to be considered transitional and what shared, in terms of present distribution. This pattern of transition of families is reinforced and made far more complex by numbers of transitional genera and species.

Among mammals, too, there is extensive and complex transition of the existing Nearctic and Neotropical faunas, formed by overlapping of families, genera, and species. The main part of the transition occurs in southern Mexico and Central America, but some Nearctic elements extend for varying distances into South America, and some (but fewer) Neotropical ones extend into parts of temperate North America. Further details of this transition of mammal faunas are given in Chapter 6 (pp. 333–335).

This has been no more than a superficial sketch of the transition of the Nearctic and Neotropical vertebrate faunas. This transition has never been adequately treated in detail by zoogeographers. It has received nothing like the attention given to Wallace's Line and Wallacea, although it is probably equally important. I have not space for a more detailed treatment, but a few generalities are worth giving in summary of it. They are offered tentatively, for consideration, rather than as final conclusions.

Transition of Nearctic and Neotropical faunas occurs in Mexico and Central America in all classes of vertebrates, but it is different in the different classes. It is simplest (involves fewest elements), is at the highest level (involves mostly different families), and is confined to the narrowest area in the classes that disperse most slowly. Among strictly fresh-water fishes, the transition is formed by few elements; the Nearctic and Neotropical components all belong to different families; and the area of overlap extends only from the Rio Grande to Guatemala. At the other extreme, the transition of Nearctic and Neotropical birds is formed by a great number of elements; most of them are only genera and species, although some different families are involved too; and the area of overlap includes much of North and South America. The complexity, level, and width of the transition in other classes of vertebrates are between these two extremes. All this is as would be expected: transition is proportional to the powers of dispersal of the different animals concerned.

Transition of the temperate North American and tropical South American vertebrate faunas has just been summarized *as a complex overlapping in a broad zone,* in terms of present distributions, but the transition can be looked at in other ways. An actual boundary line can be drawn between the Nearctic and Neotropical faunas, and transition can be considered as crossings of the boundary, still in terms of present distributions. Or either overlapping in a broad zone or crossings of the boundary may be considered in terms of past movements rather than present distributions. A writer must be clear in his own mind which of these points of view he is taking at any particular moment, and readers must know, too. I am now going to try to summarize the direction of transition *across a boundary line* between the Nearctic and Neotropical faunas, in terms of present distribution.

The fauna of Central America is much more Neotropical than

Nearctic in all classes of vertebrates, except perhaps in reptiles. This is, presumably, because Central America is tropical. Many of the Neotropical groups extend to the tropical lowlands of Mexico but no farther; it is principally these groups that determine the commonly accepted boundary between the Neotropical and Nearctic Regions, in southern Mexico about at the line between the tropics and the north-temperate zone. On the other hand, the fauna of the tableland of Mexico is mainly Nearctic, and some Nearctic groups reach southern limits in southern Mexico near the northern boundary of the Neotropical groups. The question now is whether transitional groups extend across the boundary in one direction more than in the other.

Few strictly fresh-water fishes extend across the boundary line in either direction, and none of them goes far. Of amphibians, a few groups extend across the boundary in each direction, and for roughly equal distances in each direction. Of reptiles, many extend across the boundary in each direction, but the number of Nearctic groups extending to Central but not South America is unexpectedly large. Of birds, there is more northward than southward extension of nonmigratory groups, but the difference is partly made up by migrants that breed in the north and winter southward into the edge of the tropics. Of mammals, however, a number of Nearctic groups extend across the boundary into the Neotropics, but few Neotropical ones (except bats) extend far northward.

That different classes of vertebrates seem to differ in direction of transition across the boundary of the Nearctic and Neotropical faunas suggests differences in relative dominance. Fresh-water fishes and amphibians extend across the regional boundary about as much southward as northward, which suggests that the Nearctic and Neotropical faunas of these animals are about equally dominant. Mammals (other than bats) extend more southward than northward; the Nearctic mammal fauna is evidently dominant, perhaps because so many "Nearctic" mammals really represent dominant Megagean groups, while parts of the South American fauna have been isolated through the Tertiary. Of reptiles, too, more probably extend southward (at least to Central America) than northward; this again is probably partly because some "Nearctic" reptiles really represent dominant Megagean groups. Of birds, however, more extend northward than southward; the Neotropical bird fauna seems to be relatively dominant, perhaps because it is a very large fauna and has been less isolated (because birds can fly) than the rest of the vertebrate fauna of South America. It seems, then, that the Nearctic and Neotropical faunas are about equally dominant in their fishes and amphibians; the Nearctic may be dominant in at least some reptiles as well as mammals; and the Neotropical is dominant in birds. These are tentative, oversimplified findings. I am not sure even of the situations, much less of the explanation of them. I hope that someone else will make a more thorough analysis and explanation of Nearctic-Neotropical faunal transition.

Wallace's Line, Weber's Line, and Wallacea

The faunas of the Oriental and Australian Regions are extraordinarily different. What happens where they meet in the Indo-

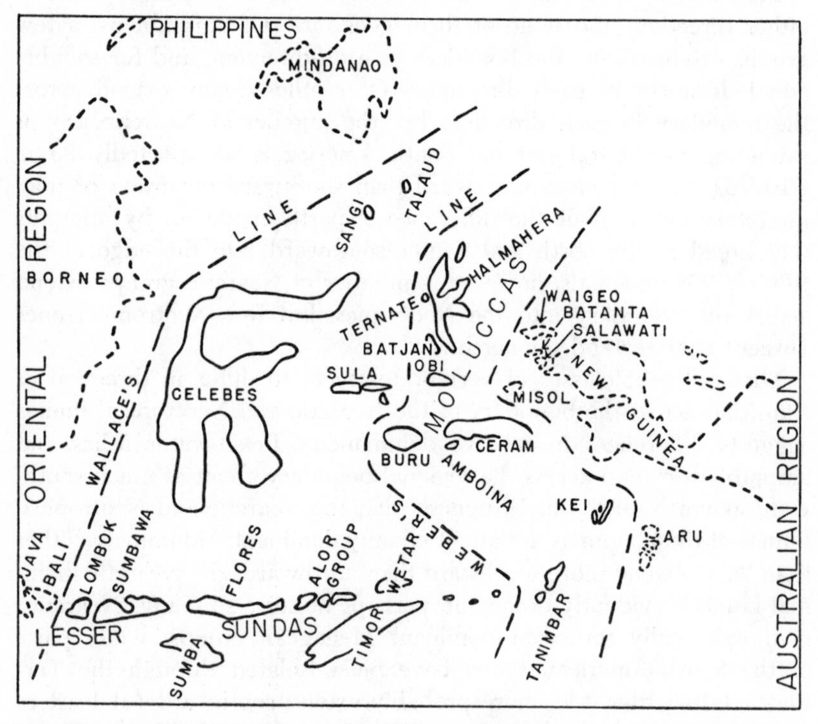

Fig. 55. Islands of Wallacea (solid lines) and adjacent land (broken lines). Lines of dashes show Wallace's Line, Weber's Line (the line of faunal balance), and the boundary of the Australian Region.

Australian Archipelago (Fig. 55) has long fascinated zoogeographers. In fact "Wallace's Line," the supposed boundary between the Oriental and Australian faunas, was the focus of generations of zoogeographers after Wallace.

As first drawn by Wallace (1860), the line ran between Bali and Lombok, between Borneo and Celebes, and between the Philippines and the Sangi and Talaud Islands (and the Moluccas). The difference between the faunas of Bali and Lombok, which are separated by a (deep) strait only about 15 miles wide, was especially emphasized. Later authors debated the position of the line, especially the northern part of it. Huxley, for example, who named Wallace's Line, thought it should split the Philippines, putting Palawan in the Orient and the rest of the Philippines in the Australian Region. Still later even the existence of the line was debated—whether there was any definite boundary between the Oriental and Australian faunas. Mayr (1944) outlines the history of the debate. I cannot give this matter the space it deserves but shall indicate how the main classes of vertebrates are distributed between the Orient and Australia, and what alternatives their distribution suggests in the way of regional boundaries or transition areas. The Philippines will not be discussed now. They lie to one side of the main line of transition and are best treated as fringing islands (pp. 500–506), although their fauna is transitional in some ways.

Of strictly fresh-water fishes, there is a great fauna in the Orient, and the main part of it reaches Sumatra, Java, and Borneo, but that is the limit of most of it. Of the many strictly fresh-water groups on Borneo, none reaches Celebes, except two or three species probably carried by man. On Java, there is some decrease of the Oriental fish fauna from west to east, although I cannot give the details of it; and no strictly fresh-water fishes except a few small cyprinids reach the Lesser Sundas, except again certain species probably carried by man. Bali does *not* have any significant fraction of the Javan fish fauna (p. 51). This is a fact that ichthyologists have been reluctant to accept. The only strictly fresh-water fishes apparently native on Bali are two or three small cyprinids of the genera *Puntius* and *Rasbora,* and both genera occur also on Lombok, and *Rasbora* extends to Sumbawa too. Secondary-division cyprinodonts and their derivatives do reach Celebes, and the probably salt-tolerant cyprinodont genus *Aplocheilus* extends also to Lombok and Timor. At the other end of the archipelago, the Australian Region is almost completely without primary-division fishes, and the one or two that are there (the lungfish and osteoglossid) are localized. However, certain small Australian fishes of the atherinid subfamily Melanotaeniinae are supposed to be zoogeographically significant. They are derived from marine or salt-tolerant ancestors, but they are thought to be confined to fresh water now. They occur from the warmer part of Australia to New Guinea, the Aru Islands, and Waigeo (Beaufort 1926, p. 103, map) and so are apparently confined to the limits of the Sahul shelf. However, their distribution and salt tolerance are perhaps not yet fully known.

Of amphibians, caecilians are not known east of Java and Borneo (disregarding the Philippines). Of Oriental frogs, the pelobatid *Megophrys* and the brevicipitid *Microhyla* extend to Bali; *Bufo,* to Celebes and to Bali and Lombok; the brevicipitid *Kaloula,* to Celebes and along the Lesser Sundas to Flores; *Rhacophorus,* to Celebes and to Timor; and ranids, including *Rana* in a broad sense, to New Guinea, the Solomons, etc., and northern Australia. Of frogs of the Australian Region, leptodactylids reach New Guinea and the Aru Islands; the brevicipitid subfamily Asterophryinae extends to the Moluccas; Australian forms of *Hyla* extend in decreasing numbers to the Moluccas, Timor, and the eastern Lesser Sundas, to Sumba; and one genus of sphenophryine brevicipitids (*Oreophryne*) extends from New Guinea to the Moluccas, Celebes (some Philippines), and the Lesser Sundas from Flores to and including Bali. The distributions of genera and especially of species of frogs are really more complex than this summary suggests, and some of the details given may be wrong, but the general pattern is certainly correct. A noteworthy detail is that 10 species of frogs are considered the same in Borneo and Celebes, on opposite sides of Wallace's Line, but some of them may have been carried by man. For example, Parker (1934, pp. 85–86) thinks *Kaloula pulchra* may have been introduced into Celebes, although another Oriental species of the same genus (*K. baleata*) is supposedly native on Celebes.

The distribution of reptiles between the Oriental and Australian Regions is too complex to describe in detail. Examples have been given in Chapter 4 (pp. 192–193) of families, genera, and even species that range continuously, so far as the land allows, from Asia to Australia; of genera that occur in both regions but are absent on some intervening islands; and of genera confined to the islands. There are also many Oriental groups of reptiles which do not cross Wallace's Line, and many others which extend for varying distances toward the Australian Region but do not reach it. In short, a considerable part, but only part, of the Oriental reptile fauna extends toward Australia, with progressive subtractions increasing with distance from the Orient, but with many groups actually reaching New Guinea and fewer reaching Australia; but the pattern is complicated by shared groups, by discontinuities in some cases, and by genera localized within the transition area. On the other hand, there is comparatively little extension of reptiles of Australian groups toward the Orient. Only a few Australian snakes (elapids) extend west even to the Moluccas. I have not analyzed the distribution of genera in the widely distributed lizard families Gekkonidae, Agamidae, and Scincidae, but a leafing through of these families in Rooij (1915) suggests that the Australian-New Guinean elements usually do not extend far westward. The gradual change in percentages of western (Oriental) and eastern (Australian) reptiles along the Lesser Sundas etc. to the Kei and Aru Islands is shown in Table 12.

TABLE 12. PERCENTAGES OF WESTERN (ORIENTAL) AND EASTERN (AUSTRALIAN) REPTILES ALONG THE LESSER SUNDAS ETC. TO THE KEI AND ARU ISLANDS (AFTER MERTENS 1934, P. 41)

	Western, %	Eastern, %
Bali	93.9	6.1
Lombok	85	15
Sumbawa	87.2	12.8
Flores	78.1	21.9
Alor	75	25
Wetar	65.2	34.8
Tanimbar	40.6	59.4
Kei	32.5	67.5
Aru	22.2	77.8

The distribution of birds between the Oriental and Australian Regions is even more complex than that of reptiles. There are many shared groups, many transitional ones (families, genera, and species), and many which occur in one region but not the other and which do not enter into the transition. The transition itself is very complex and covers a very wide area, for some Oriental birds extend to New Guinea or northeastern Australia, and some (but fewer) Australian ones extend into the edge of the Oriental Region, across Wallace's Line (see pp. 262–263). I shall not try to give all the details here. Some of them are given by Mayr (1944) and by authors cited by him. Ornithologists have stressed the difference between the birds of Bali and of Lombok. Statistically, there is a change of only 14.5 per cent of the birds from Oriental to Australian types from Bali to Lombok, and almost three-fourths of the birds of Lombok are still Oriental, but the dominant, conspicuous species are said to be mostly Oriental on Bali, mostly Australian on Lombok. The actual percentages of western and eastern birds along the Lesser Sundas from Bali to the Alor Group are given in Table 13.

TABLE 13. PERCENTAGES OF WESTERN AND EASTERN BIRDS ON LESSER SUNDA ISLANDS (AFTER MAYR 1944, P. 3)

	Western, %	Eastern, %
Bali	87	13
Lombok	72.5	27.5
Sumbawa	68	32
Flores	63	37
Alor	57.5	42.5

Of mammals, many Oriental families, genera, and species reach Borneo and Java, but only a few extend farther east. The Bornean (Oriental) groups of flightless land mammals (other than murid rodents) that reach Celebes are shrews (*Crocidura* and *Suncus*), tarsiers, monkeys (*Macaca* and the endemic genus *Cynopithecus*), a porcupine (*Hystrix*), squirrels, viverrids (4 genera, one of which is endemic), pigs (including the endemic genus *Babirussa*), deer (*Cervus*), and cattle (the endemic Anoa); and a small elephant occurred on Celebes in the Pleistocene (Hooijer 1951). Of these mammals on Celebes, the following extend still farther east, to some of the Moluccas: shrews, *Cynopithecus* monkeys (Batjan only, apparently introduced), 2 civets (*Viverra tangalunga* and *Paradoxurus hermaphroditus*), the Babirussa (only to Buru), pigs, and deer; but they are apparently not at all or not much differentiated on the Moluccas, and some or all may have been introduced there. *Paradoxurus hermaphroditus* is recorded even on the Kei and Aru Islands but is presumably introduced there. Java lacks some Oriental mammals that occur on Borneo. This fact was known to Wallace and is shown by comparison of the lists of mammals of Borneo and Java given by Carter, Hill, and Tate (1945)—these authors also give separate lists of the mammals of many other islands in the Indo-Australian Archipelago, although some of the lists are not quite accurate. Nevertheless, Java still has a good mammal fauna. Many groups of mammals that reach Java do not reach Bali: *e.g.,* hairy hedgehogs (gymnures), flying lemurs, lorises, gibbons, flying squirrels (4 genera), some other genera of rodents, the Asiatic Wild Dog (*Cuon*), mustelids (weasels, badgers, and otters), the Leopard, a rhinoceros, chevrotains, and probably wild buffalo. Bali does *not* share the main part of the Javan mammal fauna (*cf.* Carter, Hill, and Tate's lists). Mammals that do extend to Bali but not to Lombok include a tree shrew, 2 monkeys, a scaly anteater, 2 squirrels, the Leopard Cat, and the Tiger. Very few flightless land mammals extend east of Bali along the Lesser Sundas: porcupines

(*Hystrix*) extend to Lombok, Sumbawa, and perhaps Flores; and shrews (*Crocidura* and *Suncus*), the Crab-eating Monkey, a civet (*Paradoxurus hermaphroditus*), pigs, and *Cervus* deer occur on some or all of the Lesser Sundas to Timor. This list of mammals that reach Timor is nearly the same as the list of Oriental mammals that reach the Moluccas. It is a question whether some of them are native in either place; in fact there might be some doubt about all of them. Porcupines are good food ("like pork") and are presumably carried about. At least one of the shrews (*Suncus*) often lives in houses and "is carried about on native boats like the house mouse" (Carter, Hill, and Tate). Tame monkeys and other animals are often carried about by roaming Malays (Wallace), and the Crab-eating Monkey in particular is often carried (Wallace 1869, Vol. 1, p. 294, speaking particularly of the presence of this monkey near Timor). Some civets are carried and sold for their musk and may even be released on islands after their musk is taken (Wallace 1869, Vol. 2, p. 140). Pigs are staple food. And deer are often tamed and petted, esteemed for their flesh, and probably deliberately introduced when remote islands are settled (Wallace, *loc. cit.*). Some of these mammals on the Moluccas or Lesser Sundas are supposed to be distinguishable subspecies, but there may be a doubt how distinct they are, and a further doubt whether, even if distinct, they need be native. Some mammalogists apparently share these doubts. Laurie and Hill (1954, p. 86), for example, list endemic subspecies of pigs from Flores and Timor but consider them "feral descendants of pigs carried from island to island by human agency." I do not pretend to know just what species are native where, but I think it would be wise to doubt the natural occurrence of any Oriental mammals, except murid rodents and bats, east of Celebes and Bali.

Of Australian mammals, few extend much west of New Guinea. Monotremes (spiny anteaters) reach Salawati Island off the Bird's Head. Of marsupials, the Aru Islands have at least the following: 3 dasyurids, a bandicoot, several phalangerids, a *Thylogale* wallaby, and possibly a tree kangaroo. The Kei Islands have at least a bandicoot, *Phalanger* and a small flying phalanger (*Petaurus*), *Thylogale*, and possibly a tree kangaroo. *Pseudocheirus* phalangers reach Salawati, and *Dorcopsis* wallabies reach Waigeo etc., off the west end of New Guinea, as do *Phalanger, Petaurus*, and perhaps some other marsupials. A bandicoot and a *Dorcopsis* wallaby extend to Misol. The only marsupials that occur beyond this (so far as I know) are a bandicoot on Ceram (an endemic genus); the small flying phalanger

(*Petaurus*), which reaches Misol and the northern Moluccas (Halmahera, Ternate, and Batjan); and *Phalanger*, which extends through the Moluccas to Celebes (2 endemic species on latter) and to Wetar and Timor. Whether some of these marsupials may be introduced on some islands I do not know. *Phalanger* is certainly native west to Celebes.

As usual in cases like this, the details just given may be incomplete or partly wrong. Many details are probably still unknown; some of the records given may be wrong (and some doubtful records not given may be right); and there is doubt where some of the species are native and where introduced. But the pattern as a whole is certainly essentially correct.

Murid rodents complicate this simple pattern. In derivation they are Oriental elements in the Australian Region, but in present distribution they are shared and include both Oriental and Australian groups. The murid subfamily Hydromyinae is almost completely confined to the Australian Region, extending west only to the Aru and Kei Islands and Waigeo (*Hydromys*, Tate 1951, p. 232, map), except that 2 doubtfully related (relict?) genera are in northern Luzon in the Philippines. *Rattus* is a shared genus, with both Oriental and Australian species groups. The distribution and relationships of other murids between the Oriental and Australian Regions are too complex and in some cases too doubtful to be discussed here, but it is likely that native murids occur on all the habitable islands of the transition area.

Bats are very numerous in the Indo-Australian Archipelago, and the pattern of their distribution is complex. I have not tried to analyze it in detail. It is clear from Tate (1946) that many groups are shared; others occur in one region but not the other; many Oriental genera and species extend for varying distances toward Australia; some (fewer) Australian ones extend toward the Orient; and some genera and species are endemic within the transition area.

Of mammals as a whole, it can be said that most of the main Oriental mammal fauna, excepting murid rodents and bats, stops on Borneo and Java; rather small fractions of it reach Celebes and Bali; and little or none of it reaches the Moluccas or the outer Lesser Sundas. The main Australian mammal fauna stops in New Guinea; rather small fractions of it reach the Aru and Kei Islands etc.; a still smaller fraction reaches some Moluccas; and only *Phalanger* reaches Celebes and Timor. These Oriental and Australian mammals overlap very little. If the Oriental species on the Moluccas and Timor are introduced, the only natural overlap is on Celebes, where two species of *Phalanger* occur with a small variety of surely native Oriental mammals. The doubtful details are not very important. The important fact is that, excepting murids and bats, the mammals of the Moluccas and outer Lesser Sundas are very few and, for the most part, not much differentiated and probably recent. There is no old fauna of non-murid terrestrial mammals on these islands. Why? Murids seem to be more numerous and more endemic (older?) there; bats, still more numerous.

In summary of the transition of Oriental and Australian vertebrates, it might be said that fresh-water fishes form no transition—the strictly fresh-water fishes of the Orient and Australia do not meet at all. Non-murid terrestrial mammals form only a very slight transition. If shrews, monkeys, civets, pigs, and deer are native on the Moluccas and/or the outer Lesser Sundas, then a small fraction of the Oriental mammal fauna extends for a considerable distance across Wallace's Line and meets and overlaps a still smaller fraction of the Australian fauna on the Moluccas and on Timor as well as on Celebes. If these Oriental mammals are not native on the Moluccas and outer Lesser Sundas—and there is real doubt about it, based on what Wallace and others have said of some of the particular species on the particular islands in question—then the non-murid terrestrial placentals of the Orient and the marsupials of Australia meet only on Celebes, and further transition is made only by murids and bats. Of amphibians, there is considerable overlapping of Oriental and Australian groups, mostly different families, at least from Bali and Celebes to Timor, New Guinea, and northern Australia. Of reptiles and birds, there is overlapping not only of some families but also of many genera and species in a still wider area, from within the edge of the Oriental Region to northern and eastern Australia.

The complexity, level, and width of the transition of Oriental and Australian vertebrates vary with the different classes' probable powers of crossing salt-water barriers. This seems to me to be an obvious but very important fact derived from the preceding summary. Moreover, in at least part of the transition area, on the Moluccas and outer Lesser Sundas, the level of endemism of different vertebrates seems to vary with the power of crossing water. The *best* water crossers include the most distinct endemics. The terrestrial mammals (except murids) on these islands are surprisingly little differentiated. The frogs include no endemic genera and few endemic species (so far as I can judge from van Kampen 1923). The reptiles, birds, murids, and bats seem to include more distinct endemics. This comparison is partly subjective, but there can be no doubt about the low level of endemism of the non-murid terrestrial mammals on the Moluccas and outer Lesser Sundas. Even in the other classes, although there is some more endemism, there is a notable lack of really distinct, relict vertebrates. As to direction of transition, in all the groups that seem to show direction (amphibians, reptiles, birds, bats), there seems to be more extension of Oriental groups toward Australia than the reverse.

(I have tried to describe this situation entirely in terms of present distributions, but it implies things about the past. The composition and distribution of the transitional fauna imply that the islands between Borneo and Java on the one side and New Guinea and Australia on the other have been populated or repopulated by vertebrates rather recently, by dispersal across water gaps, and that the dispersing animals have come more from the Oriental side than from the Australian.)

The Oriental and Australian vertebrate faunas are, then, separated by a broad zone of subtraction and transition. The subtraction is a very important part of the pattern. The absence of strictly fresh-water fishes and the very poor representation of terrestrial mammals in the transition zone are as significant as the gradual change from Oriental to Australian animals from west to east in other classes.

The limits of this subtraction-transition zone are rather well defined. The western limit is the line between Borneo and Celebes and between Java and the Lesser Sunda Islands. This is virtually Wallace's Line, the line which Wallace thought separated the Oriental and Australian faunas but which in fact separates the nearly full-scale Oriental fauna from a strikingly depauperate fauna and the beginning of obvious transition. The only question here is whether Bali should go with Java or with the other Lesser Sundas, and this is perhaps a matter of preference. I have followed custom and put Bali with Java in the Oriental Region, in spite of the fact that the vertebrate fauna of Bali is unbalanced and limited, especially by lack of most fresh-water fishes and of many mammals.

The eastern limit of the subtraction-transition zone lies just west of Australia and New Guinea but includes certain small, closely associated islands: the Aru (and perhaps the Kei) Islands, and Salawati, Batanta, Waigeo (and perhaps Misol) just west of the Bird's Head of New Guinea. The islands not in parentheses have an almost pure New Guinean fauna, which is depauperate, but perhaps only in proportion to the size and nature of the islands. The Kei Islands and

Misol lack some significant groups which some of the other listed islands have, including melanotaeniine fishes and certain mammals. Whether they should be included in the Australian Region proper or in the subtraction-transition zone is a matter of preference. I have followed Mayr (1944) in putting Misol with New Guinea and the Kei Islands in the transition zone.

Between these limits, the vertebrate fauna changes progressively from west to east, from predominantly Oriental to predominantly Australian, although it is still not possible to give an exact description of the whole transition. The fauna of Celebes is at least three-fourths Oriental (Mayr 1944, p. 3, from authors cited); and Oriental elements, though reduced, are apparently still in the majority in the (more depauperate) faunas of the Talaud and Sula Islands. But on Buru, Australian (New Guinean) elements are in the majority, and they become overwhelmingly so on Ceram and the northern Moluccas. Along the line of the Lesser Sundas, the transition is more gradual and better documented. Oriental elements comprise something like 90 per cent of the vertebrate fauna of Bali; the proportion falls by steps to something like 60 per cent or 70 per cent on the Alor Group; and Oriental elements are still in the majority on Timor and even on the Tanimbar Islands, or at least on the western ones, although the fauna of these islands is very depauperate in some groups. On the basis of these facts, a line can be drawn which separates the islands with a majority of Oriental animals from those with a majority of Australian ones. This is the line of faunal balance, or Weber's Line (Fig. 55). It is discussed in more detail by Mayr (1944, pp. 10–12), who notes that there is still difficulty in fixing the southern end of it. Mayr notes that Weber's Line is nearer the Australian than the Oriental side of the transition zone, which is another way of saying what I have said above, that there is more extension of Oriental than of Australian groups into the transition area.

In bounding the Oriental and Australian Regions, zoogeographers may follow either of two reasonable courses. The eastern boundary of the Oriental Region may be drawn at or near Wallace's Line, and the western boundary of the Australian Region may be drawn just west of New Guinea (to include certain small islands named above); the intervening area may then be considered a separate subtraction-transition area, not part of either region. This area has been named "Wallacea" (by Dickerson *et al.* 1928, p. 101), and this is an appropriate and useful term in any case. The other course is to draw a common boundary between the Oriental and Australian Regions along

Weber's Line or near it. Both these courses have advantages and disadvantages. Which is taken does not matter much, if the underlying situation is understood. Perhaps recognition of a subtraction-transition area, Wallacea, is more likely to make the situation clear.

Summary of faunal transition

The general natures, relative distinctness, and arrangement of the regional faunas and faunal regions have already been discussed and are summarized in Tables 10 and 11, and Figure 45, but I want to say a little more about how the regions are separated.

Adjacent regional faunas are always connected by transitions, and the transitions usually occur in places that are either inaccessible or inhospitable to some animals, so that the transitional subfaunas are usually depauperate. Adjacent regional faunas are therefore usually separated by what I have called subtraction-transition areas. Such areas form partial barriers between most major regional faunas. Arabia and adjacent arid, not-quite-tropical southwestern Asia form a subtraction-transition area and partial barrier between the tropical Ethiopian and Oriental faunas. The Sahara Desert is an extreme subtraction-transition area and partial barrier between the Ethiopian and Palearctic faunas. The seasonally cold arctic forms a subtraction-transition area and partial barrier between the Palearctic and Nearctic faunas; climate more than the recent water gap has limited exchange of animals through the Bering region recently. Central America and Mexico form a subtraction-transition area and partial barrier between the Nearctic and Neotropical faunas; the present barrier effect is due partly to difficulty of access and limitation of area of Central America, and partly to the aridity of parts of Mexico. The islands and water gaps of Wallacea form a subtraction-transition area and partial barrier between the Oriental and Australian faunas. Only in eastern Asia, between the Oriental and Palearctic faunas, is there full-scale transition without intermediate subtraction between major faunas.

The local factors that now cause subtraction and act as partial barriers between regional faunas in some cases are and in other cases are not the same as the factors that caused or permitted the differentiation of the faunas in the past. In southwestern Asia, the present local factors probably have been the main effective ones: some degree of aridity and a not-quite-tropical climate have probably separated the Ethiopian and Oriental faunas in the past, as now, and allowed their differentiation. Seasonal cold, sometimes supplemented by a

Bering water gap, has probably been the main barrier between the Palearctic and Nearctic faunas in the past, as now. And in Wallacea, water barriers more or less like the present ones are evidently what have allowed the great differentiation of the Oriental and Australian faunas. But the Sahara is not primarily responsible for the differentiation of the Ethiopian and Palearctic faunas. The main factor has evidently been difference of climate. The Sahara is a very recent desert which has been less of a barrier in the past than now. Fossils show that many tropical Ethiopian animals have been able to reach Europe in the past, but most of them have not survived there, presumably because the temperate climate of Europe was unsuitable for them. In eastern Asia, the Himalayas form an impressive barrier between the Oriental and Palearctic faunas, but the mountains are less important than they seem. There is a very broad meeting of faunas east of them. That the Oriental and Palearctic faunas are different is evidently due mainly to the zonal climatic difference and not to the mountain barriers. In Central America, existing barriers are again not the main ones that permitted differentiation of (Nearctic and Neotropical) faunas in the past. The main barriers were Tertiary water gaps. Even now the difficulty of dispersing through Central America is probably not the main thing that keeps the Nearctic and Neotropical faunas distinct, except perhaps in fishes. The two faunas have mixed extensively in the last few million years. They remain distinct mainly (I think) because North and South America lie in different climatic zones.

Without pursuing this matter further, it may be said that all the main regional faunas are separated from each other either by the approximate line between the tropics and north-temperate zone or by salt water, and it follows that zonal climate and salt-water barriers are the most important inorganic factors which have caused or permitted differentiation of regional faunas. Other barriers, such as deserts and mountain ranges, are secondary, although they may sharpen faunal boundaries or displace them a little. The Sahara and the Himalayas are often considered major barriers by zoogeographers, and of course they are barriers, but the distribution of animals across or around them now and in the past shows that in the long run they have been less important than they seem.

As to direction of transitions, the direction of that between the Ethiopian and Oriental faunas is doubtful: that is, it is doubtful whether more African groups now extend toward the Orient than the reverse. The direction of transition of both the Ethiopian and Oriental faunas with the Palearctic is northward: many more tropical groups extend into the north-temperate zone than the reverse. Transition between the Oriental and Australian faunas is toward Australia. The direction of transition between the existing Palearctic and Nearctic faunas is doubtful. And transition between the Nearctic and Neotropical faunas is mixed in direction: among reptiles and mammals, there seems to be more extension of northern groups into the edge of the Neotropical Region than of Neotropical groups northward, but among birds there seems to be more extension of tropical groups northward.

Direction of transition in these cases cannot be canceled by redrawing regional boundaries at "lines of faunal balance." The limits of regional faunas are usually determined not by the balance of transitional groups but by other groups peculiar to one fauna or the other which stop at approximately common boundaries (see Fig. 53). For example, in eastern Asia the northern boundary of the Oriental Region is determined by many tropical groups that stop in southern China, even though many other tropical groups cross the boundary northward. In America, the northern boundary of the Neotropical Region is set by the many non-transitional groups that stop in the tropical lowlands of Mexico, even though some groups extend farther northward. And in the East Indies, the boundaries of the main Oriental and Australian faunas are well defined, even though there is transition in a broad area between the boundaries.

Of course all this, like so much else here, is much too simple in many ways, but it is the best I can do within the limits imposed by space and time.

REFERENCES

Anonymous. 1893. The Nearctic Region and its mammals. *Natural Science,* **3,** 288–292.

Beaufort, L. F. de. 1926. *Zoögeographie van den Indischen Archipel.* Haarlem, Dr. Erven F. Bohn.

Blanford, W. T. 1890. . . . Anniversary address . . . . *Proc. Annual Meeting Geological Soc. London,* **1890,** 13–80.

Carter, T. D., J. E. Hill, and G. H. H. Tate. 1945. *Mammals of the Pacific world.* New York, Macmillan.

Dickerson, R. E., *et al.* 1928. *Distribution of life in the Philippines.* Philippine Bureau Sci. Monograph 21.

Dunn, E. R. 1922. A suggestion to zoogeographers. *Science,* **56,** 336–338.

Fernald, M. L. 1929. Some relationships of the floras of the Northern Hemisphere. *Proc. International Congress Plant Sci.,* **2,** 1487–1507.

Gadow, H. 1913. *The wanderings of animals.* Cambridge, England, Cambridge U. Press; New York, Putnam's.

Heilprin, A. 1887. *The geographical and geological distribution of animals.* New York, Appleton.

Hooijer, D. A. 1951. Pygmy elephant and giant tortoise. *Sci. Monthly,* **72,** 3–8

Hora, S. L. 1952. Recent advances in fish geography of India. *J. Bombay Nat Hist. Soc.,* **51,** 170–188.

———. 1953. The Satpura hypothesis. *Science Progress,* **41,** 245–255.

Huxley, T. H. 1868. On the classification and distribution of the *Alectoromorphae* and *Heteromorphae. Proc. Zool. Soc. London,* **1868,** 294–319.

Laurie, E. M. O., and J. E. Hill. 1954. *List of land mammals of New Guinea, Celebes, and adjacent islands . . . .* London, British Mus.

Li, H.-L. 1952. Floristic relationships between eastern Asia and eastern North America. *Tr. American Philosophical Soc.,* New Ser., **42,** 369–429.

Lydekker, R. 1896. *A geographical history of mammals.* Cambridge, England, Cambridge U. Press.

L(ydekker), R. 1911. Zoological distribution. *Encyclopaedia Britannica,* 11th ed., **28,** 1002–1018.

Mayr, E. 1944. Wallace's Line in the light of recent zoogeographic studies. *Quarterly Review Biol.,* **19,** 1–14.

Merriam, C. H. 1892. The geographic distribution of life in North America . . . . *Proc. Biol. Soc. Washington,* **7,** 1–64.

Mertens, R. 1934. Die Insel-Reptilien . . . . *Zoologica* (Stuttgart), **32,** G. Lieferung, 1–209.

Parker, H. W. 1934. *A monograph of the frogs of the family Microhylidae.* London, British Mus.

Rooij, N. de. 1915. *The reptiles of the Indo-Australian Archipelago,* Part 1. Leiden, E. J. Brill.

Schmidt, K. P. 1954. Faunal realms, regions, and provinces. *Quarterly Review Biol.,* **29,** 322–331.

Sclater, P. L. 1858. On the general geographical distribution of the members of the class Aves. *J. of Proc. Linnean Soc. (Zool.),* **2,** 130–145.

Scrivenor, J. B., *et al.* 1943. A discussion on the biogeographic division of the Indo-Australian Archipelago . . . . *Proc. Linnean Soc. London,* 154th Session, 120–165.

Tate, G. H. H. 1946. Geographical distribution of the bats in the Australasian Archipelago. *American Mus. Novitates,* No. 1323.

———. 1951. The rodents of Australia and New Guinea. *Bull. American Mus. Nat. Hist.,* **97,** 183–430.

van Kampen, P. N. 1923. *The Amphibia of the Indo-Australian Archipelago.* Leiden, E. J. Brill.

Wallace, A. R. 1860. On the zoological geography of the Malay Archipelago. *J. of Proc. Linnean Soc. London,* **4,** 172–184.

———. 1869. *The Malay Archipelago . . . .* London, Macmillan, 2 vols.

———. 1876. *The geographical distribution of animals.* London, Macmillan, 2 vols.

chapter 8

Island patterns

This chapter is concerned with the vertebrates of islands, with their nature, distribution, and significance. The classic and still useful book on island life is Wallace's (1880 and later editions), although study of island animals from a modern point of view actually began before this, with Darwin's investigation of the birds and animals of the Galapagos Archipelago. There are excellent articles by Gulick (1932) on the biological peculiarities of oceanic islands and by Mertens (1934) on the reptiles of the islands of the world. And much special information is available in local works referred to below. Probably the most important of them is *Fauna Hawaiiensis*, the introduction of which (Perkins 1913) ought to be read by all zoogeographers.

I had planned to begin this chapter with a special consideration of the vertebrates of the West Indies (the islands I know best), using them to illustrate the principles of distribution of animals on islands, but I have changed this plan. The West Indies and their plants and animals have been discussed by a small group of specialists in a series of meetings at Harvard in the spring of 1954. The results of the discussion are being edited by Dr. E.nest Williams and will, I hope, be published soon. They cover the West Indian fauna far better than I could do single-handed. I have therefore abridged my present treatment of the West Indies and include them among other islands without special emphasis. I should add that I owe a good deal to the West Indian discussion group and especially to Dr. Williams, who has read and criticized this whole chapter. My plan now is to begin with a preliminary discussion of the nature of islands and island faunas, stressing the things to look for; then to discuss the vertebrate faunas of the principal islands and archipelagos of the world; and then to attempt a brief synthesis and conclusions.

Many doubts and difficulties are encountered in dealing with islands and island faunas.

Geographers still disagree even about the areas of some islands and the widths of some water gaps.

Geologists are doubtful or disagree among themselves about the histories of some islands and have been seriously misleading in some cases. For example, Schuchert (1935) insisted so strongly that the Greater Antilles (the four large islands of the West Indies) were geologically continental islands that it did not occur to me until recently even to suspect he might be wrong. He said (p. 109), "If the latter [geologists and stratigraphers] know anything at all, it is that the present Greater Antilles are not oceanic or volcanic islands, but consist of continental rocks and are now fragments of a once greater Central America." This was unfortunate language. It now seems that the Greater Antilles *are* oceanic, volcanic islands! Obviously zoogeographers should use geological evidence with caution and should depend primarily on zoological evidence to decide the nature and history of island faunas. If the geological and zoological evidence agree, that is fine. If not, the reasons for the disagreement are worth looking for.

Zoogeographers disagree among themselves too, and the zoological evidence is incomplete or doubtful in many cases.

The faunas of many islands are still incompletely known, and, what is worse, the fossil records of island animals are usually poor and in most cases no older than the Pleistocene.

The taxonomy of some island animals is doubtful. Supposedly distinct forms sometimes turn out not to be distinct, and vice versa. Some taxonomists distinguish island populations as "species" on the slightest characters. The taxonomists should not be blamed too much for this. They often have to do the best they can with inadequate material if an island fauna is to be made known at all, and they can only report situations as they see them. Besides doubt about the reality of some island species, there is doubt about the ancestors and/or relationships of some of them.

Further doubt has been caused by the accidental or intentional introduction onto islands of many different kinds of animals, by both ancient and modern man. If an animal on an island is endemic and sufficiently distinct, it may be presumed to be native, but there are many doubtful cases in which island forms are not differentiated or only slightly so and occur in such a way as to suggest the possibility of introduction. In such cases the zoogeographer is caught between two alternatives: to risk treating as native something which may be introduced, or to risk treating as introduced something which may be native. These alternatives are equally bad. The best procedure in these cases is to suspend judgment, but even that complicates discussion of island faunas, for the doubtful species have to be mentioned and the reasons for the doubts given. Certain Oriental mammals on the Moluccas and outer Lesser Sundas (pp. 466 and 467) and the Philippines, the bush pig on Madagascar, the frogs on Fiji, and the raccoons on certain small islands in the West Indies are a few of many such cases.

A few vertebrates, especially certain lizards and rats, are very widely distributed on remote oceanic islands. They are often carried in boats, but it is sometimes argued that the same qualities which enable them to ride uninvited in loaded canoes etc. make them also particularly liable to be "rafted," so that they may have reached the most remote islands naturally. There is, I think, a way to tell whether these animals have dispersed naturally or been carried by man. If they have crossed great widths of ocean naturally, they should have been doing so for many millions of years, and distinct endemic forms of them, including endemic genera as well as species, ought to occur on remote islands. If, however, they have been carried by man, they can have been on remote islands only a few thousand years, and their level of endemism on the islands should be low, although (by bad judgment or for lack of information) taxonomists may consider some of the island populations "subspecies" or slightly defined species.

Even when an island fauna is adequately known, there may still be doubts about the significance of it, about why some animals are present and others absent. For example, the tree frog *Hyla arborea* is absent in England, although it is present on the adjacent mainland. Beaufort (1951, pp. 54–55) thinks it has been unable to reach the British Isles since the Pleistocene. But Malcolm Smith (1951, p. 17) says this frog has been introduced into England and has not survived there. Has this frog been kept out of England by the post-Pleistocene water barrier or is the climate of England unsuitable for it?

Zoogeographers are continually plagued by such difficulties and doubts in trying to deal with island life. They can only proceed as best they can, avoiding laborious arguments about details and trying to discover the principles and patterns of distribution of life on islands as well as to itemize it.

Although a distinction is usually made between continents and islands, it is an arbitrary one, as I have already said (p. 419). All the continents and islands can be arranged in one very irregular series, according to decreasing area and/or increasing isolation, although of course these two things are only partly correlated. Even the greatest continents have sometimes been isolated: Africa and Eurasia may have been separated early in the Tertiary; if so, the separation probably limited the faunas of both continents in some ways. North America is cut off from the Old World now. The smaller continents, South America and Australia, have been more isolated. And the series continues through large islands like Madagascar, to smaller ones. All land areas from the largest to the smallest have faunas that are to some extent limited by isolation from other lands, although the effect is most obvious on the smallest and most isolated islands. For this reason the basic principles of distribution of animals ought to be the same on islands as on continents. This is more than a philosophical concept. The continents and islands of the world do form one complex, discontinuous system of land over which animals have dispersed in certain definite ways, and what is learned about animals on islands may fairly be added to what has been learned about continental animals to make up the pattern of animal distribution and dispersal as a whole.

In *Island Life* Wallace made (or took from Darwin) a sharp distinction between continental islands and oceanic ones. He defined continental islands as detached fragments of continents, consisting of complex continental rocks and always inhabited by some terrestrial mammals and amphibians; and oceanic islands as having originated in the ocean, consisting of volcanic rocks and coral (limestone) and lacking terrestrial mammals and amphibians. Geologically, this distinction may be true, although the geology of islands is more varied than Wallace knew, but zoologically it is not true. Of course, some island faunas are purely continental, being obviously parts of continental faunas cut off recently by flooding of connecting land, and others are purely oceanic, but these two kinds of faunas are connected by such an array of mixtures, intermediates, and doubtful cases that to classify island faunas simply in two main categories is wrong.

Although the distinction between continents and islands is arbitrary, and although (zoogeographically) islands cannot be sharply divided into continental and oceanic ones, an arbitrary classification of islands is useful. For convenience I have made three loose categories of them: (1) obvious, recent continental islands, with (more or less reduced) continental faunas; (2) fringing archipelagos, not simply recent-continental but lying close to continents and receiving the fringes of continental faunas; and (3) single islands or isolated archipelagos not falling into either of the preceding categories, but diverse otherwise, from large islands like Madagascar to the smallest oceanic ones. The islands to be specially discussed are classified in this way in Table 14, and their positions are shown in Figure 56.

TABLE 14. ARBITRARY CLASSIFICATION OF ISLANDS TO BE DISCUSSED

- Recent Continental Islands
  - Tropical
    - Sumatra, Java, and Borneo
    - New Guinea
    - Ceylon
    - Formosa
    - Trinidad
  - Not Tropical
    - British Isles
    - Japan
    - Newfoundland, Greenland
    - Tasmania
    - Tierra del Fuego, Falklands
- Fringing Archipelagos (all tropical)
  - Philippines
  - Western Pacific islands
  - West Indies
- Other Islands and Archipelagos (not arranged by climate)
  - Celebes
  - Madagascar, Mascarenes, and Seychelles
  - New Caledonia, Lord Howe Island, Norfolk Island
  - New Zealand
  - Hawaiian Islands
  - Galapagos
  - Islands of the Atlantic Ocean
  - Antarctica

For each of the islands and archipelagos listed I shall give something of its position, size, nature, and probable history, and of the

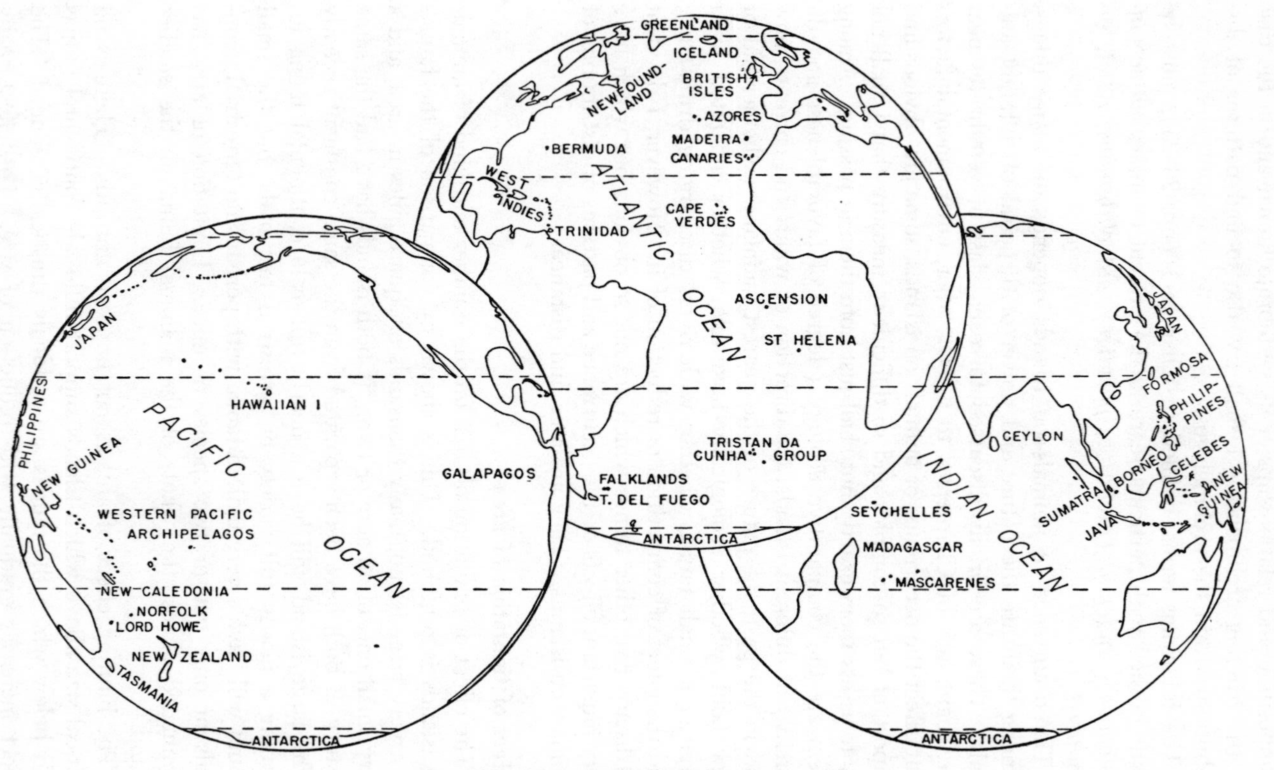

Fig. 56. Map of islands to be discussed.

composition and relationships of its vertebrate fauna; and in the case of the fringing archipelagos I shall try also to find patterns of distribution within the archipelagos.

The following terms, defined in Chapter 1 (pp. 23–25), must be distinguished with particular care in the present chapter: *native* (*indigenous*), *endemic* (*exclusive, peculiar*), *autochthonous, relict, introduced.*

The occurrence of animals on islands depends on many things. Among them are the nature and power of dispersal of different animals. These are briefly discussed for each class of vertebrates near the beginnings of Chapters 2 to 6. Of other, environmental factors that affect the occurrence of animals on islands, some are obvious and important but so complex and so difficult to measure that I shall not try to assess them exactly but shall just note them in passing. Among them are the comparative ecology (degree of favorableness and diversity) of different islands; the direction of winds and currents, now and in the past; the nature of the nearest mainland (whether old or new land, whether ecologically favorable, whether inhabited by a large or a small fauna, whether with rivers discharging drift, etc.); and the effect of competition or reduction of it. However, I do want to discuss two other things which are not so obvious but which have very important effects on the occurrence and patterns of distribution of animals on islands. They are area and distance.

Effect of limitation of area

The effect of limitation of area on the number and kind of animals on islands is profound. This is shown by comparison of the faunas of a very large island, many thousands of square miles in area, and a very small one, of an acre or two. If both islands have had the same history—if both have been separated from the same continent recently—the large island will have a nearly full-scale continental fauna including a nearly full variety of terrestrial mammals, but the small island will have a very limited fauna with perhaps no terrestrial mammals or only one or two species of mice. Limitation of area has eliminated almost the whole of the mammal fauna on the smaller island.

The limiting effect of area is not imposed abruptly. There is no critical size above which islands support full-scale continental faunas and below which the faunas are suddenly drastically reduced. The effect increases gradually with reduction of size. The effect really begins on the continents. All land masses, from the largest to the smallest, have faunas that are limited by available area, although the effect is very complex and is modified by many other factors. In fact the area of the world as a whole probably limits, in a very complex way, the number of animals that exist on it (see pp. 29–30).

Some figures are available which suggest more specifically how area limits the numbers of some animals on some islands. Table 15

TABLE 15. FROGS AND TOADS, LIZARDS, AND SNAKES OF GREATER ANTILLES AND TRINIDAD

Compiled from sources cited in reference list at end of chapter. The figures in each space are for genera: species + additional subspecies.

	Frogs and Toads	Lizards	Snakes	Totals
Cuba, about 40,000 sq. mi.	4: 24 + 5	13: 37 + 10	8: 15 + 1	25: 76 + 16
Hispaniola, 30,000 sq. mi.	4: 27 + 1	16: 41 + 4	9: 16 + 2	29: 84 + 7
Jamaica, 4,500 sq. mi.	2: 16 + 0	8: 17 + 0	5: 6 + 0	15: 39 + 0
Puerto Rico, 3,400 sq. mi.	3: 13 + 0	7: 21 + 1	4: 6 + 0	14: 40 + 1
Trinidad, 2,000 sq. mi.	12: 23	15: 19	28: 38	55: 80

shows the numbers of species of amphibians and reptiles on the four islands of the Greater Antilles, and Table 16 shows the numbers on

TABLE 16. AMPHIBIANS AND REPTILES OF FOUR ISLANDS OF DIFFERENT SIZES IN THE LESSER ANTILLES (FROM DUNN 1934, P. 108)

Montserrat, 8 by 5 miles, 3002 feet high:	9 species
Saba, 2½ miles in diameter, 2820 feet high:	5 species
Redonda, 1 mile in diameter, 1000 feet high:	3 species
Sombrero, 1 mile long, 40 feet high:	1 species

four islands of the Lesser Antilles. Inspection of these tables suggests that, within the size range of these islands (except the two smallest), division of area by ten divides the amphibian and reptile fauna by two (Table 17), but this ratio is a very rough approximation, and it might not hold in other situations.

Whatever the exact ratio, it is obvious that the limiting effect of area must be considered in comparing the faunas of islands, even of large islands. It should be added that the larger number of species on large islands is only partly due to duplication of species in differ-

TABLE 17. APPROXIMATE RELATION OF AREA TO NUMBER OF SPECIES OF AMPHIBIANS AND REPTILES ON CERTAIN WEST INDIAN ISLANDS (SEE TABLES 15 AND 16)

Approximate Area, sq. mi.	Approximate Number Species	(Actual Number Species)
40,000	80	(76–84)
4,000	40	(39–40)
(400	20)	—
40	10	(9)
4	5	(5)

ent places on single islands. More species occur *together* on large than on small islands. Of course, the effect of area is modified by many other things, including extent of high land and ecological diversity; and, on the other hand, area probably has other effects on island animals. In some cases limitation of area may limit the size of individual animals (but this effect is complex and probably not predictable—see Mertens 1934, pp. 120–123), and it probably also affects the structure of populations and the direction of evolution.

Effect of distance; immigrant and relict patterns

The effect of distance—of the width of water gaps—on the composition and patterns of distribution of island faunas is profound too. To understand it, it is necessary to understand something about the dispersal of land animals across water (see Darlington 1938).

The dispersal of terrestrial animals across water is often referred to as "accidental dispersal" or "random dispersal," but these are not good terms. The non-committal term "over-water dispersal" is preferable.

Dispersal of *individual* land animals over water is largely accidental (and accidents must occur also in dispersal over land), but in the course of time statistical probability comes into play and determines what sorts of animals cross water most often and what islands they most often reach. Even if there is only one chance in a million that a given individual of a species will get across a given water gap, out of many million individuals some may be almost sure to cross; and, other things being equal, they will be much more likely to cross narrow water gaps than wide ones. This is obvious, at least in a general way.

The effect of the width of water gaps is not just inversely proportional to distance or to the square of distance. It depends on rate of loss, on the death rate during dispersal. The death rate of most terrestrial animals during dispersal across salt water is presumably high and presumably forms a geometric progression: if only one individual in a thousand survives the crossing of a hundred miles of sea, only one in a thousand of the remainder will be expected to survive a second hundred miles, etc. In this case, if there is one chance in a thousand that a given sort of animal will reach an island 100 miles out at sea, there is only one chance in a million that it will reach an island 200 miles out, and only one in a billion that it will reach one 300 miles out.

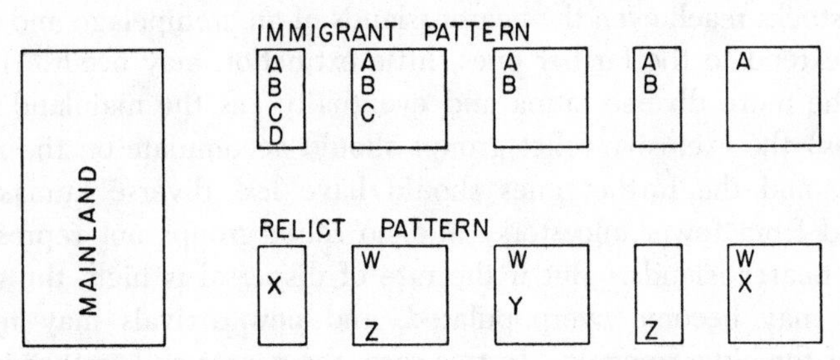

Fig. 57. Diagram to compare a simple immigrant pattern, formed by dispersal of several groups of animals from the mainland for different distances along a series of islands, and a relict pattern, formed by partial extinction of an old fauna formerly common to all the islands.

In nature, this matter is complicated by biological factors and by differences in the nature and age of different islands, the direction of winds and currents, and other things. But distance is basically important. Unless other factors are very unequal, animals dispersing from a continent to an archipelago may usually be expected to reach the nearest islands first and to spread to other islands across the narrowest water gaps. The resulting pattern of distribution should be orderly, with related forms occurring in series on adjacent islands along the routes of immigration. This is an *immigrant pattern.*

In contrast to this is the pattern that would be expected if a group of animals were once well represented on an archipelago and were then reduced in numbers and eliminated on some of the islands. This might happen if the archipelago were detached from a continent and had at first a continental fauna, much of which later became extinct. In this case the survivors would probably not form orderly series on adjacent islands but would occur irregularly, partly on the largest and most favorable islands regardless of position and partly according to chance. This would be a simple *relict pattern.*

Simple immigrant and relict patterns are contrasted in Figure 57.

Immigrant and relict patterns might be mixed. This could happen if new immigrants dispersed into an archipelago already occupied by a relict fauna, or if an old immigrant fauna were decimated by some process other than replacement from the mainland, for example by partial submergence of the islands. Replacement from the mainland should form a different sort of pattern, described below.

Immigrant patterns probably become modified in the course of time, and the pattern of modification probably varies with the rate of dispersal of the animals concerned. If the rate of dispersal is low, if few stocks reach even the nearer islands of an archipelago and if still fewer extend to the farther ones, little extinction may occur. In this case, the more diverse fauna and eventually (as the mainland fauna changes) the exclusive relict groups should accumulate on the nearer islands, and the farther ones should have less diverse faunas (descended from fewer ancestors) and no relict groups not represented on the nearer islands. But if the rate of dispersal is high, the nearer islands may become overpopulated, and new arrivals may replace some of the older animals. In this case, the nearer and farther islands should have equally diverse faunas (in proportion to area), but the farther ones will probably have the exclusive relicts, the last survivors of groups that have been replaced on the nearer islands as well as on the mainland. These different results may be reached in different groups of animals on the same archipelago. Under appropriate conditions, most terrestrial mammals and amphibians, which cross salt water with difficulty, should accumulate principally on the nearer islands of an archipelago, and their exclusive relicts should be there; other animals that disperse more freely should undergo more replacement on the nearer islands, and their exclusive relicts should be on the farther islands.

This can be illustrated by the simple case of two similar islands lying one beyond the other (Fig. 58). In this case, among animals whose rate of dispersal has been low, the most diverse fauna and most of the exclusive relicts should be on the near island. Among animals whose rate of dispersal has been high, the two islands should have about equally diverse faunas, but most of the exclusive relicts should be on the outer island. There might also be animals whose rate of dispersal and replacement would be intermediate and would produce an approximately equal number of exclusive relicts on each island; in this an immigrant pattern might simulate a relict one. But, if the islands are truly comparable in size and habitats and if enough stocks are involved, it is unlikely that a relict pattern would come to simulate an immigrant one.

Of course these different patterns would probably be more or less modified by redispersals of old island animals and in other ways. Complications would be most likely to occur among the most actively dispersing animals. Animals with low powers of dispersal, indicated

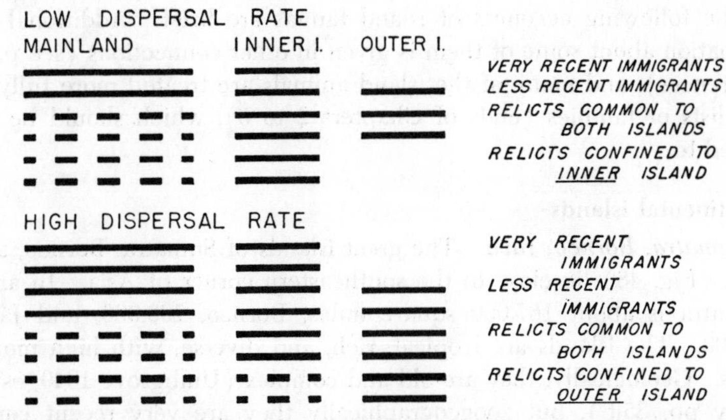

Fig. 58. Diagram of effect of dispersal rate on pattern of distribution on a series of two islands. Each island is limited (by area) to a fauna of six units. In the first case, rate of dispersal is low, the inner island is not overpopulated, and the outer island has received less than a maximum fauna; and, when groups become extinct on the mainland (broken lines), the exclusive relicts occur on the *inner* island. In the second case, rate of dispersal is high, the inner island has been overpopulated, and some older groups have become extinct there as well as on the mainland, and the outer island has received a full fauna; and the exclusive relicts occur on the *outer* island.

by simplicity of their immigrant patterns and by occurrence of localized relict-endemics mostly on islands nearest the mainland, should be the most dependable indicators of the past.

In considering the areas of islands and the widths of water gaps, the past as well as the present must be considered, especially the changes of sea level that occurred in the Pleistocene (p. 583). At times during the Pleistocene, when the ice on land reached its maximum, sea level everywhere was probably at least 100 meters lower than now. Such a lowering of the sea would increase the size of many islands and reduce the width of many water gaps. In fact it would join some islands, such as Sumatra, Java, and Borneo, to the mainland or to each other. On the other hand, when there was little or no ice on land, before the Pleistocene and perhaps at times during interglacial ages, sea level was probably 20 to 50 meters higher than now; this must have reduced the size of some islands, completely drowned others (*e.g.*, some of the Bahamas), and increased the width of some water gaps.

All that has just been said about immigrant and relict patterns, and mixed and modified ones, is preliminary and hypothetical. It remains to be seen whether the hypothetical patterns occur in fact.

The following accounts of island faunas are brief. Additional information about some of them is given in other connections (see page references), and many of the island animals are treated more fully in the lists of families (ends of Chapters 2 to 6), which should be referred to.

Continental islands

Sumatra, Borneo, Java. The great islands of Sumatra, Borneo, and Java (Fig. 48) lie close to the southeastern corner of Asia. In area, Sumatra is about 167,000 square miles; Borneo, 290,000; and Java, 49,000. The islands are tropical, rich, and diverse, with high mountains. Geologically, they are old and complex (Umbgrove 1949, especially pp. 35ff.), but zoogeographically they are very recent continental islands. They are separated from each other and from Asia by shallow water, much of it less than 50 meters deep (Fig. 59), and they were all connected together and to the mainland in the Pleistocene and perhaps later: according to Beaufort (1951, p. 84, from Smit Sibinga) the Malay Peninsula, Sumatra, and Java were connected until within early historic times.

The fauna of these islands is a nearly complete continental one, and many of the species in it are the same as those of the mainland. Fresh-water fishes, amphibians, reptiles, birds, and mammals are all well represented in it. Moreover, much of the fauna is generally distributed over all three islands, although there are inequalities. It is not possible to itemize this whole fauna here, but certain significant details of distribution of the fishes and mammals are worth giving. Fresh-water fishes and terrestrial mammals are stressed here and later in this chapter because they tell more than any other animals about land in the past.

The fresh-water fishes are not distributed according to existing land. Those of Sumatra and western Borneo are more alike than those of the western and eastern sides of Borneo. The explanation (and the distribution of the fishes is part of the evidence of it) is that, when the islands were connected with each other and with the mainland, the great land area they formed was drained by several separate rivers, especially one flowing north between what are now Sumatra (and the mainland) and Borneo and one flowing east between what are now Java and Borneo (Fig. 59), and that the fishes are still distributed according to the old river systems. However, the history of these rivers and of their fishes was apparently not simple. Java shares

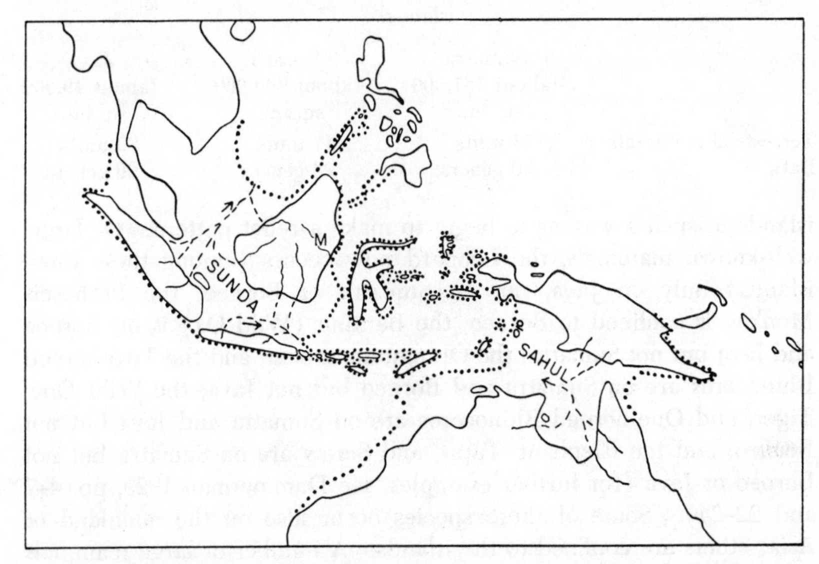

Fig. 59. The Indo-Australian Archipelago, with probable extensions of land during Pleistocene lowering of sea level (dotted lines). Former drainage of the Sunda shelf indicated by broken lines connecting existing rivers (simplified, after Kuenen 1950, pp. 482–483, Fig. 203); drainage of the Sahul shelf diagrammatic, details imaginary. M, Mahakam River. *Cf.* map of Wallacea, Figure 55.

more fresh-water fishes with Sumatra than with Borneo, as if exchange between Java and Borneo was interrupted earlier. Moreover, Java shares more fishes with the Mahakam River of *eastern* Borneo (see Fig. 59) than with southern Borneo, and this apparently can be accounted for only by change of drainage patterns or succession of fish faunas (Dammerman 1929, pp. 17 and 24–25, from Beaufort).

A rich Oriental mammal fauna extends to all three islands but it already shows what seem to be the beginnings of the effects of isolation and limitation of area. The smallest island, Java, has the fewest mammals (Table 18). (But Sumatra, though smaller than Borneo, has more groups of mammals than the latter—an effect of distance or time?) Moreover, extinctions seem to have occurred on different

TABLE 18. COMPARISON OF THE MAMMAL FAUNAS OF SUMATRA, BORNEO, AND JAVA

The figures are from Carter, Hill, and Tate's (1945) lists. These lists are composed partly of species and partly (in groups in which the species have not been fully worked out) of genera, so the figures given represent neither species nor genera but an arbitrary compromise. The Elephant, probably introduced on Borneo, and a rabbit, introduced on Java, have not been counted on those islands.

	Sumatra (about 167,000 sq. mi.)	Borneo (about 290,000 sq. mi.)	Java (about 49,000 sq. mi.)
Terrestrial mammals	55 units	47 units	33 units
Bats	30 genera	28 genera	26 genera

islands in such a way as to begin to make a relict pattern. Of large, well-known mammals, the Leopard now occurs (among these three islands) only on Java, not on Sumatra or Borneo; the Proboscis Monkey is confined to Borneo; the Banting (Wild Ox) is on Borneo and Java but not Sumatra; the Orangutan, a bear, and the Two-horned Rhinoceros are on Sumatra and Borneo but not Java; the Wild Dog, Tiger, and One-horned Rhinoceros are on Sumatra and Java but not Borneo; and the Elephant, Tapir, and Serow are on Sumatra but not Borneo or Java (for further examples, see Dammerman 1929, pp. 4–7 and 22–24). Some of these species occur also on the mainland of Asia; others are confined to the islands. A number of large mammals now absent on Java were there in the Pleistocene, including the Orangutan, several large felids, a hyaena, bears, several genera of elephants, *Hippopotamus,* tapirs, antelopes, and more than one species of wild cattle (Koenigswald 1939), so the fauna has clearly been reduced and was not originally depauperate. However, some details of distribution and some of the reduction may date from before the island's isolation; limitation of area may have an effect on peninsulas as well as on islands. This whole subject is much too complex to discuss further here. Dammerman (1929) gives additional details and compares the fauna of Java with that of the other islands.

New Guinea. New Guinea has the same geographical relationship to Australia that Sumatra etc. have to Asia. It lies on the continental shelf of Australia, separated from the continent by a narrow and shallow water gap, and it was broadly connected to Australia in the Pleistocene. The island is slightly more than 300,000 square miles in area (the largest habitable island in the world), and is tropical, rich, and diverse, with very high mountains.

The vertebrate fauna of New Guinea has been summarized in Chapter 7 as part of the fauna of the Australian Region. New Guinea, like Australia, lacks or has very few true fresh-water fishes. New Guinea has the same families of frogs as Australia, but different representations of them (*i.e.*, more brevicipitids, fewer leptodactylids); much the same reptile fauna, although again different families are differently represented; and basically similar bird and mammal faunas, although details differ. In short, the vertebrate fauna of New Guinea is basically the same as that of Australia but different in proportions and details. The differences are probably due more to the difference of climate (New Guinea mainly tropical and forested, Australia mainly temperate and open) than to the water barrier.

The rivers of southern New Guinea and of the north coast of Australia share some species of fishes and other aquatic vertebrates. This is a clear indication that in the Pleistocene the land connection between Australia and New Guinea was drained by a river, now mostly under the sea, with headwaters stretching from what is now southern New Guinea to northern Australia (Fig. 59).

Ceylon. Ceylon lies southeast of the tip of India, is about 25,000 square miles in area, and is a diverse, tropical island with some mountains. Regardless of its history otherwise, zoogeographically it is a recent continental island. It is on the continental shelf of India, and the gap of 30 miles or so between the island and the mainland is crossed by "Adam's Bridge," a chain of shoals. The island was therefore presumably connected to India by land late in the Pleistocene. Its fauna is consistent with this.

The fauna of Ceylon is moderately rich in all the main classes of vertebrates, but the fishes and mammals are, as usual, most significant. Of fishes, Ceylon has a variety of strictly fresh-water cyprinoids, catfishes, and labyrinth fishes (Munro 1955). Of mammals (Table 6, p. 328), Ceylon has 32 non-gnawing terrestrial species, 23 lagomorphs and rodents, and 28 bats. Many of the species of mammals and other vertebrates are the same as those of southern India, but others are endemic. However, Ceylon lacks some significant groups. For example, certain groups of fresh-water fishes that live in mountain torrents in southern India are absent on Ceylon (p. 63). Among the large (and therefore well-known) mammals, Ceylon lacks the Wolf and Wild Dog but has a jackal; has the Sloth-bear; lacks the Tiger but has the Leopard; lacks rhinoceroses but has the Elephant; and has a chevrotain and three deer but lacks all the numerous bovids of India except possibly the Water Buffalo, which may or may not be native.

Different animals are probably absent from Ceylon for different reasons. The torrent fishes and perhaps other specialized animals may have been unable to reach Ceylon even when a (low) land connection existed. But large mammals could and did reach Ceylon. Pleistocene fossils known there include a giraffe, two or three bovids, a hippopotamus, two rhinoceroses, two or three elephants, and a lion (Deraniyagala 1949). They and very likely others became extinct there later perhaps partly because of limitation of area.

Although most of the vertebrates of Ceylon are Indian or related to Indian ones, a few (*e.g.*, some lizards and snakes, p. 191, and some birds, p. 261) have their closest relatives in the Indochinese or Malayan Subregions. They are geographical relicts. However, such relicts are probably no more numerous in Ceylon than in southern India.

Formosa. Formosa lies about 90 miles off the coast of China, on the continental shelf, just on the edge of the tropics. Its area is about 14,000 square miles, and it is an ecologically rich and diverse island, with high mountains. It has evidently been connected to the mainland recently, for its fauna is mainly that of southern China but less diverse.

The fauna of Formosa is moderately rich in all the main classes of vertebrates. As usual, they cannot all be itemized here, but something can be said of the fishes and mammals. The strictly fresh-water fishes of Formosa include at least 42 cyprinids and cyprinoids, 8 catfishes, and 4 labyrinth fishes (Mori 1936). This is a good representation of dominant Oriental groups of fresh-water fishes. Of mammals, Carter, Hill, and Tate list five genera of insectivores, including a mole; one monkey; eleven carnivores, including the Asiatic Black Bear and Clouded Leopard; a scaly anteater; five ungulates, including a pig, three deer, and the Serow, which is the only native bovid; a hare and a number of rodents; and bats. This mammal fauna is only a little smaller than that of Ceylon and is equally characterized by absence of most large mammals including most bovids. It might be added that the birds of Formosa have been covered by Hachisuka and Udagawa (1950–1951), who also describe the island and its vegetation. Birds are numerous on Formosa but less so than in comparable areas of the tropical mainland of Asia.

Trinidad. Trinidad lies off the north coast of South America, off Venezuela. Its area is less than 2000 square miles, but it is a fairly diverse, tropical island, with extensive lowlands, and with some mountains up to about 4000 feet high. It is on the continental shelf and is separated from the mainland by water only about 10 miles wide and

15 meters deep, and it has evidently been connected with the mainland very recently. Geographically and politically, Trinidad is sometimes treated as one of the West Indies, but zoogeographically it is very different, for it has a moderately rich continental fauna of all the main classes of vertebrates, and many of the species are the same as those of the mainland.

The fresh-water fishes of Trinidad (Eigenmann 1909, pp. 327–328; and other authors to Inger 1949) include a variety of the characins and fresh-water catfishes which are dominant in South America but which do not reach the true (zoogeographically speaking) West Indies. The amphibians and reptiles of Trinidad are more numerous and generically more diverse than those of the West Indies proper (Table 15, p. 483), and include relatively more snakes and fewer lizards, as the continental fauna does. The birds of Trinidad are far more numerous than those of the West Indies proper: Bond (1936, p. XV) says that if the birds of Trinidad and the associated smaller island of Tobago were added to the West Indian fauna, they would nearly double the number of genera and would add 14 Neotropical families not otherwise represented in the West Indies. The native mammals of Trinidad (Vesey-FitzGerald 1936) include at least 4 opossums; 2 monkeys; 2 anteaters and an armadillo; a felid (the Ocelot), 2 mustelids, and a procyonid; a deer and a peccary; a squirrel, 9 murid, and 6 hystricomorph rodents; and 34 bats. This is a good mammal fauna for so small an island. This list may be incomplete, but the absence from it of certain large mammals, including the Jaguar, Puma, and tapirs, is probably significant.

The British Isles. The British Isles (Fig. 60) consist of two main islands lying one beyond the other off the coast of Europe: Great Britain, about 89,000, and Ireland, about 32,000 square miles. They are cold-temperate, but the severity of the climate is lessened by the Gulf Stream. Within the limits of climate, the islands are ecologically diverse, with low mountains as well as lowlands. They lie on the continental shelf and were connected to the mainland not long ago: Great Britain was separated from the mainland probably only about 7000 years ago, but Ireland was evidently separated from Great Britain somewhat earlier. Some leading references to works on the vertebrates of the British Isles are Hinton *et al.* (1935, list of British vertebrates); M. Smith (1951, British amphibians and reptiles); Matthews (1952, British mammals, including a chapter on the origin of the British mammal fauna); and Praeger (1950, natural history of Ireland).

The British fauna includes representatives of all the main classes of vertebrates. All the species are European or are closely related to European ones. The whole of the island fauna, at least of terrestrial vertebrates, occurs in Britain; only part of it extends to Ireland. Of strictly fresh-water fishes, Britain has the Pike (which extends to Ireland), eleven cyprinids (five extend to Ireland), two cobitids (one extends to Ireland), and two percids (one extends to Ireland). Of amphibians, Britain has three salamanders (*Triturus*), two toads, and one native and two probably introduced frogs (*Rana*),

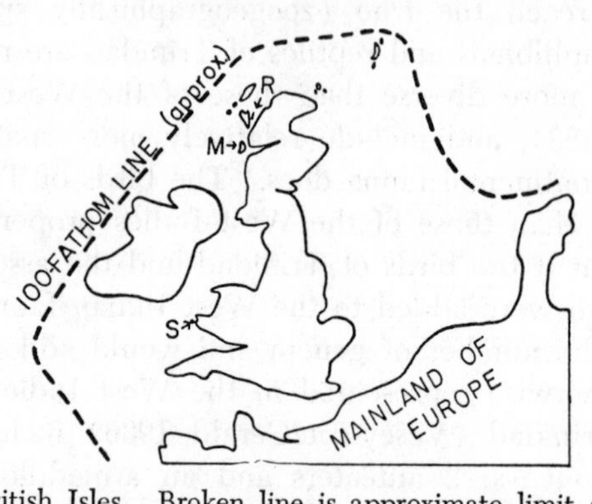

Fig. 60. The British Isles. Broken line is approximate limit of the continental shelf, much or all of which was land at times during the Pleistocene. R, Raasay; M, Mull; S, Skomer.

but of the six native amphibians, perhaps only one (a salamander) occurs naturally in Ireland. Of reptiles, Britain has three lizards (*Anguis fragilis* and two species of *Lacerta*) and three snakes (including the European Viper), but of these six native reptiles, only one (*Lacerta vivipara*) occurs in Ireland. Ireland has no snakes and perhaps no native frogs (the Common Frog is probably introduced and so may be the Natterjack Toad, of which a colony is localized in southwestern Ireland), and both Britain and Ireland are now without turtles. Of birds, Britain and Ireland have too many to discuss in detail. Of mammals, Britain has a hedgehog, four shrews, and a mole (only the hedgehog and one shrew reach Ireland); the Red Fox (also in Ireland), six mustelids (also in Ireland), and the Wild Cat (not in Ireland); two native deer (one in Ireland); three native hares and rabbits (two also naturally in Ireland); and eleven rodents and twelve bats (some also in Ireland, some not). Besides these, the Wolf, Brown Bear, Wild Boar, and Beaver existed in Britain within historic times, and the first three occurred also in Ireland.

British vertebrates had a very complex history during the Pleistocene. Successive faunas reached the islands (which were part of the mainland) in successive glacial and interglacial periods but were successively reduced or destroyed. Matthews (1952, pp. 364ff.) traces something of this history for the mammals. However, except for a few subarctic relicts, the existing fauna is probably a new one composed entirely of post-Pleistocene immigrants. A part (but only a part) of the post-Pleistocene fauna of western Europe has reached Britain; and a part (but only a part) of the British fauna has reached Ireland, which was isolated before Britain was. The resulting main pattern of distribution is a rather simple immigrant one. However, the details of it are more complex than described. The fauna diminishes northward in Britain, and many species are otherwise localized. If these islands should remain isolated and should retain their fauna while the mainland fauna was replaced, it is Britain, the nearer island, which would have the exclusive relicts, including such things as the legless anguid lizard and the mole. Ireland has no terrestrial vertebrates not represented in Britain too. However, among the small rodents, there has been some local replacement. Two forms of the Bank-vole, *Clethrionomys glareolus,* have apparently invaded Britain successively. The *nageri* group apparently came first. It was too late to reach Ireland, but it did reach small islets just off the west coast of Britain which may still have been connected to the main island. The *glareolus* group proper came later. It entirely replaced the *nageri* group on the main island of Britain, but representatives of *nageri* still survive on the small west-coastal islands of Mull, Raasay, and Skomer. Matthews (1952, pp. 147ff.) gives the distributional and fossil evidence of this and suggests other replacements among British field mice (*Microtus*). These are small examples of how successive immigrations may displace relicts from the more to the less accessible islands of an archipelago.

Japan. The Japanese archipelago is a chain of islands paralleling the east coast of Asia. The largest island, Honshu, is about 87,000 square miles, and to the south but closely associated with it are Kyushu, about 16,000 square miles, and Shikoku, about 7000. These three islands form the main part of what is sometimes called Japan proper. Hokkaido, about 30,000 square miles, lies to the north and has a colder climate and a more limited, almost purely Palearctic fauna. Japan proper is warm-temperate and is ecologically diverse, with lowlands and high mountains. These islands are connected with the continental shelf of Asia, and they were presumably connected with Asia by land in the Pleistocene. Their fauna is consistent with this.

The strictly fresh-water fishes of Japan proper are forty-odd cyprinoids and three catfishes and are a mixture of northern-Asiatic and (especially southward) southern-Chinese groups, with some endemic genera of cyprinoids (Mori 1936, pp. 10–14; see also Okada 1955). Japanese amphibians include salamanders (three families) and frogs (three families, five genera, about ten species in Japan proper—Okada 1931, p. 8) and are a mixture of northern, southern, and endemic species. Japanese reptiles include an endemic emydine turtle and a non-endemic *Trionyx;* a few gekkonid, scincid, and lacertid lizards; and several colubrid snakes and a pit viper (*Agkistrodon*). Japanese birds are numerous. Japanese mammals are six genera of insectivores, including a mole; a macaque monkey; three canids, two bears, and six mustelids; the Wild Boar, a deer, and the Serow; three rabbits etc. and ten genera of rodents; and eleven genera of bats, including *Pteropus* (flying foxes).

The main island of Japan is about the size of Great Britain, and the archipelago is comparable in some ways with the British Isles, but Japan is warmer and has not been buried under ice. The Japanese fauna is correspondingly more diverse, older (more differentiated), and more mixed in its geographical relationships. Some of the differentiation probably antedates the breaking of land connections, for the mountains of Japan probably formed isolated habitats before that. The mixed (northern and southern) relationships of the Japanese fauna suggest changes in the distribution of some groups on the mainland since the island fauna was formed.

Newfoundland and Greenland. Newfoundland lies off the northeastern corner of North America, off the Gulf of St. Lawrence. It is about 42,000 square miles in area. It is south of the latitude of Britain but is much colder, cold-temperate to subarctic, with extensive coniferous forest and tundra. It is connected with the continental shelf of North America on the Labrador side, where the present water gap is only about 10 miles wide. However, the island was more or less covered by ice at times in the Pleistocene, and it is doubtful whether there has been a land connection since then. The fauna probably requires none; the terrestrial mammals may, I suppose, have come across ice. That Newfoundland has received more large than small mammals suggests derivation of the fauna across ice.

The fauna of Newfoundland is very limited. There are no strictly fresh-water fishes, probably no native amphibians (at least one frog has been introduced), and no reptiles. Birds are numerous. Native mammals (Bangs 1913) include the Wolf, Black Bear, and five smaller carnivores (and the Polar Bear and Arctic Fox come in occasionally on ice floes); the Caribou; the Arctic Hare, Beaver, Muskrat, and a field mouse (*Microtus*); and a bat. However, many mammals that occur on the Labrador peninsula (including 16 genera and subgenera listed by Bangs) are absent on Newfoundland, which has no Moose, no shrews, no Porcupine, no squirrels, and only one mouse, which (excepting the bat) is the only really small native mammal—if it is really native.

Greenland (Clark 1943) is far north, reaching closer to the north pole than any other land. It is an immense island, about 840,000 square miles in area, but all except about 130,000 square miles is now covered by ice. Only some edges of the island are habitable. Geologically it is a continental island, but its existing terrestrial fauna has evidently reached it recently across ice, not land. Greenland of course has no strictly fresh-water fishes, amphibians, and reptiles. Breeding land birds are few (Salomonsen and Gitz-Johansen 1950–1951): the Rock Ptarmigan (American form), two falcons, White-tailed Eagle (Eurasian), Short-eared Owl (doubtful if breeds), Snowy Owl, Wheatear (Eurasian), Fieldfare (Eurasian), two pipits (one American, the other Eurasian), White Wagtail (Eurasian), red-polls (two subspecies), Lapland Longspur, Snow-bunting, and Raven. The birds not otherwise designated in this list are circumpolar species—I have not tried to determine the relationships of the Greenland populations. The Fieldfare is a north-European thrush (*Turdus*) which apparently established itself in southern Greenland after a trans-Atlantic "irruption" in January 1937 (Salomonsen 1951). A number of additional land birds have been recorded from Greenland as occasional visitors, and many water birds breed there. The land mammals of Greenland are the Arctic Fox and Wolf, Polar Bear, Short-tailed Weasel, Caribou and Musk Ox, and the Arctic Hare and a lemming. They are all derived from arctic North America.

Tasmania. Tasmania is just south of Australia. Its area is about 26,000 square miles; its climate is cool-temperate; and it is rather diverse ecologically, with low mountains. It is on the continental shelf of Australia and was connected by land with the continent recently, as the fauna shows. Tasmanian vertebrates are listed by Lord and Scott (1924).

Tasmania, like most of Australia, lacks strictly fresh-water fishes, but galaxiids etc. are numerous. Of frogs, three genera of leptodactylids and *Hyla* reach the island. So do several lizards and three elapid snakes, but no turtles. There are numerous birds. Tasmanian mammals include the Echidna and Platypus, a variety of marsupials (dasyurids, bandicoots, phalangerids, wombats, and kangaroos etc.), several genera of murid rodents including *Hydromys,* and six species of bats. These animals are all Australian species, or are related to Australian ones, or are species which occurred in Australia once. Two of the most striking of them, the Marsupial Wolf and the Tasmanian Devil (good-sized carnivorous marsupials), do not occur in Australia now but are fossil there in the Pleistocene.

Tierra del Fuego and the Falkland Islands. Tierra del Fuego is at the southern end of South America, separated from the continent by a very narrow strait. The main island is about 18,000 square miles in area; adjacent smaller islands raise the total land area south of the Straits of Magellan to about 27,000 square miles. The climate is cold-temperate but varies from wet to dry in different places, so that the islands have both forest and open habitats.

The fishes of Tierra del Fuego are like those of the southern tip of the mainland: salt-tolerant peripheral forms including galaxiids. The amphibians that reach the islands are only one or two small frogs; the reptiles, two or three small iguanid lizards. The land birds include at least hawks and owls, a parrot, a woodpecker, flycatchers, and icterids and a few other songbirds, but Tierra del Fuego of course lacks many families of birds characteristic of the tropical rain forest and apparently lacks also some more-southern South American birds such as rheas, tinamous, and antbirds. The southern limits of these groups are presumably determined by climate and not by water barriers, although water must be a barrier to the rheas. The mammals of Tierra del Fuego are a fox, two otters, the Llama, several cricetid rodents, one hystricomorph (*Ctemomys*), and a vespertilionid bat. There are no marsupials, edentates, or monkeys, and only the one hystricomorph. So far as I know, there is nothing in the vertebrate fauna of Tierra del Fuego which could be considered a significant relict.

The Falkland Islands lie about 300 miles east of the southern end of South America. The two main islands are about 2500 and 2000 square miles in area. The highest point is about 2000 feet. The islands are cool-temperate in climate and are ecologically limited, without trees but with low vegetation. These islands lack strictly fresh-water fishes, amphibians, and reptiles. They have hawks and owls and a few small land birds as well as water birds. The only certainly native land mammal is (or was) the "Falkland Wolf," a large endemic species of fox related to the South American foxes (*Dusicyon*). Darwin (1839, 1952 reprint, p. 249, footnote) had "reason to believe there is likewise a field-mouse," but so far as I know this has not been confirmed. Cattle, horses, pigs, hares, rabbits, *Rattus*, and *Mus* have been introduced. The endemic fox was incurably unsuspicious of man and became extinct about 1875. It apparently fed on birds and perhaps seals and other shore animals. How it reached the Falklands is unknown. The islands are on the continental shelf and may have been connected with South America by land, but it is strange if only a fox used the connection. Perhaps the small area and limited habitats of the islands may have limited the fauna to begin with, and the fox may have eliminated any smaller mammals that got there. This may have been the next-to-last stage in disappearance of a mammal fauna on too small and too unfavorable an island. Or the fox may have reached the Falklands on drifting ice during the Pleistocene, as the Arctic Fox probably reached Iceland and still occasionally reaches Newfoundland.

Summary of continental islands. Although the continental islands discussed above are diverse, their faunas have certain characteristics in common. Some of the characteristics are obvious and expected; others are perhaps more subtle.

The composition of the faunas of these recent continental islands is of course not limited by animals' powers of crossing salt water. The faunas usually include representatives of all the main classes of vertebrates consistent with climate, although Pleistocene rather than existing climate has limited some of them.

Distance (width of water gaps) is relatively unimportant on continental islands. Ceylon, 30 miles out, and Formosa, 90 miles out from Asia, have rather similar faunas. In the cases where distance is correlated with distribution, other factors are probably responsible. For example, the fact that Ireland has a smaller fauna than Great Britain is probably due not simply to distance but to the fact that Ireland was isolated first and is a less diverse island.

The effect of area in limiting and modifying the faunas of continental islands is visible or is at least suspected in many of the cases discussed. There seems to be a general tendency toward loss of large mammals from island faunas, even on large islands. The effect of area probably increases with time. It may be a very long time before limitation of area has its full effect on such large islands as Sumatra,

Borneo, and Java. The effect may be felt more quickly on smaller and simpler islands like the Falklands.

Effects of time on the faunas of continental islands are obvious in some cases. Comparison of the very recent British fauna with the older Japanese one suggests that time has not only eliminated some large mammals on Japan but has increased endemism and increased the diversity of relationships of the island fauna, as distributions have changed on the mainland.

The effect of zonal climate on the faunas of continental islands is basically the same as on continents: the colder the climate, the less diverse the fauna. Cold may limit the composition of a fauna in unexpected ways. For example, cold Tierra del Fuego not only lacks most amphibians and reptiles but has a highly selected mammal fauna, consisting almost entirely of representatives of groups which reached South America comparatively recently from the north. The only old South American mammal on Tierra del Fuego is one hystricomorph rodent; otherwise the whole of the characteristic, older half of the South American mammal fauna is lacking. Under other circumstances, a situation like this might lead to doubts whether an island fauna were really derived from an adjacent continent.

As to patterns on continental archipelagos, Great Britain and Ireland form a series of two islands, of which the outer has been less accessible than the inner, and terrestrial vertebrates have formed a fairly simple immigrant pattern on them since the Pleistocene. If these islands were permanently isolated, and if the mainland fauna changed, the inner island (Britain) would have the principal exclusive relicts. However, even in this case some details of distribution of small mammals suggest the beginnings of replacement on the inner island, with relicts isolated on small coastal islands west of Britain. Sumatra, Borneo, and Java show what seem to be the beginnings of a relict pattern produced by partial extinction of an originally full-scale continental fauna. All this agrees well enough with theory as set forth at the beginning of this chapter.

Fringing archipelagos

The Philippines. The Philippine Islands (Fig. 61) lie between South China and the Greater Sunda Islands, but zoogeographically they are a fringing archipelago extending northeast from Borneo for nearly a thousand miles. The largest island, Luzon, about 40,000 square miles, is the outermost main island. Mindanao, about 35,000 square miles, is at the southeastern end of the archipelago. Palawan, nearest Borneo, is only about 4500 square miles. The other principal islands, each of about 4000 or 5000 square miles or less, are closely spaced between Mindanao and Luzon. The islands are tropical and are ecologically diverse, some of them with high mountains. Accord-

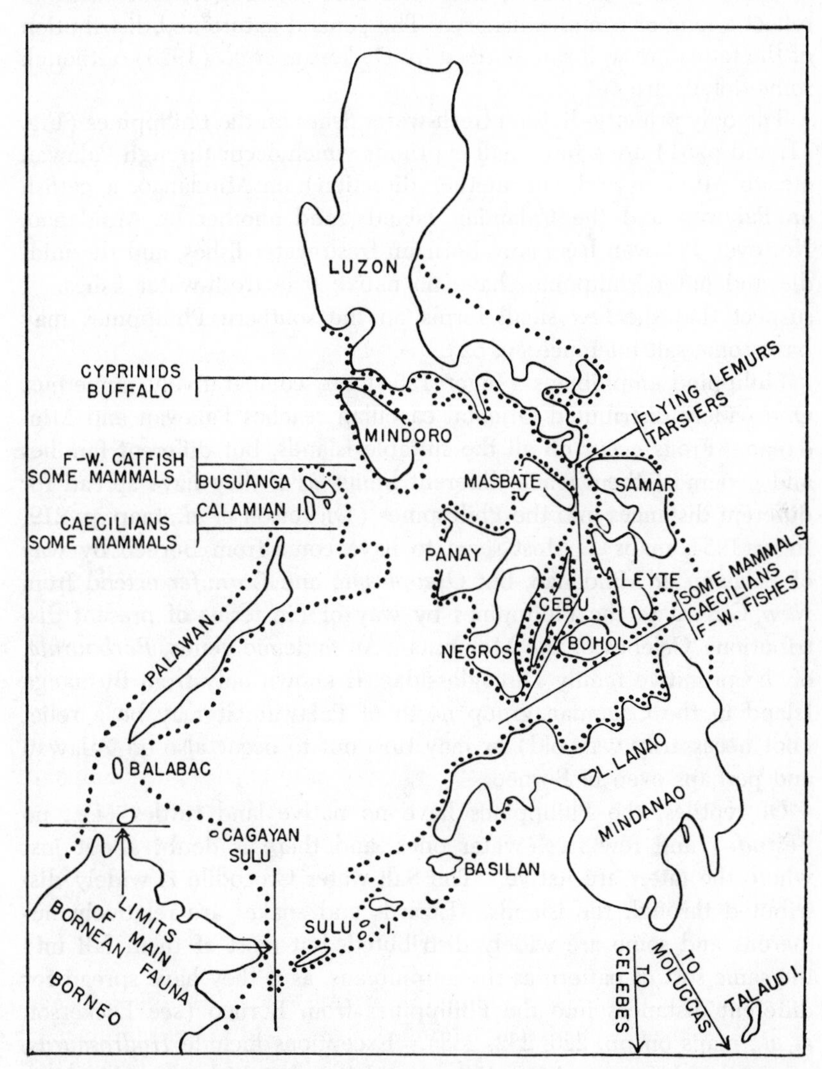

Fig. 61. The Philippine Islands, with possible extensions of land (dotted lines) during Pleistocene lowering of sea level (*cf.* Inger 1954, p. 455, Fig. 81). Some of the connections shown are doubtful. Ruled lines show limits of extension into the Philippines (from Borneo) of the animals named.

ing to the fourteenth edition of the *Encyclopaedia Britannica,* the Philippines are on the crumpled edge of the continental platform of Asia, are geologically complex, but are largely or partly volcanic. Inger (1954, pp. 449–456) reviews the apparent history of the islands in more detail. However, the vertebrate fauna does not seem to reflect a long or complex history. The general nature and distribution of the fauna are well summarized by Dickerson *et al.* (1928), although some details are out of date.

The only primary-division fresh-water fishes on the Philippines (Fig. 11, and p. 51) are a few small cyprinids which occur through Palawan etc. to Mindoro and (in another direction) on Mindanao, a catfish on Palawan and the Calamian Islands, and another on Mindanao. Not even Palawan has many Bornean fresh-water fishes, and the middle and outer Philippines have no native true fresh-water fishes. I suspect that the few small forms on the southern Philippines may have some salt tolerance (p. 52).

Philippine amphibians (Inger 1954) are comparatively numerous. One widely distributed Oriental caecilian reaches Palawan and Mindanao. Frogs occur on all the suitable islands, but different families and genera of them reach different limits, as if they have spread for different distances into the Philippines (Dickerson *et al.,* map, p. 219; Inger 1954, maps). Most seem to have come from Borneo by way of Palawan or Mindanao, but *Oreophryne* and *Cornufer* extend from New Guinea to the Philippines by way of (in terms of present distribution) Celebes or the Moluccas. An endemic genus, *Barbourula,* of the primitive family Discoglossidae, is known only from Busuanga Island in the Calamian group north of Palawan; it may be a relict (not necessarily very old) or may turn out to occur also on Palawan and perhaps even in Borneo.

Of reptiles, the Philippines have no native land turtles (*i.e.,* no *Testudo*) and few fresh-water ones, and there is doubt about just where the latter are native. The Salt-water Crocodile is widely distributed through the islands. Lizards and snakes are relatively numerous and some are widely distributed, but most of them fall into the same sort of pattern as the amphibians, as if they have spread for different distances into the Philippines from Borneo (see Dickerson *et al.,* maps on pp. 226, 232, 238). Exceptions include *Hydrosaurus,* a genus of large, semi-aquatic, agamid lizards which occurs in New Guinea, the Moluccas, Celebes, and the Philippines to Luzon (but not Palawan?), and *Steganotus,* a genus of terrestrial colubrid snakes known from northeastern Australia, New Guinea, the Moluccas, and the eastern Philippines to Luzon.

Philippine birds (Delacour and Mayr 1946) are numerous. Most of them have come from or through Borneo, via either Palawan or the Sulu Islands and Mindanao. A few have come from Asia etc. across the sea. And a few, including megapodes, cockatoos, lories, some pigeons, and a few others, are distributed as if they have come from the Australian Region, Moluccas, and Celebes. There are many endemic species and some endemic genera, especially in the eastern Philippines (Mindanao to Luzon), but (I think) no striking relict birds and no flightless ones. Palawan with the Calamian Islands etc. has an even higher proportion of Bornean birds than the other Philippines.

Philippine mammals are treated by Taylor (1934) and briefly by Carter, Hill, and Tate (1945, pp. 216 and 217), and Dr. Karl Koopman has given me some additional information about them. They include a moderate variety of small forms, especially rodents and bats, but very few large ones. Most of them, *excepting the murid rodents,* form a pattern of apparent, rather recent immigration from Borneo. Many do not extend beyond Palawan etc. and/or Mindanao, and very few reach Luzon. Of insectivores, there is an endemic genus (*Podogymnura*) of gymnures on Mindanao, and shrews of the widely distributed genera *Crocidura* and *Suncus* are widespread in the Philippines. The Flying Lemur reaches Basilan, Mindanao, Bohol, Leyte, and Samar. One genus of tree shrews reaches Palawan, and another (*Urogale*) is endemic to Mindanao. Tarsiers reach Mindanao, Bohol, Leyte, and Samar. A loris reaches Mindanao. The Crab-eating Monkey apparently extends to all the main islands and is the only primate (except man) on some of them, including Palawan and Luzon. *Manis* reaches Palawan. There are no native canids on the Philippines, and the only native felid is the Leopard Cat (the common small wildcat of southern Asia), which reaches Palawan, the Calamianes, Panay, Negros, and Cebu. An otter reaches Palawan, and a badger, Palawan and the Calamianes. The Binturong and a mongoose reach Palawan (the mongoose, the Calamianes too), and two other genera of viverrids (*Viverra* and *Paradoxurus*) extend to all the main Philippines. Of non-murid rodents, three genera of squirrels extend to various southern islands, but none reaches Luzon; and a porcupine reaches Palawan and the Calamianes. Of ungulates, a chevrotain reaches at least Balabac (just south of Palawan), and a small buffalo is endemic on Mindoro, but only pigs and small samba deer (*Cervus*) are widely distributed in the Philippines. Both fruit and insectivorous bats are numerous throughout the Philippines. Most are related toward the Orient, fewer toward Celebes, the Moluccas, the Australian Region, etc. (Tate 1946, lists on pp. 4–7).

Of all the mammals thus far listed, the only ones (besides bats) that reach Luzon are *Crocidura* and *Suncus* shrews, the Crab-eating Monkey, *Viverra* and *Paradoxurus* civets, pigs, and deer. This is almost exactly the same assemblage that reaches the Moluccas and outer Lesser Sundas! Most or all of these animals are associated with man in one way or another (p. 467) and (I think) some or all of them may be introduced on Luzon. If these mammals are really native on Luzon, it is surprising that there are no distinct endemic genera of them there.

The murid rodents are differently distributed. Murids are numerous throughout the Philippines. I shall not try to detail their distributions but shall stress that endemic genera of them are widely scattered on the Philippines (except Palawan) *and are especially numerous on northern Luzon.* Dickerson *et al.* (p. 276) quote Thomas on the discovery of six new genera of murids on Luzon, "a proportion of novelty that has perhaps never been equaled in the history of Mammal-collecting." Some of the genera are so peculiar that their relationships are doubtful, but some seem to be related toward Celebes and Australia: *e.g.*, two of the genera in the mountains of northern Luzon seem to be related to the Hydromyinae of the Australian Region.

The usual explanations of the distribution of mammals on the Philippines—complicated land bridges or rafting from several directions, partly across wide water gaps—are not very convincing. A better explanation seems to be that mammals have reached and dispersed across the Philippines wholly or mainly in one direction (northeastward from Borneo), across a series of water gaps; that murids dispersed sooner or faster than the other terrestrial mammals, reaching Luzon before the others; and that there has been time since then for the evolution of murid genera on Luzon, and for replacement of some of the older groups of murids on the Sunda Islands, with a consequent eastward shifting of distribution patterns. Murids (rodents) do cross water gaps more often than most mammals. They alone of flightless terrestrial placentals have reached Australia. Some of the very stocks that reached Australia may also have spread over the Philippines. If all this is correct, Luzon once had a mammal fauna consisting only of murid rodents and bats—unless by some chance there were relict marsupials there.

This explanation implies certain things about the history of the Philippines: that they have never been very accessible islands except perhaps briefly in the Pleistocene, and that they may have been much smaller and more isolated than now during most of their history. If, as Simpson (1945, p. 205) thinks, murids did not rise until late in the Miocene, that may have been the time of the beginning of the existing Philippine fauna. It remains to be seen whether the geological history of the island is consistent with all this. Philippine vertebrates as a whole do seem consistent. There is a notable lack of relict vertebrates, except of murids, especially on the outer Philippines. There is some endemism in all classes of vertebrates, but more of species than of genera, and in most cases more on the near islands than on the far ones. There is a conspicuous pattern of recent immigration, mostly from Borneo, running through almost the whole of the vertebrate fauna. The few cases in which Philippine vertebrates are related toward Celebes or the Australian Region (at least two frogs, a lizard, a snake, and a few birds and bats, as well as some murids) can be accounted for partly by eastward shifts in the main distributions of the groups, *i.e.*, by extinctions on Borneo etc. Only in a few exceptional cases (if any) need there have been direct dispersal across water from Celebes or the Moluccas to the Philippines.

Whether or how the Philippines were connected with Borneo in the Pleistocene is a question. The contours of the sea bottom (Fig. 61) suggest that Palawan and the Calamian Islands were connected and Mindanao not, but Palawan does not have a decisively continental fauna. Both Palawan and Mindanao have received a few (very few!) fresh-water fishes, Palawan no more than Mindanao; and cyprinids extend beyond Palawan to Mindoro, beyond the supposed land bridge. Mindanao as well as Palawan seems to have received a fair variety of amphibians from Borneo. Both islands have received mammals, and each has got significant groups which the other has not. The mammals of Palawan may be less differentiated, those of Mindanao more so; and Palawan has apparently received more carnivorous mammals, Mindanao more non-carnivorous (and more arboreal?) ones. This suggests that the mammals of Mindanao (as compared with those of Palawan) are older and may have had a little more difficulty reaching the Philippines. The endemic buffalo on Mindoro is beyond the limits of the supposed Borneo-Palawan-Calamian bridge. The most significant parts of the Philippine fauna, therefore, do not clearly show the existence or limits of a Palawan land bridge. Access to Palawan has apparently been only slightly, not much, easier than to Mindanao. Does the relatively small size of Palawan account for all that it lacks?

Whether or not all of the Philippines, or just Palawan, or none of them are included in the Oriental Region is unimportant, provided the situation is understood. They are fringing islands, with a fringing, progressively depauperate, chiefly Oriental vertebrate fauna.

Islands of the tropical western Pacific. Here will be considered together, as one fringing archipelago, all the islands of the tropical

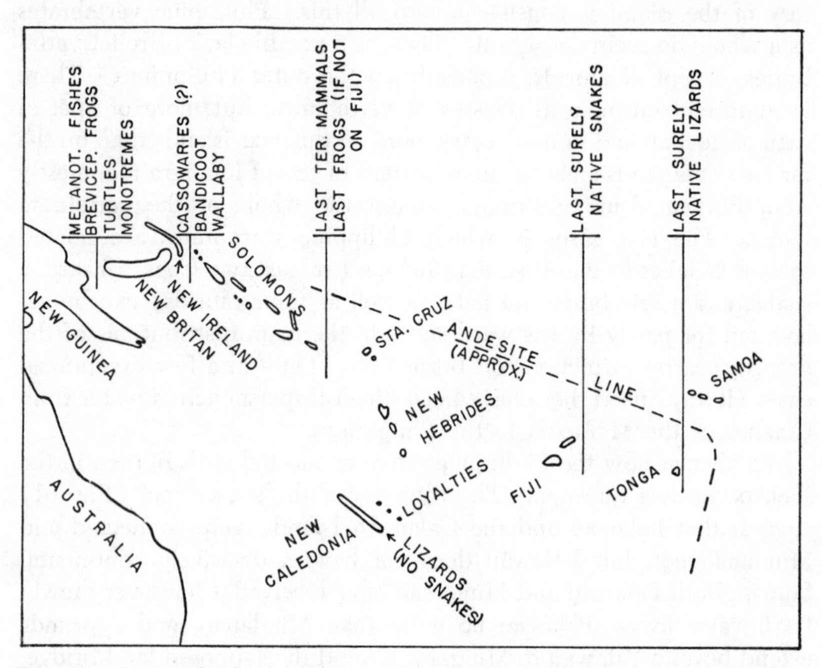

Fig. 62. Principal islands of the western tropical Pacific. Ruled lines show limits of extension into the Pacific (from New Guinea) of the animals named.

western Pacific east of Australia and New Guinea, and as far out into the ocean as terrestrial vertebrates go (Fig. 62). The purpose now is to find the main pattern of distribution in the area as a whole rather than the peculiarities of single island faunas. Most of the islands in this area are small and widely spaced, but certain groups of them, though still widely spaced, form an irregular series of steppingstones eastward from New Guinea: the Bismarcks, Solomons, Santa Cruz and New Hebrides, Fijis, Tongas, and Samoa. Many Pacific islands are low atolls, but some, including some in each group named above, are higher and ecologically more diverse. Geologically, the islands vary somewhat. Outside the Andesite Line (see map, Fig. 62), all the islands are of simple, basaltic, volcanic rock (plus coral). Inside the line, the rocks are more complex, but many of the islands are still volcanic, and I cannot see that the line has much to do with the present limits of vertebrates. None of these islands is connected with the continental shelf. Even the Bismarcks and Solomons are narrowly separated from it and from each other. Zoogeographically, all these islands should be treated as of unknown geological history, and their faunas should be allowed to speak for themselves.

True fresh-water fishes are lacking on all of these islands. The Melanotaeniinae, supposedly secondarily confined to fresh water, are unknown east of New Guinea.

Of frogs, leptodactylids and brevicipitids (the former poorly but the latter well represented on New Guinea) apparently do not reach even the Bismarcks or Solomons. *Hyla* and ranids (especially Cornuferinae) do reach the Bismarcks and Solomons: the Solomons have two species of *Hyla* and eight genera (three endemic) and fourteen species of ranids (Brown 1952), as well as the introduced *Bufo marinus*. Beyond that, frogs are unknown on the Santa Cruz Islands, New Hebrides, etc., but two ranids occur on the Fijis. They are species of *Cornufer* and *Platymantis,* both genera occurring also on the Solomons and elsewhere. The Fijian species are considered endemic (Brown and Myers 1949, Brown 1952), but they are not very strongly characterized. *Cornufer vitiensis* of Fiji, for example, apparently differs from *C. guppyi* of the Solomons only in being smaller, with smaller finger disks (Brown 1952, p. 32). These frogs on Fiji seem to me to be too far out in the Pacific, too little differentiated, too little known, and too closely associated with man (the Fijians eat frogs and presumably carry them about) to be accepted as surely native. A species of *Cornufer* (*pelewensis*) occurs on the Palau Islands, but Inger (letter of March 16, 1955) writes that, on comparison of specimens, he cannot distinguish it from *C. "rubristriatus"* of New Guinea. It may be introduced on the Palaus.

Of reptiles, turtles (excepting marine ones) apparently do not extend east of New Guinea (but a meiolaniid was on Walpole Island southeast of New Caledonia in the Pleistocene). The Salt-water Crocodile extends far into the Pacific; its limits are not determined. Several groups of snakes reach the Solomons, but few go farther. A small boid, *Candoia* (*Enygrus*), extends to the Tongas and Samoa, but it may have been carried to the outer islands by man. It is arboreal and likes to hide quietly in dense clumps of leaves; in New Guinea, I have had specimens of the genus fall from such places into my insect-beating net. It might easily be carried by accident in bunches of bananas or thatch. *Typhlops* too may be accidentally carried by man, so that its outer limits are doubtful. A small elapid, *Ogmodon*, a genus endemic to Fiji but related to New Guinean ones, probably marks the outer limit of surely native snakes. Snakes apparently do *not* occur on New Caledonia (Mertens 1934, p. 39) in spite of several erroneous reports of them there, and they do not occur on New Zealand. Lizards go farther. Native, endemic lizards occur east, in decreasing numbers, at least to the Tongas, beyond Fiji (but apparently not to Samoa), and on New Caledonia and New Zealand. Most are geckos and skinks, but a fine, relict, endemic genus of iguanids is on Fiji and the Tongas. Certain geckos and skinks extend still farther, occurring on almost every island across the warmer part of the Pacific almost to the coast of South America. Endemic species and subspecies of them have been described on some of the most remote islands, and it is commonly thought that they have spread across the ocean naturally, on drift. But these lizards are not really much differentiated on the more remote islands. On the Hawaiian Islands, for example (Oliver and Shaw 1953), no species of them is really endemic. The Hawaiian Islands are old enough to have a highly endemic fauna of insects, land mollusks, and birds. If lizards have really spread across the Pacific naturally, why did none reach Hawaii or other remote islands long enough ago to have evolved a comparable degree of endemism?

Of birds, cassowaries reach New Britain (unless they are introduced there) but not the Solomons. Other land birds are fairly numerous on the Solomons—less so than on New Guinea, but the Solomons are smaller islands. East of the Solomons, birds decrease in numbers in proportion to distance (see Fig. 32, p. 252), although of course the bird fauna varies with the size and nature of the different islands too. A few Polynesian (Old World) birds have reached the Hawaiian Islands, as have also some American ones. Easter Island now lacks land birds; whether there were ever any there is unknown. Old World land birds that have spread far eastward across the Pacific include, in order of numbers, pigeons, Old World flycatchers, parrots, honey eaters, and others. Groups that have failed to spread far include cockatoos, bee eaters, rollers, hornbills, pittas, drongos, sunbirds, and flowerpeckers (Mayr 1939).

Of mammals excepting bats, only a small fraction of the rather rich

New Guinean fauna of marsupials and rodents extends eastward. New Britain has apparently only a bandicoot, two species of *Phalanger* and a *Petaurus,* a wallaby (*Thylogale*), and about four genera of rats, including *Hydromys* and *Rattus,* but no monotremes and no marsupial carnivores. New Ireland has the bandicoot, two species of *Phalanger,* the wallaby, and *Rattus* (and other rats?). Only one *Phalanger* and about three genera of rats extend to the Solomons; some of the rats are very distinct endemic species if not genera there. This is probably the limit of native flightless land mammals in this direction. However, mice and rats associated with man occur on even the most remote islands of the tropical Pacific. Besides the world-wide commensal species, members of the Oriental *Rattus concolor* or *exulans* group, which often live in native houses, occur east to Hawaii and the Marquesas and south to New Zealand. Slightly differentiated forms of this group are supposedly endemic on remote islands, but it is unlikely that they are native there. Old World fruit and insectivorous bats extend east to Samoa but decrease rapidly in numbers with distance from the mainland. Certain insectivorous bats reach also to the Caroline and Marshall Islands, and New Caledonia and New Zealand. Only the American *Lasiurus* reaches the Galapagos and Hawaii. Bats are apparently unknown on the Society Islands, Tuamotus, etc.

The apparent eastern limits of different groups of vertebrates along the main axis of dispersal into the tropical Pacific are, then, these: strictly fresh-water fishes and non-marine turtles do not extend east of New Guinea; frogs and flightless land mammals reach the Solomons; snakes, Fiji; lizards, Fiji and the Tongas; bats, Samoa; and land birds, the farthest Tuamotus. These limits seem proportional to the probable powers of dispersal of the different animals. The only obvious, striking relict in the vertebrate fauna of the more remote islands is the iguanid lizard on Fiji; it need not be very ancient. The rest of the vertebrates, although they include endemics, form a well-marked immigrant pattern of spread across the islands. The main pattern is almost wholly one of differential spread across water gaps into a fringing archipelago, and not one of survival from ancient land masses.

Certain frogs, snakes, lizards, and rats occur beyond the limits given; and some of them on remote islands have been supposed to be native, endemic forms. Perhaps some of them are. But they are all too far out, too little differentiated, in some cases too little known, and too much associated with man to be accepted without question.

They are all eaten by man, or eat his food, or live in his houses, or are likely to be carried by him accidentally. Some or all of these animals have probably been carried beyond their natural limits by man within the last few thousand years. To what extent these island populations are really differentiated, and what the differentiation means, is still to be determined in most cases.

The West Indies. The West Indies (Fig. 63) lie east of Central America, between North and South America, mostly within the edge

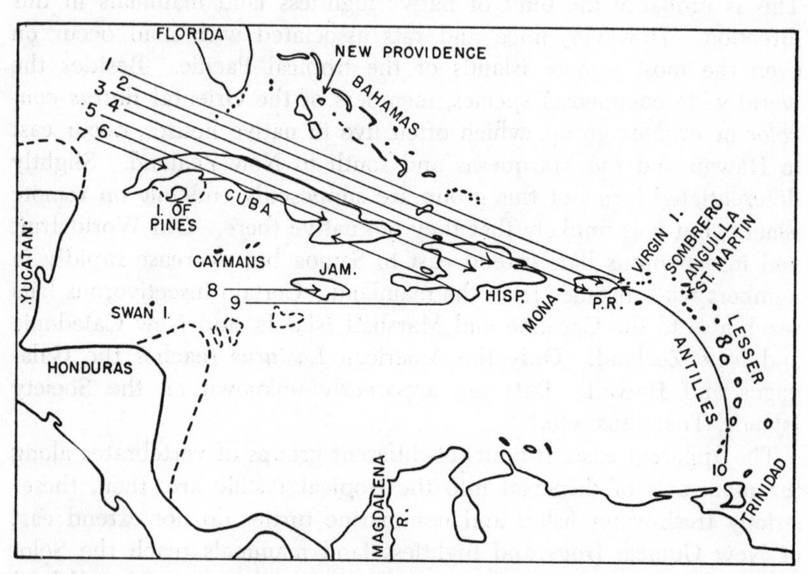

Fig. 63. The West Indies. Broken lines show banks off Central America that may have been land in the past. Numbered arrows show distance and (apparent) direction of dispersal on the Greater Antilles of the following animals: 1, *Bufo* (toads); 2, ground sloths; 3, *Nesophontes* (small insectivores); 4, cichlid fishes; 5, *Solenodon* (larger insectivores); 6, atelopodid frogs; 7, gar pikes; 8, monkeys; 9, *Hyla* (frogs); 10, *Leptodactylus* (frogs). This list includes all secondary-division fishes of the Greater Antilles except Cyprinodontes, all amphibians except *Eleutherodactylus*, and all flightless land mammals except rodents. The arrows therefore illustrate the limited diversity of the Greater Antillean fauna, its orderliness, and the fact that the most significant parts of it form a simple pattern of apparent immigration, mostly from Central America, with Cuba the most important port of entry.

of the tropics. The largest, probably oldest, and richest of the islands are the four Greater Antilles: Cuba, about 40,000 square miles; Hispaniola (Haiti and/or Santo Domingo), about 30,000, but with more high land than Cuba; Jamaica, about 4500; and Puerto Rico, about

3400 square miles. The Lesser Antilles form an irregular chain between the Greater Antilles and Trinidad (which faunally is part of South America) and are relatively small but ecologically varied islands. The Bahamas, north of the Greater Antilles and east of Florida, are very low and rather dry as well as small. All these islands will be treated in detail by Dr. Williams (see beginning of this chapter). Here I shall deal, briefly, mainly with the Greater Antilles, to put them in their place among the other important islands of the world.

The Greater Antilles, formerly supposed to be blocks of continental rock, are now considered geologically oceanic: they are made up (complexly) of volcanic rocks and marine limestone. Different islands and parts of islands have changed level at different times during the Tertiary, but whether the islands have actually been connected together by land is unknown. They are not connected to any continental shelf but are nearest to that of Central America, especially to the Honduras-Nicaraguan bank. However, in this as in other similar cases, it seems best to discount geology and see what the fauna of the islands tells.

Strictly fresh-water fishes do not reach the Greater Antilles. Of secondary-division, salt-tolerant fishes, which can disperse through the sea but do so with difficulty, a North American gar pike reaches Cuba; endemic species of a South and Central American genus of cichlids are on Cuba and Hispaniola; and both oviparous and viviparous Cyprinodontes, some North and some Central or South American in origin, are on Cuba, Jamaica, and Hispaniola but are most numerous and most differentiated on Cuba.

Of amphibians, neither (primarily tropical) caecilians nor (primarily northern) salamanders reach the West Indies, and only five genera of frogs do so. The single (tropical American) genus *Eleutherodactylus* makes up the greater part of the frog fauna of the islands; I cannot trace its history. The (tropical American) genus *Leptodactylus* reaches the Lesser Antilles, Puerto Rico, and Hispaniola, but not Cuba or Jamaica. *Hyla* reaches Jamaica (four species), Hispaniola (four species), and Cuba (one species), but not Puerto Rico. *Bufo* reaches Cuba (three species), Hispaniola (one species), Puerto Rico (one species), and the Virgin Islands (one species), but not Jamaica. And *Sminthillus* (one species) occurs only on Cuba, where it is widely distributed at low altitudes; it is apparently an endemic genus of the family Atelopodidae, a tropical American family unknown north of Costa Rica on the mainland. This Greater Antillian amphibian fauna is very limited, derived from a small fraction of the mainland fauna, and distributed on the islands in a very orderly way (Fig. 63).

Of reptiles, the West Indies have a few turtles, two crocodiles (but no alligator or caimans), some snakes, and many lizards. The only native non-marine turtles now on the Greater Antilles are slightly defined species of the fresh-water emydine genus *Pseudemys,* on all the Greater Antilles and on the Caymans and some Bahamas; they are closely related to Central American species. Large land tortoises (*Testudo*) are subfossil on Cuba, Mona Island (between Hispaniola and Puerto Rico), and Sombrero and presumably occurred on other islands. The common American crocodile reaches Cuba, Hispaniola, and Jamaica, and a larger species is endemic to Cuba and the Isle of Pines. There are too many lizards and snakes to list, but something can be said of them in general. On the mainland and on the continental island of Trinidad, snakes outnumber lizards; but on the West Indies proper, lizards far outnumber snakes (Table 15, p. 483). The West Indies lack poisonous snakes, except for the Fer-de-lance (possibly introduced) on some of the Lesser Antilles. Of the lizards, many have wide, orderly distributions, but others are localized. The small family Xantusiidae of Central and southwestern North America is represented on the West Indies by an endemic genus and species localized in eastern Cuba. Burrowing amphisbaenids, now apparently absent in most of Central America, are represented on the West Indies by three genera, including *Cadea,* which is endemic to Cuba and the Isle of Pines, and which represents an old northern group of the family.

Birds are fairly numerous on the West Indies, but many mainland families, including many dominant South American ones that reach Trinidad, are absent on the West Indies proper (pp. 240 and 493). Many species, some genera, and one slightly defined small family (Todidae) of birds are endemic to the West Indies, but there are no old relicts and no flightless birds there, unless a subfossil rail on Puerto Rico was flightless. (The Zapata Rail, on Cuba, is now known to be able to fly.) Different land birds have evidently reached the West Indies from different directions: from North, Central, and South America. Most of the resident land birds on the Greater Antilles have (I believe) probably come from Central America; the endemic family Todidae is closely related to the primarily Central American motmots. Many migratory land birds that breed in North America winter in the West Indies, and some of them (especially some wood warblers, p. 282) have established resident forms on the islands. See list of bird families (end of Chapter 5) for further details and for indication of all the American families that do not reach the West Indies.

West Indian land mammals other than bats (which are numerous) are very few, and most are now extinct. All known fossil as well as living ones are included in the following summary. The oldest fossils are no older than the (probably late) Pleistocene, so that the fossil and living species form(ed) a single fauna. There are two groups of West Indian insectivores. One, *Solenodon,* is known only on Cuba and Hispaniola and still exists very locally on both islands. Its relatives are in the Oligocene of North America. The other, *Nesophontes,* is subfossil on Cuba, Hispaniola, and Puerto Rico; it represents another old stock which also presumably came from North America, for insectivores probably did not reach South America until recently. One monkey, an endemic genus, is fossil on Jamaica. One or two stocks of ground sloths reached the West Indies and evolved endemic genera there; if there were two stocks, one reached Cuba, Hispaniola, and Puerto Rico; the other, apparently only Cuba and (perhaps) Hispaniola; both were of course originally derived from South America. All other known Greater Antillean land mammals, excepting bats, are rodents, mostly hystricomorphs. The latter include representatives of four families, all presumably derived originally from South America. Living and/or extinct capromyids (four existing and three extinct West Indian genera) are most diverse on Cuba and Hispaniola, but single genera extend(ed) to Jamaica, Puerto Rico (introduced?), the Caymans, Swan Island, and some Bahamas. Echimyids (five genera, all extinct at least on the West Indies) occurred on Cuba, Hispaniola, and Puerto Rico. Heptaxodontids (seven extinct genera) occurred on Jamaica, Hispaniola, Puerto Rico, and Anguilla and St. Martin east of Puerto Rico. And octodontids (one extinct endemic genus) occurred only on Jamaica. The distributions of the West Indian hystricomorphs are given in somewhat more detail in the list of mammal families (Chapter 6). The relationships of some of these animals are still doubtful. Of non-hystricomorph rodents, the only ones apparently native on the West Indies are rice rats (*Oryzomys* etc.), which have reached Jamaica as well as some of the Lesser Antilles, and which belong to the (in America) originally northern family Cricetidae. Besides these, two opossums, an armadillo, agoutis, and raccoons are on some of the Lesser Antilles, and a raccoon is on New Providence Island in the Bahamas, but these animals are not much differentiated and some of them may be introduced. It will be seen that the flightless land mammals of the West Indies are very limited, with no large forms (the ancestors of the West Indian ground sloths were apparently small), no carnivores except doubtfully native raccoons on a few small islands, and no ungulates. Moreover their pattern of distribution is orderly. On the Greater Antilles, five or six main groups (the two of insectivores, one or two of ground sloths, and the first two of rodents) are distributed as if they entered the West Indies by way of Cuba and extended to Hispaniola and in some cases other islands; and four groups (the monkey and three groups of rodents), as if they entered the West Indies by way of Jamaica, only one of them (one of the rodents) extending farther. As compared with other Greater Antillean mammals, the rodents are (or were) much more differentiated on the outer islands, with endemic genera on Puerto Rico and one (*Amblyrhiza,* a very large rodent) even on Anguilla and St. Martin.

The West Indian vertebrate fauna as a whole is made up of different classes in proportion to their probable powers of dispersal across salt water. Strictly fresh-water fishes are absent, but slightly salt-tolerant ones are present on some islands. Amphibians are represented by few genera, although one of them has radiated extensively on the islands. Non-marine turtles are few, and snakes are outnumbered by lizards, which are numerous. Birds and bats are numerous. But flightless land mammals are few and small; the majority are rodents; and (at least on the large islands) there are no native carnivores.

The most significant parts of this fauna, the groups that probably have most difficulty dispersing across salt water, are distributed on the Greater Antilles in an orderly way, in what seems to be an immigrant pattern, as if most of them have reached the islands from the west, from Central America, through Cuba more than through Jamaica. Secondary-division fishes are most numerous on Cuba, fewer on Hispaniola and Jamaica, and absent on Puerto Rico. Of frogs, excluding unanalyzed *Eleutherodactylus,* two genera seem to have come through or to Cuba; one, through Jamaica; and one, possibly through the Lesser Antilles to Puerto Rico and Hispaniola. And of flightless land mammals, five or six groups seem to have come through Cuba, four through or to Jamaica.

This fauna is a mixture of originally North American and originally South American groups, but the northern and southern ones apparently did *not* reach the Greater Antilles in different ways. Cuba was apparently the port of entry for the southern cichlid as well as for the northern gar pike, for the southern atelopodid frog as well as for the northern *Bufo,* and for southern ground sloths and some hystricomorph rodents as well as for northern insectivores; and Jamaica was apparently the port of entry for an originally northern cricetid (the rice rat) as well as for some southern groups.

The distribution of the most distinct relict-endemic West Indian vertebrates is significant. Cuba has the atelopodid frog, the xantusiid lizard, and *Cadea.* Jamaica has (or had) the only monkey and only octodont rodent. Each of these animals apparently represents a separate invasion of the West Indies by a fairly old stock unknown elsewhere in the West Indies. They seem to have survived on port-of-entry islands beyond which they have not spread. *Hispaniola has no comparable exclusive groups.* It is a large island, much larger than Jamaica. It is not so large as Cuba, but it has more high land and is probably capable of supporting an equal fauna (see Table 15, p. 483). That Hispaniola apparently has no exclusive relict-endemic vertebrates, of groups not represented on other islands, suggests that it has not been an important port of entry for land vertebrates and suggests also that the dispersal of these animals into the West Indies has been fairly simple. If it had been complex, with successive invasions and replacements, Hispaniola should have more than its share of exclusive relict-endemics. The presence of relict-endemics on Cuba and Jamaica is significant, but their absence on Hispaniola may be even more so, for Hispaniola should be the inner keep for an older fauna, if there was an older fauna.

To recapitulate, different kinds of vertebrates are represented on the West Indies in proportion to their probable powers of crossing salt water. The most significant groups of them (those that have most difficulty crossing salt water) are few and are distributed on the Greater Antilles in what seems to be an immigrant pattern, as if most of them have come from the west, by way of Cuba or (less often) Jamaica. The Greater Antillean fauna is a mixture of both North and South American elements, which seem to have reached the islands by the same, not different, routes. And the exclusive relict-endemics occur on Cuba and Jamaica, not Hispaniola, which again suggests that Cuba and Jamaica have been the ports of entry, and which suggests further (as does the orderly pattern of distribution) that the dispersal of the most significant vertebrate groups over the Greater Antilles has been relatively simple, with not much succession and replacement, although there must have been some. This is the kind of fairly simple, orderly immigrant pattern, with exclusive relict-endemics on islands nearest the mainland (Central America), that is most likely to give dependable clues to the past (p. 487).

The simple explanation of all this is that terrestrial vertebrates have reached the Greater Antilles across salt water; that the existing and late-fossil forms are an original accumulation, mostly distributed along original routes of immigration; and that most of the significant Greater Antillean groups have come from Central America by way of Cuba or (less often) Jamaica at a time when the Central American fauna was a mixture of both North and South American elements.

There are two further comments to be made about this. Simpson (1956, p. 13), in a recent paper on zoogeography of West Indian land mammals, notes that air-borne and raft-borne animals have different means of dispersal (which is true), and he implies that they have reached the Greater Antilles from different directions, but this does not necessarily follow. So far as I can see, the pattern of distribution on the islands of presumably air-borne invertebrates (*e.g.*, the winged carabid beetles that I study) is fundamentally the same as the pattern of water-borne vertebrates. It is hard to believe that these patterns would be the same if the animals really reached the islands from different directions. The other comment is this. It might be supposed that the Gulf Stream would prevent drift from reaching the West Indies from Central America and might favor drift transport from South America in spite of the greater distance. But this plausible idea is not supported by facts. It is particularly significant that the secondary-division fishes that have reached the islands seem to have come mainly from Central and North America, across the Gulf Stream. Most of these fishes are small. They cannot swim long distances but must be carried almost passively by surface currents. If currents really favored transport from northern South America rather than from Central America, the West Indies should have a pimarily South American fish fauna, which is not the case. Other water-borne animals which seem to have reached the Greater Antilles from Central America include fresh-water turtles (Williams 1956) and the ancestor of the endemic Cuban crocodile, which is related to a Central American species. The only satisfactory explanation of all this seems to me to be that air-borne and water-borne (including drift-borne) animals have reached the Greater Antilles from the same direction, mostly from Central America.

There is a suggestive similarity of pattern between the Philippine and West Indian vertebrate faunas. Each shows what seems to be a fairly simple immigrant pattern produced by spread into and across the islands from the nearest mainland; each has exclusive relict-endemics chiefly on the islands nearest the mainland; but, in each case, rodents show more differentiation on the outer islands than the other flightless land vertebrates do. It seems likely that these patterns have been produced by similar processes, by spread across water gaps chiefly in one direction, outward from the nearest mainland, from Borneo in one case and Central America in the other. More complicated explanations of the West Indian pattern are possible, and I hope Dr. Williams will consider them, but this simple one seems to me satisfactory now, although there must have been complications and exceptions in some cases.

When the geological history of Central America and the fossil record of its fauna are better known, they may settle the question of the origin of the Greater Antillean fauna. But there are times, and this may be one of them, when the distribution of animals now (or very recently) tells certain things about the past better than geology or paleontology can. The history of Central America is not well known. North and South America were certainly separated during most of the Tertiary, and there is geological evidence of a water gap at the lower end of Central America, separating it from South America, but geologists do not yet know when or whether Central America was separated from North America too. The composition and especially the distribution of the Greater Antillean vertebrate fauna suggest that Central America was separated; that it was an island, not a peninsula of North America, through at least part of the Tertiary; that it had a mixed northern and southern fauna of small mammals and other vertebrates; and that most Greater Antillean vertebrates were derived from it.

Summary of fringing archipelagos. The Philippines, the islands of the western Pacific, and the West Indies are the principal archipelagos of the world which, though not simply recent continental, lie close to continents and receive the fringes of continental faunas and are extensive enough, with enough comparable islands, to show informative patterns of distribution. The archipelago of Wallacea (Chapter 7, pp. 462ff.) might be added to this list. It lies between the continental shelves of Asia and Australia and receives the fringes of the faunas of both continents.

All the main islands of the Philippines, the nearer western Pacific islands (the Bismarcks and Solomons), the West Indies, and Wallacea have some flightless land mammals and frogs and so would be continental islands by Wallace's criteria. Some zoogeographers still consider some of these islands continental and explain their faunas by systems of land connections. But on all these islands different classes of vertebrates are represented in approximate proportion to their probable powers of crossing salt water: true fresh-water fishes are absent, except a very few on the southern Philippines and western Lesser Sunda Islands; amphibians and flightless land mammals are limited in basic stocks, and rodents make up more than the usual proportion of the mammals; lizards tend to be disproportionately numerous among the reptiles; and bats and birds are numerous. Still more significant, the main parts of each fauna are distributed in an orderly way, along what seem to be original routes of immigration into the archipelagos, and old relicts are the exception, not the rule. Most of what relicts and well-marked endemics there are occur on islands nearest the mainland, except that among certain animals, notably rodents, which have more than usual powers of crossing salt water, the more distinct relicts or endemics tend to occur farther from the mainland: endemic genera or very distinct groups of rodents occur or have occurred not only on the outer Philippines and outer West Indies (Puerto Rico etc.) but also on the Solomons (at the limit of distribution of terrestrial mammals on islands in the western Pacific) and on Ceram and Flores (in the depauperate zone of Wallacea).

All this is evidence that vertebrates have dispersed from the continents onto fringing archipelagos, across water gaps, much more extensively and in some cases much farther than Wallace guessed. This is in itself an important fact, regardless of what it implies about the history of islands and island faunas.

Other islands and archipelagos

Celebes. Celebes is in the Indo-Australian Archipelago, across Wallace's Line from Borneo. It is about 70,000 square miles in area and is a tropical, diverse, evidently old island. It is separated from Borneo by Makassar Strait, now about 65 miles wide, but the narrowest deep water is only about 25 miles wide (Umbgrove 1949, p. 3, Fig. 4), and the strait was probably no wider than this at times in the Pleistocene. Celebes has already been discussed as part of Wallacea but deserves separate attention.

Celebes lacks strictly fresh-water fishes but has a few secondary-division Cyprinodontes. Of amphibians, Celebes lacks caecilians but has many frogs. Most are Oriental, but *Oreophryne* is Australian. Ten of the frogs are supposed to be undifferentiated Oriental species, but some of them may be introduced on Celebes (p. 464). Reptiles are numerous on Celebes and are mostly Oriental in their relationships. They include a *Testudo;* and a much larger *Testudo* was on Celebes in the Pleistocene (Hooijer 1951). Birds are numerous and are a mixture of (more) Oriental and (fewer) Australian groups. Of mammals, Celebes has two marsupials representing two groups of *Phalanger;* shrews (*Crocidura* and *Suncus*); *Tarsius,* a monkey of an endemic genus, and an endemic species of *Macaca;* four genera (one endemic) of civets; an endemic genus of pig (*Babirussa*), one of small buffalo (*Anoa*), and more ordinary pigs, and deer; three genera of squirrels (with some endemic species), a porcupine, and various murids, including five endemic genera; and bats. All these mammals except *Phalanger* and perhaps some of the murids and bats have been derived from the Orient. Some Pleistocene fossil mammals have been found on Celebes. Most belong to groups still represented in the existing fauna, but there was also a small elephant, standing perhaps 6 feet high at the shoulder (Hooijer).

This fauna is unique in completely lacking fishes of strictly fresh-water families while having several large mammals including formerly an elephant. How the larger mammals reached Celebes is still debated. Beaufort (1951, pp. 177–178) and Hooijer think they came over a land bridge via the Philippines, but there is no trace of the most significant groups on Mindanao, and Mindanao does not have a continental vertebrate fauna. Moreover, Celebes itself does not have a continental vertebrate fauna. I think that the larger mammals of Celebes, including the elephant, probably crossed the 25-mile barrier of Makassar Strait during the Pleistocene.

Madagascar, Mascarenes, and Seychelles. Madagascar (Fig. 64) lies east of Africa, on the southern edge of the tropics. Its area is about 240,000 square miles. It is an old, geologically and ecologically diverse island, probably once more extensively forested than now. Its central axis is an ancient, granitic plateau 3000 to 5000 feet above sea level, with scattered volcanic peaks rising to over 9000 feet. The lower, coastal portions of the island, especially on the west, toward Africa, are composed of sedimentary rocks of Triassic, Jurassic, Cretaceous, and Tertiary ages, the most recent rocks being nearest the sea, and this suggests that the channel between the island and Africa has existed at least in part since the Triassic. The channel is now more than 1000 meters deep for a width of 100 kilometers or more for its whole length (Millot 1952, p. 17). It is doubtful when or whether Madagascar has had a land connection with either Africa or Asia. The existing fauna probably requires none. However, there were dinosaurs on Madagascar in the Mesozoic, and if they required a land connection, there was one (see Chapter 10). The narrowest water gap, toward Africa, is now about 260 miles wide but was probably somewhat narrower at times in the Pleistocene, and the Comoro Islands may have acted as steppingstones for some animals, although the water gaps are fairly wide there too. Rand (1936) gives an excellent description of Madagascar as well as of its bird fauna, and Millot (1952) briefly reviews its whole fauna.

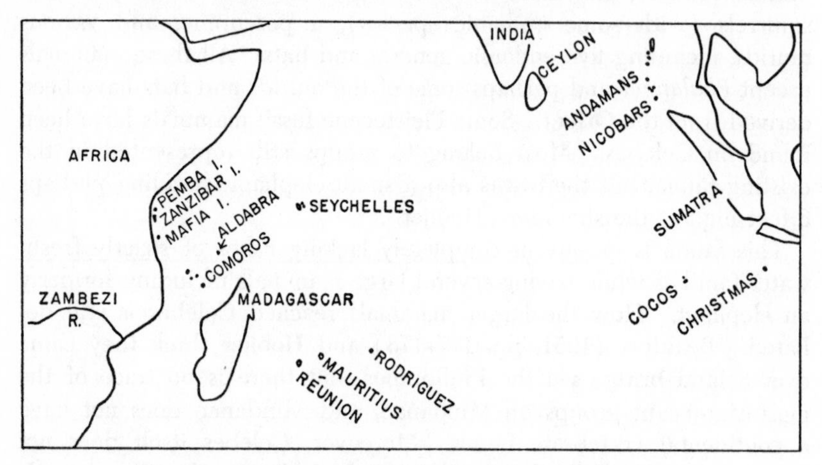

Fig. 64. Madagascar and islands of the Indian Ocean.

Madagascar lacks strictly fresh-water fishes but has a few secondary-division cichlids and cyprinodonts, probably derived from Africa, and various peripheral fishes.

Of amphibians, Madagascar lacks caecilians and some continental families of frogs but has other frogs, especially rhacophorids and brevicipitids. Most of the 150-odd species and some of the genera are endemic, and so is one brevicipitid subfamily, in which 8 genera, 21 species have apparently evolved from one ancestor on Madagascar. Some Madagascan frogs are related toward Africa; others, toward tropical Asia. For some further details see page 147.

Of reptiles (p. 190), Madagascar lacks many important continental groups including trionychid and emydid turtles; agamid, lacertid, amphisbaenid, and cordylid lizards, and *Varanus;* and leptotyphlopid worm snakes, pythons, elapids, and viperids—Madagascar, like the large islands of the West Indies, has no poisonous snakes. But the island has a good representation of other reptiles: fresh-water turtles (pelomedusids), land turtles (*Testudo* and relatives); lizards, including many geckos, relict iguanids, many chameleons, many skinks, and a few gerrhosaurids; and snakes, including *Typhlops,* specialized boids, and colubrids. A number of genera of reptiles are endemic, and one of the genera of geckos (*Uroplates,* with about five species) is sometimes considered an endemic family. Most Madagascan reptiles are African in their relationships, but a few are Oriental, and the (fresh-water) pelomedusid turtle *Podocnemis* and the (terrestrial) iguanids apparently have their closest living relatives in America.

Of birds, Madagascar has a very distinct fauna, summarized by Rand (1936), including representatives of about 51 families. Gulls, herons, ducks, rails, hawks, cuckoos, sylviids, and vangas are most numerous. This fauna includes a large endemic fraction, a large African fraction, and a smaller Oriental fraction. There are a few wholly endemic families and subfamilies: extinct, giant elephant birds (two genera, several species); flightless, rail-like mesoenatids (two genera, three species); two subfamilies of rollers (with together four genera, six species); philapittas (two genera, four species); and vangas (nine genera, eleven species). Although many families of birds that occur in Africa are represented on Madagascar, many others are not: secretary birds, ostriches, guinea fowls (except one species which may be introduced), plantain eaters, cranes, tree hoopoes, hornbills, mousebirds, barbets, woodpeckers, honey guides, broadbills, jewel thrushes, river martins, true shrikes, helmet shrikes, finches, buffalo weavers, widow birds, tickbirds, etc. are all lacking on Madagascar. The Madagascan bird fauna is therefore much less rich than the African one. See list of bird families (end of Chapter 5) for further details. The Oriental fraction of the Madagascan bird fauna includes four tropical Asiatic genera not represented in Africa and a few other birds with apparent Oriental relationships (Rand, p. 298).

Of mammals, Madagascar has a very limited, highly endemic fauna consisting of one endemic family of insectivores (Tenrecidae), with ten or more genera, perhaps thirty species; three endemic families of lemurs, with about ten existing genera, twenty species, all perhaps derived from one ancestor; one supposed endemic subfamily of cricetine rodents, with about seven genera, twelve species; and about seven endemic genera, twelve species of viverrids, representing one or more ancestor(s). Additional fossil or subfossil forms of these groups are known, and also a pygmy hippopotamus, the largest mammal known to have reached Madagascar. It was abundant there in the Pleistocene, and it was not the same as the hippo now existing on the adjacent African mainland. There are also on Madagascar, but probably introduced, one or two shrews, an Indian civet, rats and mice, and a bush pig of the African genus *Potamochoerus,* which does not occur in the Pleistocene fossil deposits, where remains of the small hippo are numerous. Finally Madagascar has many bats with African and some with Oriental relationships, the latter including several groups of *Pteropus.*

That this Madagascan vertebrate fauna has been derived across salt water is indicated by the absence of strictly fresh-water fishes and by the nature of the rest of the fauna, especially the terrestrial mammals, which are apparently derived from a few, small ancestors. The fewness of rodents is a discordant detail but may be due to chance; the absence of hystricomorphs is noteworthy. Most of the Madagascan vertebrates apparently came from Africa, fewer (at least some birds and bats) from tropical Asia. Some Madagascan frogs and reptiles have Oriental relationships, which does not necessarily signify Oriental origin. What has been said of frogs (pp. 149–150) is true of many other vertebrates: their pattern of distribution around and across the Indian Ocean is full of discontinuities and mixed relationships which indicate that the fauna is partly a relict one and that existing geographical relationships are partly the result of extinction and survival rather than of direct dispersal. The relationship of some Madagascan reptiles to tropical American ones is a result of still more extensive extinctions.

The Mascarene Islands are small islands far out in the Indian Ocean east of Madagascar. Réunion (or Bourbon) is about 400 miles from Madagascar, nearly 1000 square miles in area, and reaches an altitude of about 10,000 feet. Mauritius (or Maurice, formerly called Île de France) is more than 100 miles northeast of Réunion and more than 500 miles from Madagascar, about 700 square miles, and 2500 feet in altitude. Rodriguez is about 350 miles east of Mauritius, 40 square miles, and 1300 feet in altitude. These are all volcanic, oceanic islands, rising from deep water. They are tropical, with moderate to heavy rainfall, and were originally forested, although now they are largely stripped.

The Mascarenes have no strictly fresh-water fishes, amphibians, or land mammals except bats. Their reptiles, however, are remarkable. All the islands possessed giant land tortoises (*Testudo*) which have become extinct within historic times. (An African-Madagascan pelomedusid is recorded from Mauritius, but I assume it is introduced.)

Snakes are known, living or subfossil, only on Mauritius and adjacent, small Round Island, but they are two endemic genera of boids (or possibly primitive colubrids). Whether they occurred on the other Mascarenes is unknown. Of native lizards, Réunion and Mauritius have both geckos and skinks, but Rodriguez apparently has only one large, endemic gecko (a *Phelsuma*). Mascarene birds too are (or were) remarkable. No part of the world had a more extraordinary bird fauna, and nowhere has destruction of native birds been more rapid or more nearly complete. Hachisuka (1953, pp. 6–7) lists the known living and extinct indigenous birds of the three islands. (One or two of the birds listed are based on old, inadequate accounts of travelers and may be fictitious.) Notable Mascarene birds included the Dodo on Mauritius and its relatives on the other islands: the Solitaire on Réunion and the "White Dodo" on Rodriguez; these birds were all destroyed by man 200 or more years ago. They were descended from a ground-feeding pigeon which lost its power of flight presumably after it reached the different islands. Other extinct Mascarene birds included one or more flightless rails on each island, a supposedly flightless heron on Rodriguez, an "almost flightless" one on Réunion, and a supposedly flightless parrot on Mauritius. And the Mascarenes have or had also various water birds, and hawks, owls, several pigeons, several parrots, and a small variety of passerines. Large fruit bats (*Pteropus*) occur on all the islands, and an insectivorous bat (*Scotophilus*) is recorded from Mauritius.

The Seychelles are a group of about 45 islands, about 600 miles northeast of Madagascar, somewhat farther from the coast of Africa, and much farther from India. The largest and highest island, Mahé, has an area of about 55 square miles and an altitude of about 3000 feet. The total land area of the Seychelles is less than 150 square miles, but the islands are all connected by the 100-meter line and are surrounded by a shallow bank some 12,000 square miles in area; there are additional large banks in the vicinity, so the existing islands may be the remnants of an extensive, low archipelago. The underlying rock of the islands is said to be old granite like that of the central plateau of Madagascar. The Seychelles are tropical and well watered and were once heavily forested, but most of the forest has been cut. The fauna of the Seychelles is described in reports of the Percy Sladen Trust Expedition, summarized by Gardiner (1936).

The Seychelles lack strictly fresh-water fishes but have amphibians. Caecilians are represented there by an endemic genus with six species, all presumably derived from one ancestor which Parker (1941) seems to imply was probably African rather than Indian—but caecilians do not occur on Madagascar. The frogs include an endemic subfamily (two genera, three species) somewhat doubtfully assigned to the Pelobatidae, a family not known in either Africa or peninsular India but occurring in the eastern Orient and the north-temperate zone. The Seychelles have also an endemic species of the African-Madagascan rhacophorid frog genus *Megalixalus* and a possibly introduced species of *Rana*. The absence of *Rhacophorus,* which occurs in the Oriental Region and Madagascar, and of brevicipitid frogs, so numerous in Madagascar, is noteworthy. Of reptiles, the Seychelles have or had giant land tortoises (*Testudo*), now extinct, and an undifferentiated fresh-water pelomedusid, which (I suspect) may be introduced; several lizards (geckos, skinks, and a chameleon), some African and some apparently Oriental in origin; and two colubrid snakes, one an endemic genus. The Seychelles have or had fourteen endemic land birds (Vezey-FitzGerald and Betts 1940), most of which probably still exist. Some are said to be African, others Oriental in relationships. They include a small falcon, two pigeons, two parrots, and passerines. The Seychelles do not have, and so far as is known have never had, any flightless birds. There are no flightless land mammals there, but fruit bats (*Pteropus*) and insectivorous bats (two species of *Coleura*) occur, both with Oriental affinities.

The Mascarenes and the Seychelles are small, remote islands, with very limited faunas, but their faunas differ in unexpected ways. The Mascarenes lack amphibians and have or have had several groups of flightless birds. The Seychelles have endemic, probably rather old groups of amphibians, but no flightless birds. These differences suggest that the islands have had very different histories. The Mascarenes have probably always been remote specks of land and have accumulated only a very limited fauna of exceptionally good water crossers. The Seychelles may be the remnants of an extensive archipelago of larger islands which extended near enough to Africa or India to receive a considerable vertebrate fauna (parts of which apparently did not come through Madagascar), most of which became extinct when the islands were reduced in size. Such a fauna might have included terrestrial mammals, and they might have retarded the evolution of flightless birds. This history would be consistent with the geological nature of the Seychelles and with the form of the sea bottom around them. But it need not require complete connection of the Seychelles with any other land.

New Caledonia, Lord Howe, and Norfolk Island. New Caledonia lies about 800 miles east of Australia but closer to the New Hebrides etc., in the edge of the tropics. It is about 6500 square miles in area, with mountains over 5000 feet, and is an ecologically diverse, geologically complex, old island, with highly endemic invertebrates. The vertebrate fauna, however, is very limited. There are no true fresh-water fishes; of salt-tolerant ones, the most notable is probably a galaxiid, the most northern representative of its family. There are no amphibians, no native mammals except bats, and no non-marine turtles, although a meiolaniid occurred on Walpole Island, southeast of New Caledonia, in the Pleistocene. Snakes are absent (Mertens 1934, p. 39), although they have been mistakenly recorded several times. The only native, originally flightless land vertebrates on New Caledonia are therefore lizards, and they are only geckos (including endemic genera) and skinks. New Caledonian birds (Mayr 1945, pp. 149–175) include about 68 native land and fresh-water species of which 16 are endemic or shared only with the Loyalty Islands, and a few of the genera and one family are endemic. The native birds include hawks, owls, a button quail, rails, the Kagu, pigeons, parrots, cuckoos, an owlet nightjar, a nightjar, swifts, a kingfisher, and 24 songbirds. The endemic family (Rhinochetidae) contains only the Kagu, a large, rather crane-like bird which still has well-developed wings but does not fly. A New Caledonian rail (a species of *Tricholimnas*) may have become flightless too. However, New Caledonia lacks more strikingly flightless birds. Of mammals, rats are present but are presumably introduced, and fruit bats (*Pteropus*) and insectivorous bats occur. There is nothing in this vertebrate fauna which may not reasonably be supposed to have reached New Caledonia across water.

Lord Howe is a small volcanic island about 300 miles east of Australia and about twice as far south (and a little west) of New Caledonia, 5 square miles in area and nearly 3000 feet in greatest altitude. Much of the island is forested. Its native vertebrates are limited to an apparently endemic skink (a species of *Lygosoma* or *Leiolopisma*) and non-endemic geckos (the latter perhaps brought by man), a few land birds, and a bat (*Scotophilus*); and a meiolaniid tortoise was on the island in the Pleistocene. The birds (Hindwood 1940) formerly included about fifteen native, resident land species (but eight of them are now extinct): one or two pigeons, a parrot, a rail and a gallinule (both apparently flightless), an owl, an insect-eating kingfisher, and eight passerines. Of these fifteen species, two or three seem to have come from Australia, three from New Caledonia, one from New Zealand, and the geographical origin of the rest is doubtful. Additional land birds and some fresh-water ones visit the island regularly or irregularly, and (as usual on remote islands) various sea birds breed there.

Norfolk Island is a small volcanic island about 800 miles east of Australia and midway between New Caledonia and New Zealand, about 13 square miles in area, and 1000 feet high. The climate is favorable. The plants are said to show relationships to those of New Zealand and include the Norfolk Island Pine, a fine araucarian. There are probably no native flightless land vertebrates; one or two widely distributed geckos occur on the island but are probably introduced. Norfolk Island land birds (Mathews 1928–1936) are an endemic pigeon (extinct), two parrots (one extinct), an owl, an insectivorous kingfisher, and nine passerines, most or all endemic. No rail is known for certain, but a medium-small one is doubtfully recorded from the island, which surely should have had rails originally.

New Zealand. New Zealand is about 1000 miles east-southeast of Australia. Its two main islands have a combined area of more than 100,000 square miles. The islands are geologically complex, old, with high mountains, and are ecologically diverse within the limits of their temperate climate. New Zealand vertebrates are listed and described by Hutton and Drummond (1923) and are listed by Powell (1949, with references). Some of them are discussed elsewhere in this book or are described in more detail in the lists of families.

New Zealand lacks strictly fresh-water fishes but has peripheral ones, including galaxiids. Amphibians are represented by the frog *Leiopelma* (two or three species), a relict which has survived on New Zealand while its relatives have disappeared from most of the rest of the world. Of reptiles, New Zealand has the Tuatara (*Sphenodon*), an even more striking relict, the last survivor of an order that disappeared from the fossil record a hundred million years ago—but the Tuatara need not have been on New Zealand that long. New Zealand also has geckos (two endemic genera, the only ovoviviparous geckos) and skinks (endemic, ovoviviparous species of *Lygosoma*), but no turtles or snakes except marine ones. New Zealand has (or had) a unique assemblage of flightless birds: huge, extinct moas (two families), kiwis, a variety of thoroughly flightless rails, an extinct flightless goose, a flightless parrot, and perhaps (if it was flightless) the only flightless perching bird, *Traversia*. Of flying birds, New Zealand has small endemic families (Xenicidae, Callaeidae), endemic genera and species of other families (including hawks and owls, an extinct quail, cuckoos, parrots and pigeons, a kingfisher, and various passerines), and some non-endemics, including recent arrivals and stragglers from Australia (p. 240). Of native land mammals, New Zealand has only two insectivorous bats, one considered an old endemic family (Mystacinidae), and the other a species of *Chalinolobus* probably derived from Australia more recently.

It is not necessary to look for land connections to account for the existing New Zealand vertebrate fauna. Frogs sometimes cross at least moderate gaps of salt water, and there is no limit to the width of water one might conceivably cross, especially one which, like *Leiopelma,* can develop on land. *Sphenodon* should be a good water crosser, perhaps equal to some lizards. The New Zealand vertebrate fauna as a whole exhibits every level of endemism from an endemic order (*Sphenodon*) to non-endemic species (of birds). This is consistent with the whole existing fauna having accumulated gradually, across wide water gaps. However, New Zealand is a very old island, and I should not like to be dogmatic about its early history or the early history of its fauna.

Hawaiian Islands. The Hawaiian Islands are an archipelago about 1500 miles long in the mid-Pacific, about 2000 miles from North America and farther from Old World continents. There are "stepping-stones" toward Asia and Australia, but some of the water gaps are wide. The largest island, Hawaii, is about 4000 square miles in area. The islands are volcanic and are surrounded by deep water. They lie across the northern edge of the tropics and are, for their size, rich and diverse islands, with high mountains. They are old enough to have evolved a highly endemic fauna of invertebrates and birds, but this may not have required many million years. The main islands are apparently no older than the Pliocene. The Hawaiian fauna was catalogued in *Fauna Hawaiiensis,* a fine work for its time (1899–1913). Zimmerman (1948) has brought up to date what is known in general of the nature and history of the islands and of their flora and fauna.

There are no strictly fresh-water fishes and no native amphibians or reptiles on these islands, only birds and a bat, the latter a species of *Lasiurus* derived from America. Eight frogs and toads, one fresh-water turtle, nine lizards, and one terrestrial snake occur on the islands (Oliver and Shaw 1953), but they are all introduced (although one frog and six lizards have been mistakenly described as new forms from the Hawaiian Islands!). Of birds, the Drepaniidae have reached the level of a slightly differentiated family of nine genera, twenty-two full species, confined to the Hawaiian Islands northwest to Laysan. They are probably derived from a nectar-eating American honey creeper, but they have evolved nectar-, insect-, and fruit- or seed-eating lines on the islands, and the main lines have evolved different geographical forms on different islands. The result is an outstanding example of adaptive and geographical radiation in isolation, described in detail by Amadon (1950). The latter also briefly discusses the other native Hawaiian land and fresh-water birds. Mayr (1943) discusses them zoogeographically. He thinks there have been fourteen separate invasions of them on the Hawaiian Islands proper, showing seven levels of endemism (Table 19). Several of the endemic genera have set

TABLE 19. LAND AND FRESH-WATER BIRDS OF THE HAWAIIAN ISLANDS (EXCEPTING LAYSAN)

Modified from Mayr 1943, pp. 45–46. The 14 separate invasions are listed in order of apparent relative age (numbers 1–7).

Endemic Family

(1) Drepaniidae (probably from America)

Endemic Genera

(2) A rail (derivation unknown)
(3) Two genera of honey eaters (from Australasian ancestor)
(4) A thrush (from American ancestor)
(4) An Old World flycatcher (from Polynesian ancestor)
(4) A goose (from American ancestor)

Endemic Species

(5) A crow (probably from North America)
(5) A hawk (from America)
(5) A duck (probably North American)

Endemic Subspecies

(6) The Short-eared Owl (probably from North America)
(6) The Stilt (from America)
(6) A gallinule (from America)
(6) The Coot (from America)

Not Endemic

(7) The Black-crowned Night Heron (from America)

off species on different islands. Besides the forms listed in the table, another endemic genus of rail (perhaps related to the Hawaiian one) and an endemic species of Old World warbler (an *Acrocephalus*) are (or were) on Laysan. And besides the resident forms, the Golden

Plover and a few other non-marine birds migrate or wander to the islands. This bird fauna is a classic example of an oceanic fauna derived from a few ancestors which have gradually accumulated from different directions (the majority apparently from America) and which have partly compensated for their fewness by radiation of some of the older stocks. The Hawaiian and Laysan rails had become flightless, but no other flightless birds are known from the Hawaiian Islands.

The Galapagos. The Galapagos Islands are on the Equator about 600 miles west of South America and more than 700 miles southwest of Central America. There are five large and ten smaller islands, with a total area of about 2870 square miles; Albemarle Island accounts for about 1650 of these. Several of the larger islands rise to from 2000 to 5000 feet. Geologically, they are all volcanic, oceanic islands and are separated from the mainland by deep water. They are rather arid, but there is some forest on the higher slopes.

The Galapagos lack strictly fresh-water fishes, and amphibians. Of reptiles, they have giant land tortoises (about thirteen races of one endemic species of *Testudo*), two large iguanas (forming two related endemic genera), smaller iguanids (endemic species of the South American genus *Tropidurus*), geckos (the widely distributed genus *Phyllodactylus*), and colubrid snakes (endemic species of the tropical American genus *Leimadophis,* possibly representing two different mainland stocks of the genus). Galapagos land birds are dominated by Darwin's finches, sometimes considered a subfamily (Geospizinae), a group of four or more genera, twelve species confined to the Galapagos, except that one of the genera, with one species, is endemic to Cocos Island, between the Galapagos and Central America. This is a natural group, derived from one ancestor (probably an American finch, but just what it was is unknown) which has radiated on the Galapagos as the drepaniids have done on the Hawaiian Islands, but less extensively. Different ones are ground feeding, vegetarian, insectivorous, woodpecker-like, cactus feeding, and warbler-like, and their beaks are modified accordingly. Also they differ geographically, from island to island. Their variation on a common pattern was one of the things that suggested evolution to Darwin. Lack (1947) tells in a readable way what is now known of them and refers to more technical works on them. Other Galapagos birds include an endemic genus of mockingbirds (*Nesomimus*) with different species or subspecies on different islands; an endemic genus of pigeon (*Nesopelia*); an endemic hawk (a *Buteo*), two owls, a rail, a swallow, and three flycatchers; an endemic subspecies of the Golden Warbler; an undifferentiated cuckoo; the Barn Swallow and Bobolink, which visit the islands more or less regularly; and various wading and swimming birds. Harris' Cormorant, an endemic genus, is flightless, and there is a penguin in the Galapagos, but there are no flightless land birds there. Of mammals, the Galapagos have small, endemic cricetid rodents, including what is sometimes considered an endemic genus (*Nesoryzomys*), but Simpson puts them all in *Oryzomys,* the same genus that has reached Jamaica. There is also a bat of the genus *Lasiurus,* the same genus that has reached Hawaii.

The whole Galapagos vertebrate fauna is apparently oceanic, derived, like the Hawaiian fauna, from a few ancestors which have gradually accumulated on the islands and which have partly compensated for their fewness by radiation of some groups. The ancestors apparently included one land tortoise, three or more lizards, one or possibly two snakes, a dozen or more land birds (to produce only the resident forms), one cricetid rodent, and a bat. They evidently all came from America, but whether most of them came from South or from Central America is a question. The islands are nearer South America, but Cocos Island forms a steppingstone toward Central America, and the Galapagos may receive drift more often from that direction. Agassiz (1892, pp. 59–60 and 70ff.) found that currents off Panama, sometimes moving 75 miles a day, carried immense quantities of drift toward Cocos Island and the Galapagos, and he had some unflattering things to say about persons who built land bridges to these islands! It has been suggested that some Galapagos vertebrates, including the iguanas, are related to West Indian forms and are derived from the same source, presumably Tertiary Central America. This is an interesting possibility worth further consideration.

Islands of the Atlantic Ocean (Fig. 56). The Canaries, Madeira, and the Azores lie above the tropics at increasing distances off northwestern Africa and southwestern Europe, the nearest islands of the three groups being about 60, 350, and 800 miles off the mainland; and the Cape Verde Islands lie within the tropics about 300 miles off Africa. These islands are all volcanic and all rise from deep water. They are small islands, the largest only a few hundred square miles in area. These islands have no strictly fresh-water fishes, no native amphibians, no endemic turtles (but the widely distributed African *Pelusios subniger* occurs, perhaps introduced, on the Cape Verdes, and a *Testudo* was on the Canaries in the Pleistocene), no snakes, and no land mammals except bats. Lizards are therefore the only native flightless land vertebrates, and they occur in proportion to distance and climate: the Canaries have endemic representatives of three Mediterranean-African genera, Madeira of one, and the Azores none; the (tropical) Cape Verdes have about four or more African stocks. Some Old World land birds reach all of these islands, and bats probably do so too: a species of *Pipistrellus* is endemic even on the Azores.

Bermuda is, roughly, 600 miles off North America, above the tropics. Its present area is only about 19 square miles, but it was larger in the past (Sayles 1931). It is a volcanic island, surrounded by deep water. It has no strictly fresh-water fishes, but a salt-tolerant cyprinodont is in the mangrove swamps. Of native, flightless land vertebrates, Bermuda has only a scincid lizard (a distinct endemic species of *Eumeces*) derived from North America, but two West Indian frogs, the Giant Toad of tropical America, and a West Indian lizard have been introduced (Dunn and Conant 1937), as have the usual rats and mice. Seven land birds (Florida Gallinule, Ground Dove, Crow, Catbird, Bluebird, White-eyed Vireo, and Cardinal) are native and resident on the island. They are all North American, not West Indian, species (Bond 1936, p. XV). Many other land birds migrate or straggle to the island. Bats appear occasionally but are apparently not resident there.

More remote Atlantic islands include, in the tropics, Ascension, about 1000 miles from Africa and 38 square miles in area, and St. Helena, about 1200 miles from Africa and 47 square miles, and, in the south-temperate zone, the Tristan da Cunha group, nearly 2000 miles from the tip of South Africa and farther from South America, the largest island about 16 square miles. These are all volcanic, oceanic islands, surrounded by deep water. Their only native land vertebrates are a very few birds. Ascension had an unnamed, flightless rail, now extinct and known only from an old drawing and description. St. Helena has an endemic species of plover (a *Charadrius*), related to an African species. Before man cut their forests (Wallace 1880, p. 283), these two islands may have had other land birds but, if so, they are lost without record. The Tristan da Cunha group, though more remote, smaller in land area, and colder, has two unrelated endemic genera of flightless rails; an endemic genus of thrush (*Nesocichla*) which Ripley (1952, pp. 16–17) thinks is of American origin; and two unrelated endemic genera of finches, one (*Nesospiza,* two species, differing in size of bill) on the Tristan da Cunha group proper, the other (*Rowettia,* one species) on Gough Island, the one apparently derived from Africa and the other from America (Mayr and Amadon 1951, p. 28). However, Rand (1955) thinks all the

Tristan da Cunha land birds have probably come from South America. He notes that a South American gallinule *commonly* strays to the islands (across at least 2000 miles of sea) and that two Barn Swallows have been recorded there.

Iceland. Iceland (Clark 1943) is in the North Atlantic, about 155 miles southeast of Greenland and farther than that from Europe, and is a volcanic, oceanic island. Its area is about 40,000 square miles, but only the lower, coastal areas are habitable, and virtually the whole island was covered by ice in the Pleistocene. Its land fauna is therefore very recent and very limited. Of native land vertebrates, Iceland has only birds and mammals. The breeding land birds of Iceland are arctic or European (not primarily American) species. They include a ptarmigan, two falcons, White-tailed Eagle, Short-eared Owl (breeding only since 1928), Snowy Owl, a swallow (*Hirundo*—breeds only occasionally), Meadow Pipit, White Wagtail, Wheatear, a thrush (*Turdus*), a redpoll, the Snow Bunting, European Wren, Raven, and perhaps the Starling. (I have compiled this list from several sources, then checked it briefly against Timmermann 1938–1949.) Some other small European land birds are regular visitors; the number that reach Iceland has increased considerably in recent years, probably because the climate has become somewhat milder (Salomonsen 1948). The native land mammals of Iceland are an apparently endemic subspecies of the European Common Field Mouse (*Apodemus sylvaticus*), the Arctic Fox, and the Polar Bear. All probably came on floating ice; the Polar Bear still does so frequently. Reindeer have been introduced.

Antarctica

The Antarctic Continent is actually a huge island, and it should be mentioned here. It almost centers on the south pole and fills much of the space within the Antarctic Circle. It is nearly 1000 miles from the southern tip of South America (but there are islands in between) and farther from Australia, New Zealand, and Africa. Its area is about 5,000,000 square miles, but most of it is covered by ice, and the ice-free parts are little more than bare rock, stripped of soil by extension of the ice in the Pleistocene. What little vegetation there is is mostly lichens and mosses. Antarctic vertebrates all get their living from the sea and, excepting marine fishes, they are few: birds and marine mammals. According to Lindsey (1940), about fifteen sea birds are typical of antarctic waters; the four that breed farthest south are the Emperor and Adelie Penguins, MacCormack's Skua, and the

Snow Petrel. Their normal southern limit is the antarctic coast, but the skua, at least, wanders inland and has been recorded only 160 miles from the pole. The only non-swimming birds that reach the edge of Antarctica are sheath bills (Chionididae; see list of bird families). There are four antarctic seals, but only the Weddell Seal occurs throughout the year at the southern limit of its range, along the antarctic coast. It is the southernmost of all mammals.

Antarctica now has a fauna extremely limited by cold. But it is a very old piece of land, and it was evidently warmer than now at times in the past. There are coal seams there, and fossil wood fragments, and *Glossopteris,* and there must once have been considerable vegetation and presumably a terrestrial fauna. What the fauna was —whether it was the sort to be expected on a very large but remote island or whether it was more than that—is now unknown but may be found out some day from fossils. In the meantime, the past relation of Antarctica to other lands and the part it may have played in the dispersal of life are exciting and important questions (pp. 603ff.).

Summary of island patterns

Coldness now limits and modifies vertebrate faunas on some islands (*e.g.,* Tierra del Fuego, Antarctica). On others, Pleistocene ice as well as present cold has limited the composition or age of existing faunas (*e.g.,* Newfoundland, the British Isles as compared with Japan). Some groups of animals, notably reptiles, do not occur at all on very cold islands but reach northern and southern limits on the oceans as on continents (Mertens 1934, p. 5, map of northern and southern limits of island reptiles). On the other hand, ice bridges or drifting ice have probably carried some animals, especially some large and medium-sized mammals, to some cold islands (*e.g.,* Newfoundland, Greenland, Iceland). Cold islands are therefore special cases, outside the main pattern of island life.

The effects of area and distance on island faunas are everywhere evident and profound, although they are not easily expressed mathematically. They make or modify all the patterns discussed below.

The most important fact about island vertebrates is that most of them seem to be fairly recent immigrants, not ancient relicts.

On all the great fringing archipelagos (the Philippines, the western Pacific islands, the West Indies, and Wallacea), the main patterns of distribution are orderly immigrant patterns, as if the faunas are accumulations of animals still distributed along the routes by which they spread into the islands. No primarily relict patterns have been found on these islands. Relict patterns should be formed on archipelagos (as one is apparently being formed on Sumatra, Borneo, and Java) when continental faunas are reduced to fractions. The surviving fractions should be irregularly distributed, partly on the larger or more favorable islands regardless of position, and partly according to chance. Relicts may occur within immigrant patterns too, but their distribution should be different. They should not form relict patterns but should usually be concentrated on either the nearer or the farther islands of an archipelago, in proportion to distance and dispersal rates. Apparent examples have been found on the Philippines and West Indies. But relict *patterns* have not been found on these or any other old archipelagos, and this is the important fact.

Surprisingly few ancient relict vertebrates occur on islands. The outstanding ones are worth listing in something like their order of relict-ness. *Sphenodon* stands first. The New Zealand frog and the lungfish in Australia (which, for strictly fresh-water fishes, is a very isolated island) are extremely ancient relicts too. But these are not typical examples. The most significant thing about them may be that there are no more like them—no other such old and isolated relicts (of vertebrates) on any islands. Next on the list might come *Solenodon* and *Nesophontes* on the West Indies, and with them perhaps the tenrecids and lemurs on Madagascar, but they may not be older (on their islands) than the Oligocene or Miocene. *Podocnemis* and the iguanids on Madagascar and the iguanid on Fiji are isolated geographical relicts but need not be very old; that is, they may not have been isolated very long. There are other cases like these, but most are probably not very old, and it soon becomes difficult to draw a line between the relict and autochthonous forms.

So far as I can see, no old island has a vertebrate fauna made up chiefly of ancient relicts. Islands like Madagascar and Celebes, sometimes thought to have relict continental faunas, seem to me to have accumulated their present faunas across water. Even the island continents, Australia and (in the Tertiary) South America, may not have had relict faunas but accumulations received across water and diversified by evolution. (But Australia and South America may nevertheless have been connected to the other continents earlier, in the Mesozoic.) All islands except recent continental ones seem now to have faunas which are mainly products of accumulation and local evolution, not isolated fractions of ancient continental faunas. If there is an exception, it is New Zealand, but even there the truly ancient vertebrates are very few, and they may be old immigrants rather than fractions of an old fauna. That no old island has a fauna which is obviously a remnant of an ancient continental fauna is a fact to be accounted for.

The ages (times of beginning of accumulation and evolution) of the older existing island vertebrate faunas are unknown but can be guessed at by a series of comparisons. If geologists are correct, the Hawaiian Islands and their fauna are no older than the Pliocene. The Galapagos fauna is less differentiated and probably younger. The Philippine fauna may be no older than the Hawaiian one; most Philippine vertebrates are not much differentiated, and those that are most so, the murid rodents, belong to a family which apparently did not rise until toward the end of the Miocene. The existing fauna of Wallacea seems to be rather young too, without highly differentiated forms or striking relicts. The faunas of all of these islands may be late Tertiary and Pleistocene in origin, and this may be guessed to be the age of most other island vertebrate faunas, excepting those of New Zealand and Madagascar and perhaps New Caledonia and the Greater Antilles.

New Zealand is an old island, and *Sphenodon* and *Leiopelma* represent old groups of animals, but it is not certain when they reached New Zealand. They may have sojourned in Australia first. It is strange that they have not radiated on New Zealand. *Leiopelma* especially, as the only frog on an island apparently well suited to frogs, ought to have radiated there, but it has formed (so far as known) only two or three slightly defined species. New Zealand also has endemic genera of lizards and endemic families of birds, some of which have radiated, but how much time has been required for their evolution is unknown. This fauna is evidently old (among island faunas), but there is nothing to date it exactly. It may be (on New Zealand) wholly Tertiary, or *Sphenodon* and *Leiopelma* may be older.

Madagascar has an obviously old fauna, much differentiated, with many relicts and much radiation in some endemic groups: *e.g.*, the tenrecids (which correspond to *Solenodon* on the Greater Antilles) have evolved ten or more genera. On the other hand, the Madagascan fauna is evidently less old than the faunas of Australia and Tertiary South America. It may have begun to accumulate and evolve about the Oligocene.

New Caledonia too is supposed to be an old island, but its vertebrate fauna is not obviously old. It includes no ancient relicts. It does include endemic, presumably autochthonous genera of geckos and birds and an endemic family of birds, but they have not radiated much. The endemic bird family, Rhynochetidae, consists of a single species; it may be a relict but not necessarily an old one. This vertebrate fauna cannot be dated, but there is no obvious reason why it need be older than the mid-Tertiary.

The age of the Greater Antillean fauna is doubtful too. *Solenodon* and *Nesophontes* are old insectivores. *Solenodon* has relatives in the Oligocene. But these animals need not have reached the Antilles then. They have not radiated there (*cf.* the tenrecids on Madagascar), and they may have arrived recently, after a sojourn in Central America. The Greater Antillean fauna as a whole does not seem very old. It may be of Miocene or even more recent age.

Putting all this together, it may be guessed that (excluding Australia and South America) the oldest existing island vertebrate fauna is on New Zealand, and that it began to accumulate early in the Tertiary if not before; that the Madagascan fauna is next, and that it began (the existing fauna) about the Oligocene; that the New Caledonian and Greater Antillean faunas may be next, beginning in the mid-Tertiary; and that the vertebrate faunas of all other islands are more recent, late Tertiary or Pleistocene. These are the hypothetical ages of the existing vertebrate faunas, not the ages of the islands. Some islands are evidently older than their faunas. Madagascar, for example, existed in the Mesozoic, but its present fauna, at least the part of it that can be dated, is not Mesozoic.

The occurrence of flightless birds is consistent with this hypothetical timetable. Gigantic flightless birds, with wings nearly or completely lost, have had time to evolve on New Zealand (which has had more flightless birds than any other island) and Madagascar, but they have not evolved on other isolated islands. Less highly modified flightless birds have evolved on some other islands, especially on the Mascarenes (where other vertebrates are too few to date the fauna), but are notably few and little modified even on New Caledonia and the Greater Antilles, if any have occurred on the latter. Rails have become flightless on many islands, but this has evidently not required much time.

The occurrence of invertebrates too is at least partly correlated with this timetable. The occurrence and geographical relationships of many invertebrates are not yet well known, and it is dangerous to generalize about them from the literature, but I know that carabid beetles (which I study) are highly differentiated and apparently old on Madagascar, New Zealand, and New Caledonia, and less differentiated and apparently less old on most other islands including the Greater Antilles, Philippines, and Hawaiian Islands. Many endemic genera of carabids are listed from the Hawaiian Islands, but this is an example of how misleading published "facts" can be, for many of the supposed genera on Hawaii are artificial, and they all seem to be rather slight modifications of existing continental stocks, while some of the carabid genera of Madagascar, New Zealand, and New Caledonia are very distinct. The carabids therefore agree with the vertebrates up to a point but suggest that the New Caledonian fauna is old and the Greater Antillean fauna relatively recent. It should be added that this one family of beetles is more numerous in species than all terrestrial vertebrates put together (but only a few specialists study it) and that it is well represented on islands.

Another, separate, line of evidence bearing on the nature and origin of island faunas is the proportion of different sorts of vertebrates on islands and the limits they reach.

Strictly fresh-water fishes scarcely extend beyond the limits of the continents and recent continental islands. They reach (a very few of them, which may possibly have some salt tolerance) only the inner Philippines and inner Lesser Sundas.

Some amphibians and terrestrial mammals have reached many not-too-remote islands. Their limits are the same in some places, but terrestrial mammals and not amphibians reach the Galapagos, and amphibians and not the mammals reach the Seychelles, perhaps the Fijis, and New Zealand.

Among the terrestrial mammals, large ones rarely occur beyond the limits of continental faunas. The small elephant formerly on Celebes, the small buffaloes on Celebes and Mindoro, and the small hippopotamus formerly on Madagascar are the largest land mammals which seem to have crossed significant water gaps, and, excepting the semi-aquatic hippopotamus, they have probably crossed only very narrow ones. Smaller mammals have reached islands more often, and more remote islands, and rodents have done so especially often. Only rodents are old-endemic on the outermost Philippines; rodents are the only terrestrial placentals that have reached Australia, but several stocks of them have done so; several genera of rodents but only one of other terrestrial mammals (*Phalanger*) have reached the Solomons; rodents outnumber all other terrestrial mammals on the Greater Antilles and are the only ones that have reached some of the smaller and more isolated West Indies (the Caymans, Swan Island, and some Bahamas); and only rodents (one stock) have reached the Galapagos.

Reptiles far outstrip amphibians and terrestrial mammals on islands, but some reptiles do better than others. Fresh-water turtles are surprisingly few and surprisingly little differentiated on islands. Land tortoises, often very large ones, are famous for their occurrence on the Galapagos and on islands in the Indian Ocean and formerly occurred on some of the West Indies and Canaries; and meiolaniids occurred on Lord Howe Island and near New Caledonia in the Pleistocene. Snakes have reached most of the islands that modern tortoises have (except the Canaries and perhaps some of the Mascarenes) and extend much farther into the Pacific—throughout the Philippines, to Australia, and east to Fiji. Moreover, many more snakes than tortoises have reached islands. Lizards are the most numerous of all flightless land vertebrates on islands and they go farthest: they extend to the Tongas (beyond Fiji) in the western Pacific; they alone of flightless land vertebrates have reached New Caledonia and Lord Howe Island; at least two stocks of them have reached New Zealand; at least three, the Galapagos (against one tortoise and one or possibly two snakes); they are very numerous on the West Indies (Table 15, p. 483); and they alone of flightless land vertebrates have reached Madeira, the Cape Verdes, and Bermuda. But even lizards are probably not native on the most remote islands of the Pacific and Atlantic.

Bats occur on some suitable islands (the Hawaiian Islands and Azores) beyond the limits of lizards.

Land birds have reached all or almost all habitable islands in even the most remote parts of the oceans. They do not now occur on Easter Island or Ascension, but they may have been there in the past. Their occurrence on the Hawaiian Islands and Tristan da Cunha show that even small land birds sometimes reach oceanic islands 2000 miles or more at sea.

The order of occurrence of the main groups of vertebrates on islands is summarized in Table 20.

Plainly, the representation of different groups of vertebrates on islands, and the limits they reach, are proportional to the animals' probable powers of crossing salt water, not to their geological ages. Frogs, for example, are very old animals, fossil in the Jurassic, and some of them do well on even small islands when they get there (*e.g.*, the frogs introduced on the Hawaiian Islands and Bermuda), but they do not occur on remote islands, except New Zealand. Lizards too are an old group, but it is not their age that accounts for their numbers on islands, for most of those that occur on remote islands are

TABLE 20. MAIN GROUPS OF LAND AND FRESH-WATER VERTEBRATES IN ORDER OF THEIR OCCURRENCE ON ISLANDS

Strictly fresh-water fishes: very few; widest gaps of salt water crossed probably very few miles wide.

Terrestrial mammals

- Large mammals: very few; widest gap crossed perhaps 25 miles (elephant etc. to Celebes in Pleistocene), except semi-aquatic hippo to Madagascar across much wider gap.
- Small mammals other than rodents: not numerous; widest gap crossed possibly 200 miles (to Madagascar in past).
- Rodents: relatively numerous; widest gap crossed perhaps 600 miles or more (to Galapagos).

Amphibians: basic stocks not numerous, but some have radiated; widest gap crossed perhaps 500 (to Seychelles in past) or 1000 miles (to New Zealand, if no great difference in lie of land in past).

Reptiles

- Fresh-water turtles: few; widest gap crossed possibly 200 miles (to Madagascar in past) (but chelyids may have reached Australia from South America in the past).
- Land tortoises: few but conspicuous; widest gap crossed perhaps 600 miles or more (to Galapagos).
- Snakes: somewhat more numerous; widest gap crossed perhaps 600 miles or more (to Galapagos).
- Lizards: numerous; widest gap crossed perhaps 1000 miles (to New Zealand).

Bats: few on remote islands (ecologically restricted); widest gap crossed at least 2000 miles (America to Hawaii).

Land birds: relatively numerous; widest gaps crossed 2000 miles or more (America to Hawaii and Tristan da Cunha).

not ancient but are closely related to existing continental forms. And of mammals, it is again not the oldest ones that occur most often on islands, but rodents, some of which (the murids) are relatively recent mammals but which have reached more than their share of islands nevertheless. This pattern as a whole is plainly determined by water-crossing ability, not by age. Only the New Zealand frog seems out of place in this pattern.

The occurrence of different sorts of animals on islands is modified by other things. There is little place for fresh-water fishes and fresh-water turtles on some islands. Large mammals may not be able to exist on small islands even if they can get there. Rodents that live on seeds etc. and need little fresh water are probably better able than most mammals to live on small islands as well as to get there. Snakes are probably less able than lizards to live on small, ecologically limited islands (p. 180). And bats are obviously less adaptable than birds to island life. But these things make only minor modifications of the main pattern, which is determined primarily by distance and water-crossing ability.

To recapitulate: the immigrant patterns of the faunas of the fringing archipelagos, the fewness of truly ancient relicts on islands, the apparently rather recent age of most island faunas, and the fact that vertebrates occur on islands in proportion to probable power of crossing salt water, not in proportion to geologic age, all point to the same conclusion, that the vertebrate faunas of islands (excepting recent continental islands) are mainly accumulations derived across water, not surviving fractions of old continental faunas. There has evidently been a flow of vertebrates from the continents onto the islands of the world, the details of it depending on the widths of water gaps, the sizes and natures of islands, and the natures (not geologic ages) of different vertebrates; and there seems to have been very little return flow. Few vertebrates are distributed as if they have evolved on islands and invaded continents. That there has been such a flow —a broad, directional movement—outward from the continents onto islands is a fundamental fact of animal dispersal.

Why are there so few old vertebrates on islands? Probably partly because many islands are not old. Remote, volcanic, oceanic islands probably usually have rather short lives before they are worn down and submerged. But some islands are old. Madagascar existed in the Mesozoic and had dinosaurs then, and New Zealand and perhaps New Caledonia are very old too. Why do these islands not have Mesozoic faunas? Even New Zealand does not have a Mesozoic vertebrate fauna, whether or not *Leiopelma* and *Sphenodon* have been there since the Mesozoic. It seems likely that part of the explanation is that animals do not survive indefinitely on islands. They probably become extinct in the course of time, if not through competition with newcomers, then through some sort of deterioration in small, isolated areas. This too may be a fundamental fact, related to and partly explaining the direction of movement of vertebrates from continents onto islands.

The significance of amount of evolutionary radiation that has occurred on different islands is hard to assess. Time is involved but so are other factors, including evolutionary opportunity (in response to which the drepaniids have radiated on the Hawaiian Islands in a comparatively short time), area, and perhaps genetic state. That insectivores have radiated so little on the Greater Antilles (as compared to Madagascar) certainly suggests limitation of time. That there has been less radiation of lizards and birds on New Caledonia than on New Zealand suggests limitation of time. But why have *Sphenodon* and *Leiopelma* not radiated on New Zealand as some lizards and birds, including giant flightless birds, have done? Did they once radiate, and are they the last survivors of groups which are disappearing—perhaps even the last survivors of a Mesozoic fauna on New Zealand? Have they deteriorated genetically? Does genetic deterioration eventually stop the radiation of island animals, then lead them to almost automatic extinction? These are questions well worth careful consideration in the following chapter.

REFERENCES

Agassiz, A. 1892. Reports on the dredging operations . . . to the Galapagos . . . . *Bull. Mus. Comparative Zool.*, **23**, 1–89.

Amadon, D. 1950. The Hawaiian honeycreepers . . . . *Bull. American Mus. Nat. Hist.*, **95**, 151–262.

Bangs, O. 1913. The land mammals of Newfoundland. *Bull. Mus. Comparative Zool.*, **54**, 507–516.

Barbour, T. 1937. Third list of Antillean reptiles and amphibians. *Bull. Mus. Comparative Zool.*, **82**, 75–166.

Beaufort, L. F. de. 1951. *Zoogeography of the land and inland waters.* London, Sidgwick and Jackson; New York, Macmillan.

Bond, J. 1936. *Birds of the West Indies.* Philadelphia, Acad. Nat. Sci.

Brown, W. C. 1952. The amphibians of the Solomon Islands. *Bull. Mus. Comparative Zool.*, **107**, 1–64.

Brown, W. C., and G. S. Myers. 1949. . . . new frog . . . Solomon Islands, with notes on . . . Fijian frog fauna. *American Mus. Novitates*, No 1418.

Carter, T. D., J. E. Hill, and G. H. H. Tate. 1945. *Mammals of the Pacific World.* New York, Macmillan.

Clark, A. H. 1943. *Iceland and Greenland.* Smithsonian Inst. War Background Studies No. 15.

Dammerman, K. W. 1929. On the zoogeography of Java. *Treubia*, **11**, 1–88.

———. 1948. The fauna of Krakatau 1883–1933. *Koninklijke Nederlandsche Akademie Wetenschappen, Verhandelingen* (Tweede Sectie), **44**, 1–594.

Darlington, P. J., Jr. 1938. The origin of the fauna of the Greater Antilles, with discussion of dispersal of animals over water and through the air. *Quarterly Review Biol.*, **13**, 274–300.

Darwin, C. 1839, 1952. *Journal of researches . . . H. M. S. Beagle.* London, Henry Colburn; reprinted 1952, New York and London, Hafner.

Delacour, J., and E. Mayr. 1946. *Birds of the Philippines.* New York, Macmillan.

Deraniyagala, P. E. P. 1949. *Some vertebrate animals of Ceylon,* Vol. 1. Ceylon, Colombo: Oxford. B. H. Blackwell.

Dickerson, R. E., *et al.* 1928. *Distribution of life in the Philippines.* Philippine Bureau Sci. Monograph 21.

Dunn, E. R. 1934. Physiography and herpetology in the Lesser Antilles. *Copeia,* **1934,** 105–111.

Dunn, E. R., and R. Conant. 1937. The herpetological fauna of Bermuda. *Herpetologica,* **1,** 78–80.

Eigenmann, C. H. 1909. The fresh-water fishes of Patagonia . . . . *Rep. U. Princeton Exp. Patagonia, 1896–1899,* **3,** Part 3, 225–374.

Gardiner, J. S. 1936. Concluding remarks on the distribution of the land and marine fauna [of the Seychelles etc.] . . . . *Tr. Linnean Soc. London,* Ser. 2, **19,** 447–464.

Gulick, A. 1932. Biological peculiarities of oceanic islands. *Quarterly Review Biol.,* **7,** 405–427.

Hachisuka, M. 1953. *The Dodo and kindred birds.* London, Witherby.

Hachisuka, M., and T. Udagawa. 1950–1951. Contributions to the ornithology of Formosa. *Quarterly J. Taiwan Mus.,* **3,** 187–280; **4,** 1–180.

Hindwood, K. A. 1940. The birds of Lord Howe Island. *Emu,* **40,** 1–86.

Hinton, M. A. C., *et al.* 1935. *List of British vertebrates.* London, British Mus.

Hooijer, D. A. 1951. Pygmy elephant and giant tortoise. *Sci. Monthly,* **72,** 3–8.

Hutton, F. W., and J. Drummond. 1923. *The animals of New Zealand.* Auckland etc., Whitcombe and Tombs.

Inger, R. F. 1949. Notes on a collection of fresh-water fishes from Trinidad. *Copeia,* **1949,** 300.

———. 1954. . . . Philippine Amphibia. *Fieldiana* (Chicago Nat. Hist. Mus.), *Zool.,* **33,** 181–531.

Koenigswald, G. H. R. von. 1939. Das Pleistocän Javas. *Quartär,* **2,** 28–53.

Kuenen, P. H. 1950. *Marine geology.* New York, John Wiley; London, Chapman and Hall.

Lack, D. 1947. *Darwin's finches.* Cambridge, England, Cambridge U. Press.

Lindsey, A. A. 1940. Recent advances in antarctic bio-geography. *Quarterly Review Biol.,* **15,** 456–465.

Lord, C. E., and H. H. Scott. 1924. *A synopsis of the vertebrate animals of Tasmania.* Hobart, Oldham, Beddome and Meredith.

Mathews, G. M. 1928–1936. *The birds of Norfolk and Lord Howe Islands . . . with supplement.* London, Witherby.

Matthews, L. H. 1952. *British mammals.* London, Collins.

Mayr, E. 1939. The origin and the history of the bird fauna of Polynesia. *Proc. Sixth Pacific Sci. Congress,* **4,** 197–216.

———. 1943. The zoogeographic position of the Hawaiian Islands. *Condor,* **45,** 45–48.

———. 1945. *Birds of the Southwest Pacific.* New York, Macmillan.

Mayr, E., and D. Amadon. 1951. A classification of Recent birds. *American Mus. Novitates,* No. 1496.

Mertens, R. 1934. Die Insel-Reptilien . . . . *Zoologica* (Stuttgart), **32,** 6. Lieferung, 1–209.

Millot, J. 1952. La faune malgache et le mythe gondwanien. *Mém. Inst. Sci. Madagascar,* Ser. A, **7,** 1–36.

Mori, T. 1936. *Studies on the geographical distribution of freshwater fishes in eastern Asia.* Apparently privately published.

Munro, I. S. R. 1955. *The marine and fresh water fishes of Ceylon.* Canberra (Australia), Department of External Affairs.

Okada, Y. 1931. *The tailless batrachians of the Japanese Empire.* Tokyo, Imperial Agricultural Experiment Station.

———. 1955. *Fishes of Japan.* Tokyo, Maruzen Co.

Oliver, J. A., and C. E. Shaw. 1953. The amphibians and reptiles of the Hawaiian Islands. *Zoologica* (New York), **38,** 65–95.

Parker, H. W. 1941. The caecilians of the Seychelles. *Ann. and Mag. Nat. Hist.* (11), **7,** 1–17.

Perkins, R. C. L., *et al.* 1899–1913. *Fauna Hawaiiensis.* Cambridge, England, Cambridge U. Press, 3 vols.

Powell, A. W. B. 1949. *Native animals of New Zealand.* Auckland Mus. Handbook Zool.

Praeger, R. L. 1950. *Natural history of Ireland* . . . . London, Collins.

Rand, A. L. 1936. The distribution and habits of Madagascar birds. *Bull. American Mus. Nat. Hist.,* **72,** 143–499.

———. 1955. The origin of the land birds of Tristan da Cunha. *Fieldiana* (Chicago Nat. Hist. Mus.), *Zool.,* **37,** 139–166.

Ripley, S. D. 1952. The thrushes. *Postilla* (Yale Peabody Mus. Nat. Hist.), No. 13.

Salomonsen, F. 1948. The distribution of birds and the recent climatic change in the North Atlantic area. *Dansk Ornithologisk Forenings Tidsskrift,* **42,** 85–99.

——— (plates by Gitz-Johansen). 1950–1951. *The birds of Greenland* [3 parts]. Copenhagen, Ejnar Munksgaard.

———. 1951. The immigration and breeding of the Fieldfare . . . in Greenland. *Proc. Tenth International Ornithological Congress,* 515–526.

Sayles, R. W. 1931. Bermuda during the Ice Age. *Proc. American Acad. Arts and Sci.,* **66,** 381–467.

Schuchert, C. 1935. *Historical geology of the Antillean-Caribbean region.* New York, John Wiley.

Simpson, G. G. 1945. The principles of classification and a classification of mammals. *Bull. American Mus. Nat. Hist.,* **85.**

———. 1956. Zoogeography of West Indian land mammals. *American Mus. Novitates,* No. 1759.

Smith, M. (A.). 1951. *The British amphibians and reptiles.* London, Collins.

Tate, G. H. H. 1946. Geographical distribution of the bats in the Australasian Archipelago. *American Mus. Novitates,* No. 1323.

Taylor, E. H. 1934. *Philippine land mammals.* Philippine Bureau Sci. Monograph 30.

Timmermann, G. 1938–1949. Die Vögel Islands [Birds of Iceland]. *Vísindafélag Islendinga* (Soc. Sci. Islandica), **21, 24,** and **28** (total of 524 pages numbered consecutively).

Umbgrove, J. H. F. 1949. *Structural history of the East Indies.* Cambridge, England, Cambridge U. Press.

Vesey-FitzGerald, D. 1936. Trinidad mammals. *Tropical Agriculture,* **13,** 161–165.

Vesey-FitzGerald, D., and F. N. Betts. 1940. The birds of the Seychelles . . . . *Ibis* (14), 4, 480–504.

Wallace, A. R. 1880. *Island life.* London, Macmillan.

Whitley, G. P. 1943. The fishes of New Guinea. *Australian Mus. Mag.*, 8, 141–144.

Williams, E. 1956. *Pseudemys* . . . with a general survey of the *scripta* series. *Bull. Mus. Comparative Zool.*, 115, 145–160.

Zimmerman, E. C. 1948. *Insects of Hawaii.* Vol. 1. Introduction. Honolulu, U. of Hawaii Press.

chapter **9**

Evolution of the geographical patterns; area, climate, and evolution

Preceding chapters have dealt primarily with existing patterns of animal distribution, although the movements that have produced the patterns have been suggested. Now attention will be focused on the movements. This requires a change of method. Heretofore the method has been to present facts and draw conclusions which, though sometimes partly hypothetical, follow more or less directly from the facts. Now the conclusions of the preceding chapters will be brought together and extended into general hypotheses about the nature and pattern of animal dispersal and the factors that control it.

The main pattern of dispersal (tentative)

A main pattern of dispersal is suggested by analysis and comparison of the distributions and apparent histories of the five main classes of vertebrates (Chapters 2 to 6) and by the combined pattern of distribution of all vertebrates on continents and islands (Chapters 7 and 8).

Among existing, dominant fresh-water fishes, the most important events of dispersal seem to have been the rise of Ostariophysi in fresh water, probably in the main part of the Old World tropics, and their radiation from there, and later the rise of cypriniforms in southeastern Asia and their spread from there, accompanied by recession of some other fishes.

Among existing, dominant amphibians, the most important events seem to have been radiations of successive, dominant groups (of frogs) from the main Old World tropics, accompanied by recession of older groups.

Among existing reptiles, dispersal has been complex and is difficult to trace. However, reptiles are primarily tropical animals. They have invaded the north-temperate zone but have not often persisted there long or evolved there much. Moreover, they seem to have moved from the Old World to America and from Asia to Australia more often than the reverse. This suggests that the main center of dispersal of reptiles has been the main part of the Old World tropics, and that successive dominant groups have tended to spread from there, replacing older groups in some cases.

Among birds, dispersal has evidently been still more complex. In each main ecological group of birds successive dominant stocks have probably risen and spread over the world while successive older stocks have retreated or become extinct, but the details are mostly lost. However, what has been said of the reptiles is true also of the birds: they are most diverse in the tropics and not much differentiated in the north-temperate zone, and they seem to have moved from the Old World to North and South America and from Asia to Australia more than the reverse, so that the main part of the Old World tropics is the apparent main center of their dispersal.

Among mammals, too, dispersal has been complex, but the details of it are better known. There has been a continual, complex exchange of mammals between the Old World and North America since early in the Tertiary, but the sum of movement has been from the Old World to North America. There has been a complex exchange also between North and South America since the late Pliocene, but the sum of movement has been from North to South America. Some mammals have moved from Asia to Australia, but few have moved far in the other direction. Moreover mammals, too, are most numerous and diverse in the tropics and not much differentiated in the north-temperate zone. The main center of dispersal of mammals seems therefore to have been the main part of the Old World and especially the tropical part of it. The fossil record of mammals is full of examples of succession, of replacement of old groups by new, dominant, spreading ones.

Land and fresh-water vertebrate faunas are largest and most diverse in the great, favorable areas of the continental tropics and are reduced and limited northward and southward and outward from the large land masses. The two main regional faunas of the Old World tropics (the Ethiopian and Oriental ones) are least limited and most central, and the other continental faunas form an unsymmetrical, radial pattern of increased limitation around them (Chapter 7). Moreover, there seems to have been a flow—a broad, directional movement—of vertebrates outward from the continents onto the islands of the world (Chapter 8).

All this leads to one general conclusion. Vertebrates have moved back and forth over the world in very complex ways. But the sum of the movements seems to have been directional, from the largest and most favorable areas, especially from the main part of the Old World tropics, into smaller and less favorable areas. This is the apparent main pattern of dispersal.

Ranges, populations, and zoogeographic movements

The ranges of plants and animals are the areas occupied by populations. Changes in distributions involve changes in populations. Biologists still argue about the behavior of complex populations, but the following generalizations (inevitably too simple) seem self-evident.

Populations tend to fluctuate. The fluctuations are the result of changes in the ratio of reproduction to death. When more individuals are reproduced than die, a population increases, and when more die than are reproduced, it decreases. And when populations increase, they tend to spread, and when they decrease, they tend to lose ground.

Zoogeographers sometimes think of ranges as if they were bounded on all sides by physical barriers which prevent spreading, but this is true only in special cases: *e.g.*, aquatic animals in landlocked lakes and land animals on islands. More often, the areas occupied by populations are limited by such things as rainfall and temperature, which vary gradually from place to place and fluctuate from year to year. In such cases the areas occupied are determined not by boundaries which individuals cannot pass, and not even by boundaries beyond which conditions are unfavorable, but by equilibriums involving whole populations.

The area occupied by a population is usually not uniform but is more favorable in some parts than in others, and the ratio of reproduction to death must often vary accordingly. Conditions are likely to be most favorable near the center of the area, and there is likely to be an unfavorable marginal zone where death exceeds reproduction but where the population is maintained by excess of reproduction near the center of the area and gradual shifting of individuals toward the margins. The actual limits of range will then be determined not by the limits of favorable ground but by a constantly fluctuating equilibrium between tendency to spread at the center of the range and tendency to recede at the margins. This makes a very delicate mechanism for the spreading and receding of populations. Any change, in either the population itself or the environment, which increases the ratio of reproduction to death will cause spreading. Any change, in either the population or the environment, which decreases the reproduction/death ratio will cause receding—dying back from marginal areas. Populations should therefore fluctuate not only in numbers of individuals but in area occupied, spreading in favorable and receding in unfavorable times, and spreading when improvements occur in the populations themselves and receding when competitors make improvements. Such expansion and retraction of populations at the margins of their ranges have been observed in many animals.

The distributions of all groups of plants and animals, whether species, genera, higher groups, or whole floras and faunas, are the sums of areas occupied by populations, and the movements of the simplest and most complex groups are basically the same. All move primarily by spreading and receding of populations in something like the way described.

The movements of individual animals within the populations are essentially random, as random as the movements of molecules in a gas. The movements may be given direction in various ways, for example by winds or currents, or by compulsion to swim up a stream or to fly toward the setting sun (the movements of molecules may be directed too), but these effects are probably not very important. Populations probably usually move by random scattering of individuals, which survive in new favorable places but die in places which are, or become, unfavorable.

Dispersal cycles

The simplest pattern of dispersal would be one in which groups of plants and animals just spread steadily in all possible directions from their places of origin. The age of each group would then be proportional to area occupied, and the place of origin would be the center of the area. That this is actually the case is the "Age and area" hypothesis. It was proposed by a botanist (Willis), and its severest critic has been a botanist (Fernald 1926, with references to Willis' earlier work). Willis (1949) apparently still maintains this hypothesis in a modified form. That plants and animals tend to spread with time is (as Fernald says) a truism which is not new. But that spreading equally in all directions is the main process that has made existing distributions cannot be true. Receding has been important, too. The vertebrates that have left the best fossil records have passed through cycles of spreading and receding, and such cycles have presumably been characteristic of animals and plants in general.

Groups of animals often do pass through cycles. They originate, rise to dominance, diversify and spread, decline and retreat, and become extinct. These cycles of evolution and movement have been compared to the life cycles of individual animals, as if species and higher groups passed through periods of youth, maturity, and senescence, but, as Simpson (1949, p. 187) points out, this is a dubious analogy. Simpson (1940) has described some of the expected and observed cycles that change animal distributions—they are observed, of course, principally through the fossil record. The following account of them is partly original and partly derived from Simpson.

The simplest cycle, and the one from which more complex ones are compounded, consists of first the expansion and then the contraction of one group of animals (Fig. 65). The group may be one population or any higher group considered as a unit. If the area occupied after contraction is different from the original area, movement has occurred.

A succession or continuation of simple cycles progressing in one direction (Fig. 66) is sometimes called migration, although it has nothing to do with seasonal migration, of birds etc. Migration in the present sense may sometimes occur by concerted movement of individuals (*e.g.*, sometimes of men and lemmings), but usually it is more random. Individuals tend to disperse at random (the randomness being modified by barriers and directional influences), but only those individuals that chance to go in favorable directions gain ground; the others are lost. And ground once occupied is usually lost not by deliberate abandonment but by extinction.

The cycle of expansion and contraction of a group may be complicated in various ways. As a group contracts, it may become divided, so that its range is discontinuous (Fig. 67). Most of the patterns of discontinuity described on pages 416–419 were evidently formed in this way. If an expanding group comes against a barrier which is crossed with difficulty, a new, independent cycle may begin across the barrier. This may happen when, more or less by accident, a member of an expanding group gets across a water gap to a new land (Fig.

68). Or it may happen when a rising group, beginning in the tropics, expands far enough northward so that some member of it, because of

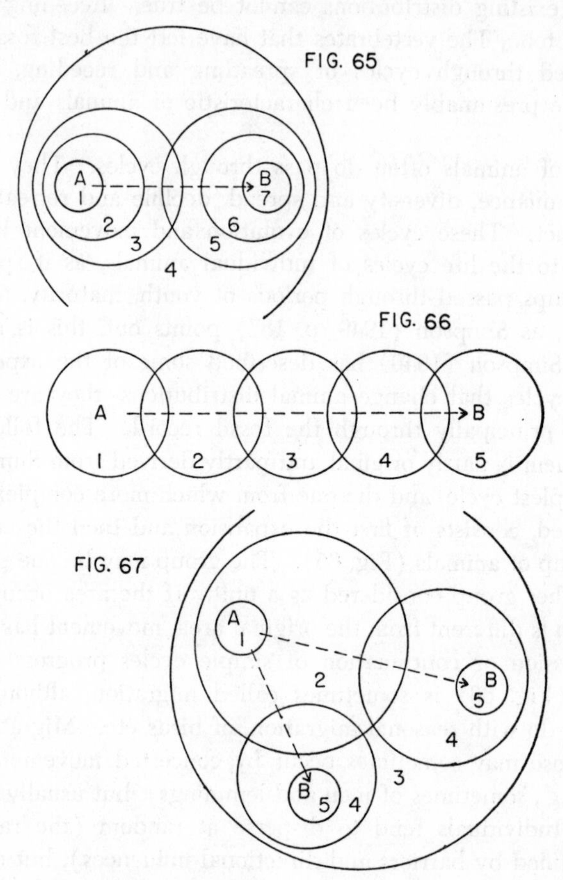

Fig. 65. Diagram of simple cycle of expansion and contraction of range, with movement. Successive stages in the cycle are numbered consecutively.

Fig. 66. Diagram of succession of simple cycles, with movement in one direction. Each circle represents a cycle like that of the preceding diagram.

Fig. 67. Diagram of expansion and contraction, ending in discontinuity.

its cold tolerance and dominance rather than by accident, crosses a northern land bridge between the Old and New Worlds (Fig. 69).

Cycles of movement are to be expected in all groups of animals and plants, and the fossil record suggests them in many groups, but only a few groups have good enough records to show the cycles in detail.

One such group is the mastodonts. Mastodonts apparently originated in Africa in the Oligocene; expanded in the main part of the Old World until, late in the Miocene, they extended northward far enough to cross into America; and then contracted, disappearing first in the

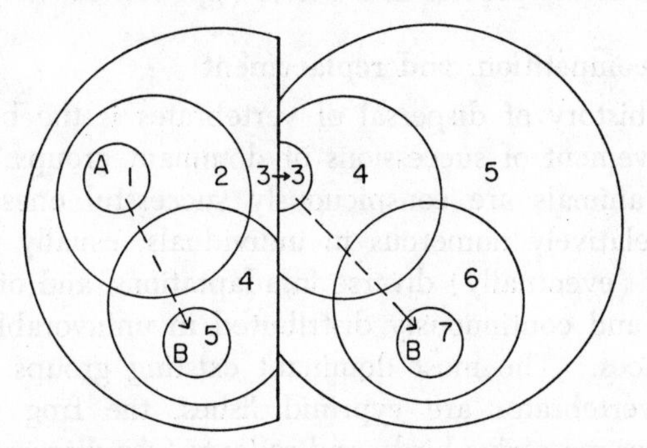

Fig. 68. Diagram of expansion and contraction, with accidental crossing of a barrier.

Old World, then in America. This cycle is diagrammed by Simpson (1940, p. 143).

The cycles of movement of single groups of animals are probably often very complex, composed of multitudes of minor cycles, with complex series of advances, retreats, and readvances. And the com-

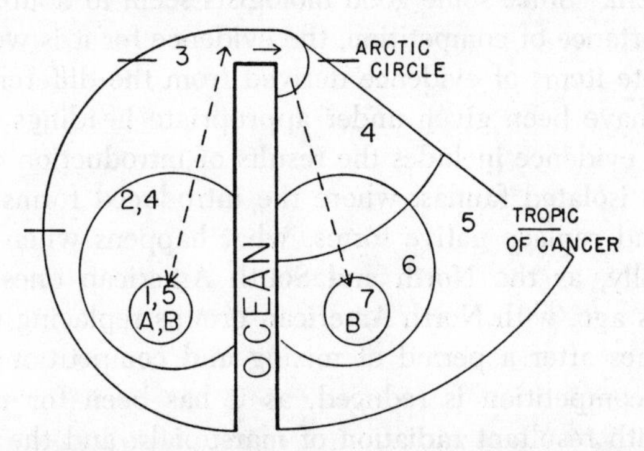

Fig. 69. Diagram of expansion and contraction, with crossing through the north followed by withdrawal southward.

plexity is multiplied when successive, dominant, competing groups of animals pass through successive cycles, each group tending to replace the ones before it. Movements of whole faunas are still more complex. When, for example, a large part of the mammal fauna of

North America moved (extended) into South America in the late Pliocene and Pleistocene, the movement began as a very complex exchange, with many movements in both directions, and it was only in the course of time that successive exchanges and selective extinctions gave direction to the process as a whole (pp. 367–368).

Dominance, competition, and replacement

The main history of dispersal of vertebrates is the history of the cycles of movement of successions of dominant groups.

Dominant animals are conspicuously successful ones. Dominant groups are relatively numerous in individuals, usually numerous in species, often (eventually) diverse in adaptations, and often (eventually) widely and continuously distributed in unfavorable as well as favorable places. The most dominant existing groups of land and fresh-water vertebrates are cyprinid fishes, the frog genus *Rana*, colubrid snakes, passerine birds, and rodents (see discussions of dominant groups in Chapters 2 to 6), and of course man. Dominance presumably reflects underlying qualities or characteristics that make for initial success and lead to the evolution of numerous, varied, successful types. The relation of dominance to evolution is discussed later in this chapter.

Dominance implies success in competition, competition being defined here as any interaction among organisms that is disadvantageous to any of them. Since some good biologists seem to doubt the fundamental importance of competition, the evidence for it is worth reviewing. Separate items of evidence derived from the different classes of vertebrates have been given under appropriate headings in Chapters 2 to 6. The evidence includes the results of introduction of dominant animals into isolated faunas, where the introduced forms often compete with and replace native forms; what happens when two faunas meet naturally, as the North and South American ones did a few million years ago, with North American groups replacing many South American ones after a period of mixing and competition; what happens when competition is reduced, as it has been for mammals in Australia, with resultant radiation of marsupials; and the many cases of replacement of one group by an apparently competing one shown by the fossil record, for example the partial replacement of some older fresh-water fishes by cyprinids and of side-necked turtles by emydines, and the many replacements in the record of mammals.

These special cases are significant, but the most significant thing of all is the general level and balance of past and present faunas.

Throughout the recorded history of vertebrates, whenever the record is good enough, the world as a whole and each main part of it has been inhabited by a vertebrate fauna which has been reasonably constant in size and adaptive structure. Neither the world nor any main part of it has been overfull of animals in one epoch and empty in the next, and no great ecological roles have been long unfilled. There have always been (except perhaps for very short periods of time) herbivores and carnivores, large and small forms, and a variety of different minor adaptations, all in reasonable proportion to each other. Existing faunas show the same balance. Every continent has a fauna reasonably proportionate to its area and climate, and each main fauna has a reasonable proportion of herbivores, carnivores, etc. This cannot be due to chance. Something holds the size and composition of vertebrate faunas within limits in spite of continual changes and successions in separate phylogenetic groups. Only competition can do this, and to do it competition must be a fundamental, ever-present force.

The spreading of dominant groups of animals and the recession of less dominant competing ones are interrelated in complex ways.

It is the dominant groups which spread. Ability to spread is one of the attributes and evidences of dominance. Dominant animals spread because the qualities that make them dominant enable them to enter new areas and new habitats, become adapted to them, and replace other animals there.

Generally speaking, animals spread to obtain advantages, not to escape disadvantages. The idea is sometimes encountered that dominant, spreading groups of animals force less dominant ones to move ahead of them, but this cannot usually be the case. Any animal normally produces surplus individuals which, constantly or occasionally, cause strong spreading pressure. The pressure may be thought of as roughly proportional to size of the surplus, or to the proportion of individuals that die because they cannot find room to live. (I know that this is an oversimplification, but it will have to do.) The arrival of a dominant competitor, causing reduction and extinction of the original species, increases the latter's spreading pressure (if it does so at all) only a little and only for a few generations. If the first species has not been able to spread before the arrival of the second, it is not likely to do so afterward. Therefore, dominant, spreading animals are not likely to push less dominant ones ahead of them but will probably overrun them and destroy them.

The spread of dominant animals and the recession of less dominant competing ones need not always be in the same direction. If a dominant group spreads across a uniformly favorable area, a less dominant competing group should recede in the same direction, as it is replaced by the invader. If, however, the dominant group spreads from a favorable into a less favorable area, for example from the tropics (where there is room for many kinds of animals) into the arctic (where there is room for few), the less dominant group may recede in the opposite direction, being replaced in the areas where there is little room and surviving in the areas where there is more room.

About the place of primitive forms in cycles of movement, there is disagreement. Some persons think that primitive forms stay at the original centers of dispersal of expanding groups of animals; others, that they are pushed to or survive at the peripheries; but both ideas probably oversimplify the matter. There are two kinds of peripheral areas: those that are peripheral because they are less favorable than the dispersal centers, and those that are peripheral because they are distant from the dispersal centers. For example for most animals the arctic is an unfavorable-peripheral area, and Australia, a distant-peripheral one. And there are two stages of spreading: the initial spread of groups of animals, and later waves of spreading of successive dominant elements within the groups. It is always the dominant elements which spread, but the patterns produced in different cases should be different. During an initial spread, the dominant elements, numerous at the dispersal center, should be the *only* representatives of the group in both kinds of peripheral areas, and non-dominant elements (if any) should occur only at the dispersal center. During the spread of later dominant elements of the same group and replacement of earlier elements, the domniant elements should again be numerous at the center; they may be relatively even more numerous in accessible unfavorable-peripheral areas, where they may overwhelm almost all other competing forms; but they will be absent in distant-peripheral areas that they have not had time to reach. In this case the older, now non-dominant elements should be numerous in distant-peripheral areas, perhaps still present in reduced numbers at the dispersal center, and least numerous or absent in unfavorable-peripheral areas. This is said of *non-dominant* elements. It would be true of primitive elements only if they were the non-dominant ones. Probably they often are, but they need not always be. To trace the geographic histories of groups of animals by picking out the primitive forms and interpreting their distributions may sometimes be possible, but it is a complicated matter. (However, it is probably a practical rule of thumb to expect most non-dominant or primitive forms to be in distant-peripheral areas.)

The place of archaic relicts (survivors of very primitive, ancient groups) in the dispersal pattern of vertebrates is worth summarizing (see under "Summary: the pattern of distribution of . . ." in Chapters 2 to 6). Of strictly fresh-water fishes, the archaic relicts are non-teleost bony fishes: bichirs in Africa, paddlefishes in China and east-central North America, the Bowfin in eastern North America, a ceratodontid lungfish isolated in Australia, and lepidosirenid lungfishes in Africa and South America. Of amphibians, caecilians and salamanders are relicts of sorts and occur principally in the tropics and north-temperate zone respectively; the three most primitive suborders of frogs occur principally in the tropics (not including Australia) and the north-temperate zone; only *Leiopelma,* on New Zealand, is strikingly isolated geographically, and it is balanced by an equally primitive frog, *Ascaphus,* in North America. Of reptiles, turtles and crocodilians are relicts of sorts, and they occur principally in the tropics, extending (turtles especially) into some temperate areas; *Sphenodon* is isolated on New Zealand, but no other archaic reptile is so isolated. Of birds, no truly archaic relicts exist; if any one part of the world has more than its share of primitive birds, it is probably South America. Of mammals, the most archaic relicts are probably the Australian monotremes; existing marsupials are relicts of a sort (but not in quite the sense that was once supposed—see p. 322), and they occur principally in the Australian Region and South America, with one species extending into cool-temperate North America; however, some placentals in the main part of the world represent stocks that may be as old as Australian and South American marsupials.

In general, archaic relicts tend to occur principally in two places in the vertebrate dispersal pattern. A few (fewer than is usually realized) occur in almost complete geographical isolation, beyond the limits of dispersal of most later competitors. But more occur in great, favorable, more or less stable areas which have been centers of evolution and dispersal: the fresh-water fishes and amphibians in the tropics and the north-temperate zone; reptiles in the tropics; and birds and mammals more in the tropics than elsewhere. The one place where archaic, relict vertebrates rarely or never occur is in unfavorable or unstable marginal areas that are not completely isolated. For example, no specially archaic reptiles are peculiar to the north-temperate zone, which is marginal for reptiles, and no specially archaic amphibians or reptiles are peculiar to Australia, which is probably marginal for them, less isolated than it is for fresh-water fishes and mammals.

The most striking relicts, the last remnants of old families and orders, survive in different ways. A few do it by means of extreme geographic isolation. Others seem to do it by means of adaptations which partly isolate them from competition: the African and South American lungfishes can breathe air and can live in seasonal or oxygen-deficient waters; and *Ascaphus* lives in cold, swift brooks out of reach of most other North American frogs. Some archaic relicts, however, compete openly and successfully with modern faunas: the African bichirs, although they can breathe air to some extent, apparently live in open competition with other African fresh-water fishes; the Bowfin, the only relict of an order dominant in the Jurassic, competes successfully with other large, predaceous fishes in eastern North America; and the North American opossum (a *primitive* marsupial, whether or not it is a relict) competes successfully with placentals and is gaining rather than losing ground. These representatives of archaic or isolated groups have retained or regained a moderate dominance.

Dispersal and climate

Much of what follows is paraphrased from my *Quarterly Review* article (1948, especially pp. 105–107), with material on birds and mammals added to that on cold-blooded vertebrates. The subject to be considered is the relation of dispersal to climate and especially to the main (tropical and north-temperate) climatic zones.

Matthew, in *Climate and Evolution* (1915, pp. 172–173; 1939, p. 3) states a thesis which can be reduced to two main propositions: (1) that the north-temperate zone, because of its variable climate, has been the principal center of evolution and dispersal of land vertebrates, and (2) that no great changes in world geography and no extraordinary land bridges are necessary to account for vertebrate distribution. Matthew's paper was an important one which helped to counteract (but did not stop!) a trend toward fancifulness of some zoogeographers and which helped to make historical zoogeography into something like a science. Nevertheless, I think that the first of his two main propositions was wrong.

Fresh-water fishes, amphibians, reptiles, birds, and mammals all seem to have dispersed from the tropics into the north-temperate zone more than the reverse. Fossils show that, in each class, some groups that have been in the north have withdrawn from there, but this does not mean that they originated there. To mistake direction of withdrawal for direction of origin and spreading is a common error. Particular details of distribution suggest northward spread of particular groups, especially of cold-blooded vertebrates (Chapters 2 to 4), but the general pattern of distribution is more significant among the higher vertebrates. Cold-blooded reptiles and warm-blooded birds and mammals are all most numerous and diverse in the tropics; various groups extend for various distances into the north-temperate zone;

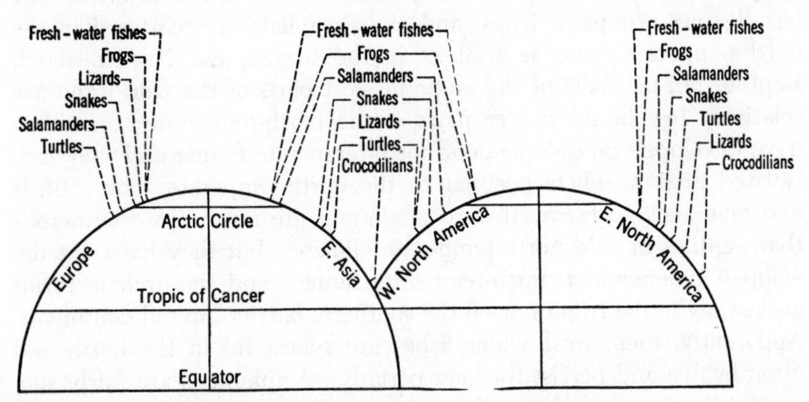

Fig. 70. Diagram of northern limits of groups of cold-blooded vertebrates. From *Quarterly Review of Biology,* 23, 1948, p. 106, Figure 3.

but very few dominant or well-defined groups of them are confined to the north-temperate zone. This is hardly consistent with the northern part of the world being the principal theater of evolution of vertebrates or the principal center of their dispersal.

North-temperate climate is characterized by two things: lower average temperature than that of the tropics and alternation of warm and cold seasons. The limiting effect of these factors on vertebrates is probably complex: the northern limits of different animals may be determined by mean temperature, or summer temperature, or winter temperature; and temperature may affect reproduction, or young, or adults, or the animals' food, or competitors. Different vertebrates do in fact reach complex northern limits, not at definite isotherms and in no fixed order (Fig. 70). Obviously the limiting effect of climate will have to be looked for, not in the simple effect of single factors on single species but in general correlations.

There is a significant correlation between northern limits, the development of independent north-temperate faunas, and the distribution of phylogenetic relicts in the different classes of vertebrates. Fresh-water fishes go far north and are numerous in cold north-temperate climates; they have developed an independent north-temperate fauna characterized by special families and even orders; and relicts of ancient groups occur in the north-temperate zone as well as in the tropics and Australia. Amphibians reach the arctic more or less around the world and are fairly numerous in the colder parts of the north-temperate zone; they have developed an independent north-temperate fauna characterized by salamanders and a few more or less distinct groups of frogs; and archaic relicts are scattered in the north-temperate zone as well as in the tropics and New Zealand. Reptiles fall far short of the arctic in most parts of the world and are relatively few in the colder parts of the north-temperate zone; they have developed no independent north-temperate fauna; and they lack isolated archaic relicts peculiar to the north-temperate zone. Birds and mammals go far northward again and are much more numerous than reptiles in cold north-temperate climates; but they have not developed independent north-temperate faunas; and the archaic forms are mainly in the tropics or on the southern, barrier-limited continents. Apparently, then, fresh-water fishes are successful in the north and often evolve and persist for long periods. Amphibians are fairly successful there and evolve and persist in some cases. Reptiles are not successful there; they invade cold northern regions but do not evolve there much or persist there long. Birds and mammals are again successful in the north but apparently do not often go through long periods of separate evolution there or persist there as relicts.

The success of cold-blooded vertebrates in the north varies with their place on the evolutionary scale. Fishes, the lowest forms, are most successful in cold places, amphibians somewhat less so, and reptiles least of all. This has something to do with reproduction and something to do with habitat. Fishes have the most primitive mode of reproduction and the one apparently least affected by cold. Amphibians have a more complicated reproduction and development, but some of them can shorten the larval period enough to complete it during the short arctic summer; their heat requirements are apparently not great. Reptiles have a still more highly organized mode of reproduction, more affected by cold. Many northern lizards and snakes, including the northernmost of all, manage to reproduce by becoming ovoviviparous: the female retains the eggs and "follows the sun" until they hatch. This is an adaptation for obtaining heat rather than for withstanding cold, and it emphasizes the fact that reptiles cannot reproduce in such cold places as amphibians. As to habitat, water is a buffer against cold; fully aquatic animals such as fishes have an advantage in cold climates which is only partly shared by amphibians and not shared by most reptiles.

Birds and mammals have reversed this trend, of retreat from cold places with increasing heat requirements and abandonment of aquatic habits. Their warm-bloodedness has enabled them to return to the north. But, probably because their warm-bloodedness makes them relatively independent of external temperature, they have not evolved major groups specially adapted and confined to the north-temperate zone as fishes and amphibians have done.

It seems (to summarize it oversimply) that many fishes and some amphibians both tolerate and become profoundly adapted to cold or seasonal climates; reptiles do neither very well; and birds and mammals tolerate cold and alternation of seasons but do not become profoundly adapted to them.

Northernmost terrestrial vertebrates belong to primarily dominant groups rather than primarily cold-adapted groups. This is a very important fact which has not been sufficiently recognized. The northernmost amphibian is a frog of the dominant, tropical and temperate genus *Rana*. It is not much different from its tropical relatives, but it extends farther north than any cold-adapted salamanders. The northernmost lizard and snake belong to genera (*Lacerta* and *Vipera*) which are widely distributed in the Old World tropics. Most northern land birds belong to widely distributed, dominant families: hawks, owls, finches, and (less dominant) phasianids. Northernmost land mammals are slightly modified hares, rodents, wolves and foxes, bears, weasels, deer, and bovids. Of course these animals are adapted to the arctic. The northernmost *Rana*, for example, has accelerated its development, and the northernmost lizard and snake are ovoviviparous and can develop rapidly by utilizing the heat of the sun. These are adaptations permitting reproduction during the short arctic summer. But they seem to be rather minor adaptations of basically dominant animals. Other minor adaptations to existence in cold places are shortening of appendages, which reduces heat loss and lessens the chance of freezing (Allen's Rule), and, among warm-blooded forms, increase of body size, which reduces heat loss (Bergmann's Rule).

The last few paragraphs suggest that terrestrial vertebrates extend northward in inverse proportion to their inherent heat requirements (especially the heat required for reproduction, which varies with place on the evolutionary scale) and in direct proportion to their inherent dominance and adaptability as well as to their minor adaptations to cold and seasonal climates.

The way in which dominant groups of vertebrates pass through the north during their dispersal cycles is exemplified by *Rana*. Different North American species of *Rana* have different development rates, which limit them northward (Moore). Only one of our *Rana, sylvatica* (or a form of it), can develop (complete its tadpole stage) during the short arctic summer, and it is the only American species which enters the arctic. A related species, *R. temporaria,* enters the arctic in Eurasia. A *Rana* of this group evidently crossed a northern land bridge between Eurasia and North America rather recently. It was not the first *Rana* to reach America, but it shows how earlier ones probably did it. *Rana sylvatica* and *temporaria* are the only species of the genus now adapted to cross a Bering bridge, if a bridge existed. If they became extinct, no existing *Rana* would have the adaptations necessary to cross an arctic bridge. But the genus has had and would still have the adaptability to do it. This is a good example of the difference between adaptations and adaptability. Ecologists can measure adaptations but they cannot measure adaptability, and that is one reason why ecology cannot give the whole answer to zoogeographic problems. *Rana* (see p. 137) has evidently originated and diversified in the Old World tropics and spread northward until, at the height of its cycle, an occasional dominant species has made the necessary minor adaptations (including shortening of larval period) to reach the extreme north and cross into North America. Many other groups of vertebrates have probably passed through the same cycle: of rise in the tropics and spread northward until an occasional, dominant, adaptable species has reached the extreme north and crossed from one half of the world to the other. Exchange of animals between the Old and New Worlds in this way is limited by the small number of species in northern faunas but facilitated by the fact that the northernmost species usually belong to dominant, adaptable groups.

Some representatives of all classes of vertebrates, at the height of their cycles of dominance and dispersal, seem to be able to pass through both warm and cold climates without profound adaptations to either, but afterward different classes behave differently. Many fresh-water fishes and some amphibians, after spreading across the climatic zones, probably gradually change from a radial to a zonal pattern of distribution by adaptation of different stocks to the tropics and to the north-temperate zone respectively. But reptiles, birds, and mammals, after radiation, apparently usually fall back into the tropics without becoming zonally differentiated. Reptiles probably do this because they are not well fitted for long existence in cool places. Birds and mammals probably do it because, although they do well in cool places, they do not become profoundly adapted to them, so that the north-temperate forms are not basically differentiated and not protected by ecological barriers against new, competing groups spreading from the tropics.

Dispersal between Old and New Worlds

In each class of vertebrates more groups seem to have moved from the Old World to North and South America than the reverse. The limitation and unbalance of the South American fresh-water fish fauna suggest that it is derived from a few immigrants which moved to South America from the Old World tropics; and ameiurid catfishes, suckers, and cyprinids have apparently moved from the Old World to North America. *Bufo, Rana,* and brevicipitid frogs have apparently moved from the Old World to North and South America, and the distributions of some other frogs (leptodactylids and *Hyla*) are consistent with an early dispersal from the Old World tropics to North and South America (and separately to Australia) followed by extinction in all or much of the main part of the Old World. Emydine turtles, *Testudo, Crocodylus,* skinks, several genera of colubrid snakes, elapids, and pit vipers all seem to have moved from the Old World to North and South America. Among birds, more movements seem to have occurred from the Old World to North America than the reverse; and many of the Old World groups have extended to South America, while no (or very few) South American birds have extended to the Old World. And many different groups of mammals have originated in the main part of the Old World and moved to North and South America, while fewer North American mammals and probably no South American ones have reached the Old World (see Chapters 2 to 6 for further details).

The separate cases listed above are, as I have said (1948, p. 107), like straws moving mostly in one direction. They suggest that, under the complex exchanges that have occurred, there is a current in the dispersal of vertebrates setting from the Old World to North and South America. The current is revealed more clearly by analysis of whole faunas, especially that of South America. South America is the critical piece of land because it is the main tropical part of the New World, to be matched against the Old World tropics, the tropics being the great reservoir and apparent main dispersal center of vertebrates.

South America was an island during most of the Tertiary. Only at the beginning of the Tertiary (if then) and again near the end of that period was the continent connected with the rest of the world so as to allow much exchange of land animals. This is proved by the history of mammals and by other evidence.

South American fresh-water fishes, it will be remembered (Chapter 2, Table 2), include many endemic families, but the orders, and the main stocks within the orders, all occur in the Old World; and there are additional, significant stocks of fresh-water fishes in the Old World that are not in South America. This suggests that part of an Old World fresh-water fish fauna moved (extended) to South America long ago, probably (if the fishes came through fresh water) before the beginning of the Tertiary.

Existing families of amphibians and reptiles in South America fall into two fairly well-defined groups (Table 21; *cf.* Schmidt 1943, p. 252). One group, marked by much generic endemism, probably dates from the earlier connection of South America with the rest of the world; the other group, marked by little or no generic endemism, probably dates from the later connection. The early South American fauna of amphibians and reptiles includes nineteen families, *all* of which occur also in the Old World, except that the Teidae is represented there by the closely related Lacertidae; but there are additional families, old or diverse in the Old World, which are not represented in the early South American fauna: of frogs, the Ascaphidae, Discoglossidae, Pelobatidae (all old), and the Bufonidae, Ranidae, and Rhacophoridae (at least diverse); of turtles, the Trionychidae (old and diverse); and of lizards, the Agamidae, Chamaeleontidae, Scincidae, and Varanidae. Other smaller or more localized families peculiar to the Old World could be added to this list. The late South American fauna of amphibians and reptiles includes ten families not present in the early fauna. Of these, nine exist or are fossil in the Old World; but there are additional families existing in the Old World which are not represented in the late South American fauna. It will be seen that the amphibians and reptiles fall into the same pattern as the fishes, but at the level of families rather than of higher groups. No (or very few) families of amphibians or reptiles are peculiar to South America; most of the South American families are represented in the Old World; and there are additional, significant families in the Old World that are not in South America. This is true of both the early and late

TABLE 21. SOUTH AMERICAN FAMILIES OF AMPHIBIANS AND REPTILES

Key: x indicates presence; 0, absence; S, families that reach only the southern edge of temperate North America.

	Arrived			
	Early	Late	Present in Old World	Present in Temperate North America
Caecilians				
Caeciliidae	x		Africa, Orient	0
Salamanders				
Plethodontidae (*Oedipus* only)		x	Europe	x
Frogs				
Pipidae	x		Africa	0
Leptodactylidae	x		Australian Region, South Africa (fossil in India)	S
Bufonidae (*Bufo* only)		x	Wide	x
Atelopodidae	x		Orient?, Africa?	0
Hylidae	x		Temperate Eurasia, Australian Region	x
Ranidae (*Rana* only)		x	Wide	x
Brevicipitidae	x		Wide	x
Crocodilians				
Crocodilidae	x		Wide	x
Turtles				
Chelydrinae (*Chelydra* only)		x	(Fossil in Europe?)	x
Kinosterninae (*Kinosternon* only)		x	0	x
Emydinae (*Geoemyda* and *Pseudemys*)		x	Eurasia	x
Testudininae (*Testudo* only)		x[a]	Wide	x
Pelomedusidae	x		Africa (fossil elsewhere)	(Fossil)
Chelyidae	x		Australian Region	0
Lizards				
Iguanidae	x		Madagascar, Fiji	x
Gekkonidae	x		Wide	S
Teidae	x		(Lacertidae)	x
Scincidae (*Mabuya* only)		x	Wide	x
Anguidae (2 genera)		x	North temperate and Orient	x
Amphisbaenidae	x		Africa etc.	x
Snakes				
Typhlopidae	x		Wide	S
Leptotyphlopidae	x		Africa etc.	x
Aniliidae	x		Orient	0
Boidae	x		Wide	x
Colubridae (s. lat.)	x		Wide	x
Elapidae	x		Wide	x
Viperidae (3 genera)		x	Wide	x

[a] Arrived in Miocene.

South American amphibian and reptile faunas. It suggests two periods of movement, during each of which part of a diverse Old World fauna reached South America.

Of birds, many may have reached South America across the Tertiary water gaps. Probably partly for this reason, partly because bird faunas have changed rapidly in other parts of the world, and partly because the fossil record of birds is very poor, I cannot divide South American birds into old and new faunas in the same way as the other vertebrates. I can only repeat what I have said above and what Mayr (1946, p. 38) has stressed, that many Old World birds have reached South America and some have radiated there, while few or no South American birds have reached the Old World.

South American mammals, even more clearly than amphibians and reptiles, divide into old and new faunas (Table 7, p. 333). The old fauna is apparently derived from only a part of the mammal fauna which existed in the main part of the world at the beginning of the Tertiary (p. 363). The new fauna represents only a part of the existing mammal fauna of the main part of the world.

For the moment I wish to draw just one conclusion from all this. Before or at the beginning of the Tertiary, immigrants representing some, but not all, of the main stocks of then-existing vertebrates somehow reached South America. And at the end of the Tertiary, additional immigrants, representing some but not all now-existing stocks of vertebrates, again reached South America. In both cases the direction of dispersal of vertebrates within the tropics, or between the tropical parts of the Old and New Worlds, seems to have been mainly from the Old World to South America.

Area, climate, and evolution

The main pattern of dispersal, tentatively stated at the beginning of this chapter, has now been elaborated and tested by consideration of the nature of zoogeographic movements; of expected and observed dispersal cycles; of the role of dominance, competition, and replacement; of the relation of dispersal to the main climatic zones; and of the direction of dispersal between the Old World and South America. All this has involved some repetition of evidence, but nevertheless the tentative pattern has been reinforced and confirmed. Dominant land and fresh-water vertebrates seem mostly to have risen in the great, favorable area of the Old World tropics and moved (spread) to the less favorable area of the north-temperate zone and the smaller area of South America. And this is just the core of the pattern, which includes movement (spreading) into still less favorable places (*e.g.*, the arctic) and still smaller areas (islands). Evolutionary theory can suggest reasons why dispersal has occurred in these directions. I am indebted to Dr. W. L. Brown for reading and usefully criticizing what I say here about evolution.

Three kinds of evolution can be distinguished in theory, although they are probably mixed in fact and may be supplemented by additional minor processes. All three have the same principal mechanism: occurrence of mutations and survival and spread of some mutations through populations.

The first kind of evolution, *differentiation of species*, is a process in which survival and spread of mutations may be partly random, not dependent on selection. It may proceed most rapidly in relatively small, isolated populations. This kind of evolution may differentiate species on islands even though the island populations are losing rather than gaining dominance.

The second kind of evolution is *adaptation to special environments*. It is not random. Mutations that happen to be advantageous under special conditions are selected. Rate of adaptation must vary with force of selection. When selection is so strong that advantageous mutations usually survive and spread, rate of adaptation will vary with size of populations, for, other things being equal, size of populations determines the number of mutations that will probably occur. Of two otherwise identical populations, one with twice as many individuals as the other, the larger is twice as likely to originate any single mutation, and in evolutionary progressions that depend on sequences of mutations the statistical advantage of large populations is great. Size of populations depends partly on the area and continuity of the environments they inhabit. It follows that adaptation should be most rapid in environments that are extensive and more or less continuous.

The third kind of evolution is *general adaptation*. It includes all the improvements of structures and functions that allow some animals to live more efficiently than others in many environments, to react more rapidly or more intelligently, or to reproduce more efficiently. It is adaptation to the general environment of the world, and it should lead to general dominance, to success over great areas and in many special environments. Like special adaptation, it should be most rapid in the largest populations, which might be expected to exist where the general conditions of life are most favorable over the largest areas. For most vertebrates this is probably in the tropics of the Old World.

Both area and climate are involved in this conception. The large habitable area of the Old World tropics gives room for large populations. The favorable, stable climate of the tropics may sometimes favor large populations and probably allows a maximum number of generations at least of cold-blooded animals, which increases the total number of individuals available for evolutionary processes in a given time.

Of course the matter is not really so simple as this. Adaptation may be most rapid in populations that not only are large but also fluctuate violently or form many small, *partly* isolated subpopulations, which only occasionally interbreed. A more serious complication is that we do not know where the largest populations really are. It is often stated by naturalists that the old, stable tropics are inhabited by enormous numbers of species with small (sparse) populations. My own experience of about five years in the tropics suggests that this is true, but that it is not the whole truth. If many rare species exist in the tropics, so do some common ones. But large populations also occur outside the tropics, perhaps especially in new or marginal areas. Some species of vertebrates are very common in great areas of the north-temperate zone, for example. Possibly the most rapid adaptation occurs in short periods when great, new, *favorable* areas first become available to life, or when an animal first becomes able to spread into great areas that are new and favorable for it (*cf.* effect of warm-bloodedness, below).

However this may be, actual patterns of distribution show that the most dominant vertebrates (cyprinid fishes, *Rana*, passerine birds, murid rodents, etc.) rise in the tropics and extend into the north-temperate zone, not (as Matthew supposed) the reverse, and this is true of the warm-blooded as well as cold-blooded groups. This is, I think, a fact that zoogeographers can contribute to evolutionary theory: *dominant vertebrates do usually evolve in the tropics, not in the north-temperate zone.*

A further consideration is that, if general adaptation proceeds by successive radiations of selected superior species, as it probably does, it should be most rapid where species are numerous as well as where populations are large. In other words, the chance of superior species appearing in an area is probably proportional to number of species in the area, just as the chance of superior individuals appearing by mutation is proportional to number of individuals. The total number of individuals of all species in whole faunas probably tends to determine the number of evolutionary improvements that occur. Moreover the mere mass and diversity of life in large, favorable areas may increase selective pressures and accelerate adaptation. The idea of evolution of dominant animals in great, densely populated areas is not new. It goes back to Darwin. I have merely restated the idea in terms of modern genetics, or at least in simple terms consistent with genetic theory.

Special adaptation and general adaptation, leading to dominance, differ only in degree. The relation between them can be illustrated. Various snakes in many parts of the world are semi-aquatic, but three groups of them have become more highly adapted than others to life in water (Chapter 4, p. 204). These groups are the Acrochordinae and Homalopsinae, fresh- and salt-water snakes derived from the Colubridae, and the Hydrophiidae, sea snakes, derived from the Elapidae. The three have originated separately, but all are confined to or center in the same place, the tropical Orient and islands to the south and east. Cyprinid fishes and probably emydine turtles did not acquire their aquatic adaptations in the Orient, but they have diversified there and apparently evolved there a dominance which has enabled them to spread widely. These three groups of snakes are probably the only land reptiles that became fully aquatic during the Tertiary, and the cyprinids are the only fresh-water fishes and the emydines the only fresh-water turtles that became dominant and spread so extensively during the Tertiary. Perhaps marine catfishes should be added to the list of aquatic animals that have risen in the tropical Orient. One family of them, Plotosidae, is confined to the Indo-Pacific region, although the other, Ariidae, occurs in warm seas around the world. These catfishes are the only Ostariophysi and perhaps the only fishes that seem to have moved successfully from fresh water into the sea recently. The convergence of clues of all these groups of animals suggests that fresh-water and estuarine habitats were very extensive in the tropical Orient during the Tertiary and were inhabited by great populations of many animals, of which some (the snakes) became specially adapted to life in the water while others (the fishes and turtles), already aquatic or amphibious, attained a general dominance which enabled them to spread into other parts of the world or to enter the sea. There is geological confirmation of this idea. The shallow, changing Tethys Sea covered much of tropical Asia well into the Tertiary and presumably offered a very large area of aquatic habitats and a very broad transition between fresh and salt water.

A special case worth a moment's consideration is what would happen, theoretically, if a group of cold-blooded vertebrates became warm-blooded, as mammals and birds have done. They should become relatively independent of temperature, should cross climatic boundaries easily, and should be relatively successful in cold places. The area of evolution of dominant groups should be increased by extension northward and might come to include the tropics plus the accessible part of the north-temperate zone. Warm-bloodedness might permit evolution of special processes. For example, more heat is apparently required for reproduction by reptiles than by amphibians, and still more heat plus control of temperature may be necessary for mammalian placental reproduction. But the most important effect of warm-bloodedness may be that, by opening areas of cool climate and reducing the importance of local climatic barriers, it increases the area in which species and higher groups of animals can evolve, increases the effective size of populations, and facilitates general adaptation. If so, warm-bloodedness may have allowed mammals not only to evolve a superior sort of reproduction but rapidly to become in every way better animals.

Another, very different thing, which tends to minimize barriers, to increase the effective size of populations, and probably to facilitate general adaptation, is ability to fly. Flight itself has obvious, direct advantages, but its indirect effect in facilitating evolution of other superior qualities may also contribute to the dominance of insects and birds.

The relation of area and climate to size of faunas, to evolution, and to direction of dispersal can now be re-examined.

Limitation of area begins by limiting number of species and also, in small areas, size of populations. This has at first nothing to do with evolution. The size of any land mass, whether a continent or an island, limits the number of animals that occur on it. This is a fact, discussed in Chapter 8 (pp. 482ff.). How area limits numbers can be suggested. Some animals, for example large mammals, simply cannot find their requirements in small areas. Others probably cannot maintain sufficient populations there to survive fluctuations. Still others probably lack room to move about in. Animals sometimes do survive only by movement and countermovement in very large areas. For example, horses (equids) probably originated in the Old World and now exist (as wild animals) only there, but they would not exist if they had not been able to move into and out of the extra area of North America. Moreover different animals affect each other, through competition, and competition may be more simple and direct in small than in large areas. Limitation of area may thus go far toward reduction of faunas before evolution comes into play.

In the course of time, evolution does come into play. The best individuals and the best species tend to evolve in the largest areas, where both individuals and species are most numerous, while evolution (of dominant forms) is retarded in small areas. And eventually the dominant animals which evolve in the large areas spread into smaller areas and replace the less dominant animals there. It is this kind of movement, spreading of dominant groups from the largest areas, which gives direction to animal dispersal as a whole.

If evolution proceeded in equal steps, and if only single steps were involved, faunas would presumably make exchanges according to size: if one fauna were twice as large as another, twice as many stocks would spread from the large to the small fauna as the reverse. But evolution usually involves series of steps. It is like a climb up a ladder of successive favorable adaptations. Some members of a small fauna may happen to take the first step as soon as any members of a large fauna do. But at each step the odds are in favor of the larger fauna, which has more climbers, and in a long climb the members of the larger fauna should have an overwhelming advantage and should become overwhelmingly dominant. This apparently does happen. The animals of continents are apparently almost always dominant over those of islands, and dispersal is almost all from continents to islands. But even in this case, although countermovements are statistically improbable, they are theoretically possible and may sometimes occur.

Besides its relative effect, evolution in too small areas may cause outright deterioration. Small, isolated populations may lose genes and may be weakened until slight fluctuations lead to their almost automatic extinction.

Limitation imposed by climate probably has the same general effect as limitation of area. Climatically unfavorable places, such as the arctic and deserts, are inhabited by few species of animals, and the climate-limited faunas are probably at the same evolutionary disadvantage as area-limited ones. Moreover they are not geographically isolated and are probably continually overrun by invaders from more favorable climates. This certainly seems to have been the case in the arctic, where the limited terrestrial vertebrate fauna now consists almost entirely of representatives of widely distributed dominant families.

The essential nature of the directional movement of animals from large to small areas, and from more to less favorable ones, is diagrammed in Figure 71.

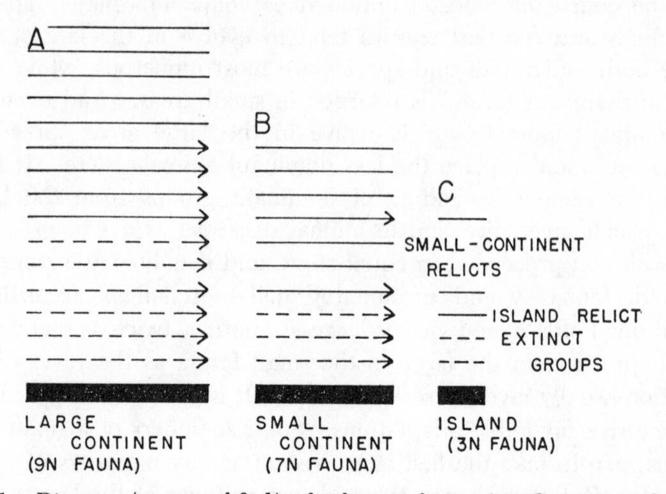

Fig. 71. Diagram (oversimplified) of relation of area (or climate) to direction of dispersal. A is a large continent (or one with a favorable climate), B a smaller continent (or one with a less favorable climate), and C an island (or very unfavorable area). Area or climate permits A to support a fauna of 9N units, B one of 7N units, and C one of 3N units. Successive dominant groups of animals evolve on A because it is the largest (or most favorable) area, spread to B, and in some cases finally reach C. When evolution of new dominant groups increases the fauna of A beyond the limit of 9N units, extinction of older groups begins; and when dispersal of the new groups to A and B increases the faunas there beyond 7N and 3N units, extinctions occur there too. There is a lag in this process. At any given time some rising, dominant groups will probably still be confined to A (top of column), and others will have reached B but not C; and some of the older groups that have become extinct on A will probably still survive as relicts on B and/or C. This main process is complicated by a proportion of back movements and by endless individual peculiarities in the evolution and dispersal of different groups.

The main pattern of dispersal

The main pattern of dispersal of vertebrates is apparently evolution of successive dominant groups in the great, favorable area of the main part of the Old World tropics and spread into smaller and/or less favorable areas, with successive replacements. Spreading can occur along three principal routes (Fig. 72).

A short route leads into temperate South Africa. The change along this route is from a larger to a smaller area and from a more to a less favorable climate. There are no great obstacles to dispersal in this direction except climate. Many tropical African vertebrates of all classes have extended into south-temperate Africa for varying distances and some have reached the Cape, although the fauna of South Africa is of much less than tropical richness.

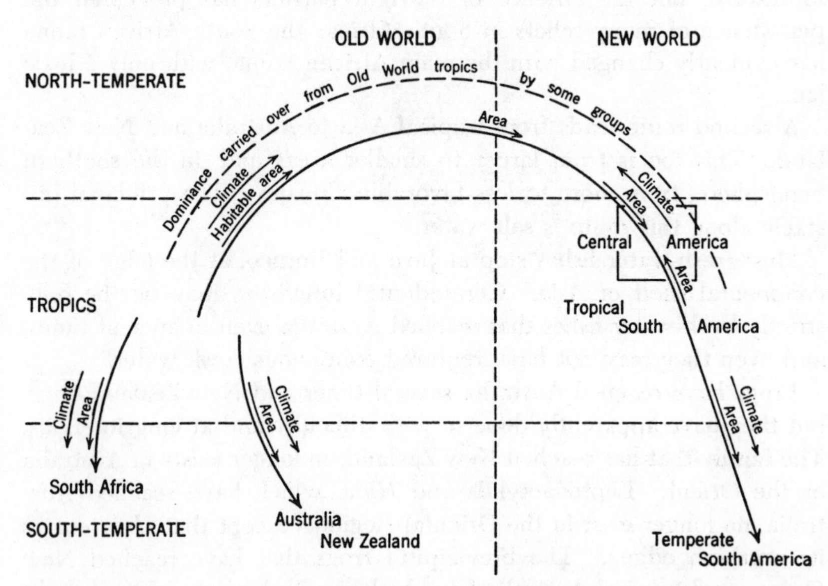

Fig. 72. Diagram of main dispersal pattern. From *Quarterly Review of Biology*, **23**, 1948, p. 112, Figure 4. Diagram is intended to suggest origin of dominant groups in main Old World tropics and dispersal along three main routes toward smaller areas and less favorable climates. Small arrows show the expected effects of area and climate on direction of dispersal at critical points.

Most South African vertebrates are not much differentiated. The most distinct of them may be an isolated genus of leptodactylid frogs. Frogs of the subfamily Brevicipitinae are confined to East and South Africa and may be retreating southward, for the East African forms have discontinuous, relict ranges. Testudinid tortoises and cordylid (zonurid) lizards are more diverse in south-temperate than in tropical Africa, but they may be adapted to steppes and deserts more than to the south-temperate climate. No families of birds are confined to South Africa; the most isolated endemic bird there may be the doubtful honey eater, *Promerops*. No families of mammals are quite confined to South Africa, but golden moles and two small families of rodents (Pedetidae and Petromyidae) occur chiefly there. But these are exceptions. Most South African vertebrates are still closely related to tropical African ones.

The South African fauna as a whole has evidently been formed by southward movement (extension) of parts of the tropical African fauna. Replacements must have occurred both in the tropics and southward, but the absence of outright barriers has prevented the persistence of many relicts in South Africa: the South African fauna has evidently changed with the main African fauna, with only a little lag.

A second route leads from tropical Asia to Australia and New Zealand. This too is from larger to smaller areas and, in the southern hemisphere, from more to less favorable climates. The principal obstacle along this route is salt water.

Most fresh-water fishes stop at Java and Borneo, at the edge of the continental shelf of Asia. Ceratodontid lungfishes may be the only strictly fresh-water fishes that reached Australia even in ancient times, and even they may not have required continuous fresh water.

Frogs have reached Australia several times and New Zealand once, but they have apparently done so with difficulty and at long intervals. The family that has reached New Zealand no longer exists in Australia or the Orient. Leptodactylids and *Hyla,* which have reached Australia, no longer exist in the Oriental Region (except that *Hyla* enters its northern edge). The brevicipitid frogs that have reached New Guinea and tropical Australia have had time to become endemic subfamilies. Only ranids are distributed as if they have reached the Australian Region recently.

Of reptiles, *Sphenodon* has reached New Zealand and survived there in isolation while its relatives have disappeared from the rest of the world. Non-marine turtles are not known to have reached New Zealand and seem to have had trouble reaching Australia. The two families, Carettochelyidae and Chelyidae, that are surely native in the Australian Region are not now in Asia; distant relatives of *Carettochelys* are fossil in Asia, but chelyids may never have been there. A trionychid may have reached New Guinea comparatively recently. Terrestrial *Testudo* has reached only to the Moluccas. Lizards seem to have reached Australia from the Orient easily and often, and gekkonid and scincid lizards have extended to New Zealand. Several terrestrial as well as aquatic snakes have reached Australia from the Orient more or less recently, but no snakes (except straggling sea snakes) have reached New Zealand.

Many birds have reached Australia, fewer New Zealand. They have apparently accumulated during a considerable period, for some are more differentiated (from Asiatic birds) than others.

Of mammals, perhaps one monotreme ancestor, at least one marsupial, perhaps five murid rodents, and many bats have reached Australia; but only two bats have reached New Zealand.

All the more recent of these vertebrates have evidently dispersed along the Asia-Australia-New Zealand arc. The earlier ones may have come the same way. There is no good evidence that any of them came from any other direction, unless the chelyid turtles (and extinct meiolaniids) came from South America through the sea. The amount of differentiation in different classes of vertebrates on Australia and New Zealand is roughly proportional to the animals' powers of crossing salt water. The poorer water crossers that have reached these isolated places at all have lagged far behind the changing Asiatic fauna or evolved divergently.

The third route of dispersal from the Old World tropics follows an arc through temperate Eurasia and North America to tropical Central and South America. From the Old World tropics into temperate Eurasia is from a more to a less favorable climate and, for many vertebrates, from a larger to a smaller available area. From temperate Eurasia to North America is from a larger to a smaller area, with no great change of climate. From North to Central America is again from a larger to a smaller area, but from a less to a more favorable climate. But from Central to South America is from a smaller to a larger area, with no great change of climate. It will be seen that theoretical forces determining direction of dispersal decrease along this arc and are reversed toward the end of it. But dominance acquired in the Old World tropics may be carried through the north-temperate zone and Central America to South America.

Different vertebrates might be expected to behave differently as they follow the arc from the Old World tropics through the north-temperate zone and into the American tropics. Strictly fresh-water fishes move slowly, from one drainage system to another, and are checked by relatively narrow barriers of salt water; they are tolerant of cold, occupy large areas in the north, and develop great, northern faunas which may block dispersal of all except the most dominant of later tropical groups. Amphibians can disperse more rapidly and get across narrow salt-water barriers somewhat more readily; they are fairly tolerant of cold and develop fairly distinct northern faunas. Reptiles can disperse still more rapidly and get across narrow ocean gaps comparatively easily; they are less tolerant of cold, occupy smaller areas in the north, and do not develop separate, northern faunas likely to block dispersal of later tropical groups. Birds disperse most rapidly of all, and (some of them) cross barriers easily; they are tolerant of cold; they do not develop important, separate north-temperate faunas, although migratory groups may tend to block dispersal of other groups through the north. Mammals disperse rapidly over land but do not cross salt-water barriers so easily; they too are tolerant of cold and do not develop important, separate north-temperate faunas. These differences should be reflected in the faunas that accumulate toward the end of the arc, in Central and South America.

The vertebrate fauna of Central America is described class by class under appropriate headings in Chapters 2 to 6. The transition of Nearctic and Neotropical (including Central American) faunas is reviewed in Chapter 7 (pp. 456ff.), but only present distributions, not movements, are considered. Now I want to review the Central American fauna and the movements that seem to have produced it.

The vertebrate fauna of Central America is significantly different from that of South America. Of true fresh-water fishes, South America has many, but very few of them have extended far into Central America, and still fewer North American ones have done so, and Central America has no important endemic groups of fishes except salt-tolerant ones. Of amphibians, Central America has fewer than South America, and the fauna is mainly transitional: a number of South American and fewer North American amphibians have extended into Central America, but there are few important endemic groups of them there. But of reptiles, Central America has 21 families plus 8 additional subfamilies against only 18 families plus 6 subfamilies in South America (Chapter 4, p. 196); some families, and many genera, are either confined to Central America or occur there but not in South America; and many of them seem to have reached Central America from the north. Of cold-blooded vertebrates, then, Central America has received few strictly fresh-water fishes but many salt-tolerant ones; a moderate number of amphibians, in reasonable proportion to the area and climate of Central America and its position between North and South America; but a disproportionately large number of reptiles; and, although a majority of the fishes and amphibians seem to have come from South America, relatively more of the reptiles seem to have come from the north. I shall try to account for this contrast among cold-blooded vertebrates before considering the warm-blooded ones.

Central America may have been an island or a series of islands, separated from North as well as South America, during part of the Tertiary. If so, this would account for the predominance of salt-tolerant fishes. It might also partly account for the diversity of reptiles, for many of the latter get across narrow salt-water barriers. But immigration across salt water leaves something unexplained. Of the few amphibians that have reached Central America recently from the north, all have entered South America too; of the many reptiles, few have entered South America. It looks as if amphibians, which disperse more slowly and may be delayed longer by barriers, and which have evolved northern, road-blocking faunas, have reached the American tropics only rarely, but always as dominant groups, while reptiles, which disperse more rapidly and are delayed less by barriers, at least of salt water, and which do not evolve road-blocking faunas in the north, have filtered through the north comparatively often, but with varying dominance. Many of the reptiles have entered and survived in the small area of Central America, but fewer of them have been able to push into the larger area of tropical South America.

Of birds and mammals, as of other vertebrates, Central America has fewer than South America. In terms of present distribution, the majority of them are South American in relationships, fewer North American; and there are few important endemic groups in Central America. In terms of past (but fairly recent) movement, however, relatively more of the birds seem to have moved from South America into Central and North America, while relatively more of the mammals have moved in the other direction from North into Central and South America. It looks as if South American birds but North American mammals are relatively dominant. This may be, as I have said (p. 461), because birds, which can fly, have been less isolated than other vertebrates in South America and have suffered less of the ill effect of limitation of area. Both birds and mammals may have dispersed so rapidly and reached Central America so often, much oftener than even reptiles have done, that recent, dominant groups have almost entirely replaced older Central American ones, while reptiles have accumulated at a rate which has added many immigrants to the old fauna but has not yet replaced it. In other words (to put it, as usual, too simply), while fishes and amphibians may have dispersed too slowly, and birds and mammals too fast, reptiles may have dispersed at just the rate necessary to cause a piling up of old and new forms in Central America without too much replacement. Whether or not this explanation is correct, a situation does exist in Central

America—a disproportionate diversity of reptiles, derived largely from North America—which must somehow be explained.

The findings of the last few paragraphs seem to agree well enough with expectation expressed in Figure 72.

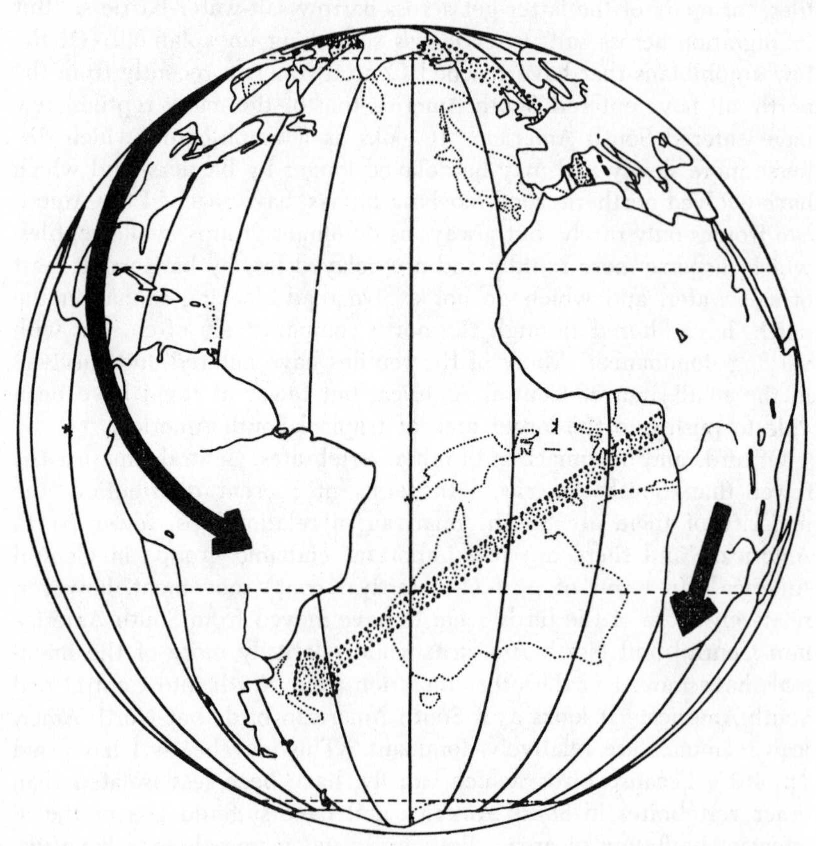

Fig. 73. The three main routes of dispersal on a double orthographic map of the world.

The nature of the directional influences diagrammed in Figures 71 and 72 is worth restating. Area and climate are not themselves directional forces. They act indirectly, through evolution. Dominant animals tend to evolve in the largest and most favorable areas and then to spread into smaller and less favorable ones. Where areas are not too unequal, there is usually not a concerted movement in one direction but an exchange, with much movement (spreading) in both directions, and it is only in the course of time that movements from larger to smaller areas predominate. As new dominant groups spread, they tend to overwhelm and replace earlier groups, and the replacements probably often (but not always) proceed from the dispersal centers outward, so that there is a continual outward shifting of distribution patterns. The dispersals of single groups may be very complex, and dispersal as a whole is much more so. Dispersal as a whole is the sum of the movements and countermovements, spreadings, competitions, extinctions, and replacements of a diversity of plants and animals over the whole of the diverse surface of the world. It is kaleidoscopic. But the kaleidoscope has main patterns which move in definite directions determined, through evolution, by area and climate.

The main pattern of dispersal described here is derived from consideration of many facts. Nevertheless I present it as an hypothesis. I think it is probably mainly true, but perhaps not wholly so. It is the best statement of the probabilities that I can make now, and it may serve as a basis for future consideration.

The three main routes of dispersal of animals over the world are shown on a double orthographic map in Figure 73.

REFERENCES

Darlington, P. J., Jr. 1948. The geographical distribution of cold-blooded vertebrates. *Quarterly Review Biol.*, **23**, 1–26, 105–123.

Fernald, M. L. 1926. The antiquity and dispersal of vascular plants. *Quarterly Review Biol.*, **1**, 212–245.

Matthew, W. D. 1915, 1939. Climate and evolution. *Ann. New York Acad. Sci.*, **24**, 171–318; reprinted (1939) as *Special Pub. New York Acad. Sci.*, **1**.

Mayr, E. 1946. History of the North American bird fauna. *Wilson Bull.*, **58**, 3–41.

Schmidt, K. P. 1943. Corollary and commentary for "Climate and Evolution." *American Midland Naturalist*, **30**, 241–253.

Simpson, G. G. 1940. Mammals and land bridges. *J. Washington Acad. Sci.*, **30**, 137–163.

———. 1949. *The meaning of evolution.* New Haven, Yale U. Press.

———. 1953. *Evolution and geography.* Eugene, Oregon, Oregon State System of Higher Education.

Willis, J. C. 1949. *The birth and spread of plants.* Geneva, Conservatoire et Jardin botaniques de la Ville.

chapter **10**

The past in the light of zoogeography

The nearly final manuscript of this chapter has been read by Professor Romer and Dr. Williams; I am indebted to both of them for many useful comments and criticisms, with the usual proviso that they are not responsible for errors that may still remain or for my conclusions.

Zoogeography, if it is to tell things about the past, should be consulted with forethought, common sense, an open mind, and a remembrance of human fallibility. Some zoogeographers urge these things on their opponents, but that is not how I mean them here. I mean simply that I shall try to practice them myself.

The history of the world's surface is known primarily from geology. Zoogeography is, in this connection, secondary. It can do no more than add details here and there to the main story that geology tells, and what it adds is sometimes not very important to geologists. The presence or absence of a land connection between two pieces of land, so important to zoogeographers, may be a matter of only a small difference of sea level in a small area for a short time, perhaps a very small detail to geologists. But zoogeography sometimes tells more general things about the nature and interrelationships of continents in the past, and then it becomes more important. Geologists have difficulty understanding zoogeography. It is better for zoogeographers to interpret their own subject, for it requires an experienced understanding of the significance of animal distributions, now and in the past, and especially an understanding of the relation between faunal differences and distance, climates, and barriers. The relation is complex, but some generalizations can be made about it.

Special cases should be treated with caution. Single, identical or closely related species or genera may occur almost without regard to distance, climates, or barriers. The Leopard and the Tolai Hare (*Lepus capensis*) extend from South Africa to Manchuria etc., although the mammal faunas of South Africa and Manchuria are very different as wholes. The Puma extends from northern British Columbia to southern South America, although the mammal faunas of these places are very different as wholes. Ordinary rats (*Rattus*) are native in Australia, in spite of the isolation of Australia and the differentiation of its mammal fauna as a whole. *Galaxias* occurs in fresh water in South Africa, southern Australia, New Zealand, and southern South America, but its distribution is (among vertebrates) unique; fortunately it is not fossil but alive and is known to enter the sea; even so, some persons suppose (I do not) that it must have dispersed over land connections in the southern hemisphere. These special cases have their own significance, but they are much less significant than whole faunas. It is whole faunas rather than special cases which should be stressed in considering the past.

Existing situations suggest that if, in the past, two contemporaneous faunas were similar as wholes and were related genetically, they occupied climatically similar, connected regions. If the genetic relationships involved the main parts of both faunas but were not close in detail—if it was a matter of related genera more than shared genera and species—the two faunas may have been far apart, connected perhaps by a long and indirect route. But if the two faunas had many genera and some species in common, they were probably closely connected. Where faunas differed, the nature of the differences may be significant. If two faunas were similar as wholes but not related in detail, they were probably separated by a barrier. If two faunas differed in size or adaptations but were in part closely related, they probably occupied climatically different but connected areas. These generalizations are easier to make than to apply. Most past faunas are incompletely known, and differences between them may be due to differences in time.

Existing animals not only give us the principles by which to interpret past distributions but also tell something of the past themselves.

For example, existing Australian mammals show conclusively that Australia has been isolated from the rest of the world at least since the beginning of the Tertiary, even though no significant fossil mammals older than the Pleistocene have been described from Australia (but the conclusiveness of this evidence depends on the fossil record of mammals in other parts of the world). Some existing animals are indicators of a more remote past than others. What matters is not the age of the animals themselves but the age of their distributions. A group of animals which dates from the Mesozoic may have redispersed many times since then, and its present distribution may be very recent, with no trace of a Mesozoic pattern left. Ages of distributions are presumably, in general, inversely proportional to rates of dispersal and powers of crossing barriers. Strictly fresh-water fishes disperse over the continents more slowly than any other vertebrates. Their present distribution may still reflect Cretaceous geography and may tell something about the arrangement of land in the Cretaceous. I doubt whether any existing animals tell anything about more ancient times.

Most of what zoogeography tells about the past comes from past distributions, which are shown by fossils but which must be judged by what we know of animal distribution now. Of the different groups of vertebrates, mammals are most significant back through the Tertiary. Dinosaurs were probably most significant through most of the Mesozoic, back to the late Triassic. Earlier in the Triassic other reptiles (perhaps especially therapsids) and stereospondyl amphibians were probably most significant. Even earlier, in the Permian and perhaps before, certain amphibians and reptiles give some clues to the state of the world, although what they tell is very scanty and very hard to read.

A zoogeographer's conclusions about the past can be determined beforehand by his premises. If he begins with the idea that animal distributions are stable and preserve old geographical patterns for hundreds of millions of years, he will find the past deceptively easy to read, and he will probably read of great land bridges or continental drift. If he thinks, as I do, that animals have great powers of movement and that they have been continually dispersing (spreading and receding) over the world, he will find the past more difficult to read. It will be very difficult to see, under the kaleidoscope of animal movement, how the continents were connected long ago and whether they too have moved. I do not mean to deny other points of view, but they ought to be clearly stated. Every writer on historical zoogeography should tell what he thinks about means and rates of dispersal of animals, the relation of recessions and extinctions to dispersal, and the age and significance of distribution patterns.

For zoogeographic purposes, the past may be divided into three parts (Fig. 6, p. 11). The most recent is the Pleistocene and post-Pleistocene, the last million years or less, about which so much is known that it can only be summarized briefly here. Before that is the Tertiary, 60 million years or so. It is in this period that zoogeography is most significant. Geological knowledge of the Tertiary is still incomplete, and the distribution of animals, especially of mammals, tells decisive things about the histories and interrelationships of the continents then. Before that is the Mesozoic and pre-Mesozoic, still longer stretches of time, when the past becomes increasingly dim, and increasingly exciting. The positions of the continents then are not known. If they have ever drifted, they may still have been doing so in the Mesozoic, and zoogeography may help to show it. For each of these three spans of time—the Pleistocene and post-Pleistocene, the Tertiary, and the Mesozoic and before—I shall try to describe first the state of the world as geologists see it, and then the contributions of zoogeography.

The Pleistocene and post-Pleistocene

The Pleistocene was the ice age, or the ice ages, for there were four of them, four successive continental glaciations in the northern parts

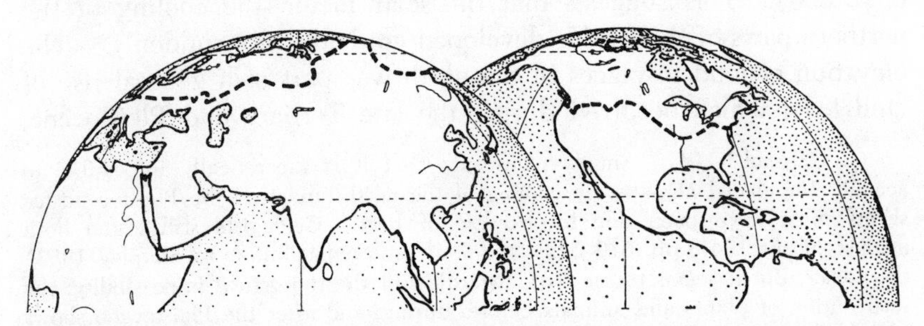

Fig. 74. Approximate southern limit of northern continental ice in the Pleistocene. Some areas north of these limits were unglaciated, in eastern Siberia, Alaska, and elsewhere.

of the world, separated by interglacial ages when the ice withdrew. The southern limits of the northern ice are shown in Figure 74. The first Pleistocene ice sheets appeared probably the better part of a million years ago. The last continental ones began to withdraw only about 10,000 years ago, according to Carbon 14 dating (Libby 1956). [I have not tried to fix the exact times of the beginning and the end of the Pleistocene. In the last few years geologists have reduced their estimates of time since the *beginning* of withdrawal of the last continental ice sheet in eastern North America from about 25,000 to about 10,000 years—it may have been only 10,000 years since the sites of Boston or Chicago were under ice. The beginning of the Pleistocene is much harder to date than the end of it, and any date given should be considered provisional.]

The causes of Pleistocene glaciation are unknown. There are many hypotheses. Flint (1947, pp. 502–512) criticizes some that seem no longer tenable. He decides (pp. 512ff.) that the glaciation was probably due to a combination of solar and topographic factors: reduction and fluctuation of the sun's heat, and elevation of land in certain places in the north. When the sun gave least heat, ice accumulated on the elevated land in northern Europe, northeastern North America, and elsewhere, and spread outward and southward from there. Biogeography can make a contribution to this hypothesis. There was a steady southward shifting of northern vegetation zones during much of the Tertiary, evidently caused by a gradual cooling of northern climates (Barghoorn 1953, p. 247). Moreover, in his analysis of exchange of mammals across the Bering bridge, Simpson found that selectivity gradually increased through much of the Tertiary, apparently as a result of a gradually increasing climatic barrier (Chapter 6, p. 366). This suggests that the solar factor—the cooling of the northern parts of the world—developed gradually, not suddenly.[1] The elevation of land in places in the north was part of a general rise of land level which occurred during the late Tertiary and Pleistocene,

[1] G. C. (not G. G.) Simpson has said that it is theoretically impossible to account for glacial climates by cooling of the north, and that the Pleistocene ice sheets must have been due to *increase* of heat. Geologists still argue both against (Flint 1947, pp. 510–512) and for (some contributors to Shapley 1953) this idea. But it seems to me that, regardless of the theoretical impossibility, the distribution of plants and animals before, during, and after the Pleistocene shows plainly that the ice was formed after a long period of cooling, when the northern parts of the world were cold. I have put this case down for future reference. Zoogeographers should respect other sciences, but they should not overrespect them, and they must not belittle their own. Later in this chapter I shall have to deal with questions—for example, the question of continental drift—which some geologists try to settle on theoretical grounds. I shall then refer to this footnote to strengthen my conviction that a zoogeographer is entitled to make his own estimate of a situation, and not to be too much bound by other people's theoretical impossibilities.

as a result of which the continents stood (and still stand) higher than they usually have during geological history.

The northern ice sheets were only the most dramatic manifestations of Pleistocene cold. Its effects were world-wide. Mountain glaciers increased on all high mountains, in the tropics and south-temperate zone as well as in the north. The amount of increase—lowering of snow lines and lengthening of glaciers—suggests that, at times during the Pleistocene, temperatures everywhere were 4° to 8° C. below present ones (Flint, pp. 455–456). Temperature probably fell synchronously everywhere, although this is not quite certain (Flint, pp. 452–453—Flint's conclusions have been questioned, but they seem reasonable to me).

Besides the lowering of temperature, there was apparently an increase of rainfall over much of the world during times of glaciation. This does not mean that all the world had heavy rain, but that the wet parts tended to be wetter and the dry parts less dry than now. There was probably also a shifting of temperate rain zones toward the tropics, into the subtropical desert zones. Lakes increased in size in southwestern North America, southwestern Asia, East Africa, South America (Lake Titicaca), and Australia; and what are now arid regions in southwestern North America, North Africa (the Sahara), Australia, etc., were evidently better watered and better vegetated than now and less of a barrier to non-desert animals.

Between the glaciations, during the interglacial ages, climates were apparently sometimes warmer and also drier than now over much of the world.

Sea level fluctuated during the Pleistocene. The water that piled up as ice on land came from the sea, and its loss caused a lowering of sea level over the whole world of the order of perhaps 100 to 120 meters at the greatest extent and 70 to 80 meters during the last advance of the ice (Kuenen 1950, p. 537), so that continents and islands which had been separated from each other by shallow water were connected together. On the other hand, when there was little or no ice on land (when the Antarctic Continent and Greenland did not have the ice caps they have now), before the Pleistocene and perhaps at times during the interglacials, sea level was 20 to 50 meters higher than now (Kuenen, p. 539), so that low edges of continents and low islands were submerged. Kuenen says that these figures must be viewed with circumspection, but that the margin of error is probably not more than 50 per cent.

It is customary to distinguish a post-Pleistocene or Recent epoch from the Pleistocene proper, but the distinction is artificial. The post-Pleistocene, which is the present, may be just another interglacial; it may end in another ice age. Ice caps still lie on Antarctica and Greenland. In the few thousand years since withdrawal of the last continental ice sheets, climate has fluctuated, being at one time somewhat warmer than now (the "climatic optimum"—Zeuner 1950, p. 67) and then somewhat colder (the "little ice age"). Now it seems to be getting warmer again. We do not know the causes of these fluctuations, and we cannot foretell how they will go in the future.

Our knowledge of the distributions and movements of animals during and after the Pleistocene is voluminous. Deevey (1949) devotes a hundred pages to what he himself says is an incomplete summary of Pleistocene (and post-Pleistocene) biogeography, with selected references. Thienemann (1950, in part) devotes still more space to the Pleistocene history of just the fresh-water fauna of Europe. Holdhaus *et al.* (1954) devote a large volume to the traces of the ice age in the fauna of Europe. Dillon (1956) makes some interesting guesses about the late Pleistocene life zones of North America. What we know is partly derived from fossils, partly deduced from present distributions and from geological and botanical evidence, *e.g.*, from pollen sequences deposited in lakes and bogs, which show cycles of vegetation and thus of climate—the pollen sequences help to show that the north was cold when the ice sheets formed. Botanists and zoologists draw heavily on geology in working out Pleistocene histories and in turn contribute something to geologists and to each other.

In the northern parts of the north-temperate zone, including much of Europe and North America, the Pleistocene was cataclysmic. It not only changed details but changed the whole underlying pattern of plant and animal distribution. In undisturbed parts of the world, organisms form a geographical pattern produced by evolution in situ, modified by movement. After northern floras and faunas had been shuffled and reshuffled four times by the advance and retreat of ice, the evolution pattern was gone, replaced by a very complex movement pattern. That is why Darwin and Wallace had to go to the tropics or the south-temperate zone to see the geographical pattern produced by evolution.

The movements of northern animals during and after the Pleistocene were almost endlessly complex, but they tended to fall into a few main patterns of movement and return movement, or spreading and receding, which often ended in geographic discontinuity.

During glacial ages, arctic animals first moved southward before the ice and then northward again as the ice withdrew, and they sometimes left behind them populations which moved upward on mountains instead of northward, became isolated, and survived at high altitudes as "stranded" arctic relicts (Fig. 75). The occurrence of arctic plants and animals isolated on high mountains in Europe (they occur also in Asia and North America) was known long ago and was properly explained by Edward Forbes (about 1846, in a publication I have not seen) and later by Darwin (1859, 1950 reprint, pp. 310ff.),

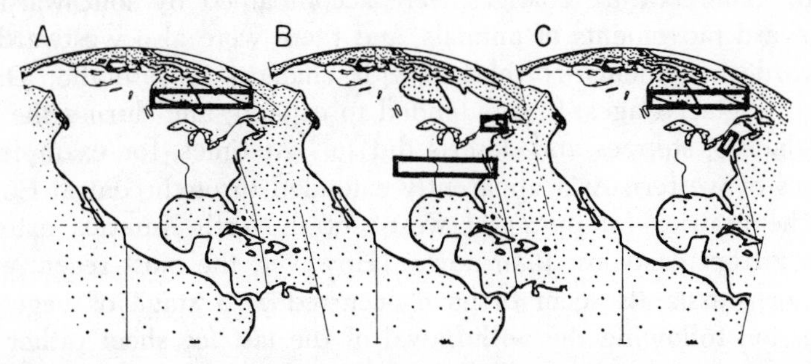

Fig. 75. Diagram of stranding of populations during advance and retreat of ice in eastern North America. A, original distribution (heavy oblong) of a subarctic species. B, distribution of the same species after advance of ice, with a population isolated in a "refugium" on Gaspé. C, distribution after retreat of ice, with a population relict on the mountains of New York and New England.

although the first explanations were too simple in detail. On the other hand, as the ice advanced, unglaciated areas were often left behind the ice front, and animals that moved southward before the ice sometimes left populations stranded in refuges cut off by ice. Some of these populations spread outward from their refuges when the ice withdrew again, and they mingled complexly with other re-advancing southern populations.

The complex spreading, readvancing, and mingling which followed the withdrawal of the last ice sheet is still going on and is the subject of intense study by northern zoogeographers. My own first zoogeographic problem was concerned with this—with repopulation of the Presidential Range of New Hampshire by carabid beetles. That the movements of our own animals during and after the Pleistocene should pre-empt our attention for a while is natural. But the post-Pleistocene readjustment of northern animals is not the essence of zoogeography any more than Wallace's Line was. Overconcentration on it can blind zoogeographers to more important things. Zoogeography should begin with some understanding of world patterns and processes, and of the cumulative effects of evolution on distribution; and the adjustment of northern animals to recent climatic changes should be recognized for what it is: a short-term, minor modification of the edge of the main thing. Overemphasis of it tends to exaggerate the apparent effect of local ecology and temporary adaptations, and to hide the importance of world situations, time, and evolution.

In the north-temperate zone south of the actually glaciated areas, the effects of the Pleistocene on animal distribution were still profound. Temperature changes were accompanied by southward and northward movements of animals, and there were also westward and eastward movements correlated with changes in rainfall. During the rainy glacial ages, forests tended to expand; and during the drier interglacials, steppes and deserts did so. At times, for example, the steppes of western Asia apparently extended through central Europe, and the prairies (steppes) of west-central North America extended much farther eastward than now. [However, the most recent extensions of grasslands seem to have occurred as a stage of vegetation succession following the withdrawal of the last ice sheet rather than as a response to aridity.] Different animals spread and receded with their habitats, and some of them eventually became stranded in detached habitat pockets (Schmidt 1938).

North-temperate fishes and other fresh-water animals were specially affected by the Pleistocene. The ice itself dammed or diverted some waters, changing drainages and allowing fishes to pass from one river system to another, and the ice gouged out thousands of depressions which became new lakes when the ice withdrew. Farther south, rainfall cycles caused expansion and contraction of lakes, which allowed fishes to disperse and then stranded them in isolated waters. The present distribution of the fishes is often evidence of the extent and connections of rivers and lakes in the Pleistocene. The fishes of the Great Basin in southwestern United States (Hubbs, Miller, *et al.*, summarized by Deevey 1949, pp. 1395–1400) are an example.

In other parts of the world, the effects of the Pleistocene on animal distributions were much less, but there were effects. Over the world as a whole, barriers were periodically reduced and dispersal was facilitated. During the glacial-pluvial ages, the dispersal of forest animals and those of other well-watered situations was favored. During the drier interglacials, steppe and desert animals were favored. At such a time, for example, arid areas may have been more nearly continuous than now between southwestern North America and southwestern South America, across the tropics, allowing an exchange of desert plants and insects which apparently did occur. Periodic lowering of sea level allowed continental faunas to reach many islands (Chapter 8). That they did reach the islands confirms that the sea was lowered. But it was not lowered beyond a certain point. A fall of sea level of 1000 meters would probably have connected Cuba to North America, and Celebes to Borneo and Asia, allowing influxes of whole continental faunas which did not occur. One of the bridges made as a result of the world-wide lowerings of sea level connected Asia and North America across Bering Strait. The connection was probably a broad plain which continued as coastal plains along the southern edges of both Asia and Alaska (see Flint, p. 529). Its climate was probably moderated by the warm Japan Current; it may have had a climate like that of the Aleutians today. It may have been covered with long grass like that which now grows on the Alaska Peninsula. Many large and small mammals might, and did, cross such a bridge. Man probably first reached America that way, late in the Pleistocene.

Within the tropics, even at low elevations, temperatures probably decreased somewhat during glacial ages, but rainfall probably increased. Lakes increased in size at least in some places, and tropical forests probably tended to expand and steppes and deserts to contract. Opposite changes probably occurred during the warmer, drier interglacials. The effect on tropical faunas may have been considerable in marginal areas but was apparently not great in general. Existing lowland tropical faunas are huge and complex, with many species of animals closely adapted to each other and to the plants around them. Such faunas are obviously old and have suffered no great disaster during the Pleistocene.

Among the special effects of the Pleistocene was a progressive withdrawal and extinction of large mammals, which is still going on. Man was probably not primarily responsible for it, but he has helped it along. Man himself evolved mainly during the Pleistocene, but not necessarily because of it.

Another special effect of the Pleistocene may have been an increase in both volume and distance of north-south migration of birds (see p. 249).

For summary of the Pleistocene, I shall paraphrase and enlarge on Flint (1947, p. 10). If it could be shown by motion picture, simplified and enormously speeded up, the scene would begin (in the Tertiary) with gradual upheaval of land and rising of high mountain ranges in many parts of the world, and (if Flint is correct) a gradual reduction and fluctuation in intensity of the sun. When these processes had gone far enough, at the beginning of the Pleistocene, the sun's fluctuations would begin to cause the waxing and waning of continental ice sheets centering on high ground in the north, as well as the waxing and waning of mountain glaciers everywhere. Accompanying these cycles would be cycles of increase and decrease of rainfall, shifting of rain zones into and out of the arid subtropics, and growth and shrinkage of pluvial lakes everywhere. At the same time the level of the sea would gradually fall and rise, making and unmaking land connections across Bering Strait and to many islands. Correlated with these inorganic cycles would be broad but irregular movements of plants and animals. Whole northern floras and faunas would move southward and northward as the climate changed, and they would change their compositions and internal patterns of distribution as they moved. Many different plants and animals would spread and recede with spreading and receding of their special habitats, and during recessions some would be stranded on mountain tops, or in ice-free refuges, or in wet, or dry, places in many parts of the world. Processions of plants and animals would cross the Bering bridge when it existed, and other processions would cross temporary bridges to islands and be cut off there, and this too would happen in many parts of the world. Large mammals would gradually become fewer; annual north-south movements of birds would increase; and man would become conspicuous and reach almost every habitable part of the world before the end of the Pleistocene.

The Tertiary

The Tertiary Period covers about 60,000,000 years (more or less) before the Pleistocene. It is the main part of the Cenozoic Era, and it is divided into five epochs (Fig. 6, p. 11). As a zoogeographer, I have long been interested in this span of time, and I have reviewed it by reading appropriate parts of Dunbar's (1949) *Historical Geology*, Romer's (1945) *Vertebrate Paleontology*, and other works, some of which are referred to below.

During the Tertiary, the continents underwent only minor changes of shape (caused mostly by incursions or withdrawals of marginal seas) but major changes of surface. All the great mountain ranges of the world were made or remade during the Tertiary. There was a general rise of land which eventually lifted the continents to their unusually high Pleistocene and present levels. And there was excessive volcanic activity in some parts of the world. The direct effect of all this on animals on continents is doubtful, but the effect on islands is obvious. In some achipelagos, islands were made and unmade as mountains were on the mainland, and where the archipelagos lay between continents, land bridges may have been made and broken repeatedly. The effect on animals was probably often very great, and the distribution of animals is often the best evidence of what happened.

Africa was apparently more stable than most continents during the Tertiary, although there was volcanic activity at least in eastern Africa.

Parts of Eurasia were very unstable. From the Mesozoic well into the Tertiary, the Tethys Sea extended across the whole of southern Europe and Asia. Europe was probably at times reduced to an archipelago. The Alps and Himalayas rose complexly from the Tethys during the Tertiary, and their rise caused complex geographical changes, including perhaps the making and unmaking of land bridges between Eurasia and Africa and between Asia and what is now peninsular India.

North America was fairly constant in shape during the Tertiary except that shallow seas overlapped parts of what are now the southeastern coastal plain and the lower Mississippi Valley. But the surface changed. The present Appalachians were raised and dissected mostly during the Tertiary, and the mountains of western North America were complexly modified.

In South America, the Andes, in their present form, are principally Tertiary and did not reach their present heights until the late Tertiary and Pleistocene. In Central America there were volcanic activity and local changes of level.

Australia was apparently relatively stable, but the islands between Asia and Australia had a very complex Tertiary history (Umbgrove 1949).

The climate of the Tertiary, at least north of the tropics, began much warmer than now and gradually cooled until the Pleistocene. Early in the Tertiary, what are now temperate types of forest extended northward into the arctic, and tropical or subtropical forest types extended far into the north-temperate zone (Barghoorn 1953). During the Tertiary, the northern zones of vegetation gradually shifted southward. Even early in the Tertiary, however, the climate was zoned, and the zones were oriented as they are now (Chaney 1940, Barghoorn). As Barghoorn says, it is unlikely that truly tropical conditions occurred in mid-latitudes; more likely, an absence of winter freezing allowed extensions northward of tropical plants. The northward extensions of tropical forests were presumably less rich and less diverse than the forests in the tropics proper.

A sudden multiplication of herbs occurred in the north-temperate zone in the mid-Tertiary (Barghoorn), perhaps in response to increasing severity of northern winters. Herbs had previously been few everywhere. They are still relatively few in the tropics. Their appearance in numbers in the north probably affected the evolution and distribution of animals, perhaps (among vertebrates) especially of seed-eating birds and rodents.

The Tertiary histories of vertebrates are summarized (hypothetically or factually) by classes in appropriate places in Chapters 2 to 6, and Romer (1945, pp. 546ff.) summarizes them in a different way, by successive epochs. Generally speaking, the present distribution of vertebrates seems to be mostly a product of Tertiary (or later) movements.

The zoogeographical evidence of the state of the world in the Tertiary is consistent with the geological and botanical evidence. For example, both fresh-water fishes (Chapter 2, p. 90) and mammals (Chapter 6, p. 366) reflect zonation of climate during the Tertiary and gradual cooling of the north, although the botanical evidence of this is more important.

The relative stability of Africa and instability of southern Asia are reflected by existing fresh-water fishes, which are old and diverse in Africa but dominated by more recent groups in tropical Asia (Chapter 2, p. 91), and this faunal difference suggests also that exchange of fishes between Africa and southern Asia has not been easy during the Tertiary. The early Tertiary record of mammals in these parts of the world is poor, but El Faiyûm mammals (in Egypt in the late Eocene and early Oligocene), which are a mixture of very distinct, previously unknown African groups and contemporaneous European ones, suggest that Africa and Eurasia were connected then but had previously been separated for a considerable time, although they had presumably been connected still earlier (Chapter 6, p. 365). The later record of mammals indicates that Africa and Eurasia were connected at least most of the time through the rest of the Tertiary.

Some persons have claimed a broad connection or union of Africa, Madagascar, and India even in the Tertiary and have named it Lemuria, but there is decisive evidence against it: *e.g.*, the absence of strictly fresh-water fishes on Madagascar and the pattern of relationships of other significant animals. Even the lemurs, for whom Lemuria was named, argue against it: the lemurs of Africa and tropical Asia are directly related to each other, but those of Madagascar belong to very distinct, separate evolutionary side lines (Fig. 76). A few relationships do exist between Madagascan and Indian vertebrates, but they are exceptions best explained by extinctions in

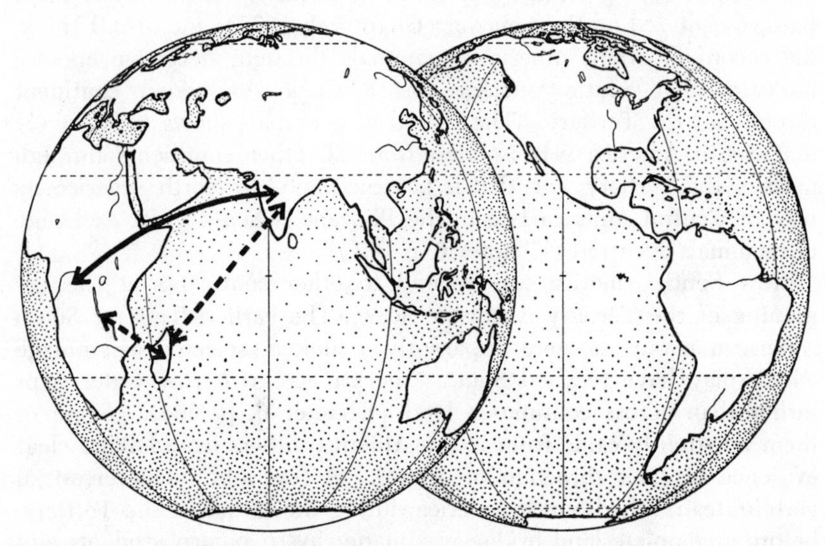

Fig. 76. Geographical relationships of lemurs. Broken arrows show relationships that should (but do not) exist if lemurs dispersed across "Lemuria." Solid arrow shows relationships that do exist between African and Oriental lemurs and that do not involve Madagascar.

Africa or (in the case of some birds and bats) dispersal across the sea (Chapter 8, p. 522). Lemuria did not exist in the Tertiary.

The times and place of land connection between Eurasia and North America during the Tertiary are plainly shown by the fossil record of mammals. Connections were apparently repeatedly made and broken across Bering Strait; the times of them are given in detail in Chapter 6 (pp. 365–366). The record of mammals does not indicate any connection across the Atlantic between Europe and North America. Up to the early Eocene, the evidence is indecisive, but it does not require an Atlantic bridge and is consistent with a North Pacific bridge. After the early Eocene, the evidence clearly favors a single, North Pacific bridge and does so increasingly in later epochs (Simpson, in papers cited in Chapter 6),

The history of South America during the Tertiary is told by the distributions of fossil and existing vertebrates. The occurrence not only of marsupials but also of placental mammals, some related to North American ones, in southern South America very early in the Tertiary indicates (if there were any doubt about it) that the continent was a continuous length of land at that time. The present distribution of strictly fresh-water fishes in South America, in one main pattern centered on the Amazon (Chapter 2, p. 72), and also, I think, the record of South American mammals through successive epochs, indicates that South America was a single, undivided continent throughout the Tertiary. The record of mammals shows conclusively that South America was isolated from all other continents through most of the Tertiary, and that it became joined to North America by way of Central America late in the Pliocene, when a great exchange of mammals occurred (Chapter 6, p. 367).

How South America was related to other continents at the beginning of the Tertiary is not so clear. The earliest Tertiary South American mammals were apparently a limited, selected assemblage which may have reached South America across narrow water gaps rather than across continuous land (Chapter 6, p. 364). Some of them apparently came from North America; others have left no clear evidence of their direction of arrival. A few additional terrestrial vertebrates reached South America during the course of the Tertiary, before a complete land bridge was made: hystricomorph rodents and monkeys came from an undetermined direction perhaps in the late Eocene and Oligocene respectively; *Testudo* and procyonids apparently came in the Miocene, the procyonids, at least, from North America. All this can be explained if South America has occupied its present position with respect to North America throughout the Tertiary, and if it has been linked to North America by a fluctuating archipelago or incomplete isthmus, which was the route of immigration of some or all of the early Tertiary mammals of South America, which allowed an occasional arrival later by "island hopping" and which finally became a complete land bridge late in the Pliocene. This partly hypothetical history is consistent within itself and with the facts so far as they go. Discovery of the right fossils in the right places will probably show whether or not it is correct. If new information supports the idea that hystricomorphs and possibly one or two other mammals reached South America directly from Africa, I should be inclined to think that they came on rafts. The Main Equatorial Current across the Atlantic is favorable (p. 17), and adequate rafts exist (p. 15).

Whether Central America was a peninsula of North America or a separate island or archipelago during most of the Tertiary is unknown, and this too is something that will probably be settled eventually by fossils. In the meantime the nature of the West Indian fauna, much of which was apparently derived from Central America during the Tertiary, seems to me to suggest that Central America was cut off from North as well as South America for a long time (see Chapter 8, p. 517). I think that during this time there accumulated and evolved on Central America a limited, highly endemic, non-dominant, insular fauna, of mixed North and South American origin, most of which (after parts of it had reached the West Indies) was overrun and destroyed by continental animals when land connections were made. There also seems to me to be evidence (mostly negative evidence) in the present distribution of birds (Chapter 5, pp. 279ff.) and past history of mammals (Chapter 6, p. 368) that no important tropical area was attached to North America during the Tertiary, and this too seems to suggest that Central America was isolated most of the time. This is, however, still an open question.

Australia has yielded almost no record of its Tertiary vertebrates. However, its present lack of native placental mammals, except bats and rather recent rodents, shows that the continent was separated from all other continents by water throughout the Tertiary. The existing Australian vertebrate fauna is complexly "stratified." It includes an endemic subclass (the monotremes); endemic superfamilies, families, and subfamilies; endemic genera of widely distributed families; endemic species groups and species of widely distributed genera; and still less differentiated and undifferentiated forms. This suggests that vertebrates arrived in Australia, in small numbers, by island hopping, continually during the Tertiary. So far as the evidence goes, all the vertebrates that reached Australia during the Tertiary came from Asia, except a few (galaxiid fishes and perhaps meiolaniid and chelyid turtles) that seem to have come from South America, perhaps by sea. This suggests that Australia occupied its present position with respect to Asia, with islands in between, throughout the Tertiary. The vertebrate (at least mammal) fauna of New Guinea seems to be less old than that of Australia (Chapter 6, p. 338), and the fauna of the islands between New Guinea and the continental shelf of Asia is still younger (Chapter 7, p. 470), with few strikingly differentiated forms, and is still distributed along apparent lines of immigration. This is consistent with the geological instability of these islands during the Tertiary, and with their older faunas having been destroyed and replaced, so that the record of early Tertiary movements across the islands to Australia has been destroyed.

As to greater land bridges and drifting continents, vertebrates show none in the Tertiary. The key to this is South America, for it is involved in most of the hypotheses of great land bridges or continental drift. Vertebrates, especially mammals, show that South America was not connected with any other continent from the beginning of the Tertiary (if then) until the late Pliocene connection with North America. *The Tertiary isolation of South America is a fact. Paleogeographers who refuse to accept it simply invalidate their own theories.* The situation at the very beginning of the Tertiary is not clear, but if South America had any connection then, it was probably with North America rather than with Africa. If there was any connection with Australia, it was evidently pre-Tertiary, before the arrival of placentals in South America.

The distribution of vertebrates adds something to what is known of Tertiary climates. Fossil floras show that North America was warmer in the Tertiary than now, but it was probably not fully tropical (pp. 589–590). The fossil record of mammals suggests this too. There seems to have been no sharp differentiation of temperate and tropical faunas in North America in the Tertiary, and when North and South America exchanged mammals at the end of the Tertiary, no third "tropical North American" fauna came to light (p. 368). If there was an endemic mammal fauna in Central America (see third paragraph above), it was apparently too small and too unaggressive to play much part in the formation of existing continental faunas. Ornithologists have, I think, exaggerated both the tropicalness of North America in the Tertiary and the importance of "tropical North America" as a center of evolution and dispersal (pp. 279ff.).

What has just been said of North America is probably true of temperate regions in general. They were warmer in the Tertiary than now but were probably not fully tropical.

The climate of the tropics during the Tertiary is harder to deduce. If the warmth of temperate climates was due to extra heat received from the sun, the tropics should have been affected too, but I am not sure just what the effects would have been. Temperature might have been kept within limits by cloud cover or atmospheric circulation. This whole subject should be worth careful consideration. All I can say about it now is that existing tropical floras and faunas are very rich and complexly integrated and seem to have suffered no great disaster for a long time.

If what is known and probable of the world during the Tertiary could be summarized in another reel of a simplified motion picture, it would have blanks and flickerings, but it would be a coherent picture nevertheless.

Even at the beginning of the Tertiary, most of the continents would be recognizable in shape, and they would lie on the world about as they do now. Asia and North America would be connected in the north. North and South America would be nearly or completely connected. Asia and Australia would be linked by an archipelago. During the Tertiary, the continents would change their surfaces but not (or very little) their positions, and the connections between them would change only in details which would be small geologically, though sometimes large in their effects on animals. Africa and Eurasia would probably be disconnected and reconnected in the first third of the Tertiary. The Bering bridge would be unmade and remade several times. North and South America would be separated (if they were connected) at the beginning and reconnected at the end of the Tertiary. The archipelago between Asia and Australia would fluctuate but would not form a complete land bridge.

Even at the beginning of the Tertiary, climates would be zoned, and the zones would be orientated about as now, and they would cause a visible zonation of vegetation. Subtropical forests would occur far into the temperate zones but would be thinner than forests in the full tropics, and there would be at least small areas of deciduous forest in the far north, where the climate would always be strongly seasonal. During the Tertiary, as the climate cooled, the northern zones of vegetation would move slowly toward the tropics, and the main movements would be complicated by minor movements of floras and parts of floras: east-west shifts, spreadings and recedings, and evolutions and extinctions.

Connected with the changes of land, climate, and vegetation would be movements of animals, endlessly complex in detail but forming a few main patterns. Within the main part of the Old World would be a succession of dispersals and redispersals, with successive dominant groups tending to spread from the tropics northward (Chapter 9). Whenever a land connection existed, there would be exchanges of animals between Asia and North America. Even early in the Tertiary, however, zonation of climate would limit exchange, blocking strictly tropical groups; and smaller and smaller fractions of faunas would be exchanged as the northern climate cooled. There would be movements in both directions, but always more from Asia to America than the reverse. In the New World, a fraction of the North American fauna would reach South America (by land?) at the beginning of the Tertiary; an occasional animal would cross the water gaps later; and at the end of the Tertiary many animals would cross the new land bridge in both directions, although the exchange would end as a great invasion of South America by animals from the main part of the world. Australia would receive an occasional immigrant from Asia across the changing water gaps. And land animals would continually or occasionally reach other islands, more or less in proportion to distance. In fact, in the motion picture, there would be a constant outward scattering of animals from continents onto islands something like water scattering from a fountain, most drops falling nearby, and fewer and fewer reaching more distant places.

The 60,000,000 years of the Tertiary would give time for much evolution, replacement, and extinction as parts of the geographical picture. Modern mammals would deploy and redeploy in the main part of the world. Great, separate faunas would radiate in South America and Australia. And these would be only the most prominent features of a kaleidoscope of change.

The Mesozoic and before

The Mesozoic Era covers about 120,000,000 years (more or less) before the Tertiary and is divided into three periods: Triassic, Jurassic, and Cretaceous (Fig. 6, p. 11).

According to Dunbar (1949), the Mesozoic began (in the Triassic) with a general emergence of the continents and with widespread aridity. In the Jurassic, the land subsided and climates tended to be wetter; and temperatures were cooler early in the Jurassic and warmer later. During the Cretaceous, the continents sank to very low levels and were extensively flooded by shallow seas. North America was temporarily cut in two by a seaway which extended northward from the Gulf of Mexico. Toward the end of the Cretaceous, the land rose again, and the Rocky Mountain system and the Andes were first formed, although both were re-formed in the Tertiary. The climate of the Cretaceous became first cooler and then warmer. Figs, tree ferns, cycads, and palms occurred in Greenland and/or Alaska. But these plants exist in some temperate regions now and do not necessarily indicate tropical conditions. I do not know how much reality there is in these supposed world-wide cycles. They may be real in a general way. However, it should be remembered that geological revolutions are not simple, and that neither temperature nor rainfall can ever have been uniformly distributed over the world (Chapter 1, p. 8).

The plants of the early Mesozoic were conifers, tree ferns, cycads, and other ancient forms. At times, especially in the mid-Mesozoic, the old floras were remarkably uniform and widely distributed. Broad-leaved trees (angiosperms) appeared in the early Cretaceous and soon became dominant everywhere.

The Mesozoic was the age of reptiles. Dinosaurs were the most conspicuous, best-known, and therefore most significant land animals. The Mesozoic saw also the endings of several ancient groups of reptiles and the beginnings of crocodiles, turtles, lizards, and snakes. There were reptiles in the sea and in the air as well as on the land. Of other vertebrates, fishes and amphibians underwent complex evolutions and replacements during the Mesozoic. Birds arose in the Jurassic. The first mammals appeared in the Triassic. But most of these groups are too unimportant or too little known during the Mesozoic to be of much zoogeographic significance.

Existing vertebrates, except perhaps some fresh-water fishes, can tell little directly about Mesozoic geography. For the most part, their distributions are Tertiary or later. Zoogeographic evidence about the Mesozoic must therefore come mostly from fossils. But the evidence of the fossils has to be interpreted by comparison with existing distributions, and interpretation is not an easy matter.

Ancient animals and plants may have had capacities and tolerances different from those of existing ones. The possibility of this increases with remoteness of time. In fairly recent times, in the Tertiary, although the significance of single species may be questioned, the significance of whole faunas and floras may be reasonably sure. For example, the occurrence not just of single plants but of subtropical plant associations far northward during the Tertiary is reasonable proof of warmer climates (Barghoorn 1953). But in more ancient times, this kind of evidence is more doubtful. Whole great groups of plants and animals may have changed their tolerances. For example, the salt tolerances of fresh-water fishes in the Mesozoic may have been very different from now (Chapter 2, pp. 45–46). The relation of reptiles to climate in the Mesozoic may have been very different from now—basically different, not just different in species' adaptations. *Sphenodon* hints at this. It lives on cool-temperate New Zealand without any apparent special adaptations for withstanding coolness or securing heat. It has not become ovoviviparous as modern lizards, both geckos and skinks, have done on New Zealand, and it has not even become diurnal but is active in the cold of the night, at temperatures uncomfortably low for clothed men (Bogert 1953, p. 169). Perhaps it and its Mesozoic contemporaries were less susceptible to cold than modern reptiles. Another hint of past difference in relation to climate comes from the American alligator, which has "surprising" tolerance for cold (Colbert *et al.* 1946). Alligators might be able to live far north of their present limits, if temperature were the only limiting factor. But there is another factor now which did not exist in the Mesozoic—the presence of many warm-blooded mammals. In the absence of effective warm-blooded competitors, a diversity of reptiles may have existed in cool northern regions, and whole faunas of them may just have suspended activity in cold weather. But the rise of warm-blooded mammals, active in cold weather, may have made this impossible.

In view of the enormous length of time (twice the Tertiary), the complexity of the world then as always, the complexity of evolution and dispersal of different vertebrates, the small amount of evidence, and the difficulty of interpreting the evidence, I shall not try to treat Mesozoic zoogeography in much detail. I shall be content to try to answer one question, although I shall try to find the answer in several different ways. The question is, was the geography of the world—the main pattern of continents and climates—in and before the Mesozoic essentially the same as now, or entirely different?

Dinosaurs

Dinosaurs are more likely than any other animals to show the geography of the world in the Mesozoic, but not too much should be expected even of them. They are discussed and classified by Romer (1945) and treated in a more popular, but authoritative, way by Colbert (1945). Dr. Colbert has read and in general approved what I say about dinosaurs here. I have urged him to write a more detailed geographical history of them himself. It would be an important contribution to zoogeography.

Dinosaurs were numerous and diverse. There were two orders of them, no more closely related to each other than to crocodiles, pterosaurs, and birds (Romer); and within each order were many different evolutionary lines. The best-known dinosaurs were giants, but there were small ones too, some as small as rabbits or smaller. The earliest dinosaurs were carnivorous and some remained so, but others, including the largest of all, became herbivorous. Dinosaurs were probably cold-blooded, but through most of their history they had no warm-blooded competitors, or at least no effective ones. They may have extended into cool climates, but probably not very cold ones. Most dinosaurs were terrestrial, but some became amphibious if not aquatic. They presumably dispersed mainly over land, but (like other land animals) they probably had some chance of crossing narrow water barriers.

Dinosaurs are divided into the following main groups.

Order Saurischia
- Suborder Theropoda
- Suborder Sauropoda

Order Ornithischia
- Four suborders (see list of reptile families, end of Chapter 4)

Theropods were terrestrial and carnivorous. They were the first dinosaurs to appear, and they soon became widely distributed and remained so until the end of dinosaur history. They reached all the main parts of the world including South America, Australia, and apparently Madagascar.

Sauropods were large, amphibious, herbivorous dinosaurs, apparently adapted to eat only soft plant material. In the mid-Mesozoic they were probably cosmopolitan, being known in North America, Europe, Africa, Madagascar, South America, and Australia. Later, in the Cretaceous, they were apparently scarce in the northern parts of the world but were widely distributed southward, in Africa, Madagascar, India, and South America.

Ornithischia, too, were large and herbivorous but were mostly terrestrial, although some became amphibious. Most of them were adapted to eat coarse vegetation. Although some are known elsewhere and earlier, they were especially numerous in the northern parts of the world, in Eurasia and North America, in the Cretaceous. One striking group of them, the Ceratopsia or horned dinosaurs, apparently originated in Asia but later evolved principally in western North America (perhaps also in Asia), perhaps being barred from eastern North America by a sea barrier (Colbert 1948).

In short: (carnivorous) theropods were widely distributed through all of dinosaur history, while (herbivorous) ornithischians and sauropods, whatever their earlier history, became zoned in the Cretaceous, the hard-plant-eating, primarily terrestrial ornithischians mostly in the north, and the soft-plant-eating, amphibious sauropods mostly farther south. This distribution has been interpreted in several different ways.

Lull, in 1910, made what is probably the best factual analysis of dinosaur distribution, but he made it too soon, before enough facts were known. To account for the facts he did know, he imagined two main, late Mesozoic continents, one northern and one southern, completely separated from each other, and he supposed that ornithischians ("orthopods") were always confined to the northern continent (North America plus most of Eurasia), that sauropods became confined to the southern one (India plus Madagascar plus Africa plus South America), and that theropods, which had dispersed earlier, occurred on both. This hypothesis was upset by the later discovery that ornithischians and sauropods were not so completely separated geographically as Lull thought.

Huene (various papers, *e.g.*, 1933, with Matley) did not demand complete separation of northern and southern faunas in the Cretaceous. On the contrary, he saw close relationships between some Cretaceous vertebrates of North and South America and also between some of those of Europe and India. But he supposed that there were long land bridges across the southern oceans, from Madagascar through India, Australia, and the Antarctic Continent to South America, and he supposed that Cretaceous sauropods dispersed that way. The best criticism of this idea is the known distribution of the dinosaurs concerned. Huene stressed three genera: *Antarctosaurus,* known from India and South America; *Laplatasaurus,* now known from East Africa, Madagascar, India, and South America; and *Titanosaurus,* from England, central Europe, Madagascar, India, and South America. If these distributions are real (much of the material on which they are based is fragmentary), they suggest incompleteness of the geographical record rather than dispersal by any definite route. There is no evidence that these particular dinosaurs ever occurred in Australia (although they may have done so), and no trace of any dinosaur has been found on the Antarctic Continent, and none on New Zealand.

Matthew (1915, pp. 275–279) thought that the distribution of dinosaurs was consistent with dispersal (1) from the north and (2) over continents arranged about as they are now. The first point seems to me unwarranted. Dinosaurs, like other cold-blooded vertebrates, probably evolved more in warm than in cool places and probably dispersed through the north more than from it. But I agree with Matthew's second point. The distribution of dinosaurs, what is known of it (which is not very much), can be accounted for well enough by dispersal over a pattern of continents like the present one, with climate zoned as now at least in the Cretaceous, when most ornithischians occurred in the north-temperate zone and most sauropods withdrew into the tropics.

The presence of dinosaurs on Madagascar is a problem. Remains of presumably terrestrial theropods have been reported there, but I do not know their nature or significance. Two genera of sauropods occurred there (see above). They were more or less aquatic animals. Some of their remains have been found in shallow-water marine deposits, and it has been suggested that they sometimes lived in shallow estuaries, walking on the bottom and feeding on water plants, and reaching to the surface for air by means of their long necks. No existing animal has quite this way of life, but the Hippopotamus comes nearest to it. Hippos are not primarily adapted for swimming, but they can swim well, and sauropods probably could do so too. A hippo did reach Madagascar in the Pleistocene, probably from Africa across the Mozambique Channel. There is geological evidence that this channel has existed since the Triassic, and I would not disregard that evidence to make a land connection for sauropods, if they came from Africa. If they came from India, they may have come by way of a series of archipelagos, with protected, shallow seas, which are suggested by a series of existing, shallow banks. I do not think the presence of sauropods on Madagascar proves a complete land connection. This is an understatement. The genera of sauropods that occurred on Madagascar were very widely distributed over the world (see second paragraph above). This is not what one would expect of large, strictly fresh-water animals even if there were land connections. Fresh-water animals move very slowly over land unless they have some special powers of dispersal. Huene thought the wide distribution of these sauropods proved land bridges. I think their wide distribution almost proves that they could disperse through the sea to some extent.

What is known of the distribution of dinosaurs does prove one main thing: their world was *not* divided into two great land masses completely separated from each other. This is not a small thing. Some persons have argued that the land was thus divided late in the Mesozoic, but plainly it was not. Moreover, the distribution of dinosaurs, so far as is known, conforms well enough with dispersal over a system of land not very different from now, with climatic zones oriented as now at least in the Cretaceous. If the dinosaurs do not prove that their world was arranged like ours, at least they do not seem to prove anything else.

At the end of the Cretaceous, dinosaurs became extinct, all of them, everywhere, apparently abruptly. Why is not known and need not concern us here. They disappeared, and mammals replaced them.

Mesozoic land bridges

During the Tertiary, land vertebrates dispersed (not in every case over continuous land) continually or occasionally between Eurasia and Africa, Asia and North America, North and South America, and from Asia to Australia, but probably not by any other regular intercontinental routes. The same routes may conceivably have been the only ones used by dinosaurs and other land animals during the Mesozoic, but the evidence is not complete, and other routes should be considered.

A land connection across the northern Atlantic, allowing dispersal between Europe and northeastern North America, has often been postulated, especially by botanists. I know of no decisive evidence either for or against its existence during the Mesozoic or early Tertiary. The record of mammals is against it later, after the early Eocene (p. 366).

There was no route of dispersal—no great land bridge—across Madagascar in the Tertiary (p. 590), and there is no good evidence of one in the Mesozoic. Madagascar did exist then and had a fauna including dinosaurs, but probably none of it survives. The present vertebrate fauna is probably wholly post-Mesozoic in origin. If parts of it are older, they cannot be dated now. If any detail of the existing Madagascan fauna is significant, it is probably the absence of strictly fresh-water fishes, which suggests that Madagascar has been cut off from all continents for a very long time. The dinosaurs that reached Madagascar in the Mesozoic may not have required continuous land. The geologic evidence, that the Mozambique Channel has existed since the Triassic, does not favor a Mesozoic connection of Madagascar with Africa.

A direct connection between Africa and South America has not existed at least since the beginning of the Tertiary. But was there a connection in the Mesozoic?

Fresh-water fishes are probably the most significant existing animals so far as Mesozoic geography is concerned. The South American fresh-water fish fauna has been analyzed in Chapter 2. It is derived from a few ancestors, extraordinarily mixed in their evolutionary levels and ecological requirements, which may have filtered along a narrow land bridge (not a broad contact) from Africa in the Cretaceous, but which may equally well have filtered through the warm edges of the northern continents. With the fishes in South America are other fresh-water animals, including pipid frogs and pelomedusid turtles, and also certain fresh-water mollusks (Pilsbry 1911; Pilsbry and Bequaert 1927, pp. 598–601), which have African relationships. Most of these animals, except the pelomedusids, have not been found fossil outside their present ranges. These animals do not answer the question of Mesozoic dispersal routes, but they suggest where the answer will come from. If so many fresh-water animals crossed a land connection between Africa and South America, still more terrestrial animals should have done so. (It is not possible to have continuity of fresh water without continuity of land too, and land animals disperse more rapidly across land connections than fresh-water animals do; if proof of this is needed, the recent movements of animals across Central America prove it.) The terrestrial groups might have changed their distributions over the world more rapidly and might no longer show old relationships which the fresh-water groups preserve, but the fossil record should show the movement of land animals between Africa and South America at the critical time. If, as seems likely, the time of dispersal of existing African-South American fresh-water animals was the Cretaceous, Cretaceous land animals should show particularly strong African-South American relationships. They do not do so. The Cretaceous reptiles of South America, for example, are at least as closely related to North American as to African ones, and there seems to be no reason why those that reached South America at that time could not have come via North America (Colbert and Romer, in Mayr *et al.* 1952, pp. 246 and 250). So far as I can see, there is at present no good evidence of an African-South American connection in the Cretaceous, although the matter is not finally settled. Whether there was a connection earlier in the Mesozoic will be considered below, under continental drift.

As to a land connection between South America and Australia, by way of the Antarctic Continent or New Zealand (see Frontispiece), there was none in the Tertiary (p. 594), and vertebrates seem to show none at any time (Simpson 1940). No land or strictly fresh-water vertebrates seem to have crossed an antarctic bridge. The fresh-water fishes of South America and Australia are wholly different, except for a few (galaxiids etc.) which have probably dispersed by sea. Leptodactylid and hylid frogs now occur chiefly in South America and Australia, but they are not on New Zealand; they have probably dispersed through the main part of the world (Chapter 3).

Meiolaniid and chelyid turtles, if they reached Australia from South America, may have done so by sea. The "ratite" birds of South America, Australia, and New Zealand (and Africa and Madagascar) are probably convergent rather than related groups (see list of bird families). Marsupials, once thought to be the best evidence of an antarctic bridge, are now thought to have reached South America and Australia separately and to have evolved parallel specializations there (p. 344). However, some plants and insects, including some carabid beetles with which I am personally familiar, do show what seem to be direct relationships between southern South America, southern Australia, and often New Zealand too. Why should the vertebrates tell one story, and the plants and insects another? There seem to be three possible explanations.

First, the plants and insects may have dispersed before the vertebrates did, by a route (not necessarily a bridge) which had ceased to be available when existing vertebrates came along. Something like this may have happened even if no bridge ever existed. Insects may have dispersed across antarctic water gaps when the climate was relatively mild, and the climate may have worsened in time to block modern land birds, which have probably dispersed recently, and which show no sign of an exchange between South America and Australia (Chapter 5, p. 257). But explanations based on time do not seem to fit some other cases. Some of the plants concerned (*e.g.*, the southern beech, *Nothophagus*) are not very ancient. There is no reason to think they are older in South America than, say, some frogs and fresh-water fishes.

Second, the different distributions of different organisms across the antarctic may be due to their different powers of dispersal. Some plants and insects cross salt water more easily than terrestrial vertebrates do, as their occurrence on remote oceanic islands proves, and they may have crossed water gaps between South America and Australia which the vertebrates could not cross. Some of them may have been carried across by wind (Chapter 1, p. 20).

The third possible explanation of different antarctic distributions is that different organisms may have dispersed in the same ways but behaved differently afterward. Some plants and insects persist for long times in small, cold areas; some of them have evidently done so on Tierra del Fuego and the southern tip of South America. Vertebrates have not persisted there. *There are no old, relict land or fresh-water vertebrates confined to the southern tip of South America.* The oldest existing South American mammals, the marsupials and edentates, do not even reach the southern tip of the continent now (Chapter 6, p. 324), although they have been in southern South America since early in the Tertiary. The same general situation exists in Australia. Certain plants and insects have evidently been isolated on Tasmania and the southern edge of Australia for a long time, but there are no old, relict vertebrates confined there—the Marsupial Wolf and the Marsupial Devil are much too recent to count in this connection. It seems that, in such cold and remote places as these, vertebrates do not tell old stories as well as some plants and invertebrates do. Mammals do show one thing about a South American-Australian bridge: there was none in the Tertiary; Australia had no land connection with any continent then (p. 593). But it seems at least possible that there was a connection between southern South America and Australia before the Tertiary; that some vertebrates as well as plants and invertebrates used it; but that only the plants and invertebrates have been able to survive upon the remnants of it.

I do not know whether there have been land connections across the antarctic or not, but I can say what I think the probabilities are. Animals have probably dispersed complexly in that part of the world, as elsewhere. Some "antarctic" plants and invertebrates, as well as most South American-Australian vertebrates, have probably dispersed through the main part of the world and then died back into their discontinuous southern ranges. Some other plants and invertebrates have probably dispersed across the antarctic when the climate was milder than now, but I think they probably did so across water gaps and not by continuous land. I doubt if any strictly terrestrial or fresh-water vertebrates made the crossing, but some salt-tolerant fishes and more or less aquatic turtles probably did. I think that, of existing, so-called "antarctic" plants and insects, some have dispersed in one way and some in the other, and I have a rule of thumb for guessing which have done which. Insects that occur in South Africa as well as southern South America and Australia, but not on New Zealand, I guess have dispersed through the main part of the world and died back into their present distributions. Insects that now occur on New Zealand as well as southern South America and Australia, but not in South Africa, I guess have dispersed directly across southern water gaps—no strictly terrestrial or fresh-water vertebrates have this distribution; none of the South American-Australian groups is represented on New Zealand. Insects that do not fit either of these patterns are more doubtful.

The final answer to the question of antarctic land bridges will probably come from the Antarctic Continent. Fossil plants and coal have already been found there. Additional fossils will probably show, eventually, whether Antarctica had a flora and fauna that may have been received over water, or whether, just possibly, it required land.

Drifting continents

Whether the continents have always stayed in their places or have moved long distances in relation to each other and to the axis of the world is a question I cannot answer but shall try to clarify. I have been interested in this subject for many years and have taken part in two symposiums which concerned it, a formal one at the meeting of the American Association for the Advancement of Science in New York, December 28 and 29, 1949 (Mayr *et al.* 1952), and an informal one at Harvard, December 13, 1951, organized by Dr. Llewellyn Price, so that what I say is at least not a hasty judgment.

The particular theory of continental drift which concerns zoogeographers is Wegener's, or Taylor's and Wegener's, modified by du Toit and others. The reference given above (Mayr *et al.* 1952) gives an entry to the literature of the subject, and a few additional leading references are listed in the chapter bibliography. I shall not review the history of the theory. The theory as it now stands is that the continents once formed a single land mass; that it broke apart; and that the pieces of it, the present continents, drifted to their present positions (Fig. 77); and that this happened shortly before and during the Mesozoic. Drift is sometimes supposed to have continued into the Tertiary, but I can see no sign of it then (p. 594).

Continental drift is sometimes dismissed with the phrase, "It makes more problems than it solves." This may be true, but it is not the point. The question is, did drift occur?

Nothing in zoogeography has brought forth more arguments or more demands for an open mind—the other man's mind—than the idea of continental drift. I have tried to keep my mind open on this subject and have made a new beginning by trying once more (as I have done before) to see if I can find any real signs of drift in the present distribution of animals. I can find none. So far as I can see, animal distribution now is fundamentally a product of movement of animals, not movement of land. Different kinds of animals have spread and receded in different ways so as to form an endless variety of different patterns which seem to be determined by the animals' evolutionary ages, powers of dispersal, relations to climate, relations to each other, etc.; and the pattern of animal distribution as a whole fits the present pattern of land, climates, and barriers of the world, not the Wegenerian pattern. Special cases emphasized by Wegenerians seem to me to be simply exceptional cases in which animal movements happen to have produced patterns which resemble Wegenerian ones. For example, special cases of bird migration happen to fit the idea of continental drift, but migration as a whole cannot be explained by drift, and

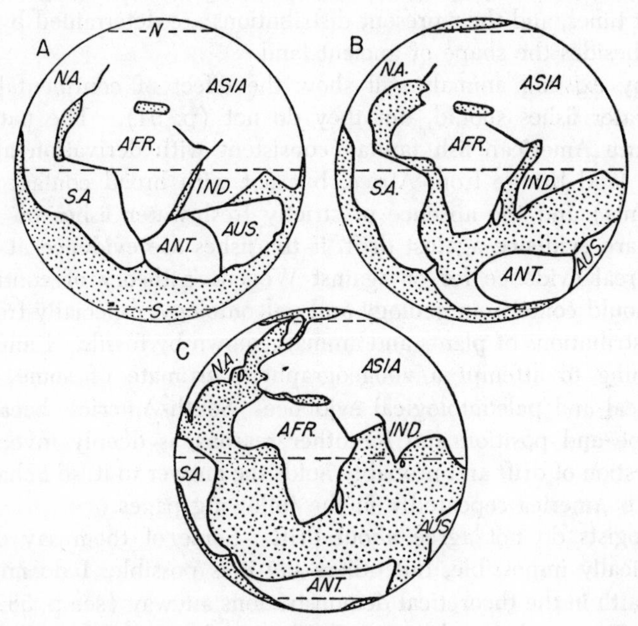

Fig. 77. Wegener's figures of continental drift, simplified and transferred to orthographic hemispheres centered on Africa at the equator.

drift does not explain even the exceptional cases very well (Chapter 5, p. 243, footnote).

Although I have made this trial as fairly as I could, I think the results were to be expected. Even if drift did occur, it was probably long ago, and existing distributions of animals and plants probably would not show it; they are probably too recent. Nevertheless Wegenerians try to use them. Their method is to try to fit existing plant and animal distributions to Wegener's hypothetical patterns of land, and when a fit is found, to claim it as evidence that the ancient land existed. For example, if a group of animals now occurs only in Africa and Brazil, it is said to date from and be evidence of a hypothetical African-Brazilian continent. This method depends on an assumption which Wegenerians usually do not put into words: that animal distributions are more permanent than land, that animals move less than continents. For an example of this sort of zoogeography, see Jeannel's (1942) book on the origin of terrestrial faunas, which I (1949) have reviewed. This method dictates its own conclusion. If animals have not moved, the continents must have done so. But animals do move. They spread and recede sometimes for long distances in short times, and their present distributions are determined by many things besides the shape of ancient land.

If any existing animals still show the effect of continental drift, fresh-water fishes should, but they do not (p. 94). The nature of the South American fish fauna (consistent with derivation along a narrow land bridge from Africa, but not with broad contact of the continents) and the absence of strictly fresh-water fishes on Madagascar are evidence against drift, if the fishes are evidence at all.

The real evidence for or against Wegener's theory of continental drift should come from geology and paleontology, especially from the past distributions of plants and animals, shown by fossils. I am therefore going to attempt a zoogeographer's estimate of some of the geological and paleontological evidence. South America, because of its shape and position and for other reasons, is deeply involved in the question of drift and probably holds the answer to it, so I shall turn to South America repeatedly in the following pages.

Geologists do not agree about drift. Some of them say drift is theoretically impossible, but others say it is possible; I do not have much faith in the theoretical demonstrations anyway (see p. 582, footnote). Too much is unknown. Geologists do not yet know how the earth originated, or how the continents originated, or why continents exist at all—why the land is not just spread out over the bottom of the oceans. With all this unknown, how can it be decided whether drift is theoretically possible or not?

It does seem sure that the continents are masses of relatively light material (sial) in approximate isostatic balance (floating) in a heavier substratum (sima). Moreover, the continents do not seem to have been pushed through the substratum by any external force. They have not plowed up the substratum, and they have not formed a symmetrical pattern (*e.g.*, a ring of land around the equator) as they would probably have done if centrifugal force, for example, had arranged them on the revolving earth. If the continents have drifted, they have probably been moved by forces inside the earth, by movements of the substratum. Such forces may exist. When mountains are folded, parts of continents are moved together for scores, perhaps hundreds, of miles. The movement is occasional, local, and occurs in a rather short time. It seems at least possible that whole continents may be moved for greater distances, occasionally and in a rather short time. But enough of this. All I want to conclude about geological theory is that it has not settled the matter of continental drift.

One of the first arguments for drift was the shapes of the continents, especially the matching shapes of Africa and South America, which look as if they might once have fitted together to form one piece of land. There are really three zigzag lines which correspond remarkably: the west edge of Africa, the east edge of America, and the Mid-Atlantic Ridge (Frontispiece). It seems to me that the correspondence of these three lines probably means something, but it does not necessarily mean drift in Wegener's sense. Long before Taylor and Wegener, George Darwin (Charles Darwin's son) and Fisher suggested that the moon broke away from the earth, from what is now the Pacific side, taking with it much of the earth's crust; and Fisher suggested that the rest of the crust then broke into pieces which were pulled apart as the earth readjusted its shape, and which became the existing continents. This was continental drift, but not Wegenerian drift. It happened, if it did happen, very early in the earth's history, long before there was life and before there was even liquid water. There may be still other possible explanations of the matching shapes of continents. Some persons think it is just chance.

South Africa and southern South America not only have matching coast lines but are similar stratigraphically, as if they had a common history before and during the Mesozoic. The degree and significance of the similarity are debated (*e.g.*, by Caster and Dunbar, in Mayr *et al.* 1952). It may be no more than can be accounted for by a broad similarity of climate, if both areas have always been (as now) in the same climatic zone.

The stratigraphic similarity includes occurrence of similar, shallow-sea, invertebrate faunas fossil in corresponding strata in South Africa and South America. The significance of this, too, has been debated. The faunas lived primarily in shallow, coastal seas, but their larvae may have dispersed across deep water: existing shallow-water invertebrates have crossed fairly wide gaps of deep water in the Pacific (Dunbar, in Mayr *et al.* 1952, p. 154); the Main Equatorial Current of the Atlantic may carry larvae from Africa to South America now (Chapter 1, p. 17); and the range of dispersal of such animals, which depends on the length of their larval periods, may have been greater in the past. The similarity of South African and South American shallow-water invertebrates in the past seems to suggest continuity of water and similarity of climate, perhaps zonation of climate, more than proximity of the continents.

Probably the strongest geological argument for continental drift is the glaciation which occurred in what are now the southern hemisphere and parts of the tropics about the beginning of the Permian, the period preceding the Mesozoic. Early Permian (formerly considered Permo-Carboniferous) glacial deposits occur in South Africa north into the tropics (signs of not one sheet but several, with separate centers of movement), on southern Madagascar, in southern South America to well within the tropics, in southern Australia, and (north of the equator) in India; and there seems to have been comparatively little glaciation elsewhere at this time. The Wegenerian explanation is that the glaciated areas formed a single continent, Gondwana, which was then at or near the south pole. However, other explanations are possible. Coleman (1926), after a comparison of recent and ancient ice ages, did not feel that continental drift was necessary to explain the ancient ones. He did not, however, decide what the causes really were, but suggested a somewhat indefinite "conjunction" of astronomic, geologic, and atmospheric conditions. Whether such an explanation is really satisfactory is still a question. The answer to it may come from better understanding of Pleistocene glaciation. There is some reason to think that climate, though always zoned to some extent, may have been abnormally strongly zoned during the Pleistocene (Barghoorn, in Shapley *et al.* 1953, p. 248). If so, reduction of the sun's heat might sometimes cause a more general cooling of the whole of the earth's surface. Even in the Pleistocene, the ice sheets in Europe and North America centered 2000 or 2500 miles from the north pole, and the southern edge of the ice in eastern North America was closer to the equator than to the pole. If cooling were less zonal, ice sheets might be formed in any climatic zone, wherever land was high enough and precipitation heavy enough.

Associated with all the main Permian ice sheets was a distinctive flora of *Glossopteris* and related plants. If these plants had limited powers of dispersal, their occurrence would be strong evidence that the glaciated areas were connected. However, the plants' seeds or spores have not been identified with certainty. Double-winged spores suitable for wind transport are associated with *Glossopteris* in some places, and if they are the spores of the plant, the latter may have dispersed over great distances.

If Africa and South America were directly connected in the Mesozoic, fossil vertebrates should show it, but they are not yet well known, and what is known is difficult to interpret. In the Cretaceous, South American vertebrates were at least as much like North American as like African ones (p. 603). In the Jurassic, too few are known to be significant. Still earlier, in the Triassic, however, they are somewhat better known. Colbert and Romer (in Mayr *et al.* 1952) make interesting analyses of what is known of the Triassic amphibians and reptiles of Africa and South America. They find that (1) during the Triassic as a whole most families of amphibians and reptiles that occurred in Africa and South America occurred also in other parts of the world, but (2) faunal relationships between Africa and South America were a little closer (Colbert), and in the mid-Triassic apparently much closer (Romer) than those between South and North America. The last finding, however, is based on scanty evidence. The mid-Triassic land vertebrates of North America are known principally from fossil *footprints* in a single formation.

As the facts now stand, it is a reasonable hypothesis that mid-Triassic vertebrates reached South America from Africa by a direct connection, but more facts are needed. The facts should be analyzed on a world-wide basis. At some time during the Triassic, land vertebrates apparently had access to all the continents. Brachyopids were cosmopolitan; Triassic dicynodonts and primitive dinosaurs are known from Africa, Europe, Asia, and North and South America; procolophonids and pseudosuchians, from Africa, Europe, and North and South America; rhynchosaurs, from Africa, Europe, Asia, and South America; and only cynodonts were confined (so far as is known) to Africa and South America in the Triassic (Colbert, in Mayr *et al.* 1952, p. 240, Table 2). Did this general accessibility of the continents exist in the mid-Triassic? Was there any land connection at all between North America and the rest of the world then? If so, why were the mid-Triassic families of Africa and South America not represented in North America—if they were not? Speaking as a zoogeographer, I should say that mere distance is not enough of an explanation. All ordinary land vertebrates can disperse over the world, even by indirect routes, in a short time, unless something stops them. Was North America in the mid-Triassic cut off from other parts of the world ecologically or by zonation of climate, if not by water? All this—the state of the whole world, and not just the relation of South America to Africa and North America—should be carefully considered before the significance of African-South American relationships can be safely interpreted.

The distribution of vertebrates before the Mesozoic still may give hints of the state of the world. Olson (1955) thinks that reptiles were able to spread, with continuity, over both Old and New Worlds not long before the Permian. Early in the Permian, however, exchange apparently stopped, and unrelated groups of reptiles evolved separate but parallel adaptive types in the Old and New Worlds. If this is correct (Olson says his interpretations are based on "certain assumptions"), it suggests a pre-Mesozoic pattern of separate continents with intermittent connections rather than Wegenerian continuity of land. This is at least the sort of evidence—comparison of faunas rather than stress of single cases—to be looked for in the remote past.

Mesosaurs occurred in South Africa and South America about the beginning of the Permian and are not known anywhere else at any time. They were small (about a yard long), active, aquatic reptiles, which lived in fresh water or perhaps sometimes in shallow seas (Caster, in Mayr *et al.* 1952, p. 130). It has sometimes been assumed that they could not cross salt water, but, comparing them with existing crocodiles, I think they may have been able to enter the sea and that no limit can be set to the width of sea they may have crossed. If mesosaurs crossed between Africa and South America through fresh water, on a land connection, land animals must have crossed too, and it will be they, when and if they are found fossil, which will prove the connection, if one existed. In the meantime, mesosaurs will probably make more arguments than they settle.

If, by way of summary, the motion-picture history of the world were begun in the Permian, more than 200,000,000 years ago, and should show only what is actually known, there would at first be no coherent picture, just small flashes on a wide screen. The first flashes would show continental ice sheets, unfamiliar plants, ancient orders of amphibians, and early reptiles. They would be seen momentarily here and there, but they would give no clear picture of the world. Later there would be no more ice sheets, and the plants and animals would change again and again. During the Permian and early Mesozoic, some of the flashes would suggest that Africa and South America had more than now in common, but that picture would be gone before the meaning of it was clear. Dinosaurs would appear in the Triassic and would soon be on all the continents, but the picture would not show how they got there. Finally, but not until the Cretaceous, the flashes would begin to suggest the world as a whole, and the main pattern of land and climates would appear, still somewhat fragmentary in the picture, but, as a whole, apparently the same as now. On this world, modern plants would suddenly be present everywhere; before the end of the Cretaceous, dinosaurs would move among maples, oaks, and beeches in the north-temperate zone. The herbivorous dinosaurs would become zoned with climate. Dinosaurs would still be present when this reel ended. They would all have disappeared and mammals would be rising in their place when the next reel, of the Tertiary, began.

Summary and requirements

What zoogeography (and some other evidence) tells of the last 200,000,000 years has already been summarized in three reels of an imaginary motion picture. Now I want to review the evidence in another way, and to outline a history of the world which seems to satisfy the requirements of zoogeography.

From the present back 60,000,000 years or so, through the Tertiary, animal distributions and other evidence show that, although details have changed, there has been no great change in the main pattern of continents and climatic zones.

Before the Tertiary, doubts begin to appear, and the doubts increase with remoteness of time. In the Cretaceous, vertebrate distributions still suggest that land and climates were arranged as now, but before that there are hints that the southern continents may have been differently arranged. This is suggested by the matching shapes of Africa and South America; by the similar stratigraphy of South Africa and South America, and their similar, fossil, coastal invertebrate faunas; by the Permian glaciations of the southern continents and India, and the associated *Glossopteris* flora; and by the similarity of South African and South American vertebrates early in the Mesozoic. Each item of this evidence can be explained away or questioned, but altogether it raises a real doubt about the arrangement of land in the southern half of the world.

But this doubt should not be overemphasized. Consider what the record of vertebrates might have shown. It might have shown that land was once divided so completely that (for example) dinosaurs evolved on one continent while mammals evolved on another. It might have shown, by fossils on Antarctica or New Zealand, that dinosaurs did pass that way. It might have shown, by identities of whole terrestrial faunas, that Africa and South America surely were in contact. But it does not show this kind of thing, at least not yet.

As far back as the record goes, in spite of all the doubts, the distribution of animals can be reconciled with a pattern of land and climates like the present one. In fact zonation of climate on the present pattern may be the best explanation of some of the situations stressed by Wegenerians, including the similar stratigraphy and shared coastal marine faunas of South Africa and South America long ago, and relationships between tropical animals on different continents at any time.

Of course the world has changed in detail. Mesozoic swamps have risen to form high mountains. Shallow seas have drained to leave dry land, and vice versa. Continents have changed their shapes and surfaces, and adjacent continents have been connected, disconnected, and reconnected. But these are minor details. The question is, was the main pattern of land and climates long ago essentially the same as now, or very different? I think, because of what the record does not show as well as because of what it shows, that probably it was the same.

I shall end this chapter with a hypothetical history of the continents which, whether true or not, seems to satisfy the requirements of animal distribution.

I assume that the moon was thrown off by the earth, from the Pacific hemisphere, very early in the earth's history (this is apparently still a respectable but not fashionable theory of the moon's origin), and that the continents were formed then, by fracture and movement in a short time (the Darwin-Fisher hypothesis), and that they have kept their identities and approximate positions since then.

Each continent (I assume) somehow affected the distribution of heat below it, so that convection currents in the substratum maintained it, continually pushing it together and keeping it from spreading out over the ocean bottom. Intermittent convection currents also pulled down geosynclines, let them fill with sediment, and then released them to rise again as mountains, often with volcanic activity; and this happened not only on, but between adjacent continents, so that all the main continents were intermittently connected in a definite pattern, mostly across the narrowest water gaps. Any continent might be cut off for a time, and the animals on it might evolve independently—but only for a time. Two continents might be cut off at the same time, and even if they were not connected together, they might preserve similar faunas while the fauna of the rest of the world changed. Something like this happened in South America and Australia in the Tertiary. In this case the faunas of the two continents were very different as wholes, although some of their components (for example, some frogs, marsupials) were similar; but in earlier cases two temporarily isolated continents may have had more similar faunas.

The climate of my hypothetical world was, of course, determined primarily by the sun's heat. Climate was always zoned by the shape and motions of the earth; and it always varied locally with height of land, relation of land to ocean, and direction of wind. Both zonation and localization of climate always had some effect on animal distribution; restriction of some groups of animals to single climatic zones on different continents sometimes counterfeited the effect of land bridges.

The exact amount of heat received from the sun was (I assume) determined by several things, including the heat of the sun itself and the composition and contents of the atmosphere. Reductions of temperature at different times may have had different causes and different results. The result may have been sometimes a general cooling of the earth's surface, with the climatic zones relatively weakly differentiated and with ice sheets forming on high ground in all latitudes, and sometimes a more zonal cooling, with ice sheets only in the colder zones. The amount of differentiation of the climatic zones may also have varied with the making and breaking of land connections between continents, and with changes in ocean circulations.

On such a world, fixed in its main pattern of land and climates but changing endlessly in detail, all that is known of animal distribution can be placed satisfactorily. At least I think so now.

REFERENCES

Barghoorn, E. S. 1953. Evidence of climatic change in the geologic record of plant life. (In) Shapley *et al., Climatic change,* pp. 235–248.

Bogert, C. M. 1953. The Tuatara: why is it a lone survivor? *Sci. Monthly,* **76,** 163–170.

Chaney, R. W. 1940. Tertiary forests and continental history. *Bull. Geological Soc. America,* **51,** 469–488.

Colbert, E. H. 1945. *The dinosaur book.* American Mus. Nat. Hist., Man and Nature Pub., Handbook 14.

———. 1948. Evolution of the horned dinosaurs. *Evolution,* **2,** 145–163.

Colbert, E. H., R. B. Cowles, and C. M. Bogert. 1946. Temperature tolerances in the American Alligator . . . . *Bull. American Mus. Nat. Hist.,* **86,** 327–374.

Coleman, A. P. 1926. *Ice ages recent and ancient.* New York, Macmillan.

Darlington, P. J., Jr. 1949. Beetles and continents. *Quarterly Review Biol.,* **24,** 342–345.

Darwin, C. 1859, 1950. *On the origin of species* . . . . London, Murray; reprinted 1950, London, Watts and Co.

Deevey, E. S., Jr. 1949. Biogeography of the Pleistocene. *Bull. Geol. Soc. America,* **60,** 1315–1416.

Dillon, L. S. 1956. Wisconsin climate and life zones in North America. *Science,* **123,** 167–176.

Dunbar, C. O. 1949. *Historical geology.* New York, John Wiley.

du Toit, A. L., and F. R. Cowper Reed. 1927. *A geological comparison of South America with South Africa.* Carnegie Inst. Washington, Pub. 381.

Flint, R. F. 1947. *Glacial geology and the Pleistocene Epoch.* New York, John Wiley.

Holdhaus, K. 1954. *Die Spuren der Eiszeit in der Tierwelt Europas.* Abhandlungen Zool.-Botanischen Gesellschaft Wien, **18.**

Huene, F. von, and C. A. Matley. 1933. The Cretaceous Saurischia and Ornithischia of the Central Provinces of India. *Mem. Geological Survey India, Palaeontologia Indica,* New Ser., **21,** No. 1.

Jeannel, R. 1942. *La genèse des faunes terrestres* . . . . Paris, Presses Universitaires de France.

Kuenen, P. H. 1950. *Marine geology.* New York, John Wiley.

Libby, W. F. 1956. Radiocarbon dating. *American Scientist,* **44,** 98–112.

Lull, R. S. 1910. Dinosaurian distribution. *American J. Sci.* (4), **29,** 1–39.

Matthew, W. D. 1915, 1939. Climate and evolution. *Ann. New York Acad. Sci.,* **24,** 171–318; reprinted (1939) as *Special Pub. New York Acad. Sci.,* **1.**

Mayr, E., *et al.* 1952. The problem of land connections across the South Atlantic, with special reference to the Mesozoic. *Bull. American Mus. Nat. Hist.,* **99,** 79–258.

Olson, E. C. 1955. Parallelism in the evolution of the Permian reptilian faunas of the Old and New Worlds. *Fieldiana* (Chicago Nat. Hist. Mus.), *Zool.,* **37,** 385–401.

Pilsbry, H. A. 1911. Non-marine Mollusca of Patagonia. *Rep. Princeton U. Exp. Patagonia, 1896–1899,* **3,** Part 5, 513–633.

Pilsbry, H. A., and J. Bequaert. 1927. The aquatic mollusks of the Belgian Congo . . . . *Bull. American Mus. Nat. Hist.,* **53,** 69–602.

Romer, A. S. 1945. *Vertebrate paleontology.* Chicago, U. of Chicago Press.

Schmidt, K. P. 1938. Herpetological evidence for the postglacial eastward extension of the steppe in North America. *Ecology,* **19,** 396–407.

Shapley, H. *et al.* 1953. *Climatic change* . . . . Cambridge, Mass., Harvard U. Press.

Simpson, G. G. 1940. Antarctica as a faunal migration route. *Proc. Sixth Pacific Sci. Congress,* **2,** 755–768.

Thienemann, A. 1950. Verbreitungsgeschichte der Süsswassertierwelt Europas. *Die Binnengewasser* (Stuttgart), **18.**

Umbgrove, J. H. F. 1949. *Structural history of the East Indies.* Cambridge, England, Cambridge U. Press.

Wegener, A. 1915. *Die Enstehung der Kontinente und Ozeane.* Brunswick, Friedr. Vieweg und Sohn.

———. 1924. *The origin of continents and oceans* [translated from 3rd German ed.]. London, Methuen.

Zeuner, F. E. 1950. *Dating the past* . . . . 2nd ed. London, Methuen.

chapter 11

Principles of zoogeography; geographical history of man

In these final pages I want to consider the principles of zoogeography and the geographical history of man.

Principles of zoogeography

Working principles of zoogeography are set forth in Chapter 1. They are premises and rules of procedure. They are listed again here to show that the premises are not simply carried over as conclusions. The working principles listed in Chapter 1 are:

1. To formulate working principles before work is begun.
2. To work with facts, so far as possible.
3. To define and limit both the work to be done and the facts to be worked with.
4. To present the facts fully and fairly.
5. To treat animals as living things which form populations capable of both spreading and receding.
6. To understand and use fairly clues to places of origin and directions of dispersal.
7. To try hypotheses when facts fail, remembering that they are hypotheses.

Derived principles of zoogeography are important general facts and probabilities found out by study of animal distribution. They are generalizations about the main pattern of animal distribution and about important conditions, factors, and processes which the pattern reveals.

The main pattern: four questions answered

The first fundamental fact of zoogeography is that animals are not distributed at random but in a definite, world-wide pattern. In the preface of this book four questions have been posed about the pattern, and answers to the questions have been developed at length in later chapters. Now I shall try to summarize the answers.

What is the main pattern of animal distribution? In general, it is a combination of several subpatterns: of limitation northward (and southward) and outward from the continents, and of further limitation and geographic complementarity of different dominant groups; of zonation with climate; of geographical radiation of dominant groups especially from the main Old World tropics; of differentiation of faunas in different main regions; and of concentration in favorable areas within the regions and limitation and often transition in unfavorable marginal areas. These subpatterns differ in detail in the different classes of vertebrates, from fishes to birds and mammals, according to the animals' requirements and powers of dispersal, and the different subpatterns are very complexly interrelated. The complexity of the whole is beyond description. However, all the subpatterns together make a simple, average pattern: concentration of the largest, most diverse, least-limited faunas in the main tropical regions of the Old World; limitation caused by climate north of the tropics; and limitation and differentiation caused by barriers in South America and Australia.

How has the pattern been formed? Apparently, by spread of successive dominant groups from the Old World tropics over much or all of the world, followed by zonation and differentiation according to climate and ocean barriers, and by retreat and replacement of old groups as new ones spread. This seems to have been the predominant pattern of movement in a diversity of other movements.

Why has the pattern been formed? Apparently, because evolution has tended to produce the most dominant animals in the largest and most favorable areas, which for most vertebrates are in the main regions of the Old World tropics. This seems to have been the predominant geographical pattern of evolution in a diversity of other evolutionary patterns.

What does animal distribution tell about ancient lands and climates? The existing distribution of vertebrates and the fossil record as far back as it is clear (geographically) tell of many small changes in the world but no great ones. The small changes have included making and breaking of land connections between Asia and North America, between North and South America, and between continents and many islands; changes in intensity and detail but not in orientation of the climatic zones; and many other small changes in the shape and surface of the land. But as far back as can be seen clearly, the main pattern of continents and climates seems to have been the same as now. It was the same through the Tertiary. It was probably the same in the Cretaceous. Before that it may have been the same, but we do not yet know for sure.

These answers have been reached by analyzing, separately, the distributions and apparent histories of the five main classes of vertebrates and finding their common pattern. It was not until I had worked over several classes separately that I saw there was a common pattern which strengthened the probabilities in the separate cases. The pattern has then been re-examined in its relation to the pattern of land and barriers of the world, and to area, climate, and evolution. The result is the system of fact, probability, and hypothesis here summarized as answers to four questions.

The complexity of animal distribution is a fact which underlies all the other facts of distribution. I have already nearly exhausted my vocabulary in stressing this. I can only repeat that existing situations are almost endlessly complex, and that they are evidently the products of exceedingly complex movements and countermovements, spreadings, recedings, disappearances, and replacements of a diversity of animals over the whole of the very complex surface of the world. The fossil records of the best-known animals hint at this complexity but cannot fully reproduce it. The whole is a kaleidoscope which cannot be seen in full detail, although main patterns can be seen.

The nature of barriers can be reduced to general principles. The present distributions and the geographical histories of vertebrates show that barriers are important but not absolute. Barriers are not impassable, just difficult to pass, the difficulty varying with the kind and width of barriers and the kinds of animals concerned. Zonal climate is one of the two most important barrier systems or limiting factors in animal distribution, and it has limited the distribution of many groups of animals for long periods. But nevertheless a succession of dominant animals of every class of vertebrates has dispersed across the climatic zones. Other ecological barriers have been less important than zonal climate in determining the main pattern of animal distribution. Oceans form the second of the two main barrier systems of the world. Their importance is strongly reflected in the main distribution pattern. But nevertheless some representatives of every class of vertebrates have crossed some widths of salt water occasionally. Mountains and deserts have been much less important, usually temporary barriers. Nothing, neither zonal climate nor ocean barriers nor anything else, has prevented the dispersal of appropriate kinds of land and fresh-water animals over all or most of the world.

Another general fact established by zoogeography is the fundamental importance of competition, broadly defined as the struggle for places in the world. Competition is shown or implied by many special situations and events of animal distribution, but especially by the distribution of past and present faunas described in Chapter 9 (pp. 552–553). Something holds the size and composition of faunas in all parts of the world and at all times within certain limits, in spite of continual changes and successions in separate phylogenetic groups. Only competition can do this, and to do it, competition must be a fundamental, ever-present force.

Connected with the fact of competition is the fact of dominance. Dominant animals are, by definition, the most successful ones. The reasons for dominance of particular animals may or may not be visible, but the fact of dominance is plain. It is a fact that in each main class of vertebrates there has been a succession of conspicuously successful, dominant forms which have multiplied, diversified, spread, and replaced other forms. Connected with this is the fact that it is usually dominant animals which spread, and that they usually move to gain advantages rather than to escape disadvantages. Less dominant animals do not usually escape by moving away from dominant competitors but are usually overrun and destroyed.

That the most dominant animals tend to evolve in and disperse from the largest and most favorable areas has been presented as a probability rather than a fact, but I think that it is a fact, necessary to explain the main pattern of animal distribution. The most important single thing in the making of the main pattern has been—not climate, or barriers, or ecological details, or ancient shapes of land, all of which are important but which have sometimes been overstressed—but evolution. Evolution with respect to area, climate, and barriers has made the broadest and deepest features of the pattern of animal distribution.

This seems to me to complete the list of important principles of zoogeography. The principles could be subdivided and minor ones could be added indefinitely, but I do not want to do it. These principles—of the existence, nature, history, causes, and significance of the pattern; of complexity; of barriers that are more or less effective but not absolute; of competition and dominance and replacement; and of the fundamental importance of evolution with respect to area, climate, and barriers—are enough.

Geographical history of man

The geographical history of man is a part of zoogeography, and although I have no special knowledge of man (I am not an anthropologist or ethnologist), I do have a knowledge of zoogeography which allows me to look at man in a special way. I am therefore going to try to trace man's geographical history. I shall try to make a coherent story of it and to tell it simply. However, I shall at the same time try to be accurate and to distinguish what is known, what is probable, and what can only be guessed at. I shall try to tell the story objectively, as if it were the story of a frog or a monkey, but I cannot always quite forget that I am talking about my own ancestors and in a way about my own history. I find it exciting in spite of myself.

Although I shall try to tell it simply, man's geographical history has not been simple. The purely geographical part is complex enough, and it is closely interwoven with the evolution of man himself and of his behavior: where man's ancestors and man have been and how they have moved over the world have depended partly on what they were like and what they were doing from time to time. I have experimented with telling the non-geographical and geographical histories entirely separately, but the results are not good. I have therefore been forced to tell the story as a whole and to separate the non-geographical and geographical parts only in a summary. The non-geographical items contribute to the geographical probabilities. And also, I think, the known geographical details and application of zoogeographical principles add something to the understanding of man's history as a whole.

Although I have tried to distinguish what is known, what is probable, and what can only be guessed at, this is not a simple thing to do. Nearly every detail of man's origin, evolution, and early history is questioned by someone or other. The literature on the subject is immense, beyond my power to read and digest. Of course, as a biologist, I know something in general of man's evolution and place in the world. To bring myself up to date I have read only three books. I have read Coon's (1954) *The Story of Man* and have been fascinated by it, although I know that in books like it there must inevitably be some errors and some failures to distinguish between fact and probability. And I have read Le Gros Clark's (1949) *History of the Primates* and (1955) *The Fossil Evidence for Human Evolution*, which are careful statements of facts and probabilities at a more technical level. In order to keep some sort of balance, I have read also

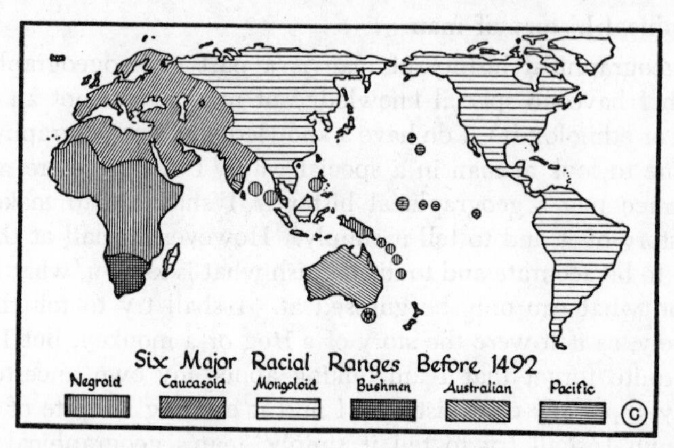

Fig. 78. Distribution of major races of man. From Coon 1954, *The Story of Man*, p. 191. © Copyright 1954 by Carleton S. Coon. Used by permission of Professor Coon and Alfred A. Knopf, Inc.

one paper critical of Le Gros Clark. It was selected almost at random and happened to be by Zuckerman (1954).

"Man" and "human" are best applied only to existing man and his immediate relatives and ancestors known to have been intelligent enough to make tools (see Le Gros Clark 1955, p. 11). Earlier or less-advanced hominids may be called "prehuman." But this distinction may be more apparent than real, as noted below. I shall also use the term "pre-man" for man's ancestors between the ape and human levels.

Le Gros Clark discusses the classification of man and his ancestors and presents a comparatively simple, sensible classification of genera and species of man, with a table (1955, p. 50) of their probable chronology during the Pleistocene. All existing men are races of one species.

The present distribution of races of man is complex (Fig. 78). It was once customary to recognize three main races, black, white, and yellow, but this was superficial and too simple. Coon (1954, pp. 190ff., with map on p. 191) considers that there are now six major races which, "before 1492," were distributed as follows. Negroids were in tropical Africa and smaller areas widely scattered across the Old World tropics: a part of southern India, the Andaman Islands, small parts of the Philippines, New Guinea and islands east to Fiji and southeast to New Caledonia, and Tasmania. Whether these scattered Negroid populations were relicts of one original, widely dispersed stock, or whether they had different origins is uncertain. Caucasoids were mainly north of the tropics in northern Africa, Europe, and western Asia, southeast into the tropics in India. Mongoloids were in northern and eastern Asia south onto the Sunda Islands, and in North and South America. These were (and are) the three main races in point of numbers and extent of distribution. Less numerous races were Bushmen in South Africa, Australians in Australia, and a Pacific Race on the more remote islands of the Pacific from Hawaii to New Zealand. These six races overlapped and mixed complexly. They can be subdivided into minor races, but that need not be considered here. I want to go on without delay to the geographical history of man.

The geographical history of man is a history of the evolutions and movements first of man's prehuman ancestors and then of genera, species, and races of man himself. Much of the earlier part of this history is still unknown. There is no unquestioned record of man or pre-man before the Pleistocene. *There is therefore a blank of probably several or many million years in the actual record of man's ancestors.* This does not weaken the evolutionary theory of man's origin, which rests on acceptance of organic evolution in general and on the evidence of comparative anatomy in particular, but it does make it difficult to fill in details. I shall begin by making some careful guesses about prehumans and about some of their activities that have geographical implications.

Prehumans evolved from a generalized anthropoid ape. This particular ape lived on the ground, but it had arboreal ancestors, and it inherited from them certain things (hands, eyes able to judge distance and see detail at close range) which preadapted it to use tools and evolve a more complex brain.

The evolution of ape toward man began with standing and walking and running on two legs. This involved not only primary changes in the legs themselves, but also, eventually, related changes from one end of the body to the other, from (among many other things) change in angle of attachment of skull to spine, to change of the "thumb" of the foot into a big toe. The more important skeletal modifications for "erect bipedalism" are listed by Le Gros Clark (1955, especially p. 110). They presumably required millions of years of evolution. Erect posture also freed the arms and hands for new uses.

Further evolution toward man was (I guess) partly correlated with prehuman behavior and the beginning of the use of tools.

Prehumans were probably omnivorous, but, in addition to other food, they may (I guess) have hunted animals from the beginning. They may have begun by eating whatever birds and smallish mammals they happened to catch (as baboons do now), and they may then gradually have evolved more systematic hunting habits. The earliest men whose habits we know were hunters, and they used tools (weapons) to kill game.

How use of tools began no one knows or can know, but reasonable guesses can be made about it. Even some monkeys and apes throw random objects at intruders. I guess that prehumans began in this way and gradually learned to hit with sticks and to throw stones both for fighting and for knocking down the birds and animals they hunted. This probably required some evolution of the hand and arm, previously adapted for grasping and pulling; and evolution (by selection) of a good hitting and throwing arm probably required time as well as use. Use of sticks and stones instead of teeth for fighting and killing probably allowed the teeth to begin to change (in general to weaken) from an ape-like toward a human condition. The nature of man's teeth early in the Pleistocene suggests that his ancestors had had some other means of fighting and killing for a considerable time. All this suggests that prehumans had a long apprenticeship in use of simple sticks and stones before man began to *make* tools. Man or pre-man may have made wooden tools, at least to the extent of breaking sticks into clubs of good length and balance, before beginning to make stone ones, although it is only the stone ones that are likely to be preserved. It is because of all this that I suspect that the distinction between man, the toolmaker, and pre-man, the tool-less, may be rather thin.

A long apprenticeship in use of sticks and stones may also have been necessary to begin the evolution of a brain which could conceive more complex tools.

An animal's brain is (in part) an organ for making simple images of the animal itself and things around it and relating them to each other. Even a simple brain not only reflects things that are happening but predicts simple things that are going to happen. When, for example, a fish rises to take a moving fly, its brain predicts where and when fish and fly will meet. The power of prediction of most animals, including most mammals and even most primates, is very limited. It is a great thing for an ape to predict that he can move a box, climb on it, and reach a bunch of bananas. Prehumans perhaps began at about this level, and rose above it when, and I think partly because, they began to use sticks and stones. To throw a stone and hit a moving animal probably requires a prediction a step above the ape-box-banana level. The next step may be to re-relate the predictions to what happens, and to make new and better predictions—to make trials and correct errors. I do not know to what extent apes can do this, but effective stone throwing probably requires more predicting and correcting than apes can do. When we pick up a handful of stones and throw them in succession, correcting toward a target, we may be doing one of the things that first began to raise the prehuman brain above the ape's level. These details are hypothetical, but the brain probably did evolve by steps something like this. Complexity of brain and complexity of behavior probably increased together, each dependent on the other. That complex behavior requires a complex brain is obvious. That the complex brain could not have evolved by itself, without complex use, seems equally certain, if not so obvious.

In the absence of pertinent fossils, it is impossible to know just when or how the different physical changes really occurred in prehuman evolution or how they were correlated with each other. Most of them probably occurred mainly before the Pleistocene, although some may have continued into that period. The evolution of the brain, or at least its increase in size as indicated by cranial capacity, apparently occurred mainly later. *Australopithecus* early in the Pleistocene still had a cranial capacity not much greater than that of an ape and only about half that of *Homo* (see Le Gros Clark 1955, pp. 119ff.). The direct line of evolution toward man may have passed through an australopithecine stage somewhat earlier than this, but nevertheless, unless the evidence is very misleading (which it may be), the brain increased in size very slowly during the prehuman stage of evolution, then rapidly in the early human stage (from australopithecine to *Homo*), and then perhaps stopped getting larger, for there seems to have been no appreciable increase of size for at least the last several tens of thousands of years. By the mid-Pleistocene the brain may have reached a size not only sufficient to the times but also sufficient for the more complex times that came later.

Man, then, is derived from an ape which came down from the trees, stood up and ran on two legs, and probably hunted game; and the evolution of structure and behavior from ape to early man probably required at least several million years. During much of this time, pre-man probably used natural sticks and stones with gradually increasing effectiveness, while he gradually perfected a hitting and throwing arm, used his teeth less and less for fighting and killing, and very gradually began to improve his brain.

Just when the evolution of ape toward man began—when the first "prehuman hominid" appeared—is of course not known. Le Gros Clark (1955, p. 171) thinks it need not have been before the early Pliocene, which would be something like 10 million years ago, but it may have been earlier. The prehuman stage of evolution probably ended—that is, man began to *make* tools—about the beginning of the Pleistocene. Small-brained *Australopithecus* may still have been using natural sticks and stones early in the Pleistocene. No made tools have been found with his remains, but many baboon skulls have been found there, the majority of them fractured in a consistent way, as if struck by well-aimed blows of an implement (Le Gros Clark 1955, p. 159). But crudely worked stones, proving the presence of man, do occur early in the Pleistocene and perhaps a little before it.

The place where prehumans evolved was somewhere in the main, warmer part of the Old World, in Africa and/or southern Eurasia. This is where man was when he and his tools first appear in the record. And no living or fossil apes or prehumans have been found in America or in the Australian Region, which probably rules out not only America and Australia but also the cooler northern part of Eurasia; for if apes or prehumans had occurred far northward in Eurasia, they would probably have reached North America.

Just where within the limits of Africa and southern Eurasia prehumans originated can only be inferred. They began their evolution by standing up and running on two legs, and this was evidently an adaptation to living in open country. Also they were probably incipient hunters. A good guess is that their place of origin was Africa, where there was probably much to be gained by running hunters: much open country to run in and much game to run after. The Pliocene mammals of Africa are still almost unknown (Hopwood 1954). If Africa is where prehumans were in the Pliocene, the lack of record of them is easily explained. But this is an incidental fact, hardly evidence.

I want to digress for a moment to ask why man's ancestors came down from the trees, left the forest, and began to pursue game in open country. It has been suggested that they were driven into the open by destruction of their forests (by a climatic change), but this seems to me unlikely. It seems to be a principle of zoogeography that movements of animals into new places are usually not escapes but spreadings of dominant groups to obtain new advantages. Prehumans probably entered open country to obtain something there, perhaps to obtain the abundant game that lived in the open. Prehumans must have had many reasons for running, but I think that in the course of time what they got by it was more important than what they escaped. If our ancestors had put safety first, they could have stayed in the trees. Some potential ancestors did so and are apes today. But from the very beginning the apes that became men probably ran after what they wanted more than away from what they feared. This is a gross oversimplification, but I think it is true, and I think it set the pattern for man's later evolution and geographical movements. The whole process has been one of gaining and utilizing successive advantages more than escaping from disasters, although escapes have sometimes been necessary too. I do not mean this philosophically, but as a plain matter of fact.

Wherever their place of origin, prehumans probably dispersed throughout the main, warmer part of the Old World, as so many other dominant groups of animals have done. Their dispersal was probably complex, as dispersals usually are, with successive spreadings, competitions, and withdrawals of successive, evolving genera and species. But we know nothing whatever about the details of this dispersal in the Pliocene except that it apparently did not extend to Australia or America and probably not to the colder northern part of Eurasia.

The complex process of evolution and dispersal of man's prehuman and almost human ancestors, which must have been going on in the Pliocene, although we have no record of it then, becomes vaguely visible early in the Pleistocene. The first actual traces of it are a few prehuman or human skulls, teeth, and bones and increasing numbers of crudely worked stones very widely scattered in the main, warm part of the Old World. The earliest of them may be late Pliocene but are more likely early Pleistocene. The worked stones show that man has existed—that is, that tools have been *made*—since the early Pleistocene if not a little before. The time since the beginning of toolmaking may be taken as something like 1,000,000 years, or 40,000 generations, supposing primitive man reproduced somewhat more rapidly than modern man. This is an arbitrary figure, but the length of time cannot be very much less than this and may have been more.

The earliest, most primitive man or almost-man who is reasonably well known is *Australopithecus* in the early Pleistocene of South Africa. Australopithecines probably stood more or less erect and had an essentially hominid dentition, but had a cranial capacity much less than that of modern man. Whether or not they had tools is uncertain. Le Gros Clark (1955, p. 160) tentatively considers *Australopithecus* prehuman but thinks it was structurally suitable to be man's direct ancestor. Australopithecines are known with certainty only from South Africa, but teeth like theirs have been found also in East Africa and Java. Australopithecines may therefore once have been widely spread in the warm part of the Old World, although this is not certain. Le Gros Clark suggests that *Australopithecus* in South Africa early in the Pleistocene may have been a slightly modified survivor of an ancestral stock from which man had already been derived in some other place.[1]

The next stage in human evolution that is reasonably well known is represented by *Pithecanthropus* (including *Sinanthropus*) in the early to middle Pleistocene in Java and China (see Le Gros Clark 1955). *Pithecanthropus* was primitive man. He apparently made tools of stone and perhaps of bone and he had fire. It is a good working hypothesis that he was directly ancestral to *Homo*, but (as Le Gros Clark emphasizes) this need not mean that the transition to *Homo* occurred in the Far East. Remains of *Pithecanthropus* are unknown elsewhere, but fossil remains of early men are so few that the negative evidence does not mean much. It is a good guess that men of the *Pithecanthropus* level of evolution were once widely distributed in the warmer part of the Old World.

The first remains that seem to represent *Homo* are from the Middle or early Upper Pleistocene (Le Gros Clark 1955, pp. 56 and 79–80). *Homo sapiens* may have existed then, although this is still doubtful. The earliest remains of *Homo* are not clearly distinguishable from *sapiens* but are too incomplete to be decisive. Le Gros Clark has reduced the described forms of fossil *Homo* to what he considers two

[1] Some persons, for example Zuckerman (1954), doubt that australopithecines represent a stage in evolution toward man. I am not qualified to decide the more obscure technical details in question, but after reading what Zuckerman says about, for example, australopithecine teeth and comparing it with what Le Gros Clark says, I am left with a strong feeling that Le Gros Clark has made a fair statement of the facts and that his conclusions are reasonable.

real species. *Homo sapiens* apparently appeared in the Middle Pleistocene as populations which, though probably not exactly like existing man, cannot be distinguished by satisfactory characters. Early *sapiens* is best known in Europe but was probably widely distributed in the main part of the Old World, and *sapiens* has continued as the main line of human evolution to the present. All existing men are, of course, *sapiens*. *Homo neanderthalensis* was apparently a specialized side line derived from *sapiens* (who was at first somewhat "neanderthaloid"), who co-existed with *sapiens* in Europe during part of the Pleistocene, and who then became extinct.

The geographical distribution of *Homo* even in the late Pleistocene is no more than suggested by his own remains, which have been found at this time, outside Europe and southwestern Asia, at only a few widely scattered localities, including southern Africa (Rhodesia) and Java. But the distribution of increasingly well-made stone tools shows that man was very widely and generally distributed in the warmer part of the Old World in the Middle and Upper Pleistocene.

Different tools or artifacts—things made by man—suggest the geographical distributions of different early men. Movius (1949, p. 409), for example, maps the distribution of two principal cultures in the Middle Pleistocene (Fig. 79). One culture, apparently the more advanced one, characterized in part by certain implements (hand axes etc.) which the other lacked, was spread all over Africa, southern Europe, and southwestern Asia to India. The other, apparently more primitive, culture was in the Far East, from southern China and northeastern India to Java. It has been suggested by Hooton and others that these two cultures belonged to different kinds of men, and it is a reasonable hypothesis that the western one belonged to early *Homo* and the eastern one to *Pithecanthropus*.

There is, however, danger of error in trying to trace the geographical history of man in further detail by artifacts. Man has probably always been acquisitive, inquisitive, and imitative. From the very beginning, men have probably taken or traded things from each other and watched and imitated each other. Man himself, his things (traded or taken), and ideas (including imitated ways of doing things) may have had partly different geographical histories since early in the Pleistocene. An exchange of the kind that may often have occurred between primitive races is described by Kipling in his story "The Knife and the Naked Chalk" (in *Rewards and Fairies*).

It is suggested by Coon (1954, p. 41—but he was probably not the first to suggest it) that the main center of evolution and dispersal of man during the Pleistocene was the region of the present Sahara Desert, East Africa, and Arabia; that western Europe was marginal; and that South Africa and eastern Asia were peripheral areas where

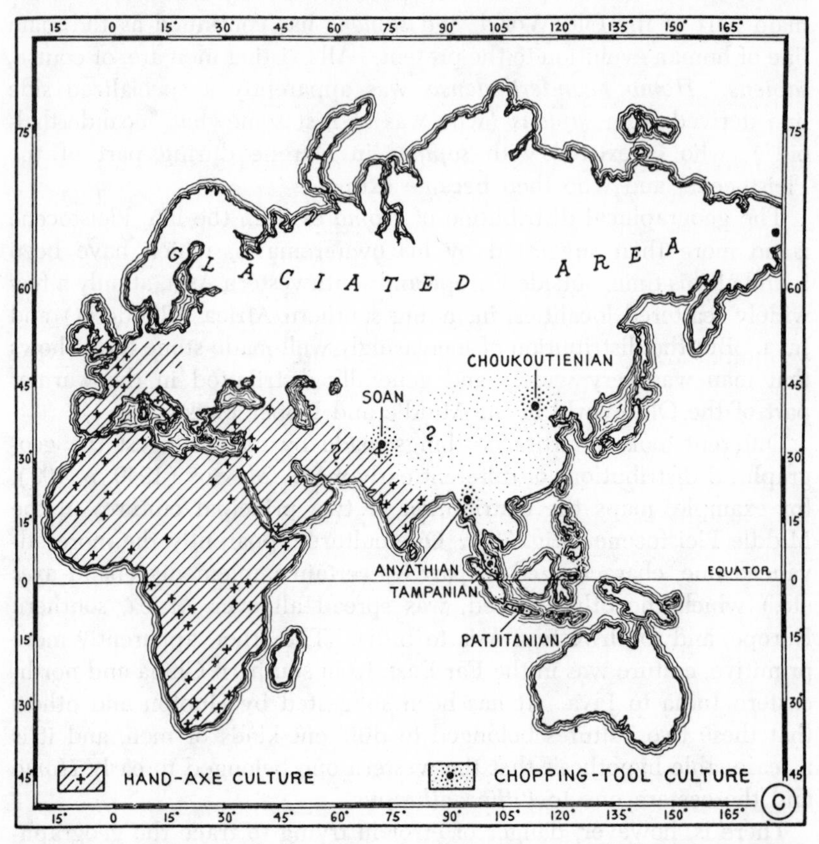

Fig. 79. Distribution of two principal cultures in the mid-Pleistocene. From Movius 1949, *The Lower Palaeolithic Cultures of Southern and Eastern Asia,* p. 409, Map 4. © Copyright American Philosophical Society. Used by permission of H. L. Movius, Jr., and the American Philosophical Society.

subhumans and archaic humans sometimes survived for a time after they had been replaced at the main center. Something like this is implied in the idea that *Australopithecus* in South Africa in the early Pleistocene may have been a survivor of a stock from which man had already been derived in some other place, and that *Pithecanthropus* may have survived in the Far East after the transition to *Homo* had occurred elsewhere. But all this is rather hypothetical.

There is, I think, no real evidence that the main course of Pleistocene evolution of man occurred in any such limited place or that man's dispersal followed any simple pattern. I should expect it to have been very complex, within the limits of Africa and southern Eurasia.

Northern Africa and southwestern Asia were probably greatly affected by Pleistocene alternations of climate. During the glacial-pluvial ages, these regions were probably comparatively well watered and probably supported large populations of man under favorable conditions. During each interglacial age these regions probably became deserts, as they are now, and men died out or left them, to return in the next age of rain. Pleistocene man may have been more affected by these changes than by southward and northward movements of the ice sheets. Some anthropologists seem to have it in the backs of their minds that the alternation of climate caused human evolution, but I doubt it. If the main center of evolution of man during the Pleistocene was in North Africa and southwestern Asia (which, as I have said, is hypothetical), it was more likely because the place was peculiarly favorable than because the climate fluctuated. Dominant animals seem usually to evolve in favorable places rather than in violently fluctuating ones. Moreover, in the case of man, the critical first steps of evolution toward an erect, tool-using, thinking animal were evidently taken *before* the first Pleistocene glaciation, and once evolution had started in so favorable a direction it was probably bound to continue and perhaps to accelerate (by progressive selection, not by any sort of orthogenesis), regardless of climatic fluctuations.

For most of his history, for more than 39,000 of the 40,000 generations I have assigned him, man made his cutting tools of broken or chipped stone, especially of flint. Flint is like an impure natural glass which breaks to a very sharp edge, although the edge is too fragile to give long service or to work hard materials (for a clear description of the use of flint see Coon, pp. 48ff.). Flint tools were improved and refined throughout the Pleistocene.

Late in the Pleistocene, probably partly because of the improvement of flint tools, man began to move out beyond his old geographical limits. Before man finished with flint, he probably reached almost the farthest parts of all the continents: Australia and Tasmania, which was connected to Australia during the ice ages; north into cold-temperate regions (but not yet the arctic); and through North and South America.

The movement northward and to America came late in the Pleistocene. Through most of that period, man was held within the warmer part of the Old World. He had fire in the mid-Pleistocene, but that alone did not enable him to live in very cold places. According to Coon, what finally broke the barrier of cold was the invention of the burin, a flint chisel which could cut bone and ivory. With the burin, man made needles; and with needles, fitted clothing of skins. Then he could and did extend northward; and then, perhaps 1000 generations ago, he finally reached America, probably during the last continental glaciation, when the sea was lowered and there was a land bridge from northeastern Asia to Alaska. There is a point to be noted here. The burin and clothing allowed man to reach America, but they were secondary things. The main thing was that man had become dominant in the warm part of the Old World, and it was his dominance which finally pushed him northward and then into the New World. This is in accordance with zoogeographic principles. It usually is primarily dominant animals which, by means of secondary, minor adaptations, extend into the cold north and spread from one-half of the world to the other.

Still later, after the Pleistocene, man learned to make still better tools by grinding (as opposed to breaking or chipping) certain stones which, though not so hard as flint, are less brittle and more durable. This was a very important event, marking the end of the Old Stone (Paleolithic) Age and the beginning of the New Stone (Neolithic) Age. With the new tools, man could fell trees and clear land, as well as make many new things of wood. The new stone tools made agriculture possible. According to Coon, the oldest known ground-stone tools, the oldest known cultivated plants, and also the oldest known domestic animals (except the dog), found together in southwestern Asia, are only about 8000 years old. Eight thousand years is only about 300 generations. A line of 300 people is not very long compared to the line of something like 40,000 people who used flint. But the invention of stone grinding began an acceleration of man's history. In something like 300 generations since then, man has passed through the New Stone, Bronze, Iron, and Steel Ages, in a crescendo of discovery, invention, increase and dispersal of populations, deforestation, consumption of resources, and increasing complexity of organizations and ideas, to the present.

Stone grinding may have been invented in several different places. At any rate ground-stone tools spread over the world rapidly, overtaking and replacing flint just as, later, iron overtook and replaced bronze (Coon). Agriculture began in several different places. It began in southwestern Asia, with wheat as the staple grain. It began in southeastern Asia, with rice as the staple. And it began, independently, in America, with corn (maize) as the staple. Other plants were cultivated too in each place, and different animals were domesticated. Only the dog had been domesticated before, while man was still a hunter.

The rise of agricultural races did not mean the immediate end of flint- and stone-armed hunters. They co-existed for a long time. In fact, primitive hunters and food gatherers still grind stones for tools and even chip flint in a few remote places including Southwest Africa, the interior of Australia, and (at least until very recently) Tierra del Fuego.

Some probable movements of New Stone Age farming races from two main centers in western and southeastern Asia are mapped by Coon (p. 126), but his map must be much too simple. There must have been very complex dispersals of man himself and of his new tools, new plants, new animals, and new ideas, all dispersing partly together and partly in different but interdependent patterns. This process gradually came into the light of history and became history as we know it, the written history of the partly combined, partly different dispersals of different races of man, of new things and new ways of doing things discovered or invented by man, of new organizations, and of new ideas. That these things have evolved and spread in patterns partly different from the evolution and spread of man himself, and different from each other, is an obvious fact. It will be further discussed and exemplified in a few pages.

Both before and after the invention of ground-stone tools, the main center from which man has dispersed over the world has been the main part of the Old World, Eurasia plus Africa.

From Asia, successive races have moved toward Australia, the later ones overrunning and replacing or mixing with the earlier ones. At least this is what seems to have occurred, although some details are still hypothetical. The earlier, more primitive races have survived in remote places toward the outer end of the dispersal line, or to one side of it on isolated islands like the Andamans and remote parts of the Philippines. The oldest or at least the outermost race we know in this succession is or was Tasmanian Man, who survived on Tasmania below Australia until about a hundred years ago. He was a spear-throwing hunter and food gatherer, who had fire and crude stone tools but not much else, not even dogs. Next are the existing Australian aborigines. They too are spear-throwing hunters and food gatherers. They probably reached Australia with flint but had learned stone grinding before the arrival of Europeans, and they had dogs. Next are the Melanesians on New Guinea etc. They make beautifully finished ground-stone tools (which they are now giving up for steel) and they have bows, unknown to Tasmanians and Australians, and some Melanesians have simple agriculture and have pigs as well as dogs, although the pigs usually run wild. I do not know whether Melanesians brought stone grinding, bows, cultivated plants, and animals with them to New Guinea or whether they received them from Asia later. Finally, behind the Melanesians, nearer Asia, are men with metal, more highly organized agriculture, and complex civilizations. Races of men are certainly now distributed along the Asia-Australia axis in the way described, but perhaps I should say again that their origin, by successive outward movements, is partly hypothetical. Their history has probably at least been more complex than indicated.

There was also a succession of movements of man from the Old World to America. The first was probably from Asia by way of the Bering bridge and Alaska during the last ice age, something like 1000 generations ago, and there may have been several later movements along the same route. From Alaska, men spread south through the whole of the New World and differentiated into minor races in different parts of it, and some of the races invented New World agriculture. Polynesians may have reached the west coast of South America, although they did not colonize it. Vikings reached the northeastern corner of North America. Finally Columbus came, but his was at best the third arrival of man in America, and he may have been much farther down the list, although his coming began a new age of history.

The last important part of the world to be reached by man was the island region of the Pacific, which was unoccupied until the coming of the Polynesians. They came from southern Asia or from islands near Asia. By just what routes they started is unknown, but it was from the Asiatic side. Anthropologists know this (Coon, pp. 161ff., with map on pp. 164–165); the Polynesians themselves know how they dispersed among the islands; and biologists know that the plants and animals that came with them were almost all Asiatic: the yams and taros and bread fruit and bananas and most other food plants, pigs and chickens, and even the rats and lizards that went along were Asiatic, not American—it is this kind of evidence that I, as a zoogeographer, am specially qualified to judge. There is no reasonable doubt that the Polynesians came from Asia, not from America, and the adventurers of the *Kon-Tiki* ought to accept this fact and be content with showing that a raft or two of Peruvian Indians may have reached Polynesia eventually and may have brought the sweet potato.[1]

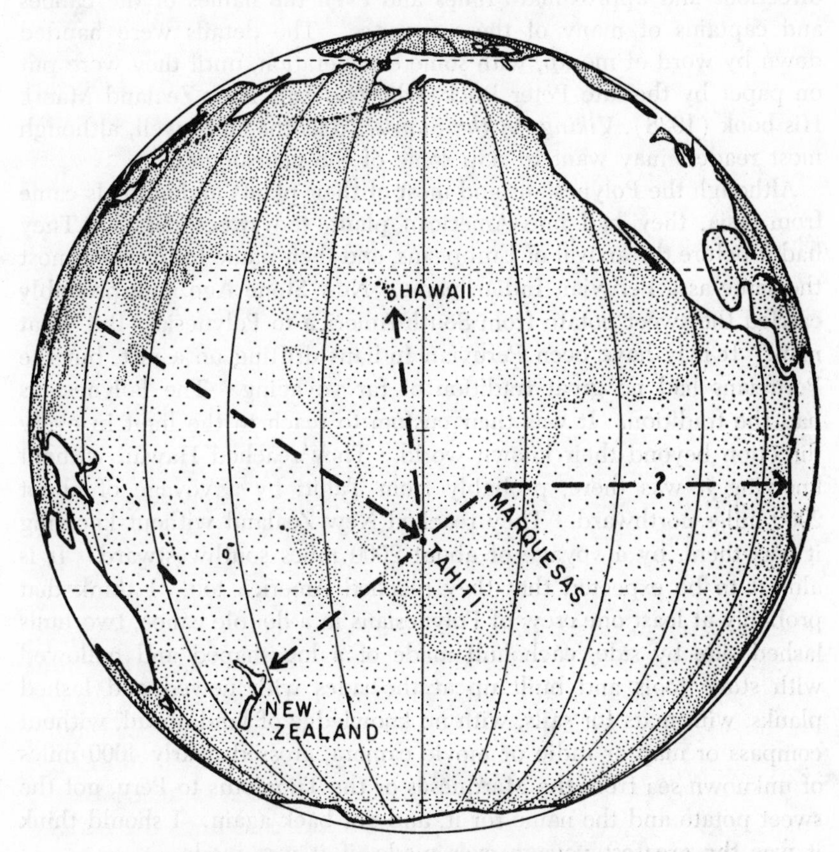

Fig. 80. Main routes of dispersal of Polynesians in the Pacific.

The Polynesians had only stone tools: finely ground axes, adzes, and chisels made of basalt. With such tools they made double canoes, fitted them with mat sails, and reached and occupied almost every habitable island in the Pacific (Fig. 80). Their first voyages through the central Pacific and to Hawaii were made only about 1500 years ago. About 1000 years ago, they reached southwest from the central

[1] I have tried to be restrained in criticism in this book. A stronger condemnation of Heyerdahl's theory that the Polynesians came from South America will be found in Merrill's (1954) *The Botany of Cook's Voyages.*

Pacific to New Zealand, and, incidentally, found and finished the huge, flightless moas there. New Zealand was not extensively settled until still later, about 600 years ago. The Polynesians still remember the directions and approximate times and even the names of the canoes and captains of many of these voyages. The details were handed down by word of mouth, with some exaggeration, until they were put on paper by the late Peter Buck, who was half New Zealand Maori. His book (1938), *Vikings of the Sunrise,* tells the story well, although most readers may want to skip some of the genealogies.

Although the Polynesians and most of their plants and animals came from Asia, they had one American plant, the sweet potato. They had it before the arrival of Europeans, and their name for it was almost the same as a Peruvian Indian name. New Stone Age man probably carried the sweet potato from South America to Polynesia. But what man? It may have been Peruvian Indians drifting on a raft, but the Peruvians had no great tradition of far voyaging. The Polynesians had the tradition. It was their custom to reach to the limit in every direction beyond their known world. They reached Hawaii without knowing it was there, probably from Tahiti by a voyage of about 2000 miles northward. They reached New Zealand without knowing it was there, by a voyage of about 2000 miles southwestward. It is almost to be expected that they reached America too. I think that probably at least one crew of Polynesians in a double canoe, two hulls lashed side by side, each hull made of a log shaped and hollowed with stone tools and built up at the sides with hewed and lashed planks, with mats for sails, with no knowledge of land ahead, without compass or map or radio or movie camera, crossed nearly 4000 miles of unknown sea from the Marquesas or the Tuomotus to Peru, got the sweet potato and the name for it, and got back again. I should think it was the greatest voyage ever made, if it was made.

In the last few pages I have tried to trace the geographical history of man up to the beginning of written history. Man's geographical history conforms to the main principles of zoogeography. It has been a complex history. Barriers have been important in it but have not been absolute; cold climate and ocean barriers limited man's distribution for a long time, but they were crossed eventually. Competition, dominance, and replacement have been inherent in it. And evolution, first of man himself and then of his tools and ideas, has been fundamental in determining the main dispersal pattern: the main dispersal pattern of man, as of so many animals, has been evolution of dominant stocks in the main, favorable part of the Old World (at first just the warmer part of it) and spreading outward from there.

I have said that, generally speaking, it has been dominant races of man which have moved and that they have moved to gain advantages more than to escape disadvantages. This needs further illustration and analysis.

One of the first great moves to gain advantage was probably from the forest into more open regions when, as a running ape, pre-man began to hunt active game on the ground. Later, each successive improvement in primitive weapons probably increased hunting efficiency and allowed man to obtain more game, and game in new places. Each improvement was probably followed by multiplication and spreading of the men who made it. The use of the dog in hunting probably had the same result. The "domestication" of fire (which was perhaps the bravest thing man ever did, as Krutch has said in *The American Scholar*) brought positive advantages, probably first the chance to extend into and use the resources of somewhat cooler regions than before, and then the ability to use food more efficiently—cooking food shortens the time of eating and leaves more time for other things. Coon (pp. 62–63) estimates that when man ate raw roots and raw meat, eating (and finding) food took most of the average day, but that when he began to cook food, eating time was reduced to about two hours a day. These advantages, too, were probably followed by multiplication and spreading of the men concerned. The invention of fitted clothing enabled man to move into still colder regions, and this movement too had the effect of gaining additional advantages more than escaping disadvantages: man did not escape from warm into cold places; he took advantage of both.

That it has usually been dominant races which have moved, and that they have moved to gain advantages more than to escape disadvantages, is an arbitrary simplification. It is partly a matter of definition, for moving to secure advantages is an evidence of dominance. The distinction between gaining advantages and escaping disadvantages is arbitrary. It is not likely that primitive man made such a distinction. The very men whose inventions and movements gained most in the end may have made them with the idea of escaping the immediate effects of cold or overcrowding. What has really happened is that there has been a very complex chain of events in which some men have been more successful than others and have multiplied and spread; their spreading has been concerned with both escapes and gains, but the gains have been more important in the long run;

the gains have been followed by new multiplications and new spreadings; which have led to new gains; etc. The process is still going on. What I want to emphasize is that at each stage some men (dominant by definition) have multiplied and spread more than others, and that in the long run positive gains have been more important than negative escapes. This is, I think, a special example of a fundamental principle of animal dispersal: that it is usually dominant animals which spread, and that they usually do so to gain advantages rather than to escape disadvantages.

I have said that man himself and his things and ideas have had partly different (but interdependent) geographical histories. There is a broad, if inexact, precedent for this in nature. Plants and animals in nature tend to form communities of more or less interdependent species, and the communities can move as wholes, but the species that compose them often have had separate, very different geographical histories of their own. (This is something that I am not sure ecological zoogeographers fully appreciate.) Animals (including man) and their parasites and diseases can have different geographical histories. (This is something that host-parasite zoogeographers sometimes do not appreciate.) For example, yellow fever was originally a disease of monkeys in tropical America and was transmitted by a tropical American jungle mosquito. Man first reached the American tropics by way of North America, on foot and bringing nothing relevant with him, and presumably became an occasional victim of the disease where it already occurred. Much later other men came across the ocean, in ships, bringing with them an Old World mosquito (*Aedes aegypti*) which habitually bred in houses and which happened to be an effective transmitter of yellow fever. The result was a series of epidemics in which the disease spread through human populations far beyond its original geographical limits, extending north at times to Philadelphia. But the disease never reached the northern limit of man in America (because *Aedes aegypti* could not stand cold) and, fortunately, never reached the Old World. It is therefore true that, although the three are sometimes closely tied together and can move together, man, yellow-fever mosquitoes, and yellow fever have very different total distributions and very different geographical histories.

A limited example of the interrelationship of advantage, dominance, and dispersal in man's history, and of the partly different dispersals of man and of his things and ideas, is worth describing. According to Coon (pp. 319–320), certain food plants (*things*) including the taro were brought to East Africa (from India) by unknown traders long ago. The *idea* of smelting iron was introduced into East Africa within 500 years of the time of Christ. The better-fed and better-armed Negroes of East and Central Africa then multiplied, became dominant, and eventually spread northward and southward at the expense of other races. The southward-moving Negroes reached South Africa about the time of White settlement and replaced most of the earlier, sparser populations of Bushmen there. Most of the Negroes in South Africa eventually outdistanced or abandoned the taro and primitive iron smelting and adopted new foods and new ideas with still different geographical histories, brought to South Africa by other races. All this culminated in the present racial crisis in South Africa. Chains of events like this have probably been going on, in an increasingly complex way, ever since man began to have possessions and ideas.

The most recent dispersals of man are recorded in written history. The general processes are probably the same ones which have been going on since the beginning of man, accelerated and multiplied in complexity. Many of the historical movements are common knowledge.

Within the theater of western Eurasia and the Mediterranean region, history records great movements of Asiatics into Europe and later of Moors across North Africa into Spain, and also very complex local movements. The Old Testament suggests the complexity of movements of man in the Near East. Early British history tells of repeated invasions and reinvasions. These are just examples. Similar local movements were probably going on all over the world. The movements resulted sometimes in replacements, sometimes in mixings and assimilations of races.

There has also been a great, complex redispersal of man outward from the main part of the Old World within the period of written history. The main movements have been made by European Caucasoids, but other races have gone or been taken along in different numbers to different places. Although Europeans have reached every part of the world, they have not occupied all parts equally. North and Central Europeans have established large populations in temperate and in some subtropical regions in North America, South Africa, Australia, and New Zealand. South Europeans have established large populations in tropical as well as south-temperate America. Europeans have been less successful in populating the Old World tropics. In some places the new European populations have replaced other races; in other places they co-exist with little mixing; in still other places there has been much mixing.

The present redispersal of man has been accompanied by an increasingly complex, constantly changing dispersal of things and ideas, which have moved partly as man has done and partly in different patterns. For example, the coffee I drank this morning may have come from three continents, but none of the men who grew it came with it; and behind this is the history of the coffee plant, which has been carried into many parts of the world where the men who first discovered coffee have not gone. For another example, the ideas that made the atom bomb came from different men in different places; some of the men carried their own ideas to the bomb; and some other ideas were used although the men who originated them did not come. These were not isolated events. If man has affected the dispersal of coffee, coffee has affected the dispersal of man by attracting people into new places to raise it; and the bomb may change the distribution of man. These are only very limited examples of the complexity of movement of men, things, and ideas over the world. The process as a whole is a network of partly separate, partly interrelated chains of events too complex to describe or even to understand.

I have repeatedly stressed that men, things, and ideas have dispersed in partly different (but interrelated) patterns probably from the beginning, with gradually increasing complexity. Within the time of written history something new has appeared in this process, with (as might be expected) a geographical history of its own. It is the system of empire or of international power. Among primitive men, conquering tribes usually destroyed or absorbed conquered ones and occupied their land, but improvements of transportation and communication gradually allowed some people to rule others without destroying them. This made empires possible (see Coon, pp. 282ff.). The first known true empire arose in southwestern Asia about 500 B.C., with the center of power in Iran. (I do not know when empires began in eastern Asia.) The center or centers of power in the West shifted successively west to Macedonia and then to Rome; then back to Byzantium (later called Constantinople or Istanbul); then (one center at least) to the Arab world; then to Spain, England, and other European countries; and now it has moved to the United States and Russia. I know that this is an absurd simplification of the movements of centers of power, but it is good enough for my purpose, which is to show that power has had its own geographical history.

The reasons why power has moved as it has are probably complex. Power probably goes wherever the requisite men, things, and ideas coincide. The first requirement is that enough people with enough intellectual capacity move to (or evolve in) a given place. A population of several tens of millions of people is probably necessary to hold international power now; additional numbers probably bring some, but perhaps not much, additional advantage. Of things important to power in recent times, iron, coal, oil, and uranium are obvious. It is probably necessary that some of these things pre-exist at the place of power, although some of them can be brought from other places. Finally it is necessary that ideas for organizing and using the people and things be brought to the place of power or originate there. Accidents have some effect. Lack of a horseshoe nail could lose a battle and move a center of power in the past, and the same sort of accident, as well as an accident of scientific discovery, can do it now. But accidents can move centers of power only within certain limits. Combinations of people, things, and ideas set the limits.

Summary of man's geographical history

Before summarizing the geographical history of man, I shall summarize his evolution and the evolution of his tools and cultures, for, as I have said, these things are closely interwoven with man's geographical history.

The evolution of structure and behavior from ape to early man probably required millions of years before the Pleistocene. During this time pre-man adapted his body in many ways to erect posture, and during much of this time he probably used natural sticks and stones with gradually increasing effectiveness, while he gradually perfected a hitting and throwing arm, used his teeth less and less for fighting and killing, and gradually began to improve his brain.

Tools made by breaking and chipping stone, mostly flint, were used and were slowly improved for perhaps a shorter time, perhaps a million years, through the Pleistocene. During this time, and mainly during the earlier part of it, man apparently evolved from the *Australopithecus* or *Pithecanthropus* level into *Homo sapiens*.

Tools made by grinding stone began to be used about 8000 years ago, and the succession of bronze, iron, and steel has occurred in the very short time since then. During this time man himself has apparently changed very little; in fact he has probably changed little since the mid-Pleistocene.

Man's geographical history began within the main, warm part of the Old World—it was actually a continuation of the geographical history of a very long line of Old World primates. The first steps toward standing, running on two legs, and entering open country may have occurred in a limited part of this area, perhaps in part of Africa. But as pre-man gradually became better adapted to standing and running, as (I think) he learned to hit and throw and to use sticks and stones more and more effectively, and as his brain gradually improved—as he gradually became more successful, more dominant—he probably spread through all the suitable, opener parts of Africa and Eurasia until he was stopped (for the time being) by cold in the north and by sea barriers in other directions. The first dispersal to these limits was probably followed by successive (but complex and irregular) redispersals of successive, evolving, increasingly dominant genera and species. Successive dispersals may have widened the outer limits of distribution a little; and, as their dominance increased, prehumans probably invaded new habitats. This is what dominant animals do: they tend to occupy relatively unfavorable as well as favorable places and to become generally distributed within their total ranges. Primitive races of *Homo sapiens* have entered almost every habitat from desert (in Australia, for example) to the heaviest jungle, and man's evolving ancestors presumably made at least partial ecological radiations too. The geographical and ecological radiations of successive, increasingly dominant prehumans were presumably accompanied by competitions and by withdrawals and replacements of earlier prehumans. No trace of all this has yet been found in the Pliocene, but this is how rising, dominant animals do disperse, and the process gradually becomes visible in the Pleistocene.

Primitive man himself dispersed throughout the main, warm part of the Old World early in the Pleistocene, if not before. Occurrence of broken-stone tools proves this. *Australopithecus*, then in South Africa, may have been early Pleistocene man or may have been a survivor of a prehuman stock which man was replacing. In either case, australopithecines were probably soon replaced. They may have been replaced by the *Pithecanthropus* stage of human evolution, which may have had a dispersal of its own in the main part of the Old World during the early and middle Pleistocene, although this is hypothetical. *Homo* apparently dispersed in the middle and late Pleistocene. The distribution of artifacts in the Middle Pleistocene suggests that *Homo* had already occupied all of the main, warmer part of the Old World except the Far East, where *Pithecanthropus* still held out temporarily.

The later history of man is mainly the history of *Homo sapiens*. His dispersals and redispersals are at first partly hypothetical, then better and better known, until they become part of written history.

The movements were at first confined to the main, warmer part of the Old World. *Pithecanthropus,* however, who had fire, had probably extended man's limits a little northward, and *Homo sapiens,* with the burin and fitted clothing, eventually extended still farther north. Late in the Pleistocene, but still in the Old Stone Age, still using flint, man made his first great dispersal outward from the main part of the Old World. He crossed a Bering bridge to North America and thence reached South America, and he spread southeast from Asia across the islands to Australia and Tasmania, probably aided by lowering of sea level which reduced the number and width of water gaps. Later, after the Pleistocene, beginning only about 8000 years ago, New Stone Age man began his own, probably complex dispersal in the main part of the Old World, and he too spread outward from there. He did not go quite as far as his predecessors in some directions, but he reached two new parts of the world: the high arctic (occupied by New Stone Age Eskimos) and the remote islands of the Pacific. I do not know whether New Stone Age man replaced Old Stone Age man (except in remote peripheral areas), or whether he mixed with him, or just converted him to New Stone tools. Probably all of these things happened in different places. Finally, within historic times, there has been still another great redispersal of man outward from the main part of the Old World, primarily of European Caucasoids, with metal and complex tools. In this case we know there has been replacement, mixing, and conversion of older races in different places.

I have summarized man's geographical history as if it occurred in separate stages, but it did not. The stages must have merged into each other. Pre-man probably continued to exist and disperse for a time after man appeared. Old Stone Age man still existed and was probably still dispersing and redispersing in remote places while New Stone Age man spread. And New Stone Age man in turn was still spreading, for example into the Pacific, during the rise and beginning of dispersal of men with metal. Moreover, dispersal during each stage was certainly very complex, and successive races of men interbred and exchanged tools extensively. Prehuman and human dispersal has therefore been a single, irregular but continuous, increasingly complex process. I have described the beginning of this process hypothetically, by applying zoogeographic principles to a few general facts and deductions about man's ancestors and their evolution. The process is continuing now in full view.

The hypothetical beginning and visible present fit together remarkably well. The present redistribution of displaced persons is part of a process of dispersal that has never stopped since prehumans left the forest and extended into open country perhaps in Africa early in the Pliocene—and that was a continuation of still earlier dispersal processes. That I must go farther now to get good fishing than I had to go when I was a boy is part of the process of destruction and following of game that began in the Old World, probably in the Pliocene. The present cutting back of forest that I have seen going on on remote mountains in the West Indies and in tropical Australia is part of a process of deforestation that began about 8000 years ago in southwestern Asia (and other places) when man began to clear land and plant crops. These present processes are the *same* ones that have been going on, irregularly but continuously and with increasing speed and complexity, since far back in human or prehuman history. A realization of this, of the continuity of the present with the past, is one of the things that should come out of study of man's geographical history.

Man's dispersal, though unique in some ways, seems in general to have followed the pattern of other predominant animals: origin in the main Old World tropics, rise to dominance there, and finally spread from there over the rest of the world.

During his evolution and dispersal, man has changed the distributions of many other animals and of many plants, by direct destruction, by accidental and intentional introduction, and most of all by deforestation, burning, and cultivation, which have changed much of the earth's surface. Some of these processes must have started in the Pleistocene, but they increased when farming began in the New Stone Age, and they have enormously accelerated within historical times. They have caused retreat or extinction of very many animals almost everywhere in the world, and increase and spreading of many other animals that happened to be preadapted to the new conditions made by man. All this is properly part of zoogeography. A distinction is often made between "natural" processes of evolution and dispersal, in which man has played no part, and "unnatural" ones, somehow affected by man. There is of course a practical difference, but it is probably one of degree more than of kind. Man has had more effect on other animals than they usually have on each other, but the effect is probably not fundamentally different otherwise. A critical study of the redistribution of plants and animals influenced by man would be an important contribution to biogeography. I have not the space or time to undertake it here. Such a study would supplement the recent volume by Thomas *et al.* (1956) on man's role in changing the face of the earth.

I want to end this book by trying to look a little way into the future, without getting involved in politics or propaganda or philosophy, but just as a zoogeographer, if this is possible.

If situations were as they used to be, if man were simply an evolving animal without tools or ideas, or if he and his tools and ideas were still evolving slowly in limited parts of the world, the future would be easy to predict. It would be probable (but not certain) that dominant men would continue to evolve in large, favorable areas especially in the main part of the Old World and to disperse outward from there, as they have done in the past. But situations are not as they used to be. The old rules of evolution still hold, but the situation is new. The evolution of man himself can play almost no part in the immediate future. There is not time. Man must make do at least for a while with the body and brain he has now. The immediate future will be determined mainly by the evolutions and movements of man's things and ideas, especially ideas. It is mostly ideas—ways of thinking and doing—which will decide who has power now, and who survives and who does not. Ideas are not limited geographically. They used to be, but they are not now; that is part of the new situation. Any large group of people can now, if they wish, draw ideas from the whole world and combine them with their own ideas to make better ones. The people who do this best, and who remember the old lesson that gains are more important than escapes (although escapes are important too), will probably hold or take power in the future. The people who cut themselves off from ideas, or who sacrifice gains to temporary security, will not hold power long. This, I think, is the plain lesson of the history of evolution and dispersal of animals and of man himself.

What might have happened to the men who first tamed fire if they had put it out through fear of it, or lost it through fear some other tribe might get it? Or did this happen? And did the other tribe finally get fire and send it downwind through long grass to destroy the men who were afraid? Or were the latter just pushed aside to survive in isolation for a while—only for a while—out of the main current of evolution of man's ideas? The greatest gains made with fire were not in war but in daily use. Cooking, as has been said, saves time in eating. The first men to eat cooked food gained leisure, and leisure let them improve their tools, teach their children, experiment, talk, and think. No men without fire could long keep pace with them.

REFERENCES

Buck, P. H. 1938. *Vikings of the sunrise.* New York, Frederick A. Stokes.

Clark, W. E. Le Gros. 1949. *History of the primates.* London, British Mus.

———. 1955. *The fossil evidence for human evolution . . . .* Chicago, U. of Chicago Press.

Coon, C. S. 1954. *The story of man.* New York, Alfred A. Knopf.

Hopwood, A. T. 1954. Notes on the Recent and fossil mammalian faunas of Africa. *Proc. Linnean Soc. London,* **165**, 46–49.

Merrill, E. D. 1954. *The botany of Cook's voyages.* Waltham, Mass., Chronica Botanica Co.

Movius, H. L., Jr. 1949. The Lower Palaeolithic cultures of southern and eastern Asia. *Tr. American Philosophical Soc.*, New Ser., 38, 329–420.

Thomas, W. L., Jr., *et al.* 1956. *Man's role in changing the face of the earth.* Chicago, U. of Chicago Press.

Zuckerman, S. 1954. Correlation of change in the evolution of higher primates. (In) Huxley, J., *et al.*, *Evolution as a process,* pp. 300–352. London, George Allen and Unwin.

Index